Medical Imaging

TECHNOLOGY AND APPLICATIONS

Devices, Circuits, and Systems

Series Editor
Krzysztof Iniewski
CMOS Emerging Technologies Research Inc.,
Vancouver, British Columbia, Canada

PUBLISHED TITLES:

Atomic Nanoscale Technology in the Nuclear Industry
Taeho Woo

Biological and Medical Sensor Technologies
Krzysztof Iniewski

Building Sensor Networks: From Design to Applications
Ioanis Nikolaidis and Krzysztof Iniewski

Circuits at the Nanoscale: Communications, Imaging, and Sensing
Krzysztof Iniewski

Electrical Solitons: Theory, Design, and Applications
David Ricketts and Donhee Ham

Electronics for Radiation Detection
Krzysztof Iniewski

Embedded and Networking Systems: Design, Software, and Implementation
Gul N. Khan and Krzysztof Iniewski

Energy Harvesting with Functional Materials and Microsystems
Madhu Bhaskaran, Sharath Sriram, and Krzysztof Iniewski

Graphene, Carbon Nanotubes, and Nanostuctures: Techniques and Applications
James E. Morris and Krzysztof Iniewski

High-Speed Photonics Interconnects
Lukas Chrostowski and Krzysztof Iniewski

Integrated Microsystems: Electronics, Photonics, and Biotechnology
Krzysztof Iniewski

Internet Networks: Wired, Wireless, and Optical Technologies
Krzysztof Iniewski

Low Power Emerging Wireless Technologies
Reza Mahmoudi and Krzysztof Iniewski

Medical Imaging: Technology and Applications
Troy Farncombe and Krzysztof Iniewski

MEMS: Fundamental Technology and Applications
Vikas Choudhary and Krzysztof Iniewski

PUBLISHED TITLES:

Nano-Semiconductors: Devices and Technology
Krzysztof Iniewski

Nanoelectronic Device Applications Handbook
James E. Morris and Krzysztof Iniewski

Nanoplasmonics: Advanced Device Applications
James W. M. Chon and Krzysztof Iniewski

Novel Advances in Microsystems Technologies and Their Applications
Laurent A. Francis and Krzysztof Iniewski

Optical, Acoustic, Magnetic, and Mechanical Sensor Technologies
Krzysztof Iniewski

Radiation Effects in Semiconductors
Krzysztof Iniewski

Semiconductor Radiation Detection Systems
Krzysztof Iniewski

Smart Sensors for Industrial Applications
Krzysztof Iniewski

Telecommunication Networks
Eugenio Iannone

Testing for Small-Delay Defects in Nanoscale CMOS Integrated Circuits
Sandeep K. Goel and Krishnendu Chakrabarty

Wireless Technologies: Circuits, Systems, and Devices
Krzysztof Iniewski

FORTHCOMING TITLES:

3 D Circuit and System Design: Multicore Architecture, Thermal Management, and Reliability
Rohit Sharma and Krzysztof Iniewski

CMOS: Front-End Electronics for Radiation Sensors
Angelo Rivetti

High Frequency Communication and Sensing: Traveling-Wave Techniques
Ahmet Tekin and Ahmed Emira

Gallium Nitride (GaN): Physics, Devices, and Technology
Farid Medjdoub and Krzysztof Iniewski

High-Speed Devices and Circuits with THz Applications
Jung Han Choi and Krzysztof Iniewski

Integrated Power Devices and TCAD Simulation
Yue Fu, Zhanming Li, Wai Tung Ng, and Johnny K.O. Sin

Labs-on-Chip: Physics, Design and Technology
Eugenio Iannone

FORTHCOMING TITLES:

Metallic Spintronic Devices
Xiaobin Wang

Microfluidics and Nanotechnology: Biosensing to the Single Molecule Limit
Eric Lagally and Krzysztof Iniewski

MIMO Power Line Communications: Narrow and Broadband Standards, EMC, and Advanced Processing
Lars Torsten Berger, Andreas Schwager, Pascal Pagani, and Daniel Schneider

Mobile Point-of-Care Monitors and Diagnostic Device Design
Walter Karlen and Krzysztof Iniewski

Nanoelectronics: Devices, Circuits, and Systems
Nikos Konofaos

Nanomaterials: A Guide to Fabrication and Applications
Gordon Harling and Krzysztof Iniewski

Nanopatterning and Nanoscale Devices for Biological Applications
Krzysztof Iniewski and Seila Selimovic

Nanoscale Semiconductor Memories: Technology and Applications
Santosh K. Kurinec and Krzysztof Iniewski

Optical Fiber Sensors and Applications
Ginu Rajan and Krzysztof Iniewski

Power Management Integrated Circuits and Technologies
Mona M. Hella and Krzysztof Iniewski

Radio Frequency Integrated Circuit Design
Sebastian Magierowski

Semiconductor Device Technology: Silicon and Materials
Tomasz Brozek and Krzysztof Iniewski

Smart Grids: Clouds, Communications, Open Source, and Automation
David Bakken and Krzysztof Iniewski

Solar Cells: Materials, Devices, and Systems
Qiquan Qiao and Krzysztof Iniewski

Soft Errors: From Particles to Circuits
Jean-Luc Autran and Daniela Munteanu

Technologies for Smart Sensors and Sensor Fusion
Kevin Yallup and Krzysztof Iniewski

VLSI: Circuits for Emerging Applications
Tomasz Wojcicki and Krzysztof Iniewski

PUBLISHED TITLES:

Nano-Semiconductors: Devices and Technology
Krzysztof Iniewski

Nanoelectronic Device Applications Handbook
James E. Morris and Krzysztof Iniewski

Nanoplasmonics: Advanced Device Applications
James W. M. Chon and Krzysztof Iniewski

Novel Advances in Microsystems Technologies and Their Applications
Laurent A. Francis and Krzysztof Iniewski

Optical, Acoustic, Magnetic, and Mechanical Sensor Technologies
Krzysztof Iniewski

Radiation Effects in Semiconductors
Krzysztof Iniewski

Semiconductor Radiation Detection Systems
Krzysztof Iniewski

Smart Sensors for Industrial Applications
Krzysztof Iniewski

Telecommunication Networks
Eugenio Iannone

Testing for Small-Delay Defects in Nanoscale CMOS Integrated Circuits
Sandeep K. Goel and Krishnendu Chakrabarty

Wireless Technologies: Circuits, Systems, and Devices
Krzysztof Iniewski

FORTHCOMING TITLES:

3 D Circuit and System Design: Multicore Architecture, Thermal Management, and Reliability
Rohit Sharma and Krzysztof Iniewski

CMOS: Front-End Electronics for Radiation Sensors
Angelo Rivetti

High Frequency Communication and Sensing: Traveling-Wave Techniques
Ahmet Tekin and Ahmed Emira

Gallium Nitride (GaN): Physics, Devices, and Technology
Farid Medjdoub and Krzysztof Iniewski

High-Speed Devices and Circuits with THz Applications
Jung Han Choi and Krzysztof Iniewski

Integrated Power Devices and TCAD Simulation
Yue Fu, Zhanming Li, Wai Tung Ng, and Johnny K.O. Sin

Labs-on-Chip: Physics, Design and Technology
Eugenio Iannone

FORTHCOMING TITLES:

Metallic Spintronic Devices
Xiaobin Wang

Microfluidics and Nanotechnology: Biosensing to the Single Molecule Limit
Eric Lagally and Krzysztof Iniewski

MIMO Power Line Communications: Narrow and Broadband Standards, EMC, and Advanced Processing
Lars Torsten Berger, Andreas Schwager, Pascal Pagani, and Daniel Schneider

Mobile Point-of-Care Monitors and Diagnostic Device Design
Walter Karlen and Krzysztof Iniewski

Nanoelectronics: Devices, Circuits, and Systems
Nikos Konofaos

Nanomaterials: A Guide to Fabrication and Applications
Gordon Harling and Krzysztof Iniewski

Nanopatterning and Nanoscale Devices for Biological Applications
Krzysztof Iniewski and Seila Selimovic

Nanoscale Semiconductor Memories: Technology and Applications
Santosh K. Kurinec and Krzysztof Iniewski

Optical Fiber Sensors and Applications
Ginu Rajan and Krzysztof Iniewski

Power Management Integrated Circuits and Technologies
Mona M. Hella and Krzysztof Iniewski

Radio Frequency Integrated Circuit Design
Sebastian Magierowski

Semiconductor Device Technology: Silicon and Materials
Tomasz Brozek and Krzysztof Iniewski

Smart Grids: Clouds, Communications, Open Source, and Automation
David Bakken and Krzysztof Iniewski

Solar Cells: Materials, Devices, and Systems
Qiquan Qiao and Krzysztof Iniewski

Soft Errors: From Particles to Circuits
Jean-Luc Autran and Daniela Munteanu

Technologies for Smart Sensors and Sensor Fusion
Kevin Yallup and Krzysztof Iniewski

VLSI: Circuits for Emerging Applications
Tomasz Wojcicki and Krzysztof Iniewski

Medical Imaging

TECHNOLOGY AND APPLICATIONS

EDITED BY
Troy Farncombe
Krzysztof Iniewski

CRC Press is an imprint of the
Taylor & Francis Group, an **informa** business

MATLAB® is a trademark of The MathWorks, Inc. and is used with permission. The MathWorks does not warrant the accuracy of the text or exercises in this book. This book's use or discussion of MATLAB® software or related products does not constitute endorsement or sponsorship by The MathWorks of a particular pedagogical approach or particular use of the MATLAB® software.

CRC Press
Taylor & Francis Group
6000 Broken Sound Parkway NW, Suite 300
Boca Raton, FL 33487-2742

First issued in paperback 2017

Version Date: 20130820

ISBN 13: 978-1-138-07228-2 (pbk)
ISBN 13: 978-1-4665-8262-0 (hbk)

Visit the Taylor & Francis Web site at
http://www.taylorandfrancis.com

and the CRC Press Web site at
http://www.crcpress.com

Contents

Preface

The field of medical imaging is particularly diverse, incorporating fields such as mechanical and electrical engineering, physics, mathematics, computer science, and, of course, medicine. Knowledge in all these fields has come together with a sole purpose: to improve patient health through noninvasive imaging.

Purpose of This Book

This book has been written with the intent of providing a snapshot of some of these fields and how they are being applied to medical imaging. The contributors to this book come from equally diverse areas, from clinical, industrial, and academic institutions. We have been fortunate to include contributions from among the foremost experts in these fields to this text.

While we consider this book to cover the state of the art in each area, it is intended as a text for graduate level students, scientists, and engineers working in the field of medical imaging. It is laced liberally with figures to aid in understanding of the rather advanced subject matter and provides abundant references for the reader seeking additional information on a given subject.

Organization

We have put together a broad-spectrum text and have tried to incorporate chapters on the major medical imaging techniques. The book is loosely divided into six different parts based on the modality. These consist of (i) optical imaging methods, (ii) x-ray imaging, (iii) ultrasound imaging, (iv) nuclear medicine and hybrid imaging, (v) magnetic resonance imaging, and (vi) software methods for image processing. Within each part, there are further chapters on specific aspects of each technique and in many cases, clinical examples and real-world experiences with these techniques.

While we realize that a single text cannot do justice to over 100 years of medical imaging, we hope that we have provided the reader with a snapshot of the state of the art and have encouraged them to further develop their own ideas in this exciting and ever-changing area.

Troy H. Farncombe
Krzysztof Iniewski

MATLAB® is a registered trademark of The MathWorks, Inc. For product information, please contact:

The MathWorks, Inc.
3 Apple Hill Drive
Natick, MA, 01760-2098 USA
Tel: 508-647-7000
Fax: 508-647-7001
E-mail: inf@mathworks.com
Web: www.mathworks.com

Editors

Troy H. Farncombe graduated from the University of British Columbia in 2000 with a PhD in physics. His graduate thesis investigated methods for performing dynamic SPECT imaging using conventional acquisition procedures. Following a postdoctoral fellowship at the University of Massachusetts Medical School, he took up a clinical role in 2002 as an imaging physicist at Hamilton Health Sciences in Hamilton, Ontario, Canada. He currently serves as an associate professor of radiology at McMaster University. In addition to clinical duties, he maintains an active research program in nuclear medicine, with a focus on investigating new types of radiation detectors for use in single photon imaging and new image reconstruction methods for quantitative SPECT imaging.

Krzysztof (Kris) Iniewski is managing R&D at Redlen Technologies Inc., a start-up company in Vancouver, British Columbia, Canada. Redlen's revolutionary production process for advanced semiconductor materials enables a new generation of more accurate, all-digital, radiation-based imaging solutions. Kris is also the president of CMOS Emerging Technologies (www.cmoset.com), an organization of high-tech events covering communications, microsystems, optoelectronics, and sensors. During the course of his career, Dr. Iniewski held numerous faculty and management positions at University of Toronto, University of Alberta, SFU, and PMC-Sierra Inc. He has published over 100 research papers in international journals and conferences and holds 18 international patents granted in the United States, Canada, France, Germany, and Japan. He is a frequent invited speaker and has consulted for multiple organizations internationally. He has also written and edited several books for Wiley, IEEE Press, CRC Press, McGraw Hill, Artech House, and Springer. His personal goal is to contribute to healthy living and sustainability through innovative engineering solutions.

Contributors

Nicola Belcari
Department of Physics
University of Pisa
Pisa, Italy

Brian G. Booth
School of Computing Science
Simon Fraser University
Burnaby, British Columbia, Canada

Albert I.H. Chen
Department of Systems Design Engineering
and
Waterloo Institute for Nanotechnology
University of Waterloo
Waterloo, Ontario, Canada

Kyusun Choi
Department of Electrical Engineering
The Pennsylvania State University
State College, Pennsylvania

Alberto Del Guerra
Department of Physics
University of Pisa
Pisa, Italy

M. Desco
Departamento de Bioingeniería e lngeniería Aeroespacial
Universidad Carlos III de Madrid
Madrid, Spain

Troy H. Farncombe
Department of Nuclear Medicine
Hamilton Health Sciences
and
Department of Radiology
McMaster University
Hamilton, Ontario, Canada

Gene Frantz
Department of Electrical and Computer Engineering
Rice University
Houston, Texas

Flavio Griggio
Department of Electrical Engineering
The Pennsylvania State University
State College, Pennsylvania

Ghassan Hamarneh
School of Computing Science
Simon Fraser University
Burnaby, British Columbia, Canada

Bjorn Heismann
Siemens Medical Systems
Erlangen, Germany

Thomas N. Jackson
Department of Electrical Engineering
The Pennsylvania State University
State College, Pennsylvania

Jin U. Kang
Department of Electrical and Computer Engineering
The Johns Hopkins University
Baltimore, Maryland

Hyunsoo Kim
Department of Electrical Engineering
The Pennsylvania State University
State College, Pennsylvania

Insoo Kim
Department of Electrical Engineering
The Pennsylvania State University
State College, Pennsylvania

Dong Soo Lee
Department of Nuclear Medicine
and
Department of Molecular Medicine and Biopharmaceutical Science
Seoul National University
Seoul, South Korea

Jae Sung Lee
Department of Nuclear Medicine
and
Department of Biomedical Sciences
and
Department of Brain and Cognitive Sciences
Seoul National University
Seoul, South Korea

Craig S. Levin
Department of Radiology
and
Department of Physics
and
Department of Electrical Engineering
and
Molecular Imaging Program at Stanford (MIPS)
Stanford University
Stanford, California

Baojun Li
Department of Radiology
Boston University Medical Center
Boston, Massachusetts

Junning Li
Laboratory of Neuro Imaging
Department of Neurology
School of Medicine
The University of California
Los Angeles, California

Aiping Liu
Department of Electrical and Computer Engineering
University of British Columbia
Vancouver, British Columbia, Canada

Chi Liu
Department of Diagnostic Radiology
and
Department of Biomedical Engineering
Yale University
New Haven, Connecticut

Angshul Majumdar
Indraprastha Institute of Information Technology
Delhi, India

Chris McIntosh
Princess Margaret Cancer Centre,
Toronto, Ontario, Canada

Martin J. McKeown
Department of Electrical and Computer Engineering
and
Pacific Parkinson's Research Centre
and
Department of Neuroscience
and
Department of Medicine (Neurology)
University of British Columbia
Vancouver, British Columbia, Canada

Greta S.P. Mok
Department of Electrical and Computer Engineering
University of Macau
Taipa, Macau, People's Republic of China

Mark Nadeski
Texas Instruments Incorporated
Dallas, Texas

Tinsu Pan
Department of Imaging Physics
MD Anderson Cancer Center
The University of Texas
Houston, Texas

Hao Peng
Department of Medical Physics
and
Department of Electrical and
Computer Engineering
and
School of Biomedical Engineering
McMaster University
Hamilton, Ontario, Canada

Jinyi Qi
Department of Biomedical
Engineering
University of California
Davis, California

Francisco E. Robles
Department of Chemistry
Duke University
Durham, North Carolina

A. Sisniega
Departamento de Bioingeniería
e lngeniería Aeroespacial
Universidad Carlos III de Madrid
Madrid, Spain

Tao Sun
Department of Electrical and
Computer Engineering
University of Macau
Taipa, Macau, People's Republic of
China

Ken Taguchi
The Russel H. Morgan
Department of Radiology and
Radiological Sciences
The Johns Hopkins University
School of Medicine
Baltimore, Maryland

Lisa Tang
School of Computing Science
Simon Fraser University
Burnaby, British Columbia, Canada

Susan Trolier-McKinstry
Department of Electrical Engineering
The Pennsylvania State University
State College, Pennsylvania

Richard L. Tutwiler
Department of Electrical Engineering
The Pennsylvania State University
State College, Pennsylvania

J.J. Vaquero
Departamento de Bioingeniería
e lngeniería Aeroespacial
Universidad Carlos III de Madrid
Madrid, Spain

Yi-Xiang J. Wang
Department of Imaging and
Interventional Radiology
The Chinese University of
Hong Kong
Shatin, Hong Kong, People's
Republic of China

Z. Jane Wang
Department of Electrical and
Computer Engineering
University of British Columbia
Vancouver, British Columbia,
Canada

Rabab Ward
Department of Electrical and Computer Engineering
University of British Columbia
Vancouver, British Columbia, Canada

Adam Wax
Department of Biomedical Engineering
Duke University
Durham, North Carolina

R. Glenn Wells
Division of Nuclear Cardiology
University of Ottawa Heart Institute
Ottawa, Ontario, Canada

Lawrence L.P. Wong
Department of Systems Design Engineering
and
Waterloo Institute for Nanotechnology
University of Waterloo
Waterloo, Ontario, Canada

John T.W. Yeow
Department of Systems Design Engineering
and
Waterloo Institute for Nanotechnology
University of Waterloo
Waterloo, Ontario, Canada

Jing Yuan
Department of Imaging and Interventional Radiology
The Chinese University of Hong Kong
Shatin, Hong Kong, People's Republic of China

Kang Zhang
GE Global Research Center
Niskayuna, New York

1

Future of Medical Imaging

Mark Nadeski and Gene Frantz

CONTENTS

1.1 Introduction

There are those who fear technology is nearly at the physical limitations of our understanding of nature, so where can we possibly go from here? But technology is not where our limits lie. Integrated circuits (ICs) have always exceeded our ability to fully utilize the capacity they make available to us, and the future will be no exception, for technology does not drive innovation. In truth, it is innovation and human imagination which drive technology.

1.2 Where Are We Going?

The broad field of medical imaging has seen some truly spectacular advances in the past half-century, many that most of us take for granted. Once a marvel only in the laboratory, advances such as real-time and Doppler ultrasonography, functional nuclear medicine, computed tomography, magnetic resonance imaging, and interventional angiography have all become available in clinical settings.

It is easy to sit back in wonder at how far the field of medical imaging has come. However, in this chapter, we glimpse into the future. Some of this future is quickly taking shape today, though some of it will not arrive for years, if not decades.

Specifically, we look at how advances in medical imaging are based on existing technology; how these technologies will provide more capacity and capabilities than we can conceivably exploit; and finally lead in to the conclusion that the future of medicine is not limited by what we know, but rather by what we can imagine.

Let us begin by looking at the edge of what is real, that wonderful place where ideas are transformed into reality.

1.2.1 EyeCam

For centuries, humanity has dreamed of being able to make the blind see. And, for as long, restoring a person's eyesight has been considered a feat commonly categorized as "a miracle."

About 10 years ago, Texas Instruments (TI) began collaborating with a medical team at Johns Hopkins, well known for its ability to make miracles happen. The team's goal was to develop a way to take the signal from a camera and turn it into electrical impulses that could then be used to excite the retina as shown in Figure 1.1. If successful, they could return some level of vision back to individuals who had lost their eyesight due to retinitis pigmatosa, a disease that affects more than 100,000 people in the United States alone.*

Now at the University of Southern California, this team continues to make significant progress. The project has evolved considerably over the years. Its initial conception consisted of mounting a camera on a pair of glasses that would require patients to rotate their heads in order to look around. Today, the team is working to actually implant a camera module *within* the eye since it is much more natural to let the eye do the moving to point the camera in the right direction. However, it is one thing to say, implanting a camera in a person's eye is more practical than mounting it

* *Source*: http://www.wrongdiagnosis.com/r/retinitis_pigmentosa/stats-country.htm

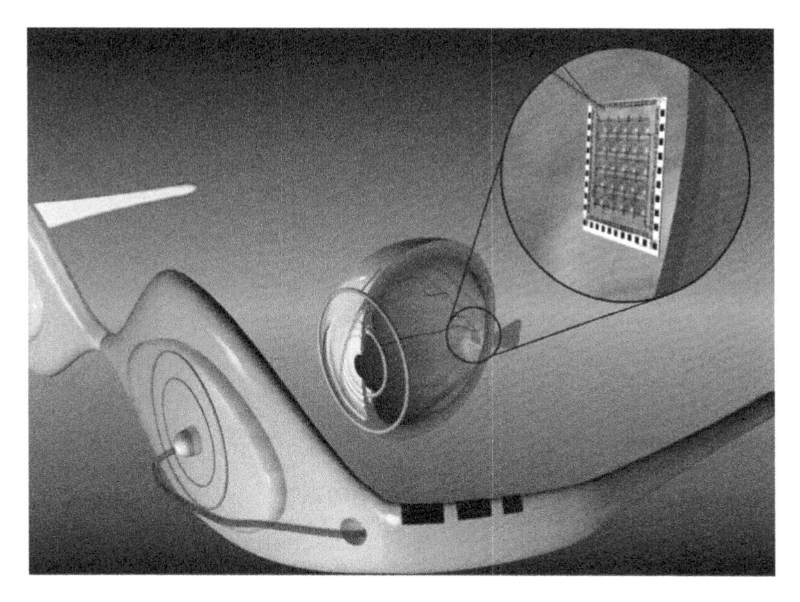

FIGURE 1.1 (See color insert.)
Example of the eyecam created and tested at the University of Southern California.

on glasses and quite another to achieve it. A number of challenges come to mind:

Size: The complete camera module has to be significantly smaller than an eyeball in order to fit.

Power: The camera must have exceptionally low power consumption. At the very least, the energy needs to be scavenged from body heat, the surrounding environment, or a yet-to-be-invented wireless power circuit.

Heat: Initial cameras may rely upon a connected power source. Even so, it is critical that the camera should not produce much heat. To be practical, the camera must be able to dissipate enough power so as not to heat the eye to the point of discomfort.

Durability: The camera must be packaged in such a way as to be protected from the fluids in the eye.

Currently working with Georgia Tech University and experts at TI, the team at USC is busy making all this happen. Is such an ambitious project even possible? Although success has yet to be seen, the team envisions a successful completion of the project. And they have good reason to be confident for they are only just pressing at the edges of possibility.

Much of what lies ahead of us in the world of medicine is the identification of technologies and devices from other parts of our world that we can apply to medical electronics. For example, Prof. Armand R. Tanguay, Jr., Principal

Investigator on the "eyecam" project, acknowledges that they have many ideas about where else in the body they could implant a camera.*

Here a camera, there a camera,
In the eye a little camera.
Old Doc Donald had a patient.
E, I, E, I, O.

Certainly, there is more than one verse to this song. The question we might ask ourselves is what do we imagine we need next?

1.3 Making Health Care More Personal

A device that can help the blind to see is a life-changing application of medical technology. Not all medical devices will have such a dramatic effect on the way we live. Most of the changes in medical care will have a much lower profile, for they will be incorporated into our daily lives. However, while their application may be more subtle, the end result will certainly still be quite profound.

The future of medicine is based upon a firm foundation of existing technologies. What is new, in many cases, is not the technology itself, but rather how the technology is applied in new ways. Consider these key technologies:

- Digital imaging
- Telecommunications
- Automated monitoring

Each of these technologies is already firmly established in a number of disparate industries. Specifically applying them to medicine will still require creativity and hard work, but doing so will enable entirely new applications. Perhaps most importantly, for health care providers and their patients, the resulting advances will help shift health care into becoming a more routine part of daily life, creating a future where medical devices help us

1. Manage our chronic conditions
2. Predict our catastrophic diseases
3. Enable us to live out our final months/years in the comfort of our homes

* Unfortunately, we cannot print any of these exciting ideas without a nondisclosure agreement in place.

1.3.1 Advances in Digital and Medical Imaging

Improving health care is the ultimate goal behind advances in medicine. As medical imaging advances, it will allow patients to have more personalized and targeted health care. Imaging, diagnosis, and treatment plans will continue to become more specialized and customized to a patient's particular needs and anatomy. We may even see therapies that are tailored to a person's specific genetics. Look at how far we have come already:

The migration to digital files: Photographic plates were once used to "catch" x-ray images. These plates gave way to film, which in turn are now giving way to digital radiography. Through the use of advanced digital signal processing, x-ray signals now can be converted to digital images at the point of acquisition while imposing no loss in image clarity. Digital files have a variety of benefits, including eliminating the time and cost of processing film, as well as being a more reliable storage medium, which can be transferred near-instantaneously across the world.

Real-time processing: The ability to render digital images in realtime expands our ability to monitor the body. Using digital x-ray machines during surgical procedures, doctors can view a precise image at the exact time of surgery. Real-time processing also increases what can be done noninvasively. For example, the Israeli company CNOGA* uses video cameras to noninvasively measure vital signs such as blood pressure, pulse rate, blood oxygen level, and carbon dioxide level simply by focusing on the person's skin. Future applications of this technology may lead to identifying biomarkers for diseases such as cancer and chronic obstructive pulmonary disease (COPD).

Evolution from slow and fuzzy to fast and highly detailed: Today's magnetic resonance imagers (MRIs) can provide higher quality images in a fraction of the time it took state-of-the-art machines just a few years ago. These digital MRIs are also highly flexible, with the ability to image, for example, the spine while it is in a natural, weight-bearing, standing position. With diffusion MRIs, researchers can use a procedure known as tractography to create brain maps that aid in studying the relationships between disparate brain regions. Functional MRIs, for their part, can rapidly scan the brain to measure signal changes due to changing neural activity. These highly detailed images provide researchers with deeper insights into how the brain works—insights that will be used to improve treatment and guide future imaging equipment.

Moving from diagnostic to therapeutic: High-intensity focused ultrasound (HIFU) is part of a trend in health care toward reducing the impact of procedures in terms of incision size, recovery time, hospital stays, and infection risk. But unlike many other parts of this trend, such as robot-assisted surgery, HIFU goes a step further to enable procedures currently done invasively to be

* www.cnoga.com

done noninvasively. Transrectal ultrasound,* for example, destroys prostate cancer cells without damaging healthy, surrounding tissue. HIFU can also be used to cauterize bleeding, making HIFU immensely valuable at disaster sites, accident scenes, and on the battlefield. Focused ultrasound even has a potential role in a wide variety of cosmetic procedures, from melting fat to promoting formation of secondary collagen to eradicate pimples.

Portability of ultrasound: Ultrasound equipment continues to become more compact. Cart-based systems increasingly are complemented and/or replaced by portable and even handheld ultrasound machines. Such portability illustrates how, for a wide variety of health care applications, medical technology can bring care to patients instead of forcing them to travel. Portable and handheld ultrasound systems have also been instrumental in bringing health care to rural and remote areas, disaster sites, patient rooms in hospitals, assisted-living facilities, and even ambulances.

Wireless connectivity: Portability can be further extended by cutting cables. Putting a transducer, integrated beamformer, and wideband wireless link into an ultrasound probe will not only enable great cost savings by removing expensive cabling from the device but will also allow greater flexibility and portability. Further reducing cost and increasing portability enables more widespread use of digital imaging technology, enabling treatment in new areas and applications. A cable-free design also complements 3D probes, which have significantly more transducer elements and thus require more cabling, something that may become prohibitively expensive using today's technology.

Fusion of multiple imaging modalities: The fusion of multiple imaging modalities—MRI, ultrasound, digital x-ray, positron emission tomography (PET), and computerized tomography (CT)—into a single device provides physicians with more real-time information to guide treatment while reducing the time that doctors must spend with patients. For example, PET and CT are increasingly being combined into a single device. While the PET scan identifies growing cancer cells, the CT scan provides a picture of the location, size, and shape of cancerous growths.

Many of the real-time imaging modalities have greatly benefitted by advances in digital signal processors (DSP), devices which specialize in efficient, real-time processing. Specifically, the ability to exponentially increase the processing capabilities in imaging machines has enabled these advances to be useful in a hospital setting. However, to drive many of these applications into more widespread usage, another order of magnitude increase in processing capability will be necessary. This is a tall challenge.

Fortunately, silicon technology companies are now turning their attention to the world of medical electronics to meet these challenges. For instance, TIs

* www.prostate-cancer.org/education/novelthr/Chinn_TransrectalHIFU.html

formed its Medical Business Unit in 2007 to address the needs of the medical industry. This type of partnership between technology companies and the medical industry will help ensure that the exciting possibilities of the future that we envision will be realized.

1.4 How Telecommunications Complements Medical Imaging

Advances in medical imaging are frequently complemented by advances in communications networks. Together, they have significantly improved patient care, while also reducing costs for health care providers and insurance companies. The question is rapidly moving importance away from where we receive treatment to how we receive treatment.

Telemedicine is the concept where a patient's medical data are transported digitally over the network to a medical professional. For example, 24/7 radiology has begun to emerge as a commonly available service. Instead of maintaining a full radiological staff overnight, a hospital emergency room can now send an x-ray via a broadband internet link to Night Hawk Radiology Services* in Sydney, Australia, or Zurich, Switzerland. Night Hawk's staff then reads the x-ray and returns a diagnosis to the ER doctors.

For some patients, the combination of imaging and communications enables diagnosis that they otherwise would not receive for reasons such as finances or distance. A prime example is the work of Dr. Devi Prasad Shetty, a cardiologist who delivers health care via broadband satellite to residents of India's remote, rural villages who otherwise would not receive it simply because of where they live.† Today, one of Dr. Shetty's clinics can handle more than 3000 x-rays every 24 h. Shetty's telemedicine program has had a major impact in India, where an average of four people have a heart attack every minute.‡

In contrast to telemedicine, telepresence§ is where a medical professional virtually visits a patient through videoconferencing. Telepresence is increasingly used in both developed and developing countries to widen the distribution of health care. Videoconferencing is often paired with medical imaging systems, such as ultrasound, to enable both telepresence and telemedicine. Such applications frequently enjoy government subsidies because they bring health care to areas where it is expensive, scarce, or both.

One example of telemedicine is the Missouri Telehealth Network,¶ whose services include teledermatology. Using this service, a patient at a rural

* www.nighthawkrad.net
† www.financialexpress.com/news/Everyone-must-have-access-to-healthcare-facilities-Devi-Prasad-Shetty/42099/
‡ www.abc.net.au/foreign/stories/s785987.htm
§ www.telepresenceworld.com/ind-medical.php
¶ www.proavmagazine.com/industry-news.asp?sectionID=0&articleID=596571

health clinic can put his or her scalp under a video camera for viewing and diagnosis by a dermatologist hundreds of miles away. Videoconferencing equipment simulates a face-to-face meeting, allowing the doctor to discuss any conditions with the patient. In the case of someone with stage 1 melanoma, early detection via telemedicine may save his or her life.

Whether patients delay a doctor's visit because of distance, cost, available resources, being too busy, or even fear, telemedicine can mean the difference between suffering with a disease or receiving treatment. Virtual house calls, where physicians use videoconferencing and home-based diagnostic equipment to bring health care to a person without necessitating a visit to the doctor, can address most of these concerns.

Virtual house calls may be particularly attractive to patients in rural areas* or those in major cities with chronic traffic jams. Virtual house calls are also a way to bring health care to patients who otherwise would not be able to see a physician, perhaps because they are bedridden, suffering from latrophobia (fear of doctors), or have limited means of transportation. Whatever the hurdle they are helping overcome, virtual house calls are yet another example of how advances in medical technology increasingly are bringing care to the patient instead of the other way around.

For example, let us take a look at how medical technology may have been able to help the nineteenth century poet, Emily Dickinson, who was reclusive to the point that she would only allow a doctor to examine her from a distance of several feet as she walked past an open door. If she were alive today, she would greatly benefit from advances in medical imaging that could accommodate her standoffishness while still diagnosing the Bright's disease that ended her life at age 55.

Future medical technology will reach even further into our lives. Imagine a bathroom mirror equipped with a retinal scanner behind the glass that looks for retinopathy and collects vital signs. In the case of Dickinson, that mirror could have noticed a gradual increase in the puffiness of her face, a symptom of Bright's disease, and alerted her physician through an integrated wireless internet connection.

One of the underlying technologies behind medical imaging—DSPs—has a lot in common with that of telemedicine. DSPs play a key role in telemedicine. For example, DSPs provide the processing power and flexibility necessary to support the variety of codecs used in videoconferencing and telepresence systems. Some of these codecs compress video to the point that a TV-quality image can be transported across low-bandwidth wired or wireless networks, an ability that can extend telemedicine to remote places where the telecom infrastructure has limited bandwidth. In the future, compression will also help extend telemedicine directly to patients' homes over cable and DSL connections.

* www.columbiamissourian.com/stories/2007/05/12/improving-care-rural-diabetics

DSPs also provide the processing power necessary to support the lossless codecs required for medical imaging, since compression could impact image quality and affect a diagnosis. Another advantage DSPs offer is their programmability, which allows them to be upgraded in the field to support new codecs as they become available, thereby providing a degree of future-proofing for hospitals and physicians.

1.5 Automated Monitoring

Consider this short list of medical devices that can operate noninvasively within our homes:

- Bathroom fixtures with embedded devices could monitor for potential problems, such as a toilet that automatically analyzes urine to identify kidney infections or the progression of chronic conditions such as diabetes and hypertension.
- A bathroom scale could track sudden changes in weight or body fat and then automatically upload this data to a patient's physician. The scale could even trigger scheduling of an appointment based on a physician's predetermined criteria.
- Diagnostic devices such as retinal scanners could be coupled with a patient's existing consumer electronics products, such as a digital camera, to provide additional diagnostic and treatment options. If the device can connect to the network, the medical data collected could automatically be made available to medical personnel.
- Sensors in the home could measure how a person is walking to determine if he or she is at risk for a medical episode such as a seizure.
- Equipment could be connected to a caregiver's network for remote monitoring. One example of such a product under development is a gyroscope-based device worn by elderly patients to detect whether they have fallen.* Near-falls could trigger alerts to caregivers while being documented and reported to a patient's physician. This device could also track extended sedentary periods, which could be a sign of a developing physical or psychological problem.
- The term "personal area network" (PAN) may come to refer to the variety of devices that work together to regularly and noninvasively monitor and record a person's vital signs. Collected data could be automatically correlated to identify more complex medical conditions.

* http://ieeexplore.ieee.org/xpl/freeabs_all.jsp?tp=&arnumber=1019448&isnumber=21925

All of these examples will change how we approach practical medicine. Passive care of this nature becomes a "round-the-clock" service rather than something that occurs infrequently and disrupts our busy schedules. Constant monitoring also enables earlier identification of health problems before conditions can become irreversible, as well as eliminates the problem of patients not being conscientious about recording information about themselves. As a result, a patient may receive care that is more thorough than if he or she found time for an office visit every week or month.

This technology could be a viable way to improve care for patients who are too busy to schedule routine doctor's visits or, as is the case with Iatrophobics, whose fear of the doctor has them putting off regular check-ups. For the rest of us, who are either too busy, too unconcerned, or too lazy to schedule regular checkups, this technology can help ensure that we do not go too long before any changes get needed care. And, as health care becomes a continuous service through automated processes, the cost of delivering care will be substantially reduced as well.

1.6 Future of Technology

This is quite an impressive list of what is just around the corner. And while many of these applications may sound like inventions from a future that we can only hope for and dream about, it is likely that the reality will be even more exciting. To understand why this is the case, let us now shift our attention to the underlying factors that enable and drive innovation.

In the decades since the invention of the transistor, the IC, the microprocessor, and the DSP, technology has significantly impacted every part of our lives. And during this time, we have seen computers shrink from filling large air-conditioned rooms until they fit into our pockets. Now we are seeing the next stage of this progression as computers move from being dedicated devices in our pockets to becoming small subsystems that are integrated into other systems or even embedded in our clothing or our bodies.

Because it is technology that has brought us where we are, it is tempting to believe that technology is the driving factor behind medical imaging and innovation. We do not believe this is the case. For example, the reason that computers became smaller was not because technology allowed them to become smaller. The reason computers became smaller was because people found compelling reasons to make them that way. It is the need or desire for portable computing, not technology, that put computers like the Blackberry and IPhone into our pockets. This point is critical:

Technology does not drive innovation. Innovation drives technology.

1.6.1 Remembering Our Focus

It will be the wants and needs of the marketplace that determine the next technology that will go in our pockets as well as the things we can expect to be embedded in our clothing and bodies. In terms of advances in medical technology, it will be the needs of the patient that dictate what comes next.

Technology is exciting, and it is easy to forget what all of these amazing advances are really all about. For whether it is a retinal scanner in a bathroom mirror or a home ultrasound machine, it is the patient who is the greatest beneficiary. These advances enable health care to become more personal by bringing patients to doctors in their offices and doctors to patients in their homes. They also increase the effectiveness of health care by providing ways to identify diseases and other conditions before they become untreatable.

At the same time, these advances also allow health care to fade into the background and become a part of daily life. Imagine being scanned each morning while brushing your teeth instead of only during an annual check-up. That would be particularly valuable for patients with chronic or end-stage diseases, because it may allow them to live their lives without having to move into a hospice.

Health care revolves around the patient, as it should. And technology in turn will help us to address their key needs, as stated earlier in this chapter, to manage their chronic conditions, predict their catastrophic diseases, and allow them to live out the last days of their lives in the comfort of their own homes.

1.7 What We Can Expect from Technology

Knowing that need drives innovation allows us to approach technology from a different perspective, one where it is more relevant to discuss what these advances will be rather than to discuss what will make them possible. For example, we could speculate on the new process technologies that will overcome current IC manufacturing limitations. However, this will tell us little about what will be built with these future ICs. Rather, it is by exploring what we can reasonably expect from technology that we can gain insight into the possible future.

Let us begin with a well-known tenant from the world of ICs, Moore's law. Moore's law [1] forecasts that the number of transistors that can be integrated on one device will double every 2–3 years. Made in 1965 by Intel co-founder, Gordon E. Moore, in 1965, this prediction has not only been adopted by the industry as a "law" but, as shown in Figure 1.2, has also accurately described the progress of ICs for the past 40 years.

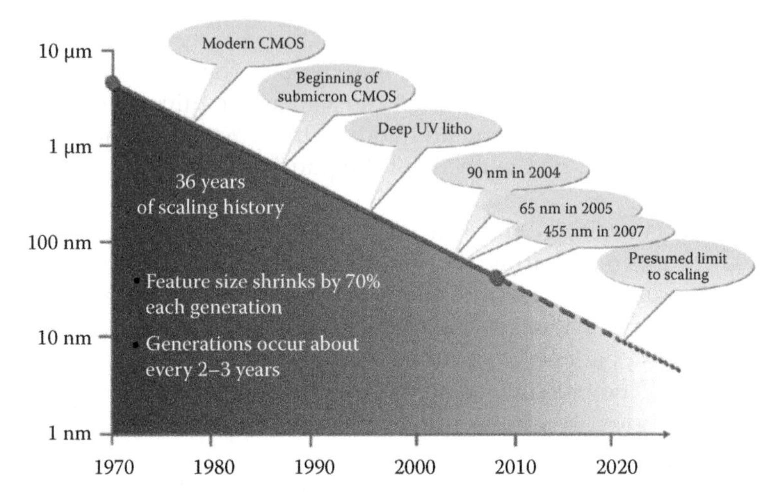

FIGURE 1.2
The trend for process technology over the past 40 years.

In practical terms, Moore's law has allowed us to reduce the price of ICs. Advances in performance and power dissipation have also affected cost, but over the past decade it has been advances in IC technology that have been primarily responsible for driving cost down.

As we look into the future, we can expect Moore's law to continue to hold true; that is, we should be able to integrate about twice the number of transistors on a piece of silicon every 2–3 years for about the same cost. What does this mean in terms of medical imaging and innovation? By the year 2020, the price of an IC could be as low as $1 for a billion transistors (a buck a billion). At that time, a high-end processor might cost $50.

Just imagine what could be done with 50 billion transistors.

Of course, this many transistors do not come without their issues:

1. The cost of developing a new IC may become prohibitively expensive.
2. Raw performance is no longer being driven by Moore's law.
3. Power dissipation must be actively managed.
4. Digital transistors are not particularly friendly to the analog world.

1.7.1 Development Cost

Back in the mid-1990s, TI introduced a new technology node for ICs. With this particular new technology node, we were able to integrate 100 million transistors on an IC. To put this in perspective, most personal computers in the 1990s had less than 10 million transistors, not counting memory. (As a short aside, advances in new process nodes are generally 0.7 times the size of previous nodes, which yields twice the number of transistors for the same die area.)

Being able to build an IC with this density does not answer the question of what will be done with them. The more difficult question, however, was what it would take to design a product with 100 million transistors. Consider the math. If we estimate that a design effort could be efficient enough such that the average time to design each transistor was about 1 h, such a project would take 50,000 staff years to complete. Clearly, starting a design of 100 million transistors from scratch would be virtually impossible. TI and other IC manufacturers solved and continue to solve this problem through intellectual property (IP) reuse. We will come back to this concept shortly.

On top of the excessive cost of an IC design, the cost of manufacturing tooling is on the order of 1 million dollars (We will overlook the billions of dollars spent on the IC wafer fab itself for the moment.). For high-volume applications, this cost spreads out to a manageable number. For example, a design with a total build of 1 million units would reduce the per-unit tooling cost to $1. For low-volume applications, however, the cost can become prohibitive: an IC design for a product anticipated to sell 10,000 units during its lifetime has a per-unit tooling cost of $100.

The obvious conclusion is that only those IC designs with extremely high volumes can be justified. Most applications, however, and not just those in the medical industry, have significantly smaller scope. Thus, these applications will not be based on ICs specifically designed for them but rather on standard programmable processors with application-specific software. And as tooling costs increase, this will be true for virtually all products developed in the future.

Because standard programmable processors allow IP to be implemented in software, the tooling cost of a programmable device can be shared among many different products, even across industries (i.e., medical imaging, high-end consumer cameras, industrial imaging, and so on). Software customizes the processor, so to speak; and the more applications a processor can serve, the lower its cost.

The trade-off of implementing IP in software is, from an engineering perspective, a fairly "sloppy" way of designing a product, meaning that the final design will require far more transistors than if an application-specific IC is used. Given the advanced state of IC process technology, however, this does not matter. Back in the 1990s, we had more processors than we knew what to do with. Today, we can build processors with more capacity than we can use. And in the year 2020, we will still have more transistors than our imaginations can exploit.

1.7.2 Performance

For years, the performance of ICs has seemed to be driven by Moore's law, just as cost was. However, if we measure raw performance—that is, the number of cycles a processor could execute—it actually drifted from Moore's law in the early 1990s. Despite this, processors have still doubled in effective performance in accordance with Moore's law through sophisticated changes to processor architectures such as deeper pipelines and multiple levels of cache memory. These changes came at their own cost—lots of transistors. Fortunately, as stated previously, we have plenty of those.

Improving performance through sophisticated changes to a processor's architecture is, in some ways, just a fancy way of saying that an architecture has been made more efficient. Deeper pipelines, for example, eliminate the inefficiencies of processing a single instruction by simultaneously processing multiple instructions. Eliminating inefficiencies only goes so far, though. Caches improved memory performance significantly, and caches for caches squeezed out a bit more performance, but having caches for caches actually begins to slow things down.

There are still many opportunities for increasing processor performance through architectural sophistication, but to achieve a major increase in performance, the industry is moving toward multiprocessing. Also referred to as multicore, the central idea is that for many applications, two processors can do the job (almost) twice as fast as one. Multiprocessing, while yielding a whole new level of performance, also introduces a whole new level of complexity to processor architectures. And as we continue to add complexity, we then must create more complex development environments to hide the complexity of the architecture from developers.

1.7.3 Multiprocessor Complexity

To understand the effect of the complexity that multiprocessing imposes on design, we need to take a look at Amdahl's law [2]. Simply stated, Amdahl's law says that sometimes using more processors to solve a problem can actually slow things down. Consider a task such as driving yourself from point A to B. There is really no way to use multiple cars to get yourself there any faster. In fact, using multiple cars along the way will likely slow you down as you stop, switch cars, and then get up to speed again, not to mention the traffic jam you would have created.

The same problem applies to an algorithm that cannot be parallelized—that is, easily distributed across multiple processors. Splitting such an algorithm across equal performance processors will slow down overall execution because of the added overhead of breaking up the task across the multiple cores. Engineers describe these types of tasks as being "Amdahl unfriendly."

Amdahl-friendly tasks are those which can be easily broken into multiple smaller tasks that can be solved in parallel. These are the types of tasks for which DSPs are well suited. Consider how a video signal can be broken into small pieces. Since each piece is relatively independent of the others, they can be processed in parallel/simultaneously.

The size of each piece depends upon the processors being used. For example, TIs developed the Serial Video Processor (SVP) [3] for the TV market. The SVP contained 1000 1 bit DSPs each simultaneously processing one pixel in

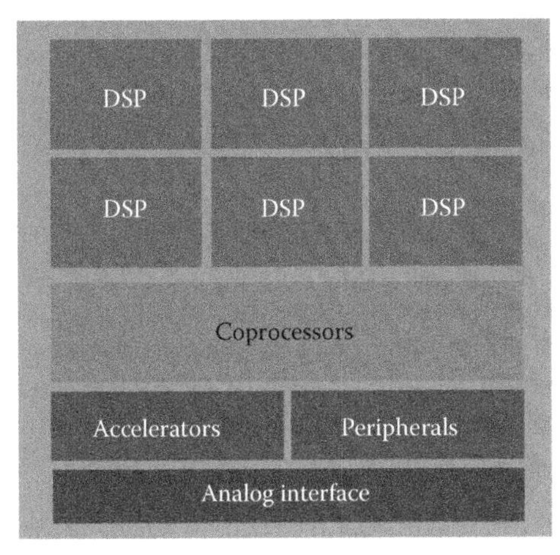

FIGURE 1.3
Higher-performance through parallelism—More DSPs + flexible coprocessors.

a horizontal line of video. Because Amdahl-friendly tasks like these can be parallelized, it is easier to determine how to architect the multiprocessing system as well as how to create the development environment with which to design applications that exploit it.

Amdahl-unfriendly tasks by far are much harder to solve as they prove difficult to divide into parts. Much of the research related to multiprocessing going on now in universities is focused on addressing how to approach such seemingly difficult problems (Figure 1.3).

1.7.3.1 Multiprocessing Elements

Even given the limitations of Amdahl's law, the best approach for increasing performance appears to be through multiprocessing. Before we can begin to consider how best to take advantage of multiple processors in the same system, however, we first need to ask, "What is a processor?" There are, after all, many different types of processing elements:

- *General-purpose processors*: Examples include ARM cores, MIPS cores, and Pentium-class processors.
- *Application-specific processors*: Examples include DSPs and graphic processing units (GPUs).
- *Programmable accelerators*: Examples include floating point units (FPUs) and video processors.
- *Configurable accelerators*: These are similar to programmable accelerators in that they can perform a range of specific tasks such as filtering or transforms.

- *Fixed-function accelerators*: These are also similar to programmable accelerators with the exception that they perform only a single task such as serving as an anti-aliasing filter for an audio signal.
- *Programmable hardware blocks*: Examples include FPGAs, PLDs, etc.

In general, the term "multiprocessing" refers to heterogeneous (different elements) multiprocessing while "multicore" refers to homogeneous (same elements) multiprocessing. The importance of this distinction is greater when talking about DSP algorithms than for more general-purpose applications. In a DSP application, there is more opportunity to align the various processing elements to the tasks that need to be accomplished. For example, an accelerator designed for audio, video/imaging, or communications can be assigned appropriate tasks to achieve the greatest efficiency. In contrast, very large, generic algorithms may be best implemented using an array of identical processing elements.

Despite all of its challenges, multiprocessing appears to be one of the next major advances that will shape IC, electronics, and medical equipment design. Today, multiprocessing is still not well understood, and progress will likely be slow as advances percolate out of university research laboratories into the real world over the next decade or so. In the meantime, we have to get used to the hit-and-miss nature of multiprocessing architectures and do our best to use them as efficiently as we can.

1.7.4 Power Dissipation

In relation to price and performance, power dissipation is the new kid on the block and where much research is starting to be focused. TI's first introduction to this important aspect of value goes back to the mid-1950s when its engineers developed the Regency radio [4] to demonstrate the value of the silicon transistor. This was the first transistor radio, and its obvious need for battery operation made it important to demonstrate low power dissipation.

Power dissipation again became important with the arrival of the calculator in the 1970s. Although most uses for these early calculators allowed for them to be plugged into a wall, sockets were not always nearby or convenient to use. The subsequent movement to LCD calculators with solar cells made low power an even more important requirement.

Lower power dissipation became a primary design constraint in the early 1990s with the arrival of the digital cellular phone. Early customers in this new market made it clear to TI that if power dissipation was not taken seriously, they would find another vendor for their components.

With this warning, TI began its now 20 year drive to reduce power dissipation in its processors. One of the results of this reduction in power is the creation of processors that have helped revolutionize the world of ultrasound

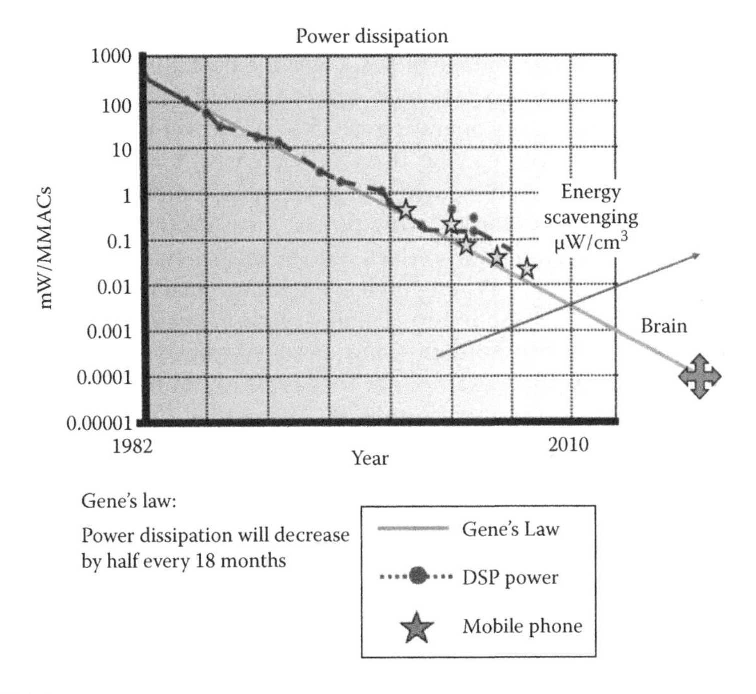

FIGURE 1.4
Gene's law. This graph shows the trend of power efficiency over time with efficiency measured as mW/MMAC. The MMAC is the base unit of DSP performance. The upward trending line on the right represents our ability to scavenge energy from the environment.

by turning the once bulky, cart-based ultrasound systems into portable and even handheld systems.

Figure 1.4 shows how power dissipation has improved over the history of DSP development. Measured in units of DSP performance—the MMAC (millions of multiplies and accumulates per second)—it shows how power dissipation has been reduced by half every 18 months. As this chart was created by Gene Frantz, Principal Fellow at TI, this trend of power efficiency over time has come to be known as Gene's law.

It should be noted that the downward efficiency trend flattened a few years ago. This occurred because of issues with IC technology where leakage power was at parity with active power. As with any problem, once understood, it was able to be resolved. Now, we are back on the downward trend of power dissipation per unit of performance.

1.7.4.1 Lower Power into the Future

As we look to the future, the question is whether IC technology will continue to follow Gene's law. There are several reasons to believe it will. The two that seem to have the most promise are lower operating voltages and the availability of additional transistors.

Much of the downward trend for power dissipation has been, in fact, due to lowering the operating voltage of ICs. Over the past 20 years, device voltage has gone from 5 to 3 to 1 V. Nor is this the end of the line. Ongoing research predicts processors operating at 0.5 V and lower [5].

Just as more transistors can be used to increase the performance of a device, they can also be used to lower the power dissipation for a given function. The simplest method for lowering power consumption is using what is known as *the father's solution*: who does not remember that loud voice resonating through the house, "When you leave a room, turn off the light!"

This wisdom is easily applied to circuit design as well—simply turn off sections of the device, especially the main processor, when they do not need to be in operation. A good example of this type of management is implemented by the MSP430 family of products. And while it does require more transistors to turn sections on and off, fortunately, it is only a few, making this an extremely efficient approach to power management.

Power can also be managed through multiprocessing, although this approach requires many more transistors. Consider that a task performed on a single processor running at 100 MHz can be performed as quickly on two of the same processors running at 50 MHz. Given the nature of power, one processor running at 100 MHz will consume the same power as two processors running at 50 MHz so long as they are operating at the same voltage.

Due to the characteristics of IC technology, however, the 50 MHz processors can operate at a lower voltage than the 100 MHz processor. Since the power dissipation of a circuit is reduced by the square of the voltage reduction, two 50 MHz processors running at a lower voltage will actually dissipate less power than their 100 MHz equivalent.

The trade-off for managing power through multiprocessing is that two slower processors require twice the number of transistors as a single, faster processor. And again, given that we can rely upon having more transistors as we move into the future, multiprocessing is a feasible method for reducing power dissipation.

1.7.4.2 Perpetual Devices

Confident that IC power dissipation will continue to go down, we can begin to think about how we might take advantage of that. One interesting corresponding area of research which is receiving a lot of attention is energy scavenging. Energy scavenging is based on the concept that there is plenty of environmental energy available to be converted into electrical energy [5]. Energy can be captured from light, walls vibrating, and variations in temperature, just to name a few examples of sources of small amounts of electrical energy.

Combining energy scavenging with ultra-low-powered devices gives us the concept of "perpetual devices." Back in Figure 1.4, there is an upsloping line that represents our ability to scavenge energy from the environment. At the point this red line crosses the power reduction curve, we will have the ability to create devices that can scavenge enough energy from the environment to operate without a traditional plug or battery power source.

Imagine the medical applications for perpetual devices. Implants once not feasible because of the need to replace batteries will be possible. Pacemakers will be able to support a wireless link to upload data, eyecams will be permanent once installed, and we may even see roving sensors that travel through our bodies monitoring our heart while cleaning our arteries.

1.8 Integration through SoC and SiP

So far we have addressed the three "P"s of value: price, performance, and power. The final aspect of IC technology that will serve as a foundation of medical imaging into the future is integration. Integration refers to our ability to implement more functionality onto a single IC as the number of transistors increases. Many in the electronics industry believe that the ultimate result of integration will be what is referred to as an SoC or "system on a chip." There is no nice way to say this: They are wrong.

To understand this position, let us look back at history. When TI began producing calculators, the initial goal was to create a "single-chip calculator." This may be surprising, but such a device has never been produced by TI or anyone else. The reason is that we have never figured out how to integrate a display, keyboard, and batteries onto an IC. What we did develop was a "subcalculator on a chip." It is important to catch the subtlety here. We did not develop the whole system, just a part of it. Perhaps, it would be more accurate to use the term subsystems on chip (SSOC).

The same is true when we look at technology today. No one creates complete systems on chip, and for many good reasons. The best, perhaps, is that in practical terms, by the time we develop an SoC, we find it has become a subsystem of a larger system. Put another way, once a technology makes sense to implement as an IC (i.e., it has passed the high-volume threshold required to reduce tooling to a reasonable per-unit cost), it has likely been found to be useful in a great variety of applications.

And this leads us to the real focal point of system integration in the future—the system in package (SiP). Figure 1.5 shows that the roadmap of component integration can be simplified to three nodes. The first node—the design is built on a printed circuit board (PCB)—is well understood by system designers. At the second node, SiP, all of the components are integrated "upward" by stacking multiple ICs into one package. It is at the third and final node that all of the components are integrated onto a single IC using SoC technology.

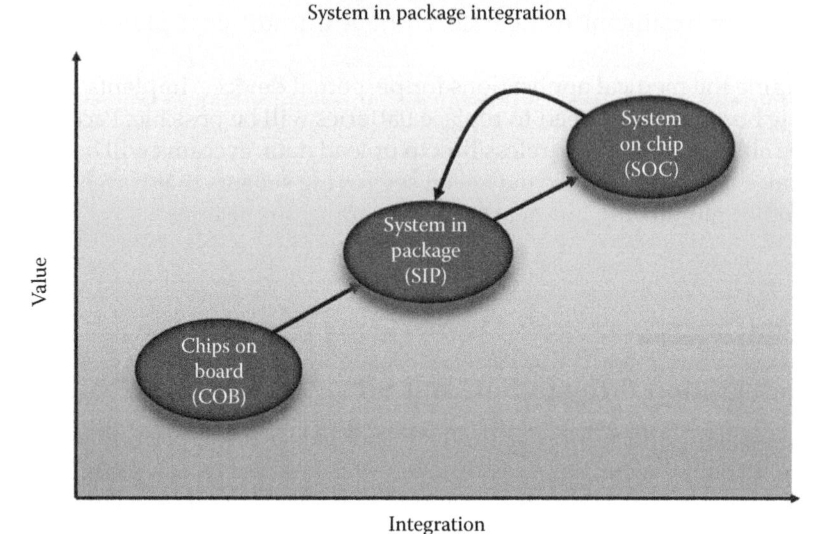

FIGURE 1.5
A simple roadmap of component integration.

Again, once a system warrants its own SoC, invariably it is being designed back into larger systems. For this reason, technology is not stable at the SoC node, and it settles back either to being placed directly onto a PCB (node 1) or into a package (node 2). As we continue to increase the number of transistors available and lower their cost, these subsystems and the IP they represent will increasingly settle into a package (SiP) and less onto a PCB.

While compelling in itself, this is not the only reason for moving away from the SoC as the ultimate way to design systems. Consider that each advance in digital IC process technology delivers more transistors and perhaps operates at a lower voltage. But the real world is not digital, nor does it follow this trend, and analog circuits, as well as RF circuits, seem to favor higher voltages.

To create a whole IC system on a single piece of silicon requires using a single process technology. However, implementing digital circuits in an analog or RF process significantly increases the relative cost of the digital circuit. Likewise, implementing analog and RF in a digital process substantially reduces signal integrity. The only way we will be able to efficiently integrate all aspects of the system into one "device" is to implement each circuit in its appropriate process technology and combine the various ICs in a single package using SiP technology.

The fact that SiP is more efficient than SoC actually turns out to be really good news for we will be able to take "off-the-shelf" devices and stack them in one package that provides virtually all of the advantages an SoC would.

The primary difference—and it is significant—is that by using SiP technology we will be able to manufacture devices within months rather than the years required to produce a new SoC design.

And, in much the same way that programmable processors can bring the cost economies of high volume to specialized applications, developers will be able to create highly optimized, application-specific SiPs as if they were standard ICs. This will give rise to a new product concept, the "boutique IC," where system designers can select from a variety of "off-the-shelf" ICs and "integrate" them into a single package for about the same cost as for the individual ICs. The resulting SiP will have the advantage of faster time-to-market, a smaller footprint, and a "living" specification where new SoCs can be integrated regularly to continually cost-reduce designs. This will certainly give new meaning to the concept of "one-stop shopping" for components.

1.9 Defining the Future

We have covered a lot of ground in our discussion of the future. When you first read about many of the medical applications we have suggested in this chapter, perhaps you thought they sounded more like science fiction than fact. However, most of these devices build upon existing, proven technologies such as digital imaging, telecommunications, and automated monitoring that will change how we approach medicine.

For the common person, and even latrophobics such as Emily Dickinson, new, noninvasive techniques that are increasingly available in the comfort of their home will make the difference in diagnosing and treating diseases before they become debilitating or life-threatening. For people who live in remote locations, telemedicine will bring doctors and patients together in new and powerful ways.

The underlying IC technology required to bring many of these devices to reality is already, or will soon be, available. Moore's law will continue to give us more transistors than we can conceivably use. More general-purpose devices based on software programming models will enable the volumes that result in these transistors being available at a reasonable cost. Using some of those extra transistors to create multiprocessing circuits will provide higher performance and lower power dissipation. Finally, SiP technology will make possible smaller, more integrated designs as well as potentially enable an entirely new way to design ICs.

As a result, technology is not the limiting factor defining the future of medical imaging. Quite the opposite, technology is more the sandbox in which we can design the creations inspired by our imagination. Innovation is driven not by the fact that we can shrink a computer to fit in our pocket or our body, but rather by the fact that we need or want to shrink that computer.

For the medical industry, it is all about the patient. And what applications will arise and how fast they will manifest depends upon what patients need. Technology, for its part, will comply by providing us everything we need to make what we envision become real. So what will the future of medical imaging bring? The future will be whatever we want to make it.

The miracles are just beginning.

References

1. G.E. Moore, Cramming more components onto integrated circuits, *Electronics*, 38(8), 4–7, April 19, 1965.
2. A. Gene, Validity of the single processor approach to achieving large-scale computing capabilities, *AFIPS Conference Proceedings*, (30), 483–485, 1967.
3. Texas Instruments, SVP for digital video signal processing, Product overview, 1994, SCJ1912, Dallas, TX.
4. A.P. Chandrakasan et al., Low-power CMOS digital design, *IEEE J. Solid-State Circ.*, 27(4), 473–484, April 1992.
5. S. Roundy, P.K. Wright, and J.M. Rabaey, *Energy Scavenging for Wireless Sensor Networks*, Kluwer Academic Press, Boston, MA, 2003.

2

Ultrahigh-Speed Real-Time Multidimensional Optical Coherence Tomography

Kang Zhang and Jin U. Kang

CONTENTS

2.1 Introduction to Optical Coherence Tomography

Optical coherence tomography (OCT) is a novel imaging modality capable of providing depth/time-resolved images of biological tissues noninvasively with micron-level resolution [1]. Since its invention in early 1990s, OCT has been viewed as an "optical analogy" of ultrasound sonogram (US) imaging, sharing many basic concepts of ultrasound such as the three fundamental imaging modes (A, B, and C scan), as illustrated in Figure 2.1.

In current clinical practice, visualization guidance during microsurgeries—such as ophthalmologic, neurological, and otolaryngologic—is realized via surgical microscopes, which limits the surgeon's field of view (FOV) and causes limited depth perception of microstructures and tissue planes beneath the surface. For example, during an internal limiting layer (ILM) peeling operation [2]—a very common type of retinal surgery—the entire thickness of human retina, consisting of more than 10 layers, is about 200–300 μm and the ILM is as thin as 1–3 μm. Therefore, such an operation is extremely challenging merely with the help of the surgical microscope, and requires rigorous and long-term training for surgeons.

As listed in Table 2.1, several well-developed imaging modalities—magnetic resonance imaging (MRI), x-ray computed tomography (CT), and ultrasound sonogram (US)—have so far been used in image-guided intervention (IGI) for many kinds of surgeries [3]. These conventional modalities,

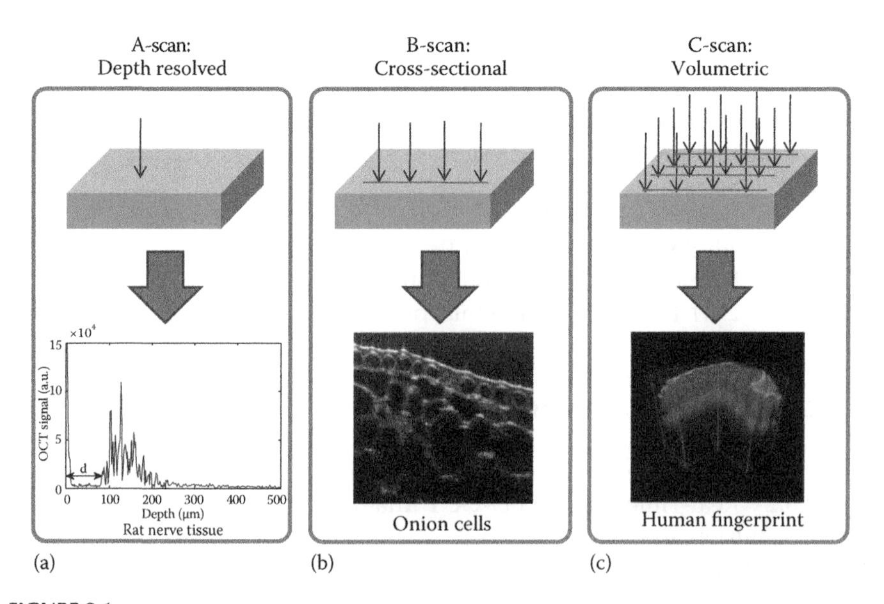

FIGURE 2.1
Three fundamental imaging modes of OCT. (a) A-scan, depth resolved; (b) B-scan, cross-sectional; and (c) C-scan, volumetric.

TABLE 2.1

Typical Parameters of 3D Biomedical Imaging Modalities

	Imaging Modality	Resolution	Imaging Depth
Conventional modality	Magnetic resonance imaging (MRI)	1–3 mm	Whole body
	X-ray computed tomography (CT)	1–3 mm	Whole body
	Ultrasound sonogram (US)	0.5 mm	10–20 cm
Optical modality	Two-photon fluorescence microscopy	<1 μm	<1 mm
	Confocal microscopy	<1 μm	<1 mm
	Optical coherence tomography (OCT)	1–10 μm	1–10 mm

however, do not work effectively for microsurgical applications due to their limited resolution, around 1 mm.

Optical imaging modalities, as listed in Table 2.1, offer microns or submicron-level resolutions and 3D imaging capability. Although two-photon fluorescence microscopy and confocal microscopy could achieve submicron-level resolutions, their imaging depth is usually limited to several hundreds of microns; moreover, their volumetric imaging speed is slow due to the scanning protocol. In comparison, OCT can achieve much deeper imaging depth of several millimeters while maintaining micron-level resolution, which makes it highly suitable for microsurgery guidance [4].

As early as the late 1990s, the concept of interventional OCT for surgical guidance was proposed using time-domain OCT (TD-OCT) at a slow imaging speed of hundreds of A-scans/s [5]. With the technological breakthroughs in Fourier domain OCT (FD-OCT) during the past decade, an ultrahigh speed OCT system is now generally available with >100,000 A-scan per second [6–14]; even mega-A-scan/second OCT was achieved [15]. Such ultrahigh acquisition speed enables time-resolved volumetric (4D) recording and reconstruction of dynamic processes such as eye blinking, papillary reaction to light stimulus [9,10], and embryonic heart beating [11–13].

For a clinical interventional imaging system, ultrahigh speed real-time imaging acquisition, reconstruction, and visualization are all essential. Several parallel processing methods have been implemented to improve image reconstruction of FD-OCT: field-programmable gate array (FPGA) has been applied to both spectrometer and swept source-based FD-OCT systems [16,17]; multicore CPU parallel processing has also been implemented [18,19]. Recently, cutting-edge general-purpose computing on graphics processing units (GPGPU) technology [20,21] has been used in FD-OCT [22–30]. Compared to FPGA and multicore processing methods, GPGPU acceleration is more cost-effective in terms of price/performance ratio and convenience of system integration.

In this chapter, the physical principles of OCT technologies will be reviewed first. Then, an ultrahigh speed FD-OCT system will be demonstrated based on our previous work, and a series of GPGPU-based FD-OCT image reconstruction and visualization technologies will be presented with imaging results.

2.2 Principles of OCT Technology

2.2.1 Optical Coherence and Interference

Optical interferometers are based on mutual coherence and interference between light beams. Figure 2.2 illustrates the principles of several commonly used optical interferometers. In this chapter, we will focus on Michelson interferometer-based OCT and introduce corresponding physical principles.

As in Figure 2.2a, with an input optical wave A_0, the optical detector measures the inference between the two output optical waves A_1 and A_2; the detected average intensity can be expressed as [31]

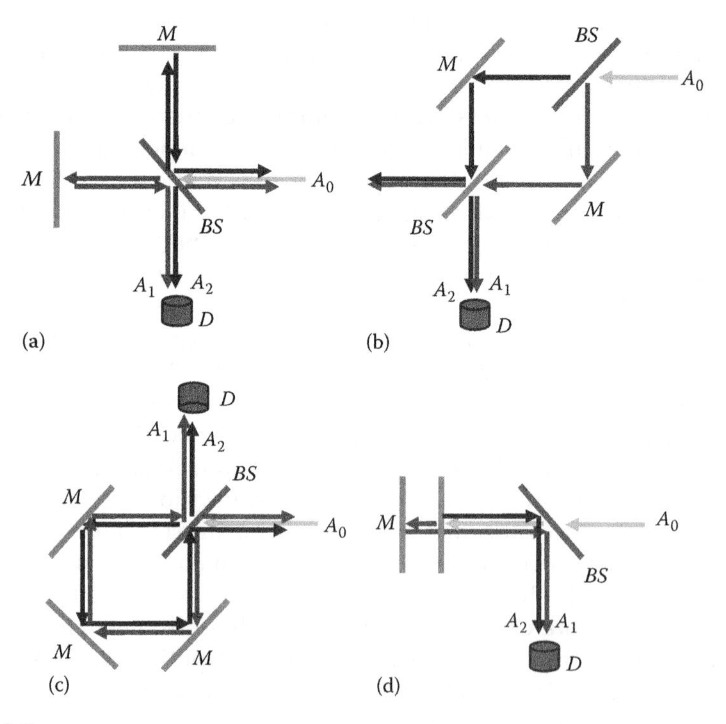

FIGURE 2.2
Several types of commonly used interferometers: (a) Michelson, (b) Mach-Zehnder, (c) Sagnac, and (d) Fizeau. D, optical detector; M, mirror; BS, beam splitter.

$$I = \left\langle \left|A_1 + A_2\right|^2 \right\rangle = \left\langle \left|A_1\right|^2 \right\rangle + \left\langle \left|A_2\right|^2 \right\rangle + \left\langle A_1^* A_2 \right\rangle + \left\langle A_1 A_2^* \right\rangle$$
$$= I_1 + I_2 + 2\mathrm{Re}\{G(\tau)\} = I_1 + I_2 + 2\left|G(\tau)\right| \cos \varphi(\tau) \tag{2.1}$$

where $G(\tau)$ is the coherence function of the light source.

2.2.2 Time Domain OCT (TD-OCT)

The time domain OCT system was first introduced by David Huang et al. from MIT in 1991 [1], based on the temporal low-coherence Michelson interferometer, as illustrated in Figure 2.3. Here, a broadband light source with a Gaussian power spectrum density $S(\nu)$ was used as the light source, and the imaging sample was modeled to be a multilayer target with reflective interfaces R_i. For each A-scan, the reference mirror moves along the axial direction and therefore interferes with each R_i within the image range ΔZ. With the time delay τ associated to the spatial displacement of a certain interface R_i relevant to the fixed reference plane, where $\tau = \Delta z / c$, Equation 2.1 can be expressed as

$$I = I_r + I_i + 2R_i \left|G(\tau - \tau_i)\right| \cos[2\pi\nu_0(\tau - \tau_i) + \varphi_a(\tau - \tau_i)]$$
$$= I_r + I_i + \frac{2R_i}{c}\left|G(\Delta z - \Delta z_i)\right| \cos[2\pi\nu_0(\Delta z - \Delta z_i) + \varphi_a(\Delta z - \Delta z_i)] \tag{2.2}$$

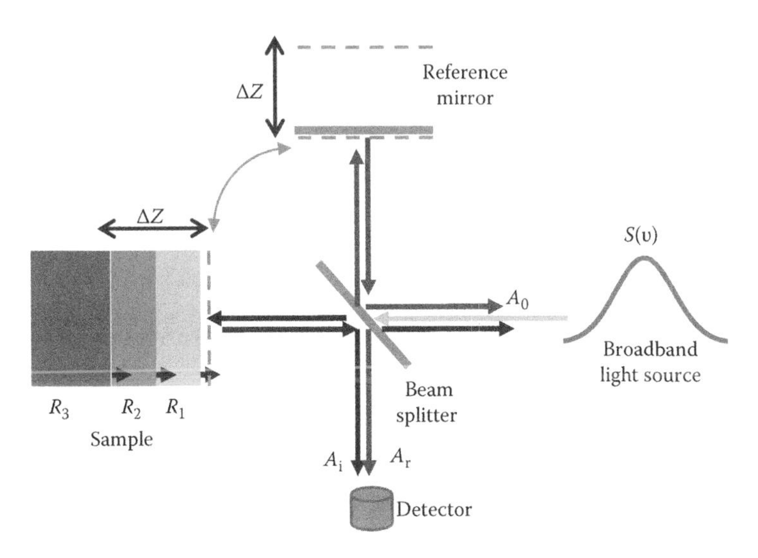

FIGURE 2.3
Principle of a time domain OCT (TD-OCT) system.

which gives the response of a single interface, or the point spread function (PSF) of the TD-OCT system. In general, for a spectrum $S(\nu)$ with Gaussian spectral shape and central frequency ν_0, according to the Fourier transform relation, $G(\tau)$ can be expressed as

$$G(\tau) = \int_0^\infty S(\nu) e^{-j2\pi\nu\tau} d\nu \tag{2.3}$$

With a normalized $S(\nu)$ as in reference [32]

$$S(\nu) = \frac{2\sqrt{\ln 2/\pi}}{\Delta\nu_c} \exp\left\{ -4\ln 2 \left(\frac{\nu - \nu_0}{\nu} \right)^2 \right\} \tag{2.4}$$

Then, $G(\tau)$ can be expressed as

$$G(\tau) = \exp\left\{ -\left(\frac{\pi\Delta\nu_c\tau}{2\sqrt{\ln 2}} \right)^2 \right\} * \exp\{-j2\pi\nu_0\tau\} \tag{2.5}$$

which gives the FWHM axial resolution as

$$\Delta z_{TD\text{-}OCT} = \frac{2\ln 2}{\pi} \frac{c}{\Delta\upsilon_c} = \frac{2\ln 2}{\pi} \frac{\lambda_0}{\Delta\lambda} \tag{2.6}$$

For multiple interfaces with reflectivity R_i, Equation 2.2 turns to

$$\begin{aligned} I = I_r + \sum_i I_i + \frac{2}{c} \sum_i R_i \left| G(\Delta z - \Delta z_i) \right| \cos\left[2\pi\nu_0 (\Delta z - \Delta z_i) + \varphi_a(\Delta z - \Delta z_i) \right] \\ + \frac{2}{c} \sum_{ij} R_i R_j \left| G(\Delta z_i - \Delta z_j) \right| \cos\left[2\pi\nu_0 (\Delta z_i - \Delta z_j) + \varphi_a(\Delta z_i - \Delta z_j) \right] \end{aligned} \tag{2.7}$$

Here, the last term in Equation 2.7 represents the autocorrelation noise.

2.2.3 Fourier Domain OCT

Fourier domain OCT (FD-OCT) was first demonstrated by Fercher in 1995 [33], which has rapidly developed and now dominates OCT technology. In general, the sensitivity of FD-OCT exceeds that of TD-OCT by more than two orders of magnitude, as well as the imaging speed (usually measured by the A-scan rate) [34]. Depending on different light source and detection mechanism, FD-OCT can be cataloged to two types, spectrometer-based spectral domain OCT (SD-OCT) and swept-laser-based swept-source OCT (SS-OCT), as illustrated in Figure 2.4. These two types of FD-OCT follow the same principle. In this chapter, we will focus on spectrometer-based SD-OCT (simply referred as "FD-OCT" in the following text).

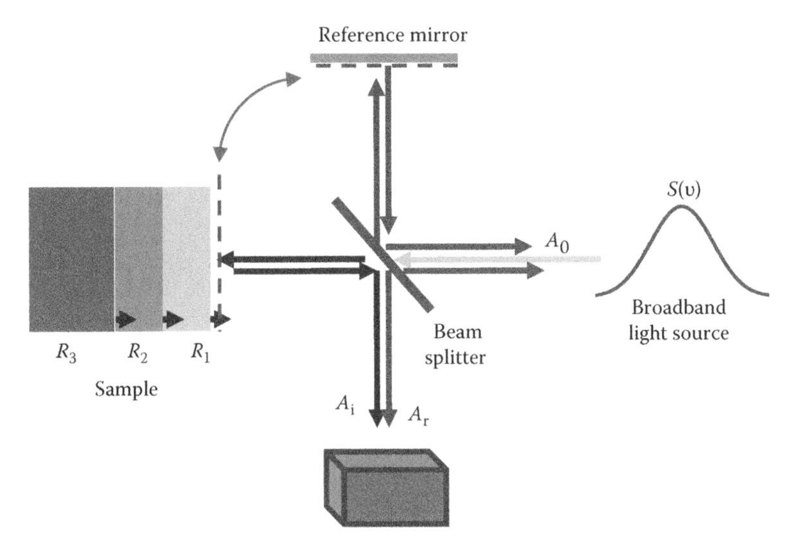

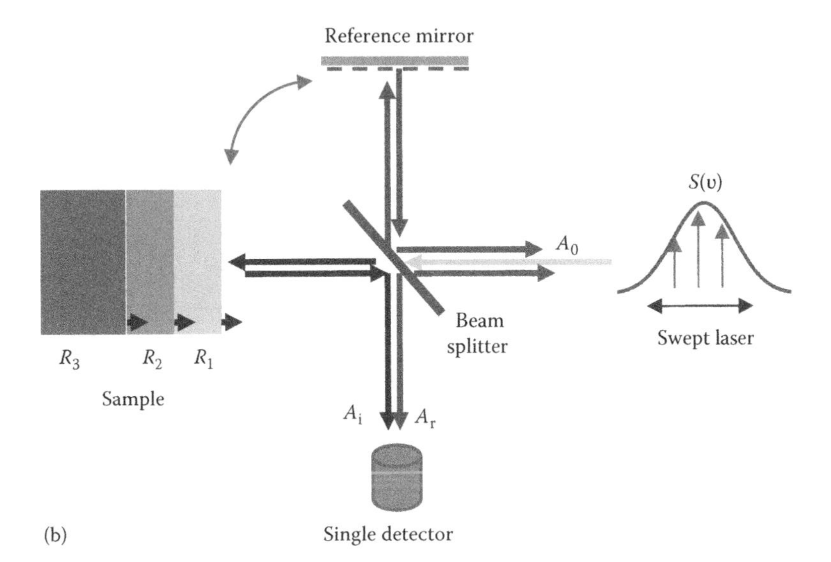

FIGURE 2.4
FD-OCT systems: (a) spectrometer-based spectral domain OCT (SD-OCT) and (b) swept-laser-based swept-source OCT (SS-OCT).

2.2.3.1 A-Scan Depth-Resolved Imaging

Figure 2.5 schematically explains the principle of an FD-OCT A-scan signal generation: each interface R_i in the multilayered sample will produce an interference pattern (interferogram) on the input broadband spectrum, noted as $S_i(\nu)$, and the spectrometer-detected spectrum is a combined interference pattern contributed by all layers. A single A-scan signal is then reconstructed by taking Fourier transform (FT) of the final combined interferogram.

Following the notations in Section 2.2.2, the final measured spectrum can be expressed as [35]

$$S_{combined}(\nu) = \rho(\nu) * S(\nu) * \left[R_r + \sum_i R_i + 2\sum_i \sqrt{R_r R_i} \cos(2\pi\nu\tau_i) + 2\sum_{i,j} \sqrt{R_i R_j} \cos(2\pi\nu(\tau_i - \tau_j)) \right] \quad (2.8)$$

where $\rho(\nu)$ is the spectral response of the spectrometer's detector array (CCD or CMOS), which is usually nonlinear with ν. Assuming a uniform $\rho(\nu)$, the A-scan signal $D(\tau)$ can be reconstructed via Fourier transformation of $S_{combined}(\nu)$ as

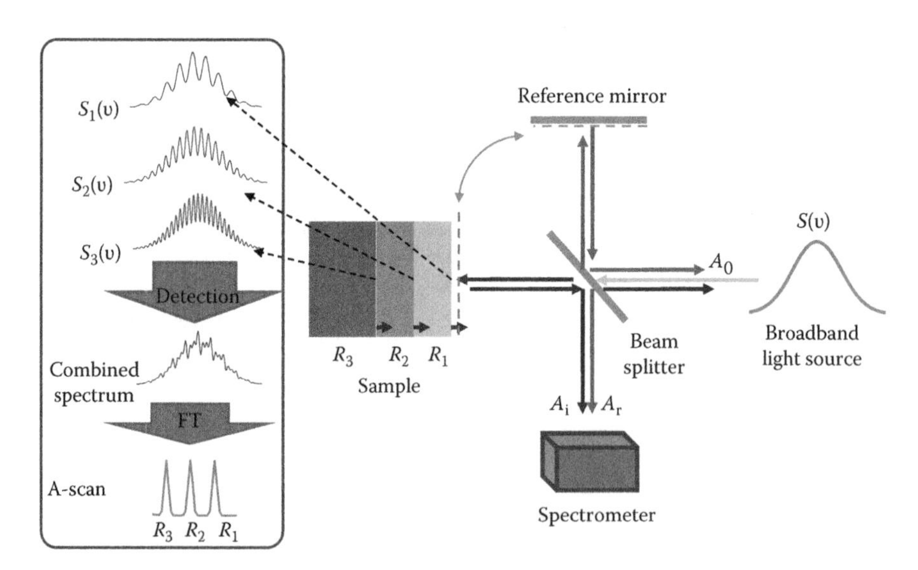

FIGURE 2.5
Working mechanism and signal processing of FD-OCT.

$$
\begin{aligned}
D(\tau) &= FT_{\nu\to\tau}\{S_{combined}(\nu)\} \\
&= \left(R_r + \sum_i R_i\right)G(\tau) + \sum_i \sqrt{R_r R_i}\,(G(\tau)\otimes\delta(\tau\pm\tau_i)) \\
&\quad + \sum_{i,j}\sqrt{R_i R_j}\,(G(\tau)\otimes\delta(\tau\pm(\tau_i-\tau_j))) \\
&= \frac{1}{c}\left(R_r + \sum_i R_i\right)G(\Delta z) \\
&\quad \text{(DC component)} \\
&\quad + \frac{1}{c}\sum_i \sqrt{R_r R_i}\,(G(\Delta z)\otimes\delta(\Delta z\pm\Delta z_i)) \\
&\quad \text{(Cross-correlation component)} \\
&\quad + \frac{1}{c}\sum_{i,j}\sqrt{R_i R_j}\,(G(\Delta z)\otimes\delta(\Delta z\pm(\Delta z_i-\Delta z_j))) \\
&\quad \text{(Auto-correlation component)}
\end{aligned} \tag{2.9}
$$

Here, the third term in Equation 2.9 stands for the desired signal for Michelson-type FD-OCT imaging while the other three terms are DC, cross-correlation, and auto-correlation noise. The function $\delta(\Delta z \pm \Delta z_i)$ models the sample's multilayered reflectivity R_i.

2.2.3.2 Imaging Resolution

From Equation 2.9, the A-scan signal is actually the convolution of the system's PSF $G(\Delta z)$ and the sample's structure function $\delta(\Delta z \pm \Delta z_i)$; therefore, the axial resolution of FD-OCT is the same as TD-OCT according to Equation 2.6, as

$$
\Delta z_{FD-OCT} = \frac{2\ln 2}{\pi}\frac{\lambda_0}{\Delta\lambda} \tag{2.10}
$$

It is also necessary to mention that the lateral resolution of both TD-OCT and FD-OCT is the same as determined by the numerical aperture (N.A.) of the imaging lens, as

$$
\Delta x_{OCT} = \frac{2\sqrt{\ln 2}}{\pi}\frac{\lambda_0}{N.A.} \tag{2.11}
$$

2.2.3.3 *Imaging Depth*

Here, we only discuss the theoretical maximum imaging depth of FD-OCT, excluding the sample's scattering and absorption effect, which will be discussed separately in Section 2.2.4.

For TD-OCT, the imaging depth is mainly determined by the scanning range of the reference mirror and coherence length of the light source. For FD-OCT, however, the imaging depth is mainly determined by the spectral resolution, which for the spectrometer-based FD-OCT, is mainly determined by the inter-pixel wavenumber spacing $\delta_s k$ of the detector array. For simplicity, assuming a linear wavenumber spectrometer with constant $\delta_s k$, the maximum imaging range for FD-OCT is given by [36]

$$\pm z_{\max} = \pm \frac{\pi}{2 * \delta_s k} \tag{2.12}$$

Here, "±" means imaging for both sides of the zero-delay plane.

2.2.4 Tissue Optics of OCT

As mentioned in Section 2.2.3.3, the actual imaging depth of OCT is usually smaller than the theoretical maximum imaging range due to the loss of optical energy inside the tissue, mainly caused by absorption and scattering.

2.2.4.1 *Tissue Absorption*

Tissue absorption originates from photon interaction with atoms and molecules of the tissue, and is closely dependent on the tissue's biological and chemical composition. The absorption coefficient μ_{ab} can be defined by Beer's law as

$$\frac{I_{output}}{I_{input}} = \exp(-\mu_{ab} * \Delta L) \tag{2.13}$$

Here, I_{input} and I_{output} are the light intensities entering and exiting the absorption path length ΔL, respectively.

Figure 2.6 illustrates the absorption coefficient spectra of several important biomaterials, which compose the absorption of most biological tissues [37]. In particular, the HbO_2 (oxygenated hemoglobin) and Hb (deoxygenated hemoglobin) present the light absorption behavior of blood. As one can observe, for most tissues, the optical absorptions are generally weak in the 600–1300 nm range; this spectrum range is known as the "therapeutic window," which means easier penetration of light waves and therefore larger imaging range. Hemoglobin dominates the absorption band of 600–1000 nm, and water absorption is the major factor from about 1000 nm to infrared.

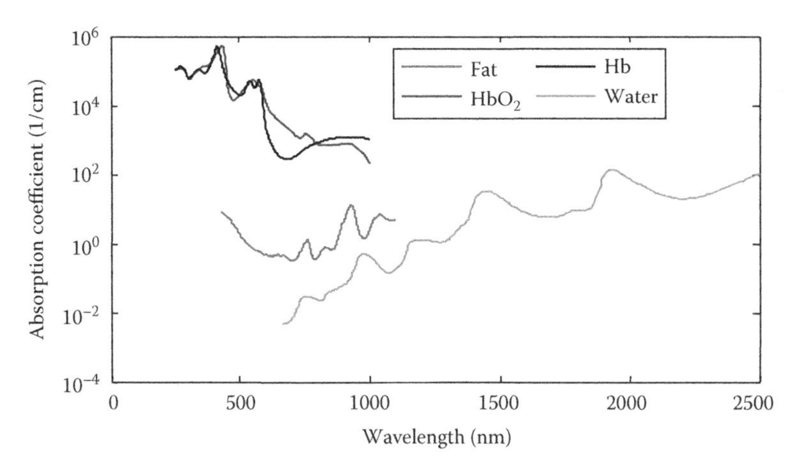

FIGURE 2.6
Absorption coefficient spectra of biomaterials: HbO_2, oxygenated hemoglobin; Hb, deoxygenated hemoglobin.

2.2.4.2 Tissue Scattering

Tissue scattering originates mainly from microscopic nonuniformity of the tissue, which can be quantitatively characterized by defining the tissue scattering cross section

$$\sigma_{sc} = \frac{P_{sc}}{I_{in}} \tag{2.14}$$

Here

I_{in} is the incident optical power

P_{sc} denotes the power redirected by the scattering centers

Scattering process can be cataloged to Rayleigh scattering (when scattering center size is much smaller than incident wavelength) and Mie scattering (when scattering center size is comparable to or larger than the incident wavelength). For Rayleigh scattering, the scattering cross section meets

$$\sigma_{sc} \propto \frac{1}{\lambda^4} \tag{2.15}$$

Therefore, under Rayleigh scattering, longer wavelength OCT (such as 1300 nm band) usually has deeper penetration depth into the tissue (such as 800 nm band), since the scattered light becomes completely incoherent and does not contribute to OCT imaging.

The selection of a proper light source according to the tissue's optical property is critical for the OCT imaging quality. For example, an 800 nm band light source is usually used for high-resolution retinal imaging, while a 1300 nm band light source is usually selected for blood vessel imaging and dermatological imaging.

2.2.5 Full-Range Fourier Domain OCT

From Equation 2.9, it is clearly shown that after a direct Fourier transformation, a pair of conjugate signals is generated and distributed at both sides of the zero-delay plane. Such complex–conjugate ghost image becomes the major artifact of FD-OCT imaging, and can be removed by several kinds of phase-shift interferometry methods [38–41]. In particular, here we introduce a very simple and effective method—the "simultaneous B-M-mode scanning method"—by applying a linear phase modulation $\varphi(t|x) = \beta^*(t|x)$ to each M-scan/B-scan's 2D interferogram frame $I(t|x,\lambda)$ [42]. Here, an M-scan means accumulation of the A-scans in the time domain at a fixed spatial point; $t|x$ indicates time/position index of A-scans in each M-scan/B-scan frame. By applying Fourier transformation along the $t|x$ direction, the following equation can be obtained:

$$\begin{aligned} F_{t|x\to f|u}\left[I(t|x,\lambda)\right] &= |E_r|^2\,\delta(f|u) + \Gamma_{f|u}\left\{F_{t|x\to f|u}\left[E_s(t|x,\lambda)\right]\right\} \\ &+ F_{t|x\to f|u}\left[E_r^*(t|x,\lambda)E_r(\lambda)\right]\otimes\delta\left(f|u+\beta\right) \\ &+ F_{t|x\to f|u}\left[E_s(t|x,\lambda)E_r^*(\lambda)\right]\otimes\delta\left(f|u-\beta\right) \end{aligned} \quad (2.16)$$

where

$E_s(t|x,\lambda)$ and $E_r(\lambda)$ are the electrical fields from the sample and reference arms, respectively

$\Gamma_{f|u}\{\}$ is the correlation operator

The first three terms on the right hand of Equation 2.16 present the DC noise, autocorrelation noise, and complex–conjugate noise, respectively. The last term can be filtered out by applying a proper band-pass filter in the $f|u$ domain. The resulting complex signal can then be converted back to the $t|x$ domain by applying an inverse Fourier transformation along the $f|u$ direction. Therefore, the FD-OCT can be extended to full-range FD-OCT free of complex–conjugate image, DC noise, and autocorrelation noise. Figure 2.7 compares the images of a multilayer phantom by half-range FD-OCT and full-range FD-OCT.

2.2.6 Optical Chromatic Dispersion and Compensation

For a Michelson-type OCT system, chromatic dispersion mismatch is a serious issue—especially for an ultrahigh resolution FD-OCT system with

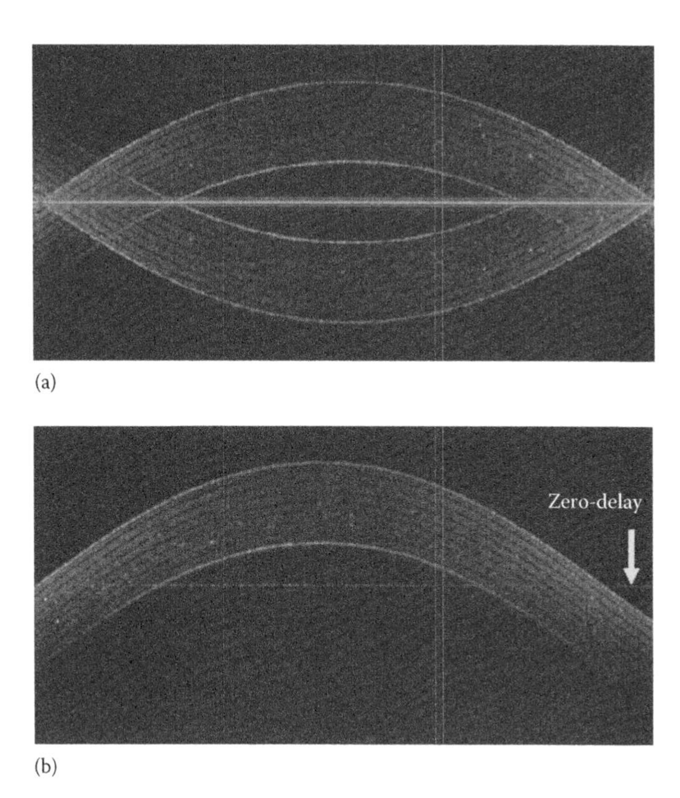

FIGURE 2.7
Imaging of a multilayer phantom: (a) half-range FD-OCT and (b) full-range FD-OCT.

an extremely broadband light source. The dispersion arises from wavelength-dependent phase velocity described by the propagation constant $\beta(\omega)$, which relates to optical material and guided wave properties. The dispersion mismatch usually results from unbalancing between optical components in the sample arm and reference arm. For some cases like retinal OCT, substantial dispersion mismatch also comes from the imaging sample itself (the vitreous humor). The $\beta(\omega)$ can be expanded as a Taylor series as

$$\beta(\omega) = \beta(\omega_0) + \left.\frac{d\beta}{d\omega}\right|_{\omega_0}(\omega - \omega_0) + \frac{1}{2}\left.\frac{d^2\beta}{d\omega^2}\right|_{\omega_0}(\omega - \omega_0)^2 + \frac{1}{6}\left.\frac{d^3\beta}{d\omega^3}\right|_{\omega_0}(\omega - \omega_0)^3 + \cdots \tag{2.17}$$

Here

$d\beta / d\omega$ is the inverse group velocity

$d^2\beta/d\omega^2$ describes the group velocity dispersion (GVD), which mainly contributes to the broadening of system PSF

The third-order dispersion $d^2\beta/d\omega^3$ brings asymmetric distortion onto the PSF. Dispersion compensation is necessary for most ultrahigh-resolution FD-OCT systems, and usually only the second and third-order dispersions are considered.

The hardware method usually matches the dispersion of the sample arms physically by putting balancing optical components on the reference arms: for example, the simplest way may be using identical optics; an alternative way may be using dispersion matching prism pairs. Both methods obviously increase the system cost for components; a perfect matching is usually difficult to realize in many cases when the dispersion comes from the sample itself (e.g., the vitreous humor dispersion for retinal imaging).

In comparison, the numerical dispersion compensation is more cost-effective and adaptable, by adding a phase correction via a Hilbert transformation of the original spectrum [43]

$$\overline{\Phi} = -a_2(\omega - \omega_0)^2 - a_3(\omega - \omega_0)^3 \tag{2.18}$$

where a_2 and a_3 were pre-optimized values according to the system properties. In many cases when the dispersion mismatching is relatively huge, a rough and low-cost hardware dispersion compensation is first implemented to approximately decrease the phase mismatch within 2π, and then a finer adjustment is given by the numerical compensation.

2.3 Ultrahigh Speed FD-OCT System Design

2.3.1 System Configuration

Figure 2.8 presents one of our ultrahigh speed FD-OCT systems incorporated with the CPU–GPU heterogeneous computing platform developed by the Photonics and Optoelectronics Laboratory of Johns Hopkins University. The high-speed spectrometer used a 12-bit dual-line CMOS line-scan camera (Sprint spL2048–140 k, Basler AG, Germany) as detector, which was set to work at the 1024 pixel mode since the spectrum only covered the array partially. Due to the property of the CMOS array sensor, the minimum line period of the camera was set to 7.8 μs, which offered a maximum line rate of 128,000 A-scan/s. A super luminescence diode (SLED) (λ_0 = 825 nm, $\Delta\lambda$ = 70 nm, Superlum, Ireland) was used as the light source, with the experimentally measured axial resolution of approximately 5.5 μm in air. The transversal resolution is approximately 40 μm, estimated with a Gaussian beam profile. The beam scanning is implemented by a pair of galvanometer mirrors and synchronized with a high-speed frame grabber (PCIE-1429, National Instruments, the United States). For the full-range FD-OCT

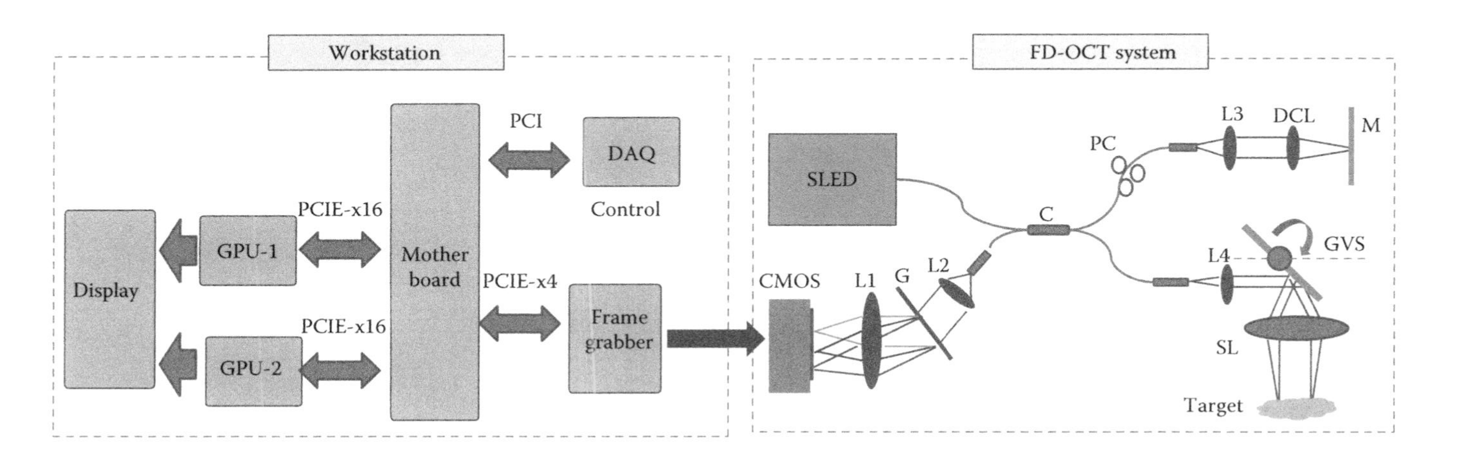

FIGURE 2.8
System configuration: CMOS, CMOS line scan camera; G, grating; L1, L2, L3, L4 achromatic collimators; C, 50:50 broadband fiber coupler; CL, camera link cable; CTRL, galvanometer control signal; GVS, galvanometer pairs (for simplicity, only the first galvanometer is illustrated); SL, scanning lens; DCL, dispersion compensation lens; M, reference mirror; PC, polarization controller.

mode, a phase modulation is applied to each B-scan's 2D interferogram frame by slightly displacing the probe beam off the first galvanometer's pivoting point (here, only the first galvanometer is illustrated) [44,45]. A quad-core Dell T7500 workstation is used to host the frame grabber (with PCIE–x4 interface), DAQ card (with PCI interface), GPU-1, and GPU-2 (both with PCIE–x16 interface). In this platform, two GPUs are utilized: the GPU-1 is dedicated for raw data processing of B-scan frames, and the GPU-2 is dedicated for volume rendering and display of the complete C-scan data processed by GPU-1. The GPUs are programmed through NVIDIA's CUDA technology [46].

2.3.2 Data Processing Flow

The CPU–GPU heterogeneous computing architecture is deployed with three major threads on the host computer, as shown in Figure 2.9: Thread 1 for raw data acquisition from the FD-OCT system; Thread 2 for GPU-based FD-OCT image reconstruction (A-scan and B-scan processing), where a different algorithm can be implemented into the "GPU-1 FD-OCT processing" block; Thread 3 for GPU-based volume rendering (C-scan processing), and the rendering methods can be applied to the "GPU-2 volume rendering" block. The three threads synchronize with each other in a pipeline mode: Thread 1 triggers Thread 2 for every B-scan, and Thread 2 triggers Thread 3 for every C-scan.

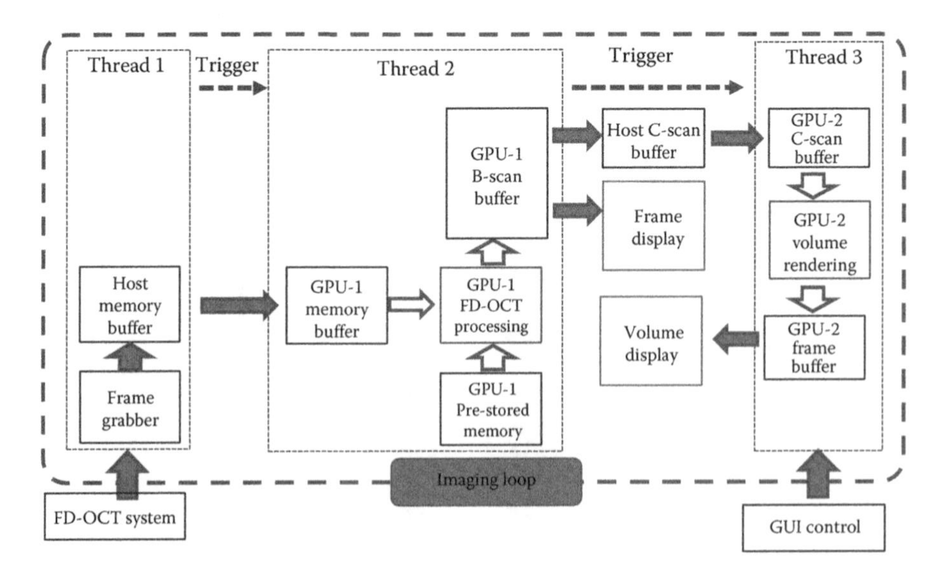

FIGURE 2.9
Data processing flowchart of the CPU–GPU heterogeneous computing architecture. Solid arrows, main data stream; hollow arrows, internal data flow of the GPU.

2.4 Ultrahigh Speed FD-OCT Data Processing

2.4.1 Numerical Interpolation Methods

In most FD-OCT systems, the spectral signal is sampled nonlinearly in k-space (e.g., by the commonly used nonlinear spectrometer), which will seriously degrade the image quality if the Fourier transformation is directly applied. Hardware solutions such as linear-k swept laser [47], linear-k spectrometer [48], and k-triggering board [49] have been developed and deployed, which however increase the system complexity and thus the development cost. A popular alternative software solution is numerical interpolation; here, we introduce the simplest linear interpolation, as well as cubic spline interpolation.

2.4.1.1 Linear Spline Interpolation

Starting from the commonly used linear spline interpolation (LSI) equation

$$S_{LSI}[j] = S[i] + \frac{S[i+1] - S[i]}{k[i+1] - k[i]}\left(k_{LN}[j] - k[i]\right) \tag{2.19}$$

where

$k[i] = 2\pi/\lambda[i]$ are the nonlinear k-space values and $\lambda[i]$ are the calibrated wavelength values of the FD-OCT system

$S[i]$ are the spectral intensity values corresponding to $k[i]$

$k_{LN}[j]$ are the linear k-space values covering the same frequency range as $k[i]$

LSI requires a proper interval $[k[i], k[i+1]]$ for each $k_{LN}[j]$, which satisfies $k[i] < k_{LN}[j] < k[i+1]$. Then, let $E[j]$ to present the lower ends for each element of $k[i]$, and then Equation 2.19 can be written as

$$S_{LSI}[j] = S\left[E[j]\right] + \left(S\left[E[j]+1\right] - S\left[E[j]\right]\right)\frac{k_{LN}[j] - k\left[E[j]\right]}{k\left[E[j]+1\right] - k\left[E[j]\right]} \tag{2.20}$$

$E[j]$ can be precalculated before interpolation based on $k[i]$ and $k_{LSI}[i]$. From Equation 2.20, one would notice that $S_{LSI}[j]$ is independent of other values in the series; therefore, this algorithm is highly suitable for the parallel computation. To facilitate the parallelization, Equation 2.20 can be further simplified to

$$S_{LSI}[j] = S\left[E[j]\right] + \left(S\left[E[j]+1\right] - S\left[E[j]\right]\right) * W[j] \tag{2.21}$$

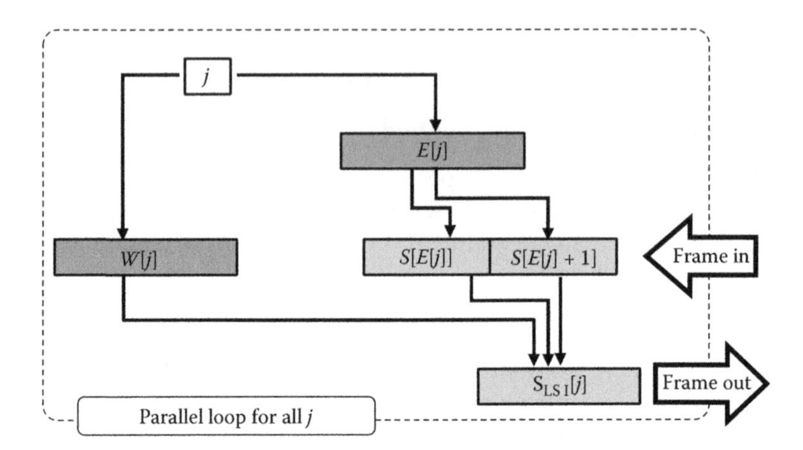

FIGURE 2.10
Parallelization of LSI. Dark gray blocks, memory for prestored data; light gray blocks, memory for real-timely refreshed data.

where $W[j] = k_{LN}[j] - k[E[j]] / (k[E[j]+1] - k[E[j]])$ can be precalculated and stored as look-up tables. Figure 2.10 shows the parallelized LSI, which can then be deployed on the GPU architecture. The values of $W[j]$ and $E[j]$ are all stored in the memory prior to interpolation, while the $S[j]$ and $S_{LSI}[j]$ are allocated in real-timely refreshed memory blocks.

2.4.1.2 Cubic Spline Interpolation

The calculation of cubic spline interpolation (CSI) is similar to LSI, given by [50]

$$S_{CSI}[j] = S\left[E[j]\right] * A[j] + S\left[E[j]+1\right] * B[j] + S''\left[E[j]\right] * C[j] + S''\left[E[j]+1\right] * D[j] \tag{2.22}$$

$$A[j] = \frac{k\left[E[j]+1\right] - k_{LN}[j]}{k\left[E[j]+1\right] - k\left[E[j]\right]} \tag{2.23}$$

$$B[j] = \frac{k_{LN}[j] - k\left[E[j]\right]}{k\left[E[j]+1\right] - k\left[E[j]\right]} \tag{2.24}$$

$$C[j] = \frac{1}{6}\left(\left(A[j]\right)^3 - A[j]\right) * \left(k\left[E[j]+1\right] - k\left[E[j]\right]\right)^2 \tag{2.25}$$

$$D[j] = \frac{1}{6}\left(\left(B[j]\right)^3 - B[j]\right) * \left(k\left[E[j]+1\right] - k\left[E[j]\right]\right)^2 \tag{2.26}$$

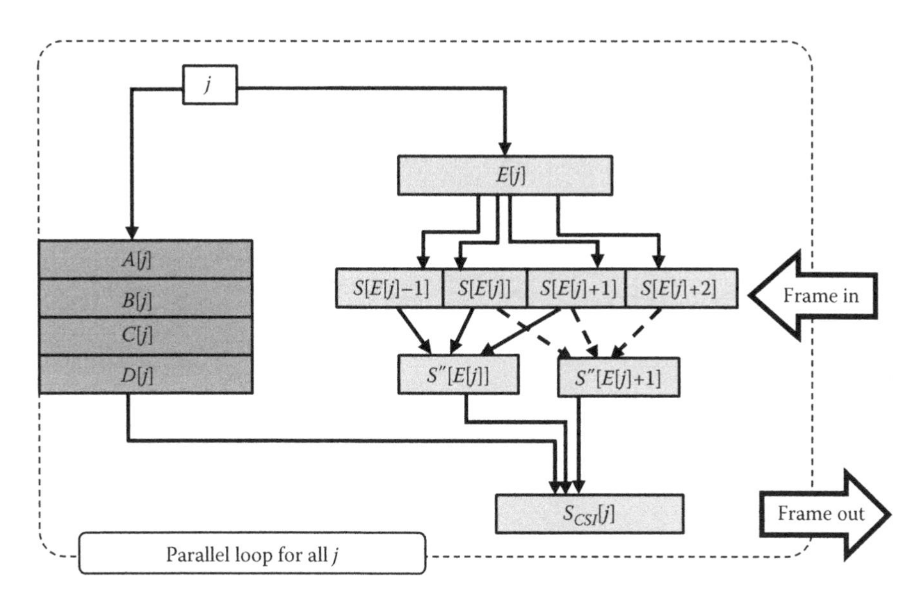

FIGURE 2.11
Parallelization of CSI. Dark gray blocks, memory for prestored data; light gray blocks, memory for real-timely refreshed data.

where $S''[E[j]]$ and $S''[E[j]+1]$ are the second derivatives besides $S_{CSI}[j]$, which can be presented in discrete forms as

$$S''[E[j]] = \left(\frac{S[E[j]+1] - S[E[j]]}{k[E[j]+1] - k[E[j]]} - \frac{S[E[j]] - S[E[j]-1]}{k[E[j]] - k[E[j]-1]} \right) \Big/ \Delta k \tag{2.27}$$

$$S''[E[j]+1] = \left(\frac{S[E[j]+2] - S[E[j]+1]}{k[E[j]+2] - k[E[j]+1]} - \frac{S[E[j]+1] - S[E[j]]}{k[E[j]+1] - k[E[j]]} \right) \Big/ \Delta k \tag{2.28}$$

where Δk is the spacing of the $k_{LN}[n]$ grid. $A[j]$, $B[j]$, $C[j]$, and $D[j]$ are all precalculated values, which are stored in the graphics memory. Figure 2.11 shows the flowchart of parallelized CSI.

2.4.2 Numerical Dispersion Compensation

As previously addressed in Section 2.2.6, as a more cost-effective and adaptable approach, numerical dispersion compensation adds a phase correction to the complex spectrum obtained via Hilbert transform $\Phi = -a_2(\omega - \omega_0)^2 - a_3(\omega - \omega_0)^3$: for the half-range FD-OCT, a Hilbert transform along k domain is realized by two FFTs and a Heaviside step filter [43]; for the full-range FD-OCT, the complex spectrum can be obtained by the modified Hilbert transform [51].

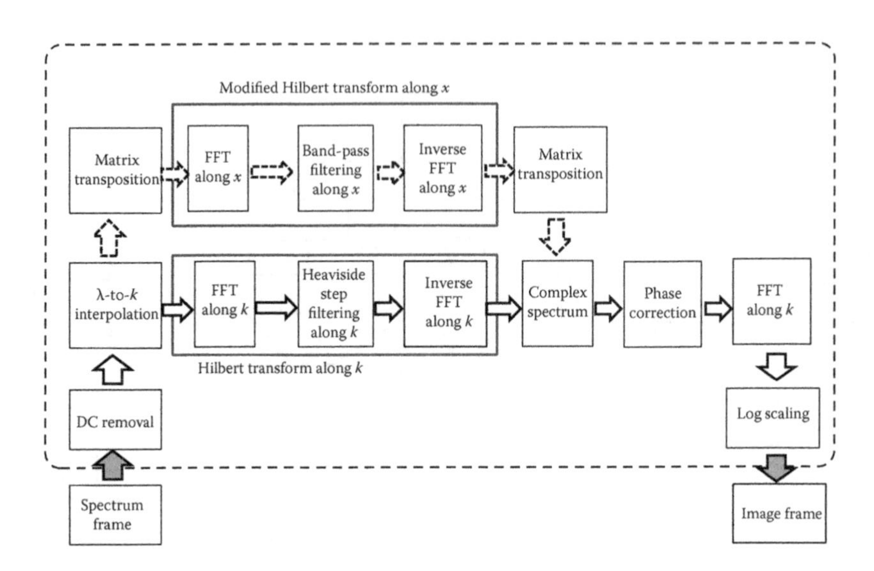

FIGURE 2.12
Numerical dispersion compensation for half-range (Hilbert Transform) and full-range (modified Hilbert transform) FD-OCT.

Both cases are illustrated in Figure 2.12, which are correspondingly implemented in the "GPU-1 FD-OCT processing" block in Figure 2.9.

2.4.3 Volume Rendering for OCT

To realize the volume rendering on GPU, here we use the ray-casting method [52–54] commonly used to visualize volumetric OCT imaging. As shown in Figure 2.13a, a 2D image frame is virtually placed at the observer's eye and

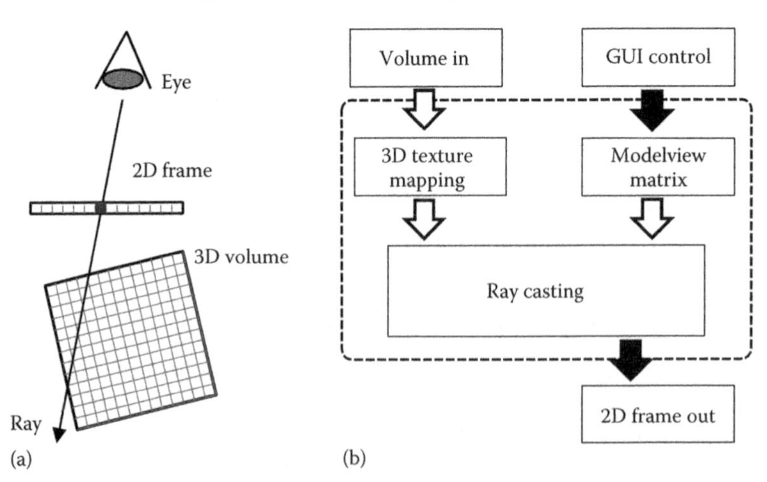

FIGURE 2.13
(a) Schematic of volume rendering via ray-casting and (b) flowchart of GPU-based volume rendering.

the 3D data volume. The relative position of the image frame and data volume is defined by a modelview matrix, which can be controlled via the GUI. Each pixel of the 2D imaging plane is the accumulation of a virtual eye-ray going through the pixels. Figure 2.13b illustrates the flowchart of GPU-based volume rendering, which corresponds to the "GPU-2 volume rendering" block in Figure 2.9. The C-scan volumetric data set is first mapped into 3D texture memory on the GPU, and then the ray-casting kernel is launched for each pixel.

2.5 Multidimensional Imaging Results

2.5.1 Real-Time 2D Cross-Sectional Imaging

Figure 2.14 presents the real-time 2D cross-sectional FD-OCT imaging results obtained by the system described in Section 2.3.1. Here, for simplicity we only demonstrate the full-range FD-OCT images, which are more representative of the system properties than the half-range images.

A NVIDIA GTX 580 GPU (512 cores, 1.59 GHz, processor clock and 1.5 GBytes graphics memory) plays the role of GPU-1 in Figure 2.9 for raw image reconstruction. Each frame consisting of 4096 A-scans (1024 pixels per A-scan) is real-timely processed with CSI combining numerical dispersion compensation as in Figure 2.12. The system runs at 31 frame/s with the full line rate of 128,000 A-scan/s.

In Figure 2.14a, an infrared (IR) card was scanned, showing clearly the plastic covering film above and the underneath fluorescence materials. Figure 2.14b presents an eight-layer polymer phantom mimicking the common case of a multilayer target such as a retina. Figure 2.14c and d shows the coronal and sagittal cross-sectional images of a human fingernail fold region, and Figure 2.14e and f presents the coronal scans of a human finger tip and finger palm regions, where the dermatological structures are clearly distinguishable.

2.5.2 Real-Time 3D (4D) FD-OCT Imaging

For real-time 3D (4D) imaging, a NVIDIA GeForce GTX 450 GPU (192 stream processors, each processor working at 1.76 GHz with 1GBytes graphics memory) was used as GPU-2 and dedicated for the C-scan volume rendering. With the scanning protocol of 100 B-scans per C-scan and 256 A-scans per B-scan, the system operates at 5.0 volume/s with full imaging rate at 128,000 A-scan/s; the system is also set to work in the full-range mode. For a 512 × 512 2D frame size, GPU-2 takes about 7 ms to render a 1024 (Z) × 256(Y) × 100(X) data volume. Therefore, multiple 2D frames can be real-timely rendered from the same volume with a different modelview matrix and displayed simultaneously.

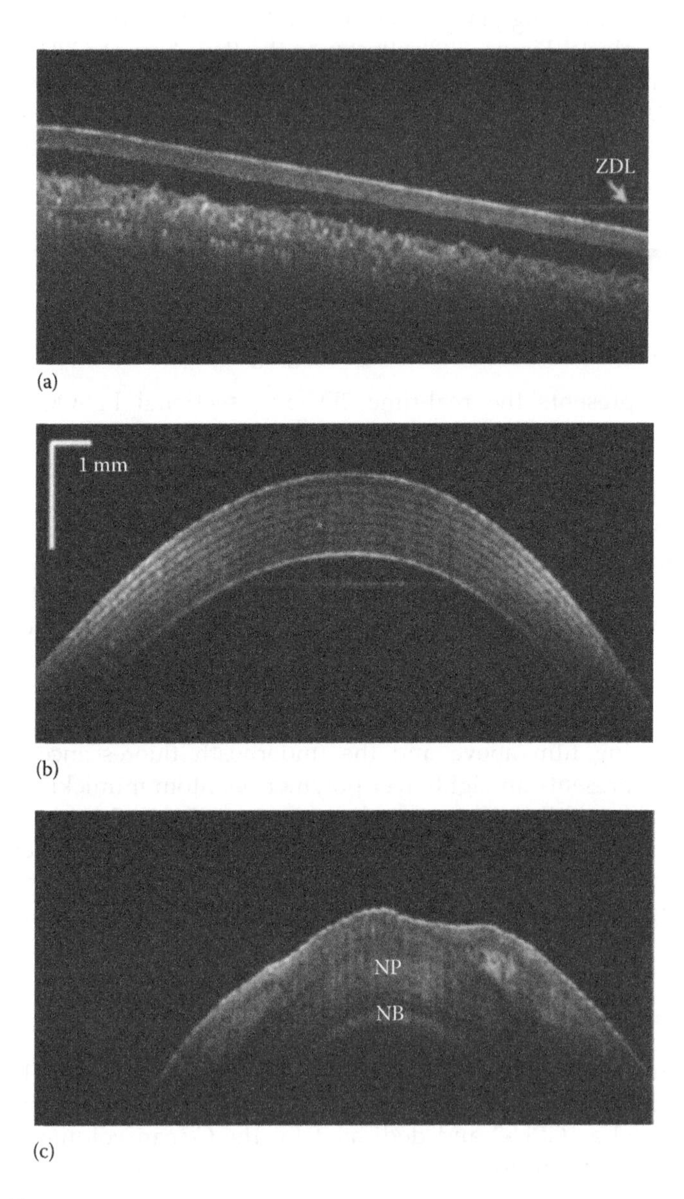

FIGURE 2.14
Real-time full-range FD-OCT images: (a) IR card, (b) eight-layer polymer phantom, (c) coronal scan of human fingernail fold.

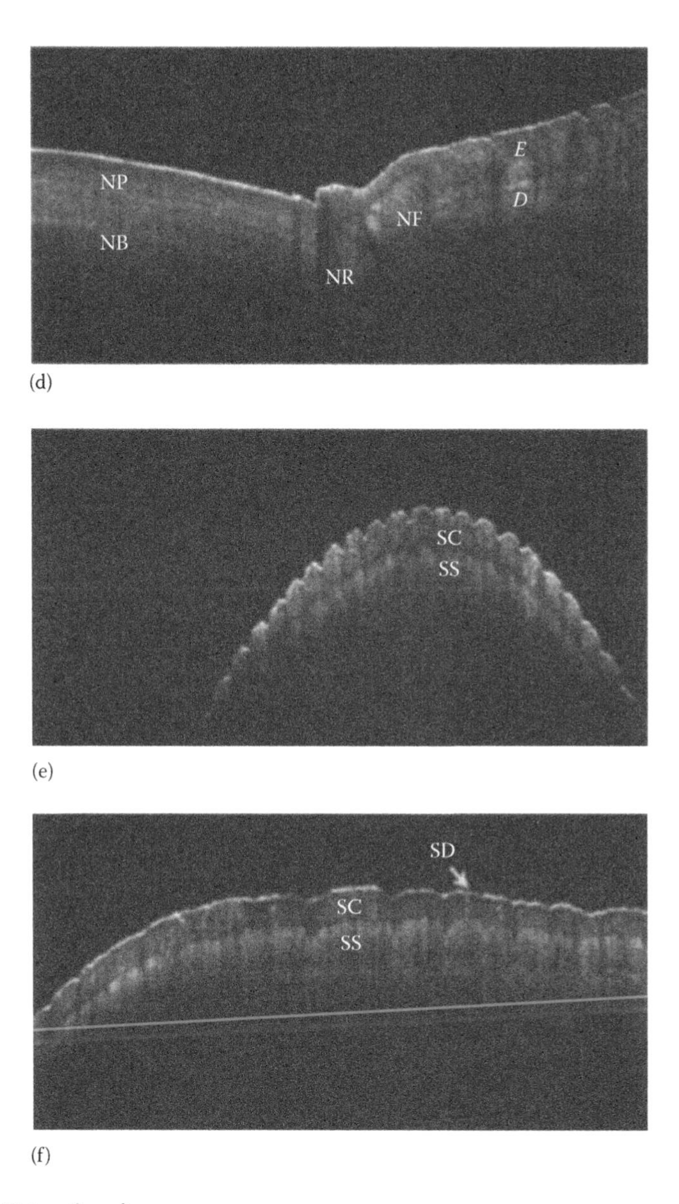

FIGURE 2.14 (continued)
Real-time full-range FD-OCT images: (d) Sagittal scan of human finger nail fold, (e) coronal scan of human finger tip, (f) coronal scan of human finger palm. SD, sweat duct; SC, stratum corneum; SS, stratum spinosum; NP, nail plate; NB, nail bed; NR, nail root; E, epidermis; D, dermis. The bars represent 1 mm in both dimensions for all images.

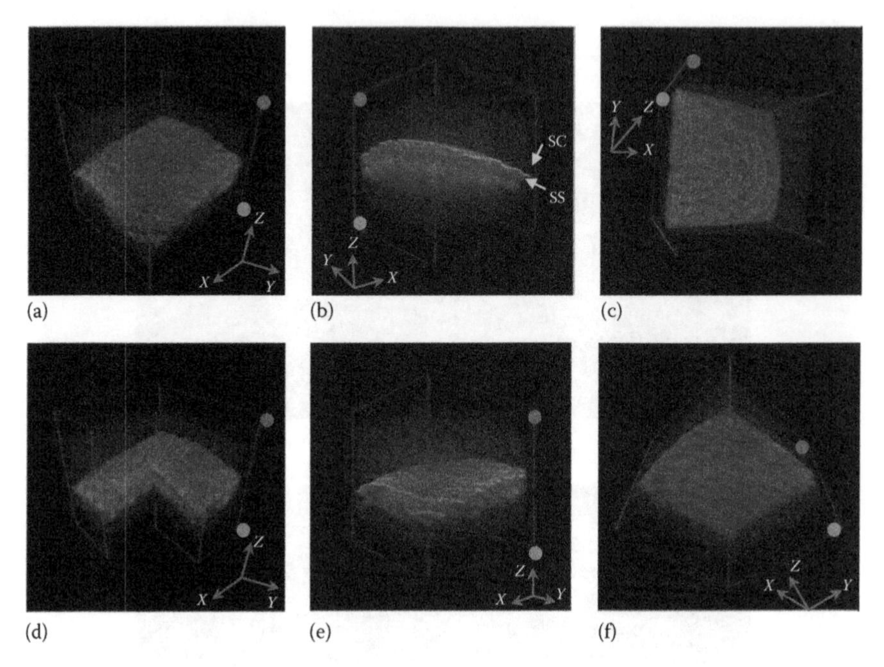

FIGURE 2.15
4D full-range FD-OCT of human finger tip. (a) through (f) are 2D frames rendered from the same data volume with different view angles. Volume size: 100 (*X*) × 256 (*Y*) × 1024 (*Z*) voxels/3.5 mm (*X*) × 3.5 mm (*Y*) × 3 mm (*Z*). The two vertexes with the big dark gray and light gray dots indicate the same edge for each rendering volume.

Figure 2.15 presents the in vivo 4D imaging of a human finger tip, where six 2D frames are rendered at one time and showing the finger print from different angles. The actual size of imaging volume is 3.5 mm (*X*) × 3.5 mm (*Y*) × 3 mm (*Z*). Figure 2.16 shows the multi-angle view of a human skin region with hairs.

2.6 Conclusion and Perspective

In this chapter, the principles of ultrahigh speed OCT were reviewed. A GPU-accelerated ultrahigh speed real-time multidimensional FD-OCT imaging platform was developed, combining the cutting edge technologies of both optical engineering and computer engineering. The imaging platform is capable of real-time data acquisition, reconstruction, and visualization. Several GPU-based algorithms were developed to accelerate the signal

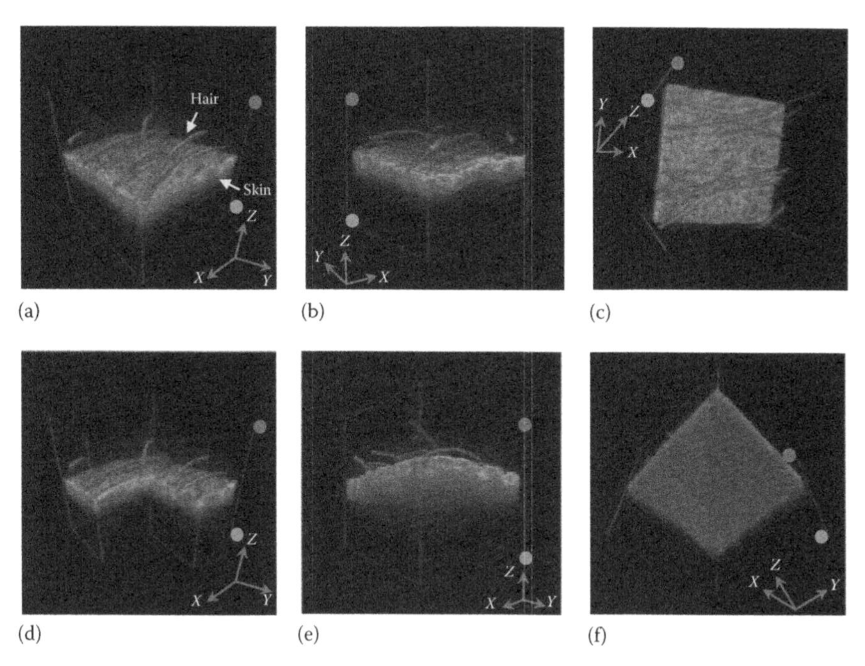

FIGURE 2.16
4D full-range FD-OCT of human skin and hair. (a) through (f) are 2D frames rendered from the same data volume with different view angles. Volume size: 100 (*X*) × 256 (*Y*) × 1024 (*Z*) voxels/3.5 mm (*X*) × 3.5 mm (*Y*) × 3 mm (*Z*). The two vertexes with the big dark gray and light gray dots indicate the same edge for each rendering volume.

processing and volume rendering. The performance of the GPU-accelerated real-time FD-OCT system was validated by in vivo human tissue imaging.

As the contemporary trend of OCT technology research, functional OCT imaging capabilities such as Doppler OCT [55,56], spectroscopic OCT [57,58], and phase-sensitive OCT [59,60] imaging are being extensively explored to provide physiological and pathological information of biological tissue. Advanced imaging analysis and computer vision technologies can also be used to realize tissue segmentation and feature reorganization. Other imaging modalities can also be combined to realize multimodal imaging OCT capability [61,62].

Acknowledgment

This work was supported by NIH grant R21 1R21NS063131-01A1.

References

1. D. Huang, E. A. Swanson, C. P. Lin, J. S. Schuman, W. G. Stinson, W. Chang, M. R. Hee et al., Optical coherence tomography, *Science* **254**, 1178–1181 (1991).
2. S. Charles, J. Calzada, and B. Wood, *Vitreous Microsurgery* (Lippincott Williams & Wilkins, Philadelphia, PA, 2010).
3. T. Peters and K. Cleary, *Image-Guided Interventions: Technology and Applications* (Springer, New York, 2008).
4. S. A. Boppart, M. E. Brezinski, and J. G. Fujimoto, Surgical guidance and intervention, in *Handbook of Optical Coherence Tomography*, B. E. Bouma and G. J. Tearney, eds. pp. 613–648 (Marcel Dekker, New York, 2001).
5. S. A. Boppart, B. E. Bouma, C. Pitris, G. J. Tearney, J. F. Southern, M. E. Brezinski, and J. G. Fujimoto, Intraoperative assessment of microsurgery with three-dimensional optical coherence tomography, *Radiology* **208**, 81–86 (1998).
6. B. Potsaid, I. Gorczynska, V. J. Srinivasan, Y. Chen, J. Jiang, A. Cable, and J. G. Fujimoto, Ultrahigh speed Spectral/Fourier domain OCT ophthalmic imaging at 70,000 to 312,500 axial scans per second, *Opt. Exp.* **16**, 15149–15169 (2008).
7. R. Huber, D. C. Adler, and J. G. Fujimoto, Buffered Fourier domain mode locking: unidirectional swept laser sources for optical coherence tomography imaging at 370,000 lines/s, *Opt. Lett.* **31**, 2975–2977 (2006).
8. W.-Y. Oh, B. J. Vakoc, M. Shishkov, G. J. Tearney, and B. E. Bouma, >400 kHz repetition rate wavelength-swept laser and application to high-speed optical frequency domain imaging, *Opt. Lett.* **35**, 2919–2921 (2010).
9. I. Grulkowski, M. Gora, M. Szkulmowski, I. Gorczynska, D. Szlag, S. Marcos, A. Kowalczyk, and M. Wojtkowski, Anterior segment imaging with Spectral OCT system using a high-speed CMOS camera, *Opt. Exp.* **17**, 4842–4858 (2009).
10. M. Gora, K. Karnowski, M. Szkulmowski, B. J. Kaluzny, R. Huber, A. Kowalczyk, and M. Wojtkowski, Ultra high-speed swept source OCT imaging of the anterior segment of human eye at 200 kHz with adjustable imaging range, *Opt. Exp.* **17**, 14880–14894 (2009).
11. M. Gargesha, M. W. Jenkins, A. M. Rollins, and D. L. Wilson, Denoising and 4D visualization of OCT images, *Opt. Exp.* **16**, 12313–12333 (2008).
12. M. Gargesha, M. W. Jenkins, D. L. Wilson, and A. M. Rollins, High temporal resolution OCT using image-based retrospective gating, *Opt. Exp.* **17**, 10786–10799 (2009).
13. M. W. Jenkins, F. Rothenberg, D. Roy, V. P. Nikolski, Z. Hu, M. Watanabe, D. L. Wilson, I. R. Efimov, and A. M. Rollins, 4D embryonic cardiography using gated optical coherence tomography, *Opt. Exp.* **14**, 736–748 (2006).
14. W. Wieser, B. R. Biedermann, T. Klein, C. M. Eigenwillig, and R. Huber, Multi-Megahertz OCT: High quality 3D imaging at 20 million A-scans and 4.5 GVoxels per second, *Opt. Exp.* **18**, 14685–14704 (2010)
15. T. Klein, W. Wieser, C. M. Eigenwillig, B. R. Biedermann, and R. Huber, Megahertz OCT for ultrawide-field retinal imaging with a 1050 nm Fourier domain mode-locked laser, *Opt. Exp.* **19**, 3044–3062 (2011).
16. T. E. Ustun, N. V. Iftimia, R. D. Ferguson, and D. X. Hammer, Real-time processing for Fourier domain optical coherence tomography using a field programmable gate array, *Rev. Sci. Instrum.* **79**, 114301 (2008).

17. A. E. Desjardins, B. J. Vakoc, M. J. Suter, S. H. Yun, G. J. Tearney, and B. E. Bouma, Real-time FPGA processing for high-speed optical frequency domain imaging, *IEEE Trans. Med. Imag.* **28**, 1468–1472 (2009).
18. G. Liu, J. Zhang, L. Yu, T. Xie, and Z. Chen, Real-time polarization-sensitive optical coherence tomography data processing with parallel computing, *Appl. Opt.* **48**, 6365–6370 (2009).
19. J. Probst, D. Hillmann, E. Lankenau, C. Winter, S. Oelckers, P. Koch, and G. Hüttmann, Optical coherence tomography with online visualization of more than seven rendered volumes per second, *J. Biomed. Opt.* **15**, 026014 (2010).
20. B. David and W. W. Hwu, *Programming Massively Parallel Processors: A Hands-on Approach* (Morgan Kaufmann, San Francisco, CA, 2010).
21. J. Sanders and E. Kandrot, *CUDA by Example: An Introduction to General-Purpose GPU Programming*, (Addison-Wesley Professional, Boston, MA, 2010).
22. Y. Watanabe and T. Itagaki, Real-time display on Fourier domain optical coherence tomography system using a graphics processing unit, *J. Biomed. Opt.* **14**, 060506 (2009).
23. K. Zhang and J. U. Kang, Real-time 4D signal processing and visualization using graphics processing unit on a regular nonlinear-k Fourier-domain OCT system, *Opt. Exp.* **18**, 11772–11784 (2010).
24. S. V. Jeught, A. Bradu, and A. G. Podoleanu, Real-time resampling in Fourier domain optical coherence tomography using a graphics processing unit, *J. Biomed. Opt.* **15**, 030511 (2010).
25. Y. Watanabe, S. Maeno, K. Aoshima, H. Hasegawa, and H. Koseki, Real-time processing for full-range Fourier-domain optical-coherence tomography with zero-filling interpolation using multiple graphic processing units, *Appl. Opt.* **49**, 4756–4762 (2010).
26. K. Zhang and J. U. Kang, Graphics processing unit accelerated non-uniform fast Fourier transform for ultrahigh-speed, real-time Fourier-domain OCT, *Opt. Exp.* **18**, 23472–23487 (2010).
27. K. Zhang and J. U. Kang, Real-time intraoperative 4D full-range FD-OCT based on the dual graphics processing units architecture for microsurgery guidance, *Biomed. Opt. Exp.* **2**, 764–770 (2011).
28. J. Rasakanthan, K. Sugden, and P. H. Tomlins, Processing and rendering of Fourier domain optical coherence tomography images at a line rate over 524 kHz using a graphics processing unit, *J. Biomed. Opt.* **16**, 020505 (2011).
29. J. Li, P. Bloch, J. Xu, M. V. Sarunic, and L. Shannon, Performance and scalability of Fourier domain optical coherence tomography acceleration using graphics processing units, *Appl. Opt.* **50**, 1832–1838 (2011).
30. K. Zhang and J. U. Kang, Real-time numerical dispersion compensation using graphics processing unit for Fourier-domain optical coherence tomography, *Electron. Lett.*, **47**, 309–310 (2011).
31. B. E. A. Saleh and M. C. Teich, *Fundamentals of Photonics*, Chapter 11 (John Wiley & Sons, New York, 2007).
32. J. M. Schmitt, Optical coherence tomography (OCT): A review, *IEEE J. Select. Topics Quant. Electron.* **5**, 1205 (1999).
33. A. Fercher, Measurement of intraocular distances by backscattering spectral interferometry, *Opt. Commun.* **117**, 43 (1995).
34. R. Leitgeb, C. Hitzenberger, and A. Fercher, Performance of Fourier domain vs. time domain optical coherence tomography, *Opt. Exp.* **11**, 889–894 (2003).

35. M. Wojtkowski, High-speed optical coherence tomography: Basics and applications, *Appl. Opt.* **49**, D30–D61 (2010).
36. W. Drexler and J. G. Fujimoto, *Optical Coherence Tomography: Technology and Applications,* Chapter 2 (Springer, New York, 2008).
37. Optical spectral data provided by Oregon Medical Laser Center (OMLC) (2001). http://omlc.ogi.edu/spectra/index.html
38. R. A. Leitgeb, C. K. Hitzenberger, A. F. Fercher, and T. Bajraszewski, Phase-shifting algorithm to achieve high-speed long-depth-range probing by frequency-domain optical coherence tomography, *Opt. Lett.* **28**, 2201–2203 (2003).
39. M. A. Choma, C. H. Yang, and J. A. Izatt, Instantaneous quadrature low-coherence interferometry with 3 x 3 fiber-optic couplers, *Opt. Lett.* **28**, 2162 (2003).
40. B. Bhushan, J. C. Wyant, and C. L. Koliopoulos, Measurement of surface topography of magnetic tapes by Mirau interferometry, *Appl. Opt.* **24**, 1489–1497 (1985).
41. A. Dubois, Phase-map measurements by interferometry with sinusoidal phase modulation and four integrating buckets, *J. Opt. Soc. Am. A* **18**, 1972–1979 (2001).
42. Y. Yasuno, S. Makita, T. Endo, G. Aoki, M. Itoh, and T. Yatagai, Simultaneous B-M-mode scanning method for real-time full-range Fourier domain optical coherence tomography, *Appl. Opt.* **45**, 1861–1865 (2006).
43. M. Wojtkowski, V. Srinivasan, T. Ko, J. G. Fujimoto, A. Kowalczyk, and J. Duker, Ultrahigh-resolution, high-speed, Fourier domain optical coherence tomography and methods for dispersion compensation, *Opt. Exp.* **12**, 2404–2422 (2004).
44. B. Baumann, M. Pircher, E. Götzinger, and C. K. Hitzenberger, Full range complex spectral domain optical coherence tomography without additional phase shifters, *Opt. Exp.* **15**, 13375–13387 (2007).
45. L. An and R. K. Wang, Use of a scanner to modulate spatial interferograms for in vivo full-range Fourier-domain optical coherence tomography, *Opt. Lett.* **32**, 3423–3425 (2007).
46. NVIDIA, NVIDIA CUDA CUFFT Library Version 4.0 (2011).
47. C. M. Eigenwillig, B. R. Biedermann, G. Palte, and R. Huber, K-space linear Fourier domain mode locked laser and applications for optical coherence tomography, *Opt. Exp.* **16**, 8916–8937 (2008).
48. Z. Hu and A. M. Rollins, Fourier domain optical coherence tomography with a linear-in-wavenumber spectrometer, *Opt. Lett.* **32**, 3525–3527 (2007).
49. D. C. Adler, Y. Chen, R. Huber, J. Schmitt, J. Connolly, and J. G. Fujimoto, Three-dimensional endomicroscopy using optical coherence tomography, *Nat. Photonics* **1**, 709–716 (2007).
50. W. H. Press, S. A. Teukolsky, W. T. Vetterling, and B. P. Flannery, *Numerical Recipes in FORTRAN* (Cambridge University Press, New York, 1992).
51. S. Makita, T. Fabritius, and Y. Yasuno, Full-range, high-speed, high-resolution 1-μm spectral-domain optical coherence tomography using BM-scan for volumetric imaging of the human posterior eye, *Opt. Exp.* **16**, 8406–8420 (2008).
52. J. Kruger and R. Westermann, Acceleration techniques for GPU-based volume rendering, in *Proceedings of the 14th IEEE Visualization Conference (VIS'03)* (IEEE Computer Society, Washington, DC, 2003), pp. 287–292.
53. A. Kaufman and K. Mueller, Overview of volume rendering, in *The Visualization Handbook*, C. Johnson and C. Hansen, eds. (Academic Press, Burlington, MA, 2005).

54. M. Levoy, Display of surfaces from volume data, *IEEE Comp. Graph. Appl.* **8**, 29–37 (1988).
55. R. Leitgeb, L. Schmetterer, W. Drexler, A. Fercher, R. Zawadzki, and T. Bajraszewski, Real-time assessment of retinal blood flow with ultrafast acquisition by color Doppler Fourier domain optical coherence tomography, *Opt. Exp.* **11**, 3116–3121 (2003).
56. S. Makita, F. Jaillon, M. Yamanari, and Y. Yasuno, Dual-beam-scan Doppler optical coherence angiography for birefringence-artifact-free vasculature imaging, *Opt. Exp.* **20**, 2681–2692 (2012).
57. U. Morgner, W. Drexler, F. X. Kärtner, X. D. Li, C. Pitris, E. P. Ippen, and J. G. Fujimoto, Spectroscopic optical coherence tomography, *Opt. Lett.* **25**, 111–113 (2000).
58. F. E. Robles, C. Wilson, G. Grant, and A. Wax, Molecular imaging true-colour spectroscopic optical coherence tomography, *Nat. Photonics* **5**, 744–747 (2011).
59. J. F. de Boer, T. E. Milner, M. J. C. van Gemert, and J. Stuart Nelson, Two-dimensional birefringence imaging in biological tissue by polarization-sensitive optical coherence tomography, *Opt. Lett.* **22**, 934–936 (1997).
60. D. C. Adler, R. Huber, and J. G. Fujimoto, Phase-sensitive optical coherence tomography at up to 370,000 lines per second using buffered Fourier domain mode-locked lasers, *Opt. Lett.* **32**, 626–628 (2007).
61. S. Jiao, Z. Xie, H. F. Zhang, and C. A. Puliafito, Simultaneous multimodal imaging with integrated photoacoustic microscopy and optical coherence tomography, *Opt. Lett.* **34**, 2961–2963 (2009).
62. E. Z. Zhang, B. Povazay, J. Laufer, A. Alex, B. Hofer, B. Pedley, C. Glittenberg et al., Multimodal photoacoustic and optical coherence tomography scanner using an all optical detection scheme for 3D morphological skin imaging, *Biomed. Opt. Exp.* **2**, 2202–2215 (2011).

3

Molecular Imaging True Color Spectroscopic (METRiCS) Optical Coherence Tomography

Adam Wax and Francisco E. Robles

CONTENTS

3.1 Introduction

Optical coherence tomography (OCT) is a cross-sectional high-resolution optical imaging modality that operates under the principles of low coherence interferometry (LCI) to enable optical sectioning, or depth resolution, within a biological sample [1]. OCT has been particularly successful for application in ophthalmology where the optical transparency of the eye, especially for wavelengths in the near-infrared region of the spectrum, allows high-resolution visualization of retinal layers for diagnosing diseases and assisting in surgeries [2]. Although originally implemented in a time domain approach where mechanical scanning of a reference mirror in an interferometry scheme enabled measurement of depth resolved reflection profiles (A-scans), recent technological advances in OCT have led to the dominance of frequency domain (FD-) OCT due to its improved acquisition speed [3] and sensitivity [4]. As the name suggests, FD-OCT measures interferometric data as a function of frequency and a Fourier transform is used to produce the A-scan, typically with an axial resolution of a few micrometers and imaging depths of 1–2 mm. To implement tomographic images, multiple A-scans are obtained using transverse scanning or a parallel acquisition scheme.

Figure 3.1a illustrates a typical FD-OCT system, where a source with a wide bandwidth, $S(\lambda)$, is split into a sample field, E_S, and reference field, E_R. The sample field E_S interacts with a sample while the reference field E_R is reflected off a reference mirror. In this example, light remitted from three different locations in the sample (m = 1, 2, and 3) is mixed with the reference field and detected using a spectrometer. Because the interferometric signal is recorded as a function of wavelength (i.e., in the frequency domain), it must be resampled to a linear wavenumber vector, $k = \lambda/2\pi$ (Figure 3.1b), to enable recovery of the A-scan (Figure 3.1c) via a fast Fourier transform (FFT).

Although the majority of clinical OCT imaging instruments measure structure by the intensity of returned light, various extensions of OCT have allowed for other parameters to be assessed: of particular relevance here is spectroscopic OCT (SOCT), which provides additional information by measuring the spatially dependent spectrum of light.

SOCT supplements the high-resolution imaging capabilities of OCT with the rich source of information available via optical spectroscopy. In this approach, the wide spectral bandwidth of the low coherence light sources required for depth sectioning is exploited to obtain spectral information. Processing of SOCT signals typically uses a short-time Fourier transform (STFT) or continuous wavelet transform (CWT) to identify the spatially varying, spectral properties of samples. Figure 3.2 illustrates STFT processing of SOCT signals. In this example, light returned from the ideal sample,

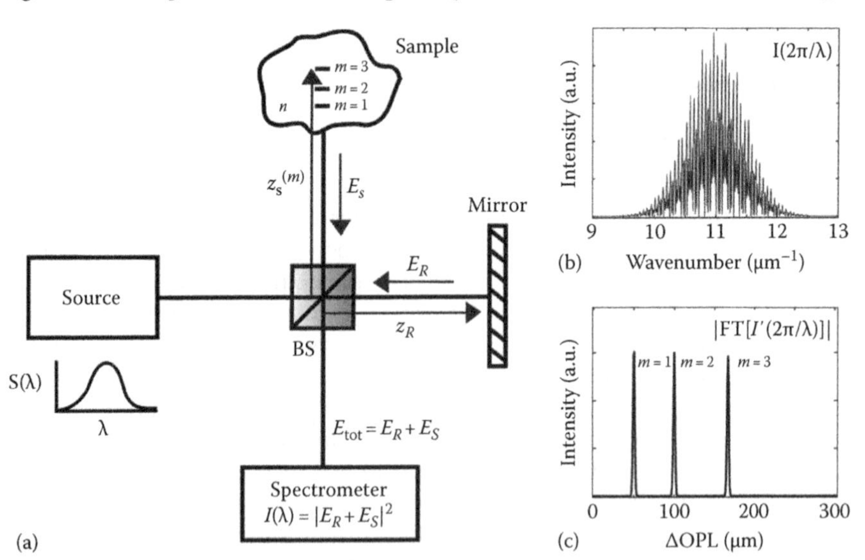

FIGURE 3.1
(a) Schematic of FD-OCT system. (b) Interferogram resampled into a linear wavenumber vector. (c) A-scan: absolute value of the Fourier transform of (b) after back ground subtraction. See text for details. (Taken from Robles, F.E., Light scattering and absorption spectroscopy in three dimensions using quantitative low coherence interferometry for biomedical applications, PhD thesis in Medical Physics, Duke University, Durham, NC, 2011.)

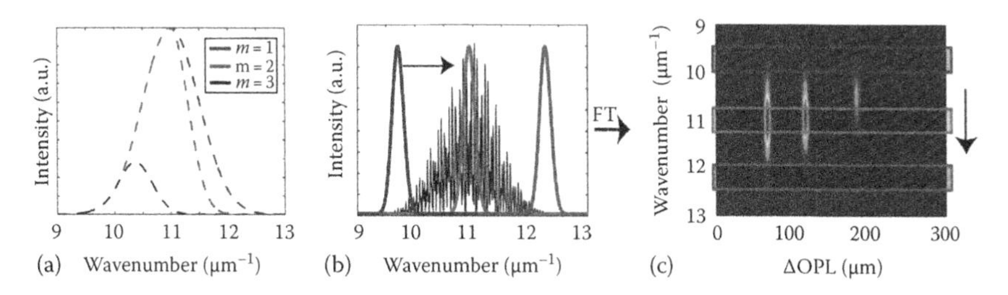

FIGURE 3.2
(a) Relative wavenumber (or wavelength) intensity returning from three simulated scatterers, as shown in Figure 3.1. (b) SOCT processing using STFT. (c) Resulting time frequency distribution (TFD). (Taken from Robles, F.E., Light scattering and absorption spectroscopy in three dimensions using quantitative low coherence interferometry for biomedical applications, PhD thesis in Medical Physics, Duke University, Durham, NC, 2011.)

simulated in Figure 3.1, undergoes attenuation, reducing the spectral intensity returned from deeper scatterers at certain wavelengths (Figure 3.2a). An STFT entails using a window that is swept across the full bandwidth with an FFT computed at each step (Figure 3.2b). The output is a time frequency distribution (TFD) map that describes the spectrum returned by each sample point. A significant limitation of the STFT is that there is a trade-off associated with the width of the window used in this process. If the window is chosen to be large in spectrum, then the TFD contains high spatial resolution, but suffers from poor spectral resolution and vice versa.

SOCT was first introduced by Morgner et al., and was used only to assess the center-of-mass of depth-resolved spectra to provide an additional mechanism of contrast [6]. However, more recent studies have explored using the full spectrum to obtain detailed functional information. For example, Faber et al. [7] and Yi et al. [8] have used SOCT to quantitatively measure the absorption coefficient of Hb. SOCT also can be used to detect the presence of exogenous contrast agents that tag specific cell receptors or trace functional mechanisms. This approach has been explored by Cang et al. who used SOCT to detect plasmonic gold nanocages in tissue phantoms [9] and by Oldenberg et al. using SOCT to detect nano rods in ex vivo human breast carcinoma [10].

To date, SOCT has not been widely applied to in vivo imaging or for preclinical studies of disease mechanisms. One view is that progress in its development has been limited by the trade-off between spatial and spectral resolution, as previously mentioned. While this trade-off is inherent in the STFT, it is not necessarily present for other methods that reconstruct TFDs; for example, bilinear distributions can avoid this limitation, although they typically have their own unique trade-offs. The widespread use of the near-infrared spectrum in OCT has also been a limiting factor in biomedical studies using SOCT, especially since biological absorbers such as hemoglobin have relatively low absorption in this range resulting in low confidence intervals [11].

In this chapter, we describe advancement of SOCT to enable depth resolved spectroscopy. By using a novel *dual window* processing method [12], the trade-off between depth and spectral resolution is effectively removed, enabling high resolution in both domains. We have advanced the DW SOCT method in two directions: Fourier domain low coherence interferometry (fLCI) for employing spectral information for analyzing scattering properties and molecular contrast true color SOCT (METRiCS OCT) for using the absorption properties of samples for true color display and to obtain functional information. The remainder of this chapter includes an overview of the DW method, discussion of its application to analyzing scattering, specifically for early cancer diagnosis, and then demonstration of its utility for analysis of absorption properties.

3.2 Dual-Window Signal Processing

Depth resolved spectroscopic information may be obtained from interferometric data to obtain functional information of a sample. As discussed earlier, a short-time Fourier transform (STFT) is typically used to access this information. This process uses a mathematical window to isolate information from a particular frequency range. When the windowed data are Fourier transformed to yield a depth resolved reflection profile, some depth resolution is lost due to the reduction of frequencies contributing to the signal. However, by sweeping the window across the full bandwidth of the source, a depth resolved reflection profile is obtained at each step, providing a map of the spectral content confined within a spatial (or depth) region (see Figures 3.2 and 3.3). The process produces a time–frequency distribution (TFD) of the signal; however, TFDs obtained using a single STFT are limited

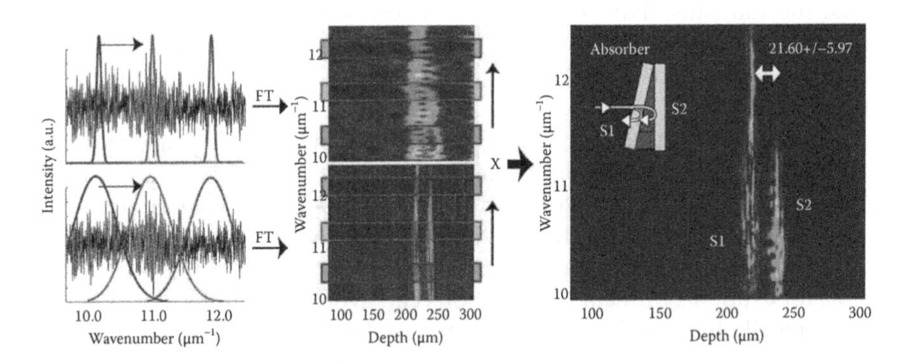

FIGURE 3.3
Procedure for calculating the DW TFD. The interferometric signal is generated from an absorbing phantom with two scattering surfaces, S1 and S2. (Adapted from Robles, F. et al., *Opt. Express*, 17(8), 6799, 2009.)

due to the inherent trade-off between the resulting spectral and the depth resolution. In order to produce distributions with high resolution in both dimensions, we have developed the dual window (DW) processing method.

The DW method is based on calculating two separate STFTs and then multiplying the results. The first STFT uses a broad spectral Gaussian window to obtain high depth resolution, while the second STFT uses a narrow spectral window to generate high spectroscopic resolution. Thus, the resulting DW TFD simultaneously achieves high resolution in both dimensions. Mathematically, this can be described as

$$\mathrm{DW}(k,z)=|E_R|^2\int 2\langle E_S(\kappa_1)\rangle\cos(\kappa_1\times\Delta\mathrm{OPL})\,e^{-(\kappa_1-k)^2/2w_1^2}e^{-i\kappa_1 z}d\kappa_1$$

$$\times\int\left(2\langle E_s(\kappa_2)\rangle\cos(\kappa_2\times\Delta\mathrm{OPL})e^{-(\kappa_2-k)^2/2w_2^2}e^{-i\kappa_2 z}\right)^* d\kappa_2, \qquad (3.1)$$

where E_S, E_R are the sample and reference fields, with the caveat that E_R is assumed to vary slowly with respect to wavenumber. $<\cdots>$ denotes an ensemble average, ΔOPL is the optical path length difference between the sample and the reference arms, z is the axial distance, and w_1 and w_2 are the widths (standard deviation of the Gaussian distribution) of the two windows. Figure 3.3 illustrates the procedure for computing the DW TFD for a signal generated by an absorbing phantom with two scattering surfaces, S1 and S2.

From the DW TFD, one can observe that the two scattering surfaces are clearly spatially resolved, and that the spectrum from S2 exhibits absorption with a loss of intensity at specific wavenumbers. To assess the fidelity of the spectral information, Figure 3.4a and b shows the spectrum from the light returned from the two coverglass surfaces, S1 and S2, respectively. Here, we note that the spectral resolution of the STFT and the DW are comparable to that of the ideal case (obtained by a transmission measurement). However, the time marginals (TFD integrated with respect to wavenumber), shown in Figure 3.4c, demonstrate the superior depth resolution of the DW method compared to that of an optimized STFT. The spatial resolution of the STFT can be improved by choosing a wider spectral window; however, this process would degrade the spectral resolution.

The depth resolved spectra from the DW TFD in Figure 3.4 also contain high-frequency oscillations. These are termed *local oscillations* and can be analyzed to reveal structural information about the sample. By taking a Fourier transform of these oscillations (e.g., blue line in Figure 3.4a), a correlation function is produced (Figure 3.4d), which contains a clear peak at a correlation distance of 21.60 ± 0.57 μm. This distance corresponds to the spacing between the two scattering surfaces and can be compared to the distance obtained by imaging, 21.6 ± 5.97 μm (see Figures 3.3 or 3.4c). By analyzing the periodicity of the local oscillations, the scattering structures in the sample can be precisely measured.

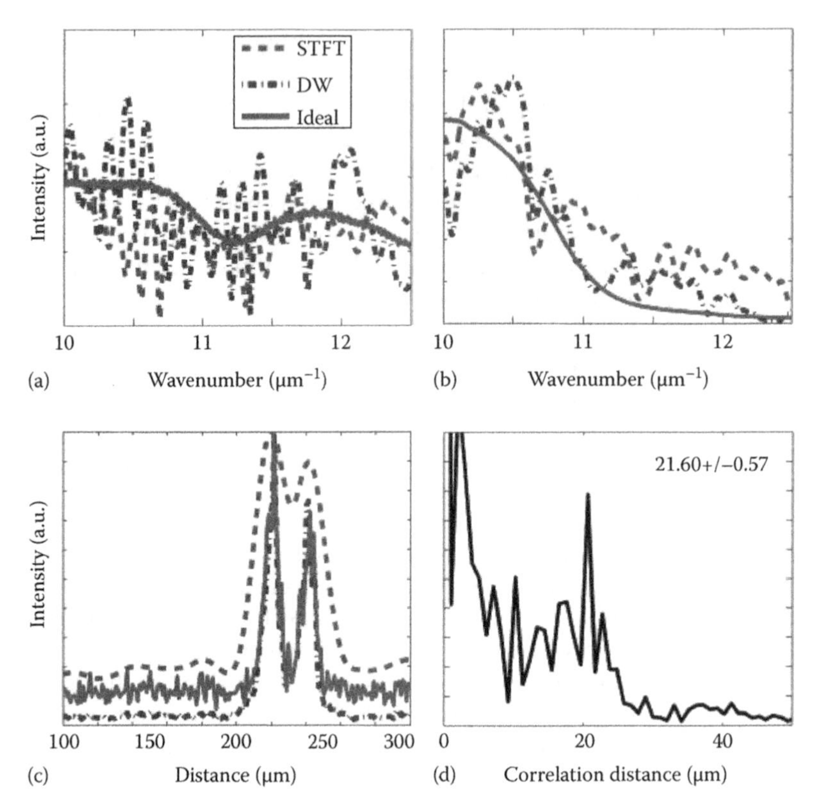

FIGURE 3.4
Depth resolved spectra from S1 (a) and S2 (b). (c) Time marginals. (d) Correlation function calculated by the Fourier transform of the DW depth resolved spectra from S1 (dotted line in (a)). The peak corresponds to the physical distance between S1 and S2. (Adapted from Robles, F. et al., *Opt. Express*, 17(8), 6799, 2009.)

This mechanism is exploited in fLCI to recover information about the cell nuclear diameter in tissue [12]. A more rigorous mathematical description on the origin of the *local oscillations* can be found in Reference [12].

3.3 Fourier Domain Low Coherence Interferometry: Analysis of Scattering Features

The fLCI approach analyzes depth-dependent, high-frequency spectral oscillations (described earlier as local oscillations) to measure the size of dominant scatterers in tissue. Through several studies of in vitro cells and tissues, the fLCI measurement has been associated with the diameter of cell nuclei [13–15]. As Figure 3.5 illustrates, light scattered by cell nuclei contains two components—one

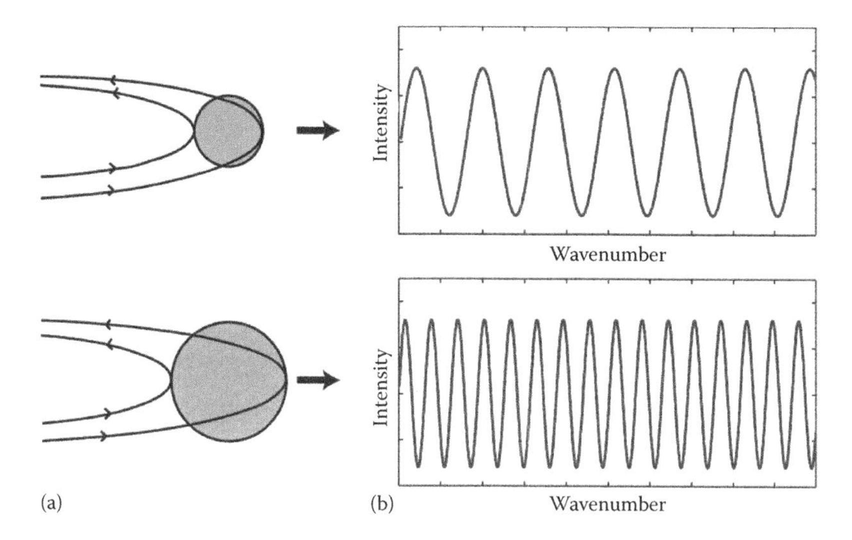

FIGURE 3.5
(a) Light is incident on cell nuclei of different sizes. (b) Localized spectra which exhibit local oscillations resulting from the induced temporal coherence of the light that is scattered from the front and back surface of the nuclei. The periodicity of the local oscillations is proportional to the size of the cell nuclei. (Adapted from Graf, R. et al., *J. Biomed. Opt.*, 14, 064030, 2009.)

originating from the front surface and the second arising from the back surface—which interact to produce periodic oscillations in the resultant local scattering spectrum. The frequency of the oscillations is proportional to the optical path length difference of the two scattered field components. An important feature of this analysis is that the local oscillations are specific to the individual scatterer, and thus high spatial and spectral resolution are required to observe them. A loss of resolution in either domain will result in a decrease in visibility of these fringes and prevent an effective analysis.

The first demonstration of the fLCI technique was accomplished using a scattering phantom consisting of dried polystyrene beads on a coverglass [17]. A subsequent study validated the ability of the approach to measure cell nuclei in vitro, confirmed via confocal microscopy [18]. The planar geometry of these samples provided two-dimensional surfaces, and thus STFT processing could be used since the loss of depth resolution was not relevant. However, the trade-off in resolution associated with STFTs precluded analysis of local oscillations in more complex, three-dimensional biological samples.

To validate the DW fLCI approach, an analysis of the scattering properties of a turbid sample containing different populations of scatterers is presented. The analysis consists of processing an OCT image with the DW method and then using fLCI and light scattering spectroscopy (LSS) to obtain structural information. LSS extracts the size of spherical scatterers by analyzing periodicity in the spectrum of singly scattered light [19]. The periodic spectral features correspond to the scattering cross section of spherical structures, and are well described by the Van de Hulst approximation [20]. These features are

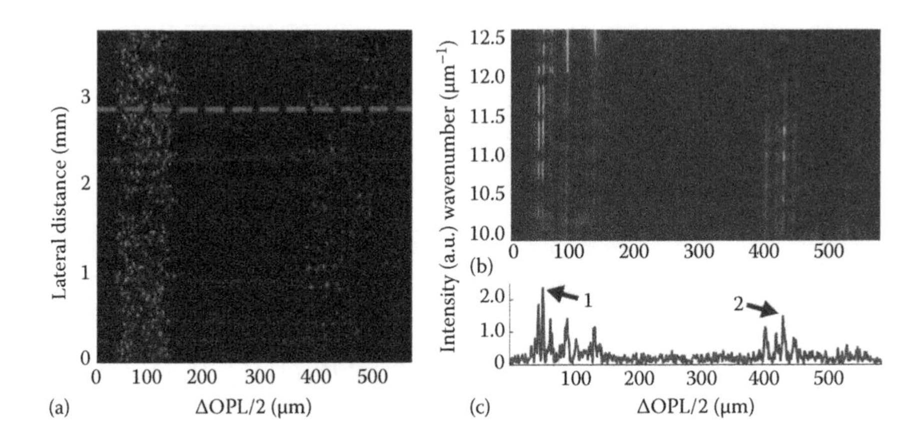

FIGURE 3.6
(a) Parallel frequency-domain OCT image of a two-layer phantom. The top layer (optical path length difference/2 [ΔOPL/2] ranging from 35 to 125 μm) contains 4.00 μm beads, while the bottom layer (400–550 μm) contains 6.98 μm beads. The dashed line corresponds to a single lateral line or A-scan (c) from which the DW TFD is generated (b). The spectra and correlation function of points 1 and 2 are analyzed later in Figure 3.7. (Adapted from Robles, F.E. and Wax, A., *Opt. Lett.*, 35(3), 360, 2010.)

distinct from the local oscillations used in fLCI and exhibit lower-frequency spectral oscillations. Thus, fLCI and LSS measurements can be obtained simultaneously by looking at the high- and low-frequency components of the depth resolved spectra, respectively.

The sample for this experiment comprised two layers, each containing a different size of polystyrene microsphere (index of refraction $n_b = 1.59$). The microspheres are suspended in a mixture of agar (2% by weight) and water ($n_a = 1.35$). The average size in the top layer is $d = 4.00 \pm 0.033$ μm, and in the bottom layer $d = 6.98 \pm 0.055$ μm. Figure 3.6a shows an OCT image of the phantom. The ability to obtain functional information from high-resolution OCT (and SOCT) images is a strength of the fLCI analysis since OCT inherently provides visual (or spatial) guidance for its functional analysis. The DW TFD is generated for each lateral line (or A-scan) of the OCT image with window sizes $w_1 = 0.0454$ μm^{-1} ($\Delta\lambda = 2.39$ nm) for high spectral resolution and $w_2 = 0.6670$ μm^{-1} ($\Delta\lambda$ = 35.1nm) for high depth resolution. The resulting DW TFD, computed from the A-scan delineated by the dotted red line in Figure 3.6a, is shown in Figure 3.6b. Figure 3.6c is the corresponding A-scan [21].

The depth resolved DW spectra contain two components that relay structural information of the sample. The low-frequency component is analyzed using the Van de Hulst approximation [20], which describes the wavelength-dependent cross section of the scatterers (LSS method). The analysis is implemented by first low-pass filtering the DW spectral profile with a hard cutoff frequency of 3.5 μm (three cycles) and the resulting distribution is fit to the theoretical Van de Hulst model using a least-squares method. The low-pass filtered data used for fitting are shown in Figure 3.7a and b, which yield sizes

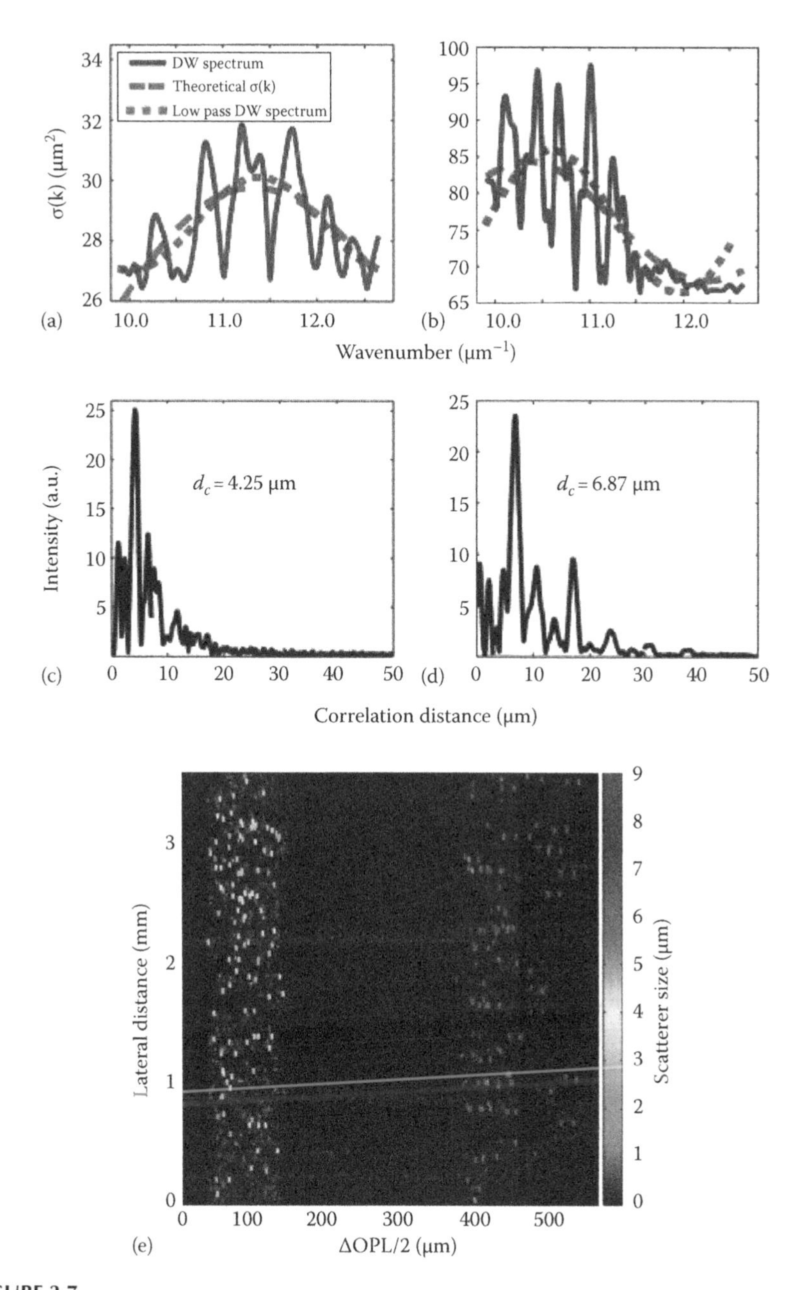

FIGURE 3.7
(a) and (b) DW spectral profiles (solid lines) and low-pass filtered spectra (dotted curves) from points 1 to 2 in Figure 3.6c, respectively. Dashed lines are the theoretical scattering cross sections for 3.97 and 6.91 μm spherical scatterers for points 1 and 2, respectively. (c) and (d) Correlation function from points 1 and 2, with correlation distances (d_c) of 4.25 and 6.87 μm, respectively. (e) Overlay of the fLCI measurements with the OCT image. (Adapted from Robles, F.E. and Wax, A., *Opt. Lett.*, 35(3), 360, 2010.)

of $d_1 = 3.97$ μm and $d_2 = 6.91$ μm for points 1 and 2 in Figure 3.7c, respectively, in good agreement with the known sizes: 4.00 ± 0.033 and 6.98 ± 0.055 μm [21].

The second component of the depth resolved DW spectra, which consists of the local oscillations (high frequencies), is analyzed using fLCI. In this case, the local oscillations arise from induced temporal coherence effects between the fields scattered by the front and back surfaces of the microspheres. To compute structural information from this component, the spectral dependence is removed by subtracting the line of best fit from the LSS analysis described earlier. The residuals are then Fourier transformed to yield a correlation function [22], where the maximum peak indicates the round-trip optical path length (ΔOPL) through the scatterer. Figure 3.7c and d plot the correlation function for points 1 and 2, from Figure 3.6c, respectively. The correlation peaks for these points are located at $d_c = \Delta OPL/(2n_b) = 4.25$ and 6.87 μm, in good agreement with both the LSS measurements and the known bead size in each layer [21].

The LSS and fLCI analyses were repeated for each A-scan and all points in the OCT image where the intensity exceeded the background level. Figure 3.7e shows the results by overlaying the fLCI measurements with the OCT image. The LSS map (not shown) yields similar results. In the top layer, the average scatterer size was 3.82 ± 0.67 and 3.68 ± 0.41 μm for the fLCI and LSS measurements, respectively (112 points). In the bottom layer, the average scatterer size was 6.55 ± 0.47 and 6.75 ± 0.42 μm for fLCI and LSS, respectively (113 points) [21].

The results of this phantom study show that the DW fLCI technique yields accurate and precise structural measurements based on spectral analysis of an OCT image. Previous efforts have only been able to recover limited details of scatterer structure, typically restricted to discriminating large and small objects. However, with the DW method, the resolution trade-off was avoided and enabled detection of local oscillations in tissue. With an eye toward early detection of cancerous lesions, Graf et al. applied fLCI to detect dysplasia using an ex vivo hamster cheek pouch carcinogenesis model. The results from this study yielded 100% sensitivity and 100% specificity, clearly demonstrating the potential of fLCI as a powerful diagnostic method [16].

To illustrate the utility of fLCI-based analysis of DW-processed OCT images, we present a study of ex vivo tissues drawn from the azoxymethane (AOM) rat carcinogenesis model, a well-characterized and established model for colon cancer research and drug development [23]. The model generates cancerous lesions in the colonic epithelium, arising from neoplastic aberrant crypt foci (ACF). In this study, we seek not only to use fLCI to identify the presence of ACF but also to identify systemic changes in the tissue characteristic of the field effect of carcinogensis. The phenomenon of the field effect describes observations that neoplastic development in one part of the colonic epithelium distorts nano- and microtissue morphology, as well as tissue function, along the entire organ. This phenomenon has been a subject of much interest since it indicates that adequate screening may be achieved by only probing certain (and more readily accessible) sections of the colon [24–26], thus suggesting a new avenue for less invasive screening for colorectal cancer (CRC). This study exploits the depth resolution of fLCI to compare the functional data of the ex vivo samples at three depths to identify changes indicative of early CRC development. The sensitivity of fLCI to the field effect is explored through measurements at two different longitudinal sections of the left colon.

All animal experimental protocols were approved by Institutional Animal Care and Use Committee of The Hamner Institute and Duke University. Forty F344 rats were randomized into groups of 10, with 30 receiving intraperitoneal (IP) injections of AOM >90% pure with a molar concentration of 13.4 M at a dose level of 15 mg/kg body weight, once per week, for two consecutive weeks (two doses per animal). The remaining 10 animals received saline by IP and served as the control group. At 4, 8, and 12 weeks after the completion of the dosing regimen, the animals (10 AOM-treated and 3 or 4 saline-treated rats per time point) were sacrificed by CO_2 asphyxiation. The colon tissues were harvested, opened longitudinally, and washed with saline. The tissue was split into —four to five different segments, each with a length of 3–4 cm, where only the two most distal segments were analyzed: the distal left colon (LC) and proximal LC. After imaging, the tissue samples were fixed in formalin and stained with methylene blue in order to be scored based on the number of ACF.

Each sample was placed on a cover glass for examination with a parallel frequency domain (pfd) OCT system. This system enables acquisition of up to 400 OCT A-scans with a single camera exposure of 20–100 ms, depending on the choice of light source and sample. This system achieved a lateral resolution of 10 μm and an axial resolution of 1.1 μm. Since the fLCI analysis retrieves optical path length rather than physical thickness, an index of refraction of n = 1.395 is used to convert the optical path length to nucleus size [27]. For the generation of the DW TFDs, the window standard deviations used were $w_1 = 0.029\ \mu m^{-1}$ and $w_2 = 0.804\ \mu m^{-1}$, resulting in TFDs with an axial resolution of 3.45 μm and spectral resolution of 1.66 nm. The spatial information provided by the OCT images is used to co-register the spectroscopic information as a function of distance from the tissue surface, allowing for consistent analysis at specific tissue depths.

Once the depth segments are aligned, regions of interest are selected for analysis and the data are averaged for that region in order to provide sufficient fringe contrast of the local oscillations. In the lateral direction, 20 DW TFDs are averaged to yield 10 different lateral segments in each OCT image. In the axial direction, we calculate the spectral averages of 25 μm depth segments from three different sections: the first at the surface (surface section 0–25 μm), a second centered about 35 μm in depth (mid-section 22.5–47.5 μm), and a third centered about 50 μm in depth (low section 37.5–62.5 μm). Figure 3.8a illustrates a typical OCT image for an AOM-treated rat tissue sample, where the dotted line delineates a region of interest for the mid-depth section. The corresponding DW TFD (after alignment and averaging laterally) is shown in Figure 3.8b.

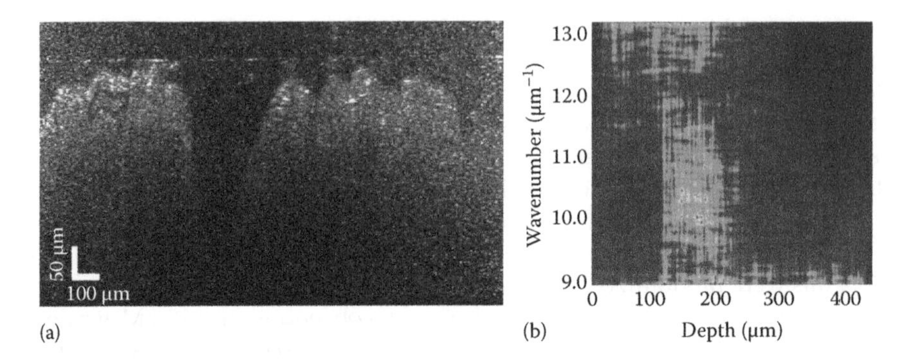

FIGURE 3.8
(a) pfdOCT image of an ex vivo rat colon sample. The line delineates an example region that is averaged across to determine the nuclear diameter. (b) DW TFD (after alignment and averaging laterally), of the laterally delineated region. (Adapted from Robles, F. et al., *Biomed. Opt. Express*, 1(2), 736, 2010.)

As described previously, the spectra from the averaged regions contain two components. The first component is the low-frequency oscillations, which may be analyzed with LSS; however, we have found that due to the lack of knowledge of the precise refractive index of the scatterer and the surrounding medium in tissue [29], the amount of useful information that can be extracted from this method is limited. Instead, the low-frequency oscillations here are isolated using a smoothing function and subtracted from the spectra. This process isolates the second component: the high-frequency components of the spectra arising from the local oscillations due to temporally coherent fields induced by the cell nuclei in the averaged region (fLCI method). Note that we assume that the cell nuclei contain a constant nuclear RI of $n_n = 1.395$ for this analysis [27]. Figure 3.9a illustrates the depth resolved spectrum (blue solid line), along with the low-frequency component (dotted black line), for the average region shown in Figure 3.8a. Figure 3.9b shows the isolated local oscillations, and Figure 3.9c, is the corresponding correlation function with a peak at 7.88 μm corresponding to the average cell nuclear diameter in the region of analysis.

Statistical analysis of the data (using a two-sided student *t*-test) revealed that the most significant layer for diagnosis using fLCI was the mid-depth section, which includes the basal layer of the epithelium. This layer has been shown to be the most diagnostically useful for discriminating dysplastic tissues using light scattering analysis [30–34]. Based on this finding, data from the two tissue segments (proximal and distal left colon) were analyzed separately only for this depth section. The measured cell nuclear diameters and number of ACF are summarized in Figure 3.10. The data show that for all the time points, and for both segments, the measured nuclear diameters for the treated groups were significantly different from the control group (p-values $< 10^{-4}$**). Furthermore, significant differences were observed for both segments after only 4 weeks post treatment. This measured increase in the

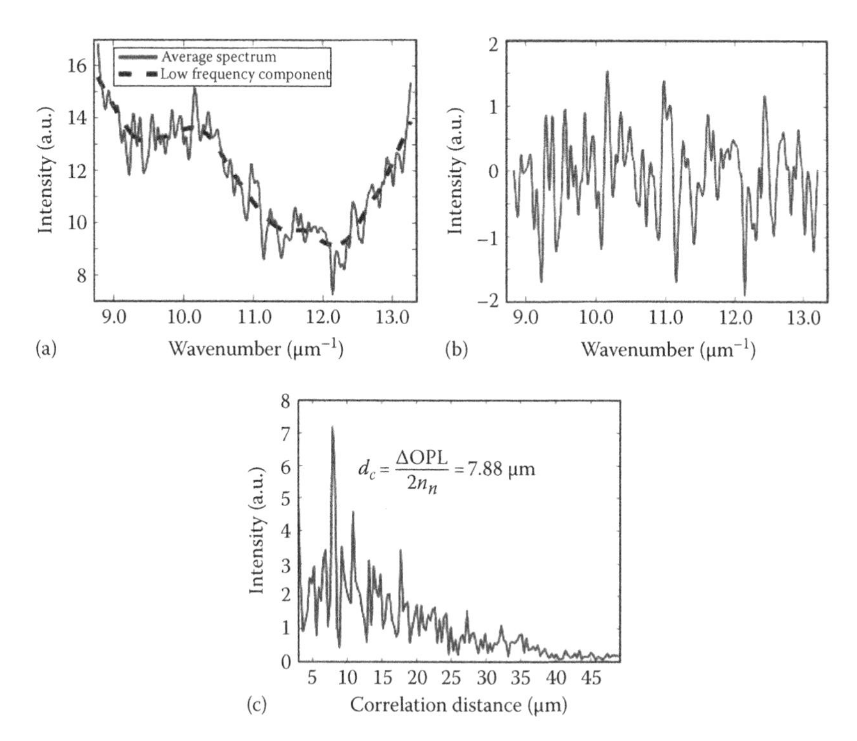

FIGURE 3.9
Average spectrum (solid line) from the delineated region in Figure 3.8, along with the low-frequency component (dotted line) (a). The low-frequency component is subtracted from the averaged spectrum to obtain the local oscillations (b). A Fourier transform yields a correlation function (c), where the peak corresponds to an average cell nuclear diameter of 7.88 μm in the region of analysis. (Taken from Robles, F. et al., *Biomed. Opt. Express*, 1(2), 736, 2010.)

nuclear diameter, however, remained relatively constant thereafter, with the exception of the last time point in the proximal LC. Here, the nuclear diameter increased dramatically from ~6.0 to ~7.2 μm. To investigate this effect further, Figure 3.10c plots the nuclear diameter as a function of the average number of ACF, which are neoplastic lesions in this model. For clarity, we also identify each point with its corresponding time period. Note that the formation of ACF was faster in the proximal LC compared to the distal LC, and that the plot shows a region of little nuclear morphological change after the initial formation of ACF. This plateau region is present in both sections and is initially independent of the number of ACF. However, once the number of ACF increased to the maximum value observed in this study (~70), the measured increase of the nuclear diameter was specific to the region manifesting more advanced neoplastic development, in contrast to the ubiquitous and relatively constant cell nuclear diameter measurements of the plateau region [28]. These data suggest the presence of a systematic change in the tissue upon formation of neoplastic lesions in this model, suggesting that a field effect has been detected.

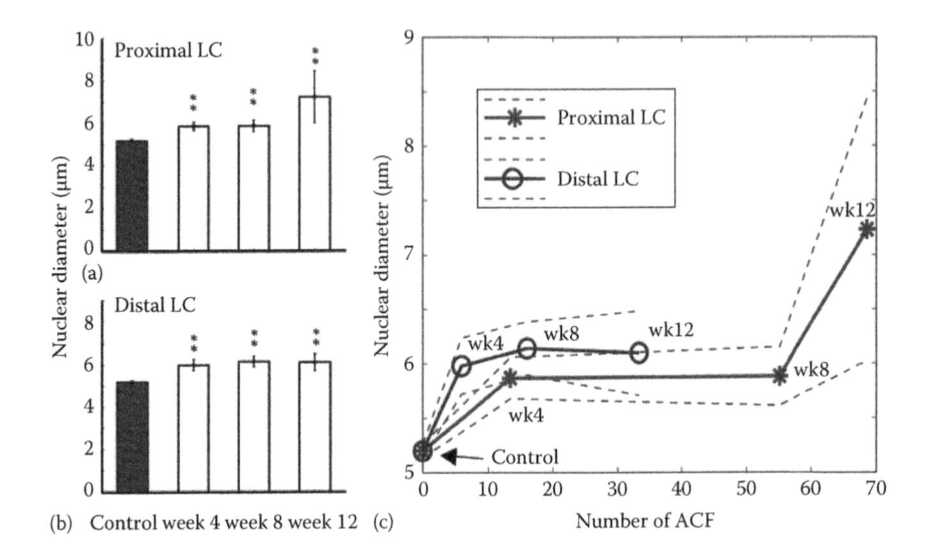

FIGURE 3.10
fLCI nuclear morphology measurements at different colon length segments. Highly statistical differences (p-value <10^{-4**}) were observed between the control group and the treated groups for the proximal LC (a) and distal LC (b). (c) Plot of the measured cell nuclear diameter as a function of the number of ACF. For clarity, the time of measurement is noted next to each point (wk = week). (Taken from Robles, F. et al., *Biomed. Opt. Express*, 1(2), 736, 2010.)

This study showed that fLCI can be used to detect structural changes associated with neoplastic development, demonstrating the potential utility of the method. Further, the measured early nuclear morphological changes were observed in both tissue segments and independently of the number of ACF, indicating detection of a ubiquitous micromorphological change of the colonic epithelium due to the field effect. However, when neoplastic development became more advanced (demarcated by the high number of ACF), the nuclear diameter increase was specific to the affected region, supporting the claim that fLCI can identify specific regions where advanced neoplastic development has occurred. Both of these aspects point to a clinical utility of fLCI for detecting CRC development and identifying its location for targeting therapy.

3.4 Molecular Imaging True Color Spectroscopic (METRiCS) OCT: Analysis of Absorption Features

Molecular imaging holds a pivotal role in medicine, owing to its ability to provide invaluable insight into disease mechanisms at the molecular and cellular level. To this end, various techniques have been developed for molecular imaging, each with its own advantages and disadvantages [35–38].

For example, fluorescence imaging achieves micrometer-scale resolution, but has low penetration depths and is mostly limited to exogenous agents. In addition to enabling analysis of scattering features, the DW method can be applied to analyze depth-resolved, absorption spectral signatures of both endogenous and exogenous chromophores, enabling a novel form of SOCT capable of molecular imaging. The compelling feature of this new imaging modality is that simultaneous structural/morphological information is provided with micrometer-scale resolution in three dimensions

A key advance which has enabled METRiCS OCT is the use of a supercontinuum laser light source in the pfdOCT system instead of the thermal light sources used in the experiments mentioned earlier. While both sources provide access to the visible region of the spectrum, the supercontinuum source avoids signal washout or loss of SNR due to coherent addition of multiple modes. With the thermal light source, this was mitigated by collimating the light incident on the sample, thereby separating these modes but resulting in a significant amount of light being discarded. Thus, imaging of biological samples could be achieved but the use of the thermal light source limited SNR to 89 dB [39], lower than most OCT systems (typically around 120 dB [4]). As an alternative source, super continuum laser sources offer a range that spans the visible spectrum yet offers a single spatial mode. This aspect enables focusing of the light to a diffraction limited point (or line) and allows for more photons to be collected, increasing SNR.

Figure 3.11 shows the pfd OCT system using the supercontinuum source (Fianium SC450–4, spectral range = 400–2000 nm, output power = 4 W,

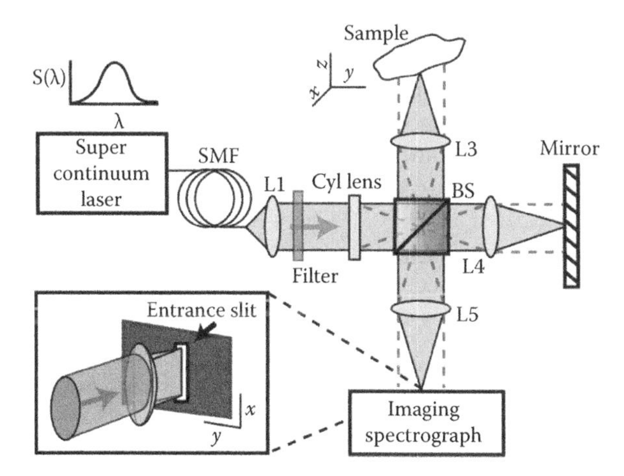

FIGURE 3.11
pfdOCT system with a super continuum laser source. The cylindrical lens delivers a line of illumination on the sample. The light path traced with the solid lines shaded in gray is in the y–z plane, where the light is focused onto the sample. The dotted line traces the light path in the x–z plane, where light is collimated onto the sample. (Taken from Robles, F.E., Light scattering and absorption spectroscopy in three dimensions using quantitative low coherence interferometry for biomedical applications, PhD thesis in Medical Physics, Duke University, Durham, NC, 2011.)

average power spectral density = 2 mW/nm). A series of filters are used to smooth the source spectrum across the detection bandwidth of 240 nm centered at 575 nm. Note that the NIR components are not used for imaging here. This system yields an experimental axial resolution of 1.2 μm (limited by the source bandwidth and system dispersion) and penetration depth of 0.4 mm (limited by the spectral resolution of the spectrograph).

In this system, the light is focused in one dimension using a cylindrical lens, producing a line of illumination at the sample and reference mirror. Light returned from the sample is mixed with the reference field at the beamsplitter, and imaged onto the entrance slit of the imaging spectrograph. The lateral resolution is 6 μm, achieved by setting the focal lengths of L3 and L4 to 100 mm and L5 to 300 mm. Here, the use of relatively long focal length lenses minimizes the effect of chromatic aberrations.

For biomedicine, molecular imaging focuses on tissue blood content and oxygen levels, which are instrumental in diagnosing diseases. To obtain this information, METRiCS OCT takes advantage of the ability to measure depth resolved spectra to analyze the very distinct absorption features of oxygenated (oxy-) and deoxygenated (deoxy-) Hb. To illustrate how absorbing signatures may be quantified with SOCT, we consider an (S)OCT signal from one scatterer or reflector:

$$I(\lambda) = 2\sqrt{I_s(\lambda) I_r(\lambda)} \cos\left(\frac{2\pi}{\lambda \times d}\right), \tag{3.2}$$

where $I_R = |E_R|^2$ is the reference field intensity, $I_S = |E_S|^2$ is the sample field intensity, and d is the round trip optical path length difference between the signal and the reference fields. In the case of a strongly absorbing sample, where scattering does not influence the spectra significantly, the modulation of the source field intensity may be described by $S'(\lambda, z_S^{(1)}) = S(\lambda) \cdot \exp(-2\mu_a(\lambda) \cdot L)$, where $\mu_a(\lambda)$ is the absorption coefficient and $L = d/(2n)$ is the distance from the surface of the absorbing sample to the scattering/reflecting point. Thus, the sample field *intensity* may be described as $I_S = R\,S(\lambda) \cdot \exp(-2\mu_a(\lambda) \cdot L)$, where R is the reflectance of the sample. Lastly, consider that that reference field intensity is equivalent to that of the source field, $I_R = S(\lambda)$, such that the OCT signal can be described as

$$I(\lambda) = 2S(\lambda) \cdot re^{-\mu_a(\lambda)L} \cos\left(\frac{2\pi}{\lambda \cdot d}\right), \tag{3.3}$$

where r is the reflectivity of the sample, $r = \sqrt{R}$. Note that the intensity depends on L and not $2L$ resulting from the square root term in Equation 3.2.

Upon processing the SOCT signal with the DW method, the spectra obtained for an optical path length $d = 2Ln$ are expected to be modulated directly by Beer's law, thus the Hb concentration (C_{Hb}) may be calculated by using

$$-\ln\left(\frac{I(\lambda, L)}{I_0}\right) = C_{Hb} L \varepsilon(\lambda) - \ln(r), \tag{3.4}$$

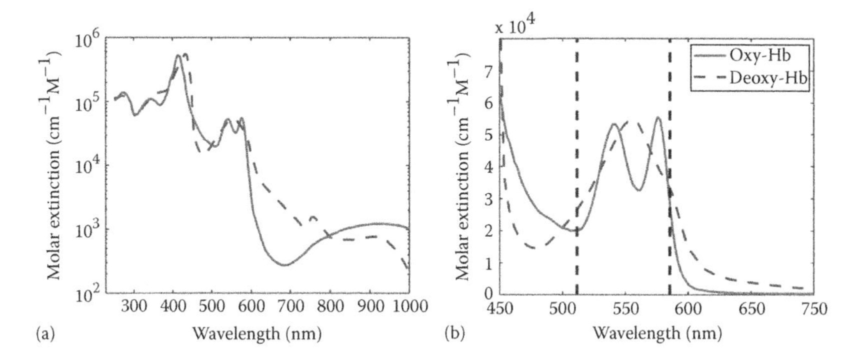

FIGURE 3.12
Molar extinction coefficients of oxy-/deoxy-Hb over a large spectral range (a), and across the visible region of the spectrum (b). The dotted black lines in (b) delineate the region where the oxy- and deoxy-Hb coefficients exhibit the greatest dissimilarity (correlation $R \sim 0$). (Taken from Prahl, S., *Optical Absorption of Hemoglobin*, 1999, http://omlc.ogi.edu/spectra/hemoglobin/; Robles, F.E. et al., *Biomed. Opt. Express*, 1(1), 310, 2010.)

where $\varepsilon(\lambda)$ is the extinction coefficient, which is independent of C_{Hb}: $\mu_a(\lambda) = C_{Hb} \cdot \varepsilon(\lambda)$. Here, the Hb molecular weight (64,500 g/mole) is used in calculating ε, shown in Figure 3.12. Figure 3.12 plots the molar extinction coefficients of oxy- and deoxy-Hb. Note that in the visible spectrum, where our pfdOCT system operates, Hb absorption is significantly stronger than in the NIR wavelength range of typical (S)OCT systems, ~800 nm and above. As a result, METRiCS OCT offers more sensitivity to low concentrations and small changes of Hb.

To illustrate this improved sensitivity, we present METRiCS OCT measurements from oxy- and deoxy-Hb phantoms prepared at varying concentrations ranging from ~5 to 70 g/L (~78 to 1085 μM). For the oxy-Hb samples, human ferrous stable Hb in lyophilized powder form was diluted in purified water until the desired concentration was achieved. To produce deoxy-Hb, a trace amount of sodium dithionite, which removes the oxygen in oxy-Hb, was introduced to a solution of the Hb powder and phosphate buffered saline (PBS). The samples were placed in a container composed of two glass slides separated by spacers of thickness $L \approx 400$ μm and the spectrum from the light reflected from the front of the rear cover glass of the sample was analyzed.

Typical measured spectra of oxy- and deoxy-Hb are shown in Figure 3.13a and b compared to the theoretical absorption curves (dotted black lines). The corresponding attenuation coefficients are shown in Figure 3.13c and d. Note that measured parameters are in excellent agreement with theoretical predictions with clear absorption peaks at 540 and 575 nm for oxy-Hb, and at 555 nm for deoxy-Hb.

To assess the sensitivity of the method, the attenuation profiles were determined for each sample and the Hb concentration was calculated by fitting the data to a line of the form $y = mx+b$, using a linear least squares method.

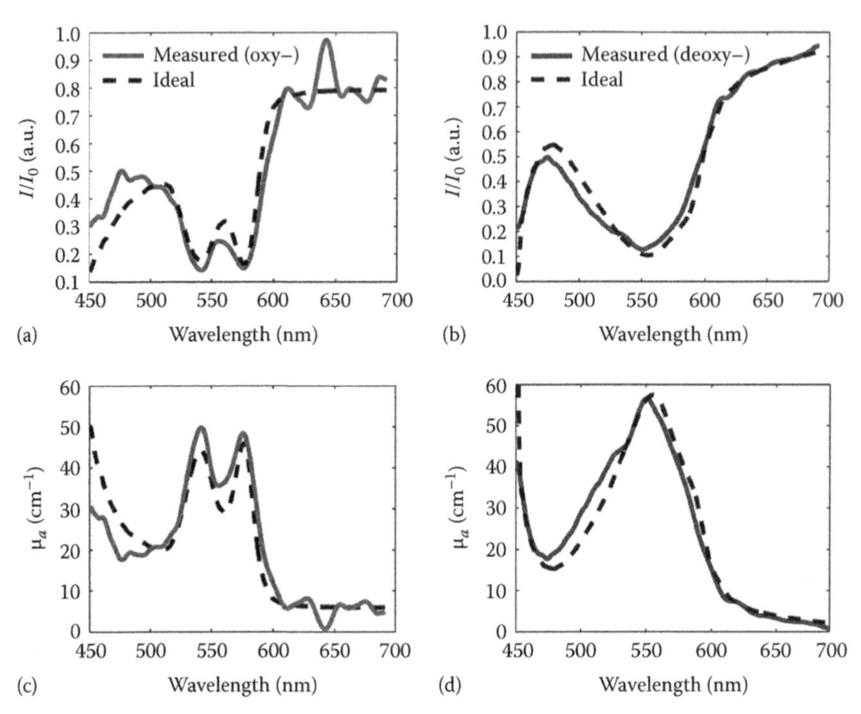

FIGURE 3.13
Measured absorption spectra for oxy-Hb (a) and deoxy-Hb (b) for concentrations of 50 and 68 g/L, respectively (solid lines). Dotted black lines are the theoretical predictions. Measured and theoretical attenuation coefficients for oxy-Hb (c) and deoxy-Hb (d). (Taken from Robles, F.E. et al., *Biomed. Opt. Express*, 1(1), 310, 2010.)

Here, x is either the oxy- or deoxy-Hb extinction coefficient and the computed $m = C_{Hb}L$. The results for oxy- and deoxy-Hb are summarized in Figure 3.14. The data show excellent agreement with the expected values producing linear fits with $R^2_{oxy} = 0.9957$ and $R^2_{deoxy} = 0.9600$. The average standard deviation for all measurements is 3.10 g/L, representing the lower limit of Hb concentration that may be detected at this depth. A more useful parameter is obtained by multiplying the minimum measurable concentration by the depth, which gives a depth-independent value of $C_{Hb-min} = 1.2$ g/L at 1 mm (noted as g/L-mm). It is important to note that the total Hb concentration in normal tissue has been reported to be ~1.8 g/L, and as much as 3.2 times higher for cancerous tissue [42].

For biological samples, another important factor that influences absorption spectra is the oxygen saturation state of Hb. When including both oxy- and deoxy-Hb absorption, the attenuation coefficient can be written as $\mu_a(\lambda) = C_{HbO_2} \cdot \varepsilon_{HbO_2}(\lambda) + C_{Hb} \cdot \varepsilon_{Hb}(\lambda)$, where ε_{HbO_2}, ε_{Hb}, C_{HbO_2}, and C_{Hb} are the molar extinction coefficients and concentrations for oxy-/deoxy-Hb,

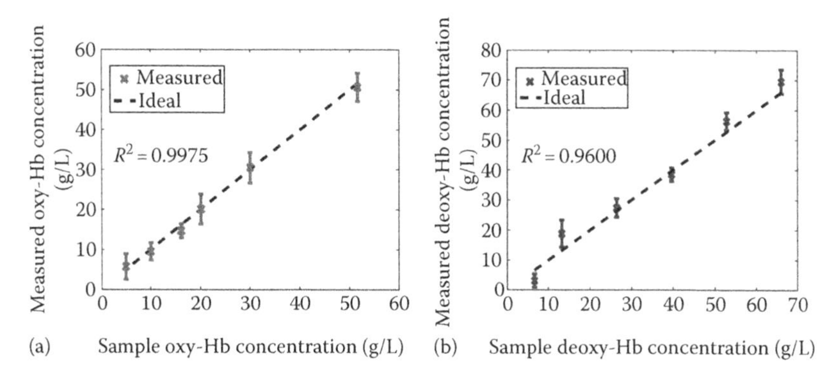

FIGURE 3.14
Measured Hb concentration for oxy- (a) and deoxy- (b) Hb samples. Error bars given by standard deviation across 25 measurements. Dotted black lines give the ideal trend with unity slope and zero intercept. (Taken from Robles, F.E. et al., *Biomed. Opt. Express*, 1(1), 310, 2010.)

respectively. Using the same formalism as with Equation 3.4, an overdetermined set of linear equations to solve for C_{HbO_2} and C_{Hb} may be written as

$$-\frac{1}{L}\ln\begin{bmatrix}\frac{I}{I_0}(\lambda_1)\\ \frac{I}{I_0}(\lambda_2)\\ \vdots\\ \frac{I}{I_0}(\lambda_n)\end{bmatrix}=\begin{bmatrix}\varepsilon_{HbO_2}(\lambda_1) & \varepsilon_{Hb}(\lambda_1) & -1/L\\ \varepsilon_{HbO_2}(\lambda_2) & \varepsilon_{Hb}(\lambda_2) & -1/L\\ \vdots & \vdots & \vdots\\ \varepsilon_{HbO_2}(\lambda_n) & \varepsilon_{Hb}(\lambda_n) & -1/L\end{bmatrix}\begin{bmatrix}C_{HbO_2}\\ C_{Hb}\\ \ln(r)\end{bmatrix} \tag{3.5}$$

where λ_i for $1 \le i \le n$ are the discretely measured wavelengths within the operating region. Finally, partial oxygen saturation (SO_2) can be calculated as

$$SO_2 = \frac{C_{HbO_2}}{C_{HbO_2} + C_{Hb}}. \tag{3.6}$$

The ability of our method to correctly measure oxygen saturation can be characterized by determining how well Equation 3.5 can be used to correlate measured data with the correct chromophore species. To determine the lowest concentration for which oxygen saturation states may be correctly measured, the correlation coefficient of the linear regression used to determine concentration in the previous analysis was calculated. Figure 3.15a shows the calculated correlation between the oxy-Hb data and the theoretical oxy- and deoxy-Hb extinction coefficients, and Figure 3.15b shows the same for the deoxy-Hb data.

Good agreement ($R > 0.65$) is observed for concentration values higher than 10 g/L if the data are compared to the correct species of Hb. A sharp cutoff can be seen below this point where the correlation coefficient drops

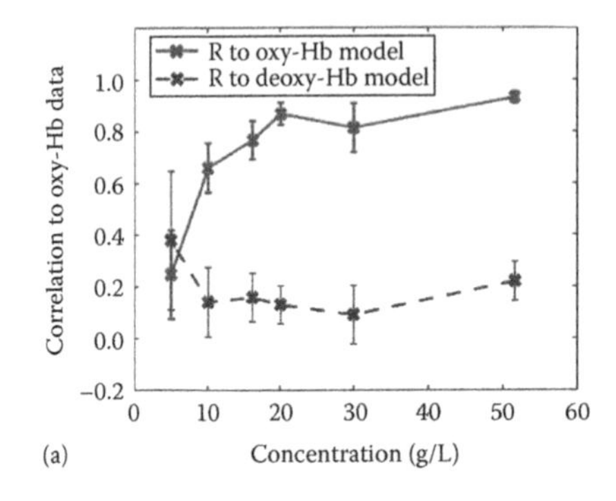

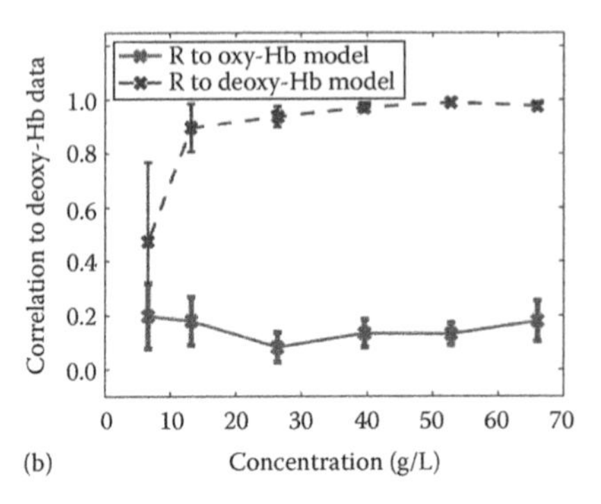

FIGURE 3.15
(a) Correlation coefficients between oxy-Hb data and ideal oxy-/deoxy-Hb extinction coefficients at varying concentrations. (b) Correlation coefficients between the deoxy-Hb data and the oxy-/deoxy-Hb extinction coefficients at varying concentrations. Defining the detectable concentration as those with correlation coefficients greater than 0.65, a threshold of 4 g/L-mm is determined for accurately assessing oxygen saturation. (Taken from Robles, F.E. et al., *Biomed. Opt. Express*, 1(1), 310, 2010.)

drastically. From this analysis, we see that while concentration values may be determined as low as 1.2 g/L-mm, oxygenation states can only be accurately determined for concentrations above 4 g/L-mm (= 10 g/L × 0.4 mm).

Molecular imaging true color spectroscopic (METRiCS) OCT incorporates the imaging capabilities of OCT with the spectral analysis described earlier. While the analysis of Hb concentration and oxygenation state points to the utility of the method, it can also be applied to achieve molecular imaging of any absorber in the visible region of the spectrum, including endogenous and exogenous chromophores. In addition, because the system employs the visible region for imaging, the sample can be presented using its true colors offering an accessible, intuitive display of the spectroscopic information.

To demonstrate true color imaging, analysis of a sample consisting of a piece of paper with three lines, red, green, and blue, is presented. Figure 3.16a shows a photograph and pfdOCT image of the piece of paper. As previously noted, pfdOCT imaging reveal the structures of the samples, which in this case includes the indentation on the paper from the pens used to draw the lines; in addition, some absorption can be observed by the loss of intensity, particularly for the red and blue lines but spectral discrimination cannot be obtained by intensity analysis alone. To obtain the spectral information, each A-scan is processed with the DW method. Figure 3.16b shows the spectra from four points, labeled R, G, B, and W with arrows in Figure 3.16a. Here, the spectra were low pass filtered using a 7th-order Butterworth filter with a cutoff of 15 cycles. The spectrum obtained for a point below the red line (point labeled R) clearly shows a peak intensity at ~630 nm, corresponding to what the human eye would perceive as red. Similarly, spectra obtained for points below the green and blue lines on the paper show peaks intensities at ~524 and ~460 nm, respectively; again, in good agreement with what the eye

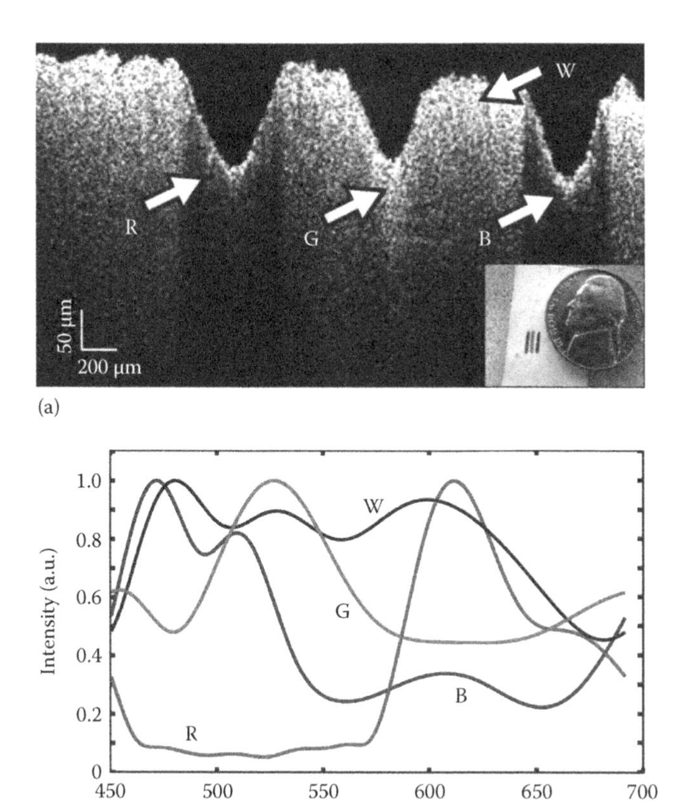

FIGURE 3.16
(a) pfdOCT image of a piece of paper with three lines drawn with colored pens, from left to right: red, green, and blue. Inset shows a photograph of the paper. (b) Spectra obtained with DW processing of OCT image from four points in the image from below the red (R), green (G), blue (B) line, and white (W) regions. (Taken from Robles, F.E., Light scattering and absorption spectroscopy in three dimensions using quantitative low coherence interferometry for biomedical applications, PhD thesis in Medical Physics, Duke University, Durham, NC, 2011.)

perceives for these colors. For completeness, a point corresponding to a white region is also plotted, and as expected, the spectrum spans all wavelengths.

The ability to extract quantitative spectral information from each point of an image is a compelling strength of the DW approach. Thus, to easily interpret this wealth of spectroscopic information, the image can be represented using the sample's true colors by dividing each spectrum into three channels using the Commission Internationale d'Eclairage (CIE) color functions [43,44], which aligns with the chromatic response of the human eye. These functions are plotted in Figure 3.17 as compared with the spectrum of the light source. For image display of the spectral data, the interferograms are not normalized before processing with the DW method, as was done previously for quantifying the spectra. Omitting this step ensures that the

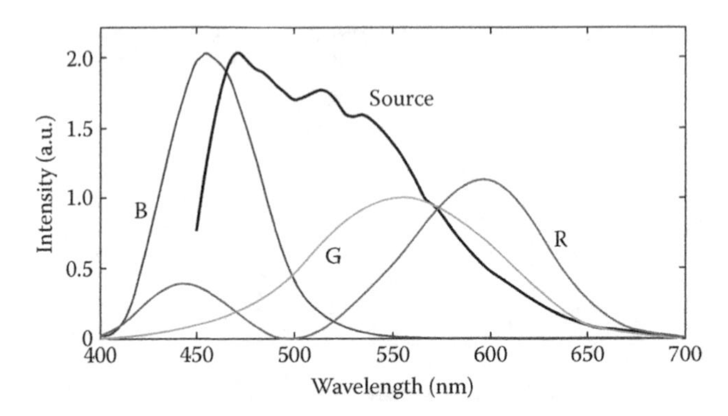

FIGURE 3.17
Commission Internationale d'Eclairage (CIE) color functions [44] and source spectrum.

RGB values are not polluted by noise, particularly in regions of low source intensity. Instead, image white balance is determined as follows [45]: First, the RGB values for each pixel are normalized. Then, the reciprocal average RGB values of a region with no expected spectral modulation are used to remove the influence (or color) of the source. Thus, a region with no absorption is assigned RGB values equal to 1. The result is a color-balanced hue map, which is then multiplied by the normalized pfdOCT image (saturation map) to yield a hue, saturation, and value (HSV) image, producing a true color representation of the sample. Figure 3.18 shows the true color image of the piece of paper produced by this process. Note that the image gives the same structural information as with pfdOCT, but now the colors provide access to the spectral information with an intuitive form of display.

To further define the imaging capabilities of METRiCS OCT, we consider a sample (Figure 3.19a) which consists of glass capillaries with ~100 μm

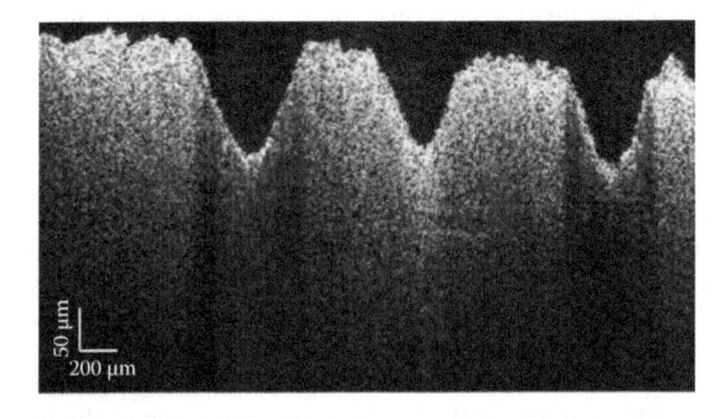

FIGURE 3.18 (See color insert.)
True color OCT image of a paper with a red, green, and blue line. (Taken from Robles, F.E., Light scattering and absorption spectroscopy in three dimensions using quantitative low coherence interferometry for biomedical applications, PhD thesis in Medical Physics, Duke University, Durham, NC, 2011.)

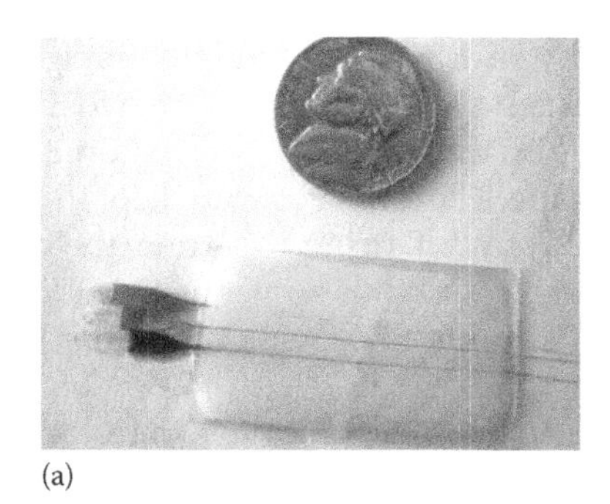

(a)

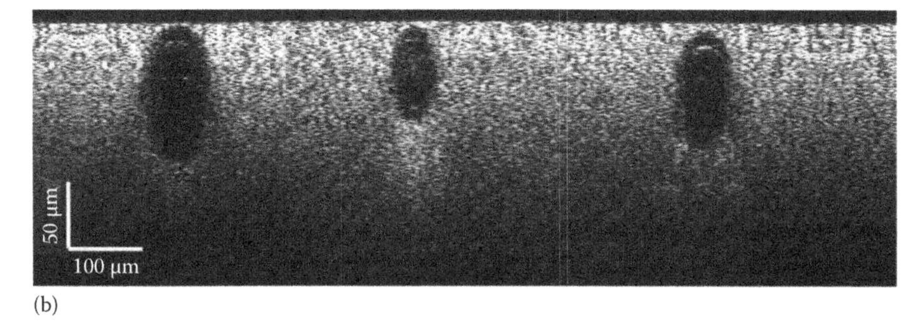

(b)

FIGURE 3.19 (See color insert.)
(a) Photograph and (b) METRiCS OCT image of tissue phantom consisting of intra-lipid and glass capillaries with food-color dyes. (Taken from Robles, F.E., Light scattering and absorption spectroscopy in three dimensions using quantitative low coherence interferometry for biomedical applications, PhD thesis in Medical Physics, Duke University, Durham, NC, 2011.)

maximum outer diameter and ~80 μm maximum inner diameter (Charles Supper, Natick, Ma), filled with red, green, and blue food-coloring dyes and immersed in agar and 2% intralipid (IL). IL is a fat emulsion that is often used as a model for scattering in tissue. Figure 3.19b shows the results of DW processing, where the absorption from the dyes in the capillaries is used to provide a true color display of the sample. A notable feature of this image is that spectral modulation is only observed from below the capillaries, where light is returned from scattering due to IL, owing to the fact that METRiCS OCT, as with conventional OCT, is only sensitive to the light returned from scatterers/reflectors. Note that no signal is present from within the capillaries, since the dyes do not scatter light. Here, spectral modulation is also observed due to the IL itself. This is due to its scattering cross section, which is proportional to $\lambda^{-2.4}$, as described by van Staveren et al. [46], producing the yellow color at deeper sample depths. It is important to point out that METRiCS OCT is sensitive to the total attenuation coefficient, which includes absorption and scattering. However, for regions immediately following a significant amount of absorption, for example, below the capillaries, the color (and hence spectrum) is dictated primarily by the absorption coefficient.

To illustrate the utility of METRiCS OCT for imaging Hb in biological samples, we present images of chick embryos in vivo. For this preparation, a small window was made on the shells of eggs containing 4-day-old embryos, and then covered with tape. The embryos were kept in an incubator at ~38°C and ~80% humidity for 7 days. During this time, the embryos' vasculature network grows considerably and thus serves as a good initial model to demonstrate METRiCS OCT's ability to provide molecular information from Hb absorption in vivo.

For the imaging experiments, the embryos were carefully removed from the shell (see Figure 3.20c) and placed on the motorized translational stage for imaging. While the pfd OCT system obtains *x*–*z*, B-mode images from a single exposure, the sample must be translated in the other (*y*) dimension to generate volumetric images. Figure 3.20a shows a B-mode, METRiCS OCT image of the chick empryo, and Figure 3.20b shows a 3D volume rendition of the entire data set. These samples contain a simple structure, consisting of the vasculature network around the yolk and the chorioallantoic membrane. As with the capillary phantom study presented previously, RBCs within the blood vessels impart a clear spectral modulation that is readily seen by the red shift in hue in locations below the vessels.

The full potential impact of this imaging method can be seen in the *en face* image, presented in Figure 3.21a. Here, several *x*–*y* planes are averaged to incorporate vessels found at different depths. This image clearly depicts

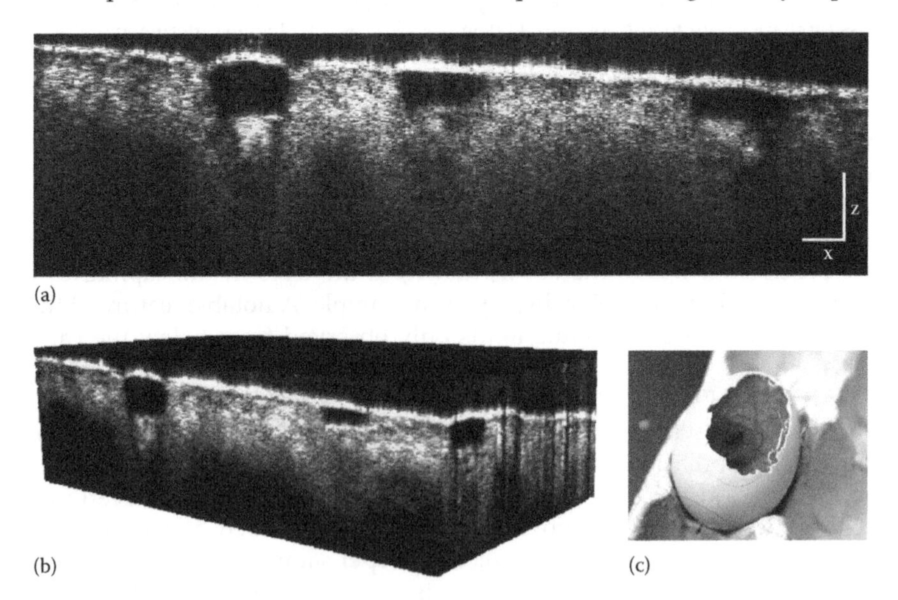

FIGURE 3.20
METRiCS OCT, B-mode image (a) and 3D volume (b) of a chick embryo. White *x* and *z* scale bars, 100 μm. Photograph of the sample (c). (Taken from Robles, F.E., Light scattering and absorption spectroscopy in three dimensions using quantitative low coherence interferometry for biomedical applications, PhD thesis in Medical Physics, Duke University, Durham, NC, 2011.)

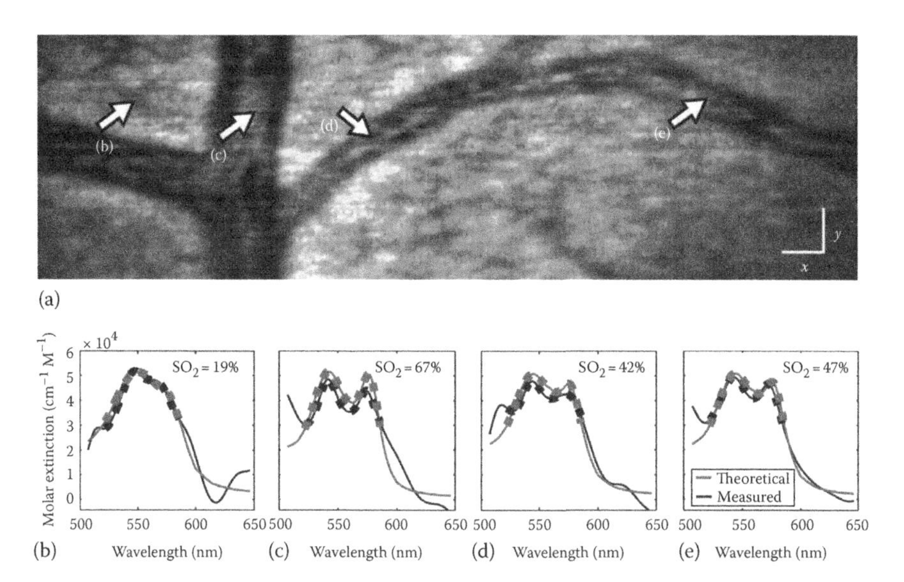

FIGURE 3.21
(a) *En face* METRiCS OCT image of chick embryo. White x and y scale bars, 100 μm. (b)–(e) Measured and theoretical spectral profiles from points (b)–(e), along with the computed Hb oxygen saturation. (Taken from Robles, F.E., Light scattering and absorption spectroscopy in three dimensions using quantitative low coherence interferometry for biomedical applications, PhD thesis in Medical Physics, Duke University, Durham, NC, 2011.)

the architecture of the vasculature network of the chick embryo, showing one major vessel, along with several branching vessels. Smaller capillaries are also seen, for example, in the upper left hand corner of Figure 3.21a. The significant advantage of METRiCS OCT is that not only is color information available but also that full detailed spectral profiles may be quantitatively analyzed. Figure 3.21b through e shows spectra from four selected points (solid black lines), along with the ideal molar extinction coefficient (red lines) for each measured Hb oxygen level. For this analysis, it is assumed that, for regions below the blood vessels, absorption is the dominant effect such that the method described earlier (Equations 3.2–3.6) may be applied to quantify the Hb oxygen levels (Figure 3.21). Note that the dotted lines are presented for only the limited bandwidth used for calculating SO_2, since this region inherently provides an orthogonal basis set. The results suggest that the major vessel in this image is an artery, where points c–d contain varying levels of oxygenated Hb. In contrast, the analysis for the smaller vessel, indicated by point (b), shows that it contains deoxygenated Hb.

As a final demonstration of the capabilities of METRiCS OCT, we present images acquired in vivo from a CD1 nu/nu normal mouse dorsal skinfold window chamber model [47]. Molecular contrast is demonstrated in this model for both endogenous and exogenous contrast. For imaging of endogenous contrast due to Hb absorption, the mouse was anesthetized and the window chamber was removed for imaging. Conventional OCT imaging of the data reveals tissue structures, as seen in Figure 3.22a. Here, several histological structures are evident, including the muscle layer, the lumen of blood vessels—including small capillaries—and the subcutaneous layer [48]. However, functional information regarding the sample is not available in this format. Using METRiCS OCT (Figure 3.22b), the spectral information reveals several interesting features: for example, the muscle layer at the surface appears relatively colorless due to low concentration of Hb, but once light traverses through the vasculature network beneath, a red shift is clearly observed due to higher concentrations of Hb. Another interesting feature is the small color variations on the top layer due to scattering from muscle and fibrous structures. Moreover, in regions below large blood vessels (>100 μm), most of the light is attenuated, thereby preventing detection of signals from below and thus creating a "shadow" effect. However, in regions without apparent blood vessels, signals are obtained throughout the full penetration depth enabled by the system.

Similar to the chick embryo results, the *en face* image, shown in Figure 3.23a, provides a more global perspective of the vasculature network. This image shows two major vessels, with the larger one on the right corresponding to a vein and the other to an artery. Several branching vessels can also be observed, along with the capillary plexus. Again, four points of interest are selected to analyze the spectra quantitatively; these points are found

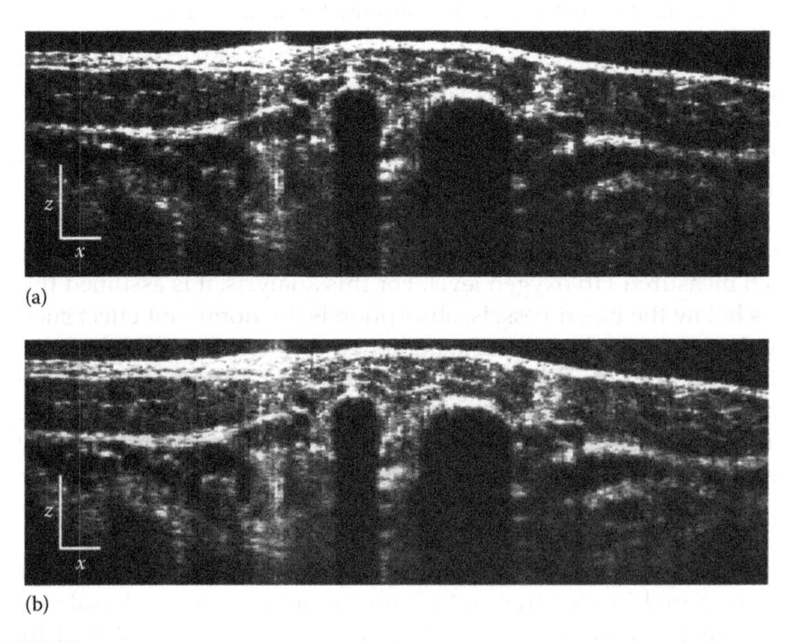

FIGURE 3.22
Tomographic (x–z) images of mouse dorsal skin flap. (a) Conventional OCT image and (b) METRiCS OCT image, demonstrating endogenous contrast molecular imaging. The white x and z scale bars are 100 μm. (Taken from Robles, F.E. et al., *Nat. Photon.*, 5(12), 744, 2011.)

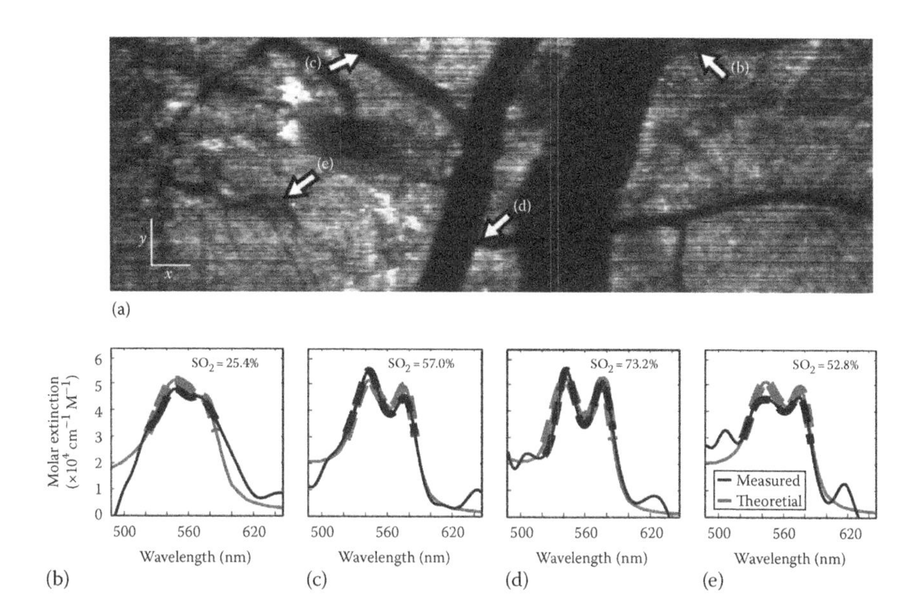

FIGURE 3.23
(a) *En face* (x–y) METRiCS OCT image using endogenous contrast, with arrows indicating points where the spectra presented in (b)–(e) are quantified. The white x and y scale bars are 100 μm. The measured spectral profiles are superposed with the theoretical Hb molar extinction coefficients. The dotted portion of the curves outlines the region used to determine SO_2 levels. All spectra were selected from depths immediately beneath each corresponding vessel. (Taken from Robles, F.E. et al., *Nat. Photon.*, 5(12), 744, 2011.)

along a vbranching vessel from the vein (Figure 3.23b), two from the artery (Figure 3.23c and d), and from the capillary network (Figure 3.23e). As Figure 3.23b through e shows, the measured spectra are well fit by the theoretical model and the calculated SO_2 levels, shown in the figure, correspond to expected values for these tissue locations in the anesthetized mouse [50].

To show the variations across individual vessels, Figure 3.24 shows a B-mode METRiCS OCT image and spectral analysis from the same data set (point b in Figure 3.23). Figure 3.24b through d presents spectra sampled across the diameter of this vessel with low oxygen levels, whereas (e–f) show spectra sampled longitudinally along a vessel with higher oxygen saturation levels. In both vessels, relatively small variations of only a few percent in saturation are observed.

Finally, imaging using an exogenous contrast agent, an area that has seen limited success in OCT to date, is presented. Here, 100 μL of sodium fluorescein (NaFS), diluted in sterile saline to 1% by weight, was introduced via a retro orbital injection, an intravenous injection which is an alternative to a tail vein injection. The images (Figures 3.25 and 3.26) clearly depict the presence of NaFS by a strong red shift in hue, which arises from stronger absorption at the lower wavelengths compared to that observed with endogenous

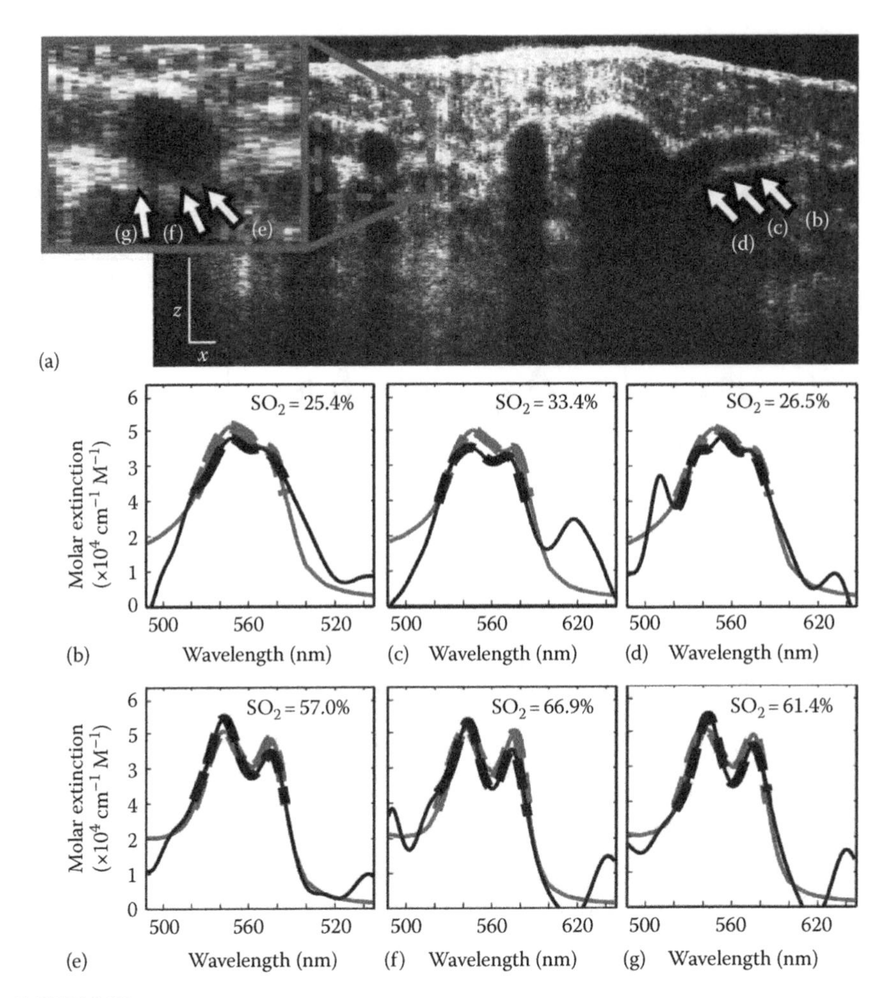

FIGURE 3.24
Tomographic (x–z) METRiCS OCT image of mouse dorsal skin flap with a close up of a vessel, and corresponding spectral profiles. (a) Cross-sectional image, acquired for point (b) in Figure 3.24. The white x and z scale bars are 100 μm. (b)–(d), Spectral profiles from points (b)–(d), sampled across the diameter of a deoxygenated vessel. (e)–(f), Spectral profiles from points (e)–(f), sampled longitudinally along a vessel with higher oxygen saturation levels. (Taken from Robles, F.E. et al., *Nat. Photon.*, 5(12), 744, 2011.)

contrast. Note that the injection results in the agent remaining confined to the vessels as it does not extravasate appreciably. Surprisingly, the agent also shows a weak increase in scattering. As a result, vessels are now characterized by the red hue of NaFS in the *en face* image (Figure 3.26a); however, large vessels still exhibit a "shadow" due to the stronger absorption.

Spectra from four points of interest are selected (Figure 3.26b through e), where the contributions of the three absorbing species (oxy-Hb, deoxy-Hb, and NaFS) are seen to affect the localized spectra to varying degrees. As shown in

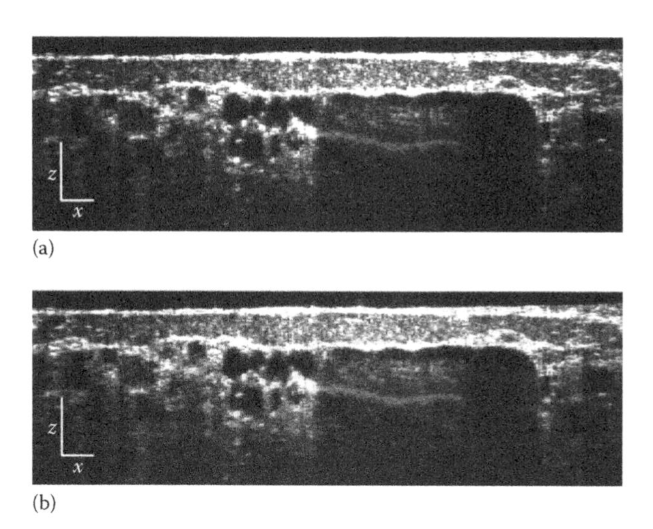

FIGURE 3.25
Tomographic (*x*–*z*) images of mouse dorsal skin flap window with exogenous contrast due to sodium fluorescein injection. (a) Conventional OCT image and (b) METRiCS OCT image, located above point (e) as shown in the *en face* (*x*–*y*) image in Figure 5.26. The white *x* and *z* scale bars are 100 μm. (Taken from Robles, F.E. et al., *Nat. Photon.*, 5(12), 744, 2011.)

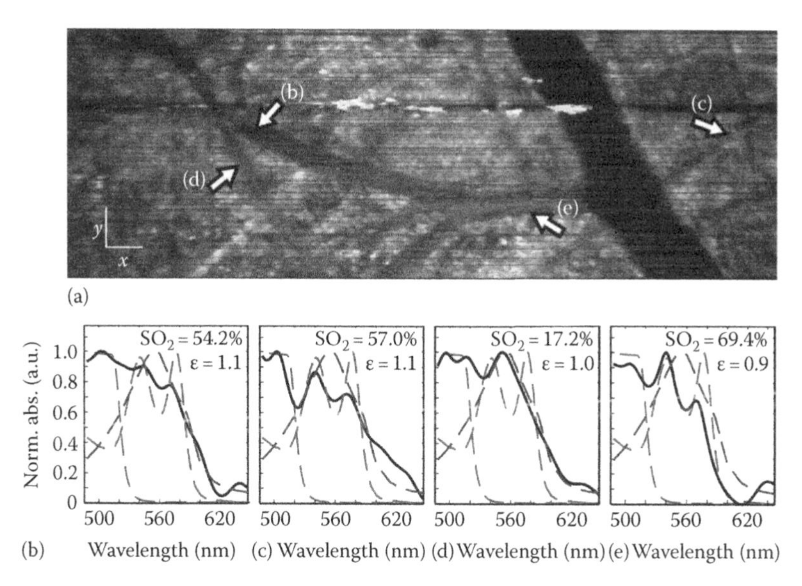

FIGURE 3.26 (See color insert.)
(a) *En face* (*x*–*y*) METRiCS OCT image of mouse dorsal skin flap window using exogenous contrast, and (b) spectral profiles from points indicated by arrows. The white *x* and *y* scale bars are 100 μm. The measured spectral profiles (black) are superposed with the theoretical oxy- (dotted red) and deoxy- (dotted blue) Hb normalized extinction coefficients, and normalized absorption of NaFS (dotted green). Also shown are the SO_2 levels and relative absorption of NaFS with respect to total Hb ($\varepsilon \equiv$ NaFS/Hb). All spectra were selected from depths immediately beneath each corresponding vessel. (Taken from Robles, F.E. et al., *Nat. Photon.*, 5(12), 744, 2011.)

Figure 3.26b through e, the presence of NaFS only contributes to the attenuation spectra in the lower wavelengths, thus assessment of SO_2 may still be accomplished as described previously. The relative amount of absorption due to NaFS with respect to total Hb ($\varepsilon \equiv$ NaFS/Hb) is also computed. The results show that this ratio is relatively constant (~1), which is expected since both NaFS and Hb are contained within the vessels and their relative contributions with respect to one another should not vary appreciably from location to location.

A final observation of this data shows an apparent effect of detector saturation due to fluorescent light from the NaFS, as the green areas of the *en face* image (Figure 3.26). The contrast agent (NaFS) absorbs light with maximum efficiency around 494 nm, producing the red hue of the vessels in the METRiCS OCT images. However, it also emits incoherent fluorescent light at a peak wavelength of 521 nm (green light). When this signal is particularly strong, green spots are seen in the image where the detector is saturated due to the fluorescent green signal.

3.5 Discussion

The development of fLCI and METRiCS OCT are a significant advance in the biomedical optics field due to the ability to provide depth resolved spectral information within an OCT image. The basis of this advance arises from the DW method, due to its ability to provide simultaneously high spectral and spatial resolution from interferometric signals. As described earlier, the DW method combines two linear time frequency distributions, each with different spatial and spectral properties, to form a new type of bilinear distribution. It has been shown mathematically [12] that the DW method probes the Wigner bilinear distribution with two orthogonal, independent windows, thus overcoming the resolution trade-off associated with linear distributions and avoiding common artifacts found in bilinear distributions.

The DW method served as a powerful tool for advance fLCI, enabling scattering signatures to be analyzed via depth resolved spectra to identify structures that were not resolved in the traditional OCT image. This method analyzes high-frequency oscillations of local spectra, resulting from induced temporal coherence effects, to determine the size of dominant scatterers. Analysis of these spectral features in complex, three-dimensional samples was not possible prior to the DW method due to the requirement for high resolution in both the spectral and the spatial dimensions. However, with the DW method, the local oscillations could be recovered and quantitative analysis was enabled. This capability was demonstrated using a phantom consisting of agar and polystyrene microspheres of different sizes in two distinct layers as a model of how cell nuclei may be distributed in different layers of tissue. Experiments with an ex vivo rat carcinogenesis model provided a powerful demonstration of fLCI's ability to detect early signs of cancerous development, with the cell nuclear diameter used as a surrogate biomarker of disease. The DW fLCI approach was shown to distinguish between normal and treated tissues, and provided evidence of the field effect of carcinogenesis.

The DW method was also used to gain access to spatially varying spectral modulations due to absorption. Unlike the local oscillations in fLCI, absorption from chromophores, such as Hb, produce a spectral modulation that is lower in frequency, thereby permitting facile separation of the two effects. With METRiCS OCT, molecular information of endogenous and exogenous chromophores is provided alongside structural information. The visible region of the spectrum affords highly sensitive measurements to biologically relevant chromophores, such as Hb and many exogenous absorbers in the visible region of the spectrum. This is important because a vast number of previously developed fluorophores, such as FDA-approved NaFS, may be leveraged for functional, tomographic imaging. METRiCS OCT not only provides contrast of these molecules but also has been shown to quantify SO_2 levels in vivo. Lastly, by having access to the full visible spectrum, an intuitive form of display using sample's true colors was achieved.

The optical imaging tools described here enable a number of different applications in biomedicine, each facing their own unique challenges in translation that will direct future work. In fLCI, an optical fiber-based system is needed to apply the method in vivo. Automated algorithms that contour the surface of tissues will facilitate analysis of large data sets corresponding to wide areas of tissue. The advantage of using fLCI is that processed data obtained from the images avoid the need to be interpreted by trained experts. The results presented here show that fLCI can assess tissue health quantitatively and provide sensitivity to precancerous states offering a potential new avenue for screening and diagnosing cancer.

Future work on METRiCS OCT will focus on application to animal models, in an effort to understand biological processes such as tumorigenesis and to identify changes to assess efficacy of therapeutic interventions. Although METRiCS OCT has only been demonstrated recently, the results show that it is a powerful tool that could enable new directions in early cancer detection, monitoring the delivery of therapeutic agents, and fundamental research, such as improving the understanding of angiogenesis and hypoxia.

Acknowledgment

This work was supported by a grant from the National Institutes of Health through the National Cancer Institute (1 R01 CA 138594).

References

1. Huang, D., E.A. Swanson, C.P. Lin, J.S. Schuman, W.G. Stinson, W. Chang, M.R. Hee, T. Flotte, K. Gregory, C.A. Puliafito, and J.G. Fujimoto, Optical coherence tomography. *Science*, 1991. **254**(5035): 1178–1181.
2. Huang, D., Y. Li, and M. Tang, Anterior eye imaging with optical coherence tomography, in *Optical Coherence Tomography: Tenchology and Applications*, W. Drexler and J. Fujimoto, eds. 2008. Springer, New York.
3. Wojtkowski, M., R. Leitgeb, A. Kowalczyk, T. Bajraszewski, and A. Fercher, *In vivo* human retinal imaging by Fourier domain optical coherence tomography. *J. Biomed. Opt.*, 2002. **7**: 457.
4. Choma, M., M. Sarunic, C. Yang, and J. Izatt, Sensitivity advantage of swept source and Fourier domain optical coherence tomography. *Opt. Express*, 2003. **11**(18): 2183–2189.
5. Robles, F.E., Light scattering and absorption spectroscopy in three dimensions using quantitative low coherence interferometry for biomedical applications, PhD thesis in Medical Physics. 2011. Duke University, Durham, NC.
6. Morgner, U., W. Drexler, F. Kärtner, X. Li, C. Pitris, E. Ippen, and J. Fujimoto, Spectroscopic optical coherence tomography. *Opt. Lett.*, 2000. **25**(2): 111–113.
7. Faber, D., E. Mik, M. Aalders, and T. van Leeuwen, Toward assessment of blood oxygen saturation by spectroscopic optical coherence tomography. *Opt. Lett.*, 2005. **30**(9): 1015–1017.
8. Yi, J. and X. Li, Estimation of oxygen saturation from erythrocytes by high-resolution spectroscopic optical coherence tomography. *Opt. Lett.*, 2010. **35**(12): 2094–2096.
9. Cang, H., T. Sun, Z. Li, J. Chen, B. Wiley, Y. Xia, and X. Li, Gold nanocages as contrast agents for spectroscopic optical coherence tomography. *Opt. Lett.*, 2005. **30**(22): 3048–3050.
10. Oldenburg, A., M. Hansen, T. Ralston, A. Wei, and S. Boppart, Imaging gold nanorods in excised human breast carcinoma by spectroscopic optical coherence tomography. *J. Mater. Chem.*, 2009. **19**: 6407.
11. Faber, D.J. and T.G.v. Leeuwen, Are quantitative attenuation measurements of blood by optical coherence tomography feasible? *Opt. Lett.*, 2009. **34**: 1–3.
12. Robles, F., R.N. Graf, and A. Wax, Dual window method for processing spectroscopic optical coherence tomography signals with simultaneously high spectral and temporal resolution. *Opt. Express*, 2009. **17**(8): 6799–6812.
13. Wang, X., B. Pogue, S. Jiang, X. Song, K. Paulsen, C. Kogel, S. Poplack, and W. Wells, Approximation of Mie scattering parameters in near-infrared tomography of normal breast tissue in vivo. *J. Biomed. Opt.*, 2005. **10**(5): 051704.
14. Drezek, R., M. Guillaud, T. Collier, I. Boiko, A. Malpica, C. Macaulay, M. Follen, and R. Richards-Kortum, Light scattering from cervical cells throughout neoplastic progression: influence of nuclear morphology, DNA content, and chromatin texture. *J. Biomed. Opt.*, 2003. **8**(1): 7.
15. Seet, K., T. Nieminen, and A. Zvyagin, Refractometry of melanocyte cell nuclei using optical scatter images recorded by digital Fourier microscopy. *J. Biomed. Opt.*, 2009. **14**(4): 044031.

16. Graf, R., F.E. Robles, X. Chen, and A. Wax, Detecting precancerous lesions in the hamster cheek pouch using spectroscopic white-light optical coherence tomography to assess nuclear morphology via spectral oscillations. *J. Biomed. Opt.*, 2009. **14**: 064030.
17. Wax, A., C. Yang, and J. Izatt, Fourier-domain low-coherence interferometry for light-scattering spectroscopy. *Opt. Lett.*, 2003. **28**(14): 1230–1232.
18. Graf, R. and A. Wax, Nuclear morphology measurements using Fourier domain low coherence interferometry. *Opt. Express*, 2005. **13**(12): 4693–4698.
19. Perelman, L.T., V. Backman, M. Wallace, G. Zonios, R. Manoharan, A. Nusrat, S. Shields, M. Seiler, C. Lima, T. Hamano, I. Itzkan, J. Van Dam, J.M. Crawford, and M.S. Feld, Observation of periodic fine structure in reflectance from biological tissue: A new technique for measuring nuclear size distribution. *Phys. Rev. Lett.*, 1998. **80**(3): 627–630.
20. van de Hulst, H.C., *Light Scattering by Small Particles*. 1981. Dover Publications, New York.
21. Robles, F.E. and A. Wax, Measuring morphological features using light-scattering spectroscopy and Fourier-domain low-coherence interferometry. *Opt. Lett.*, 2010. **35**(3): 360–362.
22. Wax, A., C.H. Yang, V. Backman, K. Badizadegan, C.W. Boone, R.R. Dasari, and M.S. Feld, Cellular organization and substructure measured using angle-resolved low-coherence interferometry. *Biophys. J.*, 2002. **82**(4): 2256–2264.
23. Reddy, B.S., Studies with the azoxymethane-rat preclinical model for assessing colon tumor development and chemoprevention. *Environ. Mol. Mutagen.*, 2004. **44**(1): 26–35.
24. Roy, H.K., A. Gomes, V. Turzhitsky, M.J. Goldberg, J. Rogers, S. Ruderman, K.L. Young et al., Spectroscopic microvascular blood detection from the endoscopically normal colonic mucosa: Biomarker for neoplasia risk. *Gastroenterology*, 2008. **135**(4): 1069–1078.
25. Kim, Y.L., V.M. Turzhitsky, Y. Liu, H.K. Roy, R.K. Wali, H. Subramanian, P. Pradhan, and V. Backman, Low-coherence enhanced backscattering: review of principles and applications for colon cancer screening. *J. Biomed. Opt.*, 2006. **11**(4): 041125.
26. Braakhuis, B.J.M., M.P. Tabor, J.A. Kummer, C.R. Leemans, and R.H. Brakenhoff, A genetic explanation of Slaughter's concept of field cancerization: Evidence and clinical implications. *Cancer Res.*, 2003. **63**(8): 1727–1730.
27. Zysk, A.M., S.G. Adie, J.J. Armstrong, M.S. Leigh, A. Paduch, D.D. Sampson, F.T. Nguyen, and S.A. Boppart, Needle-based refractive index measurement using low-coherence interferometry. *Opt. Lett.*, 2007. **32**(4): 385–387.
28. Robles, F., Y. Zhu, J. Lee, S. Sharma, and A. Wax, Fourier domain low coherence interferometry for detection of early colorectal cancer development in the azoxymethane rat carcinogenesis model. *Biomed. Opt. Express*, 2010. **1**(2): 736–745.
29. Choi, W., C. Fang-Yen, K. Badizadegan, S. Oh, N. Lue, R.R. Dasari, and M.S. Feld, Tomographic phase microscopy. *Nat. Methods*, 2007. **4**(9): 717–719.
30. Brown, W.J., J.W. Pyhtila, N.G. Terry, K.J. Chalut, T. D'Amico, T.A. Sporn, J.V. Obando, and A. Wax, Review and recent development of angle-resolved low coherence interferometry for detection of pre-cancerous cells in human esophageal epithelium. *IEEE J. Sel. Top. Quant Electron*, 2008. **14**: 88–97.

31. Terry, N.G., Y. Zhu, M.T. Rinehart, W.J. Brown, S.C. Gebhart, S. Bright, E. Carretta et al., Detection of dysplasia in Barrett's esophagus with *in vivo* depth-resolved nuclear morphology measurements. *Gastroenterology*, 2011. **140**(1): 42–50.
32. Terry, N.G., Y.Z. Zhu, J.K.M. Thacker, J. Migaly, C. Guy, C.R. Mantyh, and A. Wax, Detection of intestinal dysplasia using angle-resolved low coherence interferometry. *J. Biomed. Opt.*, 2011. **16**(10): 106002.
33. Wax, A., J.W. Pyhtila, R.N. Graf, R. Nines, C.W. Boone, R.R. Dasari, M.S. Feld, V.E. Steele, and G.D. Stoner, Prospective grading of neoplastic change in rat esophagus epithelium using angle-resolved low-coherence interferometry. *J. Biomed. Opt.*, 2005. **10**(5): 051604.
34. Wax, A., C. Yang, M.G. Müller, R. Nines, C.W. Boone, V.E. Steele, G.D. Stoner, R.R. Dasari, and M.S. Feld, In situ detection of neoplastic transformation and chemopreventive effects in rat esophagus epithelium using angle-resolved low-coherence interferometry. *Cancer Res.*, 2003. **63**(13): 3556.
35. Ntziachristos, V., C.H. Tung, C. Bremer, and R. Weissleder, Fluorescence molecular tomography resolves protease activity in vivo. *Nat. Med.*, 2002. **8**(7): 757–761.
36. Wang, X., Y. Pang, G. Ku, X. Xie, G. Stoica, and L.V. Wang, Noninvasive laser-induced photoacoustic tomography for structural and functional *in vivo* imaging of the brain. *Nat. Biotechnol.*, 2003. **21**(7): 803–806.
37. Helmchen, F. and W. Denk, Deep tissue two-photon microscopy. *Nat. Methods*, 2005. **2**(12): 932–940.
38. Boppart, S.A., A.L. Oldenburg, C. Xu, and D.L. Marks, Optical probes and techniques for molecular contrast enhancement in coherence imaging. *J. Biomed. Opt.*, 2005. **10**: 041208.
39. Graf, R., W. Brown, and A. Wax, Parallel frequency-domain optical coherence tomography scatter-mode imaging of the hamster cheek pouch using a thermal light source. *Opt. Lett.*, 2008. **33**(12): 1285–1287.
40. Prahl, S. *Optical Absorption of Hemoglobin* 1999; Available from: http://omlc.ogi.edu/spectra/hemoglobin/.
41. Robles, F.E., S. Chowdhury, and A. Wax, Assessing hemoglobin concentration using spectroscopic optical coherence tomography for feasibility of tissue diagnostics. *Biomed. Opt. Express*, 2010. **1**(1): 310–317.
42. Wang, H., J. Jiang, C. Lin, J. Lin, G. Huang, and J. Yu, Diffuse reflectance spectroscopy detects increased hemoglobin concentration and decreased oxygenation during colon carcinogenesis from normal to malignant tumors. *Opt. Express*, 2009. **17**(4): 2805–2817.
43. *Selected Colorimetric Tables*. 2011; Available from: http://www.cie.co.at/.
44. Boppart, S.A., A. Goodman, J. Libus, C. Pitris, C.A. Jesser, M.E. Brezinski, and J.G. Fujimoto, High resolution imaging of endometriosis and ovarian carcinoma with optical coherence tomography: feasibility for laparoscopic-based imaging. *Br. J. Obstet. Gynaecol.*, 1999. **106**(10): 1071–1077.
45. lves, H.E., The relation between the color of the llluminant and the color of the illuminated object. *Trans. Ilium. Eng. Soc.*, 1912. **7**: 62–72.
46. van Staveren, H., C. Moes, J. van Marie, S. Prahl, and M.J.C. van Gemert, Light scattering in Intralipid-10% in the wavelength range of 400–1100 nm. *Appl. Opt.*, 1991. **30**(31): 4507–4514.
47. Huang, Q., S. Sha, R.D. Braun, J. Lanzen, G. Anyrhambatla, G. Kong, M. Borelli, P. Corry, M.W. Dewhirst, and C.Y. Li, Noninvasive visualization of tumors in rodent dorsal skin window chambers. *Nat. Biotechnol.*, 1999. **17**(10): 1033–1035.

48. Koehl, G.E., A. Gaumann, and E.K. Geissler, Intravital microscopy of tumor angiogenesis and regression in the dorsal skin fold chamber: mechanistic insights and preclinical testing of therapeutic strategies. *Clin. Exp. Metastasis*, 2009. **26**(4): 329–344.
49. Robles, F.E., C. Wilson, G. Grant, and A. Wax, Molecular imaging true-colour spectroscopic optical coherence tomography. *Nat. Photon.*, 2011. **5**(12): 744–747.
50. Skala, M.C., A. Fontanella, L. Lan, J.A. Izatt, and M.W. Dewhirst, Longitudinal optical imaging of tumor metabolism and hemodynamics. *J. Biomed. Opt.*, 2010. **15**: 011112.

4

Spatial and Spectral Resolution of Semiconductor Detectors in Medical Imaging

Bjorn Heismann

CONTENTS

4.1 Introduction

Medical imaging devices commonly use gamma and x-ray radiation to generate internal images of the human body. Single photon emission tomography (SPECT) and positron emission tomography (PET) systems detect gamma emissions of radionuclide tracers. Computed tomography (CT), radiography, and mammography systems measure the x-ray attenuation of the human body. Figure 4.1 outlines the modes of operation of SPECT, CT, and radiography devices.

The image quality and dose usage of these systems are strongly influenced by the employed radiation detectors. From the early stages on, scintillator detectors based on materials like NaI, BGO, LSO, GOS, and CsI performed

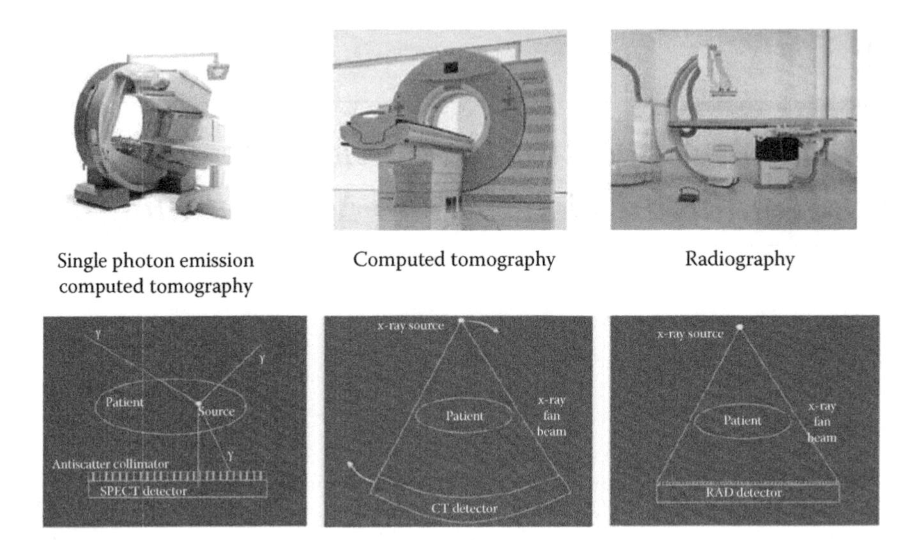

FIGURE 4.1
SPECT, CT, and radiography medical imaging devices.

the first step of radiation detection [1]. Over the past few years, an increasing number of scientific and commercial activities have used conversion semiconductor detectors for medical imaging. For lower x-ray energies, amorphous selenium detectors are routinely employed in mammography detectors. For higher x- and gamma ray energies, CdZnTe and CdTe have come into focus. SPECT prototypes in cardiology, scintimammography, and small animal imaging have been presented; see, for example, [2–4]. The authors report the improved spectral resolution and underline the potential to perform dual-isotope imaging. For CT, direct conversion counting electronics and prototype systems have been built and evaluated [5–8]. The high x-ray flux of more than 10^8 quanta per s and mm^2 is found to be a major challenge. A main reason for this is attributed to the dynamic material properties of CZT. It has been shown that defects like Te inclusions and subsequent inferior hole mobility lead to polarization in CZT detectors under medical imaging x-ray fluxes [9–12]. The main mechanism is seen in the creation of a dynamic space charge in the semiconductor bulk, degrading the charge transport properties. The potential benefits of semiconductor detectors in medical imaging rely mainly on their spatial and spectral resolution. In this chapter, we analyze the signal transport in both a scintillator and a semiconductor detector. As an application example, we focus on a CT detector. The pixel geometry, scintillator material, and thickness as well as the electronic read-out are chosen accordingly. As figures of merit, we use the modulation transfer function to quantify the spatial resolution and the detector response function D(E,E0) to analyze the spectral behavior. It should be noted that the results indicate an upper performance limit since degradations by, for example, material defects are not included.

4.2 Detector Physics

4.2.1 Indirect and Direct Conversion Detectors

The indirect conversion scintillation detector in Figure 4.2 is based on a GOS scintillator bulk material. Each pixel is enclosed by an epoxy compound filled with back-scattering TiO_2 particles. Typical pixel dimensions of around 1 mm and below are obtained. A registered photo sensor detects the secondary light photons at the bottom surface of each pixel. The primary interaction in a detector pixel is given by absorption of an incoming x-ray quantum by a gadolinium atom. The x-ray energy is converted into light photons. The energy conversion rate is around 12% [13]. Secondary light photon transport takes place. Photons which reach the photosensor contribute to the output energy signal E0. Radiography and mammography detectors follow similar designs. CsI is usually employed as a scintillator. Due to its vertical needle structure, it has the advantage of providing intrinsic light guiding properties, thus no back-scattering septa are required. This allows for an improved detector resolution at the expense of a reduced stopping power and signal speed.

Two main physical effects influence the spatial and spectral resolution in pixelized scintillator detectors. First, the primary energy deposition is not perfectly localized. For the high-Z atom gadolinium, absorption is governed by the photoelectric effect. This generates fluorescence escape photons with mean free path lengths in the order of several 100 μm. They might be reabsorbed in the pixel volume, become registered in a neighboring pixel, or leave

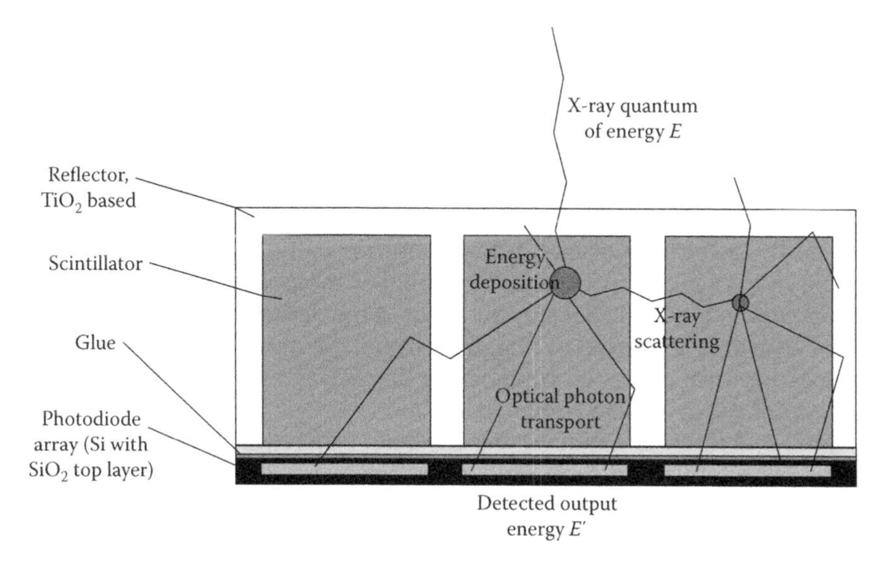

FIGURE 4.2
Schematics of a CT scintillation detector as an example for an indirect conversion detector.

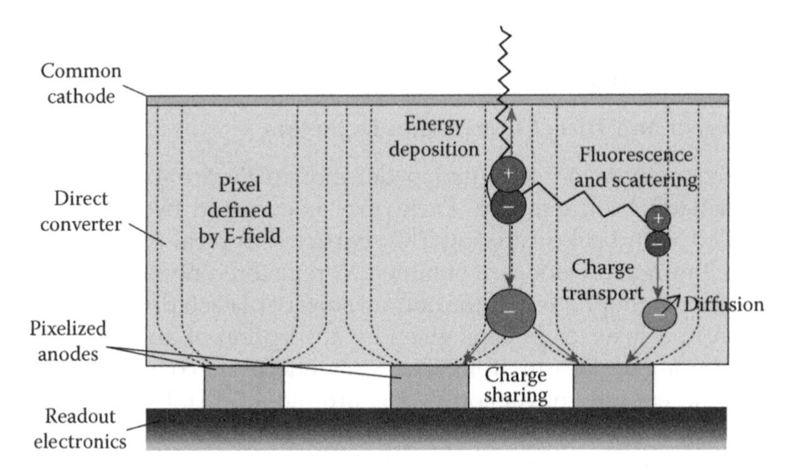

FIGURE 4.3
Schematics of a common cathode CZT direct conversion detector.

the detector volume completely. Second, light transport is affected by optical cross-talk. Septa walls are designed with a limited thickness to optimize overall dose usage and light yield. As a consequence, a significant portion of the light is transferred to adjacent "false" pixels, see Figure 4.2.

The direct conversion CZT scheme is shown in Figure 4.3. A common-cathode design with pixelized anodes on the bottom surface of the semiconductor bulk is typically used. Pixels are established by funnel-shaped electrical fields of several 100 V/mm. The physics of the primary energy deposition are comparable to the indirect conversion detector. However, the deposited energy is converted to charges instead of optical photons. The holes and electrons are separated and accelerated by the electrical field. Electrical pulse signals are induced on the electrodes. The main signal pulse is generated when the electrons follow the stronger curved electrical field in the bulk region close to the anodes.

The main signal degradation mechanisms are comparable to indirect conversion scintillator detectors. First, fluorescence scattering takes place. Due to the lower K-edge energy, the mean free path lengths of fluorescence quanta in CZT are about 100 μm. The smaller the pixel size, the more fluorescence cross-talk will affect the behavior of the detector. Second, the charge signal transport is affected by charge sharing. The moving charge cloud also induces electrical pulses on neighboring pixels [14], again mostly at the bottom part of the pixel field configuration.

Figure 4.4a through e summarizes the main difference between an indirect conversion scintillator detector and a direct conversion semiconductor detector. The scintillation detector is an optical device using light photons as intermittent information carriers. A direct conversion detector omits the conversion to light and directly generates charge carriers. It is an electrical device which employs electrons and holes to transfer the event information to the electrodes.

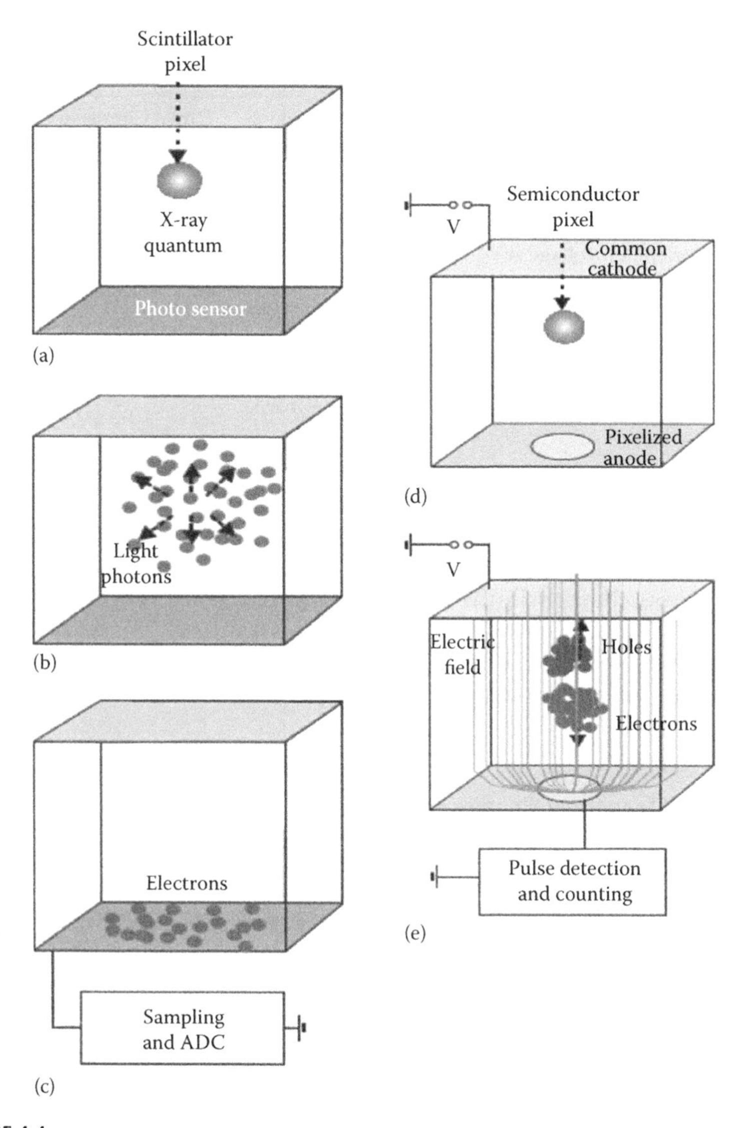

FIGURE 4.4
Signal conversion steps in an indirect conversion detector (a–c) and a direct conversion detector (d–e).

Both detector types can be operated in an integrating or a counting mode. In integrating mode, the charge information is sampled over an integration time—and converted to a digital signal. In counting mode, the total number of events is measured by counting the charge pulses. In addition to this, the energy of each absorbed quantum can be obtained by measuring the total charge or pulse amplitude of each quantum. Counting detectors thus offer spectral resolution of the input quantum field.

TABLE 4.1
Detector Simulation Parameters

	Indirect Conversion	Direct Conversion
Material	Gd_2O_2S:Pr	CdZnTe
Thickness (mm)	1.4	2.0
Pixel length (mm)	1.2	0.450
Acceleration voltage (V)	n.a.	700
Threshold noise (keV RMS)	n.a.	3

Note: We neglect electronic noise in the indirect conversion detector. It does not influence the spatial and spectral resolution comparison significantly. A threshold noise level of 3 keV (root mean square) is assumed for the direct conversion approach.

The detector parameters for our comparison are summarized in Table 4.1. For the scintillator detector, a 1.4 mm thick GOS with a pixel size of 1.2 mm has been chosen. The direct conversion detector has a 2 mm thick CZT at 700 V bias with a quadratic pixel size of (450 μm)^2. For this pixel size, fluorescence cross-talk contributes significantly. The choice reflects mostly the high-resolution case. The spatial resolution is not directly comparable to the scintillator detector. The setting is chosen to investigate if a direct conversion detector can provide improved spatial resolution at a reasonable spectral resolution.

4.2.2 Signal Transport Processes

Figure 4.5a and b outlines the cascaded system theory (CST) model of an indirect conversion integrating and a direct conversion counting detector. CST models have been applied to a number of detector evaluations, especially for flat panel radiography and mammography detectors [15].

The indirect conversion detector in Figure 4.4a has the following signal conversion steps. First, the X or gamma quantum is absorbed and its energy is converted to light photons. Second, the light photons travel through the scintillator set-up. Third, light photons are detected as an electrical current by the photo sensor. Finally, the current signal is digitized by a sampling ADC. The first step of the direct conversion detector in Figure 4.4b also consists of the x-ray energy deposition. A cloud of electrons and holes is generated. Second, the generated charges travel to the electrodes, forming current pulses. Finally, the current pulses are detected by a read-out electronic.

The signal transport can be simulated by a cascade of independent conversion steps. For the indirect conversion detector this can be done as follows [13]:

1. Primary energy deposition

   The primary energy deposition is modeled by a Monte Carlo simulation tool based on the GEANT4 particle interaction simulation framework

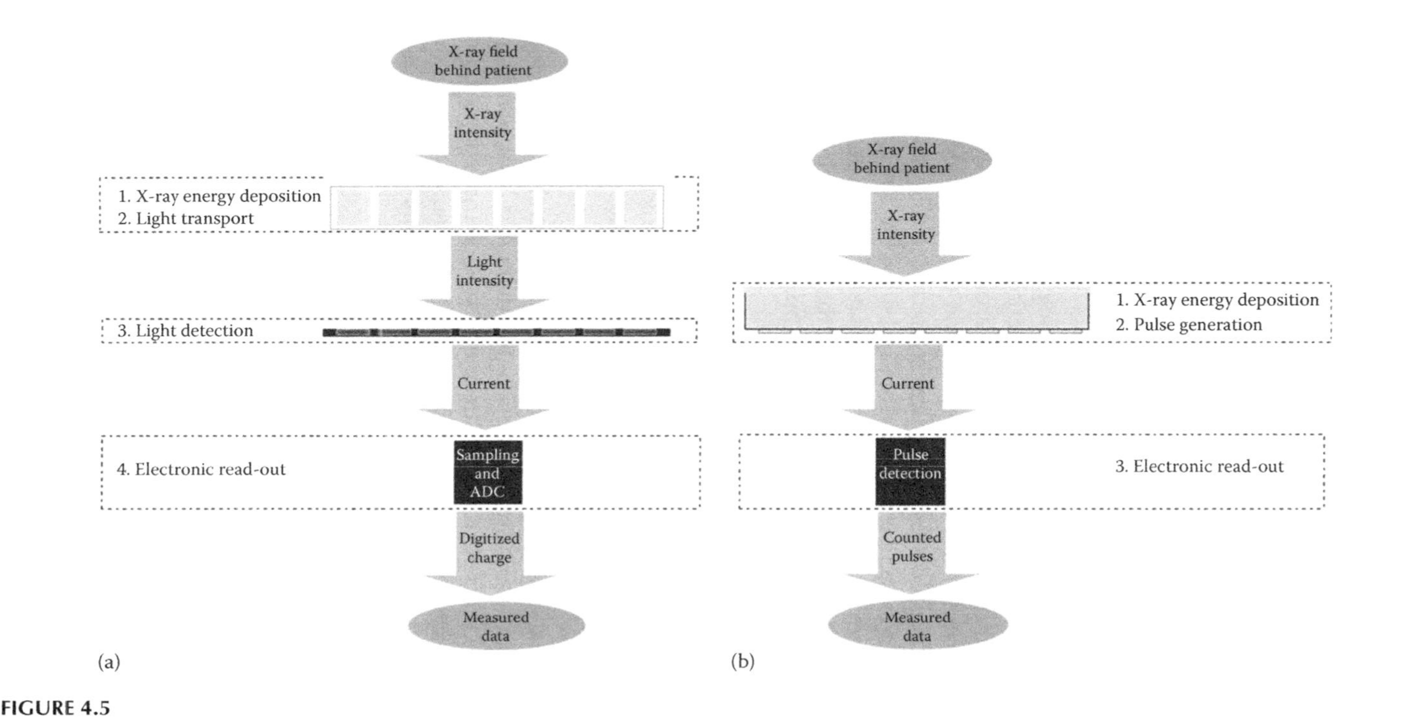

FIGURE 4.5
Cascaded detection models: (a) indirect conversion, (b) direct conversion.

[16–18]. For each incoming x-ray quantum, the spatial energy deposit is simulated on a 10 × 10 × 10 sub-voxel grid. Rayleigh and Compton scattering and escape processes generally lead to multiple deposition sites per event. Each portion of the primary energy deposit is converted into a number of optical photons, taking into account the main scintillator photon emission energies. The energy to optical photons conversion gain for $Gd_2O_2S{:}Pr$ is taken from measurements as $E_c = 0.12$ with a standard deviation of $\sigma(E_c) = 0.04$. As a result of the first step, we obtain a look-up table of individual energy deposition events. By using a high number of events (10^6 and more), systematic errors are avoided.

2. Light transport

   The second step describes the light transport to the photo sensor pixels. A photon tracking Monte Carlo simulation [13] traces the photon paths in the entire detector system until they are detected at a pixel of the photo sensor array or lost by bulk absorption or scintillator escape. Photon interaction processes like optical scattering, photon reabsorption and diffuse and specular reflection at pixel septa borders are included. The corresponding optical parameters are taken from experimental results. For a specific photon starting position, the average detection probability of an optical photon in a photo sensor pixel is obtained.

3. Light detection

   The light photons that have reached a photo sensor pixel are converted to electrical charges. The wavelength-dependent quantum efficiency $\beta(\lambda)$ of the photo sensor is taken into account. As a result of the third step, we obtain a photo current for each pixel.

4. Electronic read-out

   In the final step, the photo current is sampled to charge and digitized. For medical x-ray applications, sigma-delta ADCs are common ADC designs. Direct current measurements by a charge-coupled oscillator are also employed. The electronic read-out usually has limited linearity and additional offset noise. For the results in this chapter, nonlinearity and electronic noise do not play a role and are neglected.

For a given detector geometry, x-ray quantum input spectrum and field distribution, this scheme yields the average signal of the scintillator detector. The signal chain of the direct conversion detector in Figure 4.4b is modeled as follows [19]:

1. Primary energy deposition

   The primary energy deposition step is equivalent to the scintillator model. Instead of a GOS material, a CZT absorber is used.

2. Pulse generation

   A detailed charge transport model can be based on the work of Eskin et al. [20]. A local weighting potential allows to calculate the signal pulse shape for arbitrary charge starting positions in the detector [21]. A time-resolved pulse signal on the anode is obtained.

3. Electronic read-out

   Depending on the priority of spectral or spatial resolution, two main electronic design schemes for direct conversion detectors can be selected. Spectrally resolving detectors in SPECT and PET require a precise measurement of the energy of each quantum.

Due to this, the anode signals are usually filtered with comparably long shaping times. The signal is integrated and digitized. High-resolution detectors, on the other hand, address applications in mammography, radiography, and CT. The corresponding electronics employ shorter shaping times close to the primary pulse duration. The filtered pulse signals are usually detected by amplitude threshold triggering [5,6]. In the following we assume the second case of a high-flux x-ray detector. The threshold noise due to electronic noise contributions in the electronic read-out is included in the model.

4.3 Spatial Resolution

The spatial resolution of x-ray detectors is mainly given by pixel pitch and aperture. The pixel pitch defines the Nyquist frequency. The smaller the pixel aperture, the larger the spatial resolution will become. However, in practical imaging systems, defining the spatial resolution of a detector is a trade-off with dose usage and detector cost. In particular, scintillator detectors are often limited by the required septa walls and the cost of the required number of electronic digitization channels.

4.3.1 Definition of the Modulation Transfer Function

The Modulation Transfer Function MTF(f) is commonly used to describe the spatial resolution of pixelized detectors [15]. It is given by the normalized absolute value of the Fourier transform of the detector pixel point spread function. The MTF evaluation scheme is commonly applied to pixelated scintillator and CZT detectors, see, for example, [22,23].

4.3.2 Simulation and Measurement of the MTF

For both detector types, the MTF is determined by the "slanted slit" method. Figure 4.6 shows a slanted slit image for the indirect conversion detector. A tungsten plate with a slit of 0.1 mm width is placed on top of the scintillator array with a slit angle of approximately 3° with respect to the fundamental

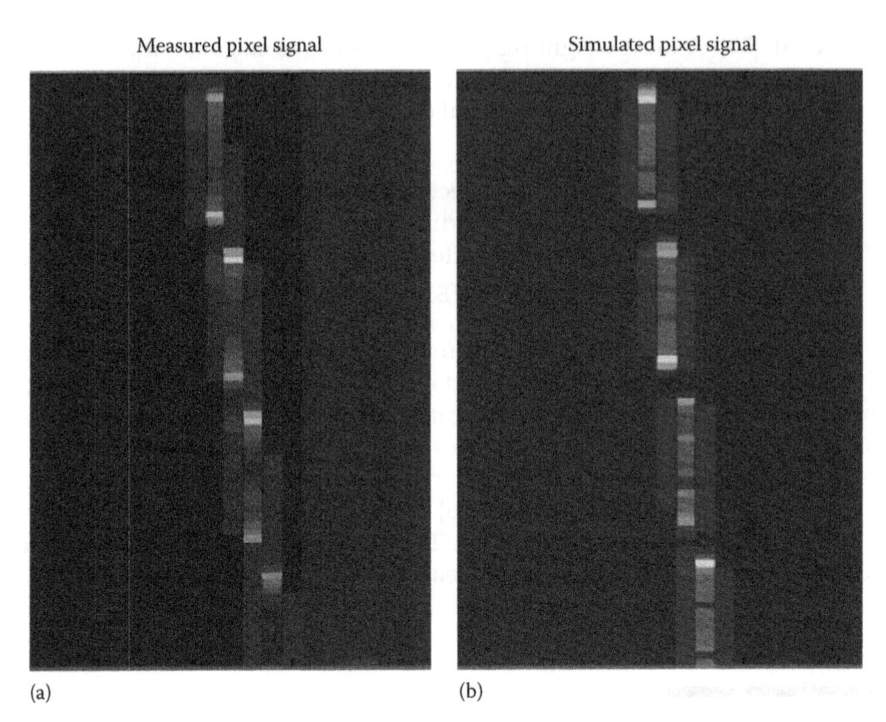

FIGURE 4.6
(a) Measured and (b) simulated slanted slit images for the indirect conversion GOS detector.

directions of the pixel lattice, here denoted as I and k. The slit is illuminated by an x-ray flat-field. Summing the image along the line direction yields the line spread function. It is an oversampled representation of the point spread function. The procedure is repeated with various angles of the slit toward the axes to obtain a two-dimensional MTF of the detector. It has been shown that measured MTFs have an excellent agreement to simulated MTFs for both indirect conversion detectors [22] and direct conversion detectors [23]. In the following, we use the simulation framework described in the previous section to obtain the slit images required for the MTF calculation.

4.3.3 Properties of the MTF

Figure 4.7 shows the MTF comparison between an indirect and a direct conversion detector. The straight lines in red and blue are simulated curves, whereas the corresponding dashed curves show the respective ideal sinc functions. The indirect conversion detector shows a mid-frequency drop in comparison to the ideal sinc function. This is mainly due to optical crosstalk, which leads to a low-pass signal filtering in the detector. In principle, the mid-frequency drop can be recovered by appropriate inverse filtering at the expense of amplified electronic noise in the signal. For high and

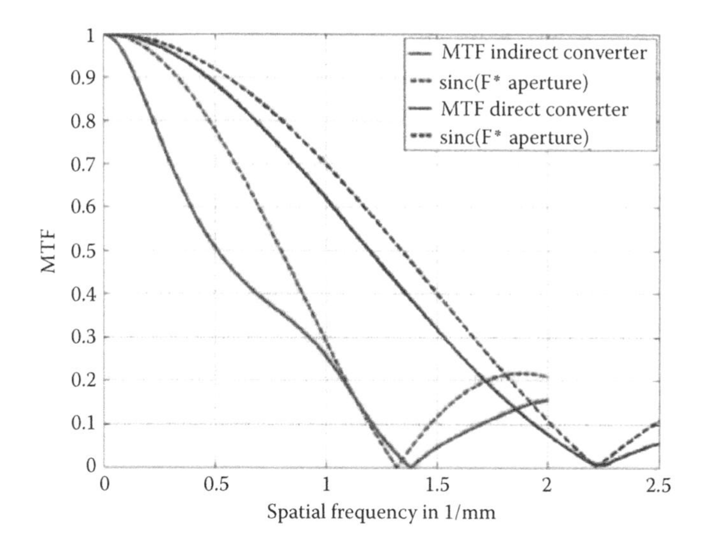

FIGURE 4.7
Modulation transfer functions for indirect conversion 1.2 mm pixel detector (dark gray) and direct conversion 450 μm detector (light gray). Dashed lines reflect the ideal sinc functions. (Reproduced from Wirth, S. et al., Simulations and measurements of the modulation transfer function of scintillator arrays, M06-257, *IEEE Nuclear Science Symposium Conference Record*, Honolulu, HI, 2008.)

medium flux medical applications, this has no major impact. The signal-to-noise ratio is mainly affected in low-flux screening applications.

In comparison to this, the direct conversion detector is close to the ideal sinc behavior. The remaining deviations are mainly due to fluorescence escapes between adjacent pixels. Despite the fact that the pixel aperture has been more than halved, charge sharing plays only a minor role compared to the effects of optical cross-talk. Note that in both detector systems, a small deviation in the zero frequency position is visible. This is due to the fact that fluorescence cross-talk leads to smaller signal contributions close to the pixel borders, effectively shrinking the pixel aperture.

4.4 Spectral Resolution

In nuclear physics and medical imaging applications like PET and SPECT, the spectral resolution of the detector is commonly described by the pulse height spectrum (PHS). A typical PHS of a CZT pixelized detector and a NaI Anger camera is shown in Figure 4.8. For x-ray applications, the detector has to register a whole range of input energies. Figure 4.9 shows a 80 and 140 kV tungsten tube spectrum. The generalization of the PHS to a range of input energies E leads to the detector response function (DRF), see [13].

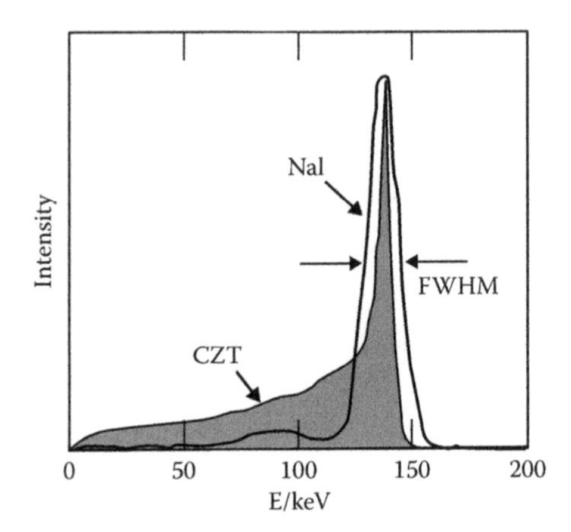

FIGURE 4.8
Pulse height spectrum (PHS) of CZT pixelized detector and a NaI anger camera. (Taken from Wernick, M.N. and Aarsvold, J.N., *Emission Tomography*, Academic Press, San Diego, CA, 2004.)

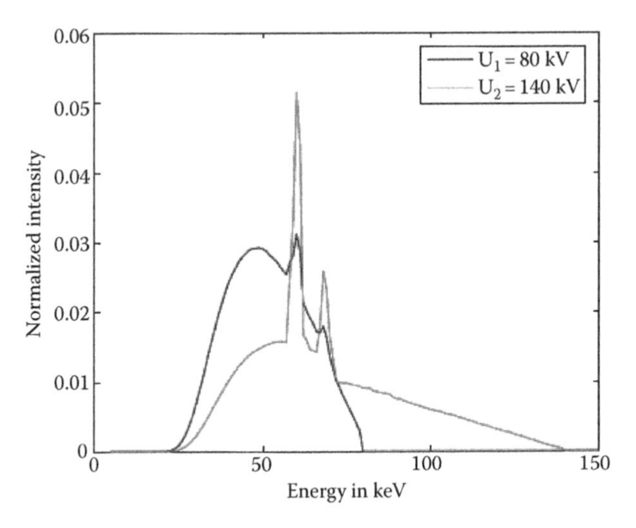

FIGURE 4.9
80 kV and 140 kV tungsten tube spectra.

4.4.1 Definition of the Detector Response Function

The DRF $D^{(i,k)}(E,E')$ yields a probability density to measure the output energy E' for an incoming quantum of energy E. The incoming quantum flux is directed at a central reference pixel. Its lateral position is equally distributed across the pixel area. The output energy is detected at a photo sensor pixel with the position (i, k). The pair (0,0) marks the center position, (1,0) the horizontal neighbors, (0,1) the vertical neighbors, etc. (see Figure 4.10).

The DRF allows us to express the statistics of the microscopic signal transport processes as a macroscopic probability function. We can simplify its

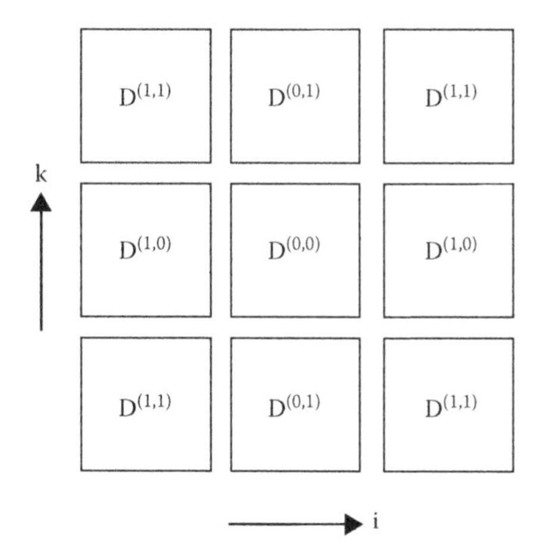

FIGURE 4.10
Spatial indices (i, k) of the detector response function. The symmetry arises for pixels unaffected by border effects.

variable dependencies for medical imaging applications. Here, the pixel-to-pixel variation of the projected anatomical input signal is usually below 1%. This is close to a flat-field irradiation of the detector. In this case, the mean signal cross-talk between pixels is symmetrical. We realize the flat-field approximation by irradiating the detector surface homogeneously. The simplified D(E,E′) function is used to describe the results.

4.4.2 Comparison of the Detector Response Functions

The DRF of the indirect and direct conversion detector set-ups are shown in Figure 4.11a and b. In both cases, the probabilities are normalized to 1 for

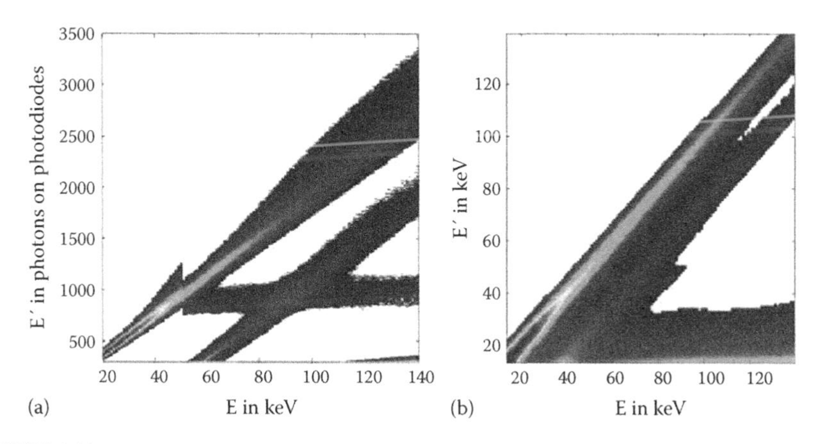

FIGURE 4.11
Detector response function for (a) indirect conversion GOS detector, (b) direct conversion CZT detector.

each input energy E. This leads to the respective color codings. Below 15 keV output energy E′, the electronic noise in the counting direct conversion detector dominates the output behavior. The respective range is omitted for better clarity.

The indirect conversion D(E,E′) in Figure 4.11a consists of the following structures:

Up to the gadolinium K-edge energy $E_K \sim 50.2$ keV, a linear branch $E \sim E'$ is visible. Its broadening is explained by the energy conversion gain variance. The output energy peak has a tail toward higher output energies E′ for increasing input energy E. This light tailing effect is due to the fact that the light transport yield increases with the interaction depth, which in turn increases with the input energy E. Above the K-edge energy, a secondary branch occurs. The events are formed by absorption of the primary energy with a fluorescence energy loss to the surroundings. The corresponding reabsorbed fluorescence events are found in the third, approximately vertical branch starting at around 50 keV output energy. Its slight inclination is again due to the increase of the interaction depth with input energy. The overall absorption probability of the quanta is reduced with increasing input energy E. The low energy output events including Compton and Rayleigh scatter depositions are not shown, see [13] for a more detailed discussion of these effects.

The direct conversion detector D(E,E′) in Figure 4.11b has a more pronounced linear branch. Its stronger relative signal content is explained by the about two times higher intrinsic conversion gain of CZT and the reduced depth-dependency due to the small pixel effect. The fluorescence branches appear at the lower Cd, Zn, and Te fluorescence energies of 23–28 keV. The differential branches are consequently closer to the main linear branch. Charge sharing events create a low-energy tail increasing toward lower output energies and overlapping with the fluorescence branches.

The spectral behavior described by D(E,E0) has consequences for both detector schemes. In the following, we consider the cases of an integrating indirect conversion detector and a counting direct conversion detector as prominent examples.

4.4.3 Integrating Indirect Conversion Detectors

For integrating indirect conversion detectors, it has been shown that the output signal variance increase leads to a Poisson excess noise [13]. Following the work by Rabbani et al. [24], a formula for the noise amplification is established as

$$f(E) = \frac{1}{\sqrt{\alpha(E)}} \frac{SNR_{out}}{SNR_{in}} = \frac{\langle E' \rangle}{\sqrt{\langle E' \rangle^2 + \sigma^2(E')}} \tag{4.1}$$

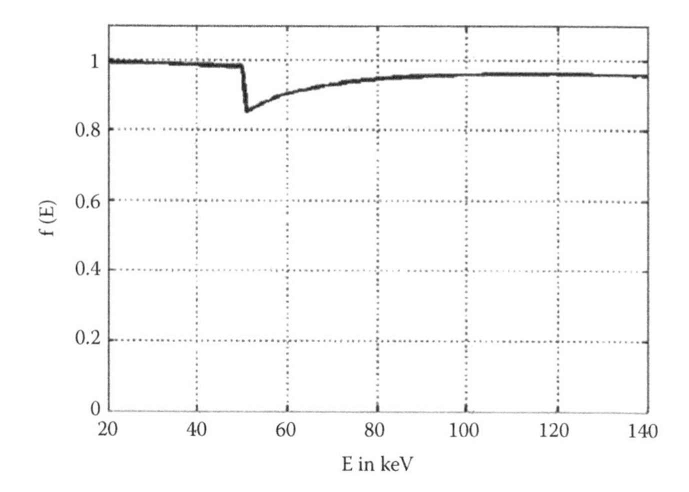

FIGURE 4.12
Generalized Swank factor f(E). (Reproduced from Heismann, B.J. et al., *Nucl. Inst. Meth. Phys. Res., A*, 591, 28, 2008.)

where
- $\alpha(E)$ is the quantum detection efficiency
- $<E'>$ is the average output energy
- $\sigma(E')$ is the output energy variance
- f(E) is a generalized energy-dependent Swank factor

The Poisson excess noise shown in Figure 4.12 is more pronounced around the K-edge. A noise increase of about 15% is visible. This is due to the fact that the output signal variance increases strongly beyond the K-edge. For continuous input spectra, a typical excess noise of 5%–10% can be estimated, depending on the input spectra and the patient attenuation.

4.4.4 Counting Direct Conversion Detectors

For a counting direct conversion detector, we can distinguish between the full energy resolution required in SPECT or PET and the binned energy resolution required for dual-energy CT or radiography. In the case of full energy resolution, D(E,E0) contains directly the normalized PHS for specific input energies.

In the following, we focus on the case of a two bin energy resolution. Like shown in Figure 4.13, this is commonly achieved by using two threshold levels in the electronic read-out. The first threshold E_{th1} discards noise events. The second threshold E_{th2} separates the output energy range into two separate bins. The diagonal rectangular sections mark the quanta events, which are correctly assigned. The lower right region contains high-energy bin primary events, which are falsely assigned to the low-energy bin.

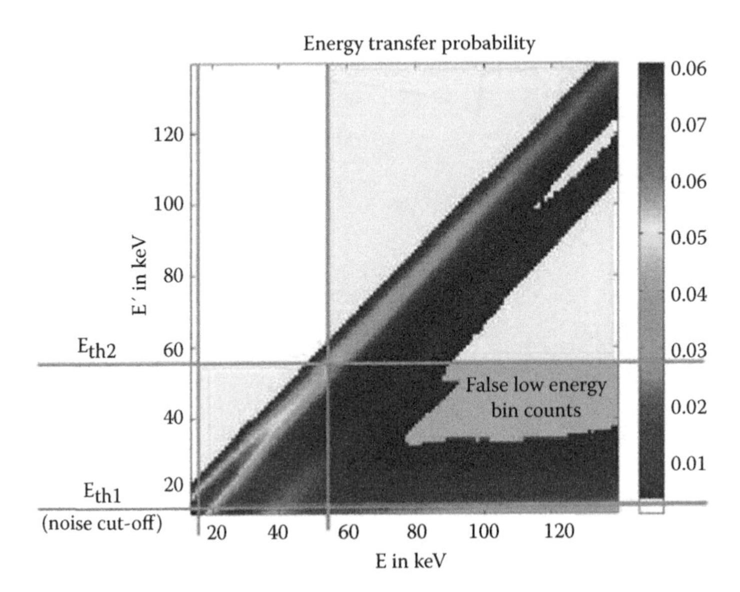

FIGURE 4.13
Schematics of energy binning for the detector response function of Figure 4.11a. Two energy threshold levels $E_{th1} = 15$ keV, $E_{th2} = 55$ keV are used.

Figure 4.14 shows the consequence of the low-energy shift. We have assumed a 140 kV tungsten x-ray tube input spectrum, see the shaded gray curve. The two detected spectra in the respective energy bins are given by the two straight curves.

Two effects are visible: First, the low-energy detected spectrum loses low-energy events. This is due to the fact that fluorescence events can carry away

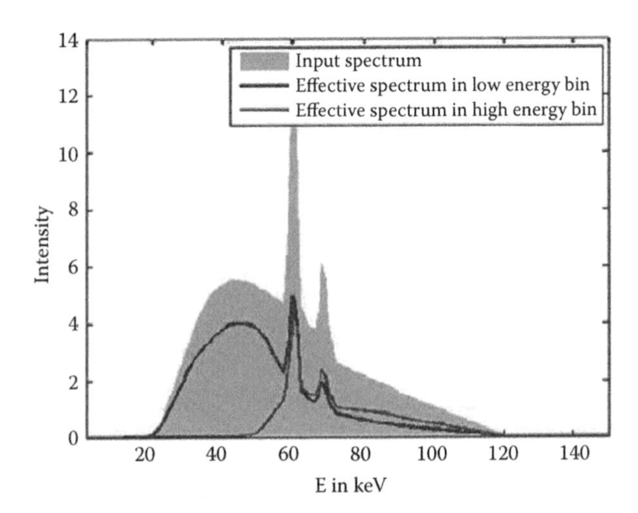

FIGURE 4.14
Detected spectra for a 140 kV tungsten input spectrum (shaded gray).

enough energy from a primary event to reduce the detected energy below the first energy threshold. Second, the two detected spectra overlap significantly. When we normalize both detected spectra to 1, we obtain the system weighting functions of the two energy bins. The overlap reaches around 60% and is thus comparably larger than the 40%–50% overlap of dual kVp or dual-source CT [25]. This indicates that the dual-energy measurement capabilities of direct conversion detectors are significantly affected by low-energy shift mechanisms due to fluorescence.

4.5 Conclusions

From the MTF results, we can deduce that the spatial resolution of semiconductor detectors is a clear potential benefit for medical imaging devices. The direct conversion of the primary x-ray field information into charge pulses omits the inter-pixel cross-talk of scintillator detectors almost completely. The direct conversion into charges demands for a strict control of electrical semiconductor defects. Low-flux screening applications in CT and other medical x-ray devices probably benefit the most from the improved spatial resolution as it requires less image filtering for the same obtained image resolution.

The intrinsic energy resolution of a counting detector read-out is a second potential benefit of a direct conversion semiconductor detector. For gamma-ray emission applications like SPECT and PET, the registered charge is a direct measure for the primary quantum energy. X-ray applications usually require only two or three energy bins defined by threshold energies. The DRF results indicate that CZT semiconductor detectors are prone to a shift of quantum detection to lower-energy bins due to interpixel fluorescence cross-talk.

The required spatial and spectral resolutions in semiconductor detectors are defined by the targeted medical device. For mammography and radiography detectors, spatial resolution is essential. CT relies on the detection of high x-ray fluxes at intermediate spatial and spectral resolution. SPECT and PET detectors mainly require a superior energy resolution. For each of these applications, detector parameters like the pixel size and the electronic read-out have to be balanced accordingly.

References

1. G. F. Knoll, *Radiation Detection and Measurement*, 3rd edn. John Wiley & Sons Inc., New York, 2000.
2. I. M. Blevis, M. K. O'Connor, Z. Keidar, A. Pansky, H. Altman, and J. W. Hugg, CZT gamma camera for scintimammography, *Phys. Med. Biol.*, 21(Suppl. 1), 56–59, 2006.

3. K. B. Parnham, S. Chowdhury, J. Li, D. J. Wagenaar, and B. E. Patt, Second-generation, tri-modality pre-clinical imaging system, M06-29, *IEEE Nuclear Science Symposium Conference Record*, Honolulu, HI, 2007.
4. D. J. Wagenaar, J. Zhang, T. Kazules, T. Vandehei, E. Bolle, S. Chowdhury, K. Parnham, and B. E. Patt, In vivo dual-isotope SPECT imaging with improved energy resolution, MR1-3, *IEEE Nuclear Science Symposium Conference Record*, Honolulu, HI, 2007.
5. E. Kraft et al., Counting and integrating readout for direct conversion X-ray imaging: Concept, realization and first prototype measurements, *IEEE Trans. Nucl. Sci.*, 54(2), 383–390, 2007.
6. D. Moraes, J. Kaplon, and E. Nygard, CERN DxCTA counting chip, *Proceedings of the 9th International Workshop on Radiation Imaging Detectors*, Erlangen, Germany, 2007.
7. Y. Onishi, T. Nakashima, A. Koike, H. Morii, Y. Neo, H. Mimura, and T. Aoki, Material discriminated x-ray CT by using conventional microfocus x-ray tube and CdTe imager, M27-2, *IEEE Nuclear Science Symposium Conference Record*, Honolulu, HI, 2007.
8. J. P. Schlomka et al., Experimental feasibility of multi-energy photon counting k-edge imaging in pre-clinical computed tomography, *Phys. Med. Biol.*, 53, 4031–4047, 2008.
9. S. A. Soldner, D. S. Bale, and C. Szeles, Dynamic lateral polarization in CdZnTe under high flux X-ray irradiation, *IEEE Trans. Nucl. Sci.*, 54(5), 1723–1727, 2007.
10. E. Bolotnikov et al., Effects of Te inclusions on the performance of CdZnTe radiation detectors, R27-2, *IEEE Nuclear Science Symposium Conference Record*, Honolulu, HI, 2007.
11. G. S. Camarda, A. E. Bolotnikov, Y. Cui, A. Hossain, S. A. Awadalla, J. Mackenzie, H. Chen, and R. B. James, Polarization studies of CdZnTe detectors using synchrotron X-ray radiation, R27-3, *IEEE Nuclear Science Symposium Conference Record*, Honolulu, HI, 2007.
12. L. Abbene, S. D. Sordo, F. Fauci, G. Gerardi, A. L. Manna, G. Raso, A. Cola, E. Perillo, A. Raulo, V. Gostilo, and S. Stumbo, Study of the spectral response of CZT multiple-electrode detectors, N24-298, *IEEE Nuclear Science Symposium Conference Record*, Honolulu, HI, 2007.
13. B. J. Heismann, K. Pham-Gia, W. Metzger, D. Niederloehner, and S. Wirth, Signal transport in computed tomography detectors, *Nucl. Inst. Meth. Phys. Res., A*, 591, 28–33, 2008.
14. T. Michel et al., A fundamental method to determine the signal-to-noise ratio (SNR) and detective quantum efficiency (DQE) for a photon counting pixel detector, *Nucl. Inst. Meth. Phys. Res., A*, 568, 799–802, 2006.
15. I. A. Cunningham, Applied linear system theory, in: *Handbook of Medical Imaging*, Vol. 1, J. Beutel, H. L. Kundel, and R. L. van Metter, Eds. SPIE, Bellingham, WA, 2000.
16. J. Giersch and J. Durst, Monte Carlo simulations in x-ray imaging, *Nucl. Inst. Meth. Phys. Res., A*, 591, 300, 2008.
17. S. Agostinelli et al., G4—a simulation toolkit, *Nucl. Inst. Meth. Phys. Res., A*, 506(3), 250–303, July 2003.
18. J. Allison et al., Geant4 developments and applications, *IEEE Trans. Nucl. Sci.*, 53(1), 270–278, February 2006.

19. B. J. Heismann, D. Henseler, D. Niederloehner, P. Hackenschmied, M. Strassburg, S. Janssen, and S. Wirth, Spectral and spatial resolution of semiconductor detectors in medical X- and gamma ray imaging, R03-1, *IEEE Room Temperature Semiconductor Workshop*, Dresden, Germany, 2008.
20. J. D. Eskin, H. H. Barrett, and H. B. Barber, Signals induced in semiconductor gamma-ray imaging detectors, *J. Appl. Phys.*, 591, 647, 1999.
21. B. Kreisler, J. Durst, T. Michel, and G. Anton, Generalised adjoint simulation of induced signals in semiconductor X-ray pixel detectors, *J. Inst.*, 3, 11, 2008.
22. S. Wirth, B. J. Heismann, D. Niederloehner, L. Baetz, W. Metzger, and K. Pham-Gia, Simulations and measurements of the modulation transfer function of scintillator arrays, M06-257, *IEEE Nuclear Science Symposium Conference Record*, Honolulu, HI, 2008.
23. T. Michel, Energy-dependent imaging properties of the Medipix2 x-ray-detector, *Proceedings of Science on the 16th International Workshop on Vertex Detectors*, Lake Placid, NY, 2007.
24. M. Rabbani, R. Shaw, and R. van Metter, Detective quantum efficiency of imaging systems with amplifying and scattering mechanisms, *J. Opt. Soc. Am. A*, 4, 895–901, 1987.
25. B. J. Heismann and S. Wirth, SNR performance comparison of dual-layer detector and dual-kVp spectral CT, *IEEE Medical Imaging Conference Record*, Honolulu, Hawaii, pp. 3280–3822, 2007.
26. M. N. Wernick and J. N. Aarsvold, *Emission Tomography*. Academic Press, San Diego, CA, 2004.

5

Design and Assessment Principles of Semiconductor Flat-Panel Detector-Based X-Ray Micro-CT Systems for Small-Animal Imaging

A. Sisniega, J.J. Vaquero, and M. Desco

CONTENTS

5.1 Introduction

In recent years, the number of animal models of human disease has increased and their use is now widespread. The need to study biological processes and morphological features in small-animal models—and to do so noninvasively so that the process can be tracked over time—has stimulated the development of high-resolution biomedical imaging devices. Nowadays, drug development relies heavily on the use of small-animal models and molecular imaging techniques, such as positron emission tomography (PET) or single photon emission computed tomography (SPECT), to provide the required functional information that characterizes the behavior of the drug. However, the results obtained are sometimes difficult to interpret due to the lack of a reliable anatomical localization of tracer uptake. To avoid this problem, registration of PET and SPECT images with accurate anatomical images [1] has proven to be an appropriate choice in new multimodality systems.

Among the different anatomical imaging techniques, x-ray microcomputed tomography (micro-CT) is the preferred complement to preclinical functional imaging modalities, due to its high-resolution capabilities and to the possibility of integration with other imaging systems.

As well as complementing the information obtained using molecular imaging techniques, micro-CT by itself is a valuable tool in small-animal imaging and is commonly used in research fields associated with the morphology of the sample (e.g., bone studies) [2–6].

In the development of x-ray micro-CT systems, most approaches make use of detectors based on x-ray image intensifiers and charge-coupled devices (CCD) to which a scintillator screen is connected either directly or using light guides (e.g., fiber optic plates) [7–9]. Recent developments in semiconductor detectors have made it possible to use new, compact devices—flat-panel detectors—for x-ray detection. These flat-panel devices can be categorized into two different groups according to the process carried out to convert the x-ray photons (primary quanta) to electric charges that are gathered and converted into a digital signal. The first approach makes use of photoconductors that directly convert the incident x-ray radiation into electric charges as secondary quanta. Devices which conform to this approach are called direct conversion flat-panel detectors. The second approach is based on scintillation screens that stop incident x-ray photons, thus producing optical photons as secondary quanta. These optical photons are then stopped by a photodiode array that provides the electric charges required by the device read-out electronics. The detectors that implement this scheme are known as indirect conversion flat-panel detectors.

Direct conversion detectors based on amorphous selenium (*a*-Se), indirect conversion detectors based on amorphous silicon (*a*-Si) coupled to scintillation screens, and indirect conversion detectors based on CCDs connected to scintillation screens using fiber tapers are compared in [10]. The authors conclude that the final reconstructed CT image quality cannot be predicted from differences in the quantum efficiency of the detectors studied, due to stability issues.

In the past 10 years, different micro-CT systems based on microfocus x-ray tubes and semiconductor-based flat-panel detectors arranged in cone-beam geometry have been developed, and their suitability has been proved [11,12]. Flat-panel-based cone-beam configurations present advantages over other configurations (e.g., pencil or fan-beam geometries) used in clinical or preclinical applications. These advantages include reduction in acquisition time, large axial field of view (FOV) with no geometrical distortions, optimization of radiated dose per time and data acquired, and a more compact, space-saving design. Additionally, indirect conversion semiconductor flat-panels are particularly interesting for small-animal imaging due to their high-resolution capability, especially when the microcolumnar scintillation screen is directly grown on the semiconductor detector. Current advances in semiconductor technology point to improved features in the future [13,14].

The main topics to be addressed during the design of micro-CT systems are reviewed later, with special emphasis on x-ray detector features and management. The performance of the system is also analyzed. This review is based on a state-of-the-art x-ray micro-CT [15] used as an add-on for small-animal PET systems [16,17]. The micro-CT system was designed to achieve an FOV that is appropriate for small rodents, a spatial resolution better than 50 μm, and a minimal radiated dose. The tomography system includes a flat-panel detector (complementary metal oxide semiconductor [CMOS] technology with a columnar cesium iodide scintillator plate) with a 50-μm pixel size, and a microfocus x-ray source with a nominal focal spot of 35 μm. Both elements are placed in a rotating gantry according to the cone-beam geometry. The magnification factor of 1.6 was obtained by applying the design specification for final resolution, FOV size, and mechanical constraints defined by the system size and radiation shields.

The performance of the flat-panel detector was evaluated to validate its suitability for the micro-CT scanner. To validate its use for preclinical in vivo imaging as an add-on system for PET/SPECT tomographs or as a standalone unit, the overall performance of the system was evaluated in terms of spatial resolution, image contrast, exposure dose, and image acquisition and reconstruction time.

5.2 Small-Animal Micro-CT Design Considerations

Small-animal micro-CT imaging systems are usually designed to provide complementary information for molecular imaging systems. The main design issues are image quality and the radiation delivered to the study animal.

Early implementations of x-ray micro-CT systems were aimed at nondestructive testing for industrial applications or at the in vitro study of biological tissues [18]. The main design goal for these systems was to achieve high image quality (resolution and contrast), regardless of the radiation dose delivered. They were not designed to facilitate the use of equipment for in vivo imaging, such as anesthesia equipment or vital sign monitoring devices. Thus, the mechanical setup was such that the x-ray source and detector were fixed while the sample was rotated between these devices. This made system calibration easier and ensured long-term stability.

In vitro micro-CT scanners can achieve very high spatial resolution values (~5 μm) using microfocus x-ray sources and area detectors such as CCDs or flat-panel detectors. However, to obtain an appropriate signal-to-noise ratio, it is necessary to perform long acquisitions at high radiation doses. Furthermore, the desired resolution in the reconstructed volume is commonly obtained in practice by using high magnification values (>3), which require the detector to be situated at some distance from the sample; therefore, such mechanical arrangements are only feasible if the x-ray source and detector are assembled on a horizontal flat surface. A typical configuration for an x-ray in vitro system is depicted in Figure 5.1.

Small-animal micro-CT scanners also offer a reasonably high spatial resolution with sufficient image quality, although they require acquisition time and radiation dose to be as low as possible. These two additional constraints reveal the need for a new approach to the implementation of the system. Therefore, a compromise must be sought between image quality, spatial resolution, acquisition time, and dose delivered. Furthermore, during the acquisition process, the animal must be kept steady while the gantry holding the

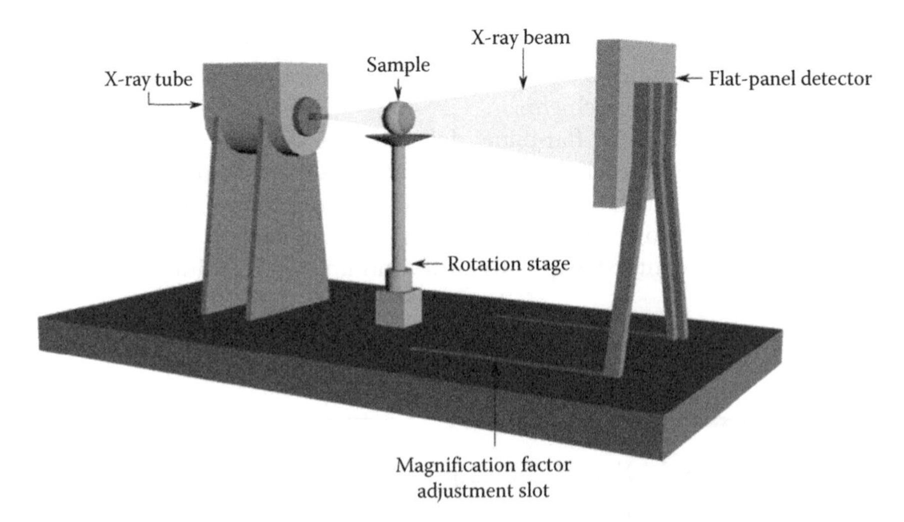

FIGURE 5.1
Typical configuration of an in vitro micro-CT scanner. The magnification factor can be adjusted by varying the distance between the sample and the flat-panel detector.

x-ray source and detector rotates around it. This is an important requirement, since it is necessary to minimize the disturbance to the animal and to avoid inconsistent projection data derived from any organ displacement that may occur if the animal is moving. Anesthesia equipment and monitoring devices (heartbeat, breathing rate) must be correctly placed, as monitoring becomes somewhat challenging if the animal is rotating.

The constraints imposed by in vivo requirements demand a more sophisticated and careful design if the system is to be compact. This requirement is even more important when space is scarce. In this setting, the mechanical features of the x-ray source and detector become more important, since these devices should be placed on a rotating gantry to keep the subject in a fixed position. To avoid instability during rotation, the weight and size of the devices should be as low as possible, and mechanical elements are necessary to counterbalance the system. Although some scanners make use of slip rings, this is not the most common solution, as they are expensive and rotation speed is not the most constraining factor. Therefore, the signal and power cables must be placed in such a way that they do not disturb the image acquisition process. This usually involves an extension of cable lengths, thus increasing signal losses and deteriorating the signal-to-noise ratio in the acquired data.

Sensor specifications are also an important issue. Fast and sensitive detectors are extremely desirable for small-animal micro-CT, given that acquisition time and dose delivered are paramount.

Since the acquired x-ray micro-CT projection data must be highly stable along the acquisition path, it is necessary to use highly stable x-ray sources and detectors. If the detector is not stable, it becomes necessary to acquire several correction datasets during a single acquisition process, thus increasing total acquisition time.

Different approaches can be adopted in the selection of the detector. Nowadays, CCDs connected optically to a scintillation screen are the most widely used detectors in small-animal micro-CT systems. These detectors show good spatial resolution (usually depending on the coupling between scintillator and sensor) and the fastest image rate. However, these devices are usually too bulky to integrate in a moving gantry. Another drawback is that the sensor can be seriously damaged by x-ray radiation [19] and must be carefully shielded, thus increasing the size and weight of the system.

Modern designs make use of the so-called flat-panel x-ray detectors (semiconductor-based light detector matrices coupled to scintillator screens [20] or direct conversion semiconductor detectors), due to their high resolution (equal to or better than that achieved by CCDs) and image quality, combined with a compact design and low weight that simplifies integration in the rotating gantry. Their main drawback is that most of them are slower than CCDs.

Some experimental developments [21–24] make use of photon-counting detectors based on cadmium telluride (CdTe) or cadmium zinc telluride (CZT) sensors. These devices are able to classify the incoming photons according to their energy, enabling accurate correction of energy-related artifacts, such as beam hardening or scatter. Much effort is being made in the development of x-ray photon-counting detectors for micro-CT systems, but these are still at an early stage. The quality and image rate of the data acquired by state-of-the-art devices is not sufficient to allow them to be used in commercial preclinical imaging systems.

5.3 Overview of Flat-Panel X-Ray Detectors for Cone-Beam Micro-CT

Flat-panel digital detectors are one of the most widely used x-ray detection devices for small-animal imaging. Several comprehensive reviews of flat-panel x-ray detector technology [20,25–27] address topics not included in the present work and could prove useful for the interested reader. This section presents a brief overview of the state-of-the-art technology of the aforementioned detectors, focusing on their suitability for small-animal cone-beam micro-CT.

Flat-panel detectors can be classified as direct conversion and indirect conversion devices (Figure 5.2). In the following, the features of flat-panel

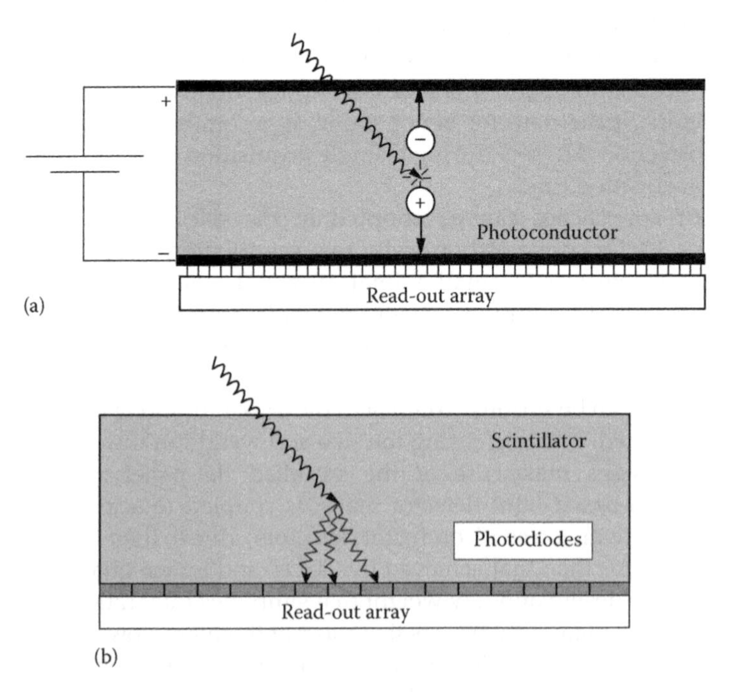

FIGURE 5.2
Scheme of direct (a) and indirect (b) detection approaches used in x-ray flat-panel detectors.

detectors are presented according to this classification. A comparative study of direct and indirect conversion flat-panel detectors can be found in [28].

Since x-ray detectors can be simply modeled using cascaded devices [29–32], for the sake of clarity, the following explanation divides signal generation into two different stages, namely, x-ray conversion and data readout. Information on the position where the x-ray photon is detected and on its deposited energy is generated at the beginning of the first stage, where the x-ray photons (primary quanta) are stopped. Subsequent steps do not add new information to form the image. Thus, it is important to have a low noise level and adequate amplification in the stages following x-ray capture to ensure that no information is lost after the x-ray photons are detected. The stage with the poorest quanta detection capability, the quantum sink, limits system performance in terms of signal-to-noise ratio. Since the information lost in the quantum sink cannot be recovered, a well-designed detector should have its quantum sink at the very first stage of image formation, namely, when x-ray photons are stopped [20].

5.3.1 Indirect Conversion Flat-Panel Detectors

5.3.1.1 X-Ray Conversion Stage

Indirect conversion flat-panel detectors detect individual x-ray photons by generating optical photons as secondary quanta. A subsequent step is needed to convert the optical photons into electric charges on each pixel. Integrated over a period of time, this charge is then amplified and digitized.

The outer layer of the detector is made of a scintillation material, which provides a variable number of optical photons per x-ray photon stopped, depending on the energy involved in the interaction.

The scintillation materials used in x-ray flat-panel detectors are usually inorganic compounds such as cesium iodide or gadolinium oxysulfide. Scintillation in such materials is based on energy transfer to the molecules of the scintillator, depending on the energy states determined by the crystal lattice of the material. In insulator or semiconductor materials, electrons remain in a discrete number of energy bands. The two main energy bands are the valence band and the conduction band. Electrons with an energy state that places them in the valence band are bound to the crystal lattice and, therefore, fixed. However, those electrons with sufficient energy to reach the conduction band can move across the crystal. The band between the conduction and valence bands is the forbidden band, where no electrons can be found in a pure crystal.

If a photon with sufficient energy reaches an electron in the valence band, it can increase its energy so as to reach the conduction band, thus leaving a hole in the valence band. The excited electron returns to the valence band by emitting a photon of energy equal to the difference between the energy levels.

This energy is usually too high and the emitted photon energy does not correspond to the visible light range.

To increase the number of visible light photons emitted, dopants are added to the crystal [33]. These materials create defects in the crystal lattice, where the normal energy band structure is modified, thus creating energy levels within the forbidden band. Excited electrons can fall first into one of these energy levels and later into the stable valence band, and emit photons with an energy that falls within the visible range.

Several scintillation materials have been used in flat-panel detectors. The most desirable characteristics of a scintillation material for this application are the possibility of implementing large-area screens, the production of a high number of optical photons per x-ray photon detected, high sensitivity to the energy spectrum of the x-ray beam, and a high degree of correlation between the direction of propagation of the incident x-ray photon and that of the optical photons generated. It is also important that the energy spectrum of the photons emitted by the scintillation material fits the reception spectrum of the photodiodes used in the second detection stage.

At present, the two scintillation materials most commonly used in flat-panel detectors are terbium-doped gadolinium oxysulfide (Gd_2O_2S:Tb) and thallium-doped cesium iodide (CsI:Tl).

Gd_2O_2S:Tb has traditionally been used to detect x-rays and can be manufactured easily and cost-effectively using well-known technology [20]. It comes in the form of a powder [34] composed of microscopic particles with a density of 7.3 g/cm^3. The particles are bound together by an acrylic binder to form a homogeneous paste that is usually applied as a coat on a glass or plastic support. A reflector can be added to increase light collection in the detector.

Gd_2O_2S:Tb has one of the highest figures of merit (defined as the best balance between high light production, fast response, and appropriate energy spectra) among powder scintillators: it is highly efficient and has a very low afterglow and an appropriate energy emission spectrum, centered at 540 nm. However, achievable spatial resolution is limited by the lateral scattering of the photons generated. Thicker scintillation screens offer a longer path for the optical photons, thus increasing scattering and worsening spatial resolution, while thinner screens stop a low fraction of the received x-ray photons, thus decreasing detector sensitivity. Therefore, a compromise must be reached between x-ray stopping power and achievable spatial resolution.

Microstructured scintillation screens based on CsI:Tl were developed to meet the main imaging needs of x-ray systems, namely, the increase in x-ray stopping power while maintaining good spatial resolution.

The manufacturing process of this type of scintillation screen is based on the deposition by thermal evaporation of long, thin, needle-shaped structures on a glass or plastic support or directly on top of the semiconductor light detectors. Each of the needle-shaped structures behaves as a light pipe,

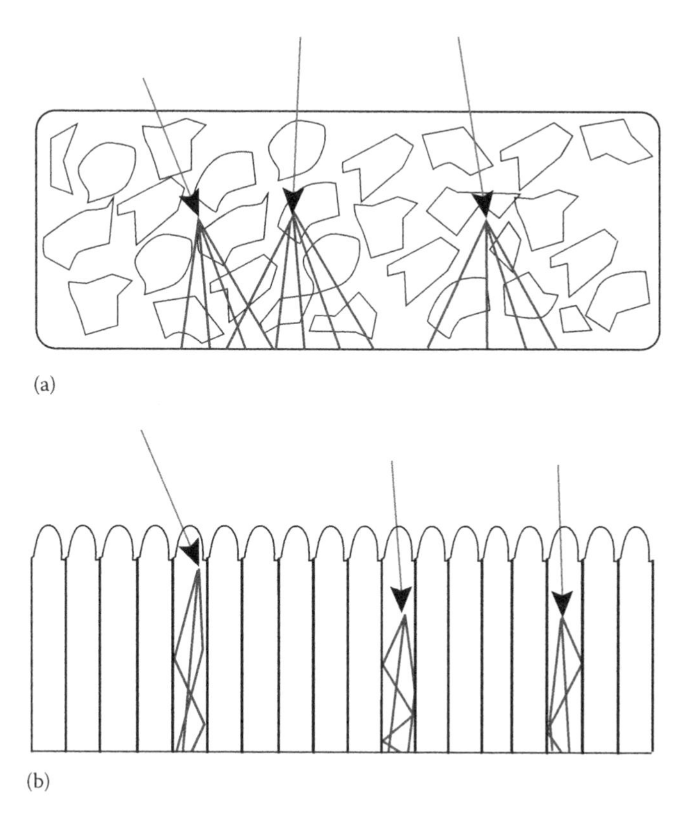

FIGURE 5.3
Simplified light transport in a granular nonstructured scintillator screen (a) and in a microcolumnar structured scintillator screen (b).

confining optical photons inside it and thus avoiding most of the lateral scattering present in nonstructured screens (Figure 5.3). This light confinement makes it possible to increase the thickness of the scintillation screen (up to 1–2 mm), while maintaining good spatial resolution. Furthermore, light production, needle stability and shape, and resolution of the scintillator screen are highly dependent on manufacturing conditions [35], namely, thallium concentration, pressure, temperature, and post-processing.

The scintillator needle layer can be deposited directly on the semiconductor detector surface without degrading the properties of the photodetector elements in the array, thus achieving optimal spatial resolution and high x-ray detection efficiency, as optical coupling agents are not necessary to glue the scintillator screen over the semiconductor surface. Besides the aforementioned advantages of this kind of scintillator, the light production per photon stopped of CsI:Tl is among the highest known, and the peak of the emitted spectrum is at 550 nm, a value that matches the spectral response of most semiconductor photodetectors currently used in the development of x-ray flat-panel detectors [36,37]. Due to its advantages over other scintillation materials, CsI:Tl is becoming the preferred material for the x-ray photon conversion stage in flat-panel detectors. However, it does have certain drawbacks, the most important being its relatively slow response time, as compared to that of Gd_2O_2S:Tb.

5.3.1.2 Secondary Quanta Detection and Read-Out

Secondary quanta conversion and data read-out in indirect flat-panel x-ray detectors have traditionally been implemented as a pixel array based on hydrogenated amorphous silicon (aSi:H) and thin-film transistor (TFT) technology. Each pixel consists of a reverse-biased photodiode and a TFT that acts as a switch. During exposure, charges are accumulated in the photodiode. After exposure, a gate pulse is generated for each row of pixels, thus switching the TFT of the given row of pixels to release the accumulated charges through the data line. The released charges are amplified and converted into a voltage using an array of charge-integrating amplifiers, with one amplifier per pixel in the row. The voltage signals from the row of pixels are then multiplexed and digitized.

The technology described has mainly been used for digital radiography detectors; however, devices using this technology show a long signal decay time and a poor fill factor due to the wide electrode width and switch size [36]. These drawbacks impair the development of sensors with an appropriate frame rate and small pixel size, thus hampering the use of this technology for micro-CT image acquisition.

A newer technology for the read-out of indirect x-ray flat-panel detectors consists of sensors based on CMOS technology. CMOS image sensors are composed of a matrix of identical pixels which have a photodiode and a MOS switch transistor [37], two scan circuits which address the different rows and columns of the sensor matrix, and an output amplifier.

The image formation process for the simplest CMOS image sensor, based on passive pixel elements, is quite similar to that of the pixel array based on aSi:H and TFT technology. First, the photodiodes are reverse biased. The incoming photons cause a decrease in the voltage of the photodiode, which is measured at the end of the imaging process. The drop in voltage gives an estimation of the number of photons that have reached the given pixel. After the pixel reading, the photodiode is reset.

Current CMOS sensors use more sophisticated pixel designs, where every pixel has an active element that acts as an individual amplifier, thus reducing the noise level of the final image. Improvements in active pixel design have led to the development of the pinned photodiode pixel, which is in use for most current CMOS image sensors. In this kind of sensor, two measurements are taken and subtracted for each image pixel (correlated double sampling, CDS). The first contains information about the number of photons reaching the pixel, while the second gives an estimation of the photodiode offset voltage. This kind of design allows a further reduction of the noise level and

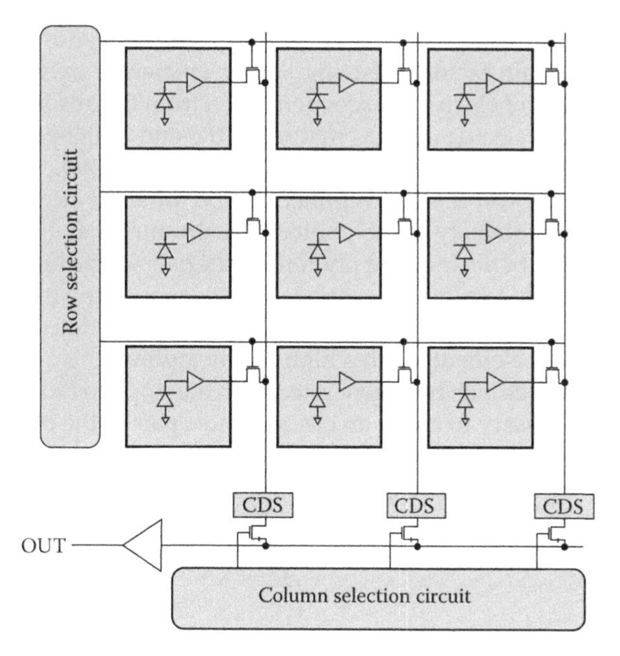

FIGURE 5.4
Sketch of a typical CMOS image sensor with active pixels and CDS.

dark current of the detector. A sketch of a CMOS sensor with active pixels and CDS circuits is shown in Figure 5.4.

5.3.2 Direct Conversion Flat-Panel Detectors

5.3.2.1 X-Ray Conversion Stage

The detection of x-ray photons to generate secondary quanta in a direct conversion flat-panel detector consists of a layer of photoconductor material. Unlike indirect conversion detectors, the secondary quanta generated are already electric charges, thus avoiding the need for an intermediate stage to convert the secondary quanta into electric charges.

The process by which electric charges are generated from stopped x-ray photons is the same for all photoconductor materials. Stopped x-rays with sufficient energy generate an electron-hole pair that drifts through the photoconductor material under the action of an externally applied electrical field. Each portion of the photoconductor layer that defines a pixel has a capacitor, which accumulates the charge generated by the stopped x-rays inside the current pixel. This accumulated charge is later measured and converted to form a pixel of the final image.

The photoconductor materials used to detect x-ray photons should be as efficient as possible in the conversion of x-ray photons to electric charges. Conversion efficiency depends on several factors associated with the photoconductor material, its manufacturing process, and the operating conditions. The main factors affecting sensor efficiency are x-ray stopping power, the number of electric charges generated from the absorbed radiation, and the number of electric charges that reach the end of the photoconductor layer and can be collected by the capacitive elements [20,26].

X-ray stopping power (i.e., the number of x-ray photons stopped) is highly associated with the density of the photoconductor material, its atomic number, and the energy of the incident photons. Materials with a high density and atomic number have a larger absorption coefficient and stop more x-ray photons. Thus, photoconductor materials for x-ray flat-panel detectors should be dense and made of elements with a high atomic number.

The number of electric charges generated from the stopped radiation depends on the energy necessary to create an electron–hole pair in the photoconductor material, that is, the ionization energy. Low ionization energy is highly desirable when attempting to generate a high number of electric charges from the incident radiation. The charge generated [26] is given by expression (5.1)

$$Q = \frac{eE}{W_{\pm}} \tag{5.1}$$

where

Q stands for the generated electric charge
e is the charge of the electron
E is the energy of the incoming radiation
$W_{\pm}$ is the ionization energy of the photoconductor

The ionization energy depends on the energy bandgap (E_g) of the material used and, in some cases, it can be modified by means of an electric field. One of the materials that allow its $W_{\pm}$ value to be modified by applying an electric field is amorphous selenium (*a*-Se), the material most widely used in the development of x-ray direct flat-panel detectors.

The fraction of the generated charge that reaches the sensor surface and can be gathered by the capacitive elements is determined by the drift mobility of the electrons and holes generated inside the photoconductor, and the mean time that electrons and holes can drift without being trapped. Taking into account the applied electric field, it is possible to define the *Schubweg*, which is the mean distance traveled by a charge carrier before being trapped [38] and is given by (5.2)

$$S = \mu\tau F \tag{5.2}$$

where

μ stands for the drift mobility of the charge carriers
τ is the mean time before being trapped
F is the applied electric field

As mentioned earlier, the most commonly used photoconductor for the development of direct x-ray flat-panel detectors is *a*-Se, despite the fact that pure *a*-Se has some undesirable properties, namely, the material is thermally unstable and crystallizes after a variable period depending on ambient conditions. To prevent crystallization and stabilize the material, some additives—usually small amounts of arsenic and a halogen (e.g., chloride)—are mixed with the original *a*-Se. The doped material, known as "stabilized *a*-Se," is more stable, thus enabling its deposition as flat screens. However, doping with a-Se can worsen the performance of the photoconductor screen. The most important drawback is the decrease in carrier mobility, which, in turn, increases the number of trapped carriers. The increase in the number of trapped carriers causes image lag and a decrease in sensor sensitivity that hamper the use of this kind of detector for the acquisition of micro-CT images. Both these effects are caused by the delayed freeing of carriers trapped during previous x-ray exposures. The image correction procedures applied to the acquired projections (see below) assume that every projection image is acquired under approximately the same conditions (dark current and x-ray conversion efficiency). Image lag prevents fulfillment of these requirements, thus leading to inconsistent datasets when image correction is performed using a correction dataset acquired before the actual acquisition process. This problem can be solved by using several correction datasets acquired during the CT image acquisition, albeit at the cost of increasing acquisition time and dose delivered.

Recent developments point to CdTe (and CdZnTe) as a photoconductor material for stopping x-ray radiation. However, in most cases, development is at an early stage and there are few manufactured devices (mainly for dental and industrial radiography).

Early studies on the properties of these photoconductor materials conclude that the sensitivity of a sensor based on CdTe or CdZnTe can be up to four times higher than that achieved using *a*-Se–based sensors [39], but they also show a strong afterglow effect [40] and poor spatial resolution [39].

Recent developments show a reduction in the afterglow and better resolution while maintaining the predicted enhancement in sensitivity. There is also a high correlation between the design of the contacts deposited over each of the material surfaces and the amount of dark current present in the detector that degrades the performance of the device, especially the achievable energy resolution [41,42]. Depending on the design of the contact plates, there are two main detector types: Ohmic detectors and Schottky detectors [41,43]. Ohmic detectors have two contacts made of platinum, while in Schottky detectors one of the contacts is made of titanium and indium, thus forming a Schottky contact that reduces the injection of holes for the same voltage bias, thus reducing the dark current of the device. The main drawback of Schottky contacts is the so-called polarization effect, which consists of a very slow increase in the number of holes trapped near the positive electron [44]. This trapping effect leads to a decrease in sensitivity and in the charge collected per detected photon [45]. There are several ways to overcome this phenomenon, the simplest and most common being to reset the bias voltage when the effect is noticeable (about 60 min after the bias voltage is applied, for modern devices) [41,42].

5.3.2.2 Signal Read-Out

In detectors based on a-Se screens, the electric charges generated within the photoconductor are gathered by capacitive elements connected to TFTs that act as switches. The reading process is almost identical to the one used for indirect x-ray flat-panel detectors based on *a*-Si:H and TFT technology. Each row of pixels is addressed using a pulse that activates the TFTs of the desired pixel row, and the signal accumulated at each pixel in the row is read and converted into a voltage. This voltage is multiplexed with the rest of the voltages coming from the different pixels in the row and the values are digitized, thus forming the image. The process is performed for every row of pixels in the sensor.

For CdTe and CdZnTe detectors, the read-out matrix is generally implemented using a CMOS or TFT application-specific integrated circuit (ASIC), although in this case, the photoconductor cannot be directly deposited over the ASIC; therefore, the ASIC is built separately and at a later stage, and the crystal is placed on top of the read-out circuit. Both are connected by flip-chip bonding using a conductive resin.

5.4 Design of an X-Ray Micro-CT System

5.4.1 Components of a Small-Animal Micro-CT System

The main components of a micro-CT system based on flat-panel detectors are the x-ray detector and the microfocus x-ray source. The constraints imposed by the application determine the type of flat-panel detector chosen by the designer.

In small-animal micro-CT systems, it is important to use a detector with an appropriate image resolution to obtain the high-quality images necessary for preclinical applications. The maximum pixel size and minimum resolution of the detector depend on the geometrical configuration of the system (see later). High sensitivity and low noise level are also highly desirable features when attempting to minimize the radiation delivered.

The output of the x-ray tube must be stable enough to ensure a constant radiation level during the acquisition process. Its focal spot size must be sufficiently small so as not to degrade system resolution. The maximum admissible size for the focal spot is determined by the detector pixel size and the geometrical configuration of the system. Furthermore, highly stable motorized devices are used in the design of small-animal micro-CT systems to place the animal within the FOV and to perform the rotational movement of the x-ray source–detector assembly around the sample. Finally, control components (e.g., computer; motion control drivers; shutter) and shield elements must be incorporated to ensure simple and safe operation.

This section summarizes the main elements of a commercial state-of-the-art in vivo micro-CT system, the Argus PET/CT (Sedecal, Madrid, Spain).

5.4.1.1 System Components

A complete description of the different components included in the design of the micro-CT system used as an example in this section can be found in [15].

The scanner design includes a computer that controls the microfocus x-ray tube and the CMOS flat-panel detector, both of which are assembled in a common rotating gantry. A linear motion stage is used to displace the sample along the FOV, thus enabling the tomograph to perform whole body scans (Figure 5.5). The assembly is enclosed in a radiation-shielded cabinet

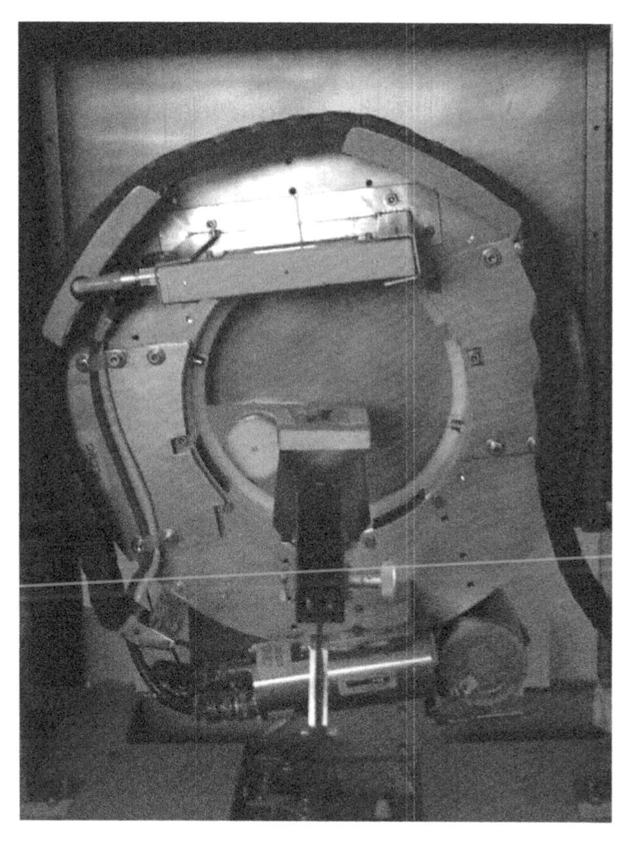

FIGURE 5.5
Small-animal CT prototype showing its components attached to the rotating gantry: microfocus cone-beam x-ray tube (bottom), object bed (center), and CMOS digital imaging sensor (top). The radiation shielding cabinet is not shown.

with openings for animal positioning, anesthesia gas lines, and physiological monitoring cables.

The computer synchronizes the gantry motions with the image integration in the detector. The data acquired from each angular position are captured by a digital frame grabber. Raw data are processed simultaneously with the acquisition, thus saving processing time during the subsequent reconstruction stage and taking advantage of the full potential offered by the system computer.

5.4.1.1.1 Flat-Panel Detector

The x-ray flat-panel detector used in the Argus PET/CT is the C7940DK-02 model (Figure 5.6) from Hamamatsu Photonics K.K. (Hamamatsu-city, Japan). A complete description and an evaluation of some of the performance parameters can be found in [36]. This is an indirect flat-panel detector based on a CsI:Tl scintillator screen and a high-fill-factor CMOS image sensor to detect the optical photons. The features of the flat-panel detector as reported by the manufacturer are summarized in Table 5.1.

The scintillator layer of the flat-panel detector consists of a matrix of 150-μm long, needle-like CsI:Tl crystal structures directly deposited over the CMOS photodiode matrix surface (Figure 5.6). A photodiode matrix with active pixel elements connected to CMOS transistor switches enables detection and read-out of the visible photons generated on the scintillation screen. The photodiode CMOS matrix has 2 × 2 and 4 × 4 binning capabilities and a high fill-factor (79%). The on-chip signal amplification channels have a low noise level, and an offset suppression circuit based on CDS is assigned to each of these channels. This design achieves a high degree of image uniformity and a low noise level. However, correction

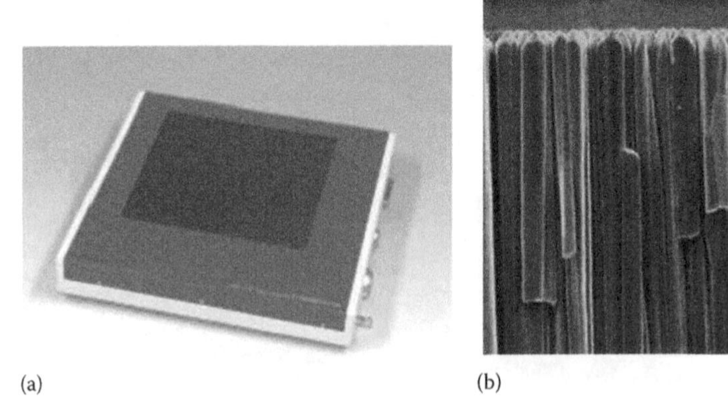

FIGURE 5.6
Hamamatsu C7940DK-02 x-ray flat-panel detector (a) and detail of the needle-like CsI:Tl crystal structures on the scintillator layer (b). (Images courtesy of Hamamatsu Photonics K.K., France, Germany.)

TABLE 5.1

X-Ray Flat-Panel Detector Features

Pixel size	50, 100, 200 μm (binning 1, 2, 4)
Photodiode area	120 × 120 mm
Number of pixels (total/active)	2400 × 2400/2240 × 2344
Frame rate	2, 4, 9 fps (binning 1, 2, 4)
Noise (rms)	1100 electrons
Saturation charge	2.2 M electrons
Dynamic range	2000
Sensitivity (at 80 kV)	25 LSB/mR

tasks must be performed after acquisition of an image in order to obtain the best achievable quality. In addition to the noise-reduction stages, every pixel in the matrix implements an overflow drain function to avoid possible blooming.

5.4.1.1.2 X-Ray Source

The scanner uses the XTG5011 Apogee microfocus x-ray source (Oxford Instruments, Scotts Valley, CA, USA). This tube has a stationary tungsten anode, a 127-μm-thick beryllium window, and a Gaussian-like focal spot measuring 46.5 × 49.1 μm, according to the manufacturer's measurements. The tube also has a maximum anode power of 75 W, limited to 50 W by the high-voltage power supply (50 kV for an anode current of 1 mA).

The working settings for this component (i.e., anode current and voltage) are managed by the control computer through an I2C interface. To reduce the radiated dose during scans and the presence of artifacts arising from the polychromatic properties of the generated x-ray beam (e.g., beam hardening), two different elements have been added to the source output window:

- A tungsten shutter to block the x-ray beam during the intervals in which the detector is not integrating a valid image, that is, when the gantry is moving. This shutter has a maximum operating frequency of 10 Hz and a minimum opening time of 20 ms. The nominal beam extinction ratio increases to 10^4 for a maximum beam energy of 30 keV, as stated by the manufacturer. Practical measurements of the beam extinction fraction for the particular settings of the current system (40 keV, 200 μA) reveal a ratio of $1{:}4{\cdot}10^3$; therefore, exposure to radiation could be considered negligible when the shutter is closed.
- An aluminum filter (thickness ranging from 0.1 to 2 mm) to filter the low-energy region of the emission spectrum, thus improving beam monochromaticity and reducing superficial dose and image artifacts.

TABLE 5.2

Rotation Stage Features

Axis diameter	78 mm
Resolution	0.001°
Accuracy	0.010°
Repeatability	0.002°
Maximum centered load	1800 N
Maximum inertia	3 kg·m^2
Maximum speed	20°/s

Minimal beam collimation is performed using a lead collimator, since most of the x-ray cone beam is used for imaging.

5.4.1.1.3 Mechanical Subsystem

The x-ray source, with its associated elements, and the flat-panel detector are placed on a circular aluminum plate fixed to a rotational motion stage. The projection dataset to be reconstructed is acquired by rotating the whole set 180° or 360° around the subject. The gantry stage is driven by a controller and a stepper motor. The main features of the rotational stage are shown in Table 5.2.

The system is provided with a linear motion stage to place the sample into the x-ray beam during the acquisition process. The sample is placed on a carbon-fiber bed with the appropriate dimensions for small animals (rats and mice). The carbon-fiber structure is attached to the linear motion stage using a metallic holder, which incorporates controls to adjust the height and lateral shift of the animal bed.

The whole mechanical system is controlled by the control computer, which interfaces with the stepper motors using motor controller circuits. These are enclosed in an electronic box containing the auxiliary electronic systems for the interlock that prevents accidental radiation leakages due to incorrect operation of the radiation-shielded cabinet elements.

5.4.2 Geometrical Configuration

Small-animal micro-CT systems based on cone-beam geometry should accomplish two basic design criteria: the resolution of the reconstructed data has to be high enough to image anatomical structures with the appropriate level of detail (~100 μm), and the transaxial FOV has to be large enough for small laboratory rodents (about 75 mm in diameter). The distances from the source to the object (D_{so}) and from the object to the detector (D_{od}) determine the size of the FOV and the magnification factor, according to the following expressions [1,46,47]

$$\text{FOV} = \frac{T_d}{M} \tag{5.3}$$

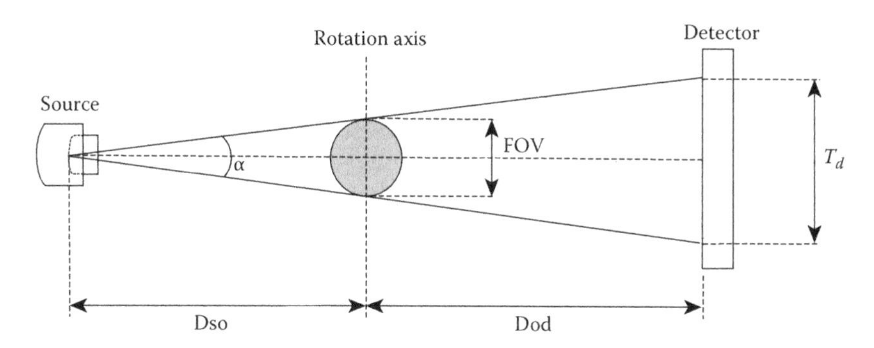

FIGURE 5.7
Geometry of the micro-CT system. T_d refers to the usable size of the x-ray CMOS detector.

$$M = \frac{D_{so} + D_{od}}{D_{so}} \tag{5.4}$$

where
T_d is the size of the flat-panel detector active area
M is the magnification factor

The useful FOV is defined as the length of the rotation axis intersected by the effective cone-beam (Figure 5.7); the "effective" section of the cone-beam is that which intersects the active area of the detector.

In order to assess the suitability of the design parameters, the theoretical system resolution can be estimated as the convolution of the effects of the finite focal spot size in the source (ρ_f) and the intrinsic resolution of the detector (ρ_d) [47].

The component of system resolution due to the detector (θ_d) can be calculated at the center of the FOV using the following expression:

$$\theta_d = \frac{\rho_d}{M} \tag{5.5}$$

If ρ_f is the focal spot size of the source, then θ_f at the center of the FOV is

$$\theta_f = \rho_f \frac{M-1}{M} \tag{5.6}$$

Assuming a Gaussian distribution for θ_f and θ_d, the resulting resolution for the reconstructed images can be estimated theoretically according to the formula

$$\theta = \sqrt{\theta_d^2 + \theta_f^2} \tag{5.7}$$

System resolution could be degraded by other factors associated with the tomographic reconstruction process and by submillimetric mechanical misalignments. Correct alignment between source and detector is critical to achieve the theoretical resolution value and avoid the presence of artifacts in the reconstructed image [48]. There are several suitable methods of experimentally estimating the differences between the geometrical parameters of the real system and those of the original design [49–52]. These methods are usually based on the acquisition of a phantom with a known geometry whose projection trajectories are fitted to the theoretical trajectories (derived from the ideal geometrical arrangement) in order to calculate the deviation between the real and the theoretical configuration parameters.

The geometrical parameters of the micro-CT subsystem of the Argus PET/CT (D_{so} = 219.8 mm and D_{od} = 131.9 mm, α = 19.4°, T_d = 120 mm) were selected to meet the design constraints, namely, resolution, FOV size, and available space). The adopted solution leads to a magnification factor (M) of 1.6 and to a theoretical resolution of 12 cycles/mm (measured as the modulation transfer function [MTF] 10%), or about 40 μm in the spatial domain.

In this system, the geometrical deviation from the original design is measured by means of an analytical procedure based on the assessment of the elliptical trajectories shown by two ball bearings placed in a soft material, such as foam, as described in [51]. Using this method it is possible to estimate the values of two of the three tilt angles of the detector, the position of the real center for the image (i.e., the projection pixel for the central and orthogonal ray), and the actual distance between the x-ray source and the flat-panel detector. The remaining angle, although not as critical as the previous two [51,53,54], should first be reduced as much as possible to be able to assume that its value is close to zero. The correct alignment in the direction of this angle is mechanically assessed. The estimated angular deviations are mechanically corrected. The offset of the central ray position is corrected online during the acquisition process.

5.4.3 Data Acquisition

Several acquisition schemes make it possible to obtain the projection data for the different angular positions. There are two main trends for the acquisition of micro-CT data. Some devices use a continuous rotation acquisition protocol, where a moving gantry performs a continuous motion while projection data are acquired. This approach has the advantage of faster acquisition times when a fast detector is used, although the effects of image lag and mechanical imperfections are more conspicuous. An alternative is the acquisition of a stack of frames for each angular projection, with the gantry steady during the acquisition (step-and-shoot). While the gantry rotates, no image is acquired and the radiation beam is blocked to minimize the effect of image lag.

The step-and-shoot approach used in the Argus PET/CT system is implemented by means of an event-driven finite-state machine with two possible states to take advantage of the maximum detector transfer rate [55] and to reduce acquisition time and delivered dose. The first state (step) is used to

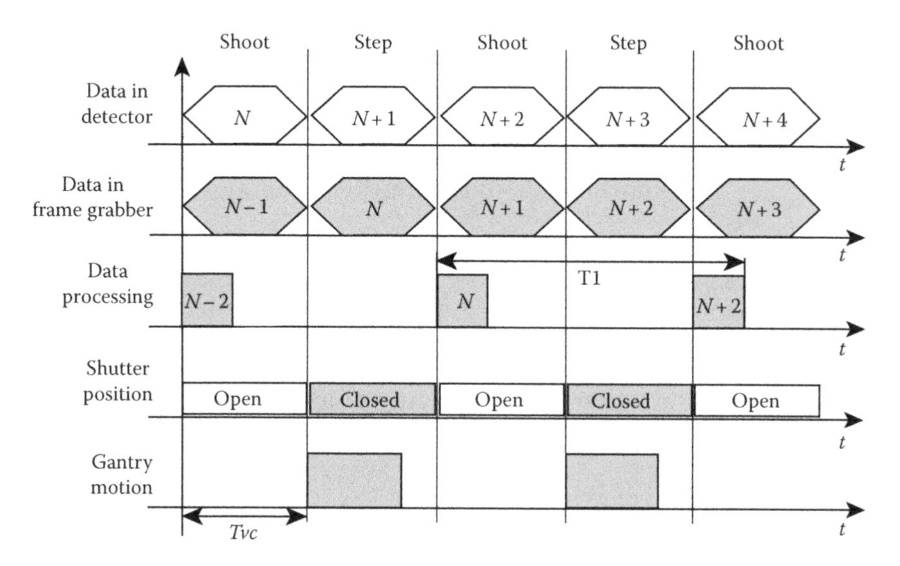

FIGURE 5.8
High-speed acquisition protocol. T_{VC} stands for the integration period in the detector and T1 represents the time elapsed between triggering of an image acquisition and storage of already processed images in RAM before they can be saved to disk. The state transitions are shown in the last row of the chronogram, and the gray rectangle indicates the time spent by the acquisition software to process the dataset acquired in the previous gantry position.

move the gantry to the next angular position and, if needed, to save data from RAM memory to disk. The second (shoot) performs two tasks:

1. Launching the acquisition of either a single frame or a frame sequence in the current angular position
2. Processing the dataset acquired in the previous gantry position

The transition between states is triggered in synchrony with the detector integration period (T_{VC}), as shown in Figure 5.8.

The processing tasks necessary to generate a projection image are performed online during the shoot state. The correction procedures are explained in the following section.

Quality of the projection images can be improved by acquiring and averaging several frames per angular position. In this case, the finite-state machine remains in shoot state for a time equal to $T_{VC}{*}N$, where N is the desired number of frames, until the last frame of the current projection arrives. Therefore, in the multiframe case, there are several "state 1" periods between two "state 2" periods. Additional pre-reconstruction processing, such as filtering, can be performed during "state 1" intervals, when the CPU is idling.

If the sample is too large to fit inside a single axial FOV, the whole volume can be acquired by performing rotations combined with axial shifts of the bed. From the parameters of the detector and the requirements of the given acquisition, it is possible to calculate the time taken by the acquisition process. This effective acquisition time (excluding warm-up and sample positioning) is given (in seconds) by the following expression:

$$T_{acq} = \left(\frac{1}{FR}\right) \cdot \left(Av_{img} + N_{loss}\right) \cdot N_p \cdot N_{FOV} \tag{5.8}$$

where

FR is the frame rate from the detector in images per second

Av_{img} is the number of averaged frames per projection image

N_{loss} is the number of frames lost due to the motion of the rotating stage per each angular projection (one in the current implementation)

N_p is the number of angular views acquired over the defined gantry rotation span

N_{FOV} is the number of axial bed positions

These parameters can be balanced to configure different acquisition protocols: for example, one option could be a high-speed, low-resolution, and low-dose scan, or alternatively a slower, high-resolution, and high-dose scan. In each case, exposure time is controlled by synchronizing the gantry rotation and shutter with the master timing from the frame grabber.

5.4.3.1 Online Correction of Raw Data

Most of the corrections to be applied to the projections can be performed online during the acquisition process, thus reducing processing time for tomographic image reconstruction.

First, despite the offset subtraction performed by the CDS circuit, every pixel has a slight offset level that must be cancelled to obtain optimal image quality. To estimate this offset level, it is necessary to acquire an image with no x-ray radiation reaching the detector (dark current image) at the same temperature and with the same binning configuration as that used for tomographic data acquisition.

After offset correction, the spatial variation of pixel sensitivity must be reduced. Pixel response can be equalized by acquiring an image with a homogeneous radiation field and no object between the x-ray source and the detector (flood-field image). Again, the acquisition setting must match the one planned for the subsequent tomographic acquisition. To calculate the equalized image, the offset corrected image is divided by the flood-field image, thus providing a flat sensitivity profile across the image pixels.

Once the image is equalized, a further correction must be performed to obtain a projection dataset free of artifacts. Due to the CMOS manufacturing process, some lines, columns, or single pixels of the pixel matrix (usually called "dead elements") do not show an appropriate response to the incoming radiation, thus giving a minimum or maximum signal value regardless

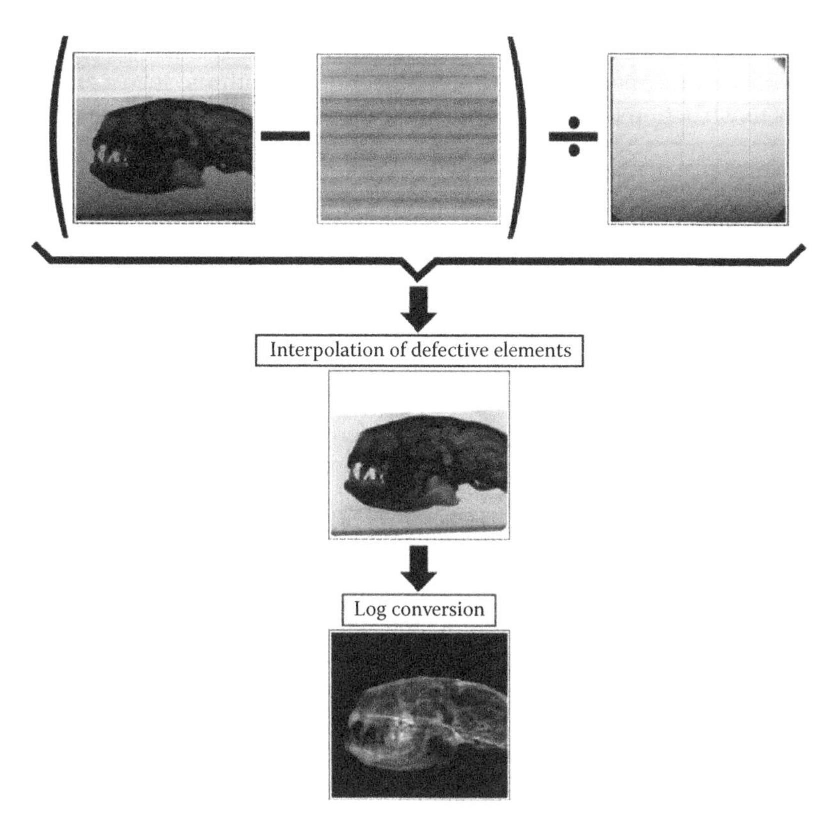

FIGURE 5.9
Raw image correction stages.

of the radiation intensity. The correction is performed by linear interpolation of the value given by the adjacent elements of the detector.

Finally, the attenuation image is generated by calculating the logarithm of the corrected projection image [56].

The different correction steps, as well as their effect on the image, are shown in Figure 5.9.

5.4.4 Tomographic Image Reconstruction

Data obtained by cone-beam micro-CT systems with a circular orbit do not fulfill Tuy's condition, which is required to guarantee consistency in an analytical reconstruction [57]. Therefore, to obtain the reconstructed volume, it is necessary to use approximated algorithms derived from filtered backprojection (FBP)—such as the one proposed by Feldkamp, Davis, and Kress, known as FDK, first described in [58]—or iterative approaches. Approximated algorithms reconstruct the image by performing a modified FBP that minimizes the effects of the inconsistency of acquired data.

Iterative and statistical algorithms, on the other hand, are intrinsically able to deal with data inconsistency. Iterative algebraic algorithms, such as the algebraic reconstruction technique (ART) [59], model the reconstruction problem as a set of algebraic equations and solve the inverse problem to obtain the attenuation value for each voxel in the reconstructed volume. One special category of iterative reconstruction methods includes those based on a statistical treatment of acquired data. These techniques treat the reconstruction as a statistical estimation problem and model the noise properties of the acquired data, achieving the best results when the original data are very noisy (e.g., at low radiation doses).

Statistical methods are now the preferred reconstruction strategy in other biomedical imaging techniques such as PET or SPECT. However, these methods present a much larger computational burden than the analytical algorithms; therefore, FDK-based methods are the preferred reconstruction technique for cone-beam micro-CT [9].

In order to provide consistent attenuation values for different materials, regardless of the scanner settings used in the experiment, the resulting volumetric data are usually represented in Hounsfield units (HU), which relate the attenuation of the different materials with that of water and are not affected by the spectral properties of the x-ray beam [18]. The conversion into HU requires a previous calibration step for different spectral configurations using several materials with a known HU value.

In the Argus PET/CT, reconstruction is performed by means of a modified FDK algorithm with a Ram-Lak filter. The algorithm is adapted to the specific geometry of the scanner, and includes beam hardening correction and HU calibration, using a phantom with seven different known materials (air, water, PMMA, Nylon, Delrin, PTFE, and aluminum). The calibration parameters are obtained for four standard settings of the x-ray beam. The reconstruction software, together with the real-time preprocessing during acquisition, achieves reasonable reconstruction times on standard personal computers (100 s to reconstruct a 512^3 voxels volume using a 2.80 GHz dual core CPU with 8 GB of RAM).

5.5 Evaluation of Micro-CT Systems

An evaluation of small-animal micro-CT performance enables us to determine the quality of the projection images and reconstructed data and to estimate the dose received by the sample.

The following sections present different techniques to evaluate the performance of small-animal micro-CT systems, based on the results obtained for the Argus PET/CT. A complete description of these results can be found in [15].

5.5.1 X-Ray Flat-Panel Detector

The features that give flat-panel detectors a greater impact on the quality of the final images are temporal stability, gain linearity, noise level, and resolution. Temporal instability increases artifacts introduced by the detector, and gain linearity affects the accuracy of tomographic image quantification and introduces ring artifacts. The noise level of the projection data is translated into noise in the reconstructed images, and the intrinsic resolution of the detector limits the final resolution that the system can achieve, as shown in Equations 3.5 through 3.7. Detective quantum efficiency for this type of detector has been evaluated in [12,60].

A possible protocol for the measurement of the different parameters and the results obtained for the C7940DK-02 are provided in the following.

- Detector stability
  One way to estimate the temporal stability of the detector involves acquisition of a set of flood-field images over a given period of time for a fixed x-ray source setting. Between consecutive acquisitions, the x-ray beam must be stopped to allow for scintillation decay in order to remove any potential afterglow contamination on the next measurement. The stability of the device can be assessed by plotting the mean pixel value as a function of time.

  To illustrate this procedure using the CA7940DK-02 system, 360 consecutive flood-field images were acquired at 30 kV and 0.4 μA, with no object between the source and the detector. Exposures were separated by 10-s intervals.

  Measurements show a peak-to-valley ratio during the experiment lower than 0.05% of the mean pixel value. This result indicates that the flat-panel detector is stable enough and the reconstructed images will not be affected by this parameter.
- Detector gain linearity
  The linearity of the detector response with the radiation received can be measured by plotting the mean pixel value as a function of the anode current. Since radiation intensity is directly proportional to the anode current [61], a linear trend is expected for the mean pixel value. For the test carried out on the C7940DK-02 system, the x-ray beam peak energy was set at 40 kV, with anode current ranging from 200 to 500 μA. A 1-mm-thick aluminum filter was placed in front of the x-ray source to reproduce a set-up commonly used for CT acquisition. The images were acquired with a homogeneous radiation field and no object between the x-ray source and detector. The result of the test shows an excellent degree of linearity ($R^2 > 0.99$) over the dynamic range of the detector.
- Detector noise level
  The noise level in the acquired projection data has a considerable impact on the final achievable quality. It can be measured by the

signal-to-noise ratio (SNR) of a set of flood-field images as a function of the anode current.

The SNR was measured as a function of anode current for the C7940DK-02. As in the previous measurement, peak x-ray beam energy was set at 40 kV, with anode current ranging from 200 to 500 μA and a 1-mm-thick aluminum filter. The trend observed was as expected.

- Detector spatial resolution

  The resolution of x-ray imaging detectors is usually expressed in terms of the presampled MTF. This describes the signal transfer characteristics as a function of spatial frequency, taking into account all the detection stages except sampling. If the sampling stage is included in the calculation, the detector spatial response may become undersampled, leading to aliasing errors in the estimated MTF.

Among the different methods proposed to estimate the presampled MTF, those based on imaging a slanted slit or a slanted edge are the most widely used due to their relatively easy implementation and accurate results.

The following paragraphs present a brief description of the procedure proposed in [62,63], which implements a variation of the slanted-edge method. A comprehensive review of the different versions of the slanted-edge approach can be found in [64]. The selected algorithm is based on the analysis of the edge response function (ERF), which is obtained by imaging a phantom consisting of an x-ray-opaque object with a polished edge. The phantom is placed directly over the detector surface at a shallow angle (1.5°–3°) with respect to the pixel matrix. The edge position is estimated with sub-pixel accuracy in the image area by using linear interpolation. This position is then fitted to a straight line by linear regression.

The slope of this line is used to determine the number of rows (N_{AV}) necessary for a 1-pixel lateral shift of the edge position in the original image (Figure 5.10).

The algorithm generates an oversampled ERF using the pixel value at the edge position for N_{AV} consecutive rows. As depicted in Figure 5.10, the value

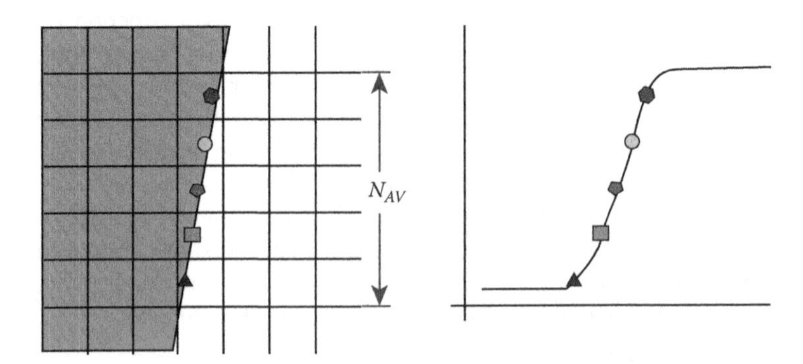

FIGURE 5.10
Estimation of the oversampled ERF from the slanted-edge images.

of the pixel on the first row (triangle mark) corresponds to the first data point of the ERF. Then, the pixel value on the second row (square mark) yields the second data point and so on. Finally, the pixel value of the N_{AV}th row (fifth row in Figure 5.10) is the last data point of the ERF.

The detector area is split up into groups of N_{AV} rows, and oversampled ERFs are estimated for each of the groups according to the process explained. These calculated ERFs are then aligned by linear regression and averaged in order to reduce noise.

The line spread function (LSF) is estimated as the 3-point derivative of the average ERF. Finally, an estimation of the MTF is obtained as the Fourier transform (FT) of the LSF. The calculated estimation is corrected for the sinc function introduced by the derivative operation, and the frequency axis is corrected for the slant angle. The result is an accurate estimation of the presampled MTF.

Another interesting parameter is the effective MTF of the system that reflects the combined effect of the detector and the finite focal spot size of the source. This can be obtained by imaging the same phantom at nominal system magnification [46] and estimating the actual resolution of the projection data of a sample placed at the center of rotation of the micro-CT system. Figure 5.11 plots the presampled MTF obtained for the flat-panel detector of the C7940DK-02. The spatial frequency where the MTF falls to 10% of the zero frequency value (MTF10%) was 8.1 lpmm, a figure compatible with the manufacturer's specifications. MTF10% measured at the nominal system

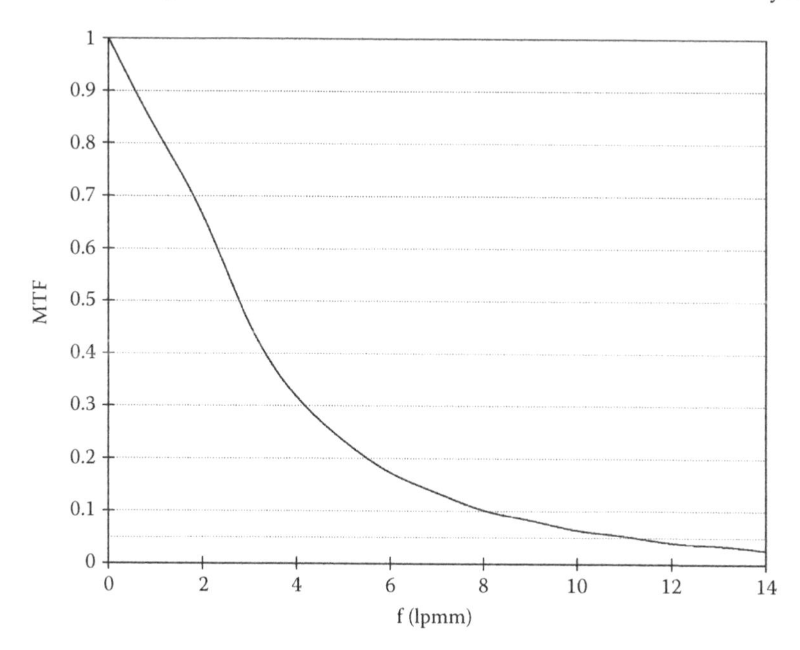

FIGURE 5.11
Presampled modulation transfer function measured with a slanted-edge method.

magnification was 11.85 cycles/mm, almost 1.6 times the intrinsic resolution, as expected due to the magnification factor used.

5.5.2 Quality Evaluation of Reconstructed Images

The quality of the reconstructed images determines their utility in preclinical studies. However, in longitudinal in vivo studies, it is necessary to reach a trade-off between image quality and accumulated radiation dose, as several CT scans are usually acquired during the experiment.

The parameters that determine the quality of the final images in preclinical applications are the noise in the reconstructed image, the contrast-to-noise ratio (CNR), spatial resolution, and HU accuracy.

Various strategies are available to measure each of the aforementioned parameters. The methods presented in [15] were used to assess the Argus PET/CT, as described later.

Noise in reconstructed images was evaluated on a homogeneous water phantom by measuring the standard deviation of the signal in HU as a function of radiated dose. The phantom was acquired six times at 25 kV and 0.6 μA at different doses (different number of averaged images for each angular position).

Measured noise level decreased proportionally to the square root of the dose, as expected, according to theoretical noise models for CT images [65]. A good soft tissue contrast is achieved for a noise level below 50 HU, which corresponded to a radiated dose of 75 mGy in our system. The curve of noise level as a function of dose can be found in Figure 5.12.

Image CNR was measured as a function of radiated dose using a contrast phantom consisting of a nylon cylinder (1.15 g/cm^3) immersed in a water container. CNR is defined as

$$\mathrm{CNR} = \frac{|\mu_n - \mu_w|}{\sqrt{\sigma_n^2 + \sigma_w^2}} \tag{5.9}$$

where μ and σ represent, respectively, the mean and standard deviation of the pixel values in the reconstructed images corresponding to water (μ_w, σ_w) and nylon (μ_n, σ_n), obtained from regions of interest created by gray-level thresholding.

Figure 5.12 shows a plot of the CNR versus the radiated dose. CNR increases almost proportionally to the square root of the dose. The CNR obtained for a dose of 75 mGy is 0.98.

The system resolution was measured following the standard test method E1695-95 [66], which is based on the examination of the CT image of a uniform disk of polycarbonate (1.18 g/cm^3) (Figure 5.13). The resolution value is derived from an analysis of the edge of the disk; in other words, the ERF is obtained and the LSF and MTF are calculated. The cut-off point where the

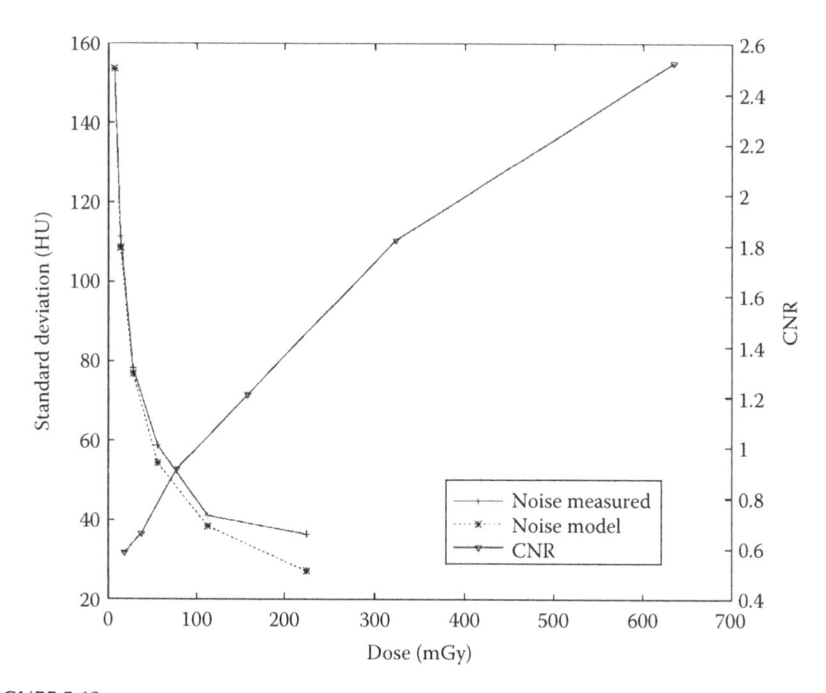

FIGURE 5.12
Noise level (standard deviation) in Hounsfield units measured and estimated using the Gaussian model and measured CNR as a function of radiated dose.

MTF decreases below 10% of its maximum value is given as the standard resolution measurement.

In the Argus PET/CT, the MTF at 10% was 11.3 cycles/mm, or 44 μm in the spatial domain (Figure 5.13). The difference between the actual resolution obtained from the reconstructed images and the resolution measured in projection data is due to the reconstruction process and to possible submillimetric misalignments [51].

The accuracy of the HU values has been assessed with the calibration phantom previously described. The phantom was previously imaged using a properly calibrated clinical CT scanner (Toshiba Aquillion 16) for two x-ray peak energy settings. The values obtained for each of the materials on the phantom were averaged over the whole area covered by them.

The same values were obtained from a third acquisition of the phantom with the small-animal micro-CT system.

Finally, the values obtained by both scanners were compared, assuming those measured by the clinical system as the gold standard.

The results of the cross-validation of the HU conversion are shown in Table 5.3. The data obtained by the micro-CT agree to a large extent with those provided by the clinical scanner, thus proving the accuracy of the HU conversion. It can be noticed that there is a strong deviation for very

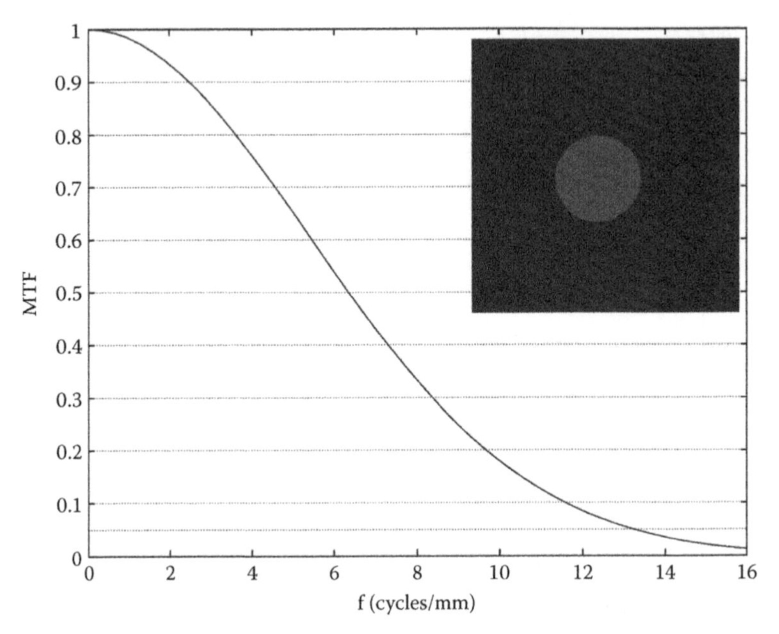

FIGURE 5.13
Modulation transfer function after tomographic reconstruction calculated using the standard protocol E1695-95. In the upper right corner of the plot, a transaxial image of the polycarbonate disk used is shown.

TABLE 5.3
Hounsfield Units Evaluation Results

Material	Toshiba Aquillion 16 (Peak Energy 120 kV)	Toshiba Aquillion 16 (Peak Energy 135 kV)	Micro-CT (Peak Energy 45 kV)
Air	−1005	−1002	−998
Water	−4	−1	1
PMMA	124	118	120
Nylon	242	290	224
Delrin	567	626	640
PTFE	885	960	937
Aluminum	4100	4070	4892

dense materials such as metals, due to the low peak energy of the spectra generated by the microfocus x-ray source.

5.5.3 Radiation Dose

Another important consideration in the design of in vivo micro-CT systems is the dose delivered to the animal. Therefore, it is necessary to provide software and hardware tools to allow the user to select the best settings for each application. Of particular importance is the resolution setting, because, if the

TABLE 5.4

X-Ray Settings and Radiation Dose for Acquisition Protocols

Voltage (kV)	Current (μA)	Time	Resolution (μm)	Radiation Dose (mGy)
25	600	6′00″	200	33.73
40	750	6′30″	100	165.2

image resolution is doubled for a given x-ray setting, voxel noise increases 4 times and the dose has to increase 16 times to maintain the same SNR [67]. For this reason, special care must be taken with x-ray settings for ultra-high-resolution protocols.

The measurement of absorbed dose may provide a more reliable assessment of in vivo biological effects than that offered by the incident radiation measurements. In the example system, thermo-luminescent dosimeters (STI, TLD-100) were introduced into representative organs in euthanized rats that had undergone standard acquisition protocols.

The dose values obtained show the potential damage to the animal for different acquisition settings.

Table 5.4 shows the results and acquisition settings for two standard protocols. The first, which is intended to provide anatomical information for PET-CT studies, does not provide high spatial resolution, as PET image resolution is usually worse than 1 mm. The second protocol represents a high-resolution setting for bone tissue, which needs higher voltage and current. The radiation doses obtained (Table 5.4) were, respectively, 0.5% and 2% of the LD50/30 (~7.3 Gy) for small rodents [67,68]. Typical results reported in the literature of delivered dose to the animal in micro-CT scans are about 100–300 mGy [69,70]. A coronal image of a large mouse using the 40 kV x-ray scan is depicted in Figure 5.14, and a maximum intensity projection image of a rat is shown in Figure 5.15.

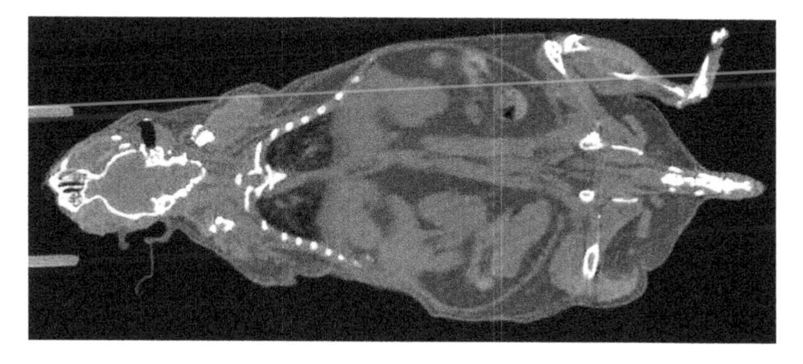

FIGURE 5.14
Coronal slice of a large mouse whole body study. The volume was acquired in two FOV positions with a 125-μm pixel size, 360 angular projections, 8 images averaged per projection, 40-kV peak x-ray beam energy, and a 200-μA x-ray source anode current.

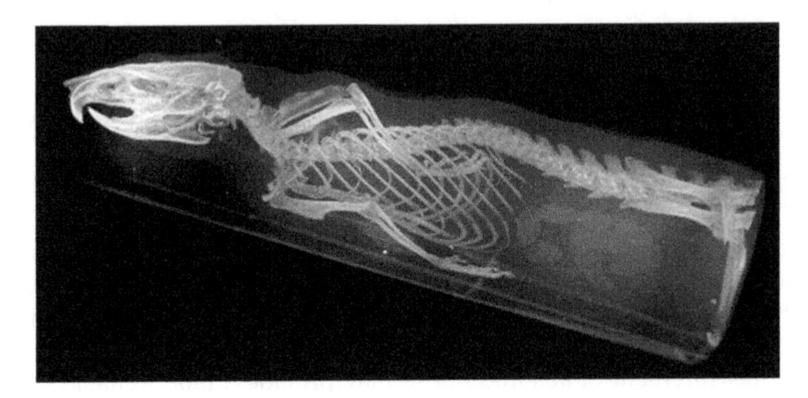

FIGURE 5.15
Volume render of a rat study, 3 beds and pixel size of 0.125 mm. The acquisition parameters were 40 kV, 200 μA, 360 angular projections, 0.125 mm pixel size, and eight images averaged per projection.

5.6 Conclusions

The design of in vivo micro-CT scanners for small-animal imaging must address a number of requirements that differ widely from those of other applications, such as in vitro micro-CT scanners or regular clinical CT scanners. The impacts of the detector device on the quality of the final image and on the mechanical constraints of the system make it necessary to select the most appropriate detector from the different commercially available options.

The general overview presented in this chapter should prevent designers with the basic aspects of the development of in vivo micro-CT systems. The example used was the Argus PET/CT. A complete assessment of its performance was provided, with emphasis on the detector component.

The size of commercial CMOS flat-panel detectors makes them a suitable choice for small-animal imaging, since they are more compact than CCDs, the most widely used small-animal imaging system.

State-of-the-art CMOS flat-panel detectors offer good results in terms of noise, contrast, and resolution, thus making it possible to optimize image quality in terms of the dose radiated.

When using low-power x-ray sources, one way to improve the SNR and, therefore, contrast of images in soft tissue, is to optimize exposure time. In this context, optimization of the acquisition protocol allows the designer to better exploit the features of the detector, thus improving system performance with regard to per-animal screening time.

The resolution of the reconstructed data should be in the tens-of-micrometer range in most multimodality preclinical experiments. This high-resolution value is achieved, thanks to the magnification effect inherent to the cone-beam geometry and to the detector and x-ray tube features. High resolution in vivo imaging may show slight quality degradation due to movements such as breathing or heartbeat. Gating the projection acquisition or performing retrospective gating over multiple exposures per projection [71] can help to recover this resolution loss at the cost of increased radiation dose and acquisition time. In addition, high-resolution imaging requires precise characterization of the system alignment; mechanical misalignments that could not be corrected have to be taken into account during the image reconstruction process, and this may prevent the use of symmetries, thus increasing reconstruction time considerably.

Equations 3.5 through 3.7 describe how the detector and source components affect the final resolution and show that the limiting factor in regular in vivo micro-CT systems is the intrinsic resolution of the detector, since the size of the finite focal spot has a negligible impact due to the magnification factor used in systems based on microfocus x-ray tubes.

References

1. M. J. Paulus, S. S. Gleason, S. J. Kennel, P. R. Hunsicker, and D. K. Johnson, High resolution X-ray computed tomography: An emerging tool for small animal cancer research, *Neoplasia*, 2, 62–70, 2000.
2. H. Ebina, J. Hatakeyama, M. Onodera, T. Honma, S. Kamakura, H. Shimauchi, and Y. Sasano, Micro-CT analysis of alveolar bone healing using a rat experimental model of critical-size defects, *Oral Diseases*, 15, 273–280, 2009.
3. N. M. Harrison, P. F. McDonnell, D. C. O'Mahoney, O. D. Kennedy, F. J. O'Brien, and P. E. McHugh, Heterogeneous linear elastic trabecular bone modelling using micro-CT attenuation data and experimentally measured heterogeneous tissue properties, *Journal of Biomechanics*, 41, 2589–2596, 2008.
4. J. U. Umoh, A. V. Sampaio, I. Welch, V. Pitelka, H. A. Goldberg, T. M. Underhill, and D. W. Holdsworth, *In vivo* micro-CT analysis of bone remodeling in a rat calvarial defect model, *Physics in Medicine and Biology*, 54, 2147–2161, 2009.
5. T. Engelhorn, I. Y. Eyupoglu, M. A. Schwarz, M. Karolczak, H. Bruenner, T. Struffert, W. Kalender, and A. Doerfler, *In vivo* micro-CT imaging of rat brain glioma: A comparison with 3 T MRI and histology, *Neuroscience Letters*, 458, 28–31, 2009.
6. H. Parameswaran, E. Bartolak-Suki, H. Hamakawa, A. Majumdar, P. G. Allen, and B. Suki, Three-dimensional measurement of alveolar airspace volumes in normal and emphysematous lungs using micro-CT, *Journal of Applied Physiology*, 107, 583–592, 2009.
7. S. C. Thacker, V. V. Nagarkar, and H. J. Liang, Characterization of a novel microCT detector for small animal computed tomogaphy (CT), *Medical Imaging 2007: Physics of Medical Imaging, Pts 1–3*, 6510, U1511–U1522, 2007.
8. V. V. Nagarkar, S. V. Tipnis, I. Shestakova, V. Gaysinskiy, B. Singh, M. J. Paulus, and G. Entine, A high-speed functional MicroCT detector for small animal studies, *IEEE Transactions on Nuclear Science*, 53, 2500–2505, 2006.

9. C. T. Badea, M. Drangova, D. W. Holdsworth, and G. A. Johnson, *In vivo* small-animal imaging using micro-CT and digital subtraction angiography, *Physics in Medicine and Biology*, 53, R319–R350, 2008.
10. A. L. Goertzen, V. Nagarkar, R. A. Street, M. J. Paulus, J. M. Boone, and S. R. Cherry, A comparison of x-ray detectors for mouse CT imaging, *Physics in Medicine and Biology*, 49, 5251–5265, 2004.
11. D. A. Jaffray and J. H. Siewerdsen, Cone-beam computed tomography with a flat-panel imager: Initial performance characterization, *Medical Physics*, 27, 1311–1323, 2000.
12. H. K. Kim, S. C. Lee, M. H. Cho, S. Y. Lee, and G. Cho, Use of a flat-panel detector for microtomography: A feasibility study for small-animal imaging, *IEEE Transactions on Nuclear Science*, 52, 193–198, 2005.
13. I. Fujieda, G. Cho, J. Drewery, T. Gee, T. Jing, S. N. Kaplan, V. Perezmendez, D. Wildermuth, and R. A. Street, X-Ray and charged-particle detection with Csi(Tl) layer coupled to a-Si-H photodiode layers, *IEEE Transactions on Nuclear Science*, 38, 255–262, 1991.
14. E. Miyata, M. Miki, N. Tawa, D. Kamiyama, and K. Miyaguchi, Development of new X-ray imaging device sensitive to 0.1–100 keV, *Nuclear Instruments & Methods in Physics Research Section A-Accelerators Spectrometers Detectors and Associated Equipment*, 525, 122–125, 2004.
15. J. J. Vaquero, S. Redondo, E. Lage, M. Abella, A. Sisniega, G. Tapias, M. L. S. Montenegro, and M. Desco, Assessment of a new high-performance small-animal X-ray tomograph, *IEEE Transactions on Nuclear Science*, 55, 898–905, 2008.
16. J. J. Vaquero, E. Lage, L. Ricon, M. Abella, E. Vicente, and M. Desco, rPET detectors design and data processing, *2005 IEEE Nuclear Science Symposium Conference Record*, 1–5, 2885–2889, 2005.
17. Y. C. Wang, J. Seidel, B. M. W. Tsui, J. J. Vaquero, and M. G. Pomper, Performance evaluation of the GE healthcare eXplore VISTA dual-ring small-animal PET scanner, *Journal of Nuclear Medicine*, 47, 1891–1900, 2006.
18. W. A. Kalender, *Computed Tomography: Fundamentals, System Technology, Image Quality, Applications*, 2nd edn., Publicis Corporate Publication, Erlangen, 2005.
19. D. Okkalides, Contrast reduction in digital images due to x-ray induced damage to a TV camera's CCD image receptor, *Physics in Medicine and Biology*, 44, N63–N68, 1999.
20. H. K. Kim, I. A. Cunningham, Z. Yin, and G. Cho, On the development of digital radiography detectors: A review, *International Journal of Precision Engineering and Manufacturing*, 9, 86–100, 2008.
21. C. Frojdh, H. Graafsma, H. E. Nilsson, and C. Ponchut, Characterization of a pixellated CdTe detector with single-photon processing readout, *Nuclear Instruments & Methods in Physics Research Section A-Accelerators Spectrometers Detectors and Associated Equipment*, 563, 128–132, 2006.
22. K. Kowase and K. Ogawa, Photon counting X-ray CT system with a semiconductor detector, *IEEE Nuclear Science Symposium Conference Record*, 5, 3119–3123, 2006.
23. Y. Onishi, T. Nakashima, A. Koike, H. Morii, Y. Neo, H. Mimura, and T. Aoki, Material discriminated X-ray CT by using conventional microfocus X-ray tube and CdTe imager, *2007 IEEE Nuclear Science Symposium Conference Record*, 1–11, 1170–1174, 2007.

24. J. P. Schlomka, E. Roessl, R. Dorscheid, S. Dill, G. Martens, T. Istel, C. Baumer, C. Herrmann, R. Steadman, G. Zeitler, A. Livne, and R. Proksa, Experimental feasibility of multi-energy photon-counting K-edge imaging in pre-clinical computed tomography, *Physics in Medicine and Biology*, 53, 4031–4047, 2008.
25. W. A. Kalender and Y. Kyriakou, Flat-detector computed tomography (FD-CT), *European Radiology*, 17, 2767–2779, 2007.
26. S. O. Kasap, M. Z. Kabir, and J. A. Rowlands, Recent advances in X-ray photoconductors for direct conversion X-ray image detectors, *Current Applied Physics*, 6, 288–292, 2006.
27. J. Yorkston, Recent developments in digital radiography detectors, *Nuclear Instruments & Methods in Physics Research Section A-Accelerators Spectrometers Detectors and Associated Equipment*, 580, 974–985, 2007.
28. T. Gomi, K. Koshida, T. Miyati, J. Miyagawa, and H. Hirano, An experimental comparison of flat-panel detector performance for direct and indirect systems (initial experiences and physical evaluation), *Journal of Digital Imaging*, 19, 362–370, 2006.
29. G. Hajdok, J. J. Battista, and I. A. Cunningham, Fundamental X-ray interaction limits in diagnostic imaging detectors: Spatial resolution, *Medical Physics*, 35, 3180–3193, 2008.
30. G. Hajdok, J. J. Battista, and I. A. Cunningham, Fundamental X-ray interaction limits in diagnostic imaging detectors: Frequency-dependent Swank noise, *Medical Physics*, 35, 3194–3204, 2008.
31. G. Hajdok, J. Yao, J. J. Battista, and I. A. Cunningham, Signal and noise transfer properties of photoelectric interactions in diagnostic X-ray imaging detectors, *Medical Physics*, 33, 3601–3620, 2006.
32. H. K. Kim, S. M. Yun, J. S. Ko, G. Cho, and T. Graeve, Cascade modeling of pixelated scintillator detectors for X-ray imaging, *IEEE Transactions on Nuclear Science*, 55, 1357–1366, 2008.
33. G. F. Knoll, *Radiation Detection and Measurement* John Wiley & Sons, New York, 2000.
34. M. Nikl, Scintillation detectors for X-rays, *Measurement Science & Technology*, 17, R37–R54, 2006.
35. B. K. Cha, J. H. Shin, J. H. Bae, C. H. Lee, S. H. Chang, H. K. Kim, C. K. Kim, and G. Cho, Scintillation characteristics and imaging performance of CsI:Tl thin films for X-ray imaging applications, *Nuclear Instruments & Methods in Physics Research Section a-Accelerators Spectrometers Detectors and Associated Equipment*, 604, 224–228, 2009.
36. H. Mori, R. Kyuushima, K. Fujita, and M. Honda, High resolution and high sensitivity CMOS PANEL SENSORS for X-ray, *2001 IEEE Nuclear Science Symposium, Conference Records*, 1–4, 29–33, 2002.
37. A. Theuwissen, CMOS image sensors: State-of-the-art and future perspectives, *ESSDERC 2007: Proceedings of the 37th European Solid-State Device Research Conference*, 21–27, Munich, 2007.
38. J. A. Rowlands and J. Yorkston, Flat panel detectors for digital radiography, in *Handbook of Medical Imaging*. Vol. 1, J. Beutel et al. (eds.), SPIE Press, Bellingham, WA, 2000, pp. 223–328.
39. Y. Izumi, O. Teranuma, T. Sato, K. Uehara, H. Okada, S. Tokuda, and T. Sato. (2001, 2009). Development of flat-panel X-ray image sensors. *Sharp Technical Journal 3*. Available: http://sharp-world.com/corporate/info/rd/tj3/3-6.html

40. S. Ricq, F. Glasser, and M. Garcin, Study of CdTe and CdZnTe detectors for X-ray computed tomography, in *11th International Workshop on Room-Temperature Semiconductor X- and Gamma-Ray Detectors and Associated Electronics*, Vienna, Austria, 1999, pp. 534–543.
41. M. Funaki, Y. Ando, R. Jinnai, A. Tachibana, and R. Ohno. Development of CdTe detectors in Acrorad, International Workshop on Semiconductor PET, 2007. Available: http://www.acrorad.co.jp/_skin/pdf/Development_of_CdTe_detectors.pdf
42. M. Tamaki, Y. Mito, Y. Shuto, T. Kiyuna, M. Yamamoto, K. Sagae, T. Kina, T. Koizumi, and R. Ohno, Development of 4-Sides buttable CdTe-ASIC hybrid module for X-ray flat panel detector, *IEEE Transactions on Nuclear Science*, 56, 1791–1794, 2009.
43. K. Kim, S. Cho, J. Suh, J. Won, J. Hong, and S. Kim, Schottky-type polycrystalline CdZnTe X-ray detectors, *Current Applied Physics*, 9, 306–310, 2009.
44. H. B. Serreze, G. Entine, R. O. Bell, and F. V. Wald, Advances in CdTe gamma-ray detectors, *IEEE Transactions on Nuclear Science*, 21, 404–407, 1974.
45. A. Cola and I. Farella, The polarization mechanism in CdTe Schottky detectors, *Applied Physics Letters*, 94, 2009, 102113.
46. S. Redondo, J. J. Vaquero, E. Lage, M. Abella, G. Tapias, A. Udias, and M. Desco, Assessment of a new CT system for small animals, in *IEEE Nuclear Science Symposiun and Medical Imaging Conference*, San Diego, CA, 2006.
47. E. Van de Casteele, Model-based approach for beam hardening correction and resolution measurements in microtomography, Department Natuurkunde Antwerpen, University Antwerpen, Antwerpen, 2004.
48. I. Vidal-Migallón, M. Abella, A. Sisniega, J. J. Vaquero, and M. Desco, Simulation of mechanical misalignments in a cone-beam micro-CT system, in *2008 IEEE Nuclear Science Symposium and Medical Imaging Conference*, Dresden, Alemania, 2008.
49. Y. Kyriakou, R. M. Lapp, L. Hillebrand, D. Ertel, and W. A. Kalender, Simultaneous misalignment correction for approximate circular cone-beam computed tomography, *Physics in Medicine and Biology*, 53, 6267–6289, 2008.
50. C. Mennessier, R. Clackdoyle, and F. Noo, Direct determination of geometric alignment parameters for cone-beam scanners, *Physics in Medicine and Biology*, 54, 1633–1660, 2009.
51. F. Noo, R. Clackdoyle, C. Mennessier, T. A. White, and T. J. Roney, Analytic method based on identification of ellipse parameters for scanner calibration in cone-beam tomography, *Physics in Medicine and Biology*, 45, 3489–3508, 2000.
52. K. Yang, A. L. C. Kwan, D. F. Miller, and J. M. Boone, A geometric calibration method for cone beam CT systems, *Medical Physics*, 33, 1695–1706, 2006.
53. G. T. Gullberg, B. M. W. Tsui, C. R. Crawford, J. G. Ballard, and J. T. Hagius, Estimation of geometrical parameters and collimator evaluation for cone beam tomography, *Medical Physics*, 17, 264–272, 1990.
54. J. Y. Li, R. J. Jaszczak, H. L. Wang, K. L. Greer, and R. E. Coleman, Determination of both mechanical and electronic shifts in cone-beam spect, *Physics in Medicine and Biology*, 38, 743–754, 1993.
55. Hamamatsu Photonics K. K., Hamamatsu Application Manual, X-ray flat panel sensor C7912 & C7942 & C7943, rev 2.10, kr1-i50006, 2003.

56. E. Lage, J. J. Vaquero, S. Redondo, M. Abella, G. Tapias, A. Udias, and M. Desco, Design and development of a high performance micro-CT system for small-animal imaging, in *IEEE Nuclear Science Symposium and Medical Imaging Conference*, San Diego, CA, 2006.
57. H. K. Tuy, An inversion-formula for cone-beam reconstruction, *Siam Journal on Applied Mathematics*, 43, 546–552, 1983.
58. L. A. Feldkamp, L. C. Davis, and J. W. Kress, Practical cone-beam algorithm, *Journal of the Optical Society of America a-Optics Image Science and Vision*, 1, 612–619, 1984.
59. R. Gordon, R. Bender, and G. T. Herman, Algebraic reconstruction techniques (Art) for 3-dimensional electron microscopy and X-ray photography, *Journal of Theoretical Biology*, 29, 471–81, 1970.
60. U. Ewert, U. Zscherpel, and K. Bavendiek. (2007, 2009). Replacement of film radiography by digital techniques and enhancement of image quality. Available: www.ndt.net/search/docs.php3?id=4516
61. J. Beutel, H. L. Kundel, and R. L. Van Metter, *Handbook of Medical Imaging, Volume 1*: SPIE Press, Bellingham, WA, 2000.
62. E. Buhr, S. Gunther-Kohfahl, and U. Neitzel, Simple method for modulation transfer function determination of digital imaging detectors from edge images, *Medical Imaging 2003: Physics of Medical Imaging, Pts 1 and 2*, 5030, 877–884, 2003.
63. E. Buhr, S. Gunther-Kohfahl, and U. Neitzel, Accuracy of a simple method for deriving the presampled modulation transfer function of a digital radiographic system from an edge image, *Medical Physics*, 30, 2323–2331, 2003.
64. E. Samei, E. Buhr, P. Granfors, D. Vandenbroucke, and X. H. Wang, Comparison of edge analysis techniques for the determination of the MTF of digital radiographic systems, *Physics in Medicine and Biology*, 50, 3613–3625, 2005.
65. H. H. Barrett, S. K. Gordon, and R. S. Hershel, Statistical limitations in transaxial tomography, *Computers in Biology and Medicine*, 6, 307–323, 1976.
66. ASTM Standard E1695, Standard Test Method for Measurement of Computed Tomography (CT) System Performance, ASTM International, West Conshohocken, PA, 2006, DoI: 10.1520/E1695-95R06E01, www.astm.org.
67. N. L. Ford, M. M. Thornton, and D. W. Holdsworth, Fundamental image quality limits for microcomputed tomography in small animals, *Medical Physics*, 30, 2869–2877, 2003.
68. A. Obenaus and A. Smith, Radiation dose in rodent tissues during micro-CT imaging, *Journal of X-Ray Science and Technology*, 12, 241–249, 2004.
69. J. M. Boone, O. Velazquez, and S. R. Cherry, Small-animal X-ray dose from micro-CT, *Molecular Imaging*, 3, 149–58, 2004.
70. M. J. Paulus, S. S. Gleason, M. E. Easterly, and C. J. Foltz, A review of high resolution X-ray computed tomography and other imaging modalities for small animal research, *Laboratory Animal*, 30, 36–45, 2001.
71. C. Chavarrias, J. J. Vaquero, A. Sisniega, A. Rodriguez-Ruano, M. L. Soto-Montenegro, P. Garcia-Barreno, and M. Desco, Extraction of the respiratory signal from small-animal CT projections for a retrospective gating method, *Physics in Medicine and Biology*, 53, 4683–4695, 2008.

6

Dual-Energy CT Imaging with Fast-kVp Switching

Baojun Li

CONTENTS

6.1 Dual-Energy X-Ray CT Imaging with Fast-kVp Switching

6.1.1 Introduction

Dual-energy CT is an imaging technique that has been known and extensively studied for many decades [1–8]. However, due to various technical challenges, only recently has dual-energy CT imaging become a reality. Recent advances in CT scanner technologies have generated a renewed interest in dual-energy CT [9–14], which has led to the commercialization of dual-energy CT systems available for routine clinical use [10,12–14].

Dual-energy CT is a special CT imaging procedure in which two CT scans of a patient are acquired in different x-ray tube potentials (and spectra) and used to perform energy- and material-selective reconstruction of the patient.

Relative to conventional single-energy (SE) CT imaging, dual-energy CT imaging offers the capability to enhance material differentiation and reduce beam-hardening artifacts.

Figure 6.1a shows a routine abdominal SE CT exam, and Figure 6.1b and c shows a corresponding dual-energy CT exam. Dual-energy CT enables the separation of kidney stone (calcium) from iodine-based intravenous contrast by decomposing the dual-energy data into iodine (atomic number of 53) and water (effective atomic number of 7.42). Furthermore, the reduction of beam-hardening artifacts near the kidney stones (circled) is readily apparent by comparing Figure 6.1a and b.

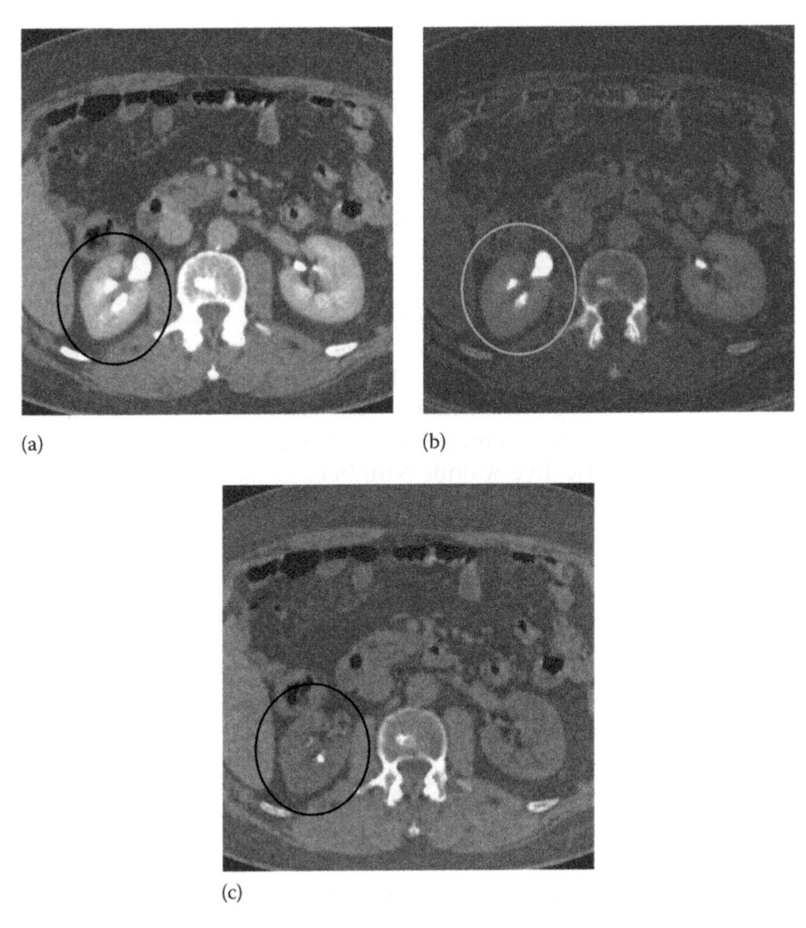

FIGURE 6.1
A comparison between a post-contrast SE abdomen CT exam and dual-energy CT exam of the same patient. (a) SE exam at 120 kVp. (b,c) Dual-energy exam using 140 and 80 kVp enables the separation of kidney stone (calcium) from iodine-based intravenous contrast by decomposing the dual-energy data into iodine (left) and water (right) material pairs. The reduction of beam-hardening artifacts near the kidney stones (circled) is readily apparent by comparing (a) and (b). (Image courtesy of Dr. Amy Hara, Mayo Clinic, Scottsdale, AZ.)

Dual-energy CT imaging has the potential to improve the efficacy of many clinical applications including abdomen [15], neurology [16], pulmonary embolism [17], and bone mineral densitometry [3]. Investigators have begun to establish the clinical efficacy of dual-energy CT for plaque characterization [18].

Conventional dual-energy CT exam is accomplished by the so-called "rotate–rotate" technique: first a low-kVp CT exam is acquired, and then the patient is translated back to the origin, followed by the acquisition of a high-kVp CT exam. Similar to digital subtraction angiograph (DSA), this technique is very sensitive to patient motion because the time interval between the two kVps is in the order of seconds. As a result, poor spatial–temporal registration between high- and low-kVp x-ray beams is a major source of image artifacts in conventional dual-energy CT imaging. Compared to SE CT, motion-induced artifacts, such as blurring of the edges and streaks centered on objects that are moving, are more severe in dual-energy CT. Furthermore, patient motion can be falsely interpreted as a change in tissue composition, which, typically manifested as light-and-dark edge effect around the moving objects, can cause inaccurate material densities and/or misdiagnosis (Figure 6.2).

To address the motion issue, a fast-kVp switching (FKS) dual-energy CT imaging method, where kVp is rapidly switched between low and high kVp in adjacent views, has recently been proposed [12–14]. Compared to

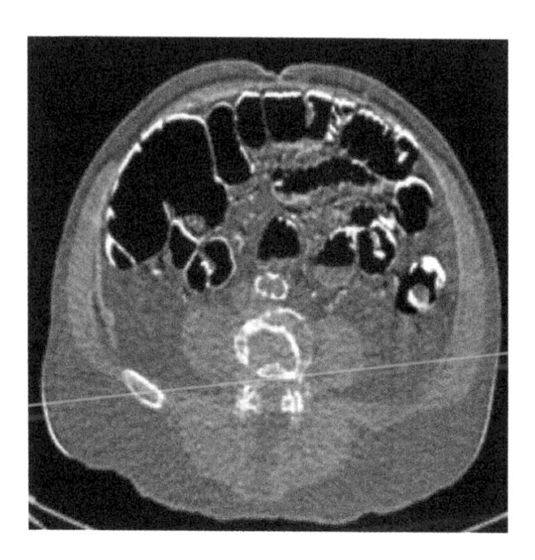

FIGURE 6.2
Patient motion is a major source of image artifacts in conventional dual-energy CT imaging due to poor spatial–temporal registration of high- and low-kVp x-ray beams. Dual-energy CT is in general more sensitive to patient motion than SE CT because material decomposition requires the two rays to pass through the same anatomy. In this example, artifacts manifested as light-and-dark edge effect near intestinal boundaries can be falsely interpreted as a change in tissue composition. (Image courtesy of Robert Beckett, GE Healthcare, Waukesha, WI.)

conventional dual-energy CT imaging, FKS dual-energy CT imaging has several benefits including precise temporal and spatial view registration, helical and axial acquisitions, and full field of view. It also presents several design challenges that warrant careful considerations.

6.1.2 Image Acquisition

6.1.2.1 X-Ray Tube/Generator

The fundamental solution to avoid the motion issue is to acquire the low- and high-kVp projections on a view-by-view basis in a single gantry rotation. This enables precise spatial–temporal registration of two different kVps, thus freezing motion and significantly reducing artifact.

The x-ray generation system must enable the rapid kVp switching to achieve sufficient energy separation and view sampling speed. The generator and tube must be capable of reliably switching between 80 and 140 kVp and have the capability to support sampling rate as quickly as every 150 μs. To ensure the signal fidelity, x-ray generator with ultralow impedance of tens of microseconds is also necessary.

Following the acquisition, correction calibrations are applied to the data. The rise and fall of the kVp waveform complicates the spectral calibration. Figure 6.3 illustrates the actual kVp waveform employed in FKS dual-energy acquisition. The nonideal kVp rise and fall makes it difficult to find a fixed kVp that matches the same spectral response as the FKS.

To estimate the effective energy, the unknown spectra of the low and high kVps are estimated as the weighted linear combination of single-kVp spectra [19]. The unknown x-ray spectra of the low- and high-kVp views are estimated as the weighted sum of several known single-kVp spectra:

$$S_p(E) = \sum_{k}^{N_k} \alpha_k S_k(E), \tag{6.1}$$

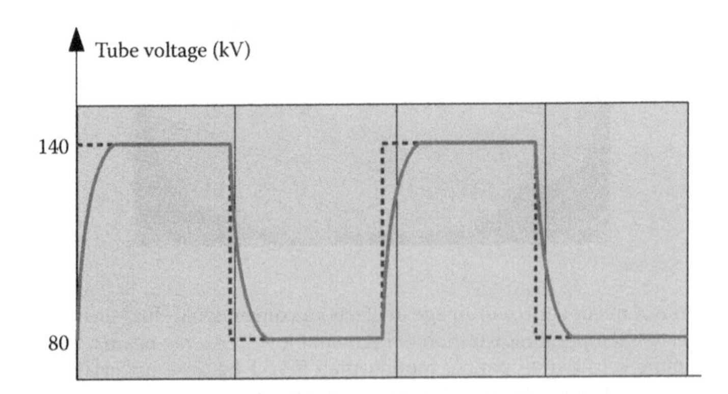

FIGURE 6.3
The rise and fall of actual kVp form complicates the spectral calibration of FKS dual-energy CT.

where

$S_k(E)$ are the basis spectra of the single kVp
N_k is the total number of the basis spectra
α_k are the weights of the basis spectra

The self-normalized detector response to this spectrum can be written as follows:

$$R(d) = \frac{\int_E \sum_k \alpha_k S_k(E) E\left[1 - e^{-\mu_d(E)t_d}\right] e^{\sum_b \mu_b(E,d) l_b(d)} dE}{\sum_d \int_k \alpha_k S_k(E) E\left[1 - e^{-\mu_d(E)t_d}\right] e^{-\sum_b \mu_b(E,d) l_b(d)} dE}, \tag{6.2}$$

in which $\mu_d(E)$ is the linear attenuation coefficient of the detectors, t_d is detector thickness, and $\mu_b(E,d)$ and $l_b(d)$ are, respectively, the linear attenuation coefficient and the thickness of bow tie material b corresponding to detector channel d. If we denote

$$G_k(d) = \int_E S_k(E) E\left[1 - e^{-\mu_d(E)t_d}\right] e^{-\sum_b \mu_b(E,d) l_b(d)} dE, \tag{6.3}$$

Equation 6.2 can be simplified to

$$R(d) = \frac{\sum_k \alpha_k G_k(d)}{\sum_d \sum_k \alpha_k G_k(d)}. \tag{6.4}$$

$R(d)$ can be measured through a fast switching air scan, and $G_k(d)$ can be calculated based on the system geometry. Hereby the weighting coefficients α_k can be solved from Equation 6.4 by least-square fitting. The overall spectrum is decomposed into a superposition of several basis spectra through the measurement of the detector response to the bow tie attenuation.

The earlier mentioned spectrum estimation technique is graphically demonstrated in Figure 6.4. $P_c(l)$ represents the equivalent x-ray attenuation of the estimated spectrum by Equation 6.4. $P(V_e,l)$ is the measured x-ray attenuation of the fast-kVp system. It is obvious that the estimated spectrum $P_c(l)$ matches the actual spectrum $P(V_e,l)$ with minimal difference.

6.1.2.2 Detector

The detector is a key contributor to FKS acquisitions through its scintillator and data acquisition system. Detector primary decay and afterglow performance are critical to avoiding spectral blurring between views. Primary

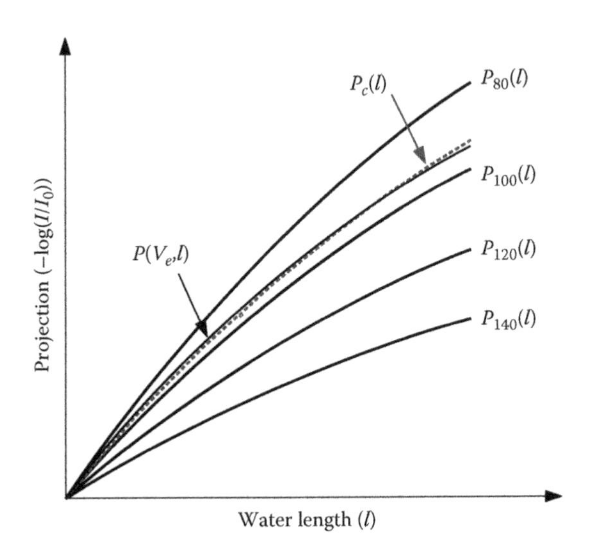

FIGURE 6.4
The estimation of effective spectrum in FKS dual-energy CT imaging. $P_c(I)$ represents the equivalent x-ray attenuation of the estimated spectrum by Equation 6.4. $P(V_e,I)$ is the measured x-ray attenuation of the fast-kVp system. It is obvious that the estimated spectrum $P_c(I)$ matches the actual spectrum $P(V_e,I)$ with minimal difference. (Image courtesy of Dr Dan Xu of GE Global Research, Schenectady, NY.)

speed and afterglow refer to the decays of light emitting from the scintillator for several to tens of milliseconds after the x-ray source is switched off. This phenomenon is analogous to the decay of light signal on the television screen after it is turned off. The residual signal from one view smears information contained in the next during a scan, thereby causing degradation of spatial resolution and undesirable cross-contamination of spectra.

Gemstone scintillator material (GE Healthcare, Waukesha, WI) is a complex rare earth-based oxide, which has a chemically replicated garnet crystal structure. This lends itself to imaging that requires high light output, fast primary speed, and very low afterglow. Gemstone has a primary speed of only 30 ns, or 100 times faster than GOS (Gd_2O_2S), while also having afterglow that is only 20% of GOS, making it ideal for fast sampling [20]. For illustration we depict in Figure 6.5 a decay curve of a Gemstone detector. The capabilities of the scintillator are also paired with an ultrafast data acquisition system (up to 7 kHz), enabling simultaneous acquisition of low- and high-kVp data at customary rotation speeds.

6.1.2.3 Flux

Compared to SE CT, the traditional flux issues are more challenging in FKS dual-energy CT imaging. There hereby needs to be a strategy for balancing the flux between the two spectra and a need for noise reduction processing (which will be discussed later in Section 6.1.4).

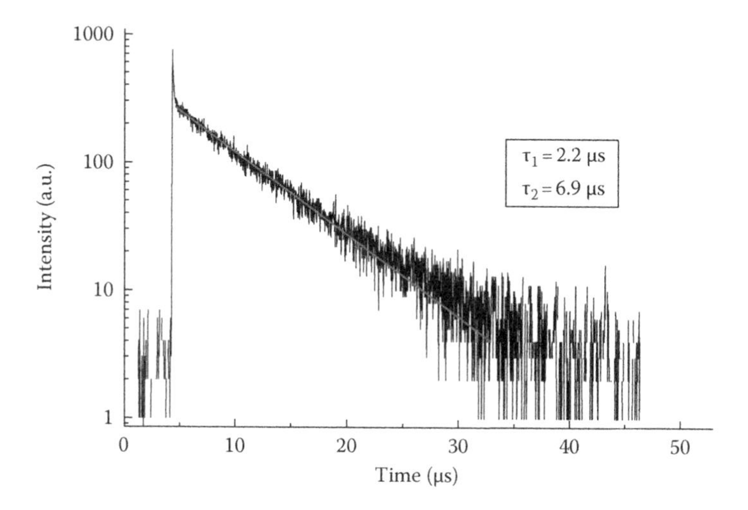

FIGURE 6.5
Gemstone scintillator material (GE Healthcare, Waukesha, WI) has a primary speed of only 30 ns, or 100 times faster than GOS (Gd_2O_2S), while also having afterglow that is only 20% of GOS, making it ideal for fast sampling. (Image courtesy of Dr Haochuan Jiang, GE Healthcare, Waukesha, WI.)

Figure 6.6 illustrates a simplified FKS dual-energy abdominal CT exam acquisition timing sequence in comparison with that of a SE abdominal CT vexam. The number of views in FKS dual-energy acquisition doubles those for SE, while the x-ray exposure time remains either the same (in abdomen cases) or similar (in head cases) as the SE acquisition. This results in much shorter view time for FKS dual-energy acquisition. As illustrated, the exposure time of a 140 kVp view in FKS dual-energy acquisition is roughly 40% of that of a single-energy view.

In dual-energy imaging, the flux ratio between the low and high kVps is in general constrained in such way that the contrast-to-noise ratio (CNR) is maximized. Multiple studies [21–23] have suggested the CNR is maximized with ~30% flux allocation, defined as the percentage of entrance skin exposure of the low kVp to that of the low and high kVp combined. That is, roughly speaking, the exposure ratio between the low and high kVps is

$$\frac{N_L}{N_H} = \frac{1}{2}. \tag{6.5}$$

The x-ray exposure is a function of x-ray energy spectrum, beam filtration, geometry, and the tube current–time product (mAs). The FKS dual-energy CT system employs identical geometry and beam filtration for the 140 and 80 kVp acquisitions, which leaves only mAs adjustable. For the x-ray tubes used in CT, the x-ray exposure of a 140 kVp beam is roughly threefold of that

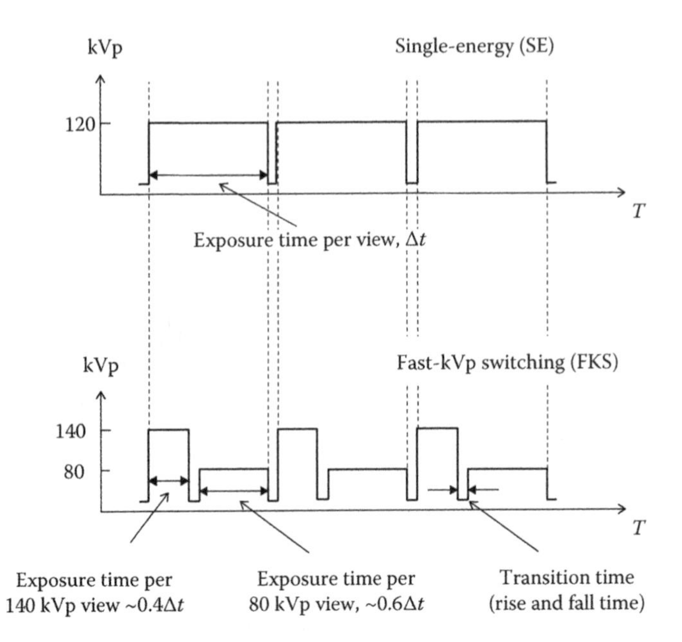

FIGURE 6.6
Comparison of the simplified acquisition timing sequence between a FKS dual-energy abdominal CT exam and a SE abdominal CT exam. The number of views in FKS dual-energy acquisition doubles those for SE, while the x-ray exposure time remains either the same (in abdomen cases) or similar (in head cases) as the SE acquisition. This results in much shorter view time for FKS dual-energy acquisition. As illustrated, the exposure time of a 140 kVp view in FKS dual-energy acquisition is roughly 40% of that of a sing-energy view. (Adapted from Li, G. et al., *Med. Phys.*, 38(5), 2595, 2011.)

of an 80 kVp beam [24]. Therefore, according to Equation 6.5, the mAs ratio between low- and high-kVp views should be kept around 3/2.

In theory, there are two potential approaches to achieve the desired flux ratio: one, modulating the tube current while maintaining constant view time (i.e., exposure time in milliseconds per view) and, two, modulating the view time while maintaining constant tube current.

Modulating the tube current is conceptually more straightforward; however, it faces an implementation dilemma. Since higher tube current is required for 80 kVp view, the filament is at a very high temperature. Upon switching to 140 kVp, which mandates a much lower tube current, it requires a mechanism to reverse the polarization between the anode and cathode, at several kilohertz, to "pinch" the flow of electrons in order to lower the current.

Hence, modulating the view time while maintaining constant tube current is the preferred approach in FKS dual-energy CT imaging for flux control [13,14]. Based on the earlier mentioned analysis, the view time ratio between the low- and high-kVp views should be 60%–40%. To demonstrate the importance of balancing flux through modulating the view time in

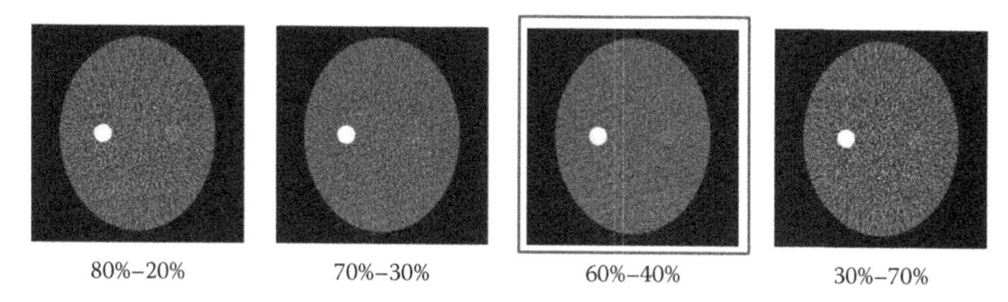

FIGURE 6.7
The reconstructed images of a simulated low-contrast lesion (2 cm in diameter) embedded in an oval phantom. The simulations were conducted under identical conditions except for the view time ratio. The view time ratio is defined as the percentage of low-kVp view to high-kVp view in each low–high-kVp pair. Comparing the results, 60%–40% view time ratio clearly offers the best CNR for the lesion detection. The result demonstrates the importance of balancing flux through modulating the view time in FKS dual-energy CT imaging.

FKS dual-energy CT imaging, we simulate a low-contrast lesion (2 cm in diameter) embedded in an oval phantom. Figure 6.7 summarizes the results from the simulation and shows the impact of various view time ratios on the image quality. Comparing the results, 60%–40% view time ratio clearly offers the best CNR for the lesion detection.

6.1.2.4 Dose

FKS dual-energy CT has been designed to minimize the additional dose relative to SE. In a recent dose and low-contrast detectability (LCD) comparison [13,14], the effectiveness of this sampling scheme with respect to dose was demonstrated by matching the LCD at a slice thickness equal to 5 mm and object size of 3 mm.

In x-ray CT imaging, dose, resolution, and noise are interdependent quantities. Thereby, it is less useful to compare dose between two systems without first equalizing the other two parameters. If the FKS dual-energy CT employs identical geometry and beam filtration as the SE CT, previous study has shown no significant difference in measured spatial resolution between the two systems, when the reconstruction kernel and display field of view are matched [25].

Therefore, for a fair comparison of dose, one only needs to equalize the amount of image noise. The LCD is a clinical-relevant image quality metric that quantifies image noise performance, making it ideal as an image quality target when measuring dose. The lower the LCD is, the less contrast a lesion of certain size (in mm) must have in order to be detected at a given confidence level (usually 95%).

Traditionally, the LCD is measured by visually detecting low-contrast objects on a specially designed phantom. An example of a popular phantom (Catphan 600, Phantom Laboratory, Salem, NY) is displayed in Figure 6.8a. The disadvantage of this approach is, not surprisingly, that the results are

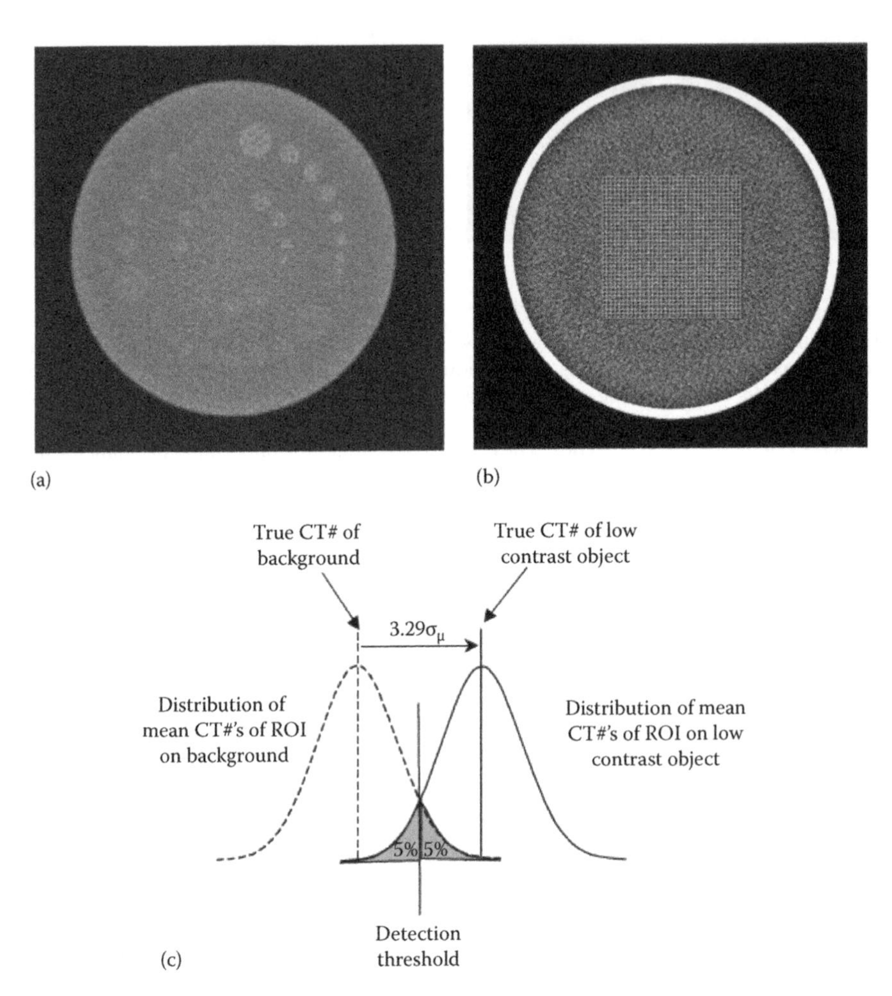

FIGURE 6.8
(a) An example of visual LCD phantom. The results are subjective to many conditions, including inter-reader variability, viewing conditions, and presence of artifacts. (b) A square array of ROIs placed in the uniform water-equivalent section of the Catphan 600 phantom. Each ROI is 6 pixels by 6 pixels that has an equivalent area to a circular object of 3 mm in diameter (DFOV = 22.7 cm, 512 × 512 image matrix). (c) By measuring the distribution of mean CT#'s of many ROIs, a prediction can be made for the necessary CT# of a low-contrast object the same size as the ROIs in order to detect it at a 95% confidence level. (Adapted from Li, G. et. al., *Med. Phys.*, 38(5), 2595, 2011.)

subjective to many conditions, including inter-reader variability, viewing conditions, and whether or not artifacts are present.

An automated statistical LCD method is thus more desirable. Chao et al. has developed one of such tools [26]. The processing begins with placing a square array of region of interest (ROI) on a uniform water-equivalent section of the Catphan 600, as depicted in Figure 6.8b. Each ROI is 6 pixels by

6 pixels, which has an area equivalent to a circular object of 3 mm in diameter for a 22.7 cm DFOV and 512 × 512 matrix sizes. As shown in Figure 6.8c, the distribution of mean CT number of the ROIs provides an estimate of the necessary contrast (in CT number) in order to detect a low-contrast object the same size as the ROIs at a 95% confidence level.

Figure 6.9 compares the measured $CTDI_{vol}$ (in mGy) between a FKS dual-energy and routine (i.e., SE) abdominal CT exams as a function of monochromatic energy (in keV, dual-energy only) and tube current (in mA, SE only). The SE measurements are denoted as "SE xxx mA," where "xxx" describes the tube current, while the dual-energy measurements are denoted as "Mono xx keV," where "xx" represents the monochromatic energy (monochromatic energy will be discussed later in Section 6.1.3.2). All other acquisition protocols were the same between the two systems compared (kVp = 120 kVp, gantry rotate speed = 1.0 s, bow tie = large body, slice thickness = 5 mm, display field of view = 22.5 cm).

From Figure 6.9, one can see that, with a tube current of 360 mA, the SE abdominal CT exam yields a LCD of 0.426% and a $CTDI_{vol}$ of 29.2 mGy for

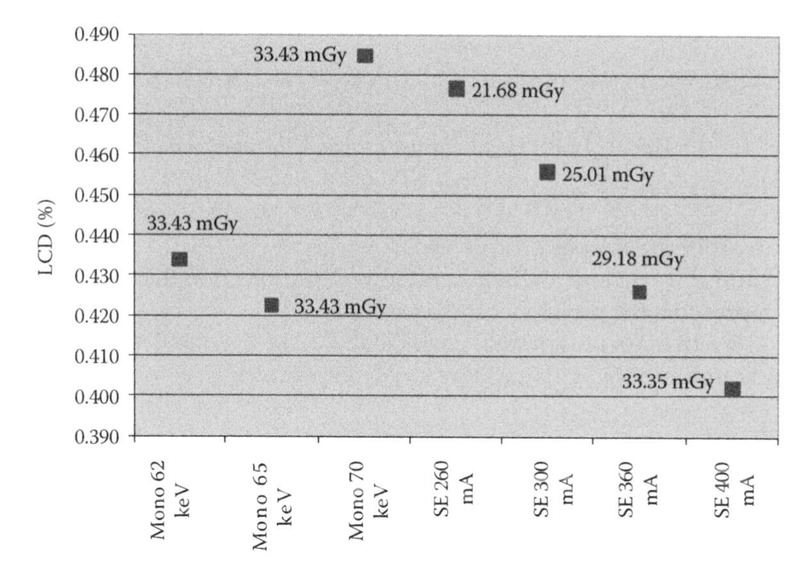

FIGURE 6.9
Comparison of the measured $CTDI_{vol}$ (in mGy) between a FKS dual-energy and routine (i.e., SE) abdominal CT exams as a function of monochromatic energy (in keV, dual-energy only) and tube current (in mA, SE only) (all other acquisition protocols were the same between the two systems compared: kVp = 120 kVp, gantry rotate speed = 1.0 s, bow tie = large body, slice thickness = 5 mm, display field of view = 22.5 cm). The SE measurements are denoted as "SE xxx mA," where "xxx" describes the tube current, while the dual-energy measurements are denoted as "Mono xx keV," where "xx" represents the monochromatic energy (monochromatic energy will be discussed later in Section 6.1.3.3). All other acquisition protocols (kVp, gantry rotation speed, bow tie, slice thickness, display field of view, etc.) were the same between the two systems compared. (Adapted from Li, G. et. al., *Med. Phys.*, 38(5), 2595, 2011.)

a 3 mm object. The FKS dual-energy abdominal CT exam (65 keV) produces a nearly identical LCD of 0.422% (0.01% = 0.1 HU) with a $CTDI_{vol}$ of 33.43 mGy for the same object size, or just 14% higher than that of a routine SE abdominal CT exam. These results were obtained using the uniform section of Catphan 600 phantom, which represents a patient with ~20 cm water-equivalent diameter.

6.1.3 Image Reconstruction

6.1.3.1 Projection-Space Material Decomposition

The mass attenuation coefficient across the x-ray spectrum is a function of two independent variables: photoelectric effect and Compton scatter [1]. Based on this principle, the low- and high-kVp projection data can be retrospectively transformed into a pair of basis materials (such as water and iodine).

Through a mathematical change of basis, one can express the energy-dependent attenuation observed in two kVp measurements in terms of two basis materials [14]:

$$
\begin{aligned}
f_1(\vec{l}) &= \alpha_1 p_H(\vec{l}) + \beta_1 p_L(\vec{l}) + \chi_1 p_H^2(\vec{l}) + \delta_1 p_L^2(\vec{l}) + \varepsilon_1 p_H(\vec{l}) p_L(\vec{l}) + \cdots \\
f_2(\vec{l}) &= \alpha_2 p_H(\vec{l}) + \beta_2 p_L(\vec{l}) + \chi_2 p_H^2(\vec{l}) + \delta_2 p_L^2(\vec{l}) + \varepsilon_2 p_H(\vec{l}) p_L(\vec{l}) + \cdots.
\end{aligned}
\tag{6.6}
$$

where

$\vec{l}$ is an arbitrary ray path,
$P_H(\vec{l})$ and $P_L(\vec{l})$ denote the high- and low-kVp projection data,
$f_i(\vec{l})$ represents the density of the basis material i,
α, β, χ, δ, ε that are polynomial coefficients,
H and L refer to the high and low kVps, respectively.

There are a couple of important facts in Equation 6.6 that should be noted here. First of all, $P_H(\vec{l})$ and $P_L(\vec{l})$ must be measured along the same ray path $\vec{l}$. This can be easily satisfied by FKS dual-energy CT imaging. The FKS mechanism ensures the low- and high-kVp projection data are spatially and temporally co-registered. (Note: Strictly speaking, the low- and high-kVp views incur a small angular offset relative to each other. To obtain truly co-registered projection pair, these views are interpolated to the same angular positions prior to material decomposition.)

Secondly, the high-order terms in Equation 6.6 are critically important to account for spectral variation over the field of view due to source spectrum, bow tie filter, detector performance, and multi-material beam-hardening effects [1]. As a consequence, projection-space material decomposition provides the opportunity for more quantitative precision than what may be achieved with SE imaging. This is the key difference between FKS

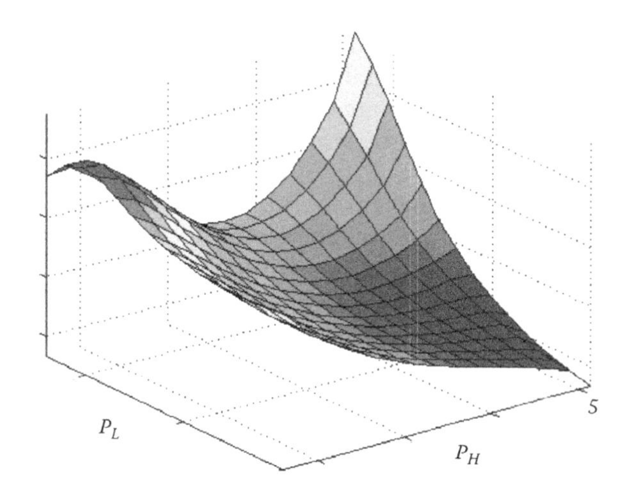

FIGURE 6.10
Equation 6.6 is overdetermined and the resolution for polynomial coefficients α, β, χ, δ, ε, etc. can be found easily through least-square fitting defined by Equation 6.7. The vertical axis represents the thickness of water.

dual-energy CT imaging and image-space dual-energy CT imaging, which usually require a post-reconstruction beam-hardening correction to recover, hopefully, the spectral information that has been lost during the image reconstruction.

To solve Equation 6.6, dual-energy projection data corresponding to different thicknesses of water and iodine are acquired. Since modern CT scanners contain a large number of detector elements, Equation 6.6 is overdetermined and the resolution for polynomial coefficients α, β, χ, δ, ε, etc. can be found easily through least-square fitting:

$$\begin{gathered}\arg\min \sum_{\vec{l}} \left\| f_1(\vec{l}) - \alpha_1 p_H(\vec{l}) - \beta_1 p_L(\vec{l}) - \chi_1 p_H^2(\vec{l}) - \delta_1 p_L^2(\vec{l}) - \varepsilon_1 p_H(\vec{l}) p_L(\vec{l}) - \cdots \right\|^2 \\ \arg\min \sum_{\vec{l}} \left\| f_2(\vec{l}) - \alpha_2 p_H(\vec{l}) - \beta_2 p_L(\vec{l}) - \chi_2 p_H^2(\vec{l}) - \delta_2 p_L^2(\vec{l}) - \varepsilon_2 p_H(\vec{l}) p_L(\vec{l}) - \cdots \right\|^2. \end{gathered} \tag{6.7}$$

Figure 6.10 shows such an example. The vertical axis represents the thickness of water.

6.1.3.2 Image Reconstruction

Through material decomposition, the energy-dependent attenuation measurements contained in kVp projections are transformed into energy-independent basis material projection data corresponding to the two basis material pair (e.g., water and iodine). Although the pair of basis material projection data (sinogram) essentially contains all useful information about the material being imaged, they are difficult to understand and interpreted by clinicians. A more useful form, which clinicians are familiar with, is the reconstructed images.

Having the identical geometry, the same reconstruction algorithm to reconstruct the SE CT images can therefore be applied to the first basis material projection data to obtain the corresponding basis material density image. The step is repeated for the second basis material as well. An example is shown in Figure 6.11. Basis material density images represent the effective density for the anatomies necessary to create the observed low- and high-kVp attenuation measurements. For instance, pure water appears as 1000 mg/cc in a water density image, and 50 mg/cc of diluted iodine is labeled as such in an iodine density image, etc. Any non-basis material is mapped

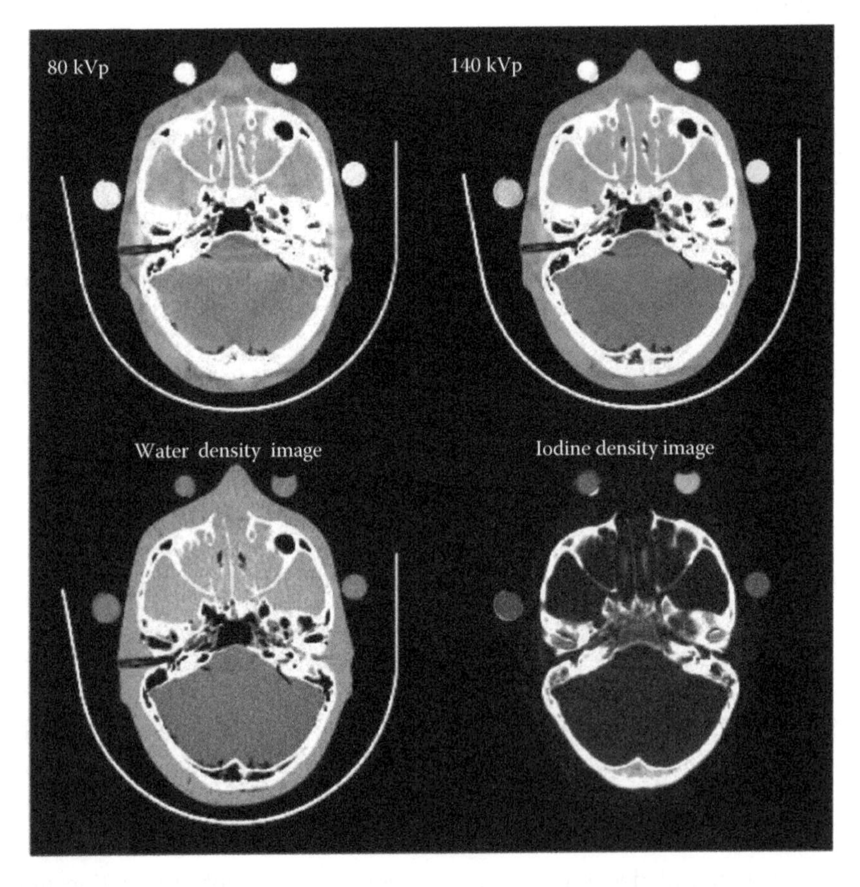

FIGURE 6.11
Exemplary reconstructed basis material density images, as well as the kVp images, from a FKS dual-energy head CT exam.

to both basis materials. For this reason, basis material density images are sometimes called "material density map."

6.1.3.3 Monochromatic Energy Image

Given the material basis density images, one can compute attenuation data that would be measured with a monoenergetic x-ray source by combining the material density images to create a monochromatic image at any specific energy level (in keV), E [14]:

$$\int u(\vec{l},E)dl = \left(\frac{\mu_1}{\rho_1}\right)(E)\int f_1(\vec{l})dl + \left(\frac{\mu_2}{\rho_2}\right)(E)\int f_2(\vec{l})dl, \tag{6.8}$$

where (μ_i/ρ_i) represents the mass attenuation coefficient for material i. For consistency with the Hounsfield unit (HU), one can normalize the attenuation measurement with respect to water. A clinical example is shown in Figure 6.12.

From Equation 6.8, the expected image noise (variance) in the monochromatic image can be expressed based on linear system theory and noise propagation principle:

$$\delta_{mono}^2 = \left[\left(\frac{\mu_1}{\rho_1}\right)(E)\right]^2 \sigma_{f1}^2 + \left[\left(\frac{\mu_2}{\rho_2}\right)(E)\right]^2 \sigma_{f2}^2. \tag{6.9}$$

Equation 6.9 implies that the noise in the generated monochromatic energy image is also energy dependent. This has been confirmed through phantom experiment. Figure 6.13 plots the measured noise (in HU) as a function of monochromatic energy level, in the range of 40–140 keV, for a 20 cm water

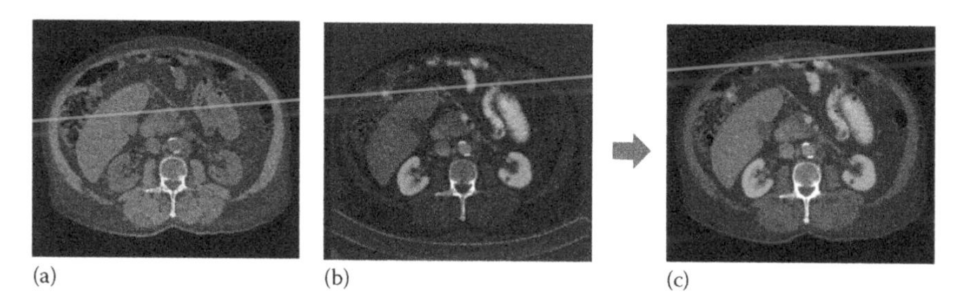

(a) (b) (c)

FIGURE 6.12
Reconstruction of monochromatic energy image from a pair of basis material density images. The monochromatic energy image, which represents attenuation data that would be measured with a monoenergetic x-ray source at any specific keV level, is computed by combining the basis material density images using Equation 6.8. (a) Water density (mg/cc), (b) iodine density (mg/cc), and (c) 65 keV monochromatic (HU).

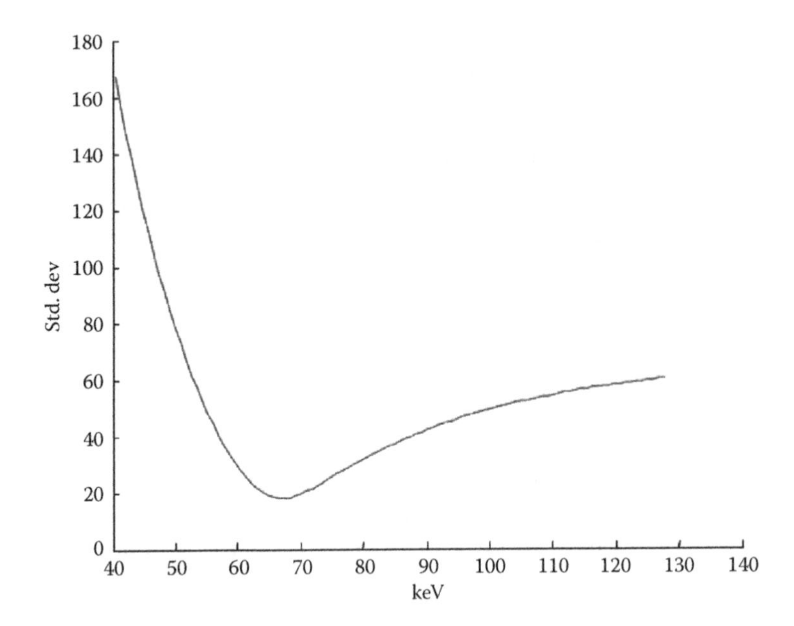

FIGURE 6.13
The noise in the generated monochromatic energy images is energy dependent. For FKS dual-energy CT imaging, the noise curve exhibits a global minimum near 65 keV for the particular acquisition protocol. (Adapted from Li, G. et. al., *Med. Phys.*, 38(5), 2595, 2011.)

phantom. The noise curve exhibits a global minimum near 65 keV for the particular acquisition protocol.

6.1.3.4 Effective Atomic Number

The x-ray linear attenuation coefficient of a periodic element, μ, can be expressed as a function of the element's material properties and E:

$$\left(\frac{\mu}{\rho}\right)(E) = \sigma(E)\frac{N_A}{A}, \tag{6.10}$$

where σ, N_A, and A are the total effective cross section, Avogadro's number, and the mass number.

The total effective cross section of x-ray radiation in function of the chemical composition of the materials can be modeled as a combination of the Compton and photoelectric effects [27]:

$$\sigma(E) = a(E)Z_{eff} + b(E)Z_{eff}^{C(E)}, \tag{6.11}$$

where Z_{eff} is the effective atomic number. The coefficients a, b, and c depend only on the energy and their values can be obtained from NIST [28].

By substituting (6.11) into (6.10), the linear attenuation can be further expressed as

$$\mu(E) = \left(a(E)Z_{eff} + b(E)Z_{eff}^{C(E)}\right)\frac{N_A\rho}{A}. \tag{6.12}$$

The simultaneous equations are formed using Equation 6.12 for two monochromatic energy images $\mu(E_1)$ and $\mu(E_2)$. Solving the simultaneous equations, Z and ρ can be obtained as follows:

$$Z_{eff} = \left[\frac{A}{N_A\rho}\cdot\frac{a(E_2)\mu(E_1) - a(E_1)\mu(E_2)}{a(E_2)b(E_1) - a(E_1)b(E_2)}\right]^{1/C(E)}. \tag{6.13}$$

Z_{eff} describes the periodic elements most closely representing its energy-dependent attenuation behavior; hence it often provides insight regarding the material's chemical composition. Knowledge of the effective atomic number of critical tissue types or specific contrast agents may be leveraged to define and import custom basis materials, allowing for the enhancement of specific anatomical structures.

Figure 6.14 shows graphically the distribution of several materials commonly encountered in diagnostic radiology in the basis material space (water and bone). Materials are clearly separated based on their chemical decomposition. The effective atomic number of each material is reflected by the angular slope. Noise is primarily responsible for the scatter of points in each material and the overlap between different materials.

6.1.4 Noise Suppression

It is well known that the basis material density images have a much lower signal-to-noise ratio (SNR) than SE CT images. This can be easily demonstrated by the following simple analysis. Let's define the SNR of iodine in a low-kVp image as

$$\mathrm{SNR}_L(x,y) \propto \frac{\mu_L^I(x,y)}{\sigma_L(x,y)}. \tag{6.14}$$

Then the SNR of iodine in a basis material density image is

$$\mathrm{SNR}_{\Delta E}(x,y) \propto \frac{\mu_L^I(x,y) - R^T\mu_H^I(x,y)}{\sqrt{\sigma_L^2(x,y) + (R^T)^2\sigma_H^2(x,y)}}. \tag{6.15}$$

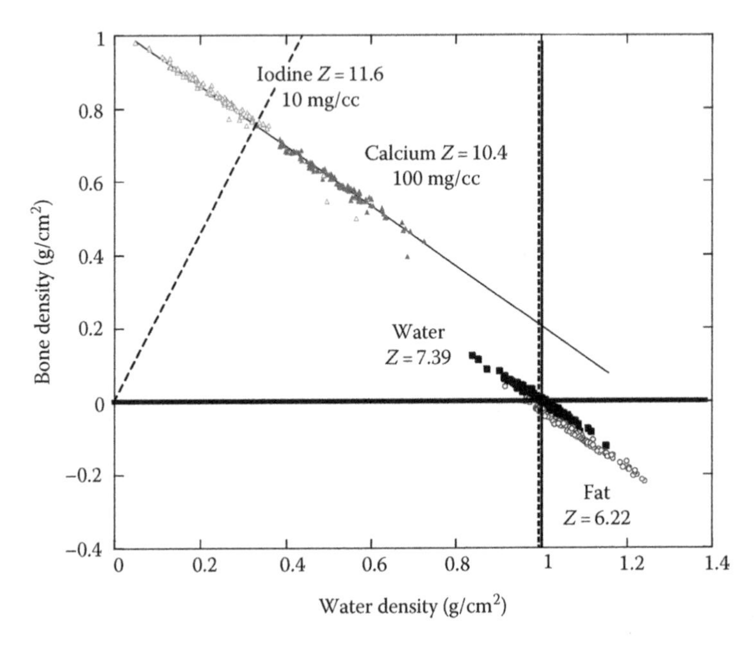

FIGURE 6.14
Graphic representation of several materials commonly encountered in diagnostic radiology in the plane of two basis materials (water and bone). Materials are clearly separated based on their chemical decomposition. The effective atomic number of each material is reflected by the angular slope. Noise is primarily responsible for the scatter of points in each material and the overlap between different materials.

By comparing (6.2) with (6.1) and using the fact that μ_L^I and μ_H^I are very close for the most of relevant energy levels, we can conclude that

$$\mathrm{SNR}_{\Delta E}(x,y) \ll \mathrm{SNR}_L. \tag{6.16}$$

A reduction of noise can be achieved by increased exposure. But based on x-ray physics, a noise reduction by a factor of n requires an exposure increase by a factor of n^2, which leads to unacceptable dose levels for most patients. Therefore, noise suppression has to be automatically applied to FKS dual-energy CT imaging and on the basis material density images in order to enhance the quality of the image while preserving the density values. This allows for a quantitative basis material density image with good image quality.

Noise reduction techniques that attempt to directly reduce noise in basis material density images have also been sought after. To this date, none of them have had success in reducing noise to satisfactory levels while minimally affecting iodine contrast without introducing artifacts.

A more significant improvement in SNR, however, results from the understanding the physical property of the noises that exist in basis material density images. It has been proven that the noises in the two resulting basis material density images are negatively correlated [2,8]. In other words,

$$COV(\sigma_{m1}, \sigma_{m2}) < 0, \tag{6.17}$$

where denote m_1 and m_2 stand for the two basis material images.

Taking advantage of the property in Equation 6.17, several noise suppression algorithms have been developed [8,29]. These algorithms subtract a weighted high-pass filtered version of the first basis material density image (e.g., water) to noise reduce the complimentary basis material density image (e.g., iodine).

These algorithms are effective in suppressing noise. However, they are at the risk of introducing a detrimental artifact. Although the high-pass filtered image is smoothed, edge structures and blood vessels full of contrast medium are added to the complimentary basis material density image, causing "cross-contamination" that changes the accuracy of the density values.

The solution, however, exists if a two-pass algorithm is employed. Upon the completion of the first pass, the noise masks from both basis material density images are generated and combined at certain energy to create a monochromatic energy image of the noises. As shown in Figure 6.15a, this image contains a large amount of edge structures and iodinated blood vessels as a result of cross-contamination. One can project these structures back to the basis material density images, where a high-pass filter is applied adaptively to both basis material density images. This second pass filter results in updated noise masks that have significantly less contamination-prune structures (Figure 6.15b and c). Reduction of contamination-prune structures is evident. The updated noise masks are then added to the complimentary basis material density image for noise suppression.

The effectiveness of the proposed two-pass algorithm is demonstrated in Figure 6.16. In this clinical example, the original water and iodine density images are very noisy. Noise-reduced water density image using a single-pass algorithm, where the cross-contaminated structures are clearly visible throughout the liver area. By comparing the complimentary noise-reduced iodine density image, one can correlate the contaminations in the water density image with the iodinated hepatic vessels in iodine density images. Using the two-pass algorithm, noise-reduced water density image is free of contamination.

6.1.5 Visualization and Clinical Applications

FKS dual-energy CT imaging provides diagnostic information beyond that found in conventional SE CT imaging, in a manner consistent with workflow, and that increases the efficacy of clinical diagnosis. The clinical application and visualization of dual-energy data is presently an area of active research.

FKS dual-energy CT images can be reviewed and quantified in a number of ways. Most commonly, it is visualized as a monochromatic energy

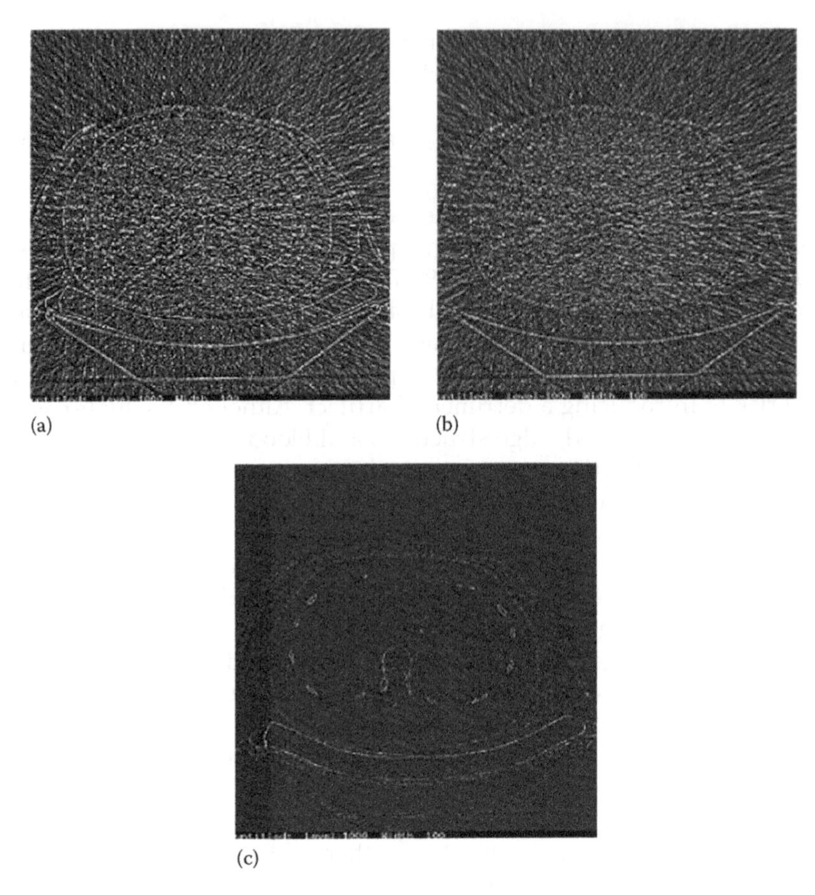

FIGURE 6.15
Example of noise mask generated from a water density image. (a) Uncorrected noise mask. Structures such as vertebrae and ribs can be clearly seen. (b) Corrected noise mask after applying the proposed algorithm. (c) The difference image between (a) and (b). Removing of the structures is evident.

image, which resembles a conventional kVp image, but with fewer beam-hardening artifacts. One may interactively change the energy level from 40 to 140 keV to best balance the needs in terms of contrast and noise. It is also common to visualize the data as a basis material density image pair. Specific material density can be directly measured from these images, which form the basis for differentiation of certain tissue of clinical interest. For example, based on the relative density of a hepatic cyst with respect to water, one can determine if the cyst is water or fat based, which may be important for making a proper diagnosis. In an effort to simplify the workflow and minimize the workload, the iodine density image can be displayed as a color overlay on the monochromatic image.

One can also visualize dual-energy data as the effective atomic number, Z_{eff}, of the materials. This information provides insight into the material's chemical composition. Alternatively, different materials can be described as histograms or scatterplots, as the example shown earlier in Figure 6.14.

To further identify trends and support additional knowledge from the data, the attenuation profile of the material versus keV may also be graphically depicted by a tissue signature curve. An example is demonstrated in Figure 6.17.

Research into FKS dual-energy CT imaging has been conducted in multiple clinical applications. A discussion of abdominal applications may be found in [15,30]. More recently, attention has been given to apply FKS dual-energy CT imaging to cardiac applications, due to its excellent temporal–spatial resolution and potential for reduced beam hardening in myocardium

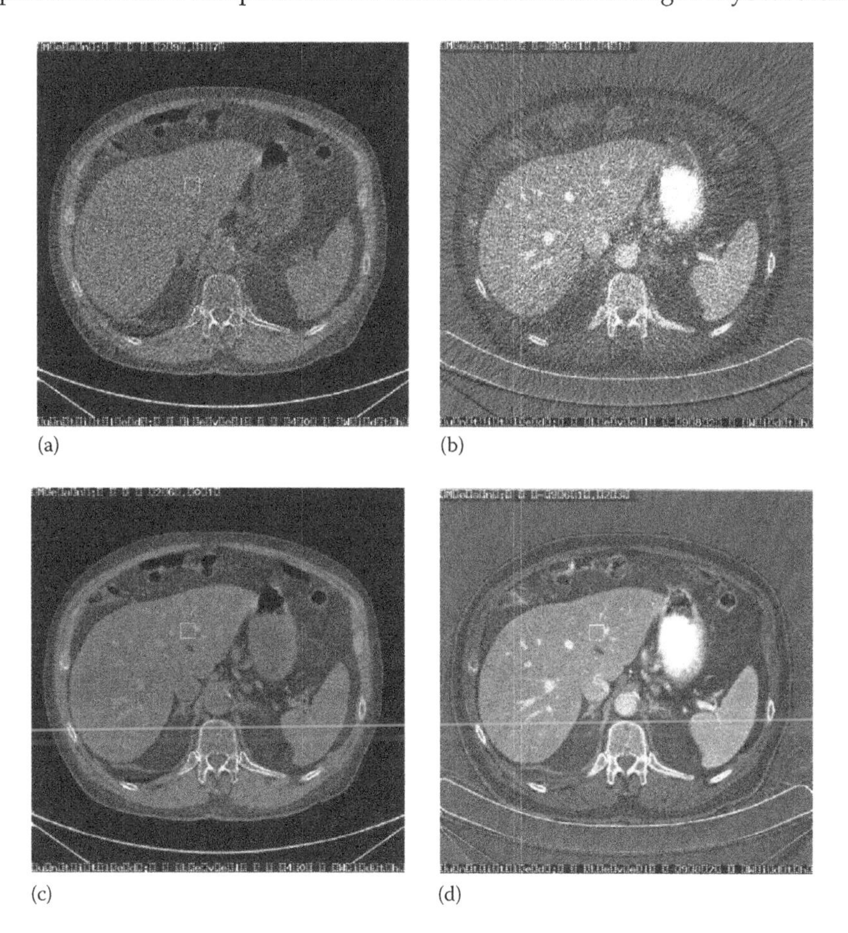

FIGURE 6.16
Example of noise suppression of a FKS dual-energy CT imaging using the two-pass algorithm. (a) Original water density image. (b) Original iodine density image. (c) Noise-reduced water density image using a single-pass algorithm, where the cross-contaminated structures are clearly visible throughout the liver area. (d) Noise-reduced iodine density image using a single-pass algorithm. The iodinated hepatic vessels are the root cause of the cross-contaminations seen in the complimentary water image in (c).

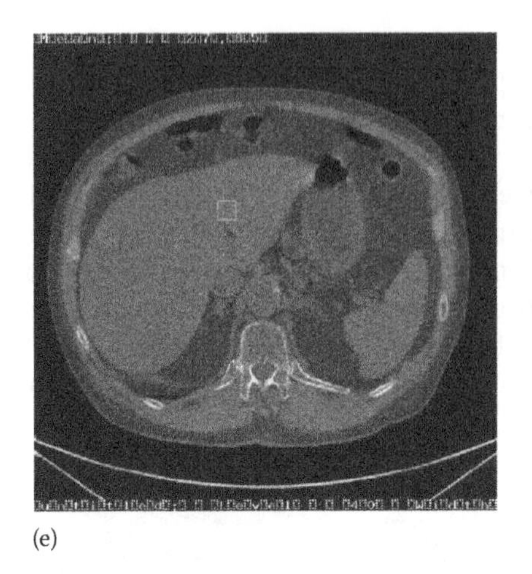

(e)

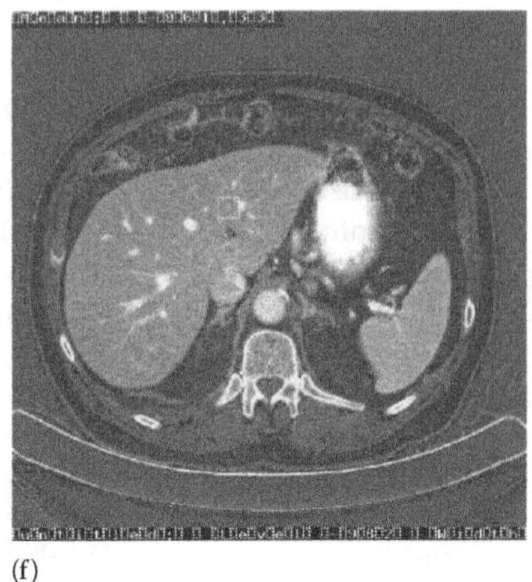

(f)

FIGURE 6.16 (continued)
Example of noise suppression of a FKS dual-energy CT imaging using the two-pass algorithm. (e) Noise-reduced water density image using the two-pass algorithm is free of contamination. (f) Noise-reduced iodine density image using the two-pass algorithm.

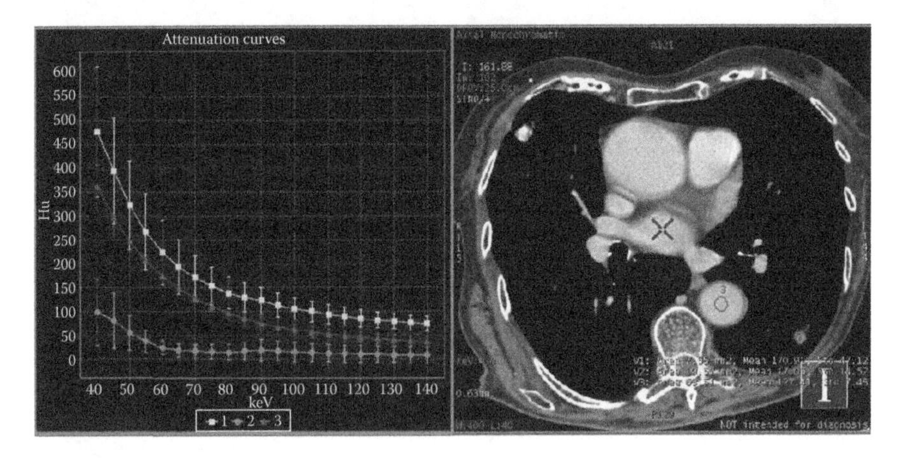

FIGURE 6.17
Tissue signature curves of two pulmonary masses, which are highlighted in the monochromatic energy image, demonstrate different behaviors in this FDK dual-energy exam. Biopsy procedures have confirmed mass 1 (upper left on image, topmost curve on graph) is a malignant nodule and mass 2 (lower right mass on image, lowest curve on graph) a calcified nodule (benign). (Image courtesy of Dr Amy Hara, Mayo Clinic, Scottsdale, AZ.)

and reduced calcium blooming. In one study, reduced calcium blooming has increased accuracy in coronary stenosis sizing through improved differentiation of calcified plaque and iodinated contrast lumen, enabling accurate measurement and risk profiles of lipid (cholesterol) plaques [18].

Thanks to its inherent beam-hardening reduction capability, FKS dual-energy CT imaging has increased accuracy in hip implant assessment

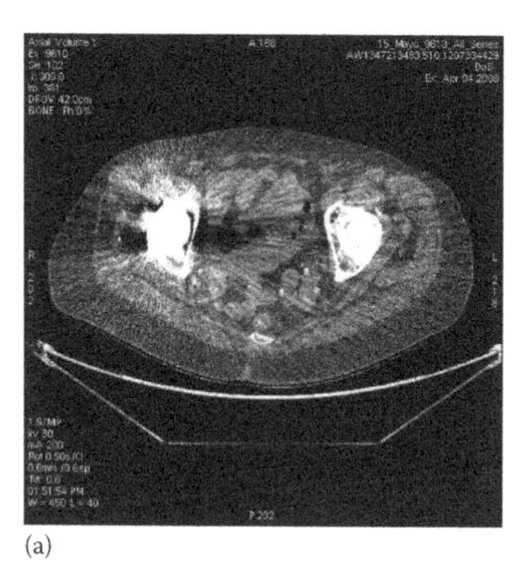

(a)

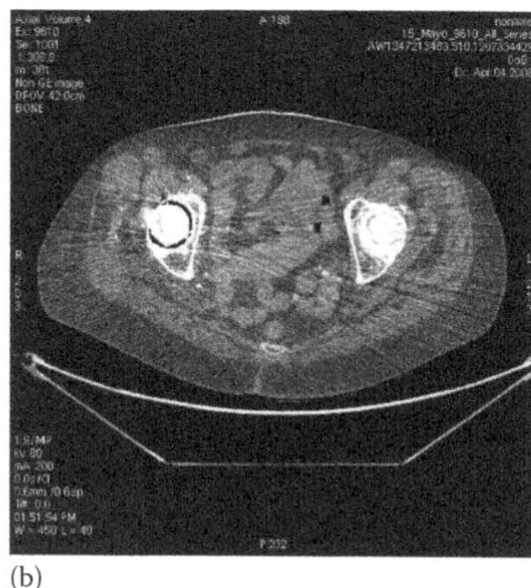

(b)

FIGURE 6.18
Example of metal artifact reduction of FKS dual-energy CT imaging. (a) SE CT exam. (b) FKS dual-energy CT exam. (Image courtesy of Robert Beckett, GE Healthcare, Waukesha, WI.)

through improved differentiation of hip bone and metallic implant, enabling accurate measurement of joint spacing (Figure 6.18).

Another excellent example of metal artifacts from pedicle screws is shown in sagittal reformatted postoperative spinal CT (Figure 6.19a). The reconstructed FKS dual-energy CT image displays significantly less streak artifacts, enabling much improved visualization of the dural sac and spinal muscles.

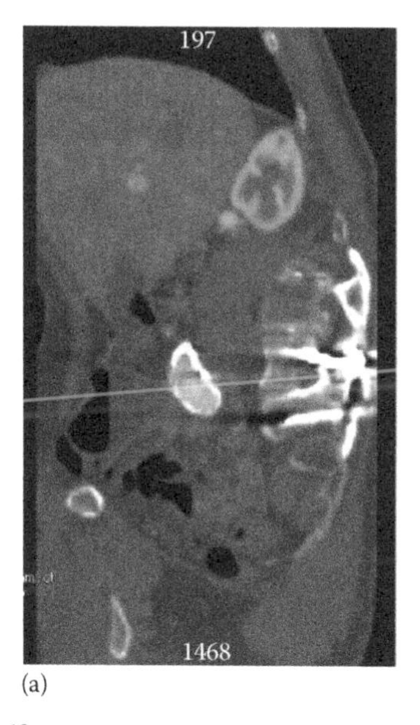

(a)

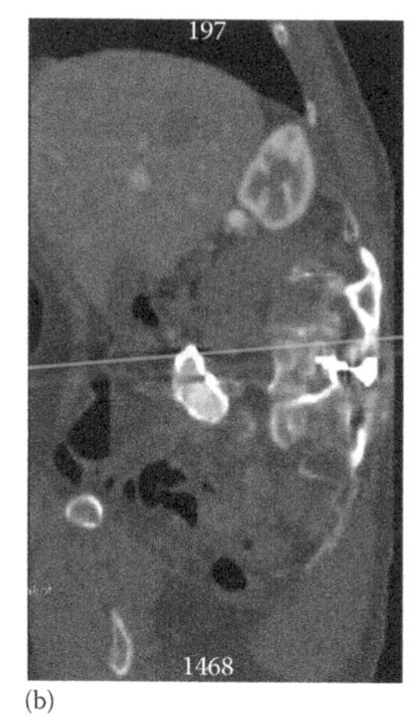

(b)

FIGURE 6.19
Example of metal artifact reduction of FKS dual-energy CT imaging. (a) SE CT exam. (b) FKS dual-energy CT exam. (Image courtesy of Dr. Amy Hara, Mayo Clinic, Scottsdale, AZ.)

References

1. R. Alvarez and A. Macovski, Energy-selective reconstructions in x-ray CT, *Phys. Med. Biol.* 21(5), 733–744, 1976.
2. R. Alvarez and E. Seppi, A comparison of noise and dose in conventional and energy selective computed tomography, *IEEE Trans. Nucl. Sci.* 26(2), 2853–2856, 1979.
3. H. Genant and D. Boyd, Quantitative bone mineral analysis using dual energy computed tomography, *Invest. Radiol.* 12, 545–551, 1977.
4. C. Cann, G. Gamsu, F. Birnberg, and W. Webb, Quantification of calcium in solitary pulmonary nodules using single- and dual-energy CT, *Radiology* 145, 493–496, 1982.
5. A. Coleman and M. Sinclair, A beam-hardening correction using dual-energy computed tomography, *Phys. Med. Biol.* 30(11), 1251–1256, 1985.
6. W. Kalender, W. Perman, J. Vetter, and E. Klotz, Evaluation of a prototype dual-energy computed tomographic apparatus. I. Phantom studies, *Med. Phys.* 13(3), 334–339, 1986.
7. J. Vetter, W. Perman, W. Kalender, R. Mazess, and J. Holden, Evaluation of a prototype dual-energy computed tomographic apparatus. II. Determination of vertebral bone mineral content, *Med. Phys.* 13(3), 340–343, 1986.
8. W. Kalender, E. Klotz, and L. Kostaridou, An algorithm for noise suppression in dual energy CT material density images, *IEEE Trans. Med. Imag.* 7(3), 218–224, 1988.
9. S. Sengupta, S. Jha, D. Walter, Y. Du, and E. Tkaczyk, Dual energy for material differentiation in coronary arteries using electron-beam CT, *Proc. SPIE Med. Imag.* 5745, 1306–1316, 2005.
10. C. Maaβ, M. Baer, and M. Kachelrieβ, Image-based dual energy CT using optimized precorrection functions: A practical new approach of material decomposition in image domain, *Med. Phys.* 36(8), 3818–3829, 2009.
11. A. Altman and R. Carmi, A double-layer detector dual-energy CT—Principles, advantages and applications, *Med. Phys.* 36(6), 2750, 2009.
12. J. Hsieh, Dual-energy CT with fast-kVp switching, *Med. Phys.* 36(6), 2749, 2009.
13. B. Li, G. Yadava, and J. Hsieh, Head and body CTDIw of dual energy x-ray CT with fast-kVp switching, *Proc. SPIE Med. Imag.* 7622, 76221Y, 2010.
14. B. Li, G. Yadava, and J. Hsieh, Head and body CTDIw of dual energy x-ray CT with fast-kVp switching, *Med. Phys.* 38(5), 2595–2601, 2011.
15. A. Silva, A. Hara, R. Paden, W. Pavlicek, and B. Morse, Dual energy (spectral) CT for routine clinical abdominal imaging, *Radiographics* 31, 1031–1046, 2011.
16. A. Kemmling, I. Noelte, J. Scharf, and C. Groden, Diagnostic advantage of dual-energy CT bone removal in cerebral CT angiography for postsurgical evaluation of standard extracranial-intracranial arterial bypass, *RSNA Scientific Assembly and Annual Program* 2009, SSE16-02, 2009.
17. S. Miura, Y. Nishimoto, S. Kitano, J. Takahama, N. Marugami, and A. Hashimoto, Lung perfused blood volume image using dual-energy CT in patients with acute and chronic pulmonary embolism, *RSNA Scientific Assembly and Annual Program 2009*, LL-CH4297-B11, 2009.

18. W. Pavlicek, P. Panse, K. A. Hara, T. Boltz, R. Paden, P. Licato, N. Chandra, D. Okerlund, R. Bhotika, and D. Langan, Initial use of fast switched dual energy CT for coronary artery disease, *Proc. SPIE Med. Imag.* 7622, 76221V, 2010.
19. D. Xu, D. Langan, X. Wu, J. Pack, T. Benson, J. Tkaczky, and A. Schmitz, Dual energy CT via fast kVp switching spectrum estimation, *Proc. SPIE Med. Imag.* 7258, 72583T, 2009.
20. H. Jiang, GE healthcare's gemstone scintillator development, *Proc. Mater. Sci. Tech.* 3, 1796–2002, 2010.
21. S. Richard and J. Siewerdsen, Optimization of dual-energy imaging system using generalized NEQ and imaging task, *Med. Phys.* 34(1), 127–139, 2007.
22. N. Shkumat, J. Siewerdsen, A. Dhanantwari, D. Williams, S. Richard, N. Paul, J. Yorkston, and R. Van Metter, Optimization of image acquisition techniques for dual-energy imaging of the chest, *Med. Phys.* 34(10), 3904–3915, 2007.
23. J. Sabol, G. Avinash, F. Nicolas, B. Claus, and J. Zhao, The development and characterization of a dual-energy subtraction imaging system for chest radiography based on CsI:Tl amorphous silicon flat-panel technology, *Proc. SPIE Med. Imag.* 4320, 399–408, 2001.
24. J. Bushberg, J. Seibert, E. Leidholdt Jr., and J. Boone, *The Essential Physics of Medical Imaging*, 2nd edn., Lippincott Williams & Wilkins, Baltimore, MD, 2002.
25. J. Fan, J. Hsieh, P. Sainath, and P. Crandal, Evaluation of the adaptive statistical iterative reconstruction technique for cardiac computed tomography imaging, *Proc. SPIE Med. Imag.* 8313, 83133F, 2012.
26. E. Chao, T. Toth, N. Bromberg, E. Williams, S. Fox, and D. Carleton, A statistical method of defining low contrast detectability, *RSNA Scientific Assembly and Annual Program* 2000, 2750–2750, 2000.
27. D. Jackson and D. Hawkes, X-ray attenuation coefficients of elements and mixture, *Phys. Rep.* 70, 169–233, 1981.
28. M. Berger and J. Hubbell, NIST XCOM: Photon cross section database, NIST Standard Reference Database 8 (XGAM) NBSIR 87-3597, 1998.
29. C. McCollough, M. Lysel, W. Peppler, and C. Mistretta, A correlated noise reduction algorithm for dual-energy digital subtraction angiography, *Med. Phys.* 16(6), 873–880, 1989.
30. M. Joshi, D. Langan, D. Sahani, A. Ramesh, S. Aluri, K. Procknow, X. Wu, R. Rahul, and D. Okerlund, Fast kV switching dual energy CT effective atomic number accuracy for kidney stone characterization, *Proc. SPIE Med. Imag.* 7622, 76223K, 2010.

7

Four-Dimensional Computed Tomography

Tinsu Pan

CONTENTS

7.1 Introduction

Respiratory motion imposes a significant challenge in radiation therapy of the lung tumor. Because most of the patients are treated in free-breathing, the information of tumor motion is critical for the radiation oncologist to delineate the target tumor. Prior to the four-dimensional computed tomography (4D-CT) imaging of the lung tumor, one or more CT scans were taken to ensure sufficient sampling of the target tumor at various phases of a respiratory cycle. This may include a free-breathing CT (FB-CT) scan plus a couple of breath-hold CT (BH-CT) scans at the end-inspiration and end-expiration phases. This practice could be problematic because a FB-CT scan can distort the shape of the target tumor. An example is shown in Figure 7.1. This is a problem even with the fast gantry rotation cycle time of less than 1 s on a modern CT scanner. A study by Underberg et al.[1] demonstrated that a single 4DCT scan could encompass a significantly larger internal target volume than 6 CT scans did in 2 of the 10 patients of stage I non-small-cell lung cancer, whose tumors exhibited the greatest mobility. Some suggested that a

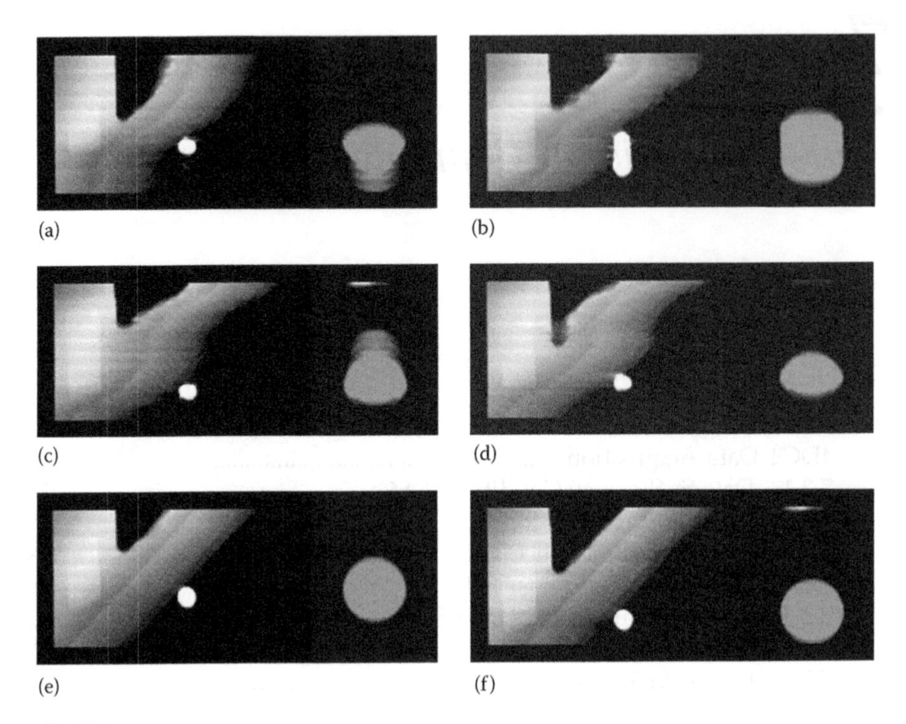

FIGURE 7.1
Images (a)–(d) of four helical CT scans of the phantom of one triangular ruler, a small ball and a large ball moving in the super-inferior direction of 2 cm over a 4-s cycle. The images (e) and (f) of the two extreme positions of the moving phantom were from 4D-CT showing the correct shape of the triangular ruler, and the two balls. The distortions in shape were noticed in all three objects in the helical CT scans (a)–(d), and each distortion is different depending on the relative motion of the helical CT scan to the phantom motion. (Reproduced from Pan T, et al., *Med Phys.*, 31(2), 333, 2004, Figures 6 and 8. With permission.)

slow CT scan of 4 s per revolution can be used to assess the tumor motion.[2] However, the 4-s scan can introduce some severe motion artifacts when compared with the average CT, averaged from the multiple phases of 4DCT as shown in Figure 7.2.

The first clinical use of 4DCT was made by commercialization of the cine 4DCT (first announced at the AAPM 2002 Annual Meeting in Montreal, Canada) on the LightSpeed MSCT scanner (GE Healthcare, Waukesha, WI).[3] The cine 4DCT utilized the cine CT scan capability already available on the LightSpeed CT scanner, which can scan at the same slice location for multiple gantry rotations. It does not require any modification on the LightSpeed CT scanner, and is equipped with an image sorting software, which correlates the same phase of CT images across the multiple table positions to a single phase of 4DCT images. The GE LightSpeed CT allows for coverage of up to over 30 cm in a single scan setup, for the application of cine 4DCT. Radiation dose in the thorax application of 4DCT is generally less than 50 mGy. Low et al.[4] proposed a similar cine 4DCT technique on a Siemens Somatom 4-slice CT using a spirometer to record the

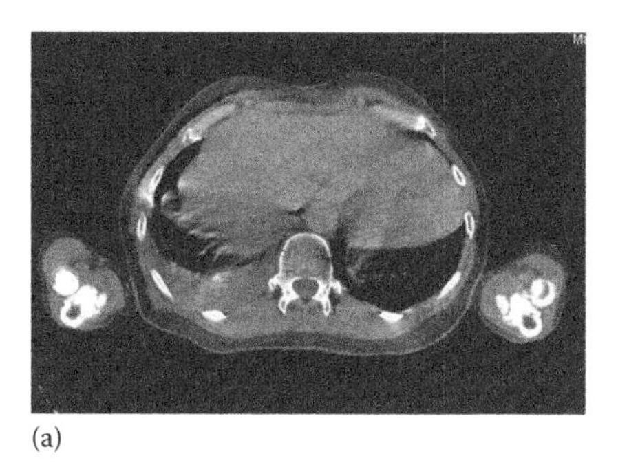

(a)

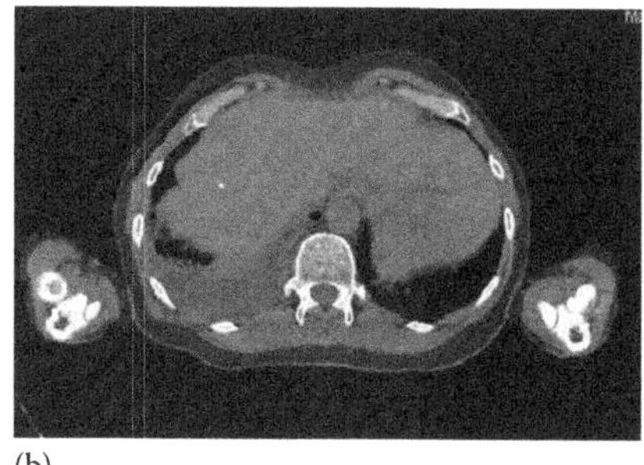

(b)

FIGURE 7.2
Slow-scan CT of the liver with a 4-s gantry rotation is in (a). The CT from averaging multiple phases of the cine CT images of 0.5-s gantry rotation over 4-s at the same location is in (b). The breath cycle of the patient was about 4 s. The artifacts in (a) were caused by the inconsistent projections from a 4-s gantry rotation, and were not present in (b). This indicated that a sub-second gantry rotation reduces the motion artifacts from reconstruction. (Reproduced from Pan T, et al., *Med Phys*, 33(10), 3931, 2006, Figure 11. With permission.)

respiratory signal. In their approach, each cine CT scan of 1 cm coverage requires a new scan setup.[4,5] It takes 15 scans of 0.5 s (7.5 s total x-ray on time) per position to scan over one respiratory cycle of data. There was an inter-scan delay of 0.25 s between two 0.5 s scans. In total, 11.25 s per step was needed by taking into account the inter-scan delay time. A pause of about 2 min after 7 scans was required for the user to reprogram another sequence of 7 scans. Although this approach could achieve 4DCT, it was not very practical for a large coverage of over 30 cm due to an inconvenience of setting up the scan protocol.

Both Phillips and Siemens adopted a low-pitch helical CT scanning mode for their 4DCT design.[6] This design was commercialized in 2006 on the MSCT scanners of at least 16 slices. In comparison, all the LightSpeed MSCT scanners of 4 slices and up, which have the cine CT scan capability that allows for a large coverage of the thorax in a simple protocol setup, are applicable for 4DCT. We will discuss the data sufficiency condition governing the 4DCT data collection; present the design of the commercial helical 4DCTs from Philips and Siemens; compare the differences between the helical 4DCT and the cine 4DCT in data acquisition, slice thickness, acquisition time, and work flow; review the respiratory monitoring devices in 4DCT; and understand the causes of image artifacts in 4DCT.

7.2 4DCT Data Acquisition

4DCT imaging needs to acquire data of a 3D volume for at least a complete breath cycle. A typical coverage is to cover the whole lungs. Since no CT detector is able to cover the whole lungs, and the conventional scan speed for diagnostic CT imaging is too fast to cover the whole lungs in a breath cycle, we need to slow down the scan speed for 4DCT imaging. Two scan modes can be utilized to achieve this goal: (1) cine CT scan and (2) low-pitch helical CT scan. Cine CT scan is a scan technique that acquires data at the same table position for more than one gantry rotation. Once the scan duration is set over one breath cycle, the need of 4DCT imaging can be achieved. Cine CT scan is similar to axial CT scan, except that the axial CT scan acquires data of either one gantry rotation for full scan reconstruction (FSR) or ½ of a gantry rotation cycle plus the fan angle of the detector for half-scan reconstruction (HSR).[7] In short, cine scan acquires data of more than one gantry rotation and axial scan acquires data of less than or equal to one gantry rotation. For helical CT, we generally need to set the pitch factor to be 0.1 or less to achieve 4DCT imaging. Special care has to be taken to ensure the scanner parameters such as pitch factor, gantry rotation speed, table speed or table translation per rotation, and detector configuration are suitable for 4DCT imaging.

7.2.1 Data Sufficiency Condition (DSC)

To achieve 4DCT imaging of an object in the presence of respiratory motion, one must acquire data at each location for the duration of one breath cycle plus the duration for one image reconstruction. This is the data sufficiency condition (DSC) for 4DCT imaging.[5] The time needed to acquire one image is equal to one gantry rotation cycle for FSR.[7] Without loss of generality, we use FSR throughout the chapter. Additional data acquisition is needed because it takes one gantry rotation of data in CT for reconstruction of an image and to ensure that there are images at both ends of a complete breath cycle. This additional data acquisition is not necessary for projection x-ray imaging such as fluoroscopy in which each projection is an image by itself.[8] Two acquisition modes of helical and cine can be used to realize the 4DCT imaging.

A helical CT scan acquires data when the table translates at a constant speed programmed by a pitch factor p, which is defined as the table travel per rotation divided by the width of the x-ray beam projected onto the rotation axis.

To satisfy the DSC with FSR, the pitch factors

$$p \leq \frac{T_g}{T_b + T_g'} \tag{7.1}$$

where T_g and T_b are the durations of the gantry rotation cycle and the breath cycle, respectively. If the breath cycle T_b is the same as the gantry rotation cycle T_g, then after one T_g (e.g., $T_b = T_g = 4$ s), the detector moves in and out of the x-ray beam in exactly one T_b. Taking into account of the extra acquisition of T_g for one image reconstruction, $p = ½$. If the breath cycle T_b is four times of T_g (i.e., $T_b = 4s$, $T_g = 1s$), then $p = 0.2$. Typically, $T_b = 4$ s and $T_g = 0.5$ s and $p = 0.11$, similar to the typical pitch factor of 0.1 or less in the helical 4DCT in Tables 7.1 and 7.2.

One important observation of the pitch factor selection is that the longer the breath cycle T_b or the shorter the gantry rotation cycle T_g, the smaller the pitch factor p becomes. In diagnostic CT imaging with patient breath-hold and without gating, p is about 1, which is 10 times faster than in a helical 4DCT scan with respiratory gating!

It is straightforward to simply increase the scan time to achieve cine 4DCT imaging of each location for the duration of one breath cycle plus the duration for one image reconstruction. This design was in place on all GE LightSpeed multi-slice CTs before the cine 4DCT was developed in 2002.[3]

TABLE 7.1

Pitch Factors p for the Philips CT Scanner and Calculated Values from Equation 7.1 at the Given Gantry Rotation Cycles T_g and Breath Cycles of 3–6s

	Philips Design		**Calculated Values with FSR**	
T_b	$T_g = 0.5$	$T_g = 0.44$	$T_g = 0.5$	$T_g = 0.44$
3	0.15	0.12	0.14	0.13
4	0.11	0.10	0.11	0.10
5	0.09	0.08	0.09	0.08
6	0.075	0.065	0.077	0.068

TABLE 7.2

Pitch Factors p for the Siemens CT Scanner and Calculated Values from Equation 7.1 at Given Gantry Rotation Cycles T_g and Breath Cycles of 3–6s

	Siemens Design		**Calculated Values with FSR**	
T_b	$T_g = 0.5$	$T_g = 1.0$	$T_g = 0.5$	$T_g = 1.0$
3	0.1	N/A	0.14	0.25
4	0.1	N/A	0.11	0.20
5	0.1	N/A	0.09	0.17
6	N/A	0.1	0.077	0.14

Only GE scanner has a user interface allowing the cine CT acquisition to be done with one scan statement. The user interfaces of Siemens and Phillips cannot be easily tailored for this purpose.[4]

7.2.2 Data Acquisition Modes

An MSCT can be characterized by the number of data channels that can be simultaneously read (SR) in the patient table direction. This number of SR is also frequently referred to as the number of slices in MSCT. An MSCT with a higher number of SR typically means a newer MSCT that can scan the same coverage faster. To acquire the cine 4DCT data with the GE scanners, the numbers of SR data channels are 4, 8, 16, and 64. To acquire the helical 4DCT data with Siemens or Philips scanners, the numbers of SR data channels are 16, 20, 40, and 64. The reason that helical 4DCT starts at 16-slice was because Siemens/Philips only introduced the low-pitch helical CT scan ($p \leq 0.1$) in their 16-slice and up CT scanners.

The number of physical data channels on an MSCT is at least the number of data channels of SR. Both GE's 4 and 8 channels have 16 physical detectors. For GE's 16 and 64 channels, there are 24 and 64, detectors, respectively. During data acquisition, the cine 4DCT utilizes the data acquisition modes of 4 × 2.5, 8 × 2.5, 8 × 2.5, and 16 × 2.5 mm on the GE's 4, 8, 16, and 64-slice CTs, respectively, to produce the images of slice thickness 2.5 mm (Figure 7.3a). In this case, a 16-slice CT may not have an advantage over an 8-slice CT in acquisition speed for the cine 4DCT because the detector size has reached 2 cm maximum for both the 8- and 16-slice GE CT.

In helical 4DCT data acquisition, the raw data before helical data interpolation are smaller than 1.5 mm. For the 16-slice, the mode of data acquisition is 16 × 1.5 mm. For the Siemens 20/40-slice, it is 20 × 1.2 mm. For the Philips 40/64-slice, it is 32 × 1.25 mm. Siemens 16, 20, and 40-slice CTs have the same detector size of 2.4 cm. Therefore, the speed of helical 4DCT is the same for the Siemens 16-, 20-, and 40-slice CTs. The Siemens 40- and 64-slice scanners use the z flying-focal spot to switch the focal spot between two positions to double-sample each angle of projection data to achieve the effect of 40 and 64-slice data even though their detector channels are 20 and 32, respectively.[9] The detector size of 16, 20, and 40-slice Siemens CTs is 2.4 cm, and it is 2.88 cm for the 64-slice Siemens CT. The Philips 40 and 64-slice CTs have 4-cm detector coverage, and their data acquisition is 32 × 1.25 mm for the helical 4DCT, which will shorten the acquisition time for a large coverage. It is important to note that slice broadening of 180% in the low-pitch helical 4DCT prevents the smallest data element in data acquisition from greater than 1.5 mm.[5]

7.2.3 Image Location, Slice Thickness, and Scan Time

Helical 4DCT scan data allow for CT image reconstruction at any location by permitting data interpolation between two neighboring detector elements, whereas cine 4DCT scan data allow for reconstructions only at the scan position.

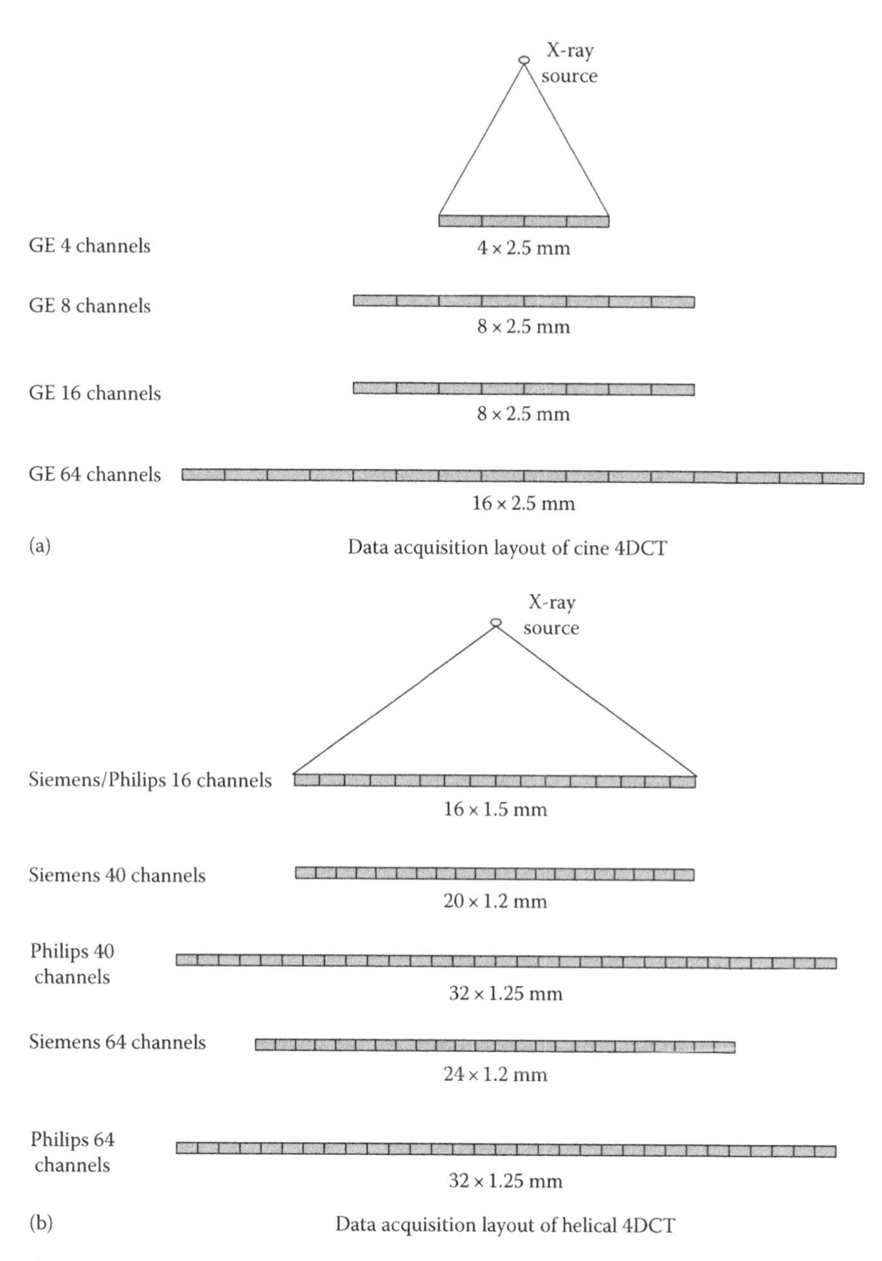

FIGURE 7.3
(a) GE detector layout for 4, 8, 16, and 64 data channels. The speed of cine 4DCT acquisitions is the same for both 8 and 16 data channels, which utilizes the maximum of 2-cm (8 × 2.5 mm) detector coverage. (b) Data acquisition layout of Siemens/Philips helical 4DCT for various data channels. The detector coverage of 4 cm is the same for the Philips 40 and 64 channels. The detector coverage of 2.4 cm is the same for the Siemens 16, 20, and 40 channels.

When using 4DCT to scan patients with lung cancer, it is important to have a complete coverage of the lungs for tumor delineation and dose calculation. The flexibility of image reconstruction at any location is not important if the images of evenly spaced slices can be reconstructed to create a 4D volume. Typical image slice thicknesses in radiation treatment planning are 2–3 mm. There is a penalty for data interpolation in helical 4DCT as it widens the slice sensitivity profile, a measure of the slice thickness of the CT image. It is important to use thin slice collimation of 16 × 1.50 mm in helical 4DCT on a 16-slice MSCT scanner because data interpolation will widen a slice thickness of 1.50 mm to almost 2.7 mm (i.e., the 1.5-mm slice will be thickened to 1.8 times of its original width). Similarly, this means that if the detector configuration is 8 × 2.50 mm, the slice thickness will become 4.5 mm in helical 4DCT. Same 8 × 2.50 mm collimation in cine 4DCT will generate the slices of 2.5 mm because no data interpolation is in the cine CT image reconstruction. In diagnostic imaging, data interpolation causes only about 20% widening when the pitch is greater than 0.5; the amount of widening reaches 180% when the pitch factor is less than 0.2.[5] To better understand this, we can take two independent slices of 2.5 mm at two neighboring locations in cine 4DCT and interpolate between them for an image located equidistance to the two slices; in such a case, interpolation would assign 50% weight to each image, and the composite image will become a 5.0-mm slice (resulting in 200% broadening of the slice thickness). Image reconstruction at pitch factors of less than 0.1 in helical 4DCT significantly thickens the CT images. Once the pitch factor p becomes 0, helical CT interpolation is no longer needed and helical 4DCT becomes cine 4DCT. That is the reason that the typical detector configuration for a 16-slice MSCT is 16 × 1.50 mm to keep the slice thickness <3 mm.

There is one additional breath cycle of acquisition time needed for the same coverage in helical 4DCT than in cine 4DCT. Scan of over half of a breath cycle time before the starting position and scan of after half of a breath cycle time after the end location to ensure each location has a complete breath cycle of data. Since the 4DCT acquisition for the lung cancer patient typically covers the whole lung, this additional acquisition of one breath cycle in the helical 4DCT is not significant when the number of breath cycles in a 4DCT is likely between 30 and 40 breath cycles. It contributes to a small amount of extra time for acquisition and a small amount of extra radiation exposure to the patient. Overall, helical 4DCT is faster than cine 4DCT because in cine 4DCT the table is paused to allow the table to move from one position to the next position; the accumulated pause time lengthens the overall acquisition time required for cine 4DCT, making helical 4DCT for the breath cycles of >6 s a faster 4DCT. The speedup (in favor of helical 4DCT) and extra radiation (in favor of cine 4DCT) are both very small.

7.2.4 Work Flow and Phase Selection Accuracy

The clinical work flow of helical 4DCT is that once the pitch factor is determined on the basis of either Table 7.1 or 7.2, collections of the helical 4DCT data and the respiratory signal have to proceed simultaneously. Reconstruction of the helical 4DCT images will not start until completion of the scan and after the respiratory signal is examined for accuracy in identification of the end-inspiration phases. This process allows for the reconstruction of 4DCT images at the specified phases in the helical 4DCT scan. Helical 4DCT tends to be slower than cine 4DCT in processing because the reconstruction is performed after (1) completion of the low-pitch helical CT scan and (2) verification of the end-inspiration phases in the respiratory signal. A suggestion to start image reconstruction before completion of data acquisition is possible. However, any change of image selection will likely result in additional image reconstruction. Because of these factors, multiple phases of helical 4DCT may not be available several minutes after the patient leaves the 4DCT acquisition session.

The clinical work flow of step and shoot cine 4DCT is that data processing is generally faster for cine 4DCT than helical 4DCT to generate the 4DCT images. Once the average duration of the respiratory signal has been determined, the cine scan duration per location is set to the average duration plus 1 s. This additional 1 s is recommended in case the patient's breath cycle becomes longer during data acquisition. Image reconstruction starts immediately when there is enough data for one image to be reconstructed. The time interval between two CT image reconstructions should be less than T_g. For example, if $T_g = 0.5$ s, then the interval can be set at 0.25 s for 50% data sharing between two image reconstructions to generate more images for data correlation with the respiratory signal, whose end-inspiration phases can be checked independently from the cine CT data collection. Increase of data sharing is to increase data sampling in cine 4DCT for data correlation so that the image selection will have a better chance of getting the images at the targeted phases. Because image reconstruction is performed independently from acquisition of the respiratory signal, more accurate correlation of the CT images to a particular phase of the breathing cycle is possible when two successive image reconstructions share more data or the time interval between two images is much shorter than T_g.

In general, phase selection accuracy is better with helical 4DCT than with cine 4DCT because the reconstruction of a particular phase is determined before image reconstruction in helical 4DCT, but not in cine 4DCT. For a long respiratory cycle of greater than 5 s, helical 4DCT may result in fewer sampling of the 4DCT images than cine 4DCT would and therefore may not fully capture the extent of tumor motion depicted in the maximum intensity projection images.[3] It is advisable to increase the number of phases in image reconstruction in helical 4DCT to improve the inclusion of the tumor motion so that the full extent of tumor motion will be included.

One important quality control measure that must be undertaken in 4DCT is to ensure that identification of the end-inspiration phases by the respiratory monitoring device is accurate. If phase calculation or identification is not accurate, both helical and cine 4DCTs will generate erroneous data,

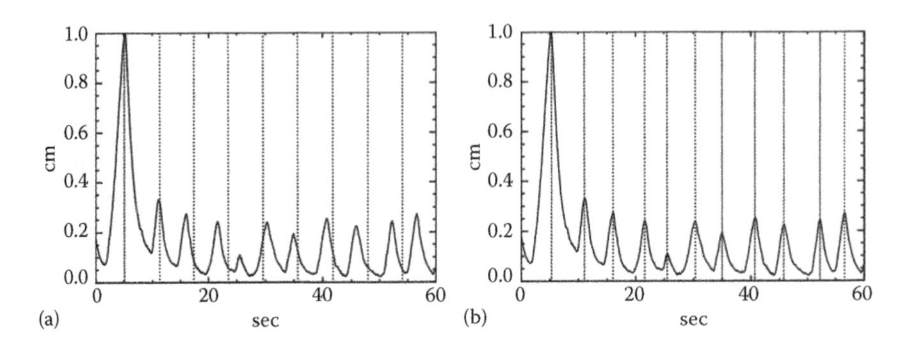

FIGURE 7.4
(a) An example of a respiratory signal over 60 s after a very irregular respiratory cycle causing an inaccurate identification of the end-inspiration phases, marked by dotted vertical lines. (b) Shows the accurate identification of the end-inspiration phases.

not representative of the 4DCT data. One example of this is illustrated in Figure 7.4. Some end-inspiration phases in (a) were incorrectly identified or missing. The correct identification of the phases is shown in (b). This step of ensuring accurate identification of end-inspiration phases is critical and should be checked in every 4DCT data processing.

7.2.5 Commercial Helical 4DCT Systems

Philips and Siemens offer helical 4DCT. Both implementations were derived from the low-pitch helical CT scan of cardiac CT, which has pitch factors p of 0.2–0.3 for a targeted heart rate of about 60 beats per minute. By lowering p to 0.1 or less, and replacing the electrocardiographic monitor with a respiratory monitor, the helical 4DCT for imaging the thorax under the condition of respiratory motion of 10–20 cycles per minute (corresponding to a 3- to 6-s respiratory cycle) becomes possible. The slower the respiratory motion, the smaller the pitch factor p has to be to meet the DSC and to avoid undersampling in helical 4DCT. An example of undersampling is shown in Figure 7.1c.

The pitch factors p of the Philips CT scanners[10] and calculated factors by Equation 7.1 for the breathing cycles of 3–6 s and gantry rotations of 0.44 and 0.5 s for FSR are listed in Table 7.1. Both factors match closely. When the gantry rotation time changes from 0.5 to 0.44 s, pitch factors p become smaller because the shorter gantry rotation time allows the scanner to cover a larger volume for the same duration at the same pitch, which may cause the scanner to scan too fast and violate the data sufficiency condition. Equation 7.1 can be used to calculate the gantry rotation cycles and the breath cycle durations other than those listed in Table 7.1.

The Siemens pitch factors[11] and calculated factors from Equation 7.1 are listed in Table 7.2. This scanner uses a single pitch factor of 0.1 and 2 gantry rotation cycles of 0.5 and 1.0 s.[12] Since the pitch factors p stay the same, the gantry rotation time has to be increased with the duration of the breath cycle.

This design is different from that of the Philips CT scanner, whose pitch factor becomes smaller when the breath cycle becomes longer. In the Siemens design, a longer gantry rotation cycle of 1.0 s is required to slow down the scan speed to accommodate for a breath cycle of ≥6 s with the same p of 0.1. One disadvantage in this design is that each CT image will have a temporal resolution of 1 s for FSR and 0.5 s for HSR. Longer breath cycles tend to have a longer duration of expiration than shorter cycles do. A decrease of the temporal resolution will increase image blurring in the CT image, particularly for the images acquired in the transition of the end-expiration phase to the end-inspiration phase, which tends to have the largest motion during respiration.

7.2.6 Respiratory Monitoring Devices

4DCT needs the timing information of the end-inspiration phase to guide its image reconstruction in helical 4DCT or image correlation in cine 4DCT. Many respiratory monitoring devices have been suggested.[8] The phase range between two end-inspiration phases is 100%, and the phase linearly increases from one end-inspiration phase to the next end-inspiration phase. Spirometer was used to measure the lung volume in inhale and exhale for the respiratory signal to correlate with the cine CT images for 4DCT.[4] The spirometer measurement is highly dependent on patient cooperation and may not be reproducible.

The most popular device for recording and monitoring the respiratory signal in 4DCT is the Real-time Position Management (RPM) respiratory gating system (Varian Medical Systems, Palo Alto, CA) shown in Figure 7.5a. The system consists of multiple infrared emitting diodes, an infrared camera, a plastic box with two or six infrared reflective markers, and the optical tracing software and computer. The infrared emitting diodes are mounted on the camera, which receive the reflected infrared signal from the reflective markers on a plastic box sitting on top of the patient's abdomen between the umbilicus and xiphoid process. The RPM device can measure the patient's respiratory pattern and range of motion and displays

(a)

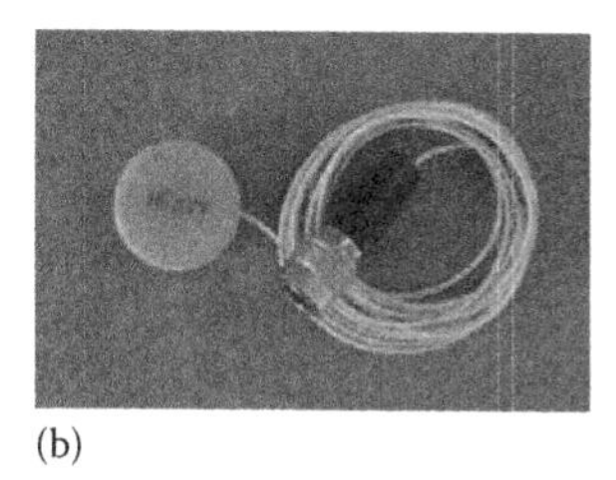

(b)

(c)

FIGURE 7.5
Respiratory monitoring devices used in 4DCT. (a) Two reflective markers on a plastic box and an infrared camera from Varian; (b) a pressure sensor from Anzai; and (c) air bellow (also a pressure sensor) from Philips.

them as a waveform. It can also send out triggers to enable prospective gating of CT. The plastic box, which is the only thing touching the patient, is almost radiation translucent, and the RPM can be integrated with some linear accelerators which use the device to turn the treatment beam on and off when the patient's breathing amplitude falls into the range of treatment. However, if the patient's abdomen or chest does not have a flat surface to allow the plastic box to sit on, placing the plastic box to the camera at the right angle may not be straightforward. The RPM system has been integrated in the GE and Philips 4DCTs.

Another device is the AZ-733V pressure sensor device (Anzai Medical Corporation, Tokyo, Japan) for recording the respiratory signal, respiratory gating and triggering. The system is comprised of chest/abdominal belt, pressure transducer (Figure 7.5b), sensor port, Wave Deck (signal processing box), laptop PC with connecting cables, and a trolley cart for storage and transport of the system. The pressure sensor can be easily attached to a patient with a Velcro belt. Measurement of the breathing signal is relative because the output of the pressure sensor is dependent on the tightness the Velcro belt fastened around the patient's chest or abdomen. One disadvantage of this system is that the sensor is radiopaque and can induce metallic artifacts. Placement of the sensor outside of the treatment field can avoid the metallic artifacts. The AZ-733V system is integrated in the Siemens 4DCT. It has been shown that there is a high correlation between the AZ-733V and RPM.[13]

A third device also used in 4DCT is the air bellow (Figure 7.5c), which is an elastic belt positioned around the abdomen that expands and contracts with the respiratory motion.[14] The device contains a pressure transducer, which converts the pressure waveform to a voltage signal, which is then digitalized and transmitted to the CT scanner. Using a respiratory bellow to monitor the breathing state has also been applied to CT-guided intervention procedure and MR imaging.

7.2.7 Image Artifacts

The basic assumption of 4DCT is that the respiratory motion is reproducible throughout the data acquisition and the duration of data acquisition at any location is at least one breath cycle plus the duration for one image reconstruction, which is one or 2/3 of a CT gantry rotation. Anything deviating from this assumption can induce artifacts manifested as (1) irregular respiratory motion, (2) misregistration at the cardiac region as the heart beating is not taken into account in 4DCT, and (3) missing data if the scan at a location is less than one breath cycle due to an underestimate of the patient's breath cycle duration. As a result, a higher helical pitch p in the Philips helical 4D or a faster CT gantry rotation in the Siemens helical 4D, or a shorter cine scan duration in GE cine 4D could introduce artifacts.[15] Figure 7.6 shows an example of these three types of artifact.

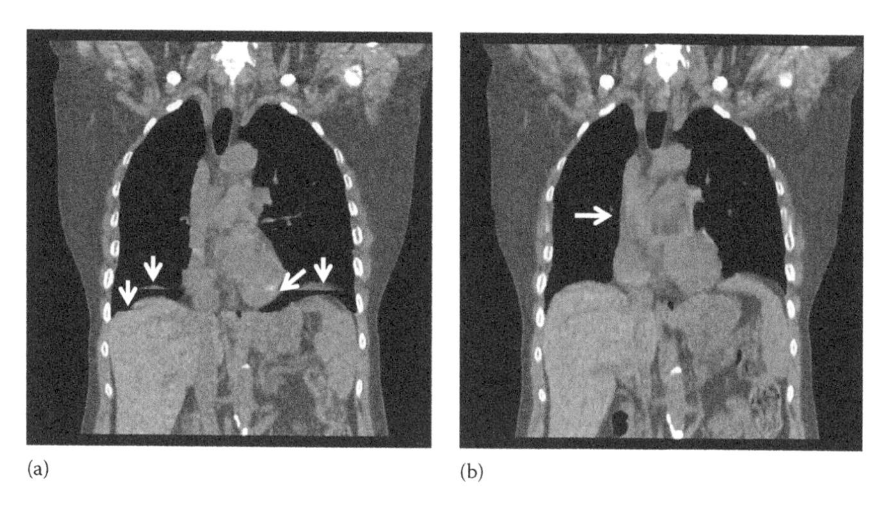

FIGURE 7.6
4DCT artifacts (pointed by arrows) due to (a) irregular respiration and misregistration in the cardiac region, not accounted for in 4DCT and (b) undersampling in a prolonged breathing cycle.

7.2.8 Average CT (ACT) and Maximum Intensity Projection (MIP) CT

ACT can be derived from averaging the 4DCT images or the cine CT images of many phases.[16,17] The conventional approach with slow-scan CT for ACT is incorrect and should be discouraged.[2,16] Figure 7.2 shows a clinic example of the same patient scanned with a slow-scan CT of 4 s and ACT of fast gantry rotation of 0.5 s. Cine-CT and low-pitch helical CT (pitch < 0.1) scans can be used to obtain ACT and both have been utilized in 4D CT imaging. ACT has been shown to be effective in dose calculation and registration with the PET data.[16,17]

MIP CT images can be derived by finding the maximum pixel value at each pixel from all the phases of 4DCT or cine CT images.[18] It has been shown that MIP CT images are effective in depicting the extent of tumor motion.[19,20] Figure 7.7 shows an example of MIP and ACT images. Both MIP and ACT images can be derived without gating.[16,18]

7.2.9 Conclusions and Future Directions

4DCT was an important development for imaging the tumors subject to the respiratory motion for radiation therapy. Its development was very closely tied to the MSCT technology, commercialized in 1998. Helical 4DCT was adopted by Siemens and Philips and commercialized in 2006, and cine 4DCT was developed by GE and commercialized in 2003. Both helical and cine 4DCT scans need to meet the requirement of data sufficiency condition to ensure there is at least one respiratory cycle of data

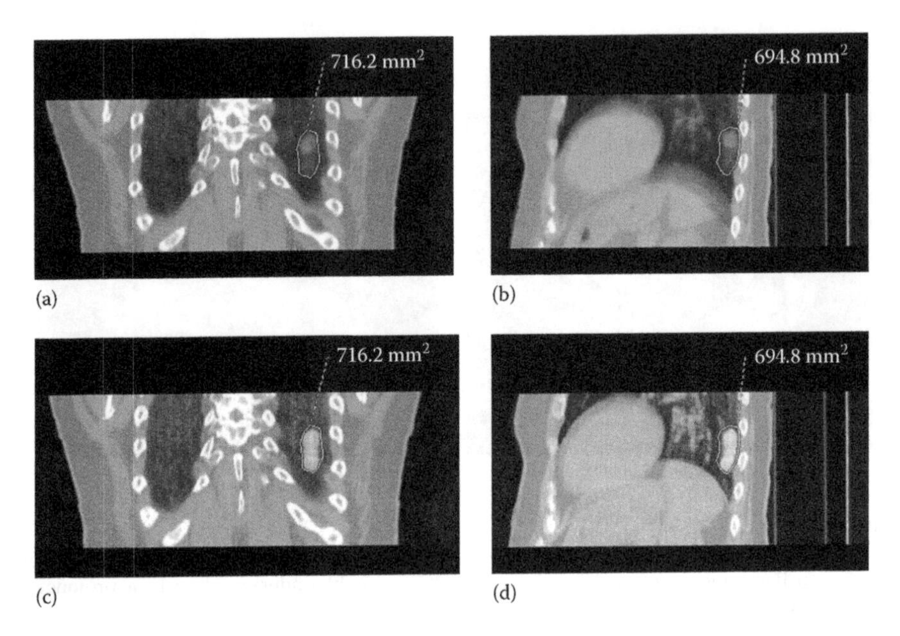

FIGURE 7.7
(a) and (b) are average CT images, (c) and (d) are the corresponding MIP CT images from the cine CT images of a patient for 4DCT. The 2D ROIs were derived from the MIP CT images.

at each location. Unlike cine 4DCT, helical 4DCT tends to result in thicker slices due to data interpolation and a longer work flow because image reconstruction cannot start until the respiratory signal of the helical 4DCT scan has been completely acquired. Commercial helical 4DCT can be performed by MSCT scanners of 16-slice or up, while cine 4DCT can be done using MSCT scanners with 4-slice or up. Three popular respiratory recording and monitoring devices are the optical device of RPM from Varian, the pressure sensor from Anzai, and the air bellow pressure sensor from Philips. To ensure a complete depiction of tumor motion, one should obtain at least 10 phases of 4DCT images in helical 4DCT, in particular for patients with the breath cycle duration of ≥6 s. The average and MIP CT images are important for registration with the PET data and tumor delineation, respectively. A very important quality control step in 4DCT is to ensure accuracy in identifying the end-inspiration phases of the respiratory signal because inaccurate identification of the end-inspiration phases will cause 4DCT to generate incorrect data. For the future development, the capability of 4DCT to cope with the irregular respiration is critical. However, it may not be possible to demand an artifact-free 4DCT due to irregular respiration when 4DCT is applied on the patient population with the lung disease which manifests itself as coughing and/or irregular respiration.

References

1. Underberg RW, Lagerwaard FJ, Cuijpers JP, Slotman BJ, van Sornsen de Koste JR, Senan S. Four-dimensional CT scans for treatment planning in stereotactic radiotherapy for stage I lung cancer. *Int J Radiat Oncol Biol Phys*. Nov 15 2004;60(4):1283–1290.
2. Keall PJ, Mageras GS, Balter JM, et al. The management of respiratory motion in radiation oncology report of AAPM Task Group 76. *Med Phys*. Oct 2006;33(10):3874–3900.
3. Pan T, Lee TY, Rietzel E, Chen GT. 4D-CT imaging of a volume influenced by respiratory motion on multi-slice CT. *Med Phys*. Feb 2004;31(2):333–340.
4. Low DA, Nystrom M, Kalinin E, et al. A method for the reconstruction of four-dimensional synchronized CT scans acquired during free breathing. *Med Phys*. Jun 2003;30(6):1254–1263.
5. Pan T. Comparison of helical and cine acquisitions for 4D-CT imaging with multislice CT. *Med Phys*. Feb 2005;32(2):627–634.
6. Keall PJ, Starkschall G, Shukla H, et al. Acquiring 4D thoracic CT scans using a multislice helical method. *Phys Med Biol*. May 21 2004;49(10):2053–2067.
7. Parker DL. Optimal short scan convolution reconstruction for fanbeam CT. *Med Phys*. Mar–Apr 1982;9(2):254–257.
8. Kubo HD, Hill BC. Respiration gated radiotherapy treatment: a technical study. *Phys Med Biol*. Jan 1996;41(1):83–91.
9. Flohr TG, Stierstorfer K, Ulzheimer S, Bruder H, Primak AN, McCollough CH. Image reconstruction and image quality evaluation for a 64-slice CT scanner with z-flying focal spot. *Med Phys*. Aug 2005;32(8):2536–2547.
10. Quick Steps for Retrospective Spiral Respiratory Correlated Imaging with Varian RPM. *for the Brilliance CT Big Bore v2.2.2 system and the Brilliance 16–64 v2.2.5*. 2007. Philips CT Product Oncology Application (453567455491).
11. Somatom Sensation Open Reference Manual. 159–177.
12. Hurkmans CW, van Lieshout M, Schuring D, et al. Quality assurance of 4D-CT scan techniques in multicenter phase III trial of surgery versus stereotactic radiotherapy (radiosurgery or surgery for operable early stage (stage 1A) non-small-cell lung cancer [ROSEL] study). *Int J Radiat Oncol Biol Phys*. Jul 1 2011;80(3):918–927.
13. Li XA, Stepaniak C, Gore E. Technical and dosimetric aspects of respiratory gating using a pressure-sensor motion monitoring system. *Med Phys*. Jan 2006;33(1):145–154.
14. Klahr P, Subramanian P, Yanof JH. Respiratory-correlated multislice CT for radiation therapy planning: imaging and visualization methods. *Medicamundi*. 2005;49(3):34–37.
15. Han D, Bayouth J, Bhatia S, Sonka M, Wu X. Characterization and identification of spatial artifacts during 4D-CT imaging. *Med Phys*. Apr 2011;38(4):2074–2087.
16. Pan T, Mawlawi O, Luo D, et al. Attenuation correction of PET cardiac data with low-dose average CT in PET/CT. *Med Phys*. Oct 2006;33(10):3931–3938.
17. Pan T, Mawlawi O, Nehmeh SA, et al. Attenuation correction of PET images with respiration-averaged CT images in PET/CT. *J Nucl Med*. Sep 2005;46(9):1481–1487.

18. Pan T, Sun X, Luo D. Improvement of the cine-CT based 4D-CT imaging. *Med Phys*. Nov 2007;34(11):4499–4503.
19. Underberg RW, Lagerwaard FJ, Slotman BJ, Cuijpers JP, Senan S. Use of maximum intensity projections (MIP) for target volume generation in 4DCT scans for lung cancer. *Int J Radiat Oncol Biol Phys*. Sep 1 2005;63(1):253–260.
20. Bradley JD, Nofal AN, El Naqa IM, et al. Comparison of helical, maximum intensity projection (MIP), and averaged intensity (AI) 4D CT imaging for stereotactic body radiation therapy (SBRT) planning in lung cancer. *Radiother Oncol*. Dec 2006;81(3):264–268.

8

Image Reconstruction Algorithms for X-Ray CT

Ken Taguchi

CONTENTS

8.1 Projection

Image reconstruction is to estimate a spatial distribution of some parameter inside an object from its projections, where projections are a set of measurements of the line integral values of the parameter. The parameter is a linear attenuation coefficient for transmission x-ray CT, while it is a radioactivity level in case of emission computed tomography (CT) such as single photon emission CT (SPECT) or positron emission tomography (PET).

Let $f(x, y)$ be the two-dimensional (2-D) distribution of linear attenuation coefficients for monochromatic x-ray CT. Using the coordinate system

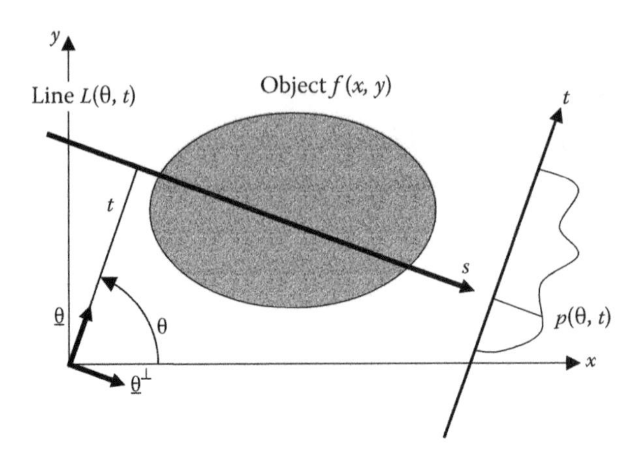

FIGURE 8.1
Parameters for parallel-beam projections.

described in Figure 8.1, projections $p(\theta, t)$, that is, line integrals of the object $f(x, y)$ along a line $L(\theta, t)$ parameterized by (θ, t), are described as

$$
\begin{aligned}
p(\theta,t) &= \int_{L(\theta,t)} f(x,y)ds \\
&= \int_{-\infty}^{\infty}\int_{-\infty}^{\infty} f(x,y)\delta(x\cos\theta + y\sin\theta - t)dxdy \\
&= \int_{-\infty}^{\infty} f(t\underline{\theta} + s\underline{\theta}^{\perp})ds
\end{aligned}
\tag{8.1}
$$

where

$\delta(\ldots)$ is the Dirac delta function

$\underline{\theta} = (\cos\theta, \sin\theta)$ and $\underline{\theta}^{\perp} = (\sin\theta, -\cos\theta)$ are two unit vectors which are orthogonal to each other

To transform a 2-D object f to its projections (or line integrals), p is called 2-D Radon transform. For reference, 3-D Radon data of a 3-D object f is plane integrals of the object.

The relationship of the intensity of two x-ray beams, one incident onto the object, I_0, and the other exiting from the object after the attenuation, I, can be expressed as

$$
I = I_0 exp\left(-\int_{L(\theta,t)} f(x,y)ds\right) \tag{8.2}
$$

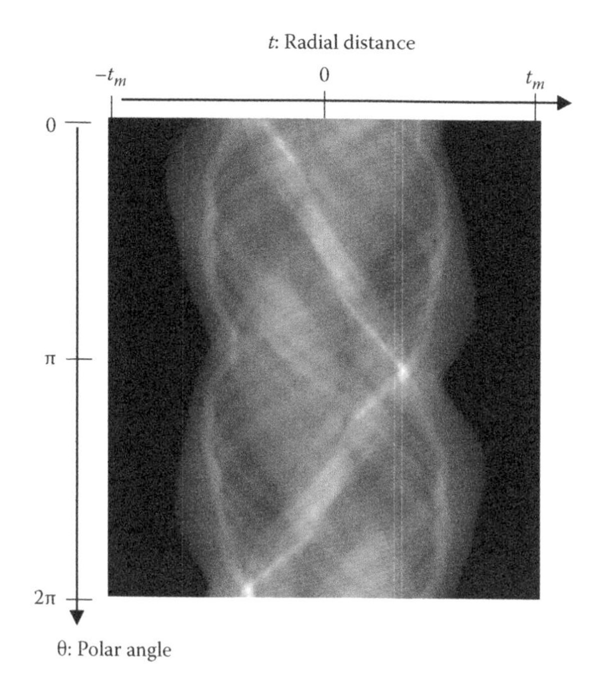

FIGURE 8.2
A sinogram $p(\theta, t)$ of a chest phantom. $|t_m|$ corresponds to the radial support of the phantom.

Thus, the projection can be obtained as

$$p(\theta,t) = -\ln\left(\frac{I}{I_0}\right) \tag{8.3}$$

An example of projections is shown in Figure 8.2. Projection data are also called a sinogram, because when a point (x, y) is projected onto a line detector, the radial distance t of the projected point can be described by a trigonometric function with the angle θ, for example, $t = t_0\sin\theta$.

8.2 Central Slice Theorem (or Fourier Slice Theorem)

There are several ways to reconstruct the object $f(x, y)$ from projections $p(\theta, t)$; however, the common foundation that relates the two functions is the central slice theorem (or Fourier slice theorem) (Figure 8.3).

$$\int_{-\infty}^{\infty} p(\theta,t)exp(-2\pi it\omega)dt = F(\omega\cos\theta, \omega\sin\theta) = F_{polar}(\theta,\omega) \tag{8.4}$$

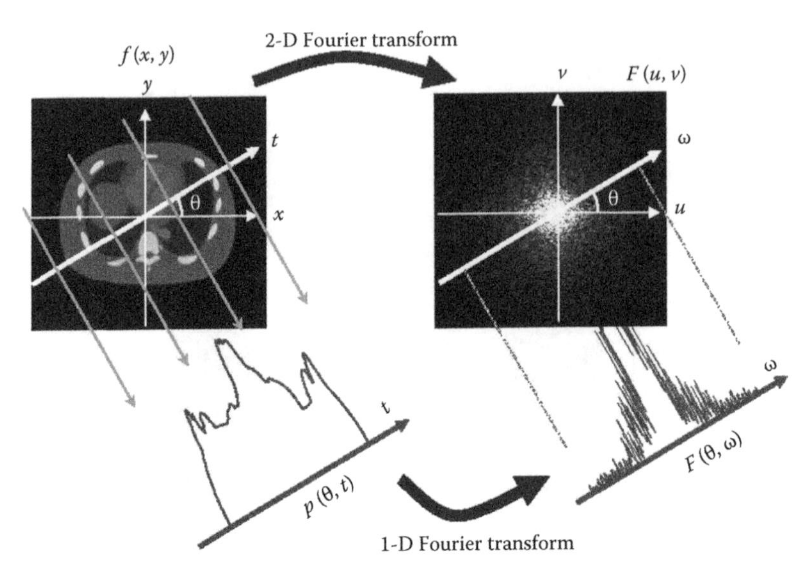

FIGURE 8.3
Central slice theorem (Fourier slice theorem).

where $F(\omega\cos\theta, \omega\sin\theta)$ and $F_{polar}(\theta, \omega)$ are the 2-D Fourier transform of the object $f(x, y)$ in the Cartesian and polar coordinate systems, respectively. The proof is shown in the following.

$$\int_{-\infty}^{\infty} p(\theta,t)exp(-2\pi it\omega)dt$$

$$= \int_{-\infty}^{\infty}\left[\int_{-\infty}^{\infty}\int_{-\infty}^{\infty} f(x,y)\delta(x\cos\theta + y\sin\theta - t)dxdy\right]exp(-2\pi it\omega)dt$$

(∵ Equation 8.1)

$$= \int_{-\infty}^{\infty}\int_{-\infty}^{\infty} f(x,y)\left[\int_{-\infty}^{\infty} \delta(x\cos\theta + y\sin\theta - t)exp(-2\pi it\omega)dt\right]dxdy$$

(Exchanging the order of integrals)

$$= \int_{-\infty}^{\infty}\int_{-\infty}^{\infty} f(x,y)exp(-2\pi i(x\omega\cos\theta + y\omega\sin\theta))dxdy$$

(∵ The property of delta function)

$$= F(u,v)\Big|_{u=\omega cos\theta,\, v=\omega sin\theta} \blacksquare \tag{8.5}$$

The central slice theorem states as follows. The one-dimensional (1-D) Fourier transform of parallel-beam projections $p(\theta, t)$ at an angle θ is equal to the Fourier transform of the object $f(x, y)$ along a line rotated by θ. Thus, once we obtain parallel-beam projections over 180° and calculate its Fourier transform, we can fill out the 2-D Fourier data space of the object, from which the object $f(x, y)$ can be reconstructed by applying inverse 2-D Fourier transform. Note that this is the method mainly used in magnetic resonance imaging (MRI), where the Fourier space is called the K-space and the component data are directly measured. Thus, MR images can also be reconstructed using filtered backprojection or other algorithms we discuss later.

8.3 Filtered Backprojection (Parallel-Beam)

In the current medical x-ray CT systems, filtered backprojection (FBP) algorithms are the standard image reconstruction method while iterative methods (discussed later) are becoming options. In SPECT and PET scanners, both FBP and fully iterative image reconstruction methods may be equally utilized options. In this section, we derive the filtered backprojection formula from the central slice theorem.

We start with the object f as the inverse Fourier transform of 2-D Fourier transform spectrum F, and then convert it to the polar coordinate system.

$$\begin{aligned} f(x,y) &= \int_{-\infty}^{\infty}\int_{-\infty}^{\infty} F(u,v)exp(2\pi i(xu+yv))dudv \\ &= \int_{0}^{2\pi}\int_{0}^{\infty} F_{polar}(\theta,\omega)exp(2\pi i\omega(x\cos\theta+y\sin\theta))\omega d\omega d\theta \end{aligned} \tag{8.6}$$

where

$u = \omega \cos \theta$

$v = \omega \sin \theta$

$du\, dv = \omega\, d\omega\, d\theta$

Splitting the integration range into two and using symmetry, $F(\theta, \omega) = F(\theta + \pi, -\omega)$, we get

$$\begin{aligned} f(x,y) &= \int_{0}^{\pi}\int_{0}^{\infty} F_{polar}(\theta,\omega)exp(2\pi i\omega(x\cos\theta+y\sin\theta))\omega d\omega d\theta \\ &\quad + \int_{\pi}^{2\pi}\int_{0}^{\infty} F_{polar}(\theta,\omega)exp(2\pi i\omega(x\cos\theta+y\sin\theta))\omega d\omega d\theta \\ &= \int_{0}^{\pi}\int_{-\infty}^{\infty} F_{polar}(\theta,\omega)exp(2\pi i\omega(x\cos\theta+y\sin\theta))|\omega| d\omega d\theta \end{aligned} \tag{8.7}$$

Substituting the central slice theorem, Equation 8.4, into $F_{polar}(\theta, \omega)$, we get

$$f(x,y) = \int_0^{\pi} \left\{ \int_{-\infty}^{\infty} \left[\int_{-\infty}^{\infty} p(\theta,t) exp(-2\pi i t \omega) dt \right] exp(2\pi i \omega (x \cos\theta + y \sin\theta)) |\omega| d\omega \right\} d\theta$$

$$= \int_0^{\pi} p^F(\theta,t) d\theta \tag{8.8}$$

where

$$p^F(\theta,t) = \int_{-\infty}^{\infty} \left[\int_{-\infty}^{\infty} p(\theta,t) exp(-2\pi i t \omega) dt \right] exp(2\pi i \omega (x \cos\theta + y \sin\theta)) |\omega| d\omega$$

$$= \int_{-\infty}^{\infty} p(\theta,t) h_R(t-t') dt' \tag{8.9}$$

and

$$h_R(t) = \int_{-\infty}^{\infty} |\omega| exp(2\pi i \omega t) d\omega \tag{8.10}$$

Equations 8.8 through 8.10 show the following two steps for image reconstruction. First, parallel projections p are convolved with a ramp filter kernel h_R along the radial direction t. The convolution can be performed in the Fourier domain for more efficient computation. Then, the filtered projection p^F are "backprojected" onto the image space over $[0, \pi)$ or 180°.

It is critical to have a clear mental picture of how images are reconstructed for investigating causes of artifacts, for example. To help readers understand the two steps in FBP intuitively, we show a result of computer simulation with a circular phantom with a diameter of 100 mm located at (0, 100 mm) in Figures 8.4 and 8.5. Figure 8.4 shows the projections p (a) and the filtered projections p^F (b), and the corresponding profiles at $\theta = \pi$. Note that there are large negative values in p^F just outside $t = (-t_0, t_0)$ where projections are nonzero, and the negative values rapidly increase toward zero.

Figure 8.5 visualizes the backprojection process over projection angle θ. It can be seen that the contour of the object becomes clearer or sharper as the angular range approaches to 180°. It can also be seen that outside the object, positive values backprojected from some angles and negative values from the other angles are cancelled out, resulting in zero. The image reconstruction is completed at 180°, and the same process is repeated from 180° to 360°.

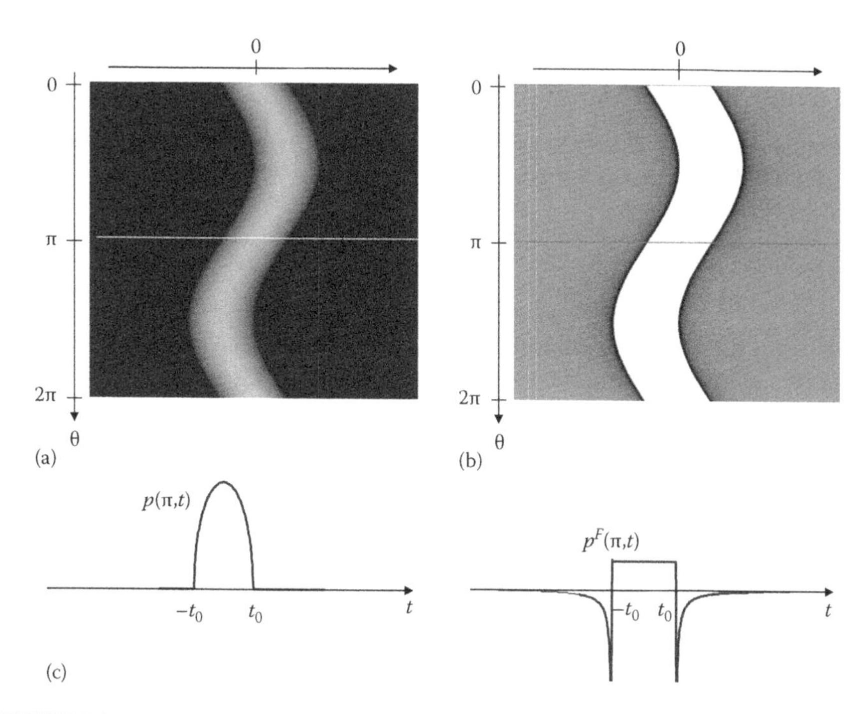

FIGURE 8.4
Projections p (a) and filtered projections p^F using Shepp–Logan filter (b) and the corresponding radial profiles at $\theta = \pi$ (c).

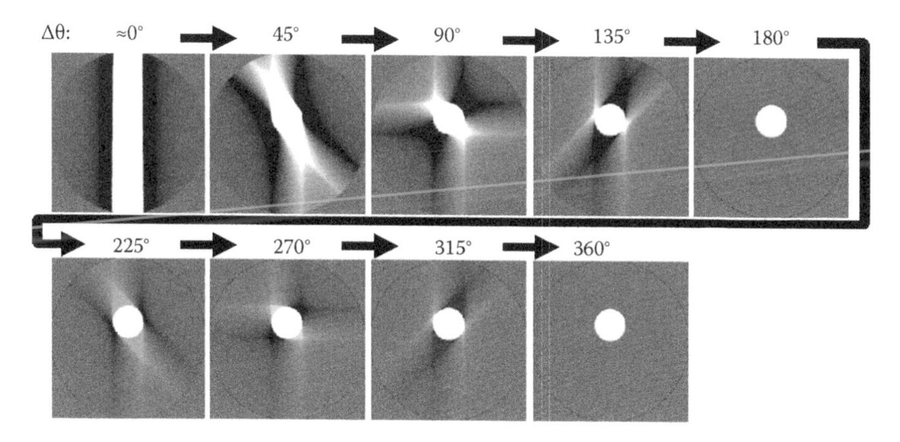

FIGURE 8.5
The process of an image being reconstructed by filtered backprojection of parallel-beam projections. Projections are in the north–south direction at $\theta = 0°$ and rotate in the counterclockwise direction.

8.4 Filter Kernels

In reality, the ramp filter kernel $|w|$ is truncated by the Nyquist frequency ω_{Nq} defined by the sampling condition

$$H_R(\omega) = \begin{cases} |\omega| & |\omega| \leq \omega_{Nq} \\ 0 & |\omega| > \omega_{Nq} \end{cases} \tag{8.11}$$

This ramp filter provides the most accurate and the sharpest image of the object $f(x, y)$ in a mathematical sense, correctly honoring all of the frequency components acquired by detectors. This filter, however, also provides the noisiest image at the presence of noise in projections. An analysis of frequency components of projections revealed that near the Nyquist frequency, there is less information from the object and more noise from quantum statistics. Shepp and Logan then proposed a modified ramp filter with an apodization window to suppress high-frequency components (i.e., "Shepp–Logan filter") shown in Figure 8.6 [1,2]. This kernel suppresses the image noise, while minimizing the loss of the spatial resolution in clinical cases.

$$H_R(\omega) = \begin{cases} \dfrac{2\omega_{Nq}}{\pi} \left| \sin\left(\dfrac{\pi\omega}{2\omega_{Nq}} \right) \right| & |\omega| \leq \omega_{Nq} \\ 0 & |\omega| > \omega_{Nq} \end{cases} \tag{8.12}$$

The Shepp–Logan filter is the standard one used for quality assurance tests in many x-ray CT scanners. But there are also other modified ramp filter

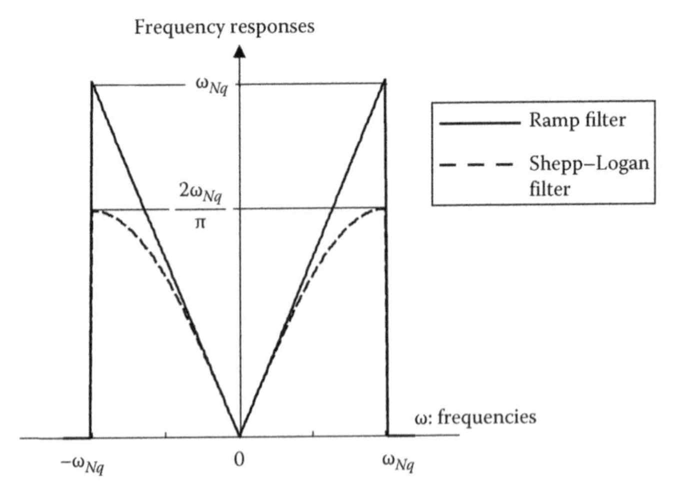

FIGURE 8.6
The frequency response of filter kernels $H_R(\omega)$.

kernels on CT scanners, which are specifically designed by enhancing and suppressing desirable frequency components for particular clinical applications such as body kernels, brain kernels, bone kernels, and lung kernels.

8.5 Filtered Backprojection (Fan-Beam)

Almost all of the x-ray CT scanners use x-ray beams divergent from the x-ray focal spot to acquire projections. Pin-hole and divergent collimators in SPECT scanners also use the divergent geometry. The geometry is called the fan-beam geometry for 2-D imaging and cone-beam geometry for 3-D imaging. Divergent beams require different FBP formula from parallel-beam geometry. In this section, we outline the FBP method for fan-beam geometry.

Fan-beam projections $g(\beta, \gamma)$ are parameterized by a projection angle β, which is an angle that the source a makes with a reference axis, and a fan angle γ, which is a locally defined angle that rotates with the source a for the ray within a fan. Line integrals of an object f can be described using a unit vector $\Theta = (\sin(\beta + \gamma), -\cos(\beta + \gamma))$

$$g(\beta,\gamma) = \int_0^\infty f(\underline{a}(\beta) + s\underline{\Theta})ds \tag{8.13}$$

Analytical methods to reconstruct images from fan-beam projections can be categorized into two groups. One is to rebin fan-beam projections into parallel-beam projections, which is called a fan-to-parallel-beam rebinning method, and the other is to reconstruct images directly from fan-beam projections, which is called a direct fan-beam FBP method. We will outline both of the methods in order.

Comparing Figure 8.7 with Figure 8.1, one may notice that the line L can be expressed using a different pair of parameters as $L(\beta, \gamma)$ or $L(\theta, t)$. Thus, the x-ray beam along L in fan-beam projections and parallel-beam projections can be related as follows.

$$\theta = \beta + \gamma, \quad t = R\sin\gamma \tag{8.14}$$

where R is the distance from the origin to the source $a = (-R \sin \beta, R \cos \beta)$. The line integral values obtained by the x-ray beam are identical; thus, one can map fan-beam projections g to parallel-beam projections p as

$$p(\theta,t) = p(\beta+\gamma, R\sin\gamma) = g\left(\theta - \sin^{-1}\left(\frac{t}{R}\right), \sin^{-1}\left(\frac{t}{R}\right)\right) \tag{8.15}$$

One can then employ the FBP method for parallel projections p to reconstruct an image f. This is the first, rebinning method.

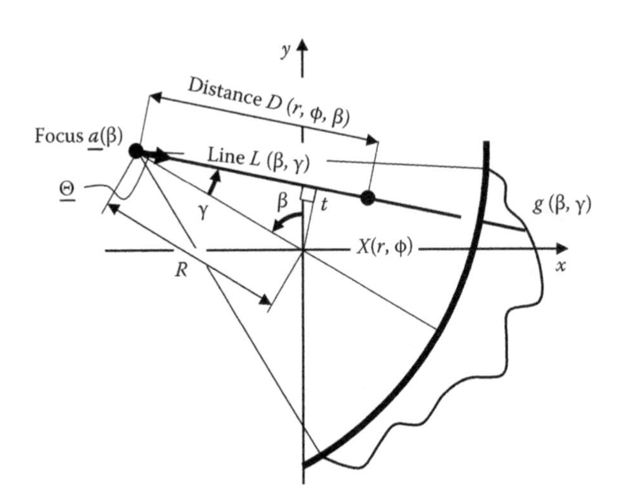

FIGURE 8.7
Parameters for fan-beam projections.

We now outline the second, direct FBP method from fan-beam projections. We start with describing the object f using the polar coordinate system (r, ϕ)

$$x = r\cos\phi,\ y = r\sin\phi \tag{8.16}$$

The FBP formula for parallel-beam geometry can then be rewritten as

$$\begin{aligned} f(r,\phi) &= \frac{1}{2}\int_0^{2\pi}\left[\int_{-t_m}^{t_m} p(\theta,t)h_R(r\cos(\theta-\phi)-t)dt\right]d\theta \\ &= \frac{1}{2}\int_0^{2\pi}\left[\int_{-\gamma_m}^{\gamma_m} p(\beta+\gamma,R\sin\gamma)h_R(r\cos(\beta+\gamma-\phi)-R\sin\gamma)R\cos\gamma d\gamma\right]d\beta \\ &\left[\because dt = R\cos\gamma d\gamma \text{ and Equations 8.14 and 8.15}\right] \end{aligned} \tag{8.17}$$

where $\gamma_m = \sin^{-1}(t_m/R)$, t_m is the radius of the support of the object f. The backprojection range is extended from π to 2π, which is normalized by multiplying 1/2.

Now we make the argument of the kernel h_R shift-invariant. Let D be the distance from the focal spot to a point-of-interest $X(r, \phi)$ as shown in Figure 8.7, and γ' be the fan angle of the ray that goes through X. Then from geometry, we have

$$D\cos\gamma' = R + r\sin(\beta-\phi) \tag{8.18}$$

$$D\sin\gamma' = r\cos(\beta-\phi) \tag{8.19}$$

and

$$D(r,\phi,\beta) = \sqrt{(R + r\sin(\beta - \phi))^2 + (r\cos(\beta - \phi))^2} \tag{8.20}$$

$$\gamma' = \tan^{-1}\frac{r\cos(\beta - \phi)}{R + r\sin(\beta - \phi)} \tag{8.21}$$

By modifying $D\sin(\gamma' - \gamma)$ using Equations 8.18 and 8.19, we get the argument of h_R in Equation 8.17 as follows.

$$\begin{aligned}
D\sin(\gamma' - \gamma) &= D\sin\gamma'\cos\gamma - D\cos\gamma'\sin\gamma \\
&\qquad \left[\because \sin(a-b) = \sin a\cos b - \cos a\sin b\right] \\
&= r\cos(\beta - \phi)\cos\gamma - (R + r\sin(\beta - \phi))\sin\gamma \\
&\qquad \left[\because \text{Equations 8.18 and 8.19}\right] \\
&= r\cos(\beta - \phi + \gamma) - R\sin\gamma \\
&\qquad \left[\because \cos(a+b) = \cos a\cos b - \sin a\sin b\right]
\end{aligned} \tag{8.22}$$

Using Equations 8.15 and 8.22, Equation 8.17 can be modified to

$$\begin{aligned}
f(r,\phi) &= \frac{1}{2}\int_{-\gamma}^{2\pi-\gamma}\left[\int_{-\gamma_m}^{\gamma_m} g(\beta,\gamma)h_R(D\sin(\gamma' - \gamma))R\cos\gamma d\gamma\right]d\beta \\
&= \frac{1}{2}\int_{0}^{2\pi}\left[\int_{-\gamma_m}^{\gamma_m} g(\beta,\gamma)h_R(D\sin(\gamma' - \gamma))R\cos\gamma d\gamma\right]d\beta
\end{aligned} \tag{8.23}$$

Here the periodicity of β over 2π is used to shift the integration range. Finally, we change the definition of the ramp filter kernel for parallel-beam projection, Equation 8.11, and obtain that for fan-beam projection

$$\begin{aligned}
h_g(\gamma) &\equiv h_R(D\sin\gamma) \\
&= \int_{-\infty}^{\infty} |\omega| exp(2\pi i\omega D\sin\gamma)d\omega \\
&= \left(\frac{\gamma}{D\sin\gamma}\right)^2 \int_{-\infty}^{\infty} |\omega'| exp(2\pi i\omega'\gamma)d\omega' \\
&= \left(\frac{\gamma}{D\sin\gamma}\right)^2 h_R(\gamma)
\end{aligned} \tag{8.24}$$

Equation 8.24 can be applied to other kernels such as Shepp–Logan filter.

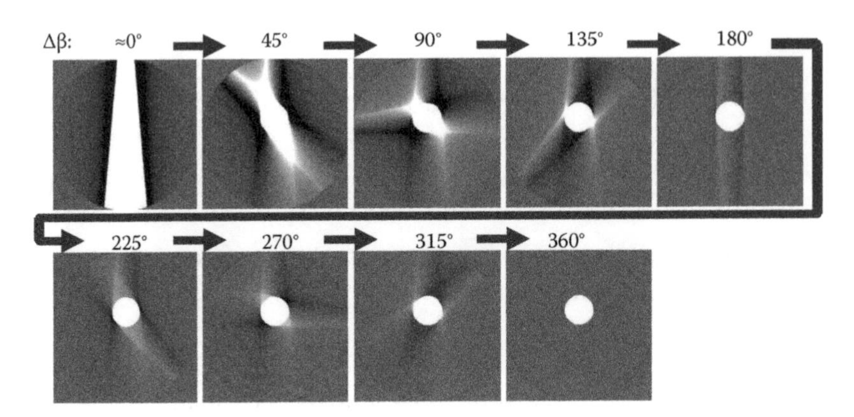

FIGURE 8.8
The process of an image being reconstructed by filtered backprojection of fan-beam projections. The focal spot is located at 12:00 at $\beta = 0°$ and rotates in the counter-clock-wise direction.

In short, the FBP method that directly employs on fan-beam projections g is as follows:

$$f(r,\phi) = \frac{1}{2}\int_{0}^{2\pi} \frac{1}{D^2(r,\phi,\beta)} \int_{-\gamma_m}^{\gamma_m} g(\beta,\gamma)h_g(\gamma' - \gamma)R\cos\gamma d\gamma d\beta \tag{8.25}$$

Similar to parallel-beam case, Equation 8.25 shows the following steps for image reconstruction. First, fan-beam projections g are weighted by cosγ. The weighted data are then convolved with a ramp filter kernel h_g with respect to the fan angle parameter γ. The filtered projections are backprojected onto the image space over $[0, 2\pi)$ or 360° with a weight calculated by the inverse square distance from the focal spot to the voxel-of-interest, $1/D^2$. Finally, the reconstructed image is scaled by $R/2$.

Similar to Figure 8.5 for parallel-beam projections, Figure 8.8 visualizes the backprojection process over projection angle β. It can be seen that the back-projected filtered data are spread from the focal spot side (on the north side when $\Delta\beta \approx 0°$) to the detector side (on the south side). Notice also that because of the divergent rays, the image reconstruction does *not* complete and artifact remains along the north–south direction when the backprojection is employed over 180°. It requires a full 360° for reconstruction to complete.

8.6 Redundancy in Projections

At the absence of scattered radiation, photon noise, and partial volume effect (and attenuation inside the object in case of SPECT and PET), projections of an object f can be calculated from line integrals of f along a ray alone, and

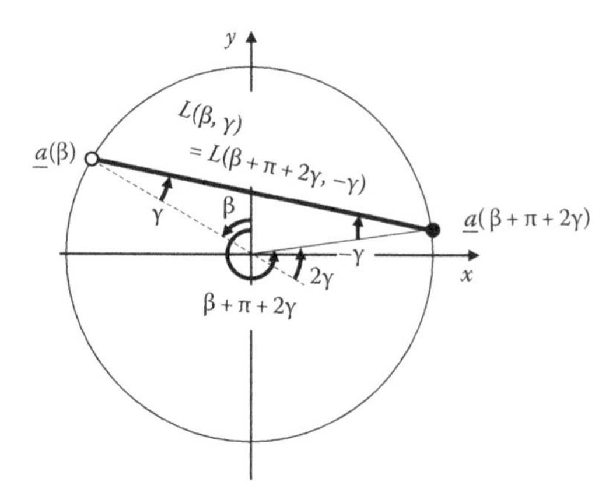

FIGURE 8.9
Conjugate (or complementary) rays in fan-beam projections.

independent of the direction of the x-ray beams. In other words, data are identical even if they are acquired when an x-ray focus and a detector swap the positions. One is called a primary ray, while the other is called a conjugate (or complementary) ray in CT community. The two rays can be related as follows (see Figure 8.9 for fan-beam projections):

$$p(\theta, t) = p(\theta + \pi, -t) \quad \left[\text{parallel-beam projections}\right] \tag{8.26}$$

$$g(\beta, \gamma) = g(\beta + \pi + 2\gamma, -\gamma) \quad \left[\text{fan-beam projections}\right] \tag{8.27}$$

Equations 8.26 and 8.27 are described using sinograms in Figure 8.10. Dashed lines indicate the primary rays at one projection angle, while solid lines

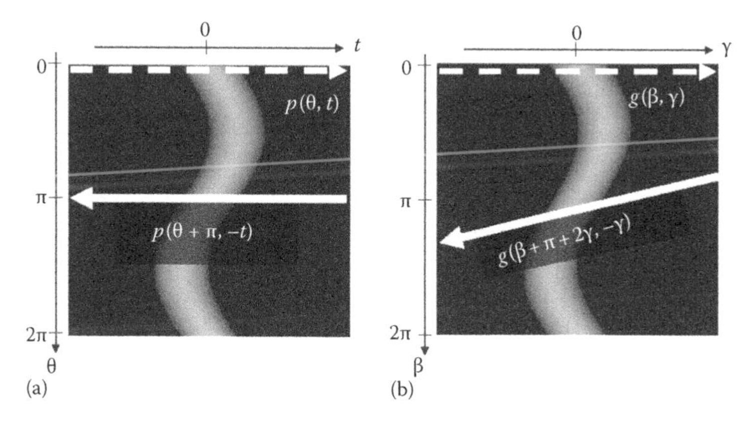

FIGURE 8.10
A primary ray at a projection angle (dashed line) and the corresponding conjugate ray (solid line) in parallel- (a) and fan-beam (b) projections. The heads and tails of the arrows indicate the corresponding order of the rays.

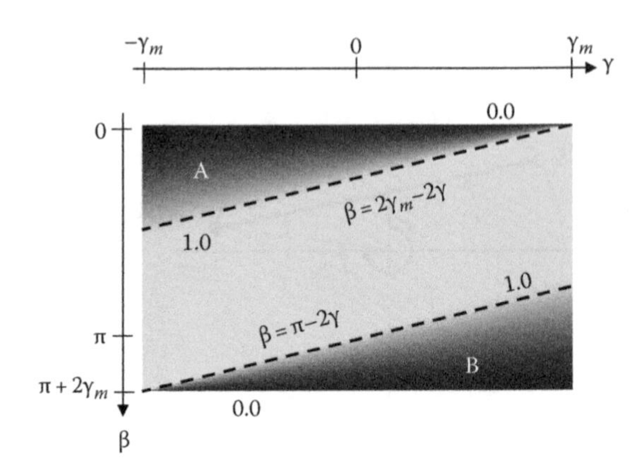

FIGURE 8.11
The redundancy weight $w(\beta, \gamma)$ of the halfscan algorithm.

indicate the corresponding conjugate rays. As shown in Equation 8.8, the angular range of projections that is necessary and sufficient for exact image reconstruction for parallel-beams is π regardless of the size of the object. The necessary angular range for fan-beams is $\pi+2\gamma_m$ (or 180° plus the full fan angle), where $2\gamma_m = 2\sin^{-1}(t_m/R)$ is the full fan angle of the detector to image an object with a circular support with a radius of t_m.

Among fan-beam projections over $\pi+2\gamma_m$, a part of projection data (indicated by A and B in Figure 8.11) are acquired twice, while the rest of the data are obtained only once. Such unbalanced redundancy has to be normalized by applying a weighting function $w(\beta, \gamma)$ to projections g prior to the convolution with h_g. With parallel-beam geometry or differentiated Hilbert transform approach, the weighting can be applied after filtering on a ray-basis or during the backprojection process on a voxel-basis.

Equation 8.25 is now generalized to

$$f(r,\phi)=\int_0^{2\pi}\frac{1}{D^2(r,\phi,\beta)}\int_{-\gamma_m}^{\gamma_m} w(\beta,\gamma)g(\beta,\gamma)h_g(\gamma'-\gamma)R\cos\gamma d\gamma d\beta \qquad (8.28)$$

Comparing Equation 8.25 with (8.28), the following observations can be made. When projections over 360° are used, all of the rays are measured exactly twice. Thus, $w(\beta, \gamma) = 1/2$ is applied to normalize the redundancy; and it is applied outside the integration because w is shift-invariant. Equation 8.25 is specifically called the fullscan.

The redundancy weighting function $w(\beta, \gamma)$ has to satisfy the following two conditions [3]: it must be continuous and smooth (twice differentiable) with respect to γ; and a sum of weights for the primary and conjugate rays is equal to 1, thus

$$w(\beta,\gamma)+w(\beta+\pi+2\gamma,-\gamma)=1 \qquad (8.29)$$

The $w(\beta, \gamma)$ is also preferred to be continuous and smooth with respect to β, but it is not necessary. For example, the following function satisfies the two conditions and is called the halfscan algorithm [3,4] (Figure 8.11). Sometimes, it is loosely called short scan or partial scan as well.

$$w(\beta,\gamma) = 3x^2(\beta,\gamma) - 2x^3(\beta,\gamma) \tag{8.30}$$

$$x(\beta,\gamma) = \begin{cases} \dfrac{\beta}{2(\gamma_m+\gamma)} & \left[0 \le \beta < 2(\gamma_m+\gamma)\right] \\ 1 & \left[2(\gamma_m+\gamma) \le \beta < \pi + 2\gamma\right] \\ \dfrac{\pi+2\gamma_m-\beta}{2(\gamma_m-\gamma)} & \left[\pi+2\gamma \le \beta \le \pi+2\gamma_m\right] \\ 0 & \left[\text{otherwise}\right] \end{cases} \tag{8.31}$$

Figure 8.12 shows fan-beam projections weighted by the halfscan weight and the filtered weighted projections. Note that projections near $\beta = 0°$ and $\pi + 2\gamma_m$ are lightly weighted and their contribution to image are decreased. Figure 8.13 shows the backprojection progresses of the halfscan algorithm. It can be seen that an image free from artifacts is reconstructed at $\Delta\beta = 225°$ (180° plus full fan angle).

This redundancy-based approach can be further generalized for projections acquired over multiple ($N > 1$) rotations as follows, which can be applied to various cases such as electrocardiogram-gated cardiac image reconstruction, compensation of detector defects, and image reconstruction of a large field of view with asymmetric detectors.

$$f(r,\phi) = \sum_{n=0}^{N-1}\int_0^{2\pi} \frac{1}{D^2(r,\phi,\beta)} \int_{-\gamma_m}^{\gamma_m} w(\beta+2\pi n,\gamma)g(\beta+2\pi n,\gamma)h_g(\gamma'-\gamma)R\cos\gamma\, d\gamma\, d\beta \tag{8.32}$$

$$\sum_{n=0}^{N-1}\left(w(\beta+2\pi n,\gamma) + w(\beta+(2n+1)\pi+2\gamma,-\gamma)\right) = 1 \tag{8.33}$$

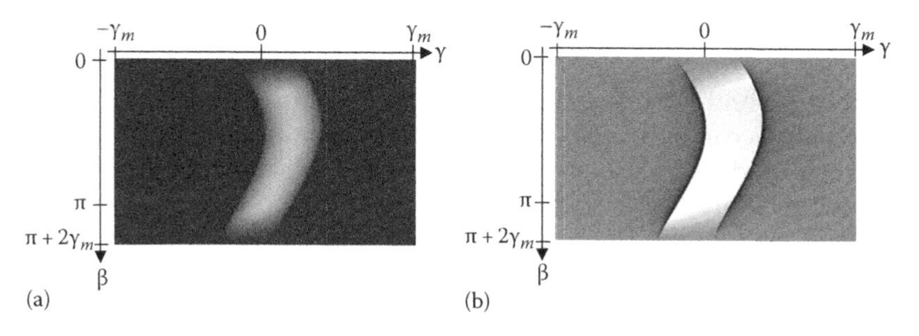

FIGURE 8.12
Weighted projections using the halfscan weight (a) and the weighted filtered projections using Shepp–Logan filter (b) of the circular phantom.

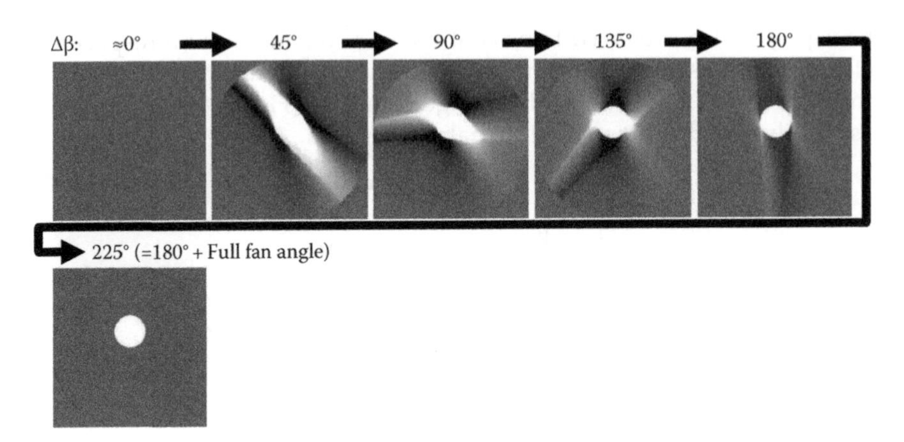

FIGURE 8.13
The process of an image being reconstructed by weighted filtered backprojection of fan-beam projections using the halfscan weight.

8.7 Cone-Beam Reconstruction

Diagnostic multidetector-row CT (MDCT) and flat-panel detector-based C-arm cone-beam CT (CBCT) acquire cone-beam projections, which are divergent with respect to z-axis (or the longitudinal direction of the object) as well as xy-axes. The angle of rays measured from the xy-plane is called the cone angle. When a number of detector rows in MDCT is small (e.g., <10 rows), such divergence is ignored and projections are treated as stacked fan-beam projections. This approach is similar to the single-slice rebinning approach used in PET and provides good image quality with no blurring along z-axis when the cone angle is small.

When the number of detector rows is large (thus the cone angle is large), however, image reconstruction algorithms have to take into account such divergence to minimize the cone angle artifacts we will discuss later. There are two methods most frequently used in CT scanners, which are extensions of the two FBP methods outlined in the fan-beam FBP section: Feldkamp algorithm [5] and cone-to-parallel fan-beam rebinning method [6–8]. The same methods can be employed with various scan modes such as an axial scan, a step-and-shoot scan, a helical or spiral scan, and a shuttle scan with appropriate redundancy weights.

The first method, Feldkamp algorithm or FDK algorithm, is an extension of the direct fan-beam FBP method and performs the following three steps which are very similar to the steps in the direct fan-beam FBP method. First, cone-beam projections $g(\beta, \gamma, \alpha)$ are weighted by $w(\beta, \gamma, \alpha)\cos\alpha\cos\gamma$, where α is the cone-angle and γ is the fan-angle, respectively, of the ray and w is the redundancy weight. Next, the weighted projections are convolved with a ramp filter kernel h_g, which is identical to the one designed for fan-beam, with respect to the fan angle parameter γ along the detector-row direction. Finally, the filtered projections are backprojected to voxels along the path the ray is acquired with the same weight used in fan-beam case, that is, an inverse square distance of the focal spot to the voxel-of-interest projected onto xy-plane (Figure 8.14).

$$f(r,\phi,z)=\int_0^{2\pi}\frac{1}{D^2(r,\phi,\beta)}\int_{-\gamma_m}^{\gamma_m} w(\beta,\gamma,\alpha)g(\beta,\gamma,\alpha)h_g(\gamma'-\gamma)R\cos\alpha\cos\gamma d\gamma d\beta \tag{8.34}$$

It can be seen that the major differences from direct fan-beam FBP method are the weighting by $\cos\alpha$, which compensates for increased path-lengths due to the cone-angle α and the cone-beam backprojection. Although this method had been considered empirical and inaccurate, later studies showed that Feldkamp algorithm is equivalent to the exact inverse 3-D Radon transform when employed with a flat-panel detector, an appropriate scan orbit, and redundancy weights [9,10]. A derivative of Feldkamp method is a hybrid filtering approach, which approximately replaces a ramp filtering by a combination of a ramp filtering *and* a differentiation step and Hilbert filtering, followed by cone-beam backprojection with an inverse distance weight [11].

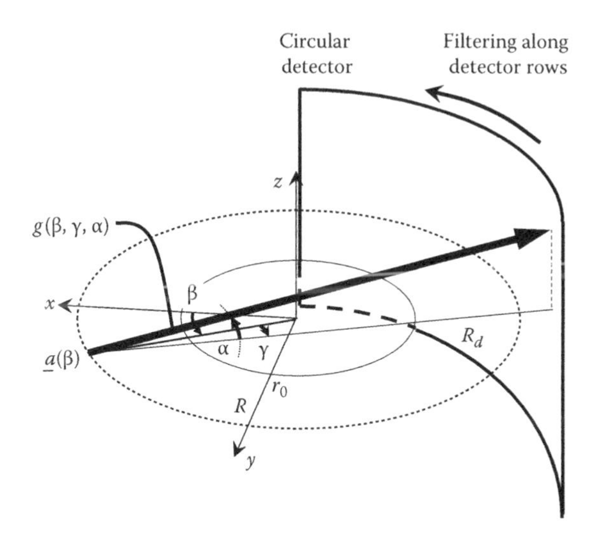

FIGURE 8.14
Cone-beam projections are filtered along detector rows and cone-beam backprojected along the path the x-ray beams are acquired.

The second method, cone-to-parallel fan-beam rebinning method, is an extension of the fan-to-parallel-beam rebinning method shown in Equations 8.14 and 8.15 and performs the following four steps. First, the fan-to-parallel-beam rebinning outlined previously is applied independently to each of cone-beam projections $g(\beta, \gamma, \alpha)$ acquired by the same detector row α as described in Equation 8.35 and Figure 8.15. This process is called cone-to-parallel fan-beam rebinning. Note that the projections from different detector rows are not mixed in this process; thus, the rebinned data $p(\theta, t, \alpha)$ maintain the divergence of beams along z-axis, while the projection of p onto the xy-plane are parallel-beams parameterized by (θ, t). Next, the rebinned data acquired by the same detector row α are convolved with the ramp filter kernel h_R, which is identical to the one designed for parallel-beam, with respect to the radial parameter t. Finally, the filtered data are backprojected to voxels along the path the ray is acquired while being weighted by $w(\theta, x, y, z)\cos\alpha$. Note that unlike direct fan- or cone-beam FBP methods, the redundancy weight optimized for each voxel can be applied during the backprojection process if desired. Because the shape of the rebinned data $p(\theta, t, \alpha)$ looks like a wedge, this algorithm is sometimes called Wedge algorithm.

$$p(\theta,t,\alpha)=p(\beta+\gamma,R\sin\gamma,\alpha)=g\left(\theta-\sin^{-1}\left(\frac{t}{R}\right),\sin^{-1}\left(\frac{t}{R}\right),\alpha\right) \quad (8.35)$$

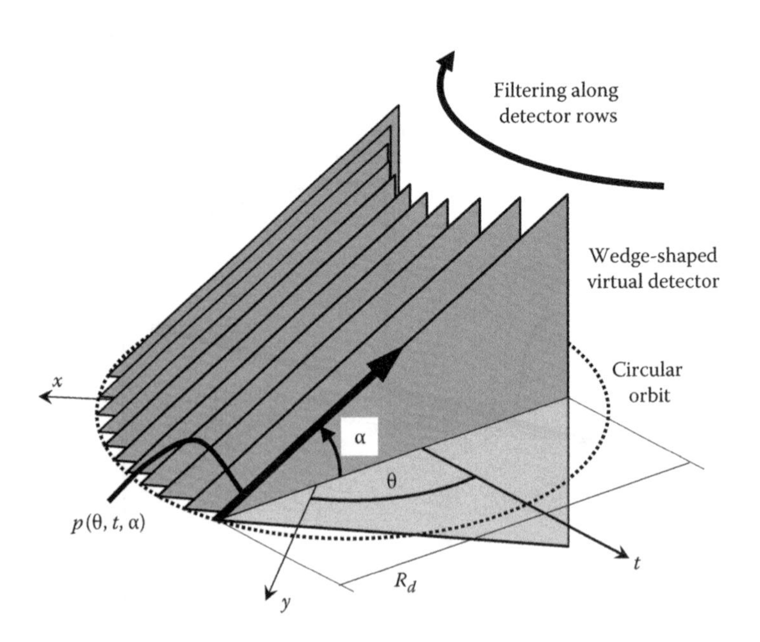

FIGURE 8.15
Cone-to-parallel fan-beam rebinning method. A part of cone-beam projections acquired at a series of focal spot positions are used to select projections that are parallel in the xy-plane, but divergent in z-axis. The rebinned projections are then filtered along the detector rows.

$$f(x,y,z)=\int_0^{\pi} w(\theta,x,y,z)\int_{-\infty}^{\infty} p(\theta,t,\alpha)h_R(t-t')\cos\alpha dt'd\theta \qquad (8.36)$$

The above two cone-beam image reconstruction methods significantly improve the image artifacts compared with the single-slice rebinning approach. The image quality is sufficient with the full cone angle up to ~5° to 10°. When the total cone angle is larger, however, there is a risk of shading artifacts and blurred edges near objects with high contrast and with a rapid change of shape in z-axis (see Figure 8.16). These are called cone-beam artifacts.

The cone-beam artifacts are generated because an axial scan along a circular orbit will not provide a complete 3-D Radon data necessary to reconstruct an exact 3-D image [5,12] (Figure 8.17). The halfscan weight designed to normalize the redundancy of 2-D Radon data may be applied for a better temporal resolution, for example. Such weights do not take into account the redundancy of 3-D Radon data correctly, and thus further degrades the image quality. A redundancy weight that balances the redundancy and the amount of the 3-D Radon data and the temporal resolution provides better image quality [12].

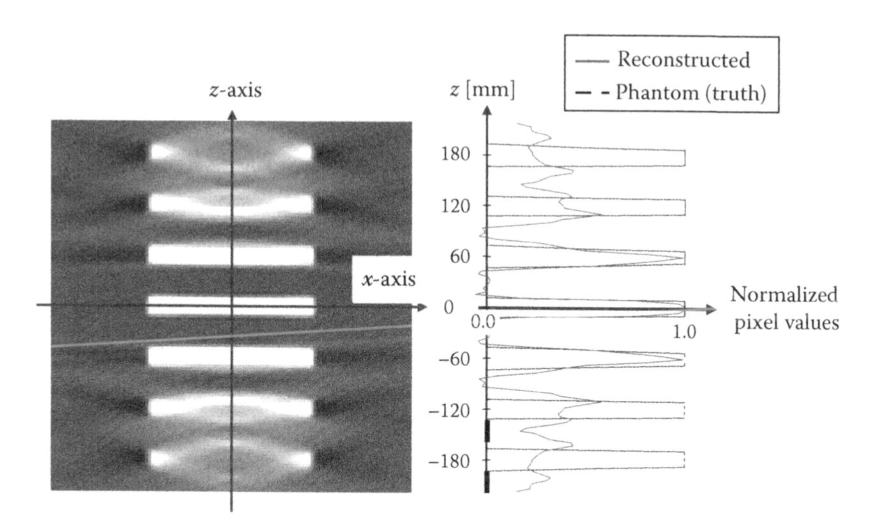

FIGURE 8.16
Cone-beam artifacts in a coronal view. A phantom with stacked disks was scanned by a circular scan at $z = 0$ mm, and the coronal image at $y = 0$ was reconstructed by Feldkamp algorithm. The focus-to-isocenter distance was 600 mm. Images near the scan plane was sharp and accurate, while images become blurry and inaccurate with shading artifacts away from the scan plane due to the effect of cone-angle.

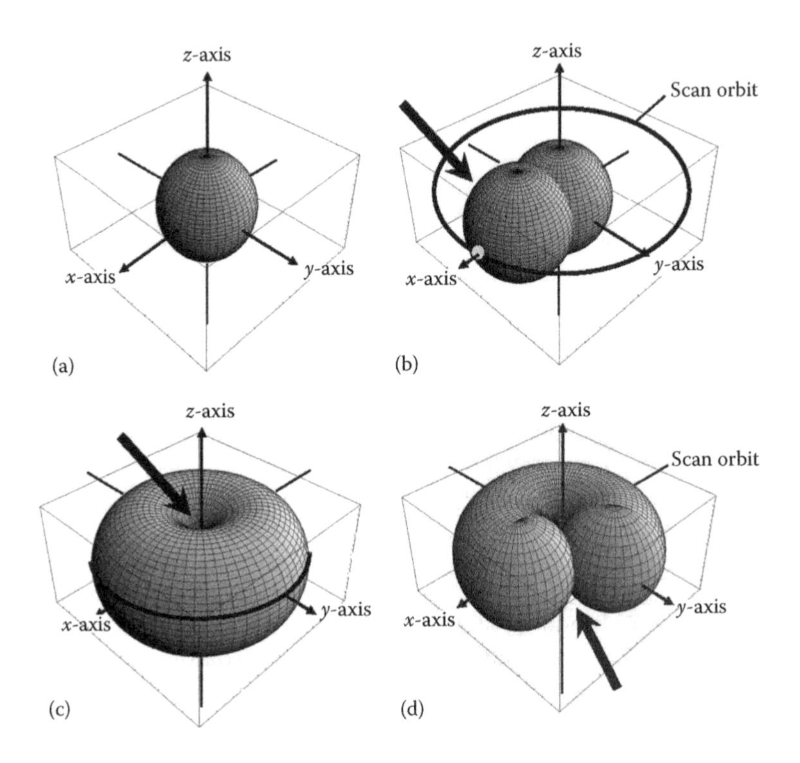

FIGURE 8.17
The necessary and obtained 3-D-Radon data (plane integrals) in 3-D-Radon space. (a) The necessary 3-D-Radon data for reconstructing an object within a ball whose radius is r_0. (b) The obtained 3-D-Radon data by a focus at β, can be described by the "surface" of the ball (arrow), whose diameter is defined by the locations of the focus and the origin. (c) Obtained 3-D-Radon data with one rotation are the surface and inside of the donut region. Comparing the obtained with the necessary 3-D Radon data (a), it can be seen that "the core of the apple" indicated by the arrow is not acquired. This is the reason of cone-beam shading artifacts away from $z = 0$ plane. (d) Fullscan Feldkamp algorithm uses the 3-D Radon data shown in (c) completely, while nonzero 3-D-Radon data used by with halfscan Feldkamp algorithm shows reduced utilization of 3-D Radon data, throwing away a part of the acquired 3-D Radon data (arrow). (From Taguchi, K., *Medical Physics*, 30, 640, 2003.)

8.8 Iterative Reconstruction Methods

8.8.1 Iterative Image Reconstruction Methods for SPECT and PET

Iterative image reconstruction (IR) methods have been investigated for many years in research community, and are now implemented and used on SPECT and PET scanners as frequently as FBP methods. The IR methods estimate the image of the object by minimizing a cost function that is defined based on a difference between the measured and the calculated projections incorporating various nonlinear physics factors such as the photon noise, the attenuation and scattering of emitted photons inside the object, and the geometrical efficiency and aperture of detectors and collimators [13]. We will outline the concept of the IR methods in this section.

When the projection process is linear, it can be described by a matrix H, which can then relate the object f and its cone-beam projection g as

$$g = Hf \tag{8.37}$$

The image reconstruction is then to obtain f from g. The FBP is an analytical, one-time method to apply the inverse of H, H^{-1}, to g. There are numerical, iterative approaches such as the algebraic reconstruction technique (ART) and simultaneous ART (SART) that use the transpose of H, H^T, and incrementally update the estimate of f to decrease the difference between the measured projection $\tilde{g}$ and the estimated projection $g = Hf^{(k)}$ from the current estimate $f^{(k)}$ at the k-th iteration. A calculation of g requires a forward projection operation $Hf^{(k)}$, while the update of the estimate is performed using the difference of the two projections as

$$f^{(k+1)} = f^{(k)} + \lambda w_H H^T(\tilde{g} - g) \tag{8.38}$$

where

H^T is the transpose of H, that is, a backprojection operation
w_H is a weighting coefficient for the contribution of voxels to a ray
λ is a parameter to control the speed of convergence

Note that one forward- and one backprojection operation is performed for one iteration process, which are required to accurately model the relation of f to g.

When the projection process is *nonlinear* due to the factors listed earlier, the relationship can be described as

$$g \sim \text{Poisson}(H_{DA}f + g_S) \tag{8.39}$$

where

H_{DA} is a modified system matrix incorporating the geometrical efficiency and detector apertures *and* the attenuation
g_S is measurements due to scattered photons

The IR methods try to minimize the cost function, for example,

$$\hat{f} = \text{argmin}_f\, LL(\tilde{g}|f) + \rho\mathcal{R}(f) \tag{8.40}$$

where

$LL(\tilde{g}|f)$ is a negative log-likelihood of $\tilde{g}$ given f
$\mathcal{R}(f)$ is a roughness penalty or *a priori* knowledge of f (e.g., log of a Gibbs prior distribution)
ρ is a parameter that balances the effect of the two terms

The update equation is more complex than ART; however, it uses one forward- and one backprojection operation per iteration.

Considering that FBP only uses one backprojection operation to reconstruct an image, performing N_{IR} iterations of IR methods in general is roughly $(2N_{IR} - 1)$ times as computationally expensive as FBP. In return, IR methods often provide more accurate and less noisy images than FBP does, thanks to accurate modeling of the nonlinear factors mentioned earlier.

8.8.2 Iterative Methods for CT

Most of iterative methods lately implemented to x-ray CT scanners seem to be different from the IR methods used in SPECT and PET. Although details are not published, some of the iterative methods (IM) may merely be an iterative image processing method, but loosely called an iterative image reconstruction method for better marketing purposes.

The current CT scanners never employ FBP alone. They perform various types of adaptive (sometimes iterative) methods to correct and process projection data prior to FBP, employ FBP, followed by various adaptive (sometimes iterative) methods on the image data to enhance the image quality. The most important nonlinear effect in x-ray CT is beam hardening effects (BHE). The BHE due to soft tissues is corrected during the adaptive pre-FBP correction process, while the BHE due to bones or contrast agents is corrected in an iterative fashion with one or a few iterations, using forward- and backprojections [14]. The adaptive post-FBP processes include an adaptive edge preserving noise reduction filtering.

The IM replaces a part of the adaptive pre-FBP processes on projections by *iterative methods* using statistical properties of projections *and* a part of the adaptive post-FBP process on image data by *iterative methods* using *a priori* knowledge and the statistical properties of the image. The iterative methods seem more aggressive than the current adaptive methods designed for similar purposes.

In contrast, true IR methods for CT may consist of a part of the adaptive pre-FBP processes, iterative image reconstruction using forward- and backprojection to iteratively estimate the image [15,16], and a part of the adaptive post-FBP processes. It is not clear at this moment which algorithms on CT scanners are IM and which are IR.

As it can be seen, both the IM and the current adaptive schemes are similar, sandwiching FBP by pre- and post-processes. If the IM models the forward imaging process, various nonlinear effects, and properties of images better than the adaptive pre- and post-FBP processes, the IM methods can provide images superior to the current CT images. The IM also has a risk of creating unrealistic images if inappropriate priors are used. The computational costs of IM and FBP may be comparable, because the most expensive step may remain the FBP process. IR methods can outperform either of the two methods, if nonlinear factors that can only be modeled in forward- and backprojection process are significant.

8.9 Other Class of Algorithms

Recently, other novel image reconstruction methods have developed. A Hilbert transform-based FBP (super short scan) method allows for reconstructing a part of the image from fan-beam projections over less than 180° [17,18]. When a detector is not large enough to cover the entire object, the *trans*-axial truncation caused strong biases and artifacts with the standard FBP method. But employing differentiated backprojection followed by inverse Hilbert transform on the image (DBP, BPF, or DBPF) accurately reconstructs an imaged region from such truncated projections.

Compressive sensing allows for exact reconstruction even from an extremely fewer number of projections that does not satisfy Nyquist–Shannon sampling theorem, when the object pixel values are piecewise constant [19–21]. The application of this method to CB-CT may be more appropriate than to diagnostic CT, because CB-CT always uses pulsed x-rays; thus, it can intentionally decrease the number of projections for some merits and use the compressive sensing technique to overcome the sampling problem. In contrast, x-rays are always on during the scan in diagnostic x-ray CT systems and it is difficult to pulse x-rays, because the time duration per projection is as short as 100–300 μs.

8.10 Artifacts (*z*-Aliasing, Cone-Angle, Scatter, Halfscan)

We outline a few artifacts that are unique to x-ray CT and related to image reconstruction.

8.10.1 Aliasing Artifacts or "Windmill" Artifacts

Windmill artifacts are black-and-white spokes (Figure 8.18) centering at high-contrast structures, which change the shape rapidly in *z*-axis such as bones at the base of skull, air pockets in intestines. The windmill artifacts rotate when we page through images along *z*-axis, which is why they are called windmill artifacts.

The windmill artifacts are caused by the aliasing along *z*-axis—the Nyquist frequency defined by a sampling pitch along *z*-axis (which is often defined by the interval of detector rows in MDCT) does not satisfy the maximum frequency component of the structure-of-interest [22,23]. It has been shown that extending the polar artifacts to the scan orbit, they cross at projection angles where detector rows that correspond to the image voxel-of-interest switch over [22,23].

There are a few methods to mitigate the problem. First one is to utilize a helical pitch that interlaces the sampling positions of primary and conjugate

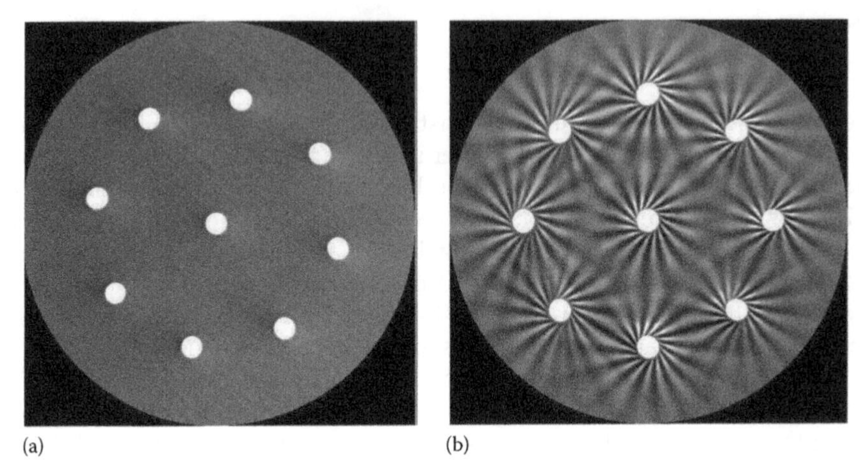

FIGURE 8.18
Cone-beam artifacts (a) and windmill artifacts (b). The phantom consists of nine balls with a diameter of 20 mm. A helical scan was performed by a MDCT with a beam width of 32 mm using a helical pitch of 1.0. Images at the edge of the ball with a thickness of 2.5 mm were reconstructed. The only difference between the two images was the configuration of the detector rows: 160 rows × 0.2 mm/row (a) and 16 rows × 2.0 mm/row (b). (From Silver, M. D. et al., *SPIE Medical Imaging: Image Processing*, San Diego, CA, 5032, 1918, 2003.)

rays in z-axis, thus decreasing the sampling pitch. The concept is similar to quarter offset detector alignment in xy-plane and can be applicable to both fan- [22,24] and cone-beam backprojection [25]. The second one is to employ cone-beam backprojection from a virtual focal spot position that is slightly offset from the actual position. This simple scheme suppresses the aliased frequency components while minimizing the loss of the original frequency components [26]. The third one is to alternate between two z-positions of the x-ray focal spot, acquiring two samples per detector row. This scheme called z flying focal spot or zFFS doubles the number of samples in the z-axis and decreases the sampling interval to the half at the iso-center [27].

8.10.2 Cone-Beam Artifacts

Cone-beam artifacts are broad artifacts in either shading or brightening spread on both sides of structures with a rapid change of shape along z-axis (see Figure 8.16).

There are two methods to mitigate the problem. First one is to use a scan mode (more specifically, a trajectory of the x-ray focal spot) that acquires more 3-D Radon data. For example, choose a helical scan over an axial scan when possible. The second one is to perform an image reconstruction algorithm and a redundancy weight that use more 3-D Radon data and are mathematically exact (or better approximate the exact solutions). An example of the redundancy weights will be discussed in the next section.

8.10.3 Halfscan Artifacts

The following phenomena are called halfscan artifacts which have been discovered relatively recently [12,28]: a larger degree of shading and brightening and a shift (or bias) of overall pixel values compared with the fullscan case, and changes of the these problems as a function of the projection angle that corresponds to the center of the halfscan range. Investigations on the artifacts are likely to continue.

Several causes may result in the artifacts; however, the main cause is considered to be a mismatch between the 3-D path of the primary ray and that of the conjugate ray, both of which cross a voxel-of-interest from the opposite direction (see Figure 8.19). The halfscan weight, derived to normalize the redundancy of 2-D Radon data, does not normalize the redundancy of 3-D Radon data. Therefore, when halfscan weight is applied to cone-beam projections, it eliminates a part of 3-D Radon data acquired by a circular, axial scan (which already is not sufficient to reconstruct exact images) (see Figure 8.17 and Reference [12] for more details). The mismatch of paths also results in discrepancies between the two rays in terms of the scattered radiation and beam hardening effects. This is the reason why image artifacts are stronger in the halfscan case than in the fullscan case and why the strengths and ranges of artifacts depend on the halfscan angle.

The other causes of halfscan artifacts include the mechanical vibration of the gantry, the effect of object motion, and the cross-scatter. The strength and direction of motion artifacts with deforming objects, such as heart, depend on the halfscan angle [29]. In dual-source CT systems, where two sets (A and B) of an x-ray tube and a detector are aligned almost perpendicularly, x-ray beams generated by one x-ray tube (A) may be scattered by the object and be detected by the other detector (B) and vice versa. The cross-scatter may leave a bias and noise even after a correction, both of which may change with the halfscan angle.

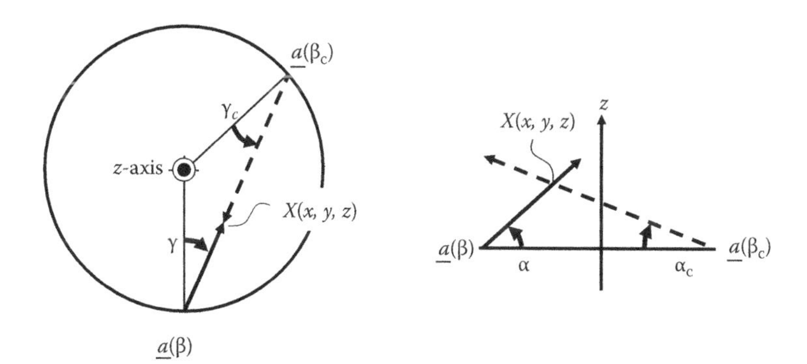

FIGURE 8.19
The paths of a primary ray and the corresponding conjugate ray do not match in cone-beam projections.

References

1. L. A. Shepp and B. F. Logan, The Fourier reconstruction of a head section, *IEEE Transactions on Nuclear Science*, NS-21, 21–43, 1974.
2. A. C. Kak and M. Slaney, *Principles of Computerized Tomographic Imaging*. New York: IEEE Press, 1987.
3. D. L. Parker, Optimal short scan convolution reconstruction for fanbeam CT, *Medical Physics*, 9, 254–257, 1982.
4. C. R. Crawford and K. F. King, Computed tomography scanning with simultaneous patient translation, *Medical Physics*, 17, 967–982, 1990.
5. L. A. Feldkamp, L. C. Davis, and J. W. Kress, Practical cone-beam algorithm, *Journal of the Optical Society of America A*, 1, 612–619, June 1984.
6. M. Grass, T. Kohler, and R. Proksa, 3D cone-beam CT reconstruction for circular trajectories, *Physics in Medicine and Biology*, 45, 329–347, 2000.
7. M. Grass, T. Kohler, and R. Proksa, Angular weighted hybrid cone-beam CT reconstruction for circular trajectories, *Physics in Medicine and Biology*, 46, 1595–610, 2001.
8. H. K. Tuy, Three-dimensional image reconstruction for helical partial cone-beam scanners, Presented at the *5th International Conference on Fully Three-Dimensional Reconstruction in Radiology and Nuclear Medicine*, Egmond Ann Zee, the Netherlands, 1999.
9. H. Kudo and T. Saito, Derivation and implementation of a cone-beam reconstruction algorithm for nonplanar orbits, *IEEE Transactions on Medical Imaging*, 13, 196–211, 1994.
10. M. Defrise and R. Clack, A cone-beam reconstruction algorithm using shift-variant filtering and cone-beam backprojection, *IEEE Transactions on Medical Imaging*, 13, 186–195, 1994.
11. A. A. Zamyatin, K. Taguchi, and M. D. Silver, Practical hybrid convolution algorithm for helical CT reconstruction, *IEEE Transactions on Nuclear Science*, 53, 167–174, 2006.
12. K. Taguchi, Temporal resolution and the evaluation of candidate algorithms for four-dimensional CT, *Medical Physics*, 30, 640–50, 2003.
13. K. Lange and R. Carson, EM reconstruction algorithms for emission and transmission tomography, *Journal of Computer Assisted Tomography*, 8, 306–316, 1984.
14. J. M. Meagher, C. D. Mote, Jr., and H. B. Skinner, CT image correction for beam hardening using simulated projection data, *IEEE Transactions on Nuclear Science*, 37, 1520–1524, 1990.
15. I. A. Elbakri and J. A. Fessler, Statistical image reconstruction for polyenergetic x-ray computed tomography, *IEEE Transactions on Medical Imaging*, 21, 89–99, 2002.
16. Y. Zhou, J.-B. Thibault, C. A. Bouman, K. D. Sauer, and J. Hsieh, Fast model-based x-ray CT reconstruction using spatially nonhomogeneous ICD optimization, *IEEE Transactions on Image Processing*, 20, 161–175, 2011.
17. F. Noo, M. Defrise, R. Clackdoyle, and H. Kudo, Image reconstruction from fan-beam projections on less than a short scan, *Physics in Medicine and Biology*, 47, 2525–2546, 2002.

18. H. Kudo, T. Rodet, F. Noo, and M. Defrise, Exact and approximate algorithms for helical cone-beam CT, *Physics in Medicine and Biology*, 49, 2913–2931, 2004.
19. E. J. Candes, J. Romberg, and T. Tao, Robust uncertainty principles: exact signal reconstruction from highly incomplete frequency information, *IEEE Transactions on Information Theory*, 52, 489–509, 2006.
20. D. L. Donoho, Compressed sensing, *IEEE Transactions on Information Theory*, 52, 1289–1306, 2006.
21. E. Y. Sidky, C.-M. Kao, and X. Pan, Accurate image reconstruction from few-views and limited-angle data in divergent-beam CT, *Journal of X-Ray Science and Technology*, 14, 119–139, 2006.
22. K. Taguchi, H. Aradate, Y. Saito, I. Zmora, K. S. Han, and M. D. Silver, The cause of the artifact in 4-slice helical computed tomography, *Medical Physics*, 31, 2033–2037, 2004.
23. M. D. Silver, K. Taguchi, I. A. Hein, B. S. Chiang, M. Kazama, and I. Mori, Windmill artifact in multi-slice helical CT, in *SPIE Medical Imaging: Image Processing*, San Diego, CA, 5032, 1918–1927, 2003.
24. K. Taguchi and H. Aradate, Algorithm for image reconstruction in multi-slice helical CT, *Medical Physics*, 25, 550–561, 1998.
25. J. Hsieh, X. Tang, J.-B. Thibault, C. Shaughnessy, R. A. Nilsen, and E. Williams, Conjugate cone-beam reconstruction algorithm, *Optical Engineering*, 46, 067001–067010, 2007.
26. I. Mori, Antialiasing backprojection for helical MDCT, *Medical Physics*, 35, 1065–1077, 2008.
27. T. G. Flohr, K. Stierstorfer, S. Ulzheimer, H. Bruder, A. N. Primak, and C. H. McCollough, Image reconstruction and image quality evaluation for a 64-slice CT scanner with z-flying focal spot, *Medical Physics*, 32, 2536–2547, 2005.
28. A. N. Primak, Y. Dong, O. P. Dzyubak, S. M. Jorgensen, C. H. McCollough, and E. L. Ritman, A technical solution to avoid partial scan artifacts in cardiac MDCT, *Medical Physics*, 34, 4726–4737, 2007.
29. K. Taguchi and A. Khaled, Artifacts in cardiac computed tomographic images, *Journal of the American College of Radiology*, 6, 590–593, 2009.

9

Portable High-Frequency Ultrasound Imaging System Design and Hardware Considerations

Insoo Kim, Hyunsoo Kim, Flavio Griggio, Richard L. Tutwiler, Thomas N. Jackson, Susan Trolier-McKinstry, and Kyusun Choi

CONTENTS

9.1 Introduction

Ultrasound techniques have wide use in various applications in numerous fields. The frequency range from 10 KHz to 1 MHz is widely used for SONAR (sound navigation and ranging), ultrasonic welding, therapeutic ultrasound, and humidifiers. A frequency range of 1–50 MHz is common in diagnostic sonography and nondestructive testing to find flaws in materials. Micron-sized silicon surface detection utilizes a frequency range of 50–200 MHz [1], and SAW (surface acoustic wave) devices use frequencies ranging from 1 to 10 GHz.

Ultrasound techniques are also common for diagnostic imaging and often supersede x-ray imaging in the medical sector [2]. The vast majority of medical ultrasound imaging occurs at frequencies between 1 and 50 MHz. For example, diagnostic imaging designed to penetrate tissues to a depth of 5–20 cm uses frequencies between 1 and 10 MHz. Recently, studies for imaging smaller organs or surfaces, such as skin, the gastrointestinal tract, and intravascular blood vessels, have emerged utilizing ultrasound with frequencies over 20 MHz. However, a need remains for better clinical ultrasound imaging for detecting skin, eye, and prostate cancers as well as many other diseases, in vivo. Imaging techniques with a resolution below 100 μm for these situations would minimize the need for biopsies [2].

Moreover, the medical community has recently expressed a desire for an ultrasound imaging system that would not only provide appropriate resolution but also would be portable [3]. For example, veterinarians would be well served by development of a portable ultrasound imaging system for onsite diagnosis of pets, zoo, and farm animals. The conventional system's size is due to the complex front-end electronics that consist of a discrete chip sets for pulsers, preamplifiers, TGCs (time gain controls), A/D (analog/digital) converters, and memory devices. In addition, the conventional design of the transducer arrays requires high-transmit-drive voltages on the order of 100 V.

During the past 20 years or so, many researchers have tried to produce miniaturized and integrated ultrasound imaging systems [4–10]. It is because integrating all electronic components into a single IC chip affords a smaller system size, higher speed, and lower power consumption than building the system with discrete chipsets. The early ultrasound ASIC chips are reported by Black et al. and Hatfiled et al. in 1994. The ASIC chips contained 16 channels of transmitters and current amplifiers with integrated transducers [4] and digital transmitters [5]. Since 2000, with the rapid development of mixed-signal IC design technology, closed-coupled ultrasound front-end electronics have emerged for high-frequency ultrasound imaging systems. Wygrant et al. developed a CMUT (capacitive micromachined ultrasound transducer) with closed-coupled electronics [6], which contain 16 receive-transmit channels. Johansson et al. introduced a portable ultrasound system using a battery-operated voltage-boosting scheme [7]. The Sonic Window, developed by Fuller et al. in 2005, included one of the most integrated ultrasound front-end IC chip concepts to date [8]. Developed for guiding needle and catheter insertion, biopsies, and other invasive procedures for which only a basic aid to diagnosis is necessary, the sonic window can also be used for C-Mode ultrasound imaging.

Kim et al. introduced a high-frequency system with a fully integrated, custom-designed ultrasound front-end IC chip, which includes both transmit and receive electronics with A/D converters and high-capacity on-chip memory [9]. The design has integrated electronics with thin-film transducers small enough to construct a portable, ultra-compact, low-power consumption ultrasound imaging system. Unlike other works mentioned earlier, which still require a significantly higher drive voltage for excitation of transducers, the transceiver chip interfaces with thin-film transducer arrays that operate below 5 V, so that the limitations of high-voltage excitation become moot.

This chapter focuses on introducing the architecture and design of a fully integrated ultrasound transceiver chip. Section 9.2 outlines the key points of ultrasound imaging fundamentals: ultrasound physics, the basics of B-mode ultrasound imaging, and beamforming architectures. Section 9.3 investigates the design consideration for ultrasound transceiver chip in portable high-frequency ultrasound imaging systems. Section 9.4 presents the Penn State portable ultrasound imaging system. The design specifications and details of the circuit components on the transceiver chip are described. In addition, the characteristics of the transceiver chip and measured signal acquisition results from the Penn State ultrasound imaging system along with the thin-film ultrasound transducer arrays are shown. Last, the summary and the conclusions are addressed in Section 9.5.

9.2 B-Mode Ultrasound Imaging System

9.2.1 Ultrasound Imaging Basics

An ultrasound wave is a longitudinal wave in which oscillations are in the same direction as propagation. In addition, the ultrasound wave is attenuated as it propagates through the medium. Several factors contribute to this attenuation.

One of the most significant factors is the absorption of ultrasound energy by the medium and its conversion into heat. The ultrasound wave loses its acoustic energy continuously as it moves through the medium. Scattering and refraction also result in some loss of energy and contribute to overall attenuation.

Reflection is an important physical phenomenon of an ultrasound wave, and it is also a key characteristic used for ultrasound imaging. A reflection occurs at any boundary between two media having different densities and/or acoustic velocities. When an ultrasound wave encounters the boundary between two media, only a portion of the wave's acoustic energy will be transmitted, and the rest of the acoustic energy will be reflected. B-Mode ultrasound imaging is based on the pulse-echo (backscattering) response of an ultrasound wave, and provides a two-dimensional, cross-sectional reflection image of the scanned object [10]. The amplitude of reflection signal is converted to the brightness information of the target object. Higher amplitude creates a brighter image, and weaker amplitude creates a dark image.

In the meantime, the ultrasound wave is attenuated as it propagates through the medium. Several factors contribute to this attenuation. One of the most significant factors is the absorption of ultrasound energy by the medium and its conversion into heat. The ultrasound wave loses its acoustic energy continuously as it moves through the medium. Scattering and refraction also result in some loss of energy and contribute to overall attenuation. A simple exponential loss of pressure amplitude expresses the attenuation of an ultrasound wave [11]:

$$p(z) = p_0 e^{-\alpha(f)\times z} \tag{9.1}$$

where

p_0 is the initial pressure amplitude

$\alpha(f)$ is the attenuation coefficient that is a function of frequency

This implies that the reflection signal from the closer medium from the observation point is stronger than the reflection signals from the far medium even if the media are the same. Therefore, time-gain compensator (TGC) is necessary in the ultrasound front-end electronics to compensate for signal attenuation as a function of depth.

Resolution is one of the most important properties to consider in designing an ultrasound imaging system. Resolution involves two factors in a B-scan: (1) resolution in the direction of the transducer motion, known as "lateral" or "transverse" resolution, and (2) resolution in the direction of acoustic pulse propagation, known as "axial" resolution.

Lateral resolution of an ultrasound imaging system (resolution in the direction of transducer motion) is the system's ability to discriminate between two closely adjacent structures placed at the same depth from the transducer surface. The ultrasound's beam width at a specific depth determines lateral resolution. The beam width varies as the wave propagates in and out of the focal region; therefore, the focusing property of the transducer is

$$L.R. = f^{\#}\lambda = \frac{f}{a}\lambda \tag{9.2}$$

where
$f^{\#}$ is the f-number
a is the aperture of the transducer
f is a focal length
λ is a wavelength

Practically, lateral resolution is proportional to 2λ [12].

Axial resolution of an ultrasound imaging system (resolution in the direction of ultrasound wave propagation) is the ability to discriminate between two closely placed structures lying along the length of the ultrasound wave. The important factor in determining axial resolution is the spatial pulse length (λ) and the numbers of pulses (N)

$$A.R. = \frac{N\lambda}{2} \tag{9.3}$$

From both (9.2) and (9.3), notably, the wavelength, and consequently the frequency, directly determines the resolution of an ultrasound system. In general, a higher-frequency ultrasound wave is more desirable for higher resolution. However, arbitrarily increasing the ultrasound wave frequency to obtain finer resolution is not desirable because the attenuation rate of ultrasound waves increase as the frequency increases. Therefore, in determining frequency, a necessary consideration is the trade-off between the resolution and the penetration distance of the ultrasound wave. Having set the frequency of an ultrasound wave, the specification for the front-end electronics, such as input buffer bandwidth and A/D conversion speed, can be determined.

9.2.2 B-Mode Imaging System Hardware

A block diagram of a general B-mode ultrasound imaging system appears in Figure 9.1. The following subsections generically describe the processing blocks.

9.2.2.1 Control Host

This grouping includes microprocessors or a host computer and post-processors. The computer or microprocessors control the entire hardware system to function in the desired modes and to provide a control interface to the front-end electronics. The post-processor performs scan conversions (i.e., imaging formation), image processing, and display.

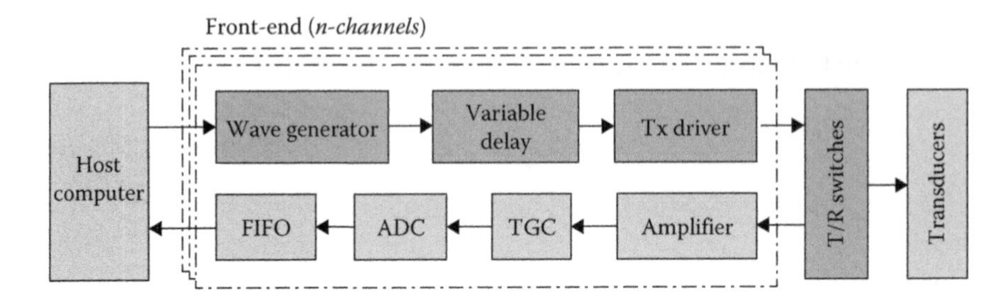

FIGURE 9.1
A block diagram of a general B-mode ultrasound imaging system.

9.2.2.2 T/R Switch

Generally, the transmit pulses use a very high voltage, typically up to 200 V, while the receiver electronics process lower voltages signals in the 10^{-3} volt range. Modern CMOS technology uses a power supply of below 5 V. Therefore, the receiver should be isolated from transmit pulses in order to protect inner circuits. The T/R switch connects transducers to the transmitter during transmit mode operation; conversely, the switch connects the transducer to the receiver during receive mode operation.

9.2.2.3 Amplifier

The amplifier is a key component in the ultrasound imaging system. It performs two functions: First, it receives the reflected signals from the transducers. This means that the dynamic range of the amplifier is crucial because attenuation of an ultrasound wave is sometimes over 100 dB. Impedance matching between the amplifier and the transducers is also important for reception. A low-noise amplifier (LNA), which has both high dynamic range and good impedance-matching properties, is preferred for the preamplifier. Second, the preamplifier enhances the received signals, but careful preamplifier gain selection avoids amplified signal saturation.

9.2.2.4 TGC

The TGC provides time-varying gain for the reflected ultrasound signal whose attenuation varies as a function of depth and the attenuation coefficient of the medium. From Equation 9.1, ultrasound waves attenuate on a logarithmic scale rather than a linear scale. This means that the TGC should express a variable linear gain range in the dB scale. The TGC can be a variable gain amplifier (VGA).

9.2.2.5 A/D Converter and Memory (FIFO)

As stated earlier, every receive channel needs one or more A/D converters for DBF. The conversion rates of three to five times the highest center frequency are necessary to reduce beamforming quantization errors [13]. As conversion speed increases, memory devices may be needed to store the digitized data and to interface the A/D converter with the receive beamformer.

9.2.3 Beamforming

Modern ultrasound imaging systems often use multichannel transducer arrays to increase beam flexibility, spatial converge, and resolution. Ultrasound pulses from each channel should have a delay in order to form a wave front that converges on a specified focal point. The transmit beamformer generates delays for each channel to focus and steer the transmit beam, and the receive beamformer performs focusing and steering of the scattered RF signals to create the B-mode images. In certain systems, the transmitter consists of delay networks. A single cycle pulse excitation signal is ideal for B-mode imaging since it yields better axial resolution. Typical systems have the flexibility to generate multiple-gated bursts of sinusoidal excitation as well as coded excitation. The delay network focuses the amount of transmitted energy into the medium and has capability for pulse and pulsed-wave Doppler transmit modes. The receive beamformer also generates delays for each channel in inverse order of transmit delays to align the amplified signals at the reference time. Then, the post processor adds the delay compensated signals and generates a large imaging signal for further image processing.

9.2.3.1 Analog Beamforming versus Digital Beamforming

Beamforming architectures are of two types: analog beamforming (ABF) and digital beamforming (DBF). ABF for ultrasound waves first appeared in the 1960s. Researchers developed DBF in the 1980s; however, not until the 1990s did DBF became feasible, because that period made available fast, high-precision A/D converters necessary for these systems. The recent development of VLSI techniques enables designing real-time digital receive beamformers, and allows amplified signals from receive channels to be coherently added after digitization [14].

The main difference between ABF and DBF, as compared in Figure 9.2, is the method of achieving beamforming. In ABF, analog delay lines for each channel delay transmit pulses, then the beam is formed. In receive mode, the amplified and delayed, reflected analog signal compensates for the transmit delays, which are subsequently accumulated to construct a large analog imaging signal. Then, an ADC digitizes the analog signal for further image processing.

Unlike the signals in an ABF system, the signals in a DBF system are sampled as close to the transducer elements as possible in receive mode, and then delayed and summed digitally. Thus, a DBF system needs an A/D converter

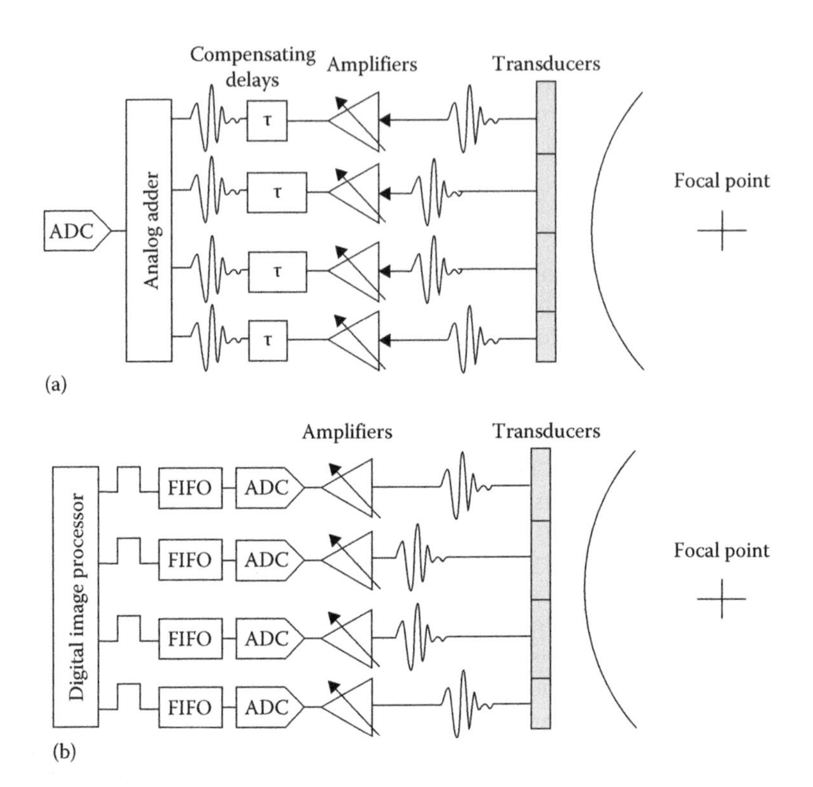

FIGURE 9.2
Simple block diagram of a typical DBF system: (a) receive analog beamforming system, (b) receive digital beamforming system.

for each channel. Since modern DBF systems use multichannels and arrays of transducers, DBF requires a large number of A/D converters. This creates a considerable disadvantage for DBF systems since ADCs consume significant power.

However, DBF systems also have considerable advantages over ABF systems. First, DBF has better control over time delay quantization errors. Analog delay lines tend to be poorly matched between channels. In DBF, synchronization with a high-frequency clock source can greatly improve delay accuracy. Second, DBF provides a finer resolution of ultrasound images. Typical analog delay accuracy is on the order of 20 ns, which constrains lateral resolution [12], but digital delay accuracy in modern digital circuit technology is on the order of a few hundred picoseconds with a few giga-hertz clock sources and PLLs (phase locked loop) [15]. Last, since the digitized data is much less susceptible to noise than analog signal, DBF systems, in stark contrast to ABF systems, can deliver clearer display images than ABF systems, which may have analog noise throughout its entire system.

9.3 Challenges to Portable High-Frequency Ultrasound Imaging System

9.3.1 Low-Voltage High-Frequency Transducer

The majority of existing transducers for medical imaging still needs a significantly higher drive voltage for excitation (above 60 V). High-voltage excitation pulses result in more complex system designs requiring protection T/R (Transmit/Receive) switches, digital controls, and charge-pump circuitry for the transmitter. Thus, the high voltage system is apt to consume high power, and consequently, is not suitable for portable systems. In addition, the existing systems interface the transducer array elements with the RF front-end by coaxial cable networks. The cables are specifically 50 ohm impedance matched to the RF front-end interface. The existence of T/R switches between the cable network and the front-end electronics makes impedance matching even more complex.

Therefore, low-voltage high-frequency ultrasound transducer arrays are crucial in portable high-frequency ultrasound imaging systems. The low-voltage transducers make the imaging system capable of integrating front-end electronics with transducers without charge-pump circuitry and RF coaxial cables. In addition, receiver protection devices are likely to be unnecessary because the transmit voltage is of the same magnitude as the CMOS logic voltage level. The integrated electronics also produces better signal integrity and noise immunity than conventional, analog, front-end electronics, which consist of discrete chipsets. Digital signal interfacing has wide acceptance for much higher signal-to-noise ratio (SNR) compliance than analog signal interfacing. In conventional systems, substantial efforts are necessary to control SNR in analog signal interfacing. However, in the integrated system, only chip-to-chip interfacing is necessary via digital signals because all analog signal processing occurs inside the chip and chips produce only digital outputs.

Researches on developing low-voltage high-frequency ultrasound transducers are in demand. First, the high-frequency PZT thin-film ultrasound transducer arrays using MEMS technology are proposed in [16]. The PZT layers can be thin, and allow reduction of the required voltages for exciting the transducer. The center frequency of the thin-film transducer is 30–70 MHz with a bandwidth of 60%–100%. Because the piezoelectric layer is a thin film (<1 μm thickness), the transducer array utilizes CMOS-compatible, low-level (below 5 V) excitation voltages. This technology enables the thin-film ultrasound transducers to be placed in close proximity to the electronics.

In addition, 20-element high-frequency ultrasound transducer array using micromolding technique is introduced in [17]. The array used piezocomposite material and thin-film Cr–Au electrodes in a mask-based process, packaged in epoxy with external connectors. The center frequency of the transducer is about 35 MHz with a bandwidth of 74% at 6 dB points.

9.3.2 Hardware Specifications

Achieving the required design specification for a high-frequency ultrasound imaging system is challenging. The linear dynamic range for an analog amplifier has a limitation of 100 dB in practical systems. However, the required dynamic range of the preamplifiers is sometimes higher than 100 dB for an ultrasound imaging system [18]. An even more critical obstacle is the requirements for the A/D converter. While the earliest commercially available digital beamformers became available in the early 1980s, they did not begin to have a significant impact until the early 1990s. Much of this delay was due to the need for A/D converters with sufficiently large numbers of bits and sufficiently high sampling rates.

Figure 9.3 shows the ADC development trends presented in the recent publications of two major international conferences: the ISSCC (International Solid State Circuits Conference) and the VLSI (Symposia on VLSI Technology and Circuits) [19]. The dynamic range and the bandwidth of the A/D converters show an inverse-proportional relationship, and the majority of the recent studies on A/D converters focused on mid-resolution (50–70 dB, i.e., 8–12 bit) and mid-frequency (10–100 MHz) ranges. The requirements for high-resolution (~10 μm minimum feature size), high-frequency (30–150 MHz center frequency of the transducers) ultrasound imaging systems are 50–70 dB dynamic range and 75–400 MHz bandwidth (gray box in the figure). As seen in the figure, a state-of-the-art A/D converter is a fundamental requirement for success of the development of portable high-frequency ultrasound imaging systems.

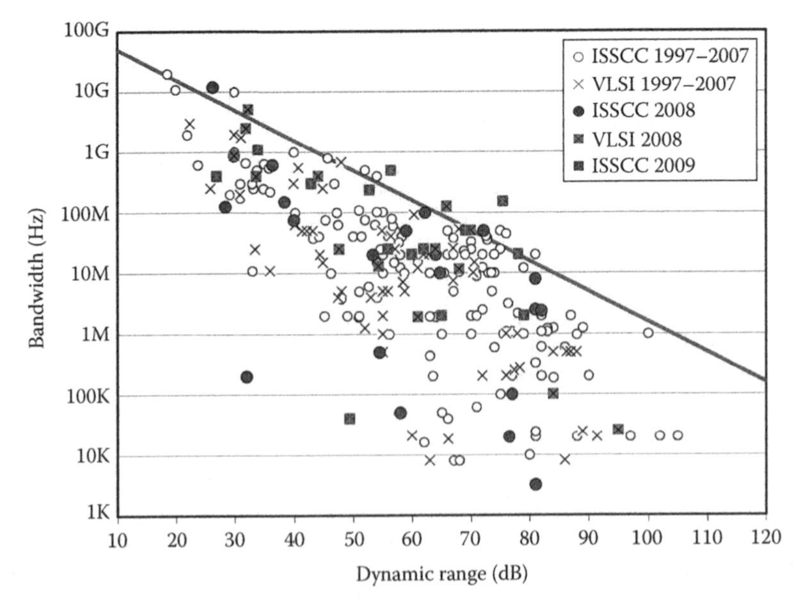

FIGURE 9.3
Dynamic range and bandwidth trends of recently published A/D converters.

9.3.3 System Architecture

Novel design architecture is essential to achieve low power consumption, while adapting DBF architecture. DBF architecture consumes considerable power because of the need for a dedicated A/D converter and memory blocks for receive channel. The need for an A/D converter per receive channel in a typical DBF system is a substantial disadvantage because of high power consumption and large system size as compared to an ABF system which delays the received signals, sums them by analog circuitry, and then converts them to a digital signal by an A/D converter.

According to Murmann [20], the power dissipation of the recently published A/D converters in ISSCC and VLSI are about 2×10^{-11} to 2×10^{-8} J at 50 dB dynamic range, as shown in Figure 9.4. For example, assuming the system has 256 receive channels (i.e., it needs 256 A/D converters with DBF architecture), the power dissipation of the A/D converter is 5×10^{-10} J (the median value), and the sampling rate is 250 MHz, the system consumes (32 W = 5×10^{-10} J × 250 MHz × 256) only for 256 A/D converters. This consumption is rarely practical or even possible considering power consumption limitations in a portable system. For example, the AD9271 chipset has a total of eight receive channels with dedicated A/D converters per each channel and consumes 1.5 W of maximum power. Thus, it needs 48 W of maximum power to build 256 receive channels using the AD9271 chips, which may not be suitable for portable devices.

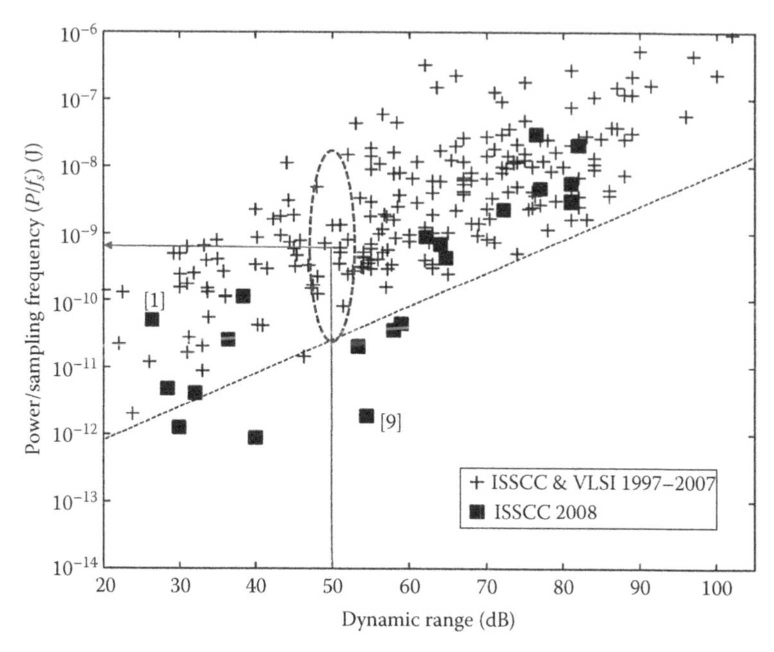

FIGURE 9.4
Power efficiency versus dynamic range in recently published A/D converters.

9.3.3.1 Shared ADC Architecture

To overcome the problem stated previously, the transceiver chip has only one A/D converter shared by the 16 receivers via a 16:1 analog multiplexer (aMUX) as proposed in [9]. Consequently, this configuration creates a DBF system but on the same order of size and power consumption as an ABF system. However, the prototype device operates 16 times slower than a conventional system because the shared ADC architecture performs 16 iterative operations accessing different channels to complete one scan. Another drawback is that the architecture requires extra digital controls.

An evaluation of the effectiveness of the shared ADC architecture, the performance of the shared ADC architecture, and the typical DBF architecture appears as a comparison in Table 9.1. In this comparison, the number of channels in the typical DFB architecture varies from 1 to 16, while the number of channels in the shared ADC architecture remains at 16, and the number of shared channels varies from 4 to 16. The sizes of one receive and transmit channel, an A/D converter, and the 3 Kb SRAM are assumed to be 0.175, 0.9, and 2.16 mm^2, respectively. These sizes are estimates based on the actual layout sizes of each component in the transceiver chip. The size of digital control circuitry, needed only for the shared ADC architecture, is expected to be 0.2 mm^2. In addition, the assumed power consumption of one receive and transmit channel, an A/D converter, 3 Kb SRAM, and the digital controls are 2, 100, 130, and 8 mW, respectively. These data were also estimated from SPICE simulation results of the each component with post-layout parameters. The time for 1-scanning is calculated based on the operational sequence and time for one complete scanning of the Penn State ultrasound imaging system described in Section 9.4.

Table 9.1 also indicates the trade-off relationship among operational speed for 1-scanning, chip size, and power consumption. Thus, to determine an optimal number of shared channels, the maximum time for 1-complete-scanning

TABLE 9.1

Performance Comparison among Shared ADC and Typical DBF Architectures

	Typical DBF Architecture			Shared ADC Architecture		
	Number of Channels			Number of Shared Channels[a]		
	1	8	16	4	8	16
Numbers of required ADC & SRAM	1	8	16	4	2	1
Chip size (mm^2)	3.24	25.9	51.8	15.2	9.1	6.1
Power consumption (mW)	240	1864	3720	960	500	270
Time for 1-scanning (μs)	50	50	50	200	400	800

[a] Total number of channels: 16.

allowed for real-time imaging is a consideration. Having established the maximum time, the total number of channels can be determined considering the chip size and power consumption specifications. For example, if the maximum allowable time for 1-scanning is 800 μs, the 16 channels can be shared. Then, if the chip size is 25 mm^2 and power consumption is 1 W, the optimal number of total channels will be 64 (16 channels of each are shared).

9.4 Penn State Portable High-Frequency Ultrasound Imaging System

9.4.1 Configuration of the Penn State Ultrasound Imaging System

Conventional front-end electronics consist of discrete chipsets for transmitters, preamplifiers, TGCs, A/D converters, and memory devices, all mounted on several PCBs (printed circuit boards). Therefore, the systems are not only large and expensive but they also have difficulty with high-speed operation. The Penn State ultrasound imaging system integrates the complete ultrasound front-end electronics onto a single IC chip with closed-coupled thin-film ultrasound transducers, and then constructs high-resolution ultrasound images. The system requires (1) a multichannel analog signal processing system including high-speed A/D converters and a transmit beamformer integrated on a single transceiver chip, and (2) auxiliary digital controls on an FPGA (field programmable gate array) chip with several discrete chipsets such as an RS232 chip, a D/A converter (DAC), and a 250 MHz crystal oscillator. These specifications enable creation of an ultra-compact, low-cost, high-speed, and high-resolution ultrasound imaging system.

The FPGA chip has been programmed to control the ultrasound transceiver chip, and helps for the transceiver chip to communicate with the imaging host via an RS-232 chip. The main control interfaces with other circuitry such as mode set, memory, and the transceiver chip according to commands received from the user. The clock buffer receives a 250 MHz clock signal from an external clock oscillator and distributes it to other circuit blocks and to the transceiver chip. The mode set consists of 1 bit serial pipeline registers that store data preset by the user, for example, channel and the preamplifier gain selection information. The 48 Kb internal memory's assignment is to store image information from the transceiver chip. The stored data, transferred to the host computer through the serial port in the FPGA chip and the RS-232 chip, allows subsequent image creation. The counter's design permits generation of digital codes that increase as an exponential function of time. The digital codes produce a pseudo-exponential analog signal for the VGA in the transceiver chip.

Figure 9.5 shows a block diagram of the transceiver chip. The receiver consists of a preamplifier and a TGC that compensates for signal attenuation as a function of depth. The TGC consists of an on-chip VGA and an external 10 bit

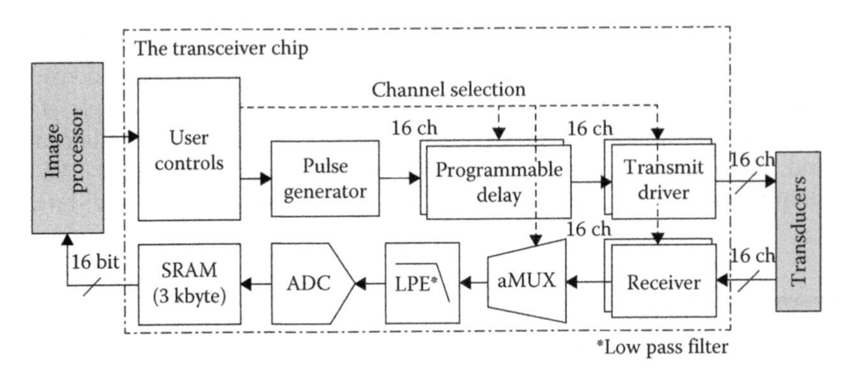

FIGURE 9.5
Simplified block diagram of the transceiver chip.

50 MHz D/A converter (AD9760, Analog Devices, Inc., MA). A time varying control signal, applied to the VGA gain control input, ensures the signal strength at the VGA output is constant over time (i.e., depth). An 8 bit A/D converter digitizes the compensated signals. The A/D converter output connects to a 3 Kb on-chip SRAM, which is large enough to store all the scanned image data for a specified depth range. In this work, the transceiver chip has 16 transmit and receive channels. The number of channels can be increased. The transceiver chip has only one A/D converter and SRAM to reduce chip area and power consumption, but the system needs an analog multiplexer (aMUX) and auxiliary digital controls for channel selection. A 16:1 analog multiplexer allows these two components to share 16 receive channels. The next subsection discusses the architectural advantage of the transceiver chip.

The transmit signal generator produces and sends a 50 MHz pulse to the thin-film ultrasound transducers through programmable delay chains to enable electronic beam focusing in the transmit mode. In a DBF system, focusing occurs by introducing delays to the transmit pulse on the elements so that emitted ultrasonic beams can be made constructively at the target of interest. Therefore, excitation pulses should be delivered to the transducer elements in an order that allows convergence of a composite wave front converges at a point. The variable delay chains in the figure determine the excitation order of the transducers.

9.4.2 Receiver

9.4.2.1 Required Specifications

The most important design specification of the system is the bandwidth of the receiver. The required bandwidth of the receiver is determined by the center frequency and bandwidth of the transducer. In this study, the target frequency of the transducers is 50 MHz with 100% bandwidth; thus, the input frequency range is 25–75 MHz [16]. Therefore, the receiver circuitry must be capable of a flat frequency response over the bandwidth of the transducer elements up to 75 MHz with a maximum gain of 20 dB.

In addition, the ratio of the A/D converter aperture and the maximum preamplifier input amplitude determines the gain of the preamplifier according to

$$\text{Gain}_{\text{preamp}} = 20 \cdot \log_{10}\left(\frac{\text{Max ADC input}}{\text{Max Premp input}}\right) \pm (\text{Design margin}) \tag{9.4}$$

In this design, the maximum input of the preamplifier is set to 0.3 V_{P-P}, and the A/D converter aperture is 1.5 V_{P-P}. Thus, the preamplifier gain is 14 dB ± (design margin). Since the thin-film transducers are currently under development, the design margin is set to ± 6 dB. Therefore, the gain of the preamplifier can be changed over a range from 8 to 20 dB at the discretion of the user.

The dynamic range of the preamplifier is another important factor in the receive circuitry. The dynamic range of the receiver determines the minimum and the maximum signal amplitudes that the system can process. Therefore, the attenuation rate and depth range of the target medium can determine the required dynamic range. Assuming the attenuation rate in soft tissue is 0.5 dB/MHz/cm [21], the total attenuation is 45 dB at 50 MHz for a penetration depth of 9 mm (i.e., the total signal path of 18 mm considering signal reflection). Adding a minimum display resolution of 30 dB, image saturation allowance of 6 dB, and noise threshold of 6 dB [22] provides a dynamic range of 87 dB [23]. Therefore, the SNR (signal-to-noise ratio) of the A/D converter needs to be greater than 87 dB, which corresponds to 15 bit resolution, according to the relationship [22]

$$\text{SNR}_{\text{ideal}} = 6.02N + 1.76(\text{dB}) \tag{9.5}$$

where N is the bit resolution of an A/D converter.

However, this dynamic range is too great for current high-speed A/D converters. The use of a VGA reduces the dynamic range requirement. Since the main purpose of the VGA is to compensate signal attenuation as a function of the depth of targets, the maximum gain of the VGA is bounded by the total signal attenuation, that is, 45 dB. The optimal gain range of the VGA is selectable according to the dynamic range of the A/D converter. For example, if the SNR of the A/D converter is 42 dB, 45 dB of variable gain range is the requirement. In this design, the target SNR of the A/D converter is 48 dB (i.e., 8 bit); thus, the required gain range of the VGA is 37 dB. Figure 9.6 illustrates the gain and dynamic range requirements.

9.4.2.2 Circuit Design Details

The receive circuitry consists of two on-chip components (the preamp and the VGA) and several off-chip components. The counter included in the FPGA chip generates time-varying digital codes that convert to the pseudo-exponential analog control signals through an external D/A converter.

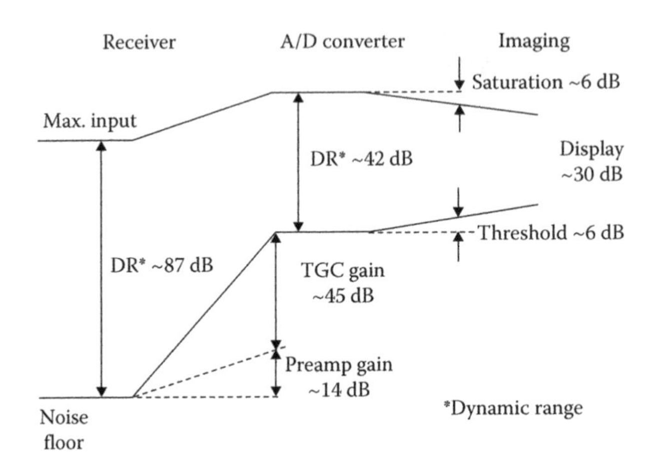

FIGURE 9.6
Transceiver chip gain and dynamic range requirements.

The VGA can produce a time-varying gain on a linear decibel scale with pseudo-exponential control signals.

Figure 9.7 shows the simplified circuit schematic of the preamplifier. The analog signals from the transducers connect to IN+ and the common ground of the transducer arrays connects to IN−. The resistance ratio of the transistors

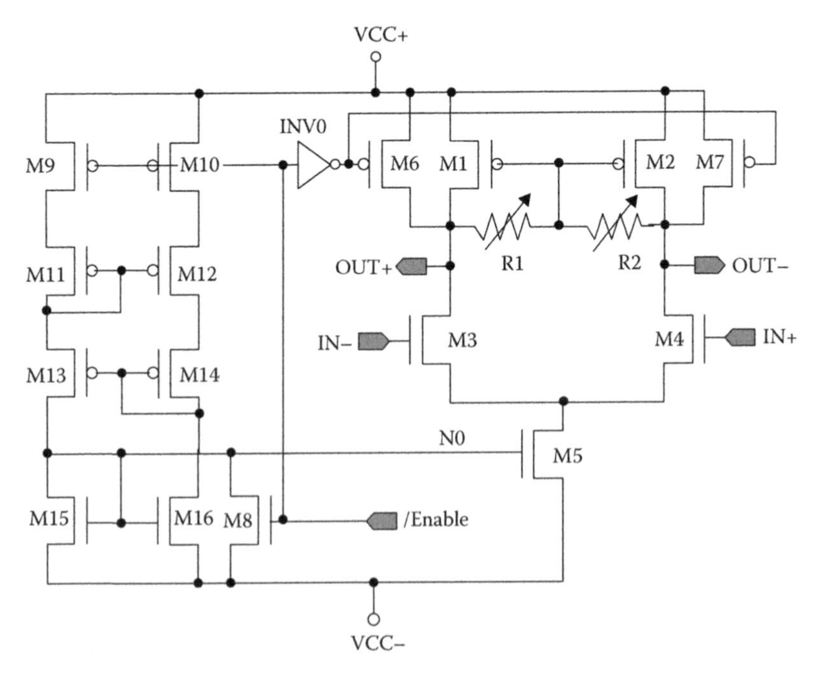

FIGURE 9.7
Preamplifier core circuit schematic.

M1 and M2 and the resistors R1 and R2 determine the gain. Expression of the voltage gain of the amplifier is

$$A_V = -g_{m1,2} \cdot \left(r_{O1,2} \| r_{O3,4} \| R_{1,2} \right) \tag{9.6}$$

where

- $g_{m1,2}$ is the transconductance of M1 and M2
- $r_{O1,2}$ is the output resistances of the transistors M1 and M2
- $r_{O3,4}$ is the output resistances of transistors M3 and M4
- R1 and R2 in the design vary the resistance value from 4 to 20 KΩ so that the user can preset the gain

The combination of transistors M6–M10 and inverter INV0 form the enable/disable switch. This internal switch eliminates external T/R switches (required in typical ultrasound transceivers) in the analog signal path. When "/Enable" is HIGH, transistor M8 turns OFF so that the voltage of the node N_0 goes to ground, and transistors M6 and M7 turn ON so that OUT+ and OUT– are tied to VCC+ regardless of the amplifier's inputs. The speed of the switch is fast enough for the amplifier to be stable before signal acquisition starts. Transistors M9–M16 generate bias voltages, independent of the power supply voltages, for the amplifier.

Since an ultrasound wave attenuates its power traveling in tissue on a decibel scale, the gain of the VGA also needs to have the capability to be changed linearly in dB [21]. Therefore, the gain of the VGA must be linear-in-dB over the linear control voltage range. To achieve this relationship, the proposed design adapts a Gilbert-type four-quadrant multiplier, whose output is equal to the product of the two inputs [24], and the generated pseudo-exponential control voltages use external circuitry with an FPGA chip and a D/A converter. Figure 9.8 depicts the folded Gilbert cell, in which the bottom differential pair of the original Gilbert cell folds without degrading performance in order to reduce the number of cascode transistors. Derivation of the analytic relationship between input and output is [25]

$$\begin{aligned} Vo &= \left(g_{m3,4} - g_{m5,6} \right) \cdot R_D \cdot (Vin^+ - Vin^-) \\ &= \sqrt{\frac{k_n(W/L)}{2I_{SS}}} \cdot g_{m1,2}(Vcp - Vcn)(Vin^+ - Vin^-), \end{aligned} \tag{9.7}$$

where

- $g_{m1,2}$, $g_{m3,4}$, and $g_{m5,6}$ are the transconductances of transistors M1 and M2, M3 and M4, and M5 and M6, respectively
- $k_N = \mu_N \cdot C_{ox}$ (μ_N is the electron mobility and C_{ox} is the gate capacitance of the NMOS transistor)
- W is the channel width of transistors M1–M4
- L is the channel length of transistors M1–M4

Transistors M9–M16 constitute a linear voltage converter.

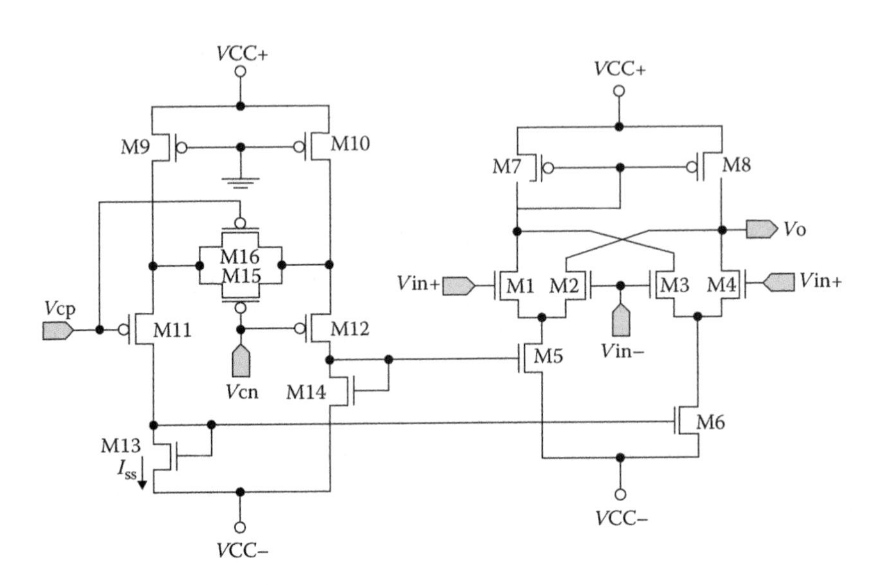

FIGURE 9.8
Folded Gilbert cell–based VGA circuit schematic.

Figure 9.9 presents the measured output of the receive circuitry including the preamplifier and the VGA. The preamplifier gain is set to 14 dB and the linear gain control range of the VGA is 23 dB with control voltage of 0.1–1.0 V, which is generated using the FPGA chip (Spartan III, Xilinx, Inc.) and a 10 bit D/A converter (AD 9760, Analog Devices, Inc.). The results demonstrate the

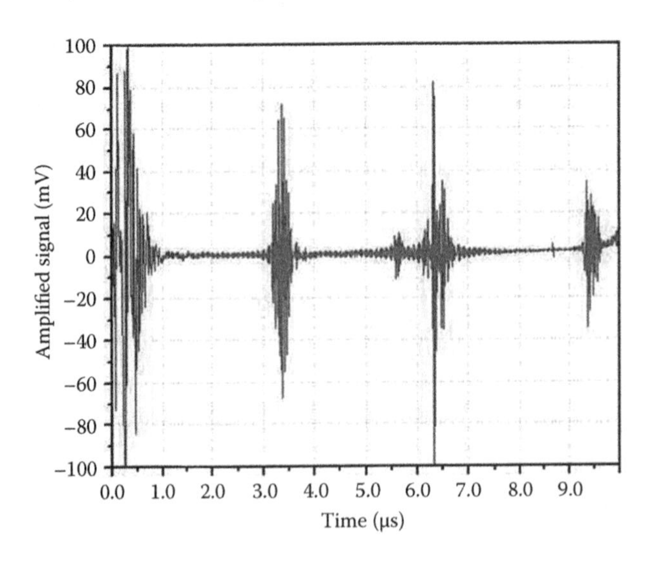

FIGURE 9.9
Amplified signals by the receiver with the VGA functioning. See the Section 4.6 about the experimental setup for this testing.

function of the receiver: the amplitude of the second and the third peaks are similar to that of the first peak due to the increasing gain over the time.

9.4.3 A/D Converter

The sampling rate of the A/D converter can be determined by the Nyquist sampling theorem [26], which states that reconstruction of a continuous-time signal from its samples is possible if the sampling frequency is greater than twice the signal bandwidth. If the center frequency of the target transducer is 50 MHz with 100% bandwidth, the bandwidth of the reflected signals will be 25–75 MHz. Given the 20 MHz design margin of the anti-aliasing filter located between the VGA and the A/D converter, the signal bandwidth will be 5–95 MHz. Therefore, the required minimum sampling rate of the A/D converter is 190 MHz according to the Nyquist theorem.

A further important determination is the effective bit resolution. This consideration depends on the characteristics of the medium. In this design, the target medium is tissue in human organs, which requires at least a 50 dB image resolution [22]. Therefore, the required effective bit resolution of the A/D converter is set to 50 dB, that is, an 8 bit resolution.

9.4.3.1 TIQ-Based A/D Converter

The design of a 190 MS/s A/D converter is a challenge in 0.35 μm CMOS technology. To achieve the A/D converter requirements, a TIQ (threshold inverter quantization) A/D converter (TIQ ADC), known for its fast conversion speed [27], has been designed. Although flash-type A/D converter has several disadvantages in terms of higher power consumption, occupying a larger area than other types of A/D converters, the shared A/D converter architecture described in Section 9.3 overcomes the drawbacks of the flash A/D converter. Since this architecture, reportedly, has the operation speed, gain, and DC offset variations of up to 18% due to process and temperature variations [27], a sampling rate of 250 MHz rather than the required sampling rate of 190 MHz is the target for the proposed design.

A TIQ comparator is one of the most important circuits in a flash A/D converter. It converts an analog input voltage into a digital logic output "1" or "0," depending on the reference voltage of the comparator. In a traditional flash A/D converter, a differential comparator, which needs a resistor-ladder circuit as an external voltage reference, is commonly used. On the other hand, the TIQ comparator, which consists of two cascaded CMOS inverters, does not need a resistor ladder circuit because it uses the built-in voltage reference of the CMOS inverters [28].

As an example of TIQ ADC operation, Figure 9.10a shows the schematic of a 2 bit TIQ ADC comprising 3 TIQ comparators, 3 gain boosters, and 1 encoder (Notably, an n bit TIQ ADC consists of $2^n - 1$ comparators and gain boosters and an n bit encoder). The TIQ comparator consists of two cascaded

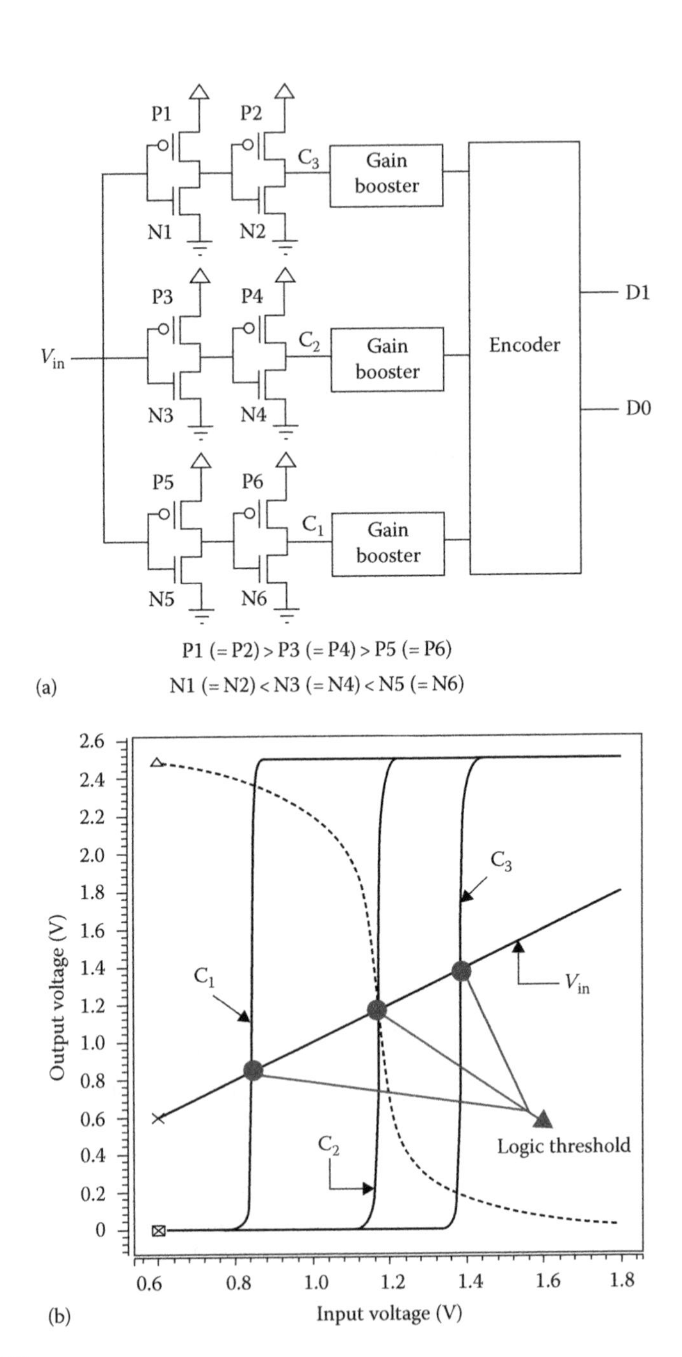

FIGURE 9.10
TIQ comparator design: (a) example of 2 bit TIQ A/D converter, (b) VTC of a TIQ comparator.

inverters: the first inverter sets the analog signal quantization according to its logic threshold, and the second inverter increases the gain of the comparator. If the analog input is higher than the logic threshold, the digital output is logic "1"; otherwise, the digital output is logic "0."

A particular output voltage (V_{out}) that is the same as input voltage (V_{in}), that is, $V_{in} = V_{out}$, determines the logic threshold of the TIQ comparator. And the logic threshold changes depending on the width ratios of the PMOS and NMOS. If P1 is the largest PMOS and P5 is the smallest PMOS, and if N1 is the smallest NMOS and N5 is the largest PMOS, as shown in Figure 9.10a, the top comparator output corresponds to C_3; the middle comparator corresponds to C_2; and the bottom comparator output corresponds to C_1 as in Figure 9.10b. Clearly, the three different ratios of the PMOS and NMOS widths result in three different logic thresholds: the logic threshold value of the top comparator is the largest and the logic threshold of the bottom comparator is the smallest.

With fixed lengths for the PMOS and NMOS transistors, increasing or decreasing the width of the PMOS or NMOS, respectively, produces the desired values for the logic thresholds. Since the TIQ comparator is used in a flash A/D converter, a $2^n - 1$ set of TIQ comparators is necessary to design an n bit flash A/D converter. Therefore, finding the exact $2^n - 1$ different logic thresholds as the reference voltages of the TIQ comparator in an input voltage range is necessary. For example, all 63 TIQ comparators for a 6 bit flash A/D converter have transistors of different sizes, so all the comparators have different logic threshold values.

9.4.3.2 Design Automation for the TIQ Comparator

The simple architecture of TIQ ADC is an advantage; however, for a TIQ ADC, the TIQ comparators' design must be precisely sized to be different from one another. Achieving the required dynamic range makes designing this feature a somewhat difficult task. Mitigating the difficulty is use of a CAD (computer aided design) tool, which automates TIQ comparator design and implementation. The systematic size variation (SSV) technique is the proposed method for easing, comparatively, the choice of needed logic thresholds from the many possible comparators [28]. The SSV technique chooses V_m from a reduced range of 3-D plots. The diagonal line drawn in the 3-D plots is the optimal line, which maintains a systematic increasing and/or decreasing order of transistor sizes. Keeping the transistor size in increasing and/or decreasing order significantly improves the linearity of the A/D converter in relationship to CMOS process variation. This method also significantly reduces the number of simulations needed for transistor size selection. The simulation is needed only along the diagonal-line region rather than on the full 3-D surface.

Using the SSV could save this design time. However, it is still time-consuming due to simulating all possible combinations of the PMOS and NMOS.

For example, according to [29], about 4 h were necessary to find 6 bit TIQ comparators using the SSV technique, and 5 Sun-Blade 2000 machines performed 28,000 simulations to find 63 TIQ comparators. Notably, the total design time increases exponentially if the bit resolution of a TIQ ADC increases.

An improved TIQ comparator design methodology has been proposed in [29]. The proposed method introduces an analytical TIQ model to overcome the drawbacks of the SSV technique. The analytical model has advantages over the SSV technique in terms of simulation time as well as accuracy.

Since the logic threshold (V_m) is a voltage in which the output voltage is equal to the input voltage ($V_{out} = V_{in} = V_m$), mathematical expression of V_m can be derived from drain current equations for both PMOS and NMOS devices. For simplicity, only a TIQ model using a Level 1 SPICE model has been derived here. More sophisticated models, such as BSIM3 (Berkeley Short-channel IGFET Model) or BSIM4 [30], may improve the model's accuracy. Also, the assumption is that the channel lengths of PMOS and NMOS devices are long enough; that is, velocity saturation does not occur. The drain currents of NMOS and PMOS, $I_{DS,\,NMOS}$, $I_{DS,\,PMOS}$, respectively, can be denoted

$$I_{DS,\,NMOS} = \frac{k_N}{2}\left(\frac{W_N}{L_N}\right)(V_m - V_{THN})^2(1+\lambda_N V_{DS}) \tag{9.8}$$

$$I_{DS,\,PMOS} = \frac{k_P}{2}\left(\frac{W_P}{L_P}\right)\left(V_{DD} - V_m - |V_{THP}|\right)^2\left(1+\lambda_P V_{DS}\right) \tag{9.9}$$

where

$k_N = \mu_N \cdot C_{OXN}$ (μ_N is the electron mobility and C_{OXN} is the gate capacitance of NMOS)

$k_P = \mu_P \cdot C_{OXP}$ (μ_P is the hole mobility and C_{OXP} is the gate capacitance of PMOS)

When the channel lengths of both transistors are the same, the logic threshold, V_m, yields

$$V_m = \frac{V_{THN} + \left(V_{DD} - |V_{THP}|\right)\cdot\Lambda\cdot\sqrt{k_P W_P / k_N W_N}}{1+\sqrt{k_P W_P / k_N W_N}} \tag{9.10}$$

where $\Lambda = \sqrt{2+\lambda_P V_{DD}/2+\lambda_N V_{DD}}$.

From Equation 9.10, the dependency of V_m on the width ratio of the PMOS and NMOS (W_P/W_N) can be derived. Figure 9.11 shows the comparison of V_m dependencies on the width ratios in analytical models and the SSV technique. A standard 0.35 μm CMOS technology is used for this graph, and the SSV technique generates a set of 63 TIQ comparators for a 6 bit TIQ ADC. The V_m

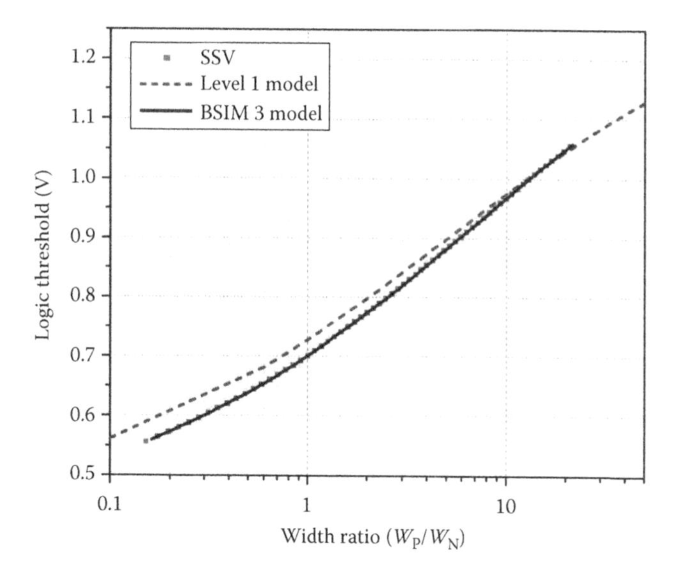

FIGURE 9.11
Accuracy comparison of various TIQ comparator models.

curve from the BSIM3 model shows excellent accuracy with an average mismatch rate of less than 1%. The maximum error is about 5%, and the average error is less than 1%. Therefore, a set of width ratios, according to ideal V_m values, can be found using the TIQ model rather than the SSV technique.

As an alternate form of Equation 9.10, a set of widths can be expressed as a function of V_m. This alternate set is

$$\left(\frac{W_P}{W_N}\right) = \frac{k_N(V_m - V_{THN})^2(2+\lambda_N V_{DD})}{k_P\left(V_{DD} - V_m - |V_{THP}|\right)^2(2+\lambda_P V_{DD})} \tag{9.11}$$

Equation 9.11 provides the width ratio for each ideal V_m. The width ratios of the TIQ comparators can be expressed as

$$\frac{W_P(i)}{W_N(i)} = R(i), \quad (i = 1 \sim 2^n - 1) \tag{9.12}$$

where

$W_P(i)$ is the i_{th} PMOS width
$W_N(i)$ is the i_{th} NMOS width
$R(i)$ is the i_{th} width ratio for the n bit TIQ comparator set

However, determining the widths of NMOs and PMOS devices cannot occur using only Equation 9.12 without another relationship between W_P and W_N.

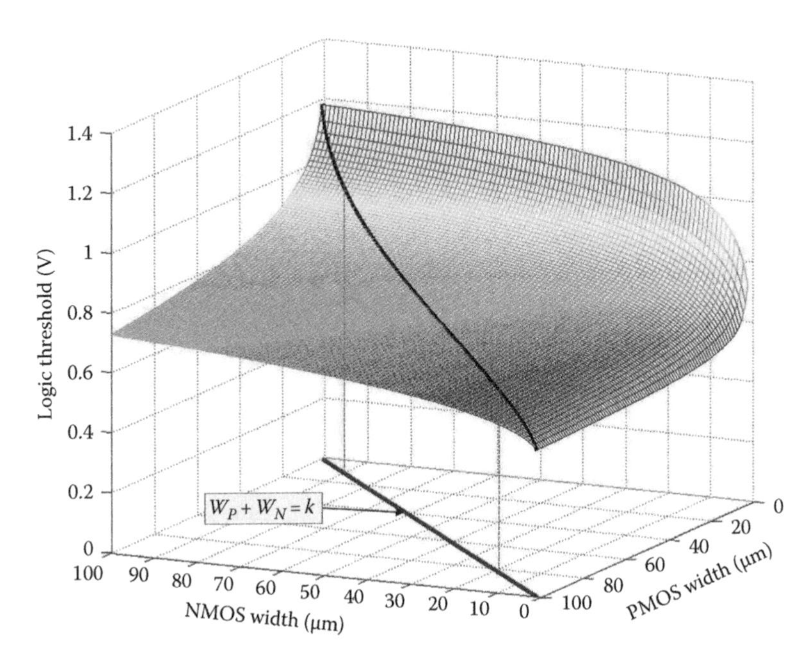

FIGURE 9.12
3-D plot of V_m and the width size relationship of the NMOS and the PMOS of the TIQ comparator analytical model.

The relationship arises from the projection of the line on the x–y plane, as illustrated in Figure 9.12. The line expression, in a general form, is

$$a \cdot W_P + b \cdot W_N = c \tag{9.13}$$

where a, b, and c are constants. In the case of process variation, shifting the NMOS and PMOS at the same ratio is desirable. Thus, the a and b can be set to 1. Therefore, (9.13) can be denoted as

$$W_P(i) + W_N(i) = k \tag{9.14}$$

From Equations 9.12 and 9.14, $W_P(i)$ and $W_N(i)$ can be derived as

$$W_P(i) = \frac{k \cdot R(i)}{1 + R(i)} \tag{9.15}$$

and

$$W_N(i) = \frac{k}{1 + R(i)} \tag{9.16}$$

The optimal transistor length parameters are derived from considering the relationship between transistor length and data conversion speed. In general, a longer transistor length in a TIQ comparator is desirable for achieving higher gain, that is, higher sensitivity, less noise, and less process variation sensitivity. However, a longer transistor yields a slower data comparison speed in the TIQ comparator. If the conversion target speed is 1 GSPS, a maximum transistor length of about 1 μm is allowed in a 0.18 μm standard CMOS technology [29]. Consequently, the optimal value of the transistor length in a TIQ comparator can be set to 1 μm for a 1-GSPS TIQ ADC.

9.4.4 SRAM

9.4.4.1 Required Specifications

The SRAM included in the design reduces the data transfer rates from the A/D converter to the imaging host processor. Since the sampling rate of the A/D converter is 250 MS/s and the bit resolution is 8 bits, the digitized image data transfers to the host at the data bandwidth of 2 GBPS. However, such high-speed data transfer is not only difficult to be achieved using less expensive circuit boards but it also generates substantial digital noise (jitters) that negatively affects the analog signal integrity of the receiver. On-chip memory devices overcome these problems. Therefore, an SRAM that can write data with a data bandwidth of 2 GBPS and that can read data with a lower data transfer rate of 250 Mbps became elements of the design for the CMOS transceiver chip.

Another key specification of the on-chip memory is capacity. The calculation of the memory capacity uses the relationship

$$\text{Capacity} = \frac{t_{\text{Travel}}}{t_{\text{sample}}} = \frac{D/c}{t_{\text{sample}}} \tag{9.17}$$

where

t_{Travel} is the total (i.e., two way) travel time of the ultrasound wave in the medium
t_{sample} is the sampling interval of the A/D converter
D is the total distance that the ultrasound wave travels in the medium
c is the speed of the ultrasound wave in the medium

Estimates indicate that a 3-Kb buffer length is necessary for $t_{\text{sample}} = 4$ ns, $D = 18$ mm, and $c = 1500$ m/s.

9.4.4.2 SRAM Design Details

Figure 9.13 shows the functional block diagram of a 3 Kb SRAM. This SRAM operates in two modes: data write mode and data read mode with auto precharge. Once the data write or data read operation begins, the operation

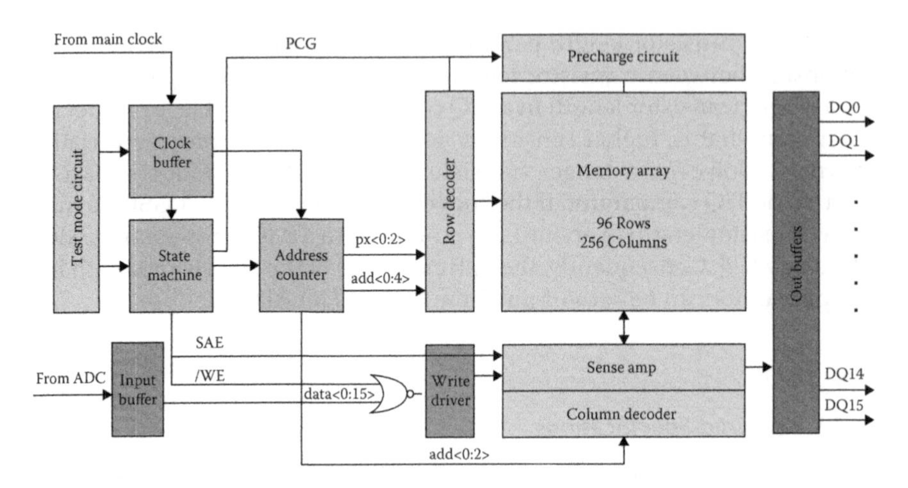

FIGURE 9.13
Functional block diagram of the 3 Kb SRAM.

continues until the all 3 Kb cells are written to or read by the address counter, which automatically, incrementally sweeps all 3000 addresses.

The SRAM design adopts asymmetric operating speed between data write and data read. As described earlier, the data write speed is faster than 125 MHz; however, a 125 MHz data readout speed could be a disaster in some inexpensive applications. Thus, this design uses two different speed clocks for data write and data read. First, a 250 MHz clock is used for data conversion and storing. Next, a 50 MHz clock is for the system I/O. Thus, the data readout uses the 50 MHz clock. The main I/O clock offers an inexpensive and easy way of transferring data.

The asymmetric operating speed between data write and data read occurs by a 2:1 multiplexer. During write operations, the SRAM shares the A/D converter clock, and the SRAM uses the main I/O clock during the readout operation. Figure 9.14 presents the SPICE simulation results of the SRAM clock transition. Until 700 ns, the SRAM writes the data in the memory cells in the two subbanks with a 250 MHz clock speed as shown in Figure 9.14b and c. Clearly, the subbanks operate one after the other as mentioned earlier. At 700 ns, the read command comes in, and the clock changes from 250 to 50 MHz. Thus, the read operation performs with the 50 MHz clock.

Figure 9.15 presents the measured outputs of the SRAM, which transferred at a rate of 400 MBPS as described earlier. In this figure, only 6 bit outputs are presented because the A/D converter shows linear characteristics with 6 bits. An 85 kHz of saw-tooth wave was generated by a function generator and fed to the A/D converter. The analog signal was sampled at a speed of 250 MS/s, then converted to digital signals. The SRAM stored the digital data at the same speed of the A/D converter's sampling rate. The test results prove the functionality of the SRAM and the A/D converter.

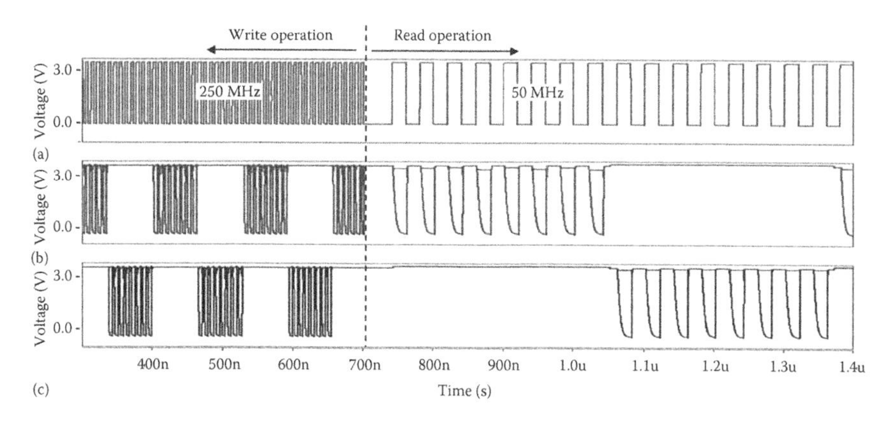

FIGURE 9.14
Simulation results of the SRAM clock modulation: (a) clock transition, (b) memory cell data in Bank 0, (c) memory cell data in Bank 1.

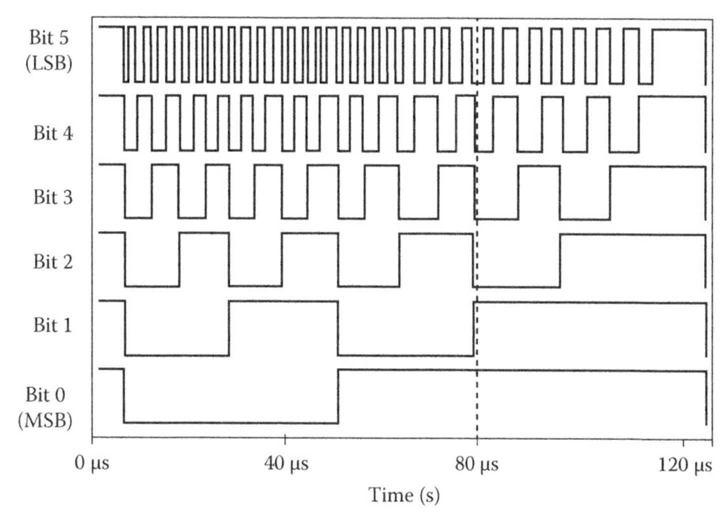

FIGURE 9.15
Measured outputs of the SRAM with the transfer rate of 400 MBPS.

9.4.5 Transmitter

The major function of the transmitter is to excite the transducer elements and to focus and/or steer the ultrasound beam. Focusing and steering the ultrasonic beam requires time delays to compensate for the acoustic signal path length differences from the transducer array to the target of interest. The expression for the focused signal $f(t)$ is

$$f(t) = \sum_{n=-N/2}^{n=N/2} X_n(t - \tau_n) \tag{9.18}$$

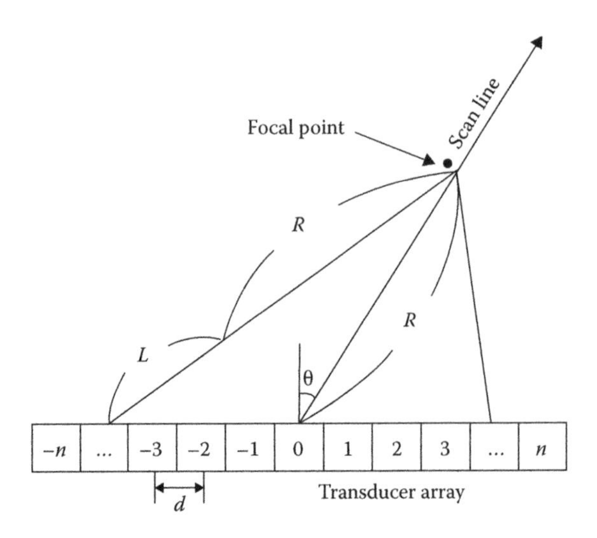

FIGURE 9.16
Dynamic focusing and steering delay.

where

N is the total number of transducer elements
X_n is the received reflected signal
τ_n is channel delay time for the nth transducer element [15]

The delay time, τ_n, derives from Figure 9.16. Assuming a signal, transmitted with the steering angle θ by exciting a transducer located at 0, a reflected signal propagates back from the focal point to the transducer array. When the distance from the focal point to the transducer center is R, the distance from the focal point to the nth transducer has the expression: $L + R$, as shown in Figure 9.16. By denoting the space between adjacent channels, d, expression of the channel delays is

$$\tau_n = \tau_n(R) = \frac{R}{c}\left[\sqrt{1+\left(\frac{nd}{R}\right)^2 + 2\left(\frac{nd}{R}\right)\sin\theta} - 1\right] \quad (9.19)$$

where c is the average propagation speed of ultrasound in the medium [31].

The design of the programmable delays provides different delay times for each of the elements in a 16 × 1 array configuration. Implementing such a fine delay step in a CMOS technology is challenging in beamformer design. The most recently designed digital clock delay schemes use PLLs (phase locked loops) [15,31]. However, a 20 ps delay step would need at least a 10-GHz clock using these methods. This is not feasible in a 0.35 μm standard CMOS technology. Therefore, this work adopts an analog delay chain method. Figure 9.17 is a block diagram of the proposed delay chain. One of five delay

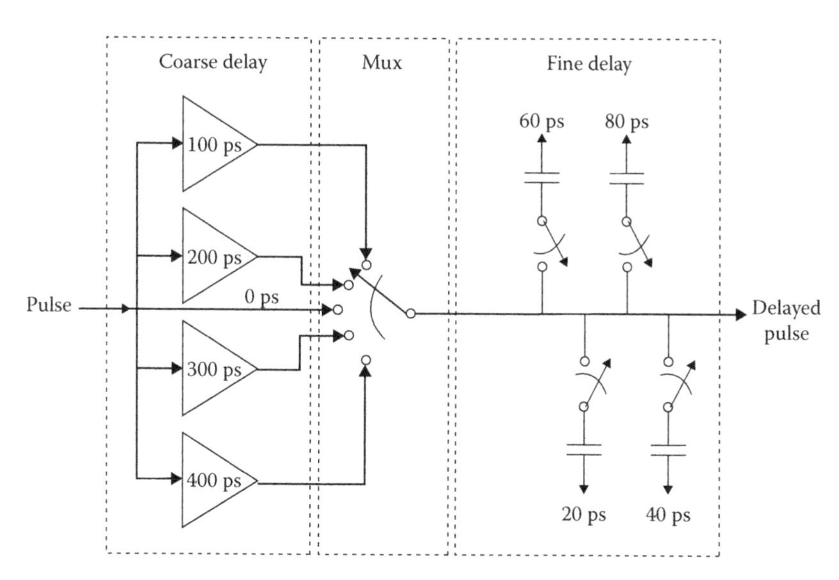

FIGURE 9.17
Circuit schematic of programmable delays for beam focusing.

buffers can be selected for coarse delay time setting with a 100-ps step. Then four different loading options can be added for the 20 ps fine delay step. Every channel on the chip has the channel delay setting circuit; therefore, a host computer can control delays between pulses by transmit beam focusing. Although required channel delays are 20–160 ps, the channel delays are programmable from 20 to 480 ps in order to compensate for the initial channel delay mismatch.

Figure 9.18 illustrates the transmit pulses generated by the transceiver chip. The transmit pulses can be delayed from 20 to 500 ps as described in the previous section. The delays by the programmable delay chain in the chip

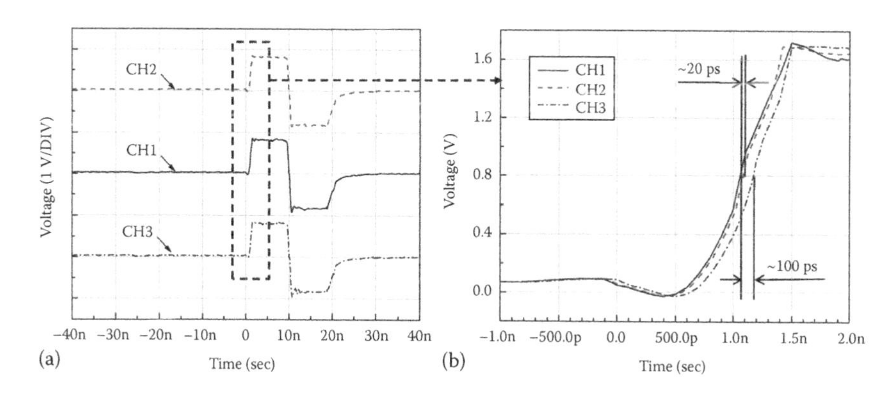

FIGURE 9.18
Measured delay of transmit channels: (a) transmit pulses with various channel delays (50 MHz), (b) magnified view of the box in blank on (a).

are measured with a high-precision digital oscilloscope (Agilent Infiniium 8000). Notably, the signals shown in the figure are calibrated to compensate for the initial channel delays due to signal line mismatches on the printed circuit board. The frequency of the pulses can be varied from 1 to 100 MHz, where 5 MHz pulses are shown in the figure.

9.4.6 Experimental Results

9.4.6.1 Prototype Experimental Board

The transceiver chip was fabricated using a TSMC 0.35 μm, double-poly, four-metal process through MOSIS. The die size was 10 mm^2. Figure 9.19 shows the mounting of the transceiver chip and the thin film transducers on test board (a), and a microphotograph of the first-generation transceiver chip (b). The average power consumption in the receive mode, consisting of 16 receiver channels, A/D converter, and SRAM, operating simultaneously, is approximately 270 mW with a 3.3 V power supply. The shared A/D converter and SRAM architecture, which reduced the number of A/D converters and SRAMs from 16 to 1, resulted in smaller chip area and lower power consumption. This size and power consumption are reasonable for a portable ultrasound imaging system using the transceiver chip; they can be decreased further if a state-of-the-art process technology, for example, a 90 nm or 0.13 μm CMOS process technology, is used. Table 9.2 summarizes the specifications for the transceiver chip.

9.4.6.2 Thin-Film Ultrasound Transducer Array

T-bar-shaped ultrasound PZT ($PbZr_{0.52}Ti_{0.48}O_3$) transducer arrays have been developed and documented in [16]. PZT layers can be thin, and allow reduction of the required voltages for exciting the transducer. Therefore, directly

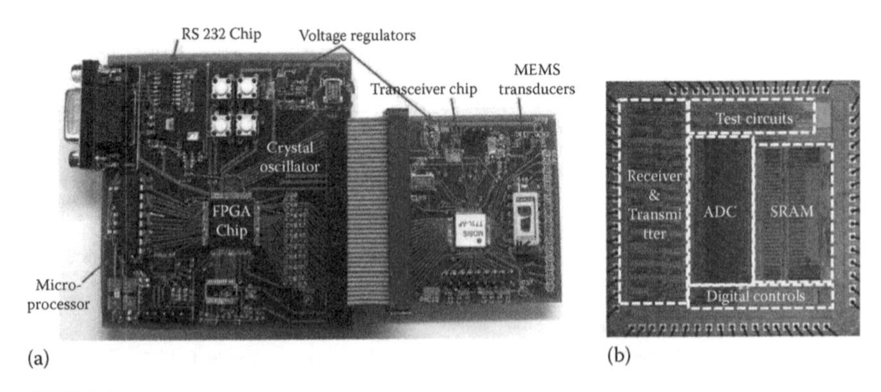

FIGURE 9.19
(a) Test board for the transceiver chip with the thin film transducer array, (b) microphotograph of the CMOS transceiver chip.

TABLE 9.2

CMOS Transceiver Chip Specifications Summary

Preamp	Gain (dB)	5–20
	Bandwidth (MHz)	>75
	Dynamic range (dB)	90
	Noise figure (dB)	10
VGA	Gain range (dB)	46
	Bandwidth (MHz)	250
	Noise figure (dB)	6–12
ADC	Resolution (bit)	6
	Conversion speed (MHz)	250
Memory	Capacity (byte)	3K
	Data bandwidth (Mbps)	250
Transmitter	Pulse frequency (MHz)	1–100
	Delay step (ps)	20
Total chip	Noise figure (dB)	9.7–14.5
	Power consumption	270 mW
	Process technology	TSMC 0.35 μm
	Chip size	10 mm^2

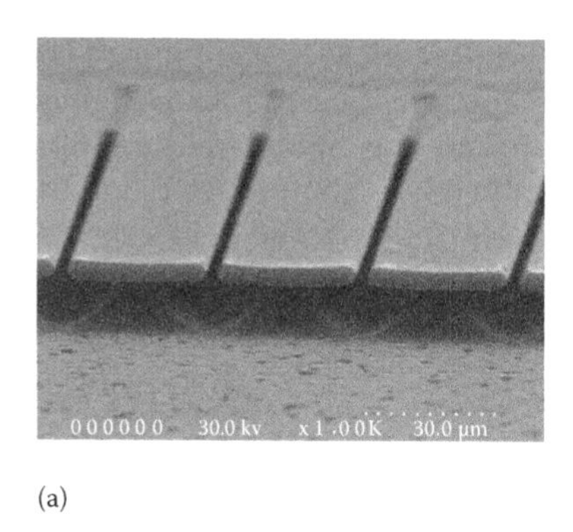

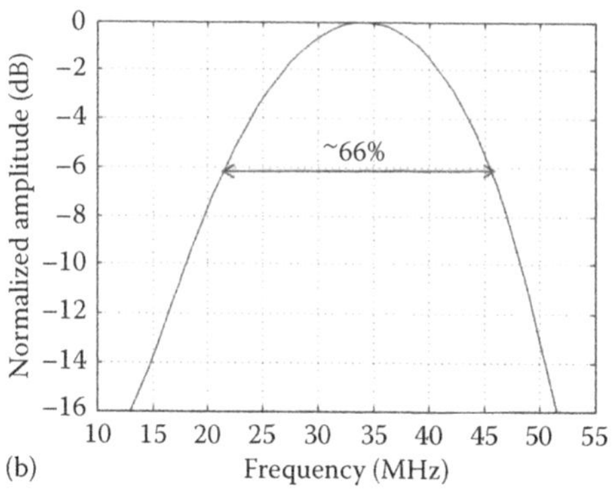

FIGURE 9.20
The SEM image of the thin-film transducer array (a) and its frequency spectrum (b).

driving the transducers with low-voltage CMOS signals is possible. The thin-film transducers' fabrication employed a Sol–gel and multilayer dry-etching process. Ti/Pt bottom electrodes were deposited on a silicon wafer on which a 300 nm thermal silicon-dioxide film was preformed. The total of 0.5–0.6 μm PZT films was deposited using the Sol–gel process over the bottom electrode; then, Pt top electrodes were formed. Finally, the T-bar shape transducer array structure was patterned by dry-etching. Figure 9.20 shows an SEM (scanning electron microscopy) image of the transducer array (a) and its frequency spectrum (b). The dimensions of a T-bar structure are 30 μm in width and 300 μm in length. Details of the thin-film transducers have reference in [16]. The target, center frequency of the thin-film transducer is 30–70 MHz, and that of the particular transducer array in these experiments is 35 MHz, with a bandwidth of 66% at 6 dB points. The capacitance of the 30 μm-wide transducer element is 145.1 pF, which indicates a dielectric constant of ~1000 at 1 kHz (as high as 1500) along with a low dielectric loss (<4%). Thus, the transducer appears to be a pure capacitance load for the preamplifier in the transceiver chip.

9.4.6.3 Pulse-Echo Experiments

The ultrasound signal acquisition experiments were conducted using the thin-film transducer with the transceiver chip. Figure 9.21 illustrates the pitch-catch mode experimental setup. The stainless steel target object was located at a distance of about 3 mm away from the transducers in the water tank. A 50 MHz transmit pulse sent to the thin-film transducers from the transceiver chip via an external pulser allowed an increase of acoustic energy. The adjacent transducer in the same array received the reflected signals; then, the transceiver chip amplified the received signals.

The gain of the preamplifier is set to 14 dB, and the gain of the VGA increased from 0 to 20 dB using varied control signals (0.15–1.0 V). Figure 9.22 shows the amplified and time-compensated ultrasound signals obtained from the VGA output using the thin-film transducer. The high-amplitude signal before 2.5 μs in the figure is from cross talk with a transmit pulse. The first peak of the detected ultrasound wave appeared around 3.5 μs, which is the first echo signal; the second peaks, around 7 μs, underwent multiple bounces. The second peaks, amplified by the VGA, compensated for the attenuation in the water. Notably, the amplitude of the second peaks is similar to that of the first peak due to the increasing gain over the time.

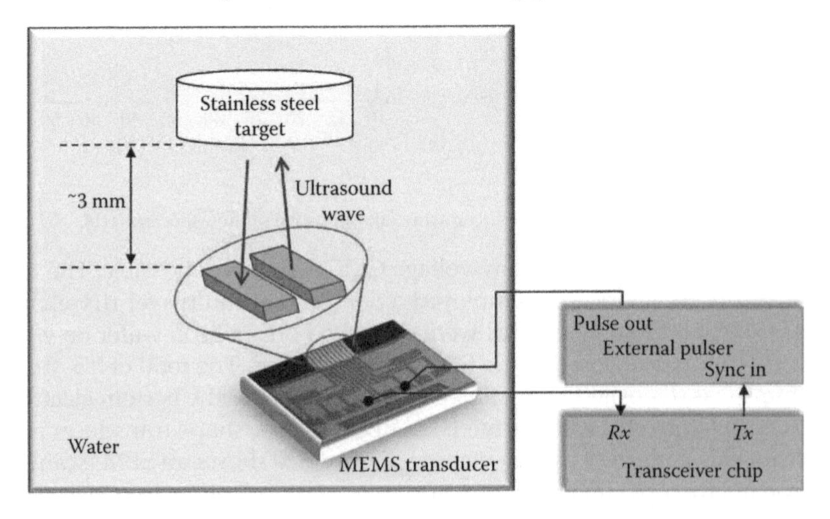

FIGURE 9.21
Pitch–Catch mode experimental setup.

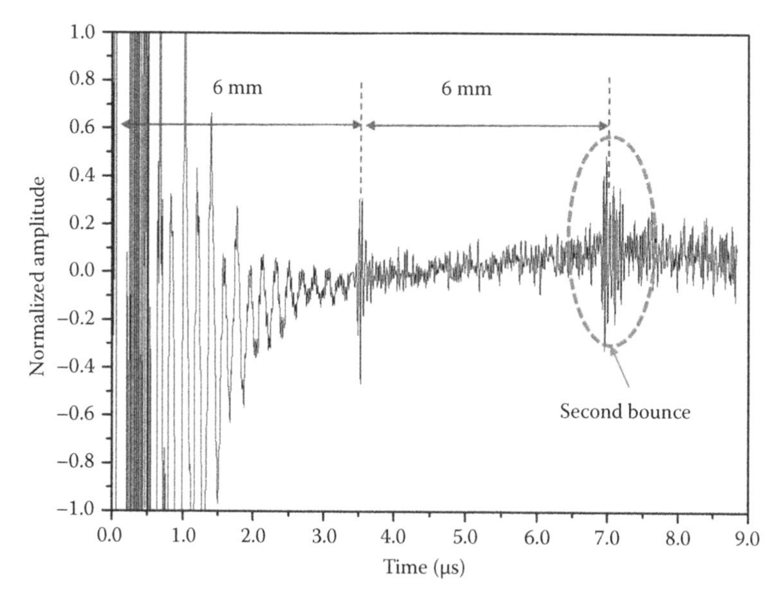

FIGURE 9.22
Thin-film transducer array pitch–catch mode test results.

9.5 Conclusion

This chapter explains the ultrasound imaging fundamentals and system considerations for developing a portable high-frequency ultrasound imaging system. The low-voltage transducer is crucial for developing a portable, ultra-compact ultrasound imaging system, because it enables the transducer to be placed with close-coupled ultrasound front-end electronics without using any expensive coaxial cables. Thus, reasonably, some cost reductions will accrue. Also, the fully integrated ultrasound transceiver chip with the shared ADC architecture is an effective solution for the portable imaging systems. It greatly decreases system size and power consumption, compared to conventional DBF ultrasound imaging systems currently marketed.

This chapter also introduces the Penn State portable ultrasound imaging system as an example of existing portable high-frequency ultrasound imaging systems. The Penn State ultrasound imaging system consists of low-voltage-operated thin-film transducer array and a fully integrated custom-designed CMOS transceiver chip. A 16 channel ultrasound receivers are shared with an A/D converter and a 3 Kb SRAM. The chip also makes it feasible for the transducers to be fabricated on the same package or board with the chip, and anticipates more cost and size reduction. Initial pulse-echo experiments using the imaging system were performed and the experimental results demonstrate the shared ADC architecture and the transceiver chip components designs.

References

1. F.S. Foster, C.J. Pavlin, J.A. Harasiewicz, D. A. Christopher, and D.H. Turnbull, Advances in ultrasound biomicroscopy, *Ultrasound in Medicine and Biology*, 26(1), 1–27, 2000.
2. P.A. Payne, J.V. Hatfield, A.D. Armitage, Q.X. Chen, P.J. Hicks, and N. Scales, Integrated ultrasound transducers, *Proceedings of the IEEE Ultrasonic Symposium*, Vol. 3, 1994, pp. 1523–1526.
3. R. Reeder and C. Petersen, The AD9271-A revolutionary solution for portable ultrasound, analog dialogue 41–07, 2007. Available: http://www.analog.com
4. William C. Black, Jr., and Doughlas N. Stephens, CMOS chip for invasive ultrasound imaging, *IEEE Journal of Solid-State Circuits*, 29, 11, 1994.
5. J.V. Hatfield, P.A. Payne, N.R. Scales, A.D. Armitage, and P.J. Hicks, Transmit and receive ASICs for an ultrasound imaging multi-element array transducer, *IEEE Colloquium on ASICs for Measurement Systems*, 1994.
6. I.O. Wygant, D.T. Yeh, X. Zhuang, S. Vaithilingam, A. Nikoozadeh, O. Oralkan, A. SanliErgun, G.G. Yaralioglu, and B.T. Khuri-Yakub, Integrated ultrasound imaging systems based on capacitive micromachined ultrasonic transducer arrays, *2005 IEEE Sensors*, 2005.
7. J. Johansson, M. Gustafsson, and J. Delsing, Ultra-low power transmit/receive ASIC for battery operated ultrasound measurement systems, *Sensors and Acturators A: Physical*, 125, 317–328, 2006.
8. M.I. Fuller, E.V. Brush, M.D.C. Eames, T.N. Blalock, J.A. Hossack, and W.F. Walker, The sonic window: Second generation prototype of low-cost, fully-integrated, pocket-sized medical ultrasound device, *2005 IEEE Ultrasonics Symposium*, 2005, pp. 273–276.
9. I. Kim, H. Kim, F. Griggio, R.L. Tutwiler, T.N. Jackson, and S. Trolier-McKinstry, CMOS Ultrasound transceiver chip for high resolution ultrasonic imaging systems, *IEEE Transactions on Biomedical Circuits and Systems*, 3(5), 293–303, 2009.
10. T.L. Szabo, *Wave Scattering and Imaging, Diagnostic Ultrasound Imaging: Inside Out*, Burlington, MA: Elsevier, 2004, pp. 213–242.
11. C.J. Pavlin and F. Stuart Foster, *Ultrasound Biomicroscopy of the Eye*, New York: Springer-Verlag, 1994.
12. M.E. Schafer and P.A. Lewin, The Influence of Front-End Hardware on Digital *Ultrasonic Imaging, IEEE Transactions on Sonics and Ultrasonics*, SU-31, 4, 295–306, 1984.
13. Wells, P.N.T. *Advances in Ultrasound Techniques and Instrumentation*, New York: Churchill Livingstone, 1993.
14. M. Karaman, A.E. Kolağasıoğlu, and A. Atalar, A VLSI receive beamformer for digital ultrasound imaging, *Proceedings of IEEE International Conference on Acoustic, Speech, Signal Processing*, San Francisco, CA, 1992, pp. V.657–V.660.
15. A. Kassem, J. Wang, A. Khouas, M. Sawan, and M. Boukadoum, Pipelined sampled-delay focusing CMOS implementation for ultrasonic digital beamforming, *Proceedings of the 3rd IEEE Workshop on SoC for Real-time Application*, 2003, pp. 247–250.

16. I.G. Mina, H. Kim, I. Kim, S.K. Park, K. Choi, T.N. Jackson, R.L. Tutwiler, and S. Trolier-McKinstry, High frequency piezoelectric MEMS ultrasound transducers, *IEEE Transactions on Ultrasonics, Ferroelectrics, and Frequency Control*, 54(12), 2422–2430, 2007.
17. S. Triger, J. Wallace, J.-F. Saillant, and S. Cochran, D.R.S. Cumming, MOSAIC: An integrated ultrasonic 2D array system, *2007 IEEE Ultrasonics Symposium*, New York, 2007.
18. R. Reeder and C. Petersen, The AD9271-A Revolutionary solution for portable ultrasound, 2007 [Online]. Available: http://www.analog.com/library/analogDialogue/archives/41–07/ultrasound.pdf
19. B. Murmann, ADC performance survey 1997–2008, 2013 [Online]. Available: http://www.stanford.edu/~murmann/adcsurvey.html
20. B. Murmann, A/D converter trends: power dissipation, scaling and digitally assisted architectures, *IEEE 2008 Custom Integrated Circuits Conference (CICC)*, San Jose, CA, September 21–24, 2008, pp.105–112.
21. A.C. Kak and K.A. Dines, Signal processing of broad-band pulsed ultrasound: Measurement of attenuation of soft biological tissues, *IEEE Transactions on Biomedical Engineering*, BME-25, 321–344, 1978.
22. H.B. Meire and P. Farrant, *Basic Ultrasound*, Chichester, England, U.K.: John Wiley & Sons, 1995.
23. R. Jacob Baker, Data converter SNR, in *CMOS Mixed-Signal Circuit Design*, Vol. 2, Piscataway, NJ: IEEE Press, 2002, pp. 63–148.
24. B. Razavi, Differential amplifiers, in *Design of Analog CMOS Integrated Circuits*, New York: McGraw-Hill Companies, Inc. 2001, pp.100–132.
25. C. Wu, C. Liu, and S. Liu, A 2GHz CMOS variable-gain amplifier with 50 dB linear-in-magnitude controlled gain range for 10GBase-LX4 ethernet, *ISSCC 2004*, Vol. 1, 2004, pp. 484–541.
26. C.E. Shannon, Communication in the presence of noise, *Proceedings Institute of Radio Engineers*, 37(1), 10–21, 1949.
27. J. Yoo, D. Lee, K. Choi, and A. Tangle, Future-ready ultrafast 8-Bit CMOS ADC for system-on-chip applications, *14th Annual IEEE International ASIC/SOC Conference*, 2001, pp. 455–459.
28. J. Yoo, K. Choi, and D. Lee, Comparator generation and selection for highly linear CMOS flash analog-to-digital converter, *Journal of Analog Integrated Circuits and Signal Processing*, 35, 179–187, 2003.
29. I. Kim, J. Yoo, J.-S. Kim, and K. Choi, Highly efficient comparator design automation for TIQ flash A/D converter, *IEICE Transactions on Fundamentals*, E91-A, 12, 3415–3422, 2008.
30. UC Berkeley Device Group, "BSIM", 2013 [Online]. Available: http://www-device.eecs. berkeley.edu/~bsim3/
31. J.H. Kim, T.K. Song, and S.B. Park, Pipeline sampled-delay focusing in ultrasound imaging systems, *Journal of Ultrasonic Imaging*, 9, 75–91, 1987.

10

Recent Advances in Capacitive Micromachined Ultrasonic Transducer Imaging Systems

Albert I. H. Chen, Lawrence L. P. Wong, and John T. W. Yeow

CONTENTS

10.1 Introduction

Ultrasound is one of the essential medical imaging modalities. The main advantage of ultrasound is its ability to penetrate tissues without the generation of harmful ionizing radiation. Therefore, it is not surprising that obstetric ultrasonography is the method of choice for prenatal care such as the monitoring of fetuses. Other benefits of using ultrasound technology include the availability of real-time results and portability of imaging units.

On the other hand, ultrasound image quality may not be comparable to other modalities such as magnetic resonance imaging or computed tomography. In addition, it is difficult for ultrasound to penetrate bones or air, making the imaging of certain organs (e.g., the heart) a challenge. To overcome those challenges, researchers have been exploring various techniques to improve performance as well as to reduce the size of ultrasonic transducers in order to effectively perform in vivo imaging.

Conventionally, ultrasonic transducers comprised of piezoelectric materials such as lead zirconate titanate (PZT). These transducers rely on the piezoelectric effect, a mechanical deformation due to the presence of an electric field and vice versa, to generate and detect ultrasound. However, as micromachining technologies mature, a new type of ultrasonic transducer has emerged. Over the past decade, research in capacitive micromachined ultrasonic transducers (CMUTs) has demonstrated promising progress (see Figure 10.1). The microelectromechanical systems (MEMS)-based transducer was first constructed by the research group from Stanford University (Haller and Khuri-Yakub, 1994). These researchers envisioned a new type of transducer, fabricated using MEMS technology, for detecting cracks on aircrafts. These transducers had demonstrated excellent sensitivity and were able to offer superior resolution than traditional PZT transducers. It was soon realized that an even higher performance resulted when the transducer was immersed (Jin et al., 1998). As the imaging resolution is generally limited by the transducer, this had led to an incredible growth of interest toward CMUT research for medical imaging.

A scanning electron microscope image of CMUT cells is shown in Figure 10.2, while the schematic of a CMUT is shown in Figure 10.3. A typical CMUT cell consists of a movable membrane, a cavity, and two electrodes. If the membrane is electrically conductive, a layer of insulator

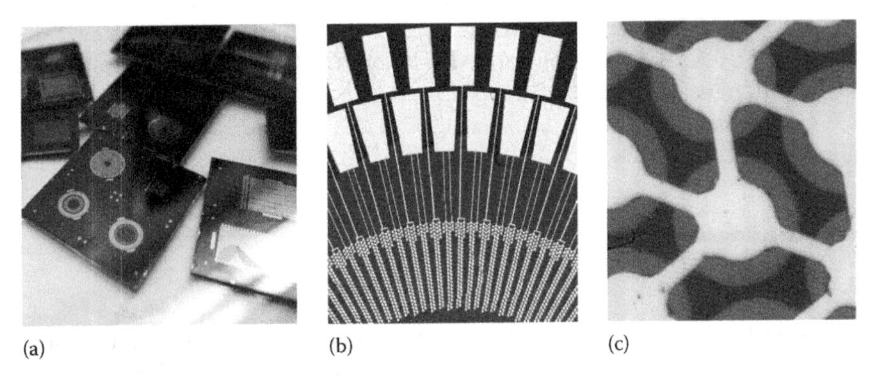

FIGURE 10.1
(a) Different CMUT designs can be made for different applications. Packing large number of ultrasonic transducers into a tiny area has never been easier! (b) Transducer array connected to electrodes demands packaging solutions. (c) Each CMUT cell is much smaller than the diameter of the human hair!

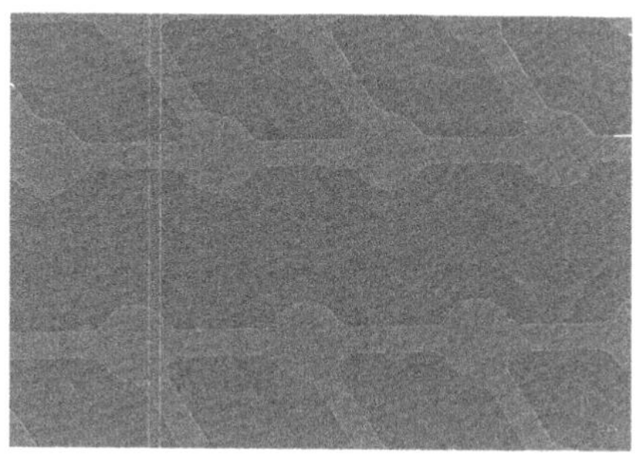

FIGURE 10.2
CMUT membrane diameter for a 10 MHz operation: 22 μm in diameter, 400 nm in membrane thickness, 11 μm in electrode diameter, and 140 nm in cavity depth. This 1D CMUT has 4140 cells per element.

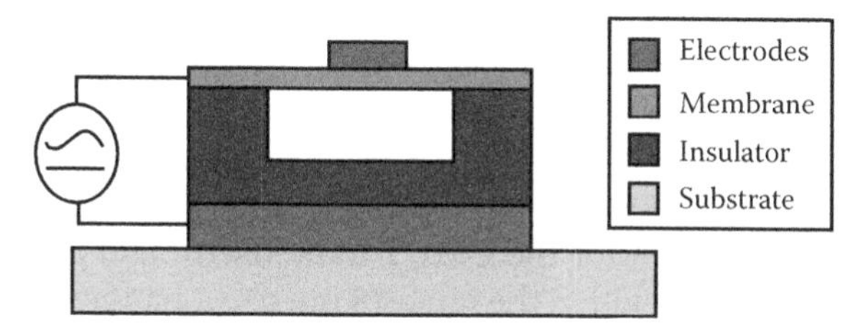

FIGURE 10.3
Cross-sectional schematic of a CMUT cell.

is also required to prevent the two electrodes from shorting. The cavity is usually vacuum-sealed to enhance the performance and the reliability of the CMUT. The CMUT structure looks similar to a parallel plate capacitor, with the top electrode and the membrane forming the top plate and the bottom electrode and the substrate forming the bottom plate. CMUTs are actuated in a fundamentally different manner than PZT transducers. They operate like a condenser microphone: when provided with a voltage excitement and a fixed bottom plate, the top plate vibrates and hence transmits sound wave. Conversely, when sound waves hit the top plate, the membrane displaces and, in turn, changes the capacitance between the plates. During this *receive* event, the voltage across the electrodes is held fixed, so as capacitance changes, electrical current is produced. From this step onward, the rest of the imaging system varies little from a conventional ultrasound imager.

CMUTs are fabricated using the same techniques as the common integrated circuit (IC) production. As a result, these novel transducers can be manufactured at a lower cost and with a higher yield. More importantly, standardized micromachining techniques allow CMUTs to be fabricated on ICs (CMUT on CMOS), producing tiny feature sizes. In other words, CMUTs are designed to be integrated with microelectronic circuits with ease—we will soon discuss why this is of major significance.

This chapter is an overview of recent advancements in CMUT systems. In the following section, we will discuss how CMUTs are fabricated and what design considerations are made. In Section 10.3, the main benefits of using CMUTs are presented. One of these benefits includes the effective integration of CMUTs with the complementary metal-oxide semiconductor (CMOS) technology—we will discuss this in Section 10.4. The applications of CMUTs are then presented in Section 10.5, while the challenges associated with the technology will be presented in Section 10.6.

10.2 Design and Fabrication

Given the similarities between a condenser microphone and a CMUT, one might wonder why the CMUT was not developed sooner. In fact, the main problem researchers faced was related to one of the main CMUT design parameters, the depth of the cavity or gap height. In order to generate an electric field large enough for the CMUT to be useful and avoid damaging the device, a small gap height in the range of micro- or nanometer is required. Thanks to the advances in micromachining technologies, such a small gap can now be easily achievable. When designing a CMUT, one should choose a small gap height to keep the bias voltage reasonable. On the other hand, the gap height should be large enough to allow room for the membrane to vibrate. Most of the time, however, the range of achievable gap height is also determined by the fabrication process.

Another important CMUT design parameter is the horizontal dimension of the CMUT membrane, or the diameter of the membrane in the case that the CMUT is circular. A bigger membrane area results in a lower resonant frequency, thus a lower ultrasound frequency. The ultrasound frequency is important because a higher frequency means a higher axial resolution in the resulting images. However, lower-frequency ultrasound has a better penetrability, so there is always a design trade-off. The size of the CMUT membrane is usually in the micrometer range for typical medical ultrasound frequency, which is in the MHz range. In order to generate sufficient energy, many CMUT cells are connected in parallel to form an element. The shape of the CMUT membrane is another key factor to be considered. Square- and hexagon-shaped membranes were reported to produce the maximum area efficiency of the device; but circular membrane is most often used in CMUTs because it generates the lowest local stress (Huang et al., 2003).

The materials used to construct the CMUTs also affect the device performance. A stiff membrane results in a higher ultrasound frequency but a lower signal amplitude because it takes more energy to vibrate. Silicon nitride and silicon oxide are two common materials for the CMUT membrane and dielectric layer (Logan and Yeow, 2009). They are used because of their mechanical properties and the fact that they can be easily grown on a silicon substrate using standard fabrication processes.

The most important CMUT fabrication step is the formation of the cavities. Different methods were proposed to form the cavities, but they can be categorized into two types: surface micromachining and bulk micromachining. Surface micromachining involves depositing the layers of the CMUT structure one by one, from the bottom to the top. The cavity is initially occupied by a sacrificial layer, which is then removed upon the completion of the structure. This step is also known as the releasing of the structure or the membrane. In order to allow etchant access to the sacrificial layer, etch holes are required on the top layers. Moreover, the sacrificial layer cannot be too thin; otherwise, the membrane may not be released properly. If a vacuum cavity is desired, the etch holes need to be sealed off. Bulk micromachining involves fabricating the bottom electrode on one wafer and the top electrode on another wafer. The wafers are bonded together after the cavities are etched on one of the wafers; therefore, the process is also known as wafer bonding. Because the cavities are formed on the top layer before being covered by another wafer, very small gap heights can be achieved using wafer bonding. Bulk micromachining is the more popular choice of CMUT fabrication because more efficient structures can be realized using wafer bonding.

Since the membrane and top electrodes are susceptible to abrasion, an encapsulation layer is required to protect the device. Furthermore, this layer should insulate the electrical components from the environment while maintaining certain biocompatibility standards. More importantly, the insulating material should minimize any adverse impact on the CMUT's acoustic performance. In its working environment, an ideal encapsulating material will have a static Young's modulus low enough to allow the CMUT elements to pull in during DC excitation as well as a dynamic Young's modulus high enough to allow acoustic matching with water at ultrasonic frequencies. Currently, polydimethylsiloxane (PDMS) is favored as the viscoelastic material for depositing on the CMUT since it meets performance specification and is well characterized for modeling (Lin et al., 2011). Ideally, a thin coating is desired as it minimizes any acoustical effect on the CMUT but is impractical as it may be easily damaged by ablative force. Through modeling and experiments, a layer in the range of 150–300 μm is considered acceptable for meeting specifications (Lin et al., 2010). Figure 10.4 shows how the encapsulation layer fits with a 1D array device. The layer can be molded into a lens for better focusing.

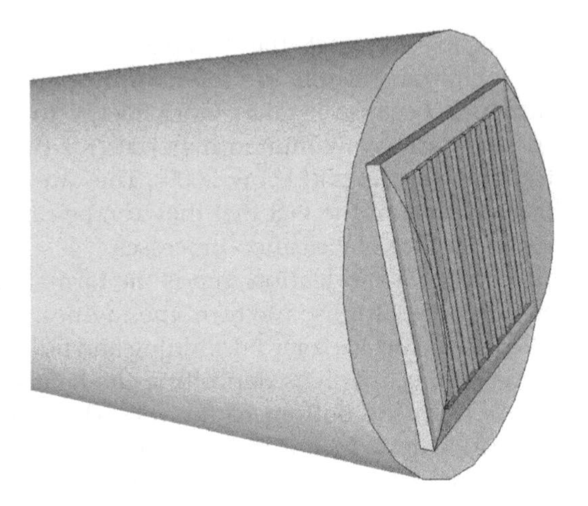

FIGURE 10.4
An encapsulation layer is typically required to insulate and protect the CMUTs. The encapsulation layer can also be molded into lens for 1D array.

10.3 Benefits

10.3.1 Wider Bandwidth

Compared to a PZT transducer, the vibrating membrane of a CMUT is substantially thinner. As a result, a CMUT experiences a much higher damping that leads to a wider bandwidth, or less *ringing*. Imagine if ringing effect was significant in our ears, we would be hearing echoes and distorted sounds constantly and be unable to appreciate a symphony and clearly distinguish between notes! In short, it is extremely important to have a wider bandwidth in order to resolve spatially.*

One reason why PZT transducers are bulky, relative to CMUTs, is that an *acoustic impedance matching* layer is required. A structure's acoustic impedance is dependent on the material composite and also geometry. For any vibration energy to transfer efficiently between mediums, the impedance of the source medium should match the impedance of the medium of interest. An impedance mismatch creates inefficiency in the acoustic energy transfer and will introduce noise and heat. Although an impedance matching layer can be designed to minimize the inefficiency, any extra thickness will contribute to additional ringing effect.†

* To be more precise, a wider bandwidth improves the *axial* resolution but not so much the *lateral* resolution.

† A great read to understand PZT transducer design can be found in Diagnostic Ultrasound Imaging by Thomas L. Szabo (Szabo, 2004).

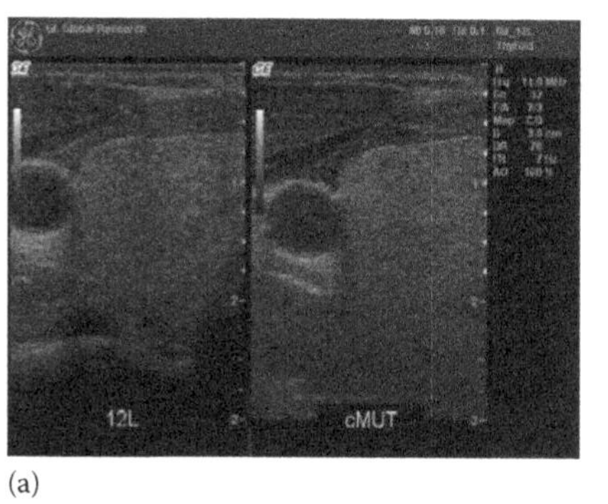

(a)

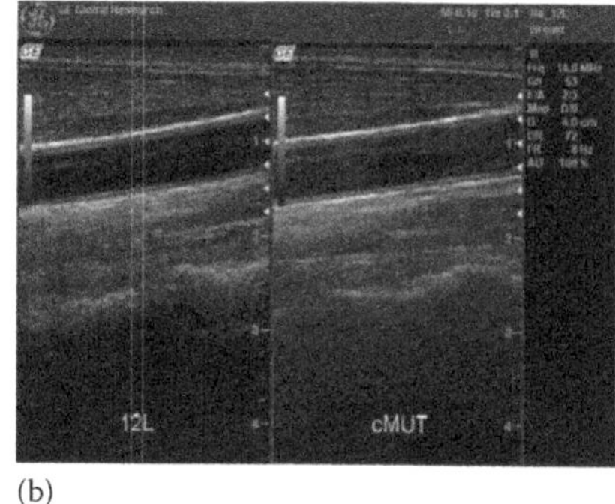

(b)

FIGURE 10.5
Comparison of imaging results between PZT array (left) and CMUT array (right). (a) Imaging of the carotid artery. (b) Imaging with the thyroid gland froma. Notice the increase in structural details of the CMUTs. (From Mills, D.M. and Smith, L.S., *IEEE Symposium on Ultrasonics*, pp. 568–571, 2003. Copyright IEEE with permission.)

In 2003, General Electric Global Research Center reported the first real-time CMUT probe with linear CMUT arrays. Compared to a benchmarking PZT probe, the CMUT images demonstrated improved anatomical border and texture information. Figure 10.5 shows the image comparison between PZT-based and CMUT-based imaging ultrasound probes (Mills and Smith, 2003). In 2009, Hitachi Medical Corporation reported the first successful commercialization of a CMUT-based array capable of 2D cross-sectional imaging (Hitachi Medical Corporation, 2009).

10.3.2 Smaller Pitch

In modern medical ultrasound imaging, it is common to focus sound with an array of transducer elements instead of a single element. Furthermore, electrical focusing, or *phased-array* focusing, is used to steer and focus the sound beam. The idea is to orchestrate pulses from each element in the array such that a maximum constructive interference can be created at a desired destination. As a result, the signal-to-noise ratio (SNR) can be increased as sound energies from multiple transducers are focused at a single point while any mechanical movement is eliminated. The latter property improves the scanning speed and accuracy when compared to traditional focusing typically requires lens and mechanical actuators for beam steering.

Eliminating mechanical movements is beneficial but phased array introduces a critical design constraint: to eliminate aliasing (also known as grating lobes in ultrasound terminology), which causes image blur, the *array pitch must be less than half of the operating wavelength*. The concept of grating lobes is illustrated in Figure 10.6. For high-frequency imaging, which is typically greater than 25 MHz, the pitch is required to be less than 30 μm. For PZT transducer fabrication, it is very difficult to achieve less than

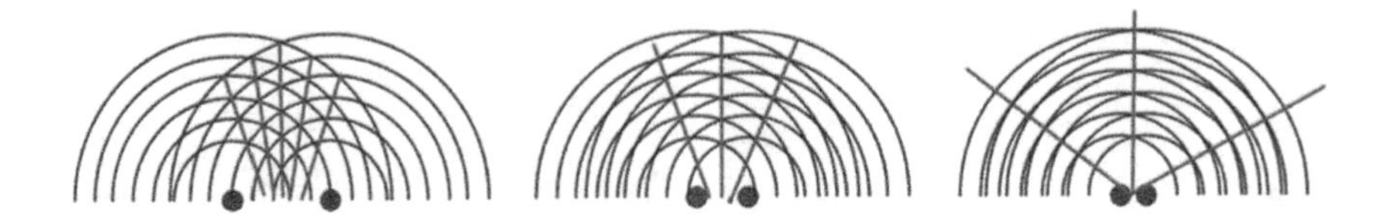

FIGURE 10.6
Ideally, the sound should focus at only one point. However, when transducers are far apart, constructive interference may occur at undesirable places and *grating lobes* occur.

40 μm in pitch due to limitations in the traditional dice-and-fill method. In contrast, from the standpoint of microfabrication (i.e., MEMS technology), micrometer-level feature size can be easily achieved. In conclusion, CMUTs are far less likely to encounter the grating lobe problem, thus better image quality.

10.3.3 IC Integration

Most medical ultrasonic imaging systems can be broken down into four parts: the ultrasonic transducers, the front-end circuitry, the cable, and the back-end processing unit. Since the transducers are typically encased in a housing that is remote to the processing unit, the raw signal received by the transducers can be significantly degraded due to cable capacitance. Therefore, front-end electronics must be incorporated with the transducers to maintain the signal integrity. Each transducer element has a dedicated transmitter and receiver circuit. The transmitter circuit typically includes a high-voltage switch to generate a large pulse, while the receiver includes an amplifier to enhance the measured signal.

It is difficult to package both the electronics and the high-density transducer array. Even though the CMOS technology can provide tiny ICs, it is still challenging to make the interconnections between individual elements and corresponding transmitter and receiver circuitry. (This is not true for CMUT technology but will be discussed shortly.) Consider a large array with 128 × 128 transducer elements; over 16,000 interconnections are needed to address each element individually; meanwhile the pitch size is constrained by the frequency. In catheter applications, although the element count is not high, interfacing the electronics is still challenging given the tiny workable area.

Traditionally, PZTs are diced into arrays with fine saws followed by the deposition of electrodes. The front-end components are fabricated separately and eventually soldered with the electrodes. However, at high densities and small surface areas, it becomes incredibly challenging to work with the traditional method. Thus, when the CMUT-on-CMOS process was demonstrated, it gave great hope to the future of ultrasound technology. The next section will discuss various CMUT and electronics integration methods.

10.4 CMUT–CMOS Integration Techniques

Although microfabrication can produce very tiny feature sizes (i.e., thin membranes and small pitches), there is a plethora of design constraints to be considered. For example, to ensure the membrane is bonded tightly onto the insulating material, high temperature is required. However, the electronics may be damaged in the high temperature. Therefore, some design trade-offs may be required. We will briefly discuss different ways of combining the CMUT with electronic circuits fabricated using the CMOS process.

10.4.1 CMUT in CMOS

CMUT and electronics are fabricated in parallel on the same wafer and under the same CMOS process (Cheng et al., 2009), as illustrated in Figure 10.7. This method reduces fabrication time and cost. However, the material type, the material properties, and the layer thickness of the CMUTs are subjected to the limitation of the CMOS process. This method is also not ideal for applications where a small transducer area is required.

10.4.2 Flip-Chip Integration

This process allows the CMUTs and the electric circuits to be optimized separately, followed by flip-chip bonding (Noble et al., 2001). This way, the high temperature for CMUT fabrication will be isolated from the CMOS fabrication. In addition, the total occupied area is minimized because both the CMUT and the CMOS circuits are stacked. However, to satisfy the requirements for flip-chip bonding, the CMUT design must have through-wafer vias (TWVs) such that both the top and bottom electrodes can be

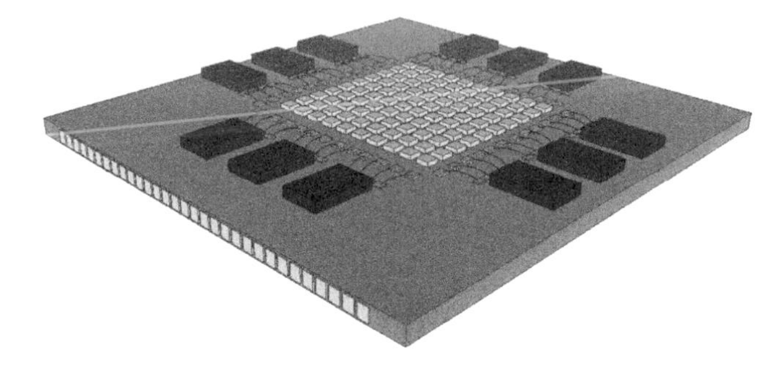

FIGURE 10.7
CMUT-in-CMOS process refers to the process where CMUTs and CMOS are fabricated on a single wafer in parallel.

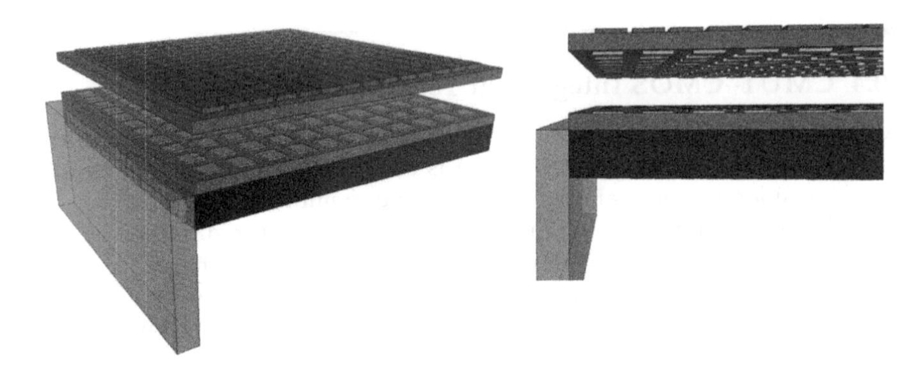

FIGURE 10.8
CMUT fabrication and CMOS process can be done separately followed by flip-chip soldering.

accessed by the bonding pads at the bottom of the wafer (Figure 10.8). As a result, this method is rather complex. Nonetheless, since flip-chip integration allows CMUT design optimization, it remains popular with current imaging research.

10.4.3 CMUT on CMOS

The most cost-effective method is the CMUT-on-CMOS process. Here, CMUT fabrication process is fabricated on top of the CMOS circuits (Gurun et al., 2011) (Figures 10.9 and 10.10), resulting in minimal in-plane area and minimal parasitic capacitance caused by electronic traces and interfaces. The main problem associated with this technique is that the fabrication temperature for the CMUTs is limited to protect the electronics. However, this process is still an active research area as steady improvements have been made on fabricating better CMUTs under low temperature.

FIGURE 10.9
CMUTs can be built on top of CMOS. This creates a very tight packaging.

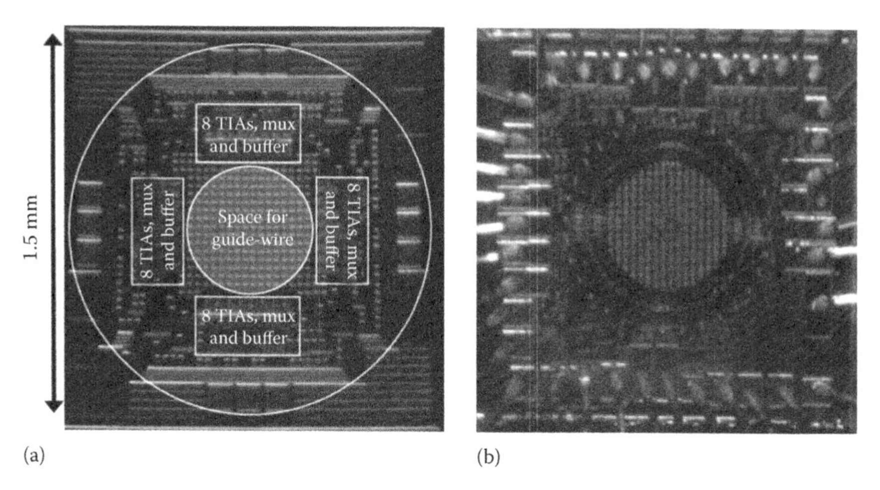

FIGURE 10.10
The application-specific integrated circuits (ASICs) can be seen (a) before the CMUT deposition (b). (From Gurun, G. et al., *IEEE Transactions on Ultrasonics, Ferroelectrics and Frequency Control*, 58–8, 1658–1668, 2011. Copyright IEEE with permission.)

10.5 Applications

The advantages of CMUT have shaped the technology to evolve into two main branches in terms of medical imaging applications. The first branch of research revolves around in vivo application where CMUTs are typically placed on the tip of catheters or endoscopes for intravascular-/intracardiac-like ultrasound imaging. Within this branch, packaging and interface engineering are the key factors in research. The second branch involves developing large and high-density 2D arrays to image large areas, typically from outside the body. Within this branch, the key factors include not only the two mentioned but also electronics simplification schemes. In this section, we will discuss the two branches, followed by the latest development in CMUT systems.

10.5.1 Small 2D Array

Intravascular ultrasound and intracardiac echocardiography (IVUS and ICE) imaging has been commercially available for over a decade. Angiography, the most widely used vascular imaging technique, cannot provide closeup anatomy of the lumen. By attaching ultrasonic transducers at the distal end of a catheter, the anatomy near the tip can be better perceived.

Traditionally, IVUS and ICE have side-looking apertures; therefore, imaging is performed radially. In recent years, engineers have been trying to implement forward-looking IVUS catheters that provide much more useful information such as plaque occluding the blood vessel. Most of the methods adopt a single piezoelectric transducer and some sort of mechanical actuators to redirect the sound beam. As discussed previously, mechanical actuation limits the frame rate and also introduces noise. In the realm of microfabrication, where millimeter spacing is considered enormous, phased-array imaging becomes a possibility because a CMUT array can be easily placed at the face of the tip instead of at the outer surface, eliminating the need for mechanical movement during a scan. In fact, several research groups have demonstrated ring array CMUT catheters where a hollow center is kept (see Figure 10.11) for fitting an ablation device capable of burning plaque occluding the vessels using high-intensity focused ultrasound (HIFU).

The rate of growth of CMUT catheters has been quite noteworthy. The Edward L. Ginzton Laboratory of Stanford University and General Electric have recently reported in vivo testing of a 9F (3 mm in diameter) CMUT catheter (Stephens et al., 2012). The device contains a linear 24-element array and has an imaging depth of penetration of around 30 mm. Using an ultrasound frequency of 8 MHz, the CMUT catheter is capable of producing high-resolution real-time 2D images as shown in Figure 10.12. Real-time volumetric imaging, with a catheter-size 64-element CMUT ring array, as illustrated in Figure 10.13, has also been demonstrated by the same team (Choe et al., 2012).

The research group from the Georgia Institute of Technology also reported a catheter-based CMUT system (Degertekin et al., 2006). Their setup also features a forward-looking ring array, but the main difference in their design is the use of CMUT-on-CMOS integration. The CMUT-on-CMOS approach reduces the parasitic capacitance of interconnects and eliminates a lot of wires, making it a very attractive option for CMUT catheter integration.

Increasing the element count and the density of the transducer will result in a better SNR and a better image resolution. However, as the element count increases, more front-end electronic components and cables are required. In the next section, we will discuss briefly several options that different research groups are investigating.

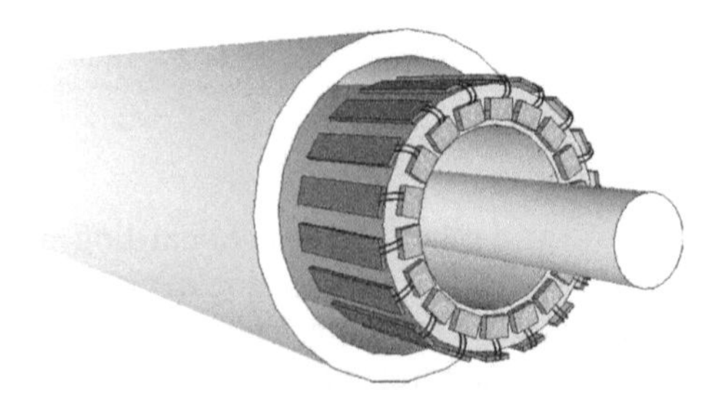

FIGURE 10.11
Ring array designs are very popular in IVUS applications since they allow a hollow center for guide wires or the possible integration of a thermal ablation tool.

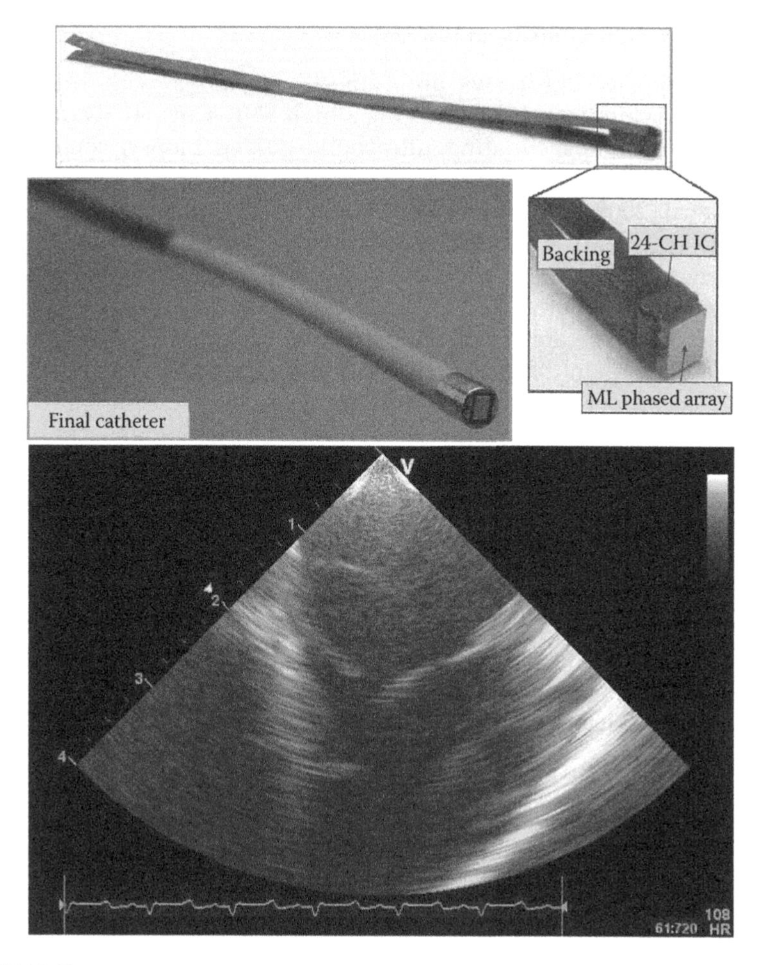

FIGURE 10.12
A 9F forward-looking microlinear catheter, in vivo in a pig heart. (From Nikoozadeh, A. et al., *IEEE International Ultrasonics Symposium Proceedings*, 2010. Copyright IEEE with permission.)

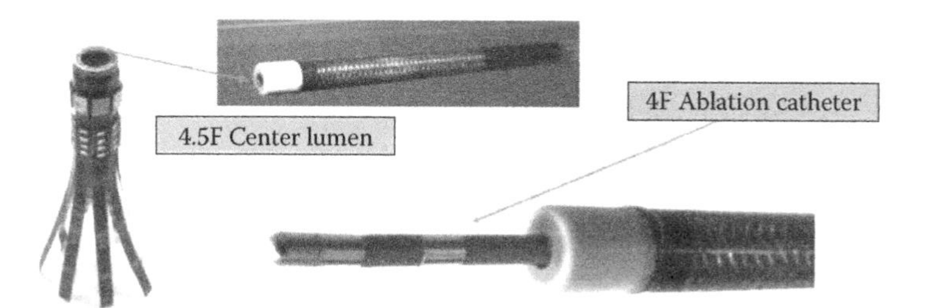

FIGURE 10.13
Real-time volumetric imaging, with a catheter-size 64-element CMUT ring array. (From Moini et al., *IEEE International Ultrasonics Symposium Proceedings*, 2011. Copyright IEEE with permission.)

10.5.2 Large/High-Density 2D Array

Large/high-density 2D arrays are typically designed with the goal of imaging large volumes while achieving a high SNR. Current 3D ultrasound applications include facilitating ultrasound-guided biopsy, screening for breast and rectal cancer, and also prenatal imaging (Leproux et al., 2010; Maruotti et al., 2009; Natarajan et al., 2011). Currently, most of the large/high-density 2D array CMUT research is conducted in parallel with solutions that simplify the interfaces. Although the actual size of the array can vary from millimeters to centimeters, the element count can vary from 1,000 to more than 1,000,000. There are two main problems that arise from this: interfacing with a large quantity of elements and ensuring the amount of data can be processed properly such that the frame rate is still considered real time. With the CMOS process, a large number of electronic circuits can be packaged into a single chip. However, at high ultrasound frequencies where the pitch size of the phased array is less than 25 μm, the area constraint placed on the electronic circuits becomes much more difficult to meet. Assuming that the area required by the electronic circuits can be small, the cable connecting the probe to the processing system can only support a small quantity of micro-coaxial cables (typically less than 128 lines) before the cable becomes a long and stiff metal rod. As a result, some kind of simplification scheme using multiplexing or simplified phased-array algorithm is necessary. Some ongoing research techniques are briefly mentioned in the following.*

10.5.2.1 Sparse Array

This simplification technique is based on an optimization problem whereby the least amount of transducer elements is used to recreate a sound beam that can otherwise be achieved by using all the elements. Depending on the pitch, size of elements, and geometries, an acoustic field, or spatial impulse responses, can be calculated. By convoluting the responses of transmit and receive aperture, sparse array researchers can come up with different sparse patterns to tackle the optimization problem. Several sparse array patterns are shown in Figure 10.14. One obvious trade-off, however, is that with fewer elements, the SNR decreases (Lockwood and Foster, 1994).

10.5.2.2 Synthetic Array

Given that transmit and receive channels or circuitry numbers may be insufficient to support the transducer, synthetic array approaches merely break down the transmit and/or receive aperture to sub-apertures at the cost of increasing the number of pulse-echo events. An analogy can be made where

* For a great comprehensive explanation of some of these techniques, please refer to "The Future of Beamforming" by K. E. Thomenius!

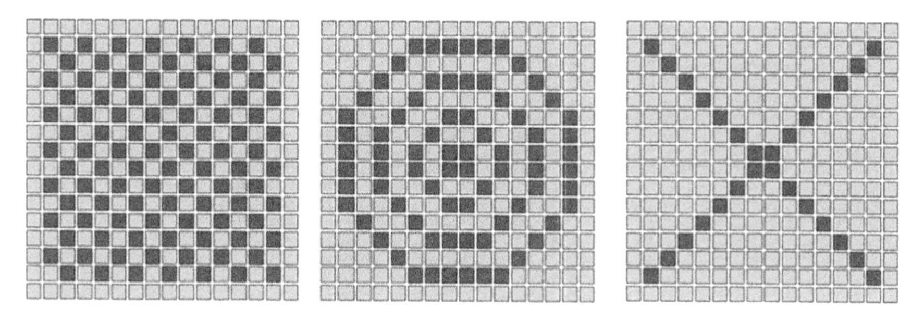

FIGURE 10.14
Some sparse array patters. Darker colored pixels represent transmit elements and lighter pixels represent receive elements.

multiple low-resolution images are taken to form a high-resolution image. However, this technique may not be suitable for events where the object moves at high speed (i.e., heart valves). Nonetheless, since image quality is typically of prime concern, synthetic array approach remains popular in 2D arrays (Thomenius, 1996).

10.5.2.3 Row–Column (Also Known as Crossed Electrodes)

Row–column (RC) focusing employs all elements to transmit and receive, thus maintaining a SNR comparable to that of an N by N full-matrix phased-array method. In addition, only 2N leads, pulsers, and amplifiers are used instead of N by N. This dramatically reduces the bulkiness of the system and makes the system extremely easy to implement. Also, the time it takes to complete a volumetric scan is comparable to a 1D array. These three properties address all the shortcomings associated with the sparse and synthetic array.

The RC method works by virtually rotating the aperture of an elongated 1D array between transmit and receive events. Imagine a 1D array overlaying another 1D array at an orthogonal rotation about the center. As depicted in the Figure 10.15, each transducer is connected to a column bus on the top while connected to a row bus on the bottom. At transmit mode, a column of beam is focused at each event, while at receive mode, the 2D information containing the plane intersecting the column of beam and the aperture is received (Figure 10.16). For more in-depth explanation, please refer to Logan et al. (2011).

Although there are many simplification schemes, there has not yet been a recognized "optimum" solution to tackle high-density arrays. It is likely that the electronics can physically scale down in the near future as the CMOS process continues to improve at an astonishing rate. With the general trend toward using GPU for 3D image reconstruction, the authors believe information processing or digital computation is likely not going to limit the system's quality or frame rate. Given that the speed of sound is ultimately the bottleneck that is limiting the frame rate and resolution, the authors believe that

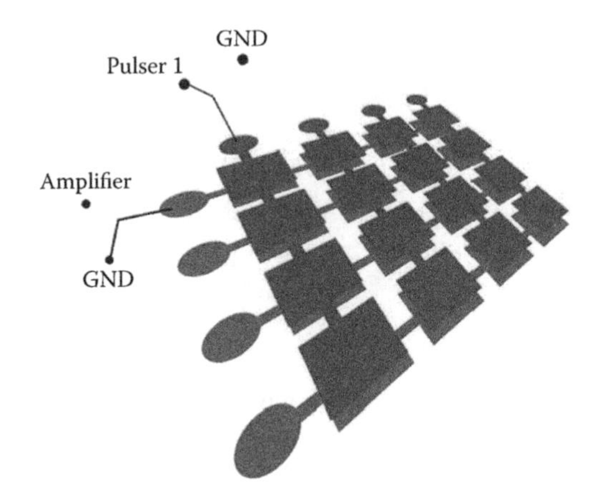

FIGURE 10.15
Row–column connection scheme. The top electrode is connected in columns, while bottom is in rows.

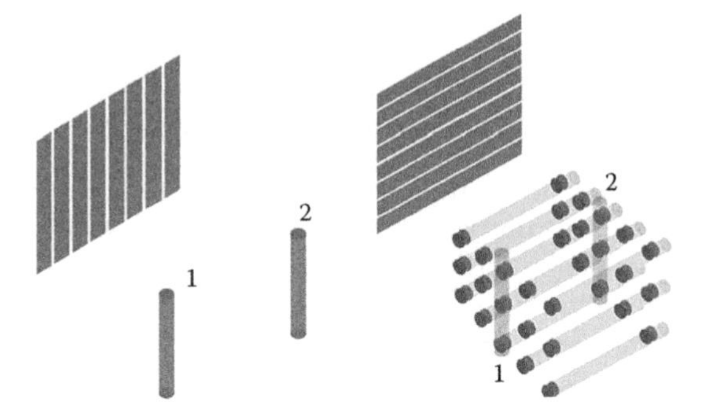

FIGURE 10.16
A single transmit column beam is formed at the transmit event. During receive event, a 2D section (dark pixels) is recorded.

an immediate problem that requires solving lies within the development of the next generation of ultrasound cable. To ensure that high-channel count CMUT arrays can be used immediately, the development of high-density yet thin and flexible cables are needed. Otherwise, perhaps the processing unit can be so small that it can be part of the probe.

10.6 Challenges

CMUT offers so many benefits over the traditional PZT transducer, yet there are several challenges that the CMUT technology needs to overcome before it can be widely adopted. The first issue is dielectric charging. When a large electric field is applied to the device, electric charges could get trapped in the insulating dielectric layer. The trapped charges create a shift in the CMUT internal electric field, causing the CMUT resonant frequency, and the resulting ultrasound frequency, to change. If different elements in a phased-array experience different frequency changes, image quality will suffer. In addition, a shift in the electric field could reduce the CMUT efficiency or cause the device to collapse. Novel CMUT structures have been proposed to mitigate the dielectric charging problem (Huang et al., 2005), but more research still needs to be done in this area.

Another limitation of the CMUT technology is the cross talk between transducer elements. Because all elements are situated on the same substrate, vibration of one element could spread to neighboring elements. The result of the cross talk is the degradation of image quality, because the unintended vibration of elements generates ultrasound that becomes noise. Models have been developed to help researchers understand the mechanism of cross talk in CMUTs (Bayram et al., 2005, 2006).

Perhaps the biggest obstacle that prevents CMUTs from achieving early success is the reliability of the CMUT device. Unlike PZT transducers that are rigid, CMUTs employ the vibration of thin membranes that can be easily damaged. Encapsulation layers have been used to protect the CMUTs, but there is always a trade-off between image quality and reliability, because while a thicker encapsulation layer provides a better protection, it reduces the efficiency of the CMUT. Fortunately, device longevity is not the biggest concern for catheter-based CMUTs because the catheter is discarded after each use. Therefore, one of the research objectives for catheter-based CMUTs is to minimize the manufacturing cost of the CMUTs and the associated front-end electronic circuits.

10.7 Conclusion

Researchers have been working on improving the CMUTs and bringing the technology to the market over the past two decades. Thanks to micromachining technologies, CMUTs offer several benefits over traditional PZT transducers such as a better axial resolution and a higher degree of integration. Because of those advantages, CMUTs are especially suitable to be used to construct small arrays in catheters or endoscopes or in high-density 2D array applications. Recent research, for example, the CMUT-integrated catheter, has demonstrated novel CMUT systems that can overcome some of the limitations of traditional medical ultrasound imaging. The capability of CMUTs can be further improved by combining high-density 2D arrays with different phased-array algorithms.

The first commercialized CMUT probe is already available, and several CMUT prototypes have been reported. Even though the CMUT technology still faces some challenges, they are tremendously outweighed by its benefits. CMUTs will unlikely replace conventional PZT transducers, but the new medical applications make them impossible to ignore. CMUTs are here to stay!

References

Bayram, B., M. Kupnik, G.G. Yaralioglu, O. Oralkan, D. Lin, X. Zhuang, A.S. Ergun, A.F. Sarioglu, S.H. Wong, and B.T. Khuri-Yakub, Characterization of cross-coupling in capacitive micromachined ultrasonic transducers. *In IEEE International Ultrasonics Symposium*, 1(2005): 601–604. doi:10.1109/ULTSYM.2005.1602924.

Bayram, B., G.G. Yaralioglu, M. Kupnik, and B.T. Khuri-Yakub, 5F-4 acoustic crosstalk reduction method for CMUT arrays. *Proceedings in International IEEE Ultrasonics Symposium*, (2006): 590–593.

Cheng, X., D.F. Lemmerhirt, O.D. Kripfgans, M. Zhang, C. Yang, C.A. Rich, and J.B. Fowlkes, CMUT-in-CMOS ultrasonic transducer arrays with on-chip electronics. *In Solid-State Sensors, Actuators and Microsystems Conference*, (2009): 1222–1225. doi:10.1109/SENSOR.2009.5285878.

Choe, J.W., O. Oralkan, A. Nikoozadeh, M. Gencel, D.N. Stephens, M. O'Donnell, D.J. Sahn, and B.T. Khuri-Yakub, Volumetric real-time imaging using a CMUT ring array. *IEEE Transactions on Ultrasonics, Ferroelectrics and Frequency Control*, 59(2012): 1201–1211.

Degertekin, F.L., R.O. Guldiken, and M. Karaman, Annular-ring CMUT arrays for forward-looking IVUS: Transducer characterization and imaging. *IEEE Transactions on Ultrasonics, Ferroelectrics and Frequency Control*, 53(2006): 474–482.

Gurun, G., P. Hasler, and F.L. Degertekin, Front-end receiver electronics for high-frequency monolithic CMUT-on-CMOS imaging arrays. *IEEE Transactions on Ultrasonics, Ferroelectrics and Frequency Control*, 58(2011): 1658–1668.

Haller, M.I. and B.T. Khuri-Yakub, A surface micromachined electrostatic ultrasonic air transducer, *Proceedings in IEEE International Ultrasonics Symposium*, 2(1994): 1241–1244.

Hitachi Medical Corporation, Development of ultrasonic transducer 'Mappie' with cMUT technology, 2009, http://www.hitachi-medical.co.jp/tech/medix/pdf/vol51/P31–34.pdf (accessed on May 21, 2013).

Huang, Y., A.S. Ergun, E. Haeggstrom, and B.T. Khuri-Yakub, New fabrication process for capacitive Micromachined ultrasonic transducers. *In IEEE the Sixteenth Annual International Conference on Micro Electro Mechanical Systems*, (2003): 522– 525. doi:10.1109/MEMSYS.2003.1189801.

Huang, Y., E.O. Haeggström, X. Zhuang, A.S. Ergun, and B.T. Khuri-Yakub, A solution to the charging problems in capacitive micromachined ultrasonic transducers. *IEEE Transactions on Ultrasonics, Ferroelectrics, and Frequency Control*, 52(2005): 578–580.

Jin, X.C., I. Ladabaum, and B.T. Khuri-Yakub, Surface micromachined capacitive ultrasonic immersion transducers. *Proceedings in the Eleventh Annual International Workshop on Micro Electro Mechanical Systems*, (1998): 649–654.

Leproux, A., M. Van Beek, U. De Vries, M. Wasser, L. Bakker, O. Cuisenaire, M. Van Der Mark, and R. Entrekin, Automated 3D whole-breast ultrasound imaging: Results of a clinical pilot study. *In SPIE Medical Imaging*, (2010): 762902–762902.

Lin, D-S., Interface engineering of capacitive micromachined ultrasonic transducers for medical applications, Doctoral Dissertation, Stanford University, (2011): 54–68.

Lockwood, G.R. and F.S. Foster, Optimizing sparse two-dimensional transducer arrays using an effective aperture approach. *Proceedings in IEEE International Ultrasonics Symposium*, 3(1994): 1497–1501.

Logan, A.S., L.L.P. Wong, A.I.H. Chen, and J.T.W. Yeow, A 32 × 32 element row-column addressed capacitive micromachined ultrasonic transducer. *IEEE Transactions on Ultrasonics, Ferroelectrics and Frequency Control*, 58(2011): 1266–1271.

Logan, A. and J.T.W. Yeow, Fabricating capacitive micromachined ultrasonic transducers with a novel silicon-nitride-based wafer bonding process. *IEEE Transactions on Ultrasonics, Ferroelectrics and Frequency Control*, 56(2009): 1074–1084.

Maruotti, G.M., D. Paladini, R. Napolitano, L.L. Mazzarelli, T. Russo, M. Quarantelli, M.R.D' Armiento, and P. Martinelli, Prenatal 2D and 3D ultrasound diagnosis of diprosopus: Case report with post-mortem magnetic resonance images (MRI) and review of the literature. *Prenatal Diagnosis*, 29(2009): 992–994.

Mills, D.M. and L.S. Smith, Real-time in-vivo imaging with capacitive micromachined ultrasound transducer (cMUT) linear arrays. *Proceedings in IEEE International Ultrasonics Symposium*, 1(2003): 568–571.

Moini et al., Volumetric intracardiac imaging using a fully integrated CMUT ring array: Recent developments, *IEEE International Ultrasonics Symposium Proceedings*, Orlando, FL, 18–21(2011): 692–695.

Natarajan, S., L.S. Marks, D.J.A. Margolis, J. Huang, M.L. Macairan, P. Lieu, and A. Fenster, Clinical application of a 3D ultrasound-guided prostate biopsy system. *In Urologic Oncology: Seminars and Original Investigations*, 29(2011): 334–342.

Nikoozadeh, A. et al., Forward-looking intracardiac imaging catheters using fully integrated CMUT arrays, *IEEE International Ultrasonics Symposium Proceedings*, San Diego, CA, (2010): 770–773.

Noble, R.A., R.R. Davies, M.M. Day, L. Koker, D.O. King, K.M. Brunson, and A.R.D. Jones, Cost-effective and manufacturable route to the fabrication of high-density 2D micromachined ultrasonic transducer arrays and (CMOS) signal conditioning electronics on the same silicon substrate. *Proceedings in IEEE International Ultrasonics Symposium*, 2(2001): 941–944. doi:10.1109/ULTSYM.2001.991874.

Stephens, D.N., U.T. Truong, A. Nikoozadeh, O. Oralkan, C. HyungSeo, J. Cannata, A. Dentinger et al., First in vivo use of a capacitive micromachined ultrasound transducer array-based imaging and ablation catheter. *Journal of Ultrasound in Medicine: Official Journal of the American Institute of Ultrasound in Medicine*, 31(2012): 247–256.

Szabo, T., *Diagnostic Ultrasound Imaging: Inside Out: Inside Out*. Academic Press, 2004.

Thomenius, K.E., Evolution of ultrasound beamformers. *Proceedings in International Ultrasonics Symposium*, 2(1996): 1615–1622.

11

PET Detectors

Alberto Del Guerra and Nicola Belcari

CONTENTS

11.1 Introduction

The objective of a positron emission tomography (PET) scan is the measurement of the local activity density $\rho(x,y,z)$ of a β^+ emitting radioisotope, distributed in a living body.

The γ-ray pair, emitted from the annihilation of the positron with an electron, is used as a source of signals to determine $\rho(x,y,z)$. Thanks to the nearly collinear (back-to-back) emission of the γ-ray pair, it is possible to define the line L along which the annihilation occurred. The line L is then usually called line-of-flight (or LOF).

The activity distribution $\rho(x,y,z)$ is measured in terms of projections ($N_{\gamma-\gamma}$) along lines L.

Each projection is obtained from the activity distribution with the line integral operator

$$N_{\gamma-\gamma} = k\int \rho(x,y,z)dl$$

where k is a factor that takes into account the probability for the γ-ray pair to be detected.

The lines L are collected by surrounding the object by detectors typically arranged in ring geometry like in Figure 11.1. Each detector is "coupled" with all the elements in an arch of opposing detectors so as to measure the LOFs. The intersection of all the so-defined sectors is called field-of-view (FOV), and represents the region of the space where the completeness of the ensemble of the LOF allows the reconstruction of the $\rho(x,y,z)$. This model assumes that each annihilation takes place on a well-defined LOF, whilst, in reality, some deviation to this model may occur due to the physics of the β^+ decay and to the effects of the interaction of the annihilation photons with the surrounding matter, such as photoelectric absorption or Compton scattering. As a consequence, this deviation ultimately limits the best achievable spatial resolution and the quantitation in the measurement of the $\rho(x,y,z)$.

The objective of the PET measurement is then to obtain the spatial coordinates of the line L where the count is detected. This can be achieved by

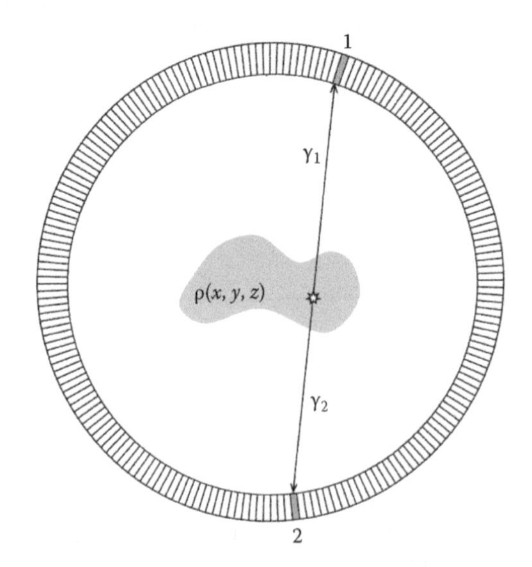

FIGURE 11.1
The collinear emission of an annihilation γ-ray pair defines the line-of-flight (LOF). The LOFs are collected by surrounding the object with a "ring" of detectors.

measuring for both γ-rays the coordinates $P(x,y,z)$ of the first interaction for each γ within the detector. In addition, the detector should also give information on the energy of the γ-rays and the time of the interaction. The accuracy and precision of energy and time measurements are relevant to reduce the noise in the final image (see Section 11.4).

11.2 PET Detectors

11.2.1 Bit of History

The simplest and most widely used solution to build a PET detector is a matrix of tiny inorganic scintillator elements (pixilated crystals) coupled to some sort of position-sensitive photodetectors able to identify the position of the pixel where the interaction occurs, the released energy, and to provide timing information.

The scintillator approach for PET detectors dates back to the early 1970s with the construction of the first positron camera by Ter-Pogossian (1975); 24 NaI scintillators, each one coupled to a photomultiplier tube (PMT), were arranged on a single hexagonal ring, providing a single transaxial image. During the 1970s, the advances in PET detector concept were mainly concerned with the addition of more detectors and rings. In those days, the one-to-one coupling between scintillator and photodetector was the limiting factor for increasing the detector granularity due to the relatively large size of the photomultipliers.

In 1986, the introduction of the "block detector" by Casey and Nutt (1986) changed the world of nuclear imaging. This invention made possible the construction of high-resolution PET tomographs at a much-reduced cost.

Still in 1986, other solutions were suggested as a PET detector. Hamamatsu proposed the first PET detector based on a position-sensitive PMT (R2486) (Uchida 1986) coupled to a matrix of bismuth germanate (BGO) crystals. The application of this system was mainly limited by the relatively high cost of the photodetector with respect to the detection improvement that could be achieved. Again in 1986, Lightstone (1986) proposed a detector module based on two $3 \times 5 \times 20$ mm^3 BGO crystals read out by one-to-one coupling with a 3×3 mm^2 silicon avalanche photodiode (APD). Although the one-to-one coupling of scintillator and photodetector would offer intrinsically better event positioning, energy resolution, timing and count rate performance, the cost and the difficulties of many readout channels strongly limited the application of this solution. Hence, almost all PET clinical tomographs built since 1986 have used some forms of the block detector. However, in the past 10 years, a renaissance of the other techniques is observed, as illustrated in Section 11.5.

11.2.2 Block Detector

In the original version of the *block detector,* a matrix of BGO crystals is attached to 4 PMTs arranged on a 2 × 2 matrix (Figure 11.2). The light from each crystal element is shared among the PMTs. To aid this process, the crystals are not optically independent, but the element-to-element separation is obtained by cuts at different depths. The closer the cut is to the side of the block, the deeper the cut is. The amount of light "seen" by each detector is linearly connected to the position of the crystal element. Information on the position of the crystal where the interaction occurred can be obtained from the PMT output signals S_A, S_B, S_C, and S_D, from PMTs *A, B, C,* and *D* arranged as in Figure 11.2, by calculating the centroid of the light distribution reaching the PMTs surface. The *x* and *y* coordinated can be derived by the simple formulae

$$x = \frac{(S_A + S_C) - (S_B + S_D)}{E}$$

$$y = \frac{(S_A + S_B) - (S_C + S_D)}{E}$$

where the energy *E* released in the interaction is given by

$$E = S_A + S_B + S_C + S_D$$

In a block detector, the precision in the determination of the position of the γ-ray interaction is given by the size of the crystal element. In PET clinical practice, the required spatial resolution along the axial direction is usually less critical than the one along the radial and tangential directions. For this reason, in some block detector rectangular pixels are used, with the longer side positioned along the axial direction.

In principle, the PMTs array and related electronics should provide a perfect discrimination of events that hit different crystal elements, that is, each element should be clearly identified as separated from the others. This process is usually called "coding." In practice, the spatial distributions of the

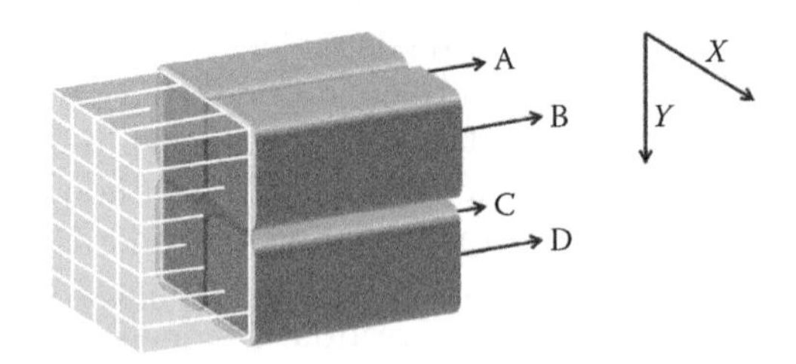

FIGURE 11.2
Scheme of a block detector. A block of scintillator is subdivided by cuts at different depths into 4 × 8 rectangular elements. The block is read out by a matrix of 2 × 2 photomultiplier tubes (outputs S_A, S_B, S_C, and S_D).

position of the events hitting different crystals are not perfectly separated and show a nonnegligible overlap. This leads to a second contribution to the degradation of the spatial resolution.

In addition, the block detector is able to measure two coordinates only (x and y) of the position of the interaction; the third coordinate (z, usually called depth-of-interaction, or DOI) can only be estimated considering the finite length of the crystal. Since its introduction, the block detector is still the preferred solution in clinical PET system.

11.2.3 Scintillators for PET

As already discussed, a PET detector should be able to (1) identify the position of the first interaction of a 511 keV γ-ray, (2) convert it to digital information while carrying information on released energy and on arrival time (coincidence measurement).

The most common solution is a scintillating material coupled to a photodetector. Scintillation detectors are widely used gamma-ray detectors that form the basis for almost all PET scanners in use today. These detectors consist of a dense crystalline scintillator material that serves as an interacting medium for γ-rays and emits visible light when energy is deposited inside them.

The intensity of the light (i.e., the number of photons) is "proportional" to the released energy. Formally, the light yield is defined as number of photons per MeV of deposited energy, usually indicated as ph/MeV or in percentage of the light yield of sodium iodide (NaI). The latter definition is useful to give a number that is approximately independent from the deposited energy.

The requirements of a scintillator to be a good candidate for PET are high electron density, high light yield, short decay time, and high effective atomic number.

Firstly, a high electron density is required so as to have a high linear attenuation coefficient at 511 keV, thus maximizing the intrinsic efficiency of the detector (for a given scintillator thickness). Other important factors are a high light yield and a short decay time.

A high light yield improves the identification of the pixels (affecting the coding factor b in the spatial resolution formula, as described in Section 11.3.2), the energy resolution (being $\Delta E/E$ proportional, excluding nonlinearity factors, to $1/\sqrt{\text{number of photons}}$ and timing performance). A short decay time helps in enhancing the timing resolution of the system, and then to reduce the width of the timing window, with a reduction of the random count rate (see Section 11.4.2) as a consequence. In addition, a very short decay time, and then a high timing resolution ($\leq$600 ps FWHM), is a fundamental requirement in time-of-flight PET.

Another important parameter is the effective atomic number (Z_{eff}), since the higher is Z_{eff} the higher is the fraction of photoelectric interaction (PE%) in the scintillator with respect to Compton scattering. In fact, as described in Section 11.4.1, the PE% at 511 keV is important for the rejection of scatter events.

TABLE 11.1

Main Characteristics of Inorganic Scintillators Suitable for PET

	NaI	BGO	GSO	LSO	LYSO	LGSO	LuAP	YAP	$LaBr_3$
Light yield 10^3 ph/MeV	38	9	8	30	32	16	12	17	60
Primary decay time	250	300	60	40	41	65	18	30	16
ΔE/E (%) at 662 keV	6	10	8	10	10	9	15	4.4	3
Density (g/cm^3)	3.67	7.13	6.71	7.35	7.19	6.5	8.34	5.5	5.08
Effective Z_{eff}	50	73	58	65	64	59	65	33	46
1/μ at 511 keV (mm)	25.9	11.2	15.0	12.3	12.6	14.3	11.0	21.3	22.3
PE (%) at 511 keV	18	44	26	34	33	28	32	4.4	14

Source: Adapted from Lecomte, R., *Eur. J. Nucl. Med. Mol. Imaging*, 36, S69, 2009.

Table 11.1 shows the main characteristics of scintillators for PET. For example, the popular BGO scintillator is very dense (7.13 g/cm^3) and has a high Z_{eff} resulting in a short attenuation length (1/μ at 511 keV = 11.2 mm) and a high PE% (44%). On the other hand, it shows a poor light yield and a long decay time, thus limiting the timing properties. In modern scanners, the BGO block is now being replaced by cerium-doped lutetium oxyorthosilicate (LSO), lutetium-yttrium oxyorthosilicate (LYSO), or gadolinium oxyorthosilicate (GSO) for their superior light yield and shorter decay times, even if the intrinsic capability of BGO in detecting photoelectric interactions remains unsurpassed. Modern block detectors now use matrices with square section and completely separated crystal elements with a size down to 3–4 mm.

11.2.4 PET Electronics

The electronics for PET can be usually described as two separate sections: one is for the energy and position measurement and the other one is for the timing. The first section is devoted to the readout of the signals from the photodetection system (e.g., the four photomultipliers in a block detector). Output signals are digitized with analog-to-digital converters (ADCs). From the combination of the digitized signals, it is then possible to estimate the position of the interaction and the total energy released in the scintillator.

A way to identify two γ-rays as an annihilation pair is to measure them in *time coincidence*, that is, the two photons are detected only if the timing difference between the two interactions is shorter than a defined value (usually few ns) called timing window or τ. This is usually done in the timing section

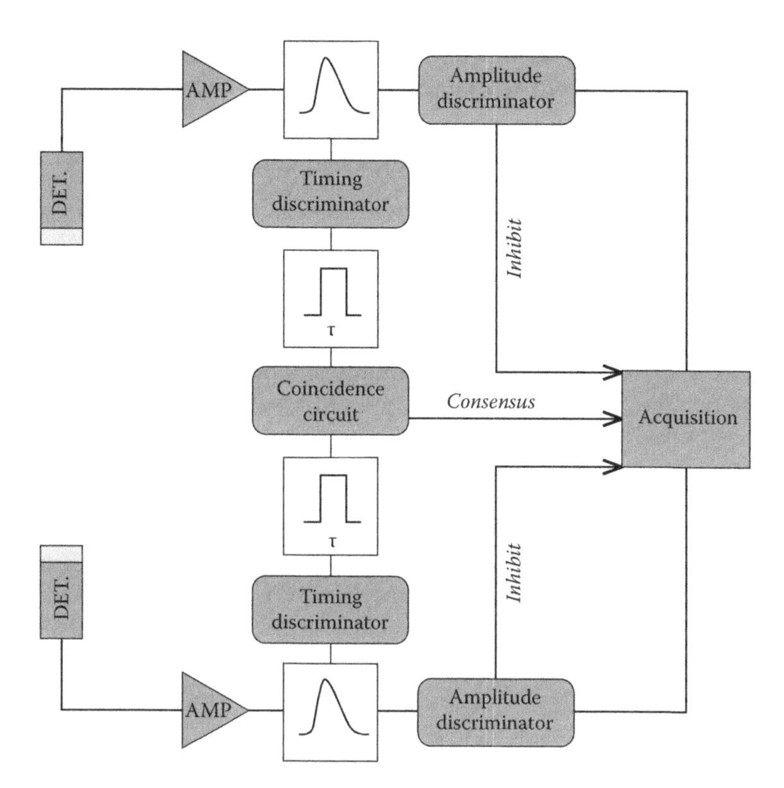

FIGURE 11.3
Simplified scheme of a PET electronics including timing and position/energy measurement sections. The PET is here represented by two opposing detectors only.

of the electronics. In its simplest form (see Figure 11.3), a timing discriminator is attached to each detector, producing a timing signal with a width equal to τ, when a single event is detected. A dedicated coincidence circuitry produces a consensus signal when two timing signals have some overlaps, enabling the acquisition of the two events, provided that the energy of both photons is above a fixed threshold (amplitude discrimination). For example, a possible solution for the simultaneous timing and amplitude discrimination is the so-called constant fraction discriminator (or CFD) (Gedcke and McDonald 1967).

11.3 Spatial Resolution Issues in PET

The best achievable spatial resolution in PET is limited due to both the physics of the β^+ decay (physics contribution) and the available technology for the position detection of two γ-rays in coincidence (technological contribution).

Both contributions are significant in PET and can be modeled to have an estimation of the expected spatial resolution of a given system.

11.3.1 Physics Contribution: Noncollinearity and Positron Range

The positron is emitted with a nonzero kinetic energy, and it is slowing down in tissue via Coulomb interactions. The energy loss continues until the positron reaches the thermal equilibrium with the surrounding tissue and annihilates with an electron. Thus, the positron range depends on tissue (water equivalent) electron density and on the positron kinetic energy (E_k). In fact, positron are emitted with a continuous energy spectrum, with a maximum energy (E_{max}) that depends on the radioisotope. The average kinetic energy is approximately one-third of E_{max}.

As a consequence, the annihilation point does not coincide with the emission point, giving rise to a nonnegligible degradation of the PET spatial resolution (Figure 11.4). Table 11.2 reports the contribution of the positron range to the spatial resolution of the scanner for various radioisotopes. It must be noted that the range distribution of the positron cannot be represented as a Gaussian function being sharply peaked at the center; see, for example, (Del Guerra 2004).

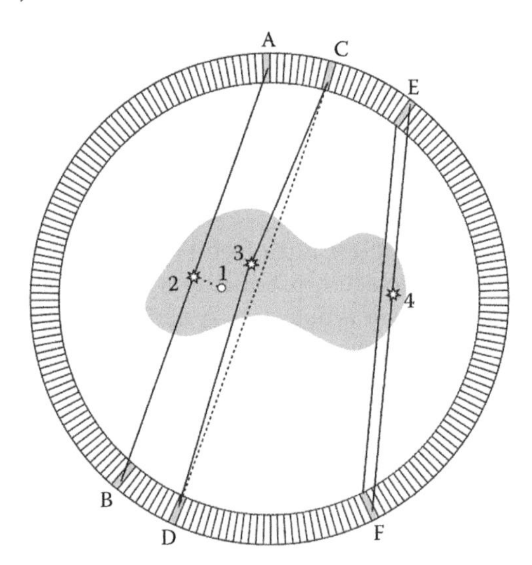

FIGURE 11.4
Effect of positron range, noncollinearity, and parallax error. A positron is emitted in 1 and annihilates in 2 due to its finite range. The two annihilation γ-rays are detected in A and B defining an LOR that does not pass by 1. An annihilation occurs in 3 and two quasicollinear γ-rays are detected in C and D. Due to the *a*-collinearity, the defined LOR does not pass by 3. An annihilation occurs at the borders of the FOV in 4 and the two γ-rays are detected in E and F. Due to the uncertainty in the measure of the depth of interaction, a parallax error occurs and the LOR defined by the two detectors does not pass by 4.

TABLE 11.2

Average Positron Kinetic Energy and Effective Range (in Water) for Different Radioisotopes

Isotope	Average E_k (MeV)	Effective Range in Water (mm)	FWHM (mm)	FWTM (mm)
^{18}F	0.242	0.54	0.10	1.03
^{11}C	0.385	0.92	0.28	1.86
^{15}O	0.735	2.4	0.50	4.14
^{68}Ga	0.740	2.8	0.58	4.83

Source: Lecomte, R., *Eur. J. Nucl. Med. Mol. Imaging*, 36, S69, 2009.

Note: The Contribution (as FWHM and FWTM) to the spatial resolution of the scanner is also reported.

In addition, the positron annihilation occurs not quite at rest. As a consequence, the two photons are not emitted, in the laboratory frame, at exactly 180°, but they have an angular deviation from noncollinearity of $\Delta\theta \pm 0.25°$ in water. In this case, the measured line-of-response (LOR) does not pass by the annihilation point, introducing an additional contribution to the degradation of the spatial resolution (Figure 11.4). The error depends on the diameter of the scanner. Larger systems are subject to a larger error, with a contribution proportional to the diameter D, that is, $\text{FWHM}_{\text{non-coll.}} = \Delta\theta \times D/4 = 0.0022D$ (Derenzo and Moses 1993).

11.3.2 Technological Contribution

The best spatial resolution achievable in PET is also limited by other factors, which are due to errors in estimation of the position of the point of the first interaction of the γ-ray in the detector. When using pixilated crystal detectors, each pair of coincidence pixels defines an LOR. Projection data are then acquired in the form of N_{ij} values that represent the number of counts measured in the LOR defined by pixels i and j.

With an ideal detector, each LOR should correspond to a well-defined LOF. However, due to the various effects that affect the detection process, such as penetration or multiple scattering within the detector itself, and limitations in the determination of the position of the interaction within the detector, a given LOF will be associated to various LOR with different probabilities. Such probability distribution is usually considered in the image reconstruction process when statistical algorithms are used.

The objective of the PET detection process is to determine an LOR; the LOR is supposed to be an accurate estimation of the LOF. On the detector, the problem is then reduced to the determination of the x, y, z coordinates of the first interaction of the annihilation photon as the best estimation of the point the LOF is passing through.

The main contributions for pixilated detector solutions can be summarized as

- Finite crystal size
- No information of the depth-of-interaction
- Inter-crystal scatter
- Crystal position readout coding

When using pixels, it is not possible to determine where exactly the γ-ray interacts within the pixel. Hence, with pixels of side d, such uncertainty contributes to the degradation of the spatial resolution with a factor $d/2$. Additional effects are due to the finite thickness of the scintillator block and in particular to the penetration and scattering of γ-rays in the scintillation material. In fact, some LOF (especially those having a high geometrical inclination) are subjected to a potentially significant parallax error, which is due to the lack of information on the depth-of-interaction (DOI) within the crystal, leading to an uncertainty in the determination of the LOR (Figure 11.4).

The effect to the spatial resolution is nonlocal with respect to the position of the annihilation in the FOV and increases toward its edges. It is also nonisotropic; the radial component of the spatial resolution along the radial direction is more affected than the tangential component. This effect is usually called *radial elongation*. The contribution to the total FWHM of the spatial resolution can be estimated as (Derenzo and Moses 1993)

$$p = \alpha \frac{r}{\sqrt{r^2 + R^2}}$$

where

r is the radial position

R is the radius of the tomograph

α is a parameter that depends on the scintillator material and thickness (e.g., it is 12.5 mm for a 30 mm thick BGO; Derenzo and Moses 1993).

When the crystal position is identified via "light sharing" technique, that is, by calculating the centroid of the light spot emerging from the crystal with a high-granularity position-sensitive photodetector, there is a nonnegligible error that is not constant on the detector surface. The average contribution is usually called *coding error*.

Another important factor contributing to the degradation of the spatial resolution is the *intercrystal scatter* (ICS). This effect happens when the first interaction in the crystal is a Compton interaction and a second interaction (either a photoelectric or a Compton interaction) occurs in a different crystal element. How and how much this event contributes to the degradation of the spatial resolution depends on the readout logic and the physical properties of the crystal. For example, block detectors are not able to distinguish the two interactions occurring in the same block, and a unique point of interaction is determined in the centroid of the two (or more) energy depositions leading to an erroneous determination of the LOR. ICS is then conceptually similar to the scattered counts concept (see Section 11.4.1), but smaller mis-placements (high frequency contribution) are observed. In this sense, the ICS contributes to the spatial resolution degradation more than to noise.

ICS can be modeled in system matrix for the iterative image reconstruction using Monte Carlo or (semi-)analytical calculations. In some cases, ICS and parallax error can be modeled together. In fact, in both cases (see Figure 11.5) there is a mispositioning error, that is, the assigned pixel (dark gray) is not the pixel where the γ-ray enters the matrix (light gray) and both effects are due to the physical properties of the interaction of the γ-ray with the scintillating material.

Considering all the contributions listed previously, the spatial resolution (FWHM) of a PET system can be estimated as (Derenzo and Moses 1993)

$$\mathrm{FWHM} = 1.2\sqrt{\left(\frac{d}{2}\right)^2 + b^2 + (0.0022D)^2 + r^2 + p^2}$$

where

- d is the pitch of the pixels of the crystal matrix
- b is the coding factor
- D is the scanner diameter
- r is the positron range contribution
- p is the contribution of the parallax effect

The ICS contribution to the spatial resolution formula is included in the coding factor b. The multiplicative factor 1.2 is the typical degradation introduced by the image reconstruction using analytical algorithm such as filtered backprojection.

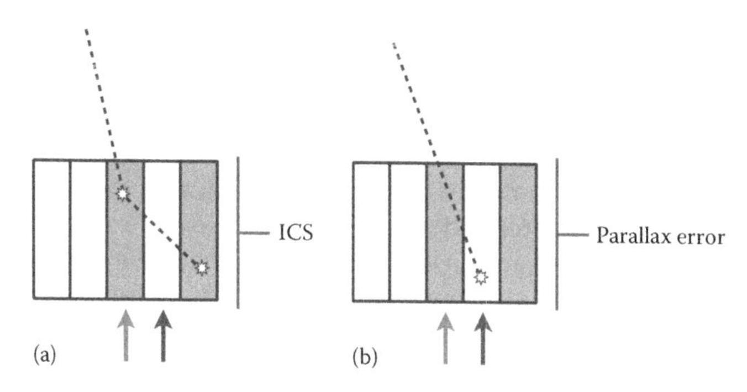

FIGURE 11.5
Schematic representation of an intercrystal scatter event (a) and a parallax error (b)

Due to the available technology, present clinical systems have a typical spatial resolution of 4–6 mm FWMH while small animal systems can offer a spatial resolution close to 1 mm FWHM at the center of the field-of-view. The resolution value in the previous equation is defined assuming infinite statistics, that is, it does not include effects from noise.

11.4 Noise in PET

Noise characteristics have important implications for quantitation and detection performance in PET imaging, especially in high-resolution scanners. The primary effect affecting noise in PET is detection efficiency.

Detection efficiency refers to the efficiency a radiation-measuring instrument has in converting emissions from the radiation source into useful signals.

Maximum detection efficiency is desirable to obtain maximum information (minimum statistical noise) with a minimum amount of radioactivity (minimum dose).

Detection efficiency is affected by several factors, including the following:

- The geometric efficiency, which depends mostly on detectors size and positioning
- The intrinsic detector efficiency, which depends mainly on absorbing material
- The recording efficiency, that is, the fraction of generated signals that are correctly recorded by the acquisition system
- The absorption and scatter of radiation within the source itself

For a given scintillating material and a given detector size, the efficiency can be improved by increasing the solid angle subtended by each detector, that is, reducing the ring diameter, or by increasing the scintillator thickness. In both cases, a more severe parallax effect is expected. Furthermore, the obvious size requirement to fit the patient or animal in the scanner should be met. For this reason, the scintillator thickness and scanner diameter are chosen so as to offer the best compromise between sensitivity and spatial resolution.

In addition to statistical noise due to the counting rate limitations of a PET system, there are several sources of noise in PET due to the physics of the γ-ray interaction with matter and to the PET technology.

Coincidence PET events are classified according to the way they are generated. A *true* coincidence is an event where the two γ-rays come from the same electron–positron annihilation and are detected in time coincidence by two opposing detectors when no interactions happen along their path. A *scatter* coincidence is an event where the two γ-rays comes from the same electron–positron annihilation, and are detected in time coincidence by two opposing

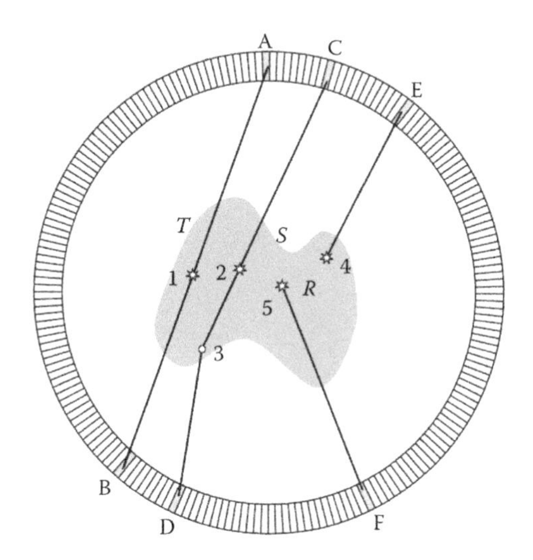

FIGURE 11.6
Representation of true (T), scatter (S), and random (R) events. A true coincidence is generated in point 1: the annihilation photons are detected in opposing crystals A and B. A scatter coincidence is generated in point 2: the one annihilation photon is detected in crystal C while the other is detected in opposing crystal D after a Compton scattering interaction in 3. A random coincidence is detected in opposing crystals E and F for two annihilations in 4 and 5 occurring with a time difference shorter than the coincidence window.

detectors where at least one of the two photons has at least one Compton interaction (scattering) along its path. A *random* coincidence is an event where the two γ-rays comes from two independent electron–positron annihilations and are detected in time coincidence by two opposing detectors. A *multiple* coincidence is an event where more than two γ-rays are detected in time coincidence opposing detectors. Scatter, random, and multiple events generate wrong LORs and thus are possible sources of noise in the PET image. The measured count rate due to true, scatter, random, and multiple coincidences are usually indicated with the letters T, S, R, and M, respectively (Figure 11.6). True and scatter coincidences are also usually called *prompt* coincidences, and the relative count rate is indicated by $P = T + S$.

11.4.1 Scatter Counts

Scatter coincidence must be rejected to reduce the noise in the PET image. Scatter counts can be identified by measuring the energy of the photons. For this reason, detectors should also be able to measure the energy of the incoming γ-ray. Hence, accepting only events where both γ-rays release 511 keV in the detectors, scatter events can be rejected. A finite energy window (EW) is applied to data, that is, only events where the released energy is within a lower and upper threshold are accepted.

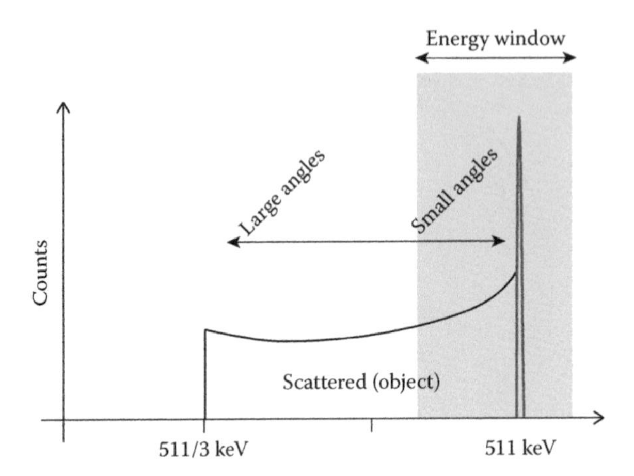

FIGURE 11.7
Energy spectrum of the 511 keV γ-rays emerging from an object. Photons subjected to Compton scattering have a reduced energy (curve in dark gray) down to 511/3 keV. Photons that emerge with no interactions are in medium gray. In this figure, the energy blurring due to the finite energy resolution is not considered.

Reducing the lower threshold of the EW increases the fraction of scattered counts and larger and larger scatter angle events are included (Figure 11.7). This is evident considering the formula for the energy of the scattered photons after a Compton interaction of a 511 keV photon

$$E_{\gamma'}(\theta) = \frac{E_\gamma}{2 - \cos\theta}$$

In addition, the photoelectric and multiple Compton components are not separated and then it is not possible to fully cut-off the scatter component from measured data. This fact is even more important considering that, in the real case, the measurement of the energy of the photons has some uncertainty due to the limited energy resolution of the detection system. The energy resolution of a detector is the accuracy in the measurement of the energy of the photon and it is usually expressed in terms of $\Delta E/E$ at 511 keV, where ΔE is the FWHM of the energy measurement distribution.

A further complication in scatter event rejection is the Compton scattering in the detector. The energy spectrum of the first interaction of a 511 keV γ-ray in a scintillator is represented in Figure 11.8. The scatter component due to intercrystal scatter is now separated from the photopeak component. In fact, what the system measures is not the γ-ray energy, but the energy of the photo—(photoelectric interaction) or recoil—electron (Compton interaction). However, an overlap between the object scatter (curve in dark gray in Figure 11.7) and ICS (curve in light gray in Figure 11.8) components is still present for $511/3 \text{ keV} < E < 2/3 \times 511 \text{ keV}$.

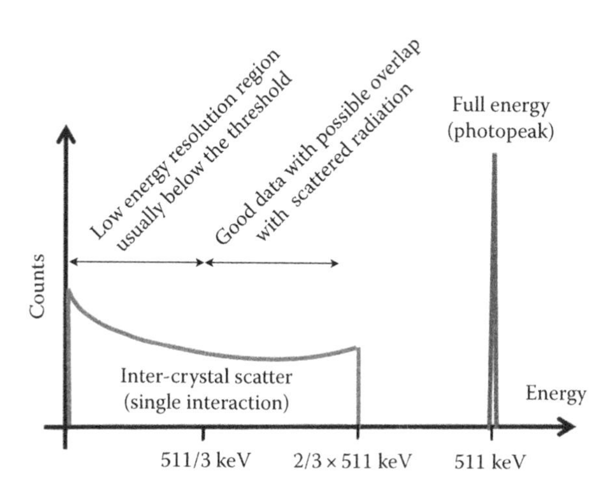

FIGURE 11.8
Energy spectrum generated by the first interaction of a 511 keV γ-ray in a scintillator. The ICS component is also considered here (in light gray). The region 511/3 keV < E < 2/3 × 511 keV has possible overlaps with object-scattered radiation component (in dark gray in Figure 11.7). The region E < 511/3 keV is difficult to use for its too low energy. In this figure, the energy blurring due to the finite energy resolution is not considered.

The situation is also complicated for the multiple interactions that may occur in the scintillating block. The ICS component and photopeak are no longer separated and the photopeak is also actually populated by events where the photons released all of its energy in multiple interaction (at least one Compton interaction and one photoelectric interaction). For this reason, the peak at 511 keV is usually called the full energy peak instead of photopeak.

For the object scattering and ICS issues, the energy resolution and the fraction of photoelectric interactions with respect to Compton interactions (usually indicated as PE% at 511 keV) are important parameters for a scintillating crystal to be suitable for PET. For example, typical values for the energy resolution of LSO readout by PMTs are 12%–20%, while the PE% at 511 keV is 34% (see Table 11.1).

Another important figure indicating the performance in scatter rejection of a PET scanner is the scatter fraction (SF). SF is the contribution of the scattered events to the reconstructed image and can be calculated as

$$SF = \frac{S}{R_{\mathrm{TOT}}}$$

where S is the count rate of scatter events while R_{TOT} is the total acquisition count rate. When the random count rate (see Section 11.4.2) is negligible, the SF is independent from the count rate and does not depend on the activity in the FOV.

In addition to the scatter rejection with the energy window, other techniques for correcting for the contribution of scattered count in the final image can be implemented in the reconstruction algorithm. These methods, called *scatter corrections*, are usually based on the estimation of the distribution of the scatter events for a given object.

11.4.2 Random Counts

The random count rate R in a LOR defined by detectors i and j can be expressed as

$$R_{ij} = C_i \times C_j \times 2\tau$$

where

C_i and C_j are the singles count rate on detector i and j and
τ is the time window

Note that the random count rate varies with the square of the singles count rate, that is, with the square of the activity.

A narrow timing window (i.e., a small τ) is then necessary to reject random counts, but a certain fraction of random counts is always present in the data. To obtain quantitative data in PET, it is necessary to estimate the rate of random coincidences in the measured data in each LOR. Methods for correcting for random counts are based either on indirect or direct estimation of R. In the first case, the method is based on the statistical estimation of R based on the measurement of the singles count rate. Hence, to estimate R_{ij} it is sufficient to measure C_i and C_j while τ is a given value. In most cases, it is not possible to estimate C_i and C_j on a pixel-by-pixel basis and hence the value of R is usually calculated at the level of the block with the assumption that the random count rate is slowly varying within the same block. This method is usually characterized by low noise for the higher single count rate with respect to coincidence counts, but also by a potentially high systematic error due to the uncertainty in the estimation of τ. A second possible approach to random count correction is to directly estimate R with the delayed window technique. This method can be simply described for a two detectors PET system. In this case, timing signals from one detector are delayed by a time significantly greater than the coincidence resolving time τ of the circuitry. Hence, the distribution in time of delayed and nondelayed timing signals are uncorrelated and the probability for a delayed coincidence to be detected is approximately equal to the probability to measure a random count. The approximation stays in the fact that, in the delayed window technique, also events generated by a coincidence could give place to a random count while this is not the case during the acquisition of prompt (i.e., generated by annihilation pairs) coincidences. A random LOR is then measured when a delayed coincidence is detected. In this case, it is also necessary to acquire the event to have an estimation of the spatial distribution of random counts. This method is usually characterized by higher noise than the indirect measurement technique, but has the advantage of not requiring the a priori knowledge of τ, thus reducing systematic errors.

The estimated random distribution can be subtracted from the prompt signal on-line, or stored as a separate sinogram or LOR-based data format for later processing (e.g., considered in the image reconstruction algorithm).

The coincidence events are detected by comparing the arrival time of all detected events to determine which ones arrived closely enough in time to be identified as an annihilation pair.

The ability of a pair of detectors to determine the time difference in the arrival of the annihilation photons is known as the timing resolution and is typically on the order of few nanoseconds.

Using a scintillator, the time resolution is ultimately limited by scintillator decay time light yield according to the formula (Hyman 1965):

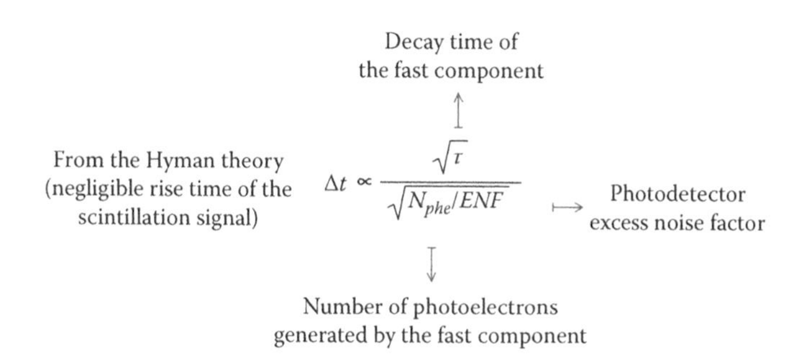

where the excess noise factor (ENF) describes the statistical noise due to the stochastic multiplication process (Musienko 2006).

Other important parameters for a scintillator to provide the best timing performance are the light transport (affecting light yield and pulse shape) and the time structure of the pulse (e.g., the rise time of the electronic signal). A typical timing window that is used in PET scanners not to accidentally reject annihilation photon pairs is typically two to three times the timing resolution, leading to values typically between 4 and 20 ns.

The noise equivalent count rate (NECR) is the figure of merit that quantifies the amount of background and statistical noise characteristic of a given PET scanner.

The formulation of NECR is as follows:

$$\text{NEC} = \frac{T^2}{R_{TOT}}$$

where R_{TOT} is the sum of true, random and scatter counts

$$R_{TOT} = T + S + kR$$

where k is a factor depending on the method used for measuring the random count rate.

NEC is an indirect measure of the noise in the data due to scatter and random events, but also due to the effect of dead time at high count rates.

Procedures for the calculation of the NEC curve are defined in the standard protocols NEMA NU2 1994 and 2001. NEC is measured for various amounts of activity (usually following the decay of the activity in a standard phantom) in the field-of-view, thus creating the typical NEC curve (Figure 11.9). This curve is characterized by a peak where both the value of the NEC and the activity (peak NEC activity) are important parameters to estimate the scanner performance. In practice, the NEC curve can be used to estimate the total amount of activity to be injected in a patient, given that an optimal scan should begin with a count rate below the value at the peak of the NEC curve.

Note the behavior of the NEC curve with the typical peak (30 cps at 500 μCi in the example).

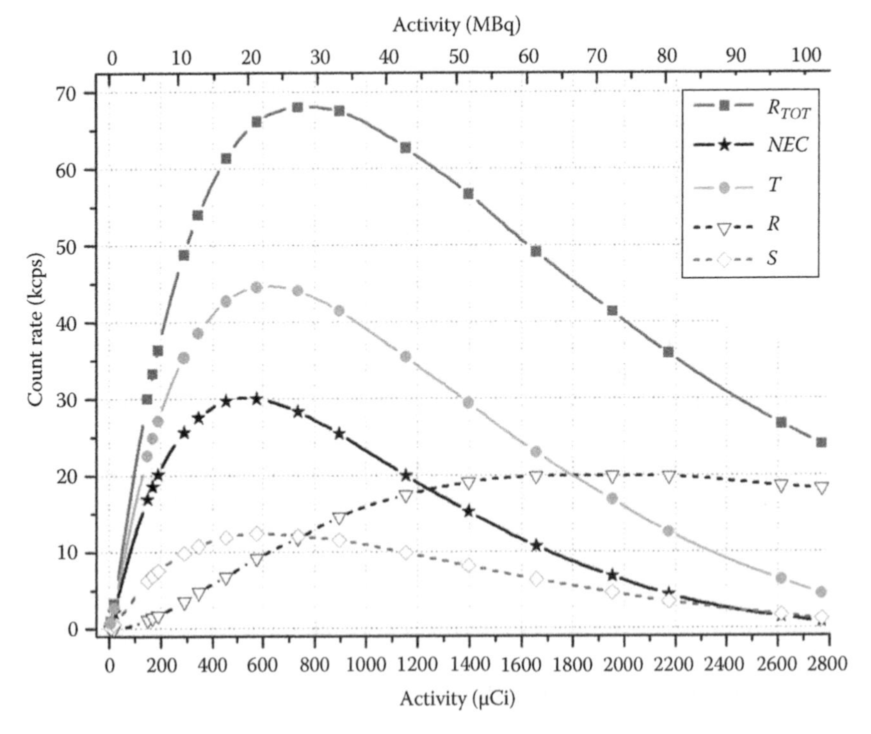

FIGURE 11.9
Example of the plot of T, R, S, R_{TOT}, and NEC for an animal scanner and a given phantom (custom mouse phantom).

11.5 Advanced PET Detectors

11.5.1 Position-Sensitive Photomultiplier Tubes

Clinical PET detectors still rely on the block detector with increasing number of crystals per block of smaller and smaller size (down to 4 mm × 4 mm). To improve the spatial resolution, smaller pixels are required. To maintain a clear identification of the pixels, the block detector is no longer adequate, that is, the coding is becoming the most limiting factor. Hence, the photodetector may take the form of a position-sensitive PMT (PS-PMT).

The PS-PMT is now at the third generation. Early PS-PMT (introduced in 1985 by Hamamatsu) was based on a typical round shape, with size up to 10 cm diameter, with a crossed wire anode structure. The second generation has a square-shaped metal channel dynode structure with a very fine anode structure made by crossed metal plates or a matrix of independent square anodes (hence they are usually called multi-anode PMT or MA-PMT). This type of PS-PMT provides a very good intrinsic spatial resolution and uniformity, but was limited in size (up to about 2 cm side). With the third generation, there has been a great improvement in the active area dimensions (up to 5 cm in side) and active-to-total area ratio (up to about 90%), still maintaining the performance as close as possible to the second-generation tubes. These larger tubes are based on the metal channel dynode structure with an anode matrix of 16 × 16 elements on a 3 mm pitch. Similar package and performances are offered by the tubes where a dual multichannel plate is used for electron multiplication instead of the common dynode structure.

MA-PMTs are now large enough to be used as a single photodetector for the construction of a modern block detector, but they can also be designed so as to be assembled in an array to cover a large detection area, with an improved effective area up to 97%. However, the cost is still much higher than PMTs commonly used in whole-body PET detectors. To fully exploit the performance of MA-PMTs, each anode should be independently acquired (multi-anode readout) like the four outputs of a block detector. However, in order to simplify the complexity of the readout system, resistive chains (Popov et al. 2001; Olcott 2005) can be used to strongly reduce the number of output channels. Using PS-PMTs, the pixel identification is still performed with light-sharing technique, that is, the γ-ray interaction position is encoded in the centroid of the light spot sampled by the PMT anode structure.

Many different combinations of pixilation and PS-PMT are presently used depending on the specific application and required FOV. For example, second- and third-generation MA-PMTs are widely used in small-animal PET instrumentation, where matrices of tiny scintillating crystals with a size close to 1 mm side are used to achieve an extremely high intrinsic spatial resolution (Figure 11.10).

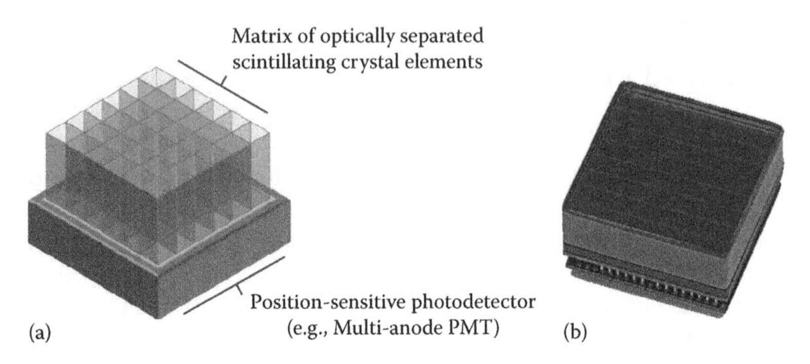

FIGURE 11.10
A possible configuration of a PET detector comprised a MA-PMT and a matrix of scintillating crystals (a). In this case, the pixel size contribution in the spatial resolution formula is $d/2$ while the coding factor is $b > 0$. (b) The popular Hamamatsu H8500 with 8 × 8 independent anodes. Its main features are minimum peripheral dead zone (1 mm) and minimal height (12 mm).

11.5.2 Semiconductor Photodetectors

The advent of semiconductor photodetectors opens new possibilities for the design of PET detectors in a wide flavor of different ways. A first attempt of such use dates back to 1986 when APDs were proposed (Lightstone 1986) as an alternative to the block detector for clinical PET.

The APDs are compact and rugged, and show properties comparable to those of traditional PMTs. Nowadays, APDs are based on well-established technologies, and they are available in a wide range of single-pixel or multipixel configuration with various layouts and pitch sizes. Small-animal PET scanners based on avalanche photodiode detectors have been proposed in recent years as an alternative to PSPMT-based systems. Most of these systems are characterized by an individual crystal readout (Lecomte et al. 1994; Pichler et al. 1998), where single elements of APDs can be used for one-to-one readout of scintillator crystals, but solutions where matrices of APD are used to readout matrices of scintillating pixels with light-sharing coding have also been proposed.

Consequently, there is no resolution loss due to light sharing or electronic coding, and the true geometric crystal resolution can be achieved. In addition, the one-to-one coupling makes the dead time and pile-up events probability negligible. The continuous slab solution coupled to an array of APD has also been investigated (Bruyndonckx et al. 2004).

Even if current APDs have very good performance (gain 10^2–10^3, rise time 2–3 ns), standard phototubes are still superior in some figure of merit such as a higher gain (~10^5), better noise and timing characteristics.

The silicon photomultiplier (SiPM, also known as multipixel photon counter or MPPC) is a newcomer in this field, but seems to be a valid candidate for a solid state–based PET system. The SiPM (Saveliev and Golovin 2000) is a densely packed matrix of small, Geiger-mode avalanche photodiode (GAPD) cells (usually called microcells) with individual quenching resistors (Figure 11.11). The microcells are multiplexed by a common metal electrode

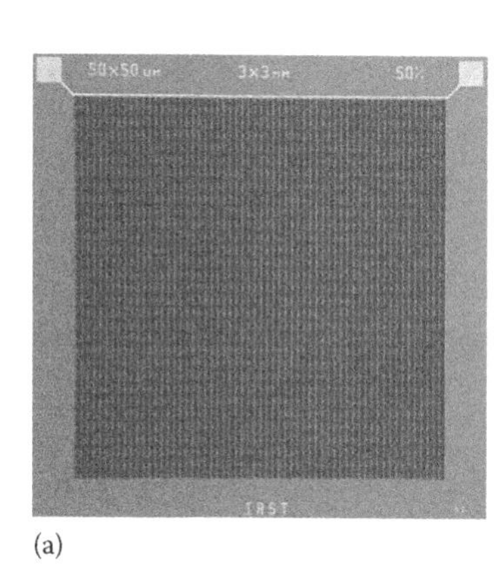

(a)

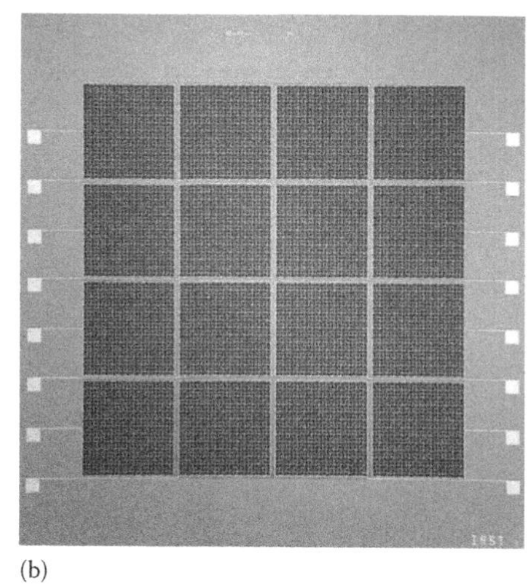

(b)

FIGURE 11.11
(a) Picture of a single SiPM of 3 × 3 mm² total area that comprises 3600 microcells of 50 × 50 μm² each with a fill factor of 50%. (b) Example of a matrix of 4 × 4 SiPM build on the same silicon substrate. (Photo courtesy of FBK-IRST, Trento, Italy.)

contact layer, preserving the proportionality of the output signal and making it suitable for spectroscopy. Thanks to the Geiger-mode operation, the SiPM exhibits a high intrinsic gain (10^5–10^6), as well as low noise properties, resulting in a dark rate below 1 kHz/mm² for a 3–4 photoelectron threshold at room temperature. These features make it possible the use of simple front-end electronics, which can even be placed far away from the detector. Compactness and versatility are important advantages of the SiPM over APD. Furthermore, the SiPM needs a much lower bias voltage (~30 to 100 V) with respect to APD. The SiPM has remarkable timing properties with an intrinsic time resolution of tens of ps (sigma) at the single photoelectron level (Collazuol 2007).

SiPMs have been developed in a wide range of microcell dimensions (from 25 to 100 μm) and pixel sizes (from 1 to 4 mm), and they are offered also in small matrices of elements grown in a common substrate. The production process on a single bulk guarantees a high uniformity in terms of breakdown voltage and gain among pixels of the same matrix.

Thanks to their larger dimensions and their high granularity, SiPM arrays can now become a real alternative to phototubes in the readout of scintillator crystal for gamma ray detection. A comparison of the performance of PMTs, APDs, and SiPM is reported in Table 11.3.

11.5.3 Alternative Photodetector–Scintillator Configurations

As an alternative method for the readout of matrices of scintillators, solutions where an element of scintillator is coupled to a single photodetector (namely, one-to-one coupling) are also used with the intent of overcoming

TABLE 11.3

Comparison of Some Characteristics of Three Types of Photodetectors Used in PET Systems

	PMT	APD	SiPM
Gain	10^5–10^7	10^2	10^5–10^6
Dynamic range	10^6	10^4	10^3/mm
Excess noise factor	0.1–0.2	>2	1.1–1.2
Rise time	<1 ns	2–3 ns	~1 ns
Dark current	<0.1 nA/cm^2	1–10 nA/mm^2	0.1–1 MHz/mm^2
QE at 420 nm	25%[a]	60%–80%	<40%[b]
Bias voltage	~800 to 2000 V	~100 to 1500 V	~30 to 50 V
Temperature coefficient	<1%/K	2%–3%/K	3%–5%/K
Magnetic susceptibility	Very high (mT)	No	No

Source: Lecomte, R., *Eur. J. Nucl. Med. Mol. Imaging*, 36, S69, 2009.
Note: The ENF appears in the timing formula in Section 4.2.
[a] High QE versions are now available.
[b] In this case, the photon detection efficiency (PDE) is reported.

the coding limitation. For example, matrices of small area APDs are used for the parallel readout of the pixilated matrices of scintillator (Figure 11.12). In this case, each crystal can be coupled to a single APD. This method, similar to those used in the early PET of the 1970s but now with a pixel pitch down to 1 mm, allows a very simple, almost "perfect" one-to-one coding (Figure 11.12). Consequently, there is no resolution loss due to light sharing or electronic coding and the true geometric crystal resolution can be achieved. An additional advantage is the negligible dead time and pile-up events probability. Although the one-to-one coupling of scintillator and photodetector

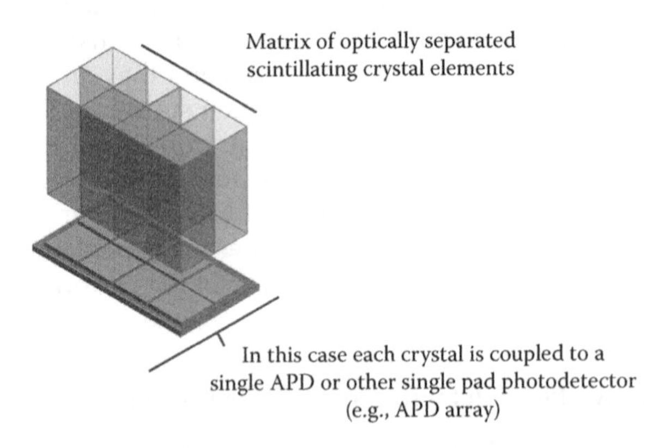

FIGURE 11.12
An example of one-to-one coupling using matrices of solid-state photodetectors. In this case, the pixel size contribution in the spatial resolution formula is $d/2$ while the coding factor is $b = 0$.

would offer intrinsically better event positioning, energy resolution, timing, and count rate performance, the cost and the difficulties of many readout channels strongly limited the application of this solution.

On the other hand, detectors based on a monolithic scintillator block coupled to high granularity position-sensitive photodetectors (Anger camera principle) have been proposed (Joung 2001; Balcerzyk et al. 2009). In this way, the limitation introduced by the finite size of the crystal elements could be overcome by measuring, with high precision, the center of gravity of the light spot. In fact, using arrays of small detector elements (matrix), the precision of localization is limited by the size d of the detector element (more precisely the detector pitch). Using a larger area detector, that is, monolithic crystals (Figure 11.13), the precision of localization is related to the positioning performance of the photodetector only, that is, the pixel size and the coding factors in the resolution formula are replaced by a single factor that takes into account the precision of the measurement of the point of interaction. Usually, this factor is not constant along the whole surface of the scintillator and degrades significantly at the borders of the detector. In addition, the dependence of the positioning precision with thickness (it becomes worse as the thickness increases) limits the possibility to use blocks with the same thickness as the more commonly used matrices of pixels.

With this solution, the spatial resolution issue is transferred to the performance of the photodetector. For this reason, finely pixilated solid-state photodetectors (such as matrices of SiPM) are the ideal solution, but the use of MA-PMTs has also been proposed.

In addition to the difficulties in deriving the position of the interaction within the scintillator, a further drawback of detectors based on monolithic crystals is related to the positioning calibration procedure, which is usually complex and time consuming.

An additional issue for the development of dedicated PET instrumentation is to provide an adequate efficiency for the detection of 511 keV photons.

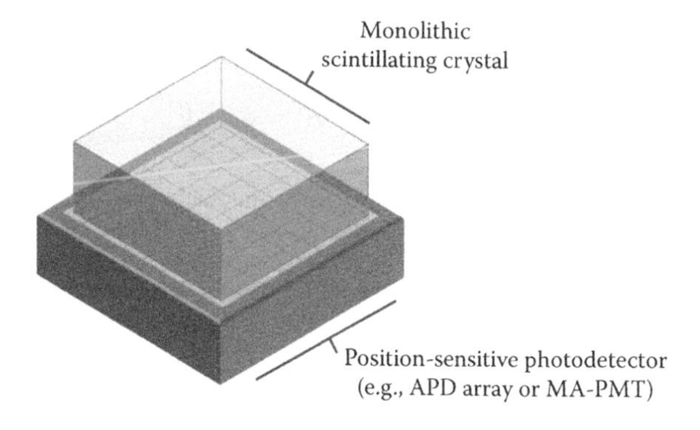

FIGURE 11.13
An example of a monolithic scintillating crystal coupled to a position-sensitive photodetector. In this case, the pixel size contribution in the spatial resolution formula is not applicable while a positioning factor >0 replaces the coding factor.

In order to maximize the efficiency of the PET system, the PET heads should be positioned close to the object and the thickness of the photon absorber should be at least one attenuation length at 511 keV. In this case, some LORs (especially those having a high geometrical inclination) are subjected to a potentially significant parallax error (see Section 11.3.1) due to the lack of information on the depth where the interaction occurs (DOI error). A number of techniques for designing detectors with DOI capability have been proposed, based either on the direct measurement of the DOI within the crystal (continuous DOI information) (Moses and Derenzo 1994; Balcerzyk et al. 2009) or by segmenting the crystal into two layers, so that the photodetection system is able to discriminate events occurring in one layer from the other(s) (discrete DOI information) (Saoudi et al. 1999; Seidel 1999) (Figure 11.14).

Latest small animal PET systems, incorporating the high granularity detectors and in some cases DOI capability offer a volumetric spatial resolution of close to 1 mm^3 that is very close to the theoretical one.

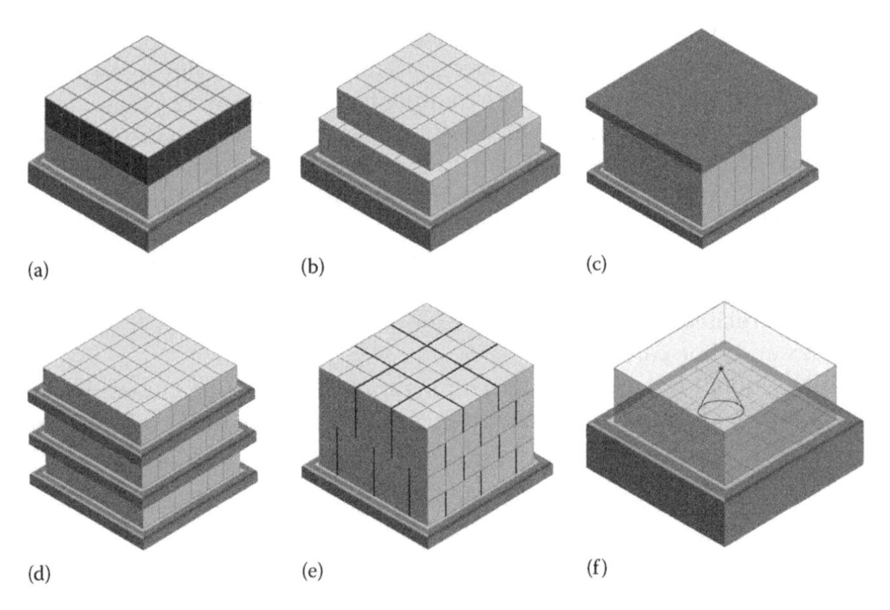

FIGURE 11.14
Various solutions for the estimation of the depth-of-interaction: (a) dual-layer phoswitch (discrete): the two layers, made of two different scintillating material, are distinguished by the different decay time of the scintillators; (b) dual-layer staggered (discrete with oversampling): pixels on the two layers are identified in different positions by the photodetector; (c) dual side readout (continuous): the DOI is evaluated by calculating the ratio of the signals measured by the detectors on both sides, which are supposed to be different due to light attenuation property of the crystal; (d) multiple photodetector readout (discrete): layers are physically separated and read out by a stack of thin solid-state photodetectors; (e) multiple layer with reflective pattern (discrete): the different reflection pattern on each layer encode the position of the pixel in each layer that will appear in a different position in the photodetector image; (f) width of the light spot in continuous scintillators (continuous): the DOI is evaluated by measuring the high light spread that is observed as the interaction point is far from the photodetector.

11.5.4 Photodetectors for PET/MRI

Nowadays, particular interest is devoted to photodetectors insensitive to magnetic fields for the development of combined magnetic resonance-PET scanners (Zaidi and Del Guerra 2011). Standard PMTs have a high magnetic susceptibility (see Table 11.3) and can work in regions where the magnetic field is at the order of mT. An early attempt to use PMTs for PET/MR was done using long optical fibers in order to transport the scintillation light outside the magnetic field region. Nowadays, PMTs are used in the tandem PET/MR configuration where the PET and the MR system, are placed back-to-back, sharing the same patient (or animal) bed, and are placed far enough to make the effect of the magnetic field negligible on PMTs. To ensure a better insulation to magnetic field effects, PMTs are usually passively shielded with metal cases.

A second approach exploits the potential of solid-state photodetectors such as APDs or SiPMs to allow the development of intrinsically magnetic field compatible PET detectors. A successful application of the solid-state technology to this purpose is the development of a fully integrated APD-based PET/MR (with a magnetic field of 3T) for clinical studies.

A key feature for the further development of PET/MR systems is the use of SiPM that can nowadays be considered the state-of-the art technology for PET/MR as demonstrated by the great interest of research groups and PET manufacturers as well as the large number of SiPM-based PET detector prototypes that are continuously proposed for both clinical and preclinical applications.

11.5.5 Direct Conversion Detectors

Semiconductors are a valid alternative to scintillators as detector material for γ-ray detection. When radiation is absorbed in a solid-state detector, ionization occurs by promoting electrons out of the valence band to the conduction band where electrons can move in the crystal lattice. When the electron is promoted to the conduction band, a positive charge (hole) in the lattice is created. If an electric field is applied across the sensitive volume of the detector, the excess of charge (both electrons and holes) is free to move. Opposite electrodes collect the charge, thus providing a signal that is proportional to the energy released in the detector. Crystalline materials that can be used in such detectors are cadmium telluride (CdTe) and cadmium zinc telluride (CdZnTe). Thanks to their characteristics (see Table 11.4), they can be used for the construction of compact γ-ray detectors (Levin 2008).

On the shadow of the successful application of CdZnTe in the construction of dedicated SPECT systems (such as in small-animal scanners), this material was also proposed in PET, thanks to the superior energy resolution with respect to scintillators together with its good photofraction (see Table 11.4). Regarding energy resolution, CdZnTe can achieve about 2% FWHM at

TABLE 11.4

Physical Properties of CdTe and $Cd_{0.9}Zn_{0.1}Te$, in Terms of Density (ρ), Effective Z (Z_{eff}), Attenuation Coefficient (μ) at 511 keV, Energy to Produce an Electron-hole Pair (E_e–h) and Photofraction (PE%)

	$Cd_{0.9}Zn_{0.1}Te$	CdTe
ρ (g/cm²)	5.78	6
Z_{eff}	49.1	50
μ (511 keV)	0.5 cm⁻²	0.51 cm⁻¹
E_{e-h}	4.64 eV	4.43 eV
PE% (511 keV)	18%	18%

511 keV, which is much better than what is usually offered by scintillation detectors. This excellent resolution can be used to improve the scatter rejection by setting a much narrower energy window, thus potentially improving the noise characteristics as described in Section 11.4.1. The intrinsic spatial resolution of these devices is dominated by the pixel size. Nowadays, devices with a pixel size of 1 mm or smaller (Vaska 2005) are available.

As of today, CdZnTe detectors have not been developed for any human PET scanners due to the challenges in timing resolution (CdZnTe has usually a worse timing resolution with respect to scintillation PET detectors, e.g., using LSO) and the relatively low PE% for 511 keV gamma rays. However, for small-animal imaging, the small size of the subjects makes the PE% a less critical requirement.

11.6 Conclusions

The performance of PET detectors has a fundamental role in image quality, affecting both spatial resolution and noise. The technologies for PET detectors are continuously being improved and new solutions are initially tested and applied in small-animal imaging instrumentation. For years, the block detector has represented the core of a clinical PET detection system. Improvements were mostly related to scintillating material and pixels size, reaching performance that are more than adequate for human imaging. The development of PET-MR instrumentation could represent the new frontier for the field of hybrid imaging. The availability of MR-compatible solid-state photodetectors like silicon photomultipliers offers new possibilities for the design of novel concepts opening a new era for PET detectors.

References

Balcerzyk M. et al. *Meas. Sci. Technol.* 20, 2009, 104011.
Bruyndonckx P. et al. *IEEE Trans. Nucl. Sci.* 51, 5, 2004, 2520–2525.
Casey M.E., Nutt R. *IEEE Trans. Nucl. Sci.* 33(1), 1986, 460–463.
Collazuol G. et al. *Nucl. Instr. Methods A* 581, 2007, 461–464.
Del Guerra A., *Ionizing Radiation Detectors for Medical Imaging*. Singapore: World Scientific, 2004.
Derenzo S., Moses W.W. *Quantification of Brain Function, Tracer Kinetics & Image Analysis in Brain PET*. De Uemura et al., Amsterdam, the Netherlands: Elsevier, 1993, pp. 25–40.
Gedcke D.A., McDonald W.J. *Nucl. Instrum. Methods* 55, 1967, 377–380.
Hyman L.G. *Rev. Sci. Instrum.* 36(3), 1965, 193–196.
Joung J. et al. *IEEE Trans. Nucl. Sci.* 48, 2001, 715–719.
Lecomte R. *Eur. J. Nucl. Med. Mol. Imaging* 36(Suppl. 1), 2009, S69–S85.
Lecomte R. et al. *IEEE Trans. Nucl. Sci.* 41(4), 1994, 1446–1452.
Levin C.S. *Proc. IEEE* 96(3), 2008, 439–467.
Lightstone A.W. et al. *IEEE Trans. Nucl. Sci.* 33(1), 1986, 456–459.
Moses W.W., Derenzo S.E. *IEEE Trans. Nucl. Sci.* 41, 1994, 1441.
Musienko Y. et al. *Nucl. Inst. Meth. Phys. Res. A* 567, 2006, 57–61.
Olcott P.D. *IEEE Trans. Nucl. Sci.* 52, 2005.
Pichler B. et al. *IEEE Trans. Nucl. Sci.* 45, 1998, 1298–1301.
Popov V. et al. *IEEE Nuclear Science Symposium Conference Records*. 2001, 1937–1940.
Saoudi A. et al. *IEEE Trans. Nucl. Sci.* 46, 1999, 462–467.
Saveliev V., Golovin V. *Nucl. Instr. Methods A* 442, 2000, 223–229.
Seidel J. et al. *IEEE Trans. Nucl. Sci.* 46, 1999, 485–490.
Ter-Pogossian M.M. et al. *Radiology* 14, 1975, 89–98.
Uchida H. et al. *IEEE Trans. Nucl. Sci.* 33(3), 1986, 464–467.
Vaska P. et al. *IEEE Nuclear Science Symposium Conference Record*. 2005, 2799–2802.
Zaidi H., Del Guerra A. *Med. Phys.* 38 (2011): 5667–5689.

12

Recent Developments of High-Performance PET Detectors

Hao Peng and Craig S. Levin

CONTENTS

12.1 Introduction

Positron emission tomography (PET) is a noninvasive medical imaging tool widely used in both clinical and preclinical research. PET has proven its value in several areas such as cancer diagnosing and staging, assessing neurological diseases, myocardium viability evaluation in cardiology, as well as radiotherapy and chemotherapy monitoring [1–4]. In recent years, PET is also being actively used in small animal research using new molecular probes labeled with positron-emitting radionuclides. Applications using small animal PET include cell trafficking as cancer cells metastasize to different organs [5–8], gene delivery and expression in living animals [9,10], and low levels of endogenous messenger ribonucleic acid (mRNA) [11,12]. Advanced instrumentation, particularly high-performance detector designs, would help PET improve its image quality and enhance its molecular sensitivity for both clinical and small animal research. This chapter will focus on four aspects of the recent developments of high-performance PET detector designs: (1) high photon detection sensitivity, (2) improved spatial resolution, (3) depth-of-interaction (DoI) design, and (4) time-of-flight (ToF) PET and silicon photomultipliers (SiPMs).

12.2 PET: Basic Principles

Here we provide a brief description of the basic principles of a PET system and some terms to be used in this chapter. First, a tracer compound labeled with a positron-emitting radionuclide is injected to the object. The radionuclide then decays, and the resulting positrons subsequently annihilate with electrons after traveling a short distance within the object. Each annihilation produces two 511 keV photons traveling approximately in opposite directions, and these photons are detected by the detectors, which usually consist of scintillation materials, photodetectors, and front-end electronics. The signal of each photon from every pair coincidence event is processed individually for spatial, energy, and time information. For a pair coincidence event, if the energy of two photons stays within a preset energy window (~20%–30% full width half maximum (FWHM) centered on the 511 keV photopeak) and the time difference stays within a preset time window (~6–12 ns), a coincidence event will be registered and constitutes a line of response (LoR) for image reconstruction. In PET, there are three types of coincidence events [1,13,14]:

- *True coincidences (T)* are the good coincident photon events where the LoR between the two coincident photons essentially passes through the points of emission.
- *Random coincidences (R)* are a source of undesirable background counts that occur when two distinct nuclei each decay nearly at the

same time and only one photon from each decay is detected within the time window. Random rates are reduced with lower detected single-photon count rate (single rate) and a narrower coincidence time window setting. It is desirable to have good coincidence time resolution so that a narrow time window may be employed to reject randoms without compromising photon sensitivity.

- *Scatter coincidences (S)* are another undesirable source of background events that occur when one (or both) annihilation photon emitted from the same nucleus or single photons emitted from two separate nuclei undergoes one or more Compton scatter (CS) interactions inside the tissue before detection. Since CS causes a photon to lose energy, its effects can be reduced through the use of a narrow energy window setting around the 511 keV photopeak. It is desirable to have good energy resolution so that a narrow energy window may be employed to reject scattered coincidences without compromising photon sensitivity.

A widely used parameter, reflecting the signal-to-noise ratio (SNR) performance of a PET system in the context of three types of coincidences and optimum settings of both energy and time windows, is the noise equivalent count (NEC) [15] shown in Formula 12.1:

$$\mathrm{NEC} = \frac{T^2}{T + S + kR} \tag{12.1}$$

where

T, *S*, and *R* are the total number of true, scatter, and random coincidences, respectively

k is the number between 1 and 2, depending on the shape of phantoms/organs being imaged and the random estimation methods being used

A higher NEC at a given injection dose implies that a PET system is able to achieve better SNR and thus contrast-to-noise ratio (CNR) performance. In the following sections, we will focus on how to improve NEC and image quality through designing high-performance PET detectors and review a number of recent developments.

12.3 High Photon Detection Sensitivity

High photon sensitivity is a critical issue in PET instrumentation as it enables high statistical quality of acquired data, which is required to realize the potential of high spatial resolution [13,14,16]. High photon sensitivity also leads to a higher NEC in Formula 12.1 (i.e., due to Poisson statistics).

For a PET system, the photon sensitivity is determined by two factors [14,17]: geometric efficiency (E_g) that reflects the solid angle coverage of the system and intrinsic coincidence detection efficiency (E_i) that is determined by the intrinsic properties of detectors (i.e., atomic number, density, thickness), crystal packing fraction, as well as the energy and time window settings. The overall system photon sensitivity (E_s) is given by Formula 12.2:

$$E_s = E_g \times E_i \tag{12.2}$$

The photon sensitivity is often quoted for a point positron source placed at the center of a PET system. For a standard clinical whole-body PET system (~700–800 mm diameter bore), E_s is normally around 0.5%–1% [18,19]; for a small animal PET system scanner (~100–200 mm diameter bore), E_s is around 1.0%–9.0% [20,21]. E_g is dependent on the total solid angle of a PET system, which can be increased either by moving detectors closer to the subject or by increasing the system coverage with more detectors. Several attempts have been made to address this issue including (1) extending the axial field of view (FoV) of a cylindrical system as deployed recently in Siemens Biography 64 TruePoint PET/CT whole-body scanner [22], (2) decreasing the system diameter, and (3) modifying the system geometry from cylindrical shape to other optimum geometries. There are several ways to increase E_i. The most straightforward way is to utilize scintillation materials with high Z and density (see Table 12.1), increasing the length of crystals, and/or packing them more tightly. However, longer crystals are associated with spatial resolution degradation toward the edge of FoV (also known as "parallax error") and

TABLE 12.1

Common Scintillators for PET Imaging

Scintillator	Density (g/cm³)	Effective Atomic Number (Z)	Linear Attenuation Coefficient (cm⁻¹)	Decay Time (ns)	Light Output (Light Photons/MeV Annihilation Photon)
NaI	3.67	50.8	0.35	230	~41,000
$Bi_4(GeO_4)_3$ (BGO)	1.06	75.2	0.96	300	~7,000
$Lu_{0.6}Y_{1.4}SiO_{0.5}$:Ce (LYSO)	7.10	54	0.5	40	~26,000
$Lu_2(SiO_4)O$:Ce (LSO)	7.40	66.4	0.86	40	~26,000
$Gd_2(SiO_4)O$:Ce (GSO)	6.71	59.4	0.70	60	~10,000
BaF_2	4.89	56	0.45	2	~2,000
$LaBr_3$	5.06	48	0.47	21	~60,000

reduced light signal to be discussed later. In addition, E_i can be increased by using monolithic scintillation crystals [23–25], by stacking semiconductor detectors slabs such as cadmium–zinc–telluride (CZT) [26–31] and by using tapered crystal arrays [31,32]. Though not being widely used in PET as those scintillation crystals shown in Table 12.1, CZT is a novel solid detector and has great potential to be used for PET detector development (density 5.61 g/cm^3; effective atomics number, 48; linear attenuation coefficient, 0.50 cm^{-1}).

Consider a dual-panel breast-dedicated PET system based on CZT detector technology as shown in Figure 12.1. A simulation study indicates that for 4 cm panel separation and a point source, the system sensitivity can reach ~32.5% at the center of FoV and can achieve >15% across the whole FoV for all three directions [33]. This is over an order of magnitude higher than the standard whole-body PET system (~1% at the isocenter).

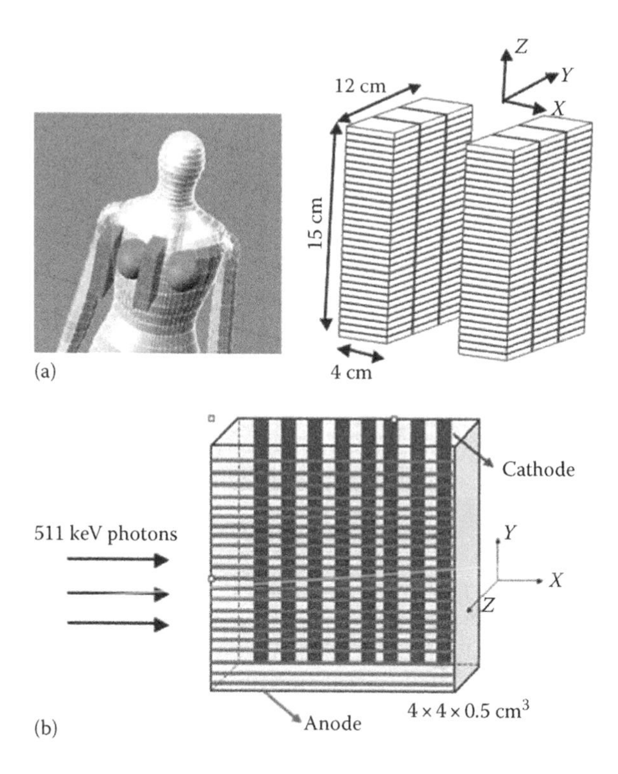

FIGURE 12.1
(a) Illustration of a dual-panel CZT-based PET system for breast cancer imaging. Each panel has dimensions of 4 × 12 × 15 cm^3. Each detector module has dimensions of 4 × 4 × 0.5 cm^3 with 25 μm inter-module spacing, giving a packing fraction of over 99%. (b) The cross-strip readout scheme for CZT detectors with sets of parallel anode and cathode strips in order to limit the number of electronic readout channels. (Adapted from Peng and Levin, *Phys. Med. Biol.*, 55, 2761, 2010.)

The increased photon sensitivity of the CZT system is attributed to the dual-panel configuration, close proximity to the breast, high packing fraction of the CZT detectors (>99%), and the detector of 4 cm thickness seen by incoming 511 keV photons.

Another group studied tapered PET detectors and compared their performance against rectangular PET detectors [32], as shown in Figure 12.2. Essentially, these efforts intend to decrease the gaps both in-between crystal pitches and between detector modules. For small-diameter animal scanners based on ring geometry, these gaps can be a significant factor limiting the sensitivity. This study investigated a small animal PET scanner of an inner diameter of 6 cm and an outer diameter of 10 cm.

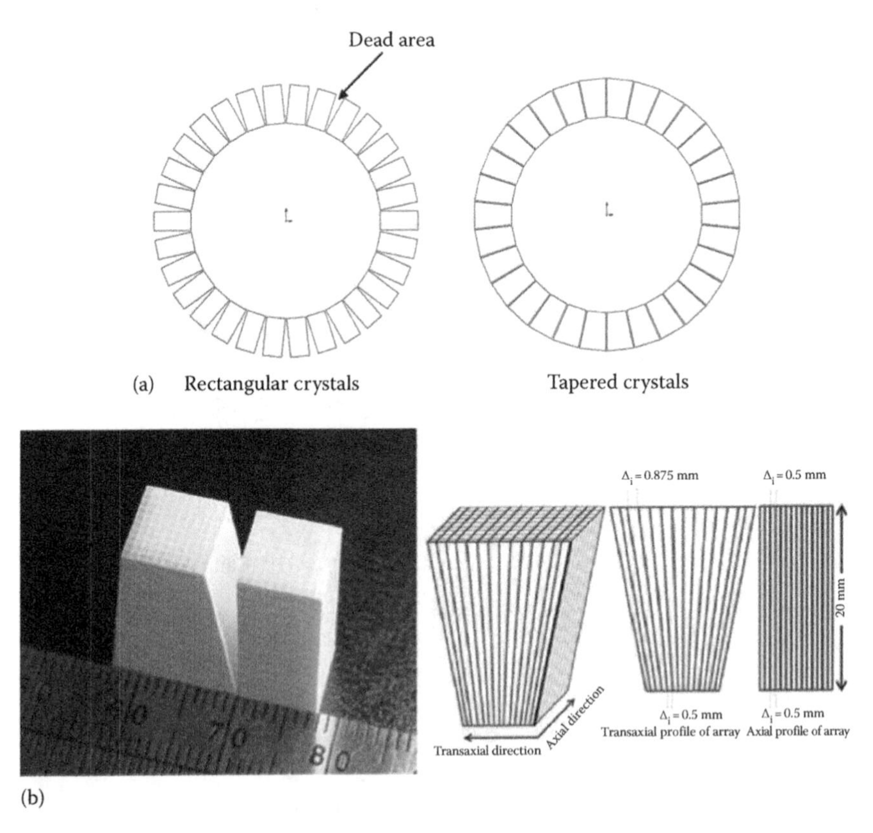

FIGURE 12.2
(a) Concept of using tapered crystals for small animal PET systems to reduce the dead area in-between detector modules (assuming a bore of 10 cm diameter, each detector module has a cross-sectional area of 1 × 1 cm^2 and crystals of 2 cm height). (b) Photograph of a tapered LSO array next to a rectangular LSO array and illustration of the dimensions of the simulated tapered LSO array. Note that there is no tapering along the axial direction. (Adapted from James et al., *Phys. Med. Biol.*, 54, 4605, 2009.)

The tapered crystals significantly reduce dead area in-between detector modules and thus increase the system packing fraction The traditional rectangular detector is a 14 × 14 array of 0.5 × 0.5 × 20 mm^3 LSO crystals, while the tapered detector is a 14 × 14 array of 0.5 × 0.875 mm^2 crystals (average height: 20 mm). The width of crystal elements is uniform in the axial direction but tapered in the transaxial dimension. Using the tapered arrays, the system achieves a photon sensitivity of ~4.2% at the center of FoV, compared to ~3.0% for the system with non-tapered crystals. Another novel approach to reducing the inter-module gaps that arise when arranging rectangular modules into a ring can be achieved by a box-shaped PET system as described in [17].

It should be emphasized that E_i is also dependent on the coincidence time (typically 2× time resolution) and energy window (typically 2× energy resolution at 511 keV photopeak) settings in a PET system as only those events staying within the windows are recorded. In general, if other conditions remain the same, a system that exhibits better energy resolution and time resolution would have a larger E_i.

12.4 Improved Spatial Resolution

A framework with regard to PET spatial resolution (i.e., "point spread function," *PSF*) has been proposed [34], based on the measurements with multiple whole-body PET scanners from different vendors [34]:

$$FWHM_{PSF} = 1.25\sqrt{\left(\frac{d}{2}\right)^2 + (0.0022D)^2 + R^2 + b^2} \tag{12.3}$$

The contributions to $FWHM_{PSF}$ near the center of FoV as shown in Formula 12.3 include crystal size, annihilation photon acollinearity, positron range, reconstruction algorithm, and crystal-decoding factor. All factors are in the unit of mm (*d*, crystal pitch; *D*, scanner diameter; *R*, effective positron range). *b is* a factor empirically derived for PET block detectors and ranges from 0 to 2.2 depending on decoding algorithms.

The spatial resolution plays a critical role for PET image quality through contrast recovery as characterized by modulation transfer function (MTF), which is obtained by calculating the Fourier transforms of the PSF [13,16,35]. MTF is widely used to quantify a PET system's ability to recover the input contrast of objects at various spatial frequencies. A system with a flat MTF curve having a value near unity would faithfully reproduce the image object with ideal contrast recovery. Furthermore, such contrast (recovery) is linked to SNR and impacts the image quality of a PET system as indicated by Formula 12.4, also known as the *Rose criterion* that has been widely used in radiology for lesion detectability [36]:

$$\mathrm{CNR} = \mathrm{contrast} \cdot \sqrt{N_{pixels}} \cdot \mathrm{SNR} \geq 4 \quad (12.4)$$

where

CNR is contrast-to-noise ratio; contrast refers to the activity concentration ratio between lesions and background depending on tumor malignance and probe specificity

N_{pixels} is the number of pixels contained by a lesion of given size (2-D case). The SNR is associated with the total counts detected, as well as the NEC value shown in Formula 12.1

The Formula 12.4 implies that using a PET system of high spatial resolution to detect smaller lesions is a challenging task. On one hand, high-resolution detector design is advantageous and can help improve contrast recovery. On the other hand, as N_{pixels} decreases for smaller lesions, SNR and photon sensitivity (Formula 12.2) need to be significantly increased to maintain sufficient CNR. Such trade-off needs to be paid enough attention when designing high-resolution PET detectors and systems.

Based on the Formula 12.3, there have been several studies trying to investigate the spatial resolution limit of PET systems. Levin et al. found that for a point source of ^{18}F in water-equivalent tissue determined by three blurring factors (positron range, photon acollinearity, and crystal pitch), 750 μm FWHM resolution is attainable with 1 mm crystal pitch and 20 cm system diameter [37]. In addition, Stefan et al. proposed a model for studying the intrinsic spatial resolution by considering the positioning uncertainty of high-energy photon inside various detector materials. The results indicate that 500 μm FWHM resolution is achievable if a 250 μm detector pixelation and 8 cm system diameter can be made [38]. Motivated by the simulation work, the group has successfully developed detectors of 500 and 750 μm LSO crystal pitches [39], as shown in Figure 12.3. Several major challenges faced by reducing the crystal pitch down to around 1 mm [1], as well as several promising solutions under investigation, are briefly summarized as follows.

12.4.1 Complex and Expensive Assembly

A major challenge of manufacturing PET detectors with smaller (e.g., <2 mm width) scintillation crystal elements is that cutting, surface treatments, and assembling crystals to make arrays are complex and expensive processes. Furthermore, optical reflectors are usually required to be inserted that typically have a minimum thickness about 50–75 μm. Assuming a single-crystal

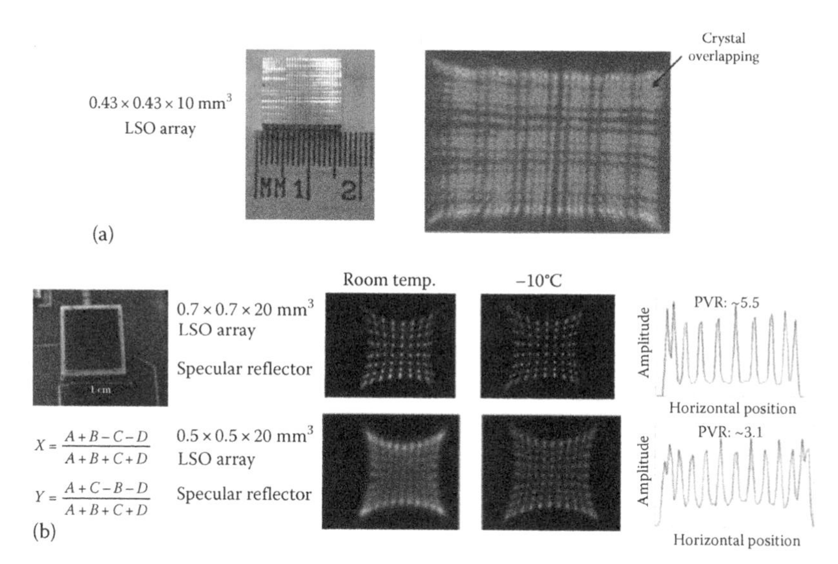

FIGURE 12.3
(a) Photograph of a 20 × 30 crystal array of 0.43 mm crystal pitch and the resulting flood histogram. The array was read out by a Hamamatsu H7546 MCPMT that has 64 individual channels arranged in an 8 × 8 grid (grid pitch: 2.25 mm). Significant crystal overlapping in the flood map is noticed. (b) Crystal flood maps for two high-resolution arrays read out by PSAPDs. Four spatial channels at four corners of each PSAPD were used to position events using Anger logic. The device has an effective area of 1.0 × 1.0 cm² and was cooled to −10°C to improve the detector' SNR and crystal separation. The 1-D horizontal profile and the average PVR are shown for the flood maps obtained with cooling. (Adapted from James et al., *Phys. Med. Biol.*, 54, 4605, 2009; Stickel et al., *J. Nucl. Med.*, 48, 115, 2007.)

pitch of 250 μm is used, the optical reflectors in-between crystal pitches would degrade the packing fraction down to ~60%–70%, significantly reducing the photon detection sensitivity.

12.4.2 Reduced Scintillation Light Output

A typical PET detector comprises long yet very crystal elements of low aspect ratio (i.e., width/height), which comprise the scintillation light collection. A weak scintillation light signal negatively impacts energy and time resolution and also limits the degree of spatial decoding. Light collection improvements can be made by proper treatment of the crystal faces, such as polishing or well-controlled etching, to enhance total internal reflection [40]. To address the issue of light loss, in 2 mm resolution PET systems built to date, relatively short (~10 mm length) crystals are used [41,42]. However, this consequently compromises intrinsic PDE with a nearly 24% reduction (~58% for a 10 mm long crystal versus ~82% for a 20 mm one), based on the attenuation coefficient of LSO shown in Table 12.1.

This challenge can be addressed by using novel CZT detectors mentioned earlier. In such semiconductor detectors, the electron–hole pairs are directly involved in the electronic signal formation, instead of the relatively inefficient process of creation/transportation of scintillation light in scintillator-based detector technology. As a result, compared to scintillator-based PET detectors, CZT detectors exhibit much better energy resolution for 511 keV photons [26–31].

12.4.3 Compton Scatter and Multiple Interactions

A 511 keV photon will interact with scintillation crystals or other type of detector materials and deposit its energy mainly through photoelectric effect (PE) and CS. If the detector elements are small, the scattered photons will usually escape into adjacent crystals before depositing their remaining energy, also known as multiple interactions. Multiple interactions occurring between adjacent crystals have been found to be a dominant source for the mispositioning of incoming 511 photons and contribute to the tails of the PSF [43,44]. This, in turn, degrades spatial resolution and contrast recovery. In general, the probability of photon scattering is higher for relatively low-Z, lower-density crystals such as GSO and CZT (see Table 12.1) and can also be significant in higher-Z crystals such as LSO, if crystal pitches are small [45]. The implication is that even though the utilization of a finer crystal pitch could improve the spatial resolution, the increased portion of multiple interactions due to the reduced crystal size might diminish that benefit while causing additional penalties such as light output loss.

12.4.4 Spatial Decoding for Block Detectors

An important question is that how the reduced light output affects the spatial decoding. To explain this, the results of several recently developed PET detectors are presented in Figures 12.3 and 12.4, including (1) an LSO crystal array of $0.43 \times 0.43 \times 10$ mm^3 coupled to a position-sensitive photomultiplier tube (PMT) whereas the crystals in the flood map cannot be clearly resolved [46], (2) two LSO crystal arrays (0.5 and 0.7 mm pitch) coupled to position-sensitive avalanche photodiode (PSAPDs) [39], and (3) an 8×8 crystal LSO arrays coupled to a 2×2 large-area APD array through light multiplexing and Anger-logic positioning [47]. In Figure 12.3, at room temperature, crystals in the flood map of the 0.5 mm array are less resolved compared to the 0.7 mm array, due to the reduced light collection (i.e., smaller crystal pitch). As PSAPDs were cooled to −10°C, the quality of both flood maps improves since the devices have higher SNR at lower temperature. The peak-to-valley ratio (PVR), an important figure-of-merit widely used for block detector evaluation, is also shown for the two profiles of the flood maps. Higher PVR indicates that crystals are better distinguished, implying that the detector is able to achieve better spatial resolution in the context of PSF in Formula 12.4 (i.e., smaller b value). A flood map for a block detector of larger crystal pitch (2.5 mm) designed for a brain PET is shown in Figure 12.4. No cooling was applied and 64 crystals can be clearly resolved. The dependency of spatial decoding on the detector's SNR is demonstrated in the simulation results [47]. In essence, when a high-resolution crystal array is used with any type of multiplexing

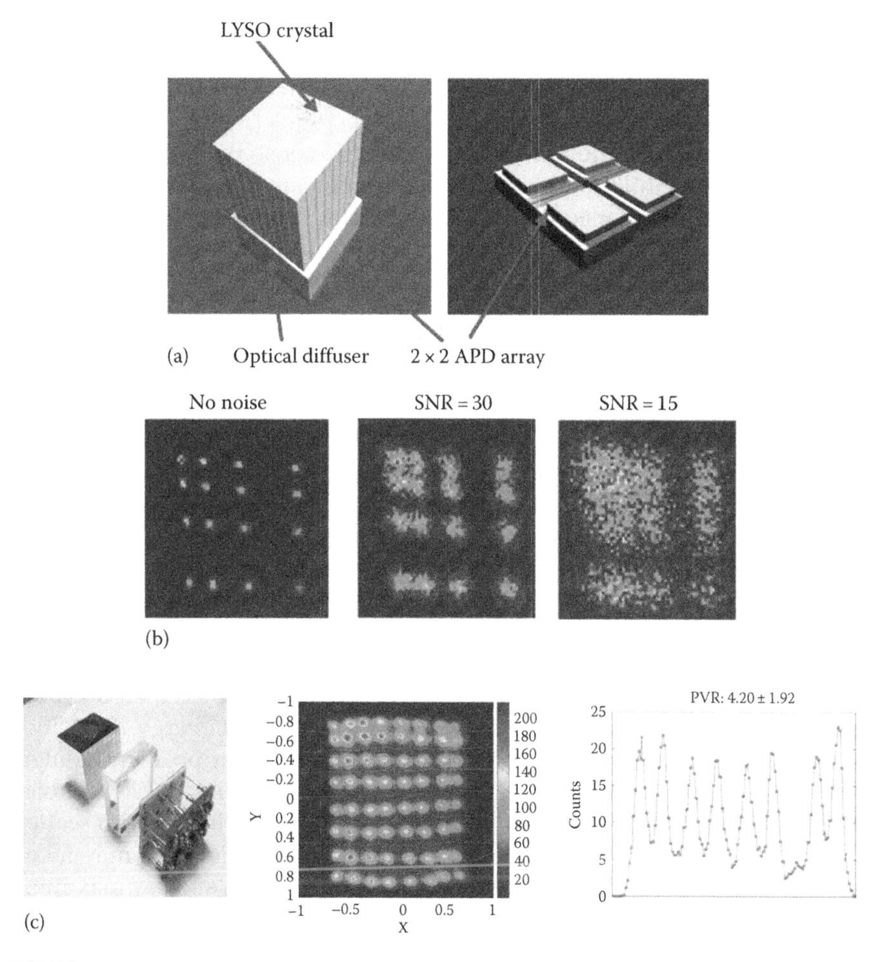

FIGURE 12.4
(See color insert.) (a) Anger-logic PET block detector based on light multiplexing. A 2 × 2 array of large-area high-gain APDs was used. (b) Simulated flood maps under different SNR conditions of photodetectors (crystal dimensions: 2.5 × 2.5 × 20.0 mm^3). Only one quadrant for the 8 × 8 array was simulated due to the symmetry. Significant degradation of the flood maps is observed as the SNR decreases. SNR refers to the ratio of the amplitude of 511 keV photopeak over the RMS noise of the detector when a single crystal is coupled directly to the photodetector. (c) The block detector prototype comprising the LYSO array (crystal dimensions: 2.75 × 3.0 × 20.0 mm^3), the optical diffuser (9 mm thickness), the APD array, and the highly compact custom readout electronics. The crystal flood map and the 1-D profile for the PVR analysis are shown. (Adapted from Peng and Levin, *Phys. Med. Biol.*, 56, 3603, 2011.)

scheme, the reduced light collection will inevitably degrade the detectors' SNR and thus the crystal separation (i.e., lower PVR), which make it less likely to achieve the theoretical resolution limit of 500 μm as predicated in [38].

12.4.5 High-Resolution PET Detector Developments

Two novel detector designs that are capable of achieving high spatial resolution while not suffering from those limitations mentioned earlier are briefly described here. One approach is to use semiconductor detectors such as CZT or cadmium telluride instead of scintillation crystals [26–31]. Unlike scintillation crystals, semiconductor detectors directly sense the ionization signal created by the annihilation photon and do not create/transport scintillation light. Such direct conversion enables it to exhibit superior energy resolution of ~1.5%–3.0% FWHM at 511 keV, compared to ~10%–20% scintillator-based detectors. In addition, fine pixelation can be set by the segmented pattern of charge-collecting electrodes deposited on the crystal slab faces, rather than relying on cutting and assembling many miniscule crystal pixels [26,29]. For instance, to achieve high intrinsic spatial resolution of 500 μm, one can deposit the electrodes with a pitch that matches the desired spatial resolution, either in a cross-strip pattern or in a fully pixelated pattern.

Another approach is to read out the larger-area side faces of the crystals, rather than the small-area ends using very thin PSAPD photodetectors (Figure 12.5) [48–52]. In this new geometry, with the crystals/photodetectors orienting edge on with respect to incoming 511 keV photons, the average light path from the crystals to the photodetectors is greatly reduced and the light collection is nearly complete (>95%). This feature helps to achieve improved energy resolution (~12%) and time resolution (~2 ns) in addition to 1 mm spatial resolution.

One promising benefit of these two novel detector designs is their 3-D positioning capability. By recording the 3-D coordinates and energy deposition for every interaction, it is possible to use more intelligent positioning algorithms to better estimate the line of entrance of an incoming 511 keV photon in order to mitigate the effect of positioning errors due to intercrystal scattering, as discussed in Section 12.4.3. Such positioning algorithms may incorporate the physics of CS and/or a probabilistic formalism such as maximum likelihood to estimate the first interaction location with good accuracy [53], as well as retain events that are normally rejected such as single (unpaired) photons, tissue-scattered coincidences, and multiple-photon coincidences [54,55].

12.4.6 High-Density Data Acquisition Systems

Despite their superior performances, these novel CZT and LSO-PSAPD detectors face challenges in readout electronics developments. For example, a dual-panel breast-dedicated detector system based on LSO and PSAPD under development [51,56,57] will have more than 10,000 individual readout channels to be processed (Figure 12.5). On the other hand, the lower-gain

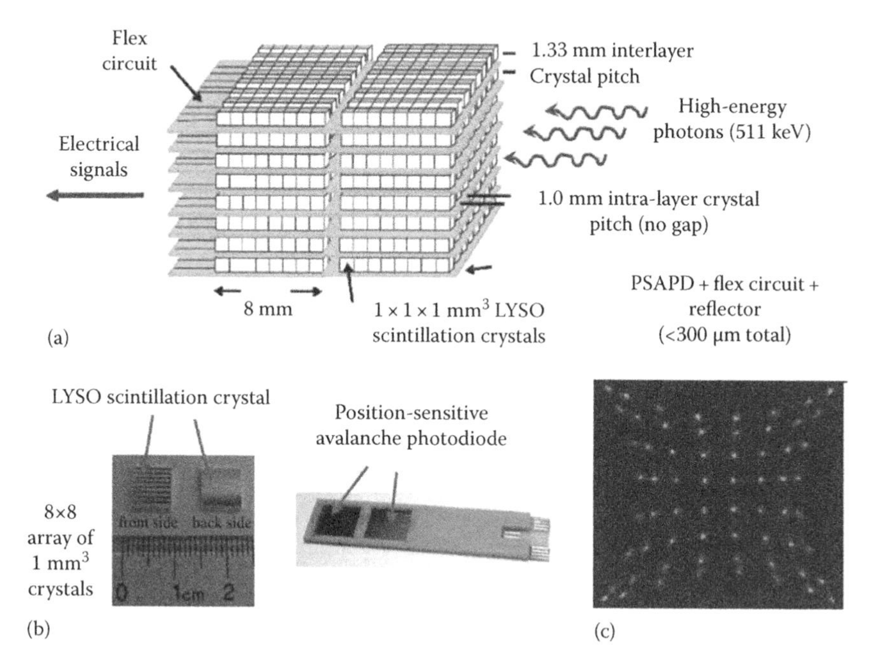

FIGURE 12.5
(a) Picture of an edge-on LYSO+PSAPD module including scintillation crystals, extra-thin PSAPD modules, and flex circuits. Each layer comprises two 200 μm thick PSAPD chips mounted on a 50 μm thick flex circuit. The flex circuit delivers bias to each PSAPD and enables readout of the four corner signals for positioning. (b) Picture of the thin module with two PSAPD chips mounted. Each dual-LYSO-PSAPD detector layer is oriented edge on so that incoming photons encounter a minimum of ~2 cm thick of LSO with directly measured photon interaction depth (~1 mm). (c) The crystal flood map of an 8 × 8 array of 1 × 1 × 1 mm³ LYSO crystals without cooling. (Adapted from Lau et al., *Phys. Med. Biol.*, 55, 7149, 2010; Vandenbroucke et al., *Phys. Med. Biol.*, 55, 5895, 2010.)

and poorer-detector SNR of CZT and APD devices (compared to PMTs) prevent extensive multiplexing to be implemented to reduce the number of readout channels, either with charge multiplexing or light multiplexing [47]. Thus, a key bottleneck in the development of a PET system built from high-resolution 3-D positioning detectors is the development of high-density application-specific integrated circuits (ASICs) that are required to read out hundreds to thousands of densely packed detector channels [1,58].

12.5 Depth-of-Interaction Design

A well-known problem in PET instrumentation is the parallax error, also known as the DoI problem, which impacts both spatial resolution and quantitative studies [59]. The framework proposed in Formula 12.2 does not take into such effect. A PET system typically comprises a circular arrangement of crystal arrays. The interaction of 511 keV photons with long crystal elements at oblique angles, as well as the penetration of 511 keV photons through multiple crystal elements, results in DoI blurring and leads to nonuniform spatial resolution.

The resolution blurring caused by the DoI effect can be mitigated if one can determine the photon interaction depth with improved accuracy. In general, DoI resolution has two types: discrete DoI and continuous DoI. The former one implies that crystals are physically segmented to independent units (similar to the pixilation through electrodes in CZT detectors). The latter implies that no physical segmentation is made and DoI is estimated/calibrated using collimated photon beams. DoI capability can be introduced for conventional scintillator–based PET detectors, and five representative DoI design schemes are summarized as follows. It is worthwhile pointing out that for those 3-D positioning detectors such as LSO+PSPAD- or CZT-based detectors examined earlier, the DoI capability has been inherently incorporated. For example, with edge-on PSAPDs configured on the sides of miniscule LSO crystal elements (Figure 5.5), the segmented crystals are able to achieve 1 mm intrinsic positioning capability along the DoI direction [26,29,48,49].

12.5.1 Multiple Crystal–Photodetector Layers

The design consists of two or more crystal layers each read out by a layer of photodetector(s) [45,51,56,60,61]. As shown in Figure 12.6a, the DoI resolution is discrete and determined by the crystal segmentation of each layer. The advantage is that the overall detector performance is not compromised, as there is no optical and electric interference between the two layers. The drawback of this design is the increased number of electronic readout channels (2x more) and increased development cost.

12.5.2 Single-Crystal Layer + Dual-Ended Photodetectors

This design employs a single-crystal layer read out by two photodetectors on both ends (Figure 12.6b). The depth information of the 511 keV photon inside the crystal is determined by the difference in the amount of light detected by the two photodetectors [62–64]. For an 8×8 array comprising $1 \times 1 \times 20$ mm^3 crystal pitches, average DoI resolution over all crystals and all depths is $\sim 3.5 \pm 0.1$ mm for the energy threshold $E > 350$ keV and $\sim 5.0 \pm 0.1$ mm for $E < 350$ keV when more CS events are included and thus degrade the photon positioning capability. Such design is also able to extract DoI resolution for even smaller arrays such as 0.5 mm and 0.75 mm pitch [38]. A minor problem about this design is that the light sharing and propagation along the long crystal (~20 mm between the two photodetectors) may result in poor detector performance in terms of energy and/or time resolution, as well as the positioning nonlinearity near the faces of two photodetectors

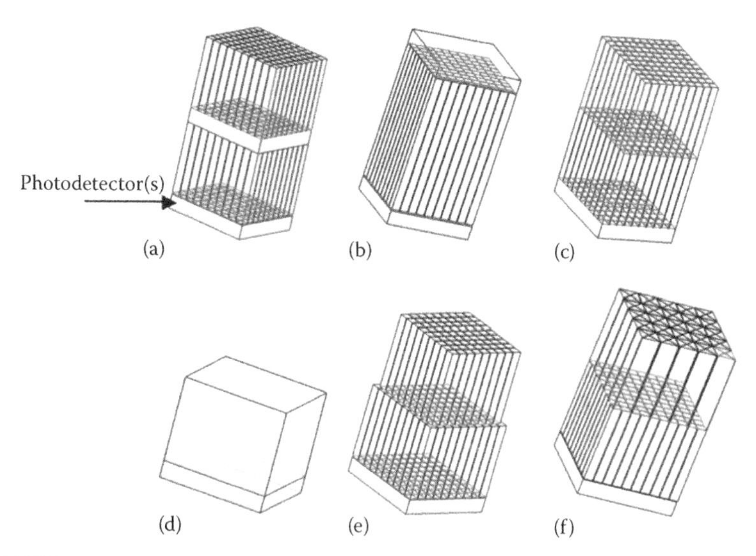

FIGURE 12.6
Illustration of several DOI detector design concepts for PET. (a) Dual-crystal photodetector(s) layers. (b) Single-crystal layer with photodetectors at each end. (c) Phoswich design with two types of scintillation materials. (d) Statistical positioning with a monolithic crystal block. (e) Dual-layer crystals with offset positions. (f) Dual-layer crystals of mixed shapes.

(i.e., nonuniform DoI resolution). In addition, as DoI resolution is directly dependent on the signal differences between the two photodetectors, the stability and calibration of detectors, particularly the gain variability between the two photodetectors, remain a challenge.

12.5.3 Phoswich Design

The phoswich detector comprises two different types of scintillation crystal materials read out by the same photodetector [65,66], as shown in Figure 12.6c. DoI capability is realized by pulse-shape discrimination based on the fact that the decay time constants of two scintillation materials are different (Table 12.1). In this design, complicated pulse-shape discrimination circuits are required to extract the timing difference between pulses originated from two types of materials. Ideally, the DoI resolution should be equal to the thickness of each crystals layer as in the design 5.1. However, several factors could degrade the DoI capability, including the limited decay time difference between the two materials, the intrinsic timing fluctuation in signal from each layer and electronics, multiple interactions involving both layers, as well as the light loss at the interface between the two crystal layers of different refractive indices. Nevertheless, such design has demonstrated good performance, and a commercial system based on such design is the GE small animal PET scanner (LYSO, 40 ns decay time, and GSO, 60 ns decay time).

12.5.4 Statistical Positioning with Monolithic Crystals

This design extracts the DoI information based on monolithic crystals and statistical modeling [67–69]. As shown in Figure 12.6d, a monolithic crystal block of relatively large size is read out by either individual photodetector(s) or position-sensitive photodetector(s), which can be coupled to either the entrance surface or the exit surface. In essence, the spatial resolution in the cross section of the crystal is obtained in the same manner as that in detectors without DoI capability. While in the DoI direction (the direction along which 511 keV photons enter the crystal), the DoI resolution is determined based on the light output intensity and light spread profile that both depend on the depth of photon interaction. Several algorithms have been developed for this design, including maximum likelihood and mean square error [67–69].

Despite the simplicity implied by the use of a single monolithic crystal layer, the complexity of this design lies mostly in the algorithms and calibrations. For instance, a set of experiments has to be done by irradiating the detector with 511 keV photon beams at a series of known positions and angles of incidence. Moreover, this design is subjected to the degeneracy of positioning near corners and edges of the crystal, where the spatial resolution is inferior compared to that near the center of the crystal. Nevertheless, such design has several attractive benefits besides the DoI capability: First, the use of monolithic crystal avoids complicated crystal cutting and assembly; second, it gets rid of dead area in-between crystal pitches and thus increases the photon detection sensitivity; and third, no extra optical interface, and therefore no light loss, is introduced as in other designs.

12.5.5 Dual-Layer Crystals of Offset Positions or Mixed Shapes

In this design, two layers of crystals (same material) are involved and are read out by only a layer of photodetectors(s). As the top and bottom layers of crystals have different light output profile, their positions can be extracted accordingly. For instance, one method is to offset the top layer by half of the crystal pitch with respect to the bottom layer as shown in Figure 12.6e [70]. For illustration only, the top layer of a 9×9 array and the bottom layer of a 10×10 array are shown. Another design consists of two crystal layers built from crystals of different shapes, and more details are provided in Figure 12.7. Each layer is built from small modules and each module comprises eight crystals (four crystals for each layer). Note that the dimension of the base of a triangular crystal is twice that of the pitch of rectangular crystals. Here the light sharing is dependent on the crystal shape. For instance, the triangular crystals in the top layer form a special light-focusing pattern that enables their differentiation from the bottom layer of square crystals.

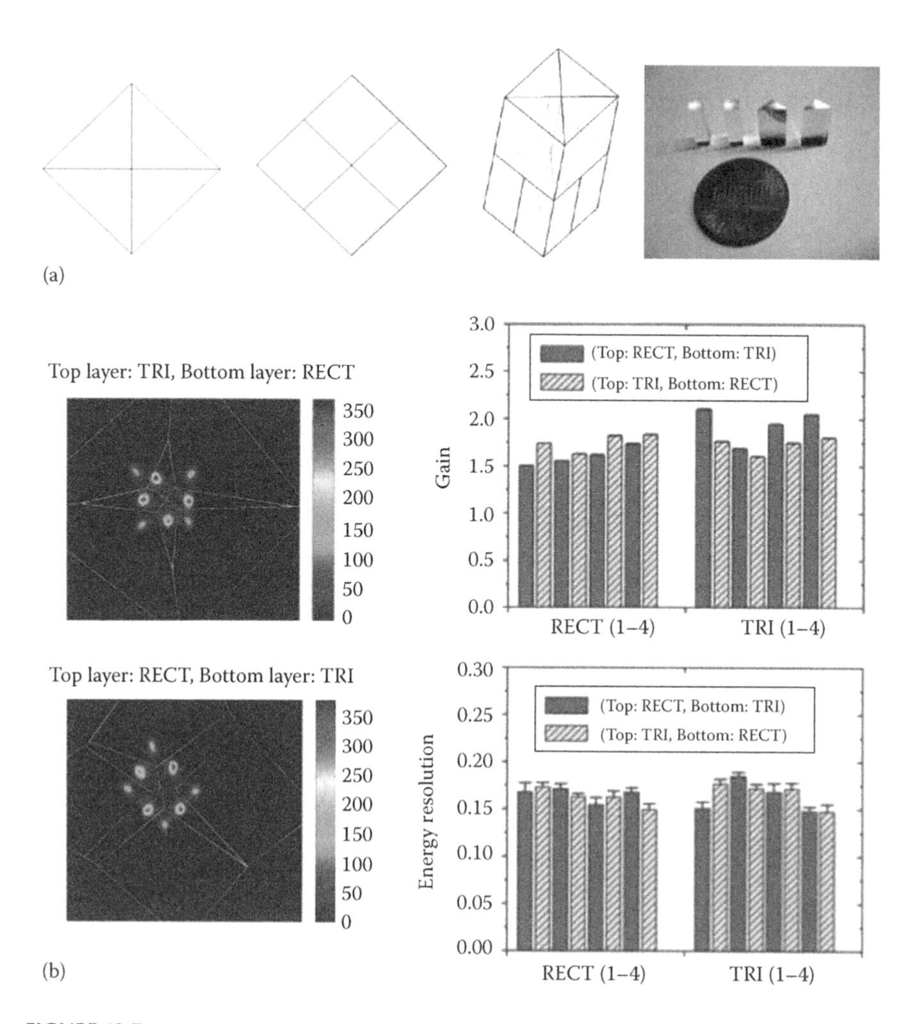

FIGURE 12.7
(a) DOI design using two layers comprising triangular (TRI) and rectangular (RECT) crystal elements. Each module consists of four triangular crystals (top layer) and four rectangular crystals (bottom layer). The picture of individual crystal segments is shown. (b) The flood histograms of a basic module (comprising eight crystals) are shown for two configurations (rectangular crystals at the top layer or at the bottom layer). The gain (in voltage) and the energy resolution FWHM after the individual crystals were segmented from the flood histograms and analyzed. No significant performance difference is observed between crystals in the two layers.

This design provides discrete DoI and does not double the number of photodetectors. Moreover, it does not require complicated readout electronics that are necessary in a phoswich design. However, one potential limitation may stay in the complexity and relatively high cost in manufacturing triangular-shape crystals.

12.6 ToF PET and SiPM Detectors

12.6.1 Time-of-Flight PET

The history and recent progress of ToF-PET instrumentation are briefly reviewed here. By accurately measuring the arrival time of two 511 keV annihilation photons, the location of their emission points can be constrained. Though this constraint is not tight enough to avoid the image reconstruction or improve the spatial resolution, it can significantly reduce the statistical noise in the reconstructed images [71–74]. The SNR improvements are due to the following reason: In a non-ToF PET, the noise from all pixels along a given LoR is correlated, while in a ToF PET, the statistical fluctuations from the data along a LoR are constrained to a reduced number of image pixels, as illustrated in Figure 12.8. The framework developed for the SNR benefit of a ToF PET is given by Formula 12.5 [75,76]:

$$\Delta x = \frac{c}{2}\Delta t \frac{\mathrm{SNR}_{\mathrm{TOF}}}{\mathrm{SNR}_{\mathrm{Non\text{-}TOF}}} = \sqrt{\frac{D}{\Delta x}} \tag{12.5}$$

where

Δx is the position uncertainty
c is the speed of light
Δt is the time difference (related to the time resolution of PET detectors)
D is the size of the object being imaged

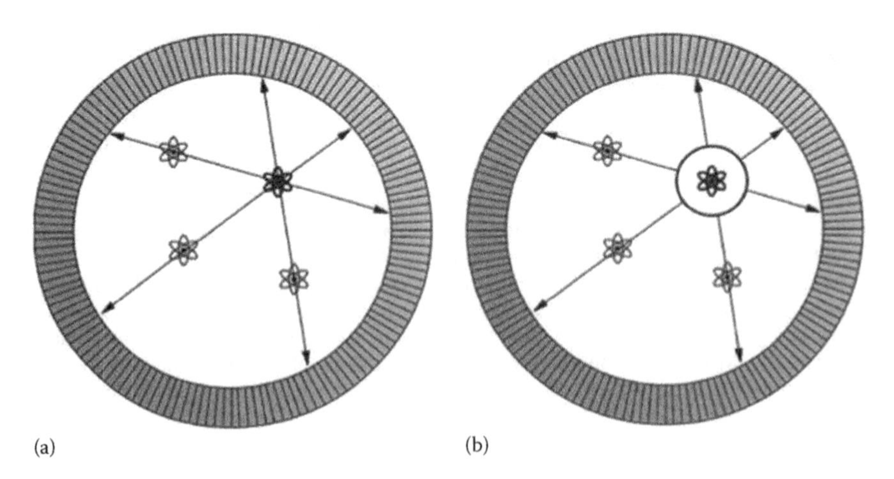

FIGURE 12.8
(a) In conventional PET, the source of the activity is localized to a LoR between a detector pair (i.e., two measured photons). (b) In ToF PET, timing information is used to restrain the source of the activity to a segment of the LoR. (Adapted from Moses, *Nucl. Instrum. Methods Phys. Res. A*, 580, 919, 2007.)

For example, for a time resolution of 500 ps that is currently achievable with fast scintillators and PMTs, Δx is ~7.5 cm long. For an imaging object of 40 cm diameter, the SNR improvement is ~2.3 and G is about 5.3. However, for small animal PET, with the coincidence time resolution of current detector technologies and small size of objects (e.g., ~2–5 cm bore diameter), no SNR improvement is expected.

Back to the time the ToF concept was introduced in the 1980s, PET systems were mainly used for brain and cardiac imaging, whereas resolving small lesions and improving SNR were not deemed critical. However, the ToF PET has regained more research and clinical attention as PET increases its role in oncology nowadays. Several simulation and experimental studies have been performed to predict the performance of ToF whole-body PET systems. Phantom studies with a LYSO-based commercial ToF-PET scanner (Philips Gemini TF PET/CT) indicate that ToF information not only increases the lesion detection ability by a SNR gain of 1.4–1.6 (10 mm diameter lesion within 35 cm diameter background) but also reduces the number of iterations required in statistical image reconstruction [77]. Clinical studies also provide promising results that a ToF PET is able to reveal/define structural details in colon cancer, abdominal cancer, and lymphoma [77]. Currently, a number of ToF-PET scanners are commercially available, including Philips (crystal: LYSO) [78], Siemens (crystal: LSO) [79], and GE (crystal: LYSO) [80]. All these systems exhibit a timing resolution of ~550–600 ps FWHM.

Very recently, the feasibility study using Cherenkov radiators for a ToF-PET system has also been made, such as PbF_2, $PbWO_4$, and lead glass [81]. For inorganic scintillation crystals such as LSO and $LaBr_3$, the intrinsic time resolution is limited by the temporal response associated with transition of electron–hole pairs between energy states (i.e., stimulation from ground states to excited states (1–2 ns) and decay from excited states back to ground states (20–40 ns)). Traditionally used in high-energy particle physics, a Cherenkov light–based detector is able to achieve the excellent timing resolution (~5–100 ps) and emerges as a promising candidate for ToF-PET systems [81]. The analytical calculation shows that the light output at 511 keV is found to be around 10–25 for high-Z Cherenkov radiators and around 50–250 for low-Z Cherenkov radiators. This poses a serious challenge for light detection and requires novel photodetectors of superior detection efficiency and low dark counts, which might be beyond the capabilities of detector technologies currently available. Besides the light yield, light photons collected by the photodetector also depend on crystal geometry as mentioned in Section 12.4.2. For standard PET crystals of smaller aspect ratio (i.e., the ratio between the width and the height), light trapping inside the crystals is significant and thus reduces light output. This implies that only $PbF_2/PbWO_4$ crystals of much larger aspect ratio should be considered for this promising application.

12.6.2 Fast and High-Z Scintillation Materials

In 1980s, a number of PET systems incorporating ToF were built based on BaF_2 or CsF as they are very fast scintillators. However, they are not suitable for PET detectors due to the low light output and low detection efficiency, as well as the lack of high-speed electronics to perform accurate timing measurements. There has been renewed interest in ToF PET over the past few years [76,82], thanks to the availability of newly developed fast scintillators, such as LSO, LYSO, and $LaBr_3$ crystals. As shown in Table 12.1, these crystals have high light output (good for energy resolution and scatter rejection), fast decay time constant (good for time resolution and random rejection), as well as good stopping power (high atomic number). Besides commercial LYSO-based ToF-PET scanners, Philips has recently developed a ToF-PET system based on $LaBr_3$ crystals [83], which have a faster decay time and higher light output than LSO or LYSO though a lower atomic number Z. Preliminary characterization indicates that an array of $4 \times 4 \times 30$ mm^3 $LaBr_3$ (5% Ce) readout with large size PMTs could achieve a time resolution of ~313 ps FWHM and an average energy resolution of 5.1% FWHM at 511 keV [84], compared to a timing resolution of ~585 ps and an energy resolution of 11.5% FWHM for LYSO crystals [78]. Furthermore, other scintillation materials such as $LaBr_3$ (Ce) with different cerium concentrations, $CeBr_3$, and LuI_3 (Ce) are also being investigated [76].

Besides the intrinsic properties of materials, other parameters are also important for detector designs for ToF-PET systems, such as crystal surface treatment and crystal height selection [85]. In conventional clinical PET detectors, relatively long crystals (e.g., $4 \times 4 \times 20$ mm^3 LSO crystal pitch) are coupled to the surfaces of PMTs through light guide. If a scintillation event occurs at the back of one detector/crystal but the front of opposing detector/crystal, the difference in prorogation time within crystals will cause additional blurring in the measured coincidence time spectrum [75]. Furthermore, the use of any light guide itself would also degrade the time resolution. In this regard, the detector design using dual-crystal layers + dual-ended photodetectors mentioned in Section 12.5.1 appears to be a promising solution, which can achieve good time resolution (i.e., less propagation time within each layer of crystal), good DoI resolution, and high sensitivity at the same time.

It should also be pointed out that there exists a potential benefit of NEC improvement due to the superior time resolution of fast scintillation crystals. The simulation of a $LaBr_3$-based ToF-PET system shows that it is able to achieve higher NEC as defined in Formula 12.1 than that of a LYSO-based ToF-PET system [83,84]. On the other hand, such system exhibits inferior performance compared to the LYSO scanner in terms of photon sensitivity. The authors provide a possible explanation for such seemingly contradictory results that the lower sensitivity of $LaBr_3$ is partially offset by its lower scatter and random fractions due to its superior energy resolution and time resolution. Nevertheless, it should be kept in mind that such NEC improvement is only possible with two modifications made: (1) The crystal thickness increases from 20 mm (LYSO) to 30 mm ($LaBr_3$), and (2) the axial FoV increases from 18 cm (LYSO) to 25 cm ($LaBr_3$).

12.6.3 SiPM Devices

12.6.3.1 Overview of SiPM Operation

Being available recently, SiPM is considered a very attractive photodetector technology for ToF-PET system development [86–92]. Each SiPM pixel, of size ranging from 1 to 3 mm, contains thousands of micro-cells. Each micro-cell has a pitch of ~20–50 μm and operates in the Geiger mode, which is also called Geiger APD and single-photon avalanche diode (SPAD). SiPMs are capable of measuring extremely low-level light signals to the extent of single photon. Compared to PMTs, they offer the "solid-state" advantages such as low operating voltages, ruggedness, compactness, and immunity to magnetic fields. Examples of SiPMs are shown in Figure 12.9.

A very brief description of how SiPM works is provided as follows, and more details about its working principles can be found in [91,92]. Each micro-cell (inside a SiPM pixel) consists of a reversely biased p-n junction and responds independently to incident photons. Electron–hole pairs are formed inside the depletion layer and drift in the presence of a high electric field. When the bias voltage increases above the breakdown voltage, each electron–hole pair is able to trigger a self-sustaining avalanche multiplication process called Geiger discharge. As a result, the number of charge carriers generated no longer reflects the intensity of incident photons. This is a fundamental difference between APDs and SiPMs, similar to that between proportional counters and Geiger tubes in gas detectors. To suppress the self-sustaining process and protect the micro-cells from excess large current, a quenching

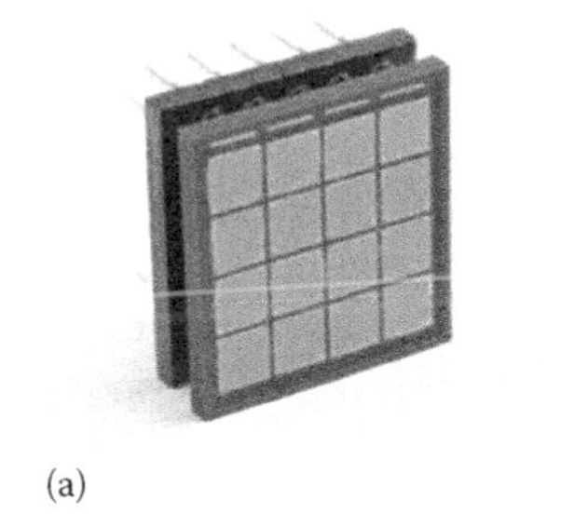
(a)

(b)

(c)

FIGURE 12.9
(a) Design of a 4 × 4 array of SiPM devices with ~3.0 mm pitch from SensL (ceramic package). (b) Design of a 4 × 4 array of SiPM devices with ~3.0 mm pitch from SensL (surface mount package). (c) Digital SiPM recently developed by Philips (a 2 × 2 array). Active area of each pixel is ~3.8 × 3.3 mm². Signal from multiple micro-cells within a pixel is digitally added. (Adapted from the technical notes from SensL and Degenhardt et al., *IEEE Nucl. Sci. Symp. Conf. Rec.*, 2383, 2009; Frach et al. *IEEE Nucl. Sci. Symp. Conf. Rec.*, 1959, 2009.)

mechanism through either passive or active circuitry is required. For the passive quenching scheme, a large resistor is connected in series with the micro-cell, and when a large current flows through, it results in a voltage drop that brings the bias across the junction below the breakdown voltage. While for the active quenching, more complex voltage and timing controlling units are to be used. After a short duration determined by the time constant of quenching circuits, the bias across the micro-cell recovers back to the breakdown voltage so that it is ready for detecting another radiation event. Behaving essentially in a binary mode, each micro-cell produces a charge output (Q_{cell}) as shown in Formula 12.6:

$$Q_{cell} = C_{cell} \times \left(V_{bias} - V_{breakdown}\right) \mathrm{PDE} = \mathrm{FF} \times P_{Geiger} \times \mathrm{QE} \tag{12.6}$$

where
C_{cell} is the micro-cell's capacitance
V_{bias} is the applied voltage
$V_{breakdown}$ is the breakdown voltage for the p-n junction (~10–70 V)

The gain of a micro-cell is approximately 10^5–10^7, very comparable to that of PMTs. Finally, a number of micro-cells are connected to a common readout line, and the summation of output of individual microcells provides the amplitude information of incoming signals.

Another important parameter of SiPM devices is photon detection efficiency *(PDE)* shown in Formula 12.6. PDE is dependent on three terms: fill factor (FF), probability of Geiger discharge (P_{Geiger}), and quantum efficiency (QE). The fill factor is the ratio of active areas of micro-cells to the total area of a single SiPM pixel. P_{Geiger} depends on the location of electron–hole pairs within the depletion layer, as well as the strength and shape of the internal electric field. QE is strongly wavelength dependent, and a higher QE in blue region is desired for PET, due to the wavelength of scintillation light photons. Typical values of PDE for commercial SiPM devices are 20%–30% [92], which are also comparable to that obtained with PMTs.

A detailed comparison among three detector technologies (PMT, APD, and SiPM) for PET can be found in [91]. In brief, SiPM provides a number of advantages for PET instrumentation. First, its small pixel size (~1–3 mm) enables detectors to achieve high spatial resolution, similar to those detector designs based on CZT and APD described earlier. Second, its high gain results in good detector SNR and eliminates the need for subsequent stages of amplification. Third, the output pulse of each micro-cell has a well-defined shape (i.e., both rising edge and amplitude), which is beneficial for accurate timing pick off [91]. A number of recent PET detector designs deploying SiPMs can be found in [91–96]. In particular, recent studies show that a time resolution of 237 and 240 ps (FWHM) is achieved for $LaBr_3$ [93] and LYSO [94] crystals of $3 \times 3 \times 20\ mm^3$ pitch, respectively.

Generally speaking, when the available space for accommodating detectors is a concern or when detectors need to work inside strong magnetic fields such as PET/MR systems, solid-state detectors such as APDs and SiPMs are preferred. On the other hand, when superior time resolution is desired as for ToF PET, PMTs and SiPMs are preferred over APDs as the latter suffer from relatively lower gain (lower SNR) but larger parasitic capacitance. However, whether SiPMs are able to achieve improved intrinsic time resolution over PMTs requires further investigation.

12.6.3.2 Digital SiPM

The SiPM operation described in the last section is also called analog SiPM, as each pixel processes analog output from many micro-cells through a common readout line and add them together. Subsequently, the output from each pixel is then digitized and processed for obtaining energy and time information. For a PET system consisting of thousands of channels, this becomes a very challenging task and requires the use of dedicated ASICs as discussed in Section 12.4.6. Such architecture faces a few potential challenges including [97,98] (1) expensive power-consuming electronics; (2) signal deterioration due to large parasitic of on-chip interconnect, bond wires, and external load; and (3) susceptibility to electronic noise and temperature variations that are typical characteristics of individual micro-cells.

To address these challenges, a novel design concept of SiPM emerges very recently, known as digital SiPM, and it is implemented in the complementary metal oxide semiconductor (CMOS) process. On one hand, CMOS sensors offer the promise of ultralow-power, low-cost, high-yield, and reliable deliverables [99–102]. On the other hand, by integrating micro-cells with the CMOS processes, it is possible to monolithically integrate the detector onto the same substrate as complex electronics required for quenching, analog-to-digital converter (ADC), time-to-digital converter (TDC), and data storage/transfer.

The architecture of a digital SiPM is illustrated in Figure 12.10. Each SiPM pixel comprises an array of micro-cells operating in Geiger mode. A logic circuit is attached to each micro-cell, allowing the micro-cell to capture, store, and output exactly one photon at a time. The CMOS inverter connected to the anode is designed to generate a high-speed asynchronous trigger and a slower synchronous data output signal, which are then transferred to other function blocks such as ADC and TDC for energy and timing information. To meet different requirements of spatial resolution, the output of ADC/TDC can be read out at either micro-cell level (i.e., for fluorescence lifetime imaging microscopy (FLIM) applications) [103,104] or pixel level (i.e., for ToF-PET applications).

The benefits of digital SiPMs over analog SiPMs are expected to be low dark count rate (DCR), high yield due to suppression of faulty micro-cells, improved timing performance, improved temperate stability, and no additional readout electronics such as ASICs required. Philips has recently demonstrated the feasibility of using digital SiPMs for PET applications [97,98], as shown in Figure 12.9. One thing worth mentioning is that the device is reported to be capable of achieving a time resolution of 170 ps FWHM, for a

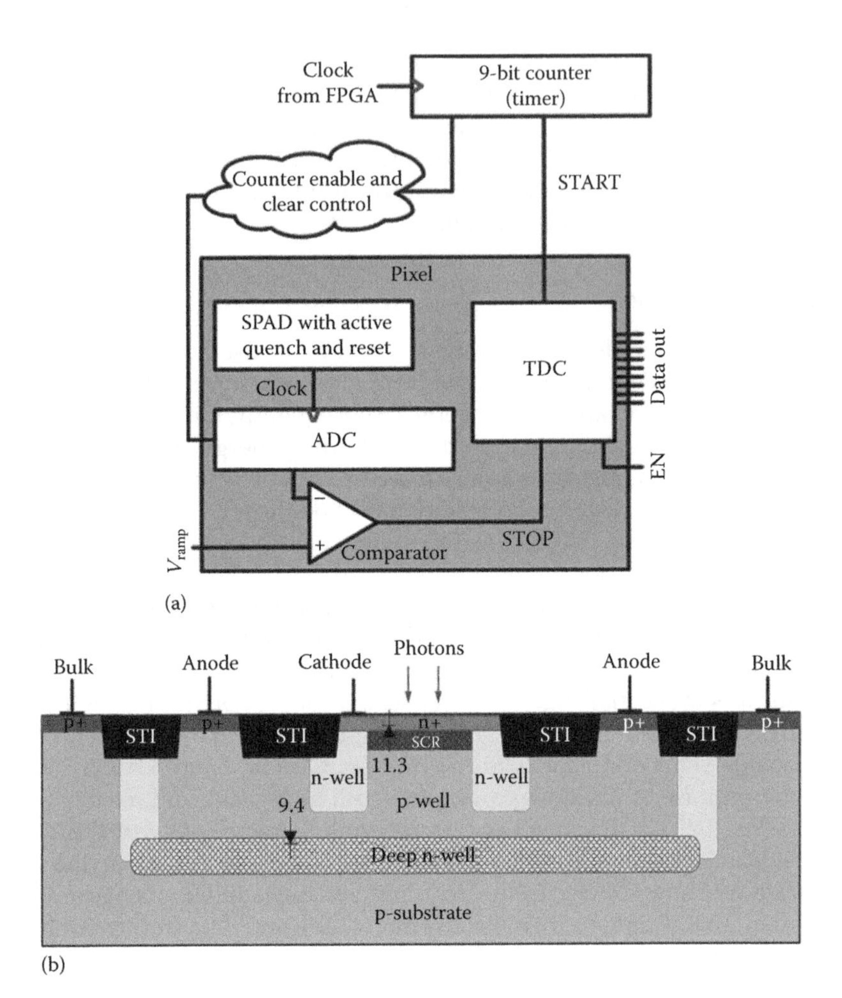

FIGURE 12.10
(a) Diagram of a digital SiPM with integrated ADC and TDC) functions. (b) Cross section of proposed SPAD structure with a novel guard ring implemented in standard 130 nm CMOS technology. SCR, space-charge region, and STI, shallow trench isolation. The novel guard ring structure is designed to suppress PEB. By violating the design rules of the standard CMOS technology and placing an n-well within the p-well, the edges of the junction are connected to two n-well guard rings that have lower doping than the n^+ layer. Furthermore, by using a deep n-well layer, a p-well can be isolated from the substrate for minimizing DCR. The breakdown voltages at two junctions (11.3 and 9.4 V) are also shown, respectively. (Adapted from Palubiak et al., *IEEE Sens. J.*, 11, 2401, 2011.)

single LYSO crystal of $3 \times 3 \times 5$ mm^3. At the time of writing, the digital SiPM device is not yet commercially available though it has stimulated great interest in the community and is deemed to become the next-generation detector technology for PET instrumentation.

In the following, the CMOS sensor fabrication and in-pixel circuitry integration are briefly discussed. Note that the micro-cell will be used interchangeably with SPAD hereafter, since the latter one has been widely used in the community of CMOS sensor development.

12.6.3.2.1 CMOS Fabrication Process

SiPMs or SPADs are traditionally designed and manufactured using proprietary processes (i.e., beveled-edge and reach-through structures), which facilitate the fabrication of stand-alone sensors with large active area (~several 100 μm in diameter) [105–107]. However, these structures and associated manufacture processes are based on ultrapure silicon wafers and nonplanar technology, which inherently limit the fabrication yield and are not suitable for monolithic integration of processing circuits.

Designing an effective guard ring is a major challenge for CMOS sensor development. Inside a SPAD, the very strong electric field within the depletion region must be uniformly distributed along the p–n junction. Therefore, a SPAD needs to have a guard ring structure in presence to avoid premature edge breakdown (PEB). Early implementations consist of a p^+ anode and n-well cathode, with a guard ring structure created using the lateral diffusion of two n-tub regions [108]. The use of relatively old 0.8 μm CMOS technology guarantees a clean fabrication process and relatively low doping levels. Driven by the needs for more complex electronics, the 0.35 μm CMOS technology was also investigated for the fabrication of SPAD arrays [109–111]. Recently, the use of more advanced deep submicron (DSM) CMOS processes and a number of novel guard ring structures were studied, and an example is provided here. As shown in Figure 12.10, by using a triple-well process and placing an n-well within the p-well (i.e., which violates standard CMOS design rules) [112], the edges of the junction are connected to two n-well guard rings that have lower doping than the n^+ layer. Furthermore, by introducing a deep n-well layer, a p-well is isolated from the substrate to help minimize DCR. Also note that the reverse breakdown of the SPAD junction occurs at 11.3 V, while the parasitic junction of the p-well/deep n-well breaks down at 9.3 V. Therefore, 9.3 V is the maximum negative voltage allowed at the anode before the breakdown of the parasitic p-n junction occurs [112].

Another challenge associated with DSM CMOS process is DCR performance, which is largely dependent on doping levels, annealing and drive-in diffusion steps, and the presence of shallow trench isolation (STI) [113–115]. The DCR could be up to several hundreds of kHz for an active area of only a few microns. For example, an implementation with lower noise (60 kHz for an active area of 10 μm in diameter and excess bias of 0.5 V) exhibits PDE of as low as 2.5% [113]. With the same technology, another study reports that a SPAD with active area of 3 μm diameter presents a ~100 Hz of DCR [114]. However, no precise characterization of the device is given, and the extremely small active area might be of no practical use. Furthermore, all these DSM CMOS implementations suffer from the tunneling effect due to high implant doping levels ($>5 \times 10^{17}$ cm^{-3}) [115].

Besides the fabrication process, another active research direction is the quenching and recharging (resetting) scheme [116,117]. An important tradeoff involved is between high speed (i.e., low dead time) and noise performance. Passive quenching and active recharging have been commonly deployed in early works. Its main advantages include the well-defined hold time before recharge and a fast quenching and recharge time, which are beneficial for after-pulsing suppression. To further reduce the dead time and suppress after-pulsing events if necessary, active quenching and active recharging circuits have also been studied [117]. Nevertheless, such tasks will pose a few additional challenges such as increased parasitic capacitance and reduced fill factor of the SPAD.

12.6.3.2.2 Digital SiPM with In-Pixel Electronics

As mentioned earlier, CMOS process can make the detector more compatible with digital circuits and minimize electronic noises. Such a benefit may consequently result in improved energy and time resolution and thus a higher NEC as shown in Formula 12.1. Furthermore, maintaining all analog processing within the detector can reduce the power consumption of data acquisition system and enable scaling up to larger arrays with high cost-effectiveness, which is especially useful for high-resolution PET systems that have thousands of channels.

A SPAD array containing 60 sensors on a chip area of 18 × 18 mm was reported using the proprietary process [118,119], where the diameter of SPADs ranges from 20 to 75 μm. The design is heavily pad limited, and most of the chip area is used for interconnecting. In addition, studies also indicate that a SPAD would suffer from increased after-pulsing and optical cross talk due to larger parasitic capacitance, if quenching electronics were not integrated into the chip [120]. The first integration of in-pixel circuits into a CMOS SPAD array was shown in [109], which consists of a 4 × 8 array containing quenching resistors and inverters using the 0.8 μm CMOS process. A larger array (32 × 32) fabricated with the same technology was also developed [121]. Unfortunately, due to the large transistor size in the old CMOS technology, very limited electronics were integrated, and only a single pixel of the whole array could be activated at any time. To address that limitation, the 0.35 μm CMOS technology was deployed to improve the performances of SPADs, allowing for the integration of large-scale on-chip electronics. For example, a 128 × 128 SPAD array with an on-chip bank of 32 TDCs for time-correlated imaging applications was presented in [122]. A 17-bit ADC has also been successfully implemented over a 2 × 7 SPAD array [123].

Despite a number of potential benefits, there exists a serious challenge for this promising detector technology: how to maintain high PDE and fill factor when in-pixel circuits are introduced to CMOS sensors. The continuous need for additional metal layers to improve interconnecting in modern CMOS process results in more complex optical stacks, which consequently decrease the PDE as shown in Formula 12.6. Meanwhile, when various designs of high-performance ADC and TDC are introduced, a significant amount of chip areas (and interconnection tracers) are to be occupied by those function blocks, which would further reduce the fill factor and PDE. All these challenges need to be taken into account when designing digital SiPMs to be deployed for PET detector development.

12.7 Conclusions

This chapter reviews current challenges of advancing PET detector technology and some recent developments. It focuses on four important aspects of PET instrumentation: high photon detection sensitivity, improved spatial resolution, DoI resolution, and ToF PET and SiPMs. Improved system geometry, novel semiconductor detectors, and tapered scintillation crystal arrays are able to enhance the photon detection sensitivity of a PET system. Novel 3-D positioning detectors are of great potential to be deployed in PET for achieving high spatial resolution better than 1 mm, such as CZT and PSAPDs. DoI capability enables a PET system to achieve uniform spatial resolution across the FoV and improve accuracy for quantitative studies. The availability of fast scintillation crystals such as $LaBr_3$ and SiPM greatly advances ToF-PET development. In particular, the development of digital SiPM using the standard CMOS process has huge potential to be used as the next-generation detector technology for PET instrumentation. All these technology advances, together with new imaging probe molecules, will ultimately enhance the molecular sensitivity of PET and increase its role in both preclinical and clinical research.

Acknowledgments

We would like to thank Canadian Breast Cancer Foundation Grant (2011R00356), NSERC Engagement Grant (537661), and NSERC Discovery Grant (596000) for their financial support. The authors would also like to thank Dr. Jamal Deen at McMaster University for his insightful discussion on CMOS sensor development.

References

1. C.S. Levin. New imaging technologies to enhance the molecular sensitivity of positron emission tomography, *Proc. IEEE*, 96, 439–467, 2008.
2. S.S. Gambhir. Molecular imaging of cancer with positron emission tomography, *Nat. Rev. Cancer*, 2, 683–693, 2002.

3. U. Nestle, H. Weber, and A.L. Grosu. Biological imaging in radiation therapy: Role of positron emission tomography, *Phys. Med. Biol.*, 54, 1–25, 2009.
4. M.E. Phelps. Positron emission tomography provides molecular imaging of biological processes, *Proc. Natl. Acad. Sci.*, 97, 9226–9233, 2000.
5. N. Adonai et al. Ex vivo cell labeling with 64Cu-pyruvaldehyde-bis (N4-methylthiosemicarbazone) for imaging cell trafficking in mice with positron emission tomography, *Proc. Natl. Acad. Sci.*, 99, 3030–3035, 2002.
6. J. Wang and L. Maurer. Positron emission tomography: Applications in drug discovery and drug development, *Curr. Top. Med. Chem.*, 5, 1053–1075, 2005.
7. J.V. Frangioni and R.J. Hajjar. In vivo tracking of stem cells for clinical trials in cardiovascular disease, *Circulation*, 110, 3378–3383, 2004.
8. F. Cao et al. In vivo visualization of embryonic stem cell survival, proliferation, and migration after cardiac delivery, *Circulation*, 113, 1005–1014, 2006.
9. K. Shah, A. Jacobs, X.O. Breakefield, and R. Weissleder. Molecular imaging of gene therapy for cancer, *Gene Ther.*, 11, 1175–1187, 2004.
10. G.D. Luker. Special conference of the American association for cancer research on molecular imaging in cancer: Linking biology, function, and clinical applications in vivo, *Cancer Res.*, 62, 2195–2198, 2002.
11. D.J. Hnatowich. Observations on the role of nuclear medicine in molecular imaging, *J. Cell. Biochem.*, 39, 18–24, 2002.
12. R. Pither. PET and the role of in vivo molecular imaging in personalized medicine, *Expert Rev. Mol. Diagnost.*, 3, 703–713, 2003.
13. S.R. Cherry. *Physics in Nuclear Medicine*, 3rd edn. Philadelphia, PA: Saunders, 2003.
14. National Electrical Manufacturers Association. NEMA Standards Publication NU 2-2001. Performance Measurements of Positron Emission Tomographs, Rosslyn, VA: National Electrical Manufacturers Association, 2001.
15. S.C. Strother, M.E. Casey, and E.J. Hoffman. Measuring PET scanner sensitivity: Relating count rates to image signal-to-noise ratios using noise equivalent counts, *IEEE Trans. Nucl. Sci.*, 37, 783–788, 1990.
16. M.E. Phelps, S.C. Huang, and E.J. Hoffman. An analysis of signal amplification using small detectors in positron emission tomography, *J. Comput. Assist. Tomogr.*, 6, 551–565, 1982.
17. F. Habte, A.M. Foudray, P.D. Olcott, and C.S. Levin. Effects of system geometry and other physical factors on photon sensitivity of high-resolution positron emission tomography, *Phys. Med. Biol.*, 52, 3753–3772, 2007.
18. J.L. Humm, A. Rosenfeld, and A. Guerra. From PET detectors to PET scanners, *Eur. J. Nucl. Med. Mol. Imaging*, 30, 1574–1597, 2004.
19. A. Alessio, P. Kinahan, P. Cheng, H. Vesselle, and J. Karp. PET/CT scanner instrumentation, challenges, and solutions, *Radiol. Clin. North Am.*, 42, 1017–1032, 2004.
20. Q. Bao, D. Newport, M. Chen, D.B. Stout, and A.F. Chatziioannou. Performance evaluation of the Inveon dedicated PET preclinical tomography based on the NEMA NU-4 standards, *J. Nucl. Med.*, 50, 401–408, 2009.
21. M.C. Huisman, S. Reder, A.W. Weber, and S.I. Ziegler. Performance evaluation of the Philips MOSAIC small animal PET scanner, *Eur. J. Nucl. Med. Mol. Imaging*, 34, 532–540, 2007.
22. D.W. Townsend, B. Jakoby, and M.J. Long. Performance and clinical workflow of a new combined PET/CT scanner, *J. Nucl. Med.*, 48(supplement 2), 437, 2007.

23. J. Joung, R.S. Miyaoka, and T.K. Lewellen. cMice: A high resolution animal PET using continuous LSO with a statistics based positioning scheme, *Nucl. Instrum. Methods Phys. Res. A.*, 489, 584–598, 2002.
24. M.C. Maas et al. Monolithic scintillator PET detectors with intrinsic depth-of-interaction correction, *Phys. Med. Biol.*, 54, 1893–1908, 2009.
25. T. Ling, K. Lee, and R.S. Miyaoka. Performance comparisons of continuous miniature crystal element (cMiCE) detectors, *IEEE Trans. Nucl. Sci.*, 53, 2513–2518, 2006.
26. J.L. Matteson, M.R. Pelling, and R.T. Skelton. CZT detectors with 3D readout for gamma-ray spectroscopy and imaging, *Proc. SPIE*, 4784, 1–13, 2002.
27. P. Vaska et al. Studies of CZT for PET applications, *IEEE Nucl. Sci. Symp. Conf. Rec.* pp. 2799–2802, 2005.
28. F. Zhang, H. Zhong, D. Xu, and L.J. Meng. Feasibility study of using two 3D position sensitive CZT detectors for small animal PET, *IEEE Nucl. Sci. Symp. Conf. Rec.* pp. 1582–1585, 2005.
29. Y. Gu et al. Study of a high resolution, 3-D positioning cross-strip cadmium zinc telluride detector for PET, *Phys. Med. Biol.*, 56, 1563, 2011.
30. G.S. Mitchell et al. CdTe strip detector characterization for high resolution small animal PET, *IEEE Trans. Nucl. Sci.*, 55, 870–876, 2008.
31. Y.F. Yang et al. Tapered LSO arrays for small animal PET, *Phys. Med. Biol.*, 56, 139–153, 2011.
32. S.S. James and S.R. Cherry. Simulation study of spatial resolution and sensitivity for the tapered depth of interaction PET detectors for small animal imaging, *Phys. Med. Biol.*, 55, 63–74, 2010.
33. H. Peng and C.S. Levin. Design study of a high-resolution breast-dedicated PET system built from cadmium zinc telluride detectors, *Phys. Med. Biol.*, 55, 2761–2788, 2010.
34. W.W. Moses and S.E. Derenzo. Empirical observation of resolution degradation in positron emission tomographs utilizing block detectors, *J. Nucl. Med.*, 34, 101, 1993.
35. H. Peng and C.S. Levin. Study of PET intrinsic spatial resolution and contrast recovery improvement for PET/MRI systems, *Phys. Med. Biol.*, 57, 101–115, 2012.
36. A. Rose. *Vision: Human and Electronic*, New York: Plenum, pp. 21–23, 1973.
37. C.S. Levin and E.J. Hoffman. Calculation of positron range and its effect on the fundamental limit of positron emission tomography system spatial resolution, *Phys. Med. Biol.*, 44, 781–799, 1999.
38. J.R. Stickel and S.R. Cherry. High-resolution PET detector design: Modeling components of intrinsic spatial resolution, *Phys. Med. Biol.*, 50, 179–195, 2005.
39. S.S. James et al. Experimental characterization and system simulations of depth of interaction PET detectors using 0.5 mm and 0.7 mm LSO arrays, *Phys. Med. Biol.*, 54, 4605–4615, 2009.
40. S.R. Cherry et al. Collection of scintillation light from small BGO crystals, *IEEE Trans. Nucl. Sci.*, 42, 1058–1063, 1995.
41. S. Surti et al. Design evaluation of A-PET: A high sensitivity animal PET camera, *IEEE Trans. Nucl. Sci.*, 50, 1357–1363, 2003.
42. J. Uribe et al. Signal characteristics of individual crystals in high resolution BGO detector designs using PMT-quadrant sharing, *IEEE Trans. Nucl. Sci.*, 50, 355–361, 2003.

43. C.S. Levin et al. Compton scatter and x-ray crosstalk and the use of very thin intercrystal septa in high resolution PET detectors, *IEEE Trans. Nucl. Sci.*, 44, 218–224, 1997.
44. R. Lecomte, D. Schmitt, and G. Lamoureux. Geometry study of a high resolution PET detection system using small detectors, *IEEE Trans. Nucl. Sci.*, 31, 556–561, 1984.
45. C.S. Levin. Design of a high resolution and high sensitivity scintillation crystal array for PET with nearly complete light collection, *IEEE Trans. Nucl. Sci.*, 49, 2236–2243, 2002.
46. J.R. Stickel, J. Qi, and S.R. Cherry. Fabrication and characterization of a 0.5-mm lutetium oxyorthosilicate detector array for high-resolution PET applications, *J. Nucl. Med.*, 48, 115–121, 2007.
47. H. Peng and C.S. Levin. Investigation of a clinical PET detector module design that employs large-area avalanche photodetectors, *Phys. Med. Biol.*, 56, 3603–3627, 2011.
48. C.S. Levin et al. Investigation of position sensitive avalanche photodiodes for a new high-resolution PET detector design, *IEEE Trans. Nucl. Sci.*, 51, 805–810, 2004.
49. J. Zhang et al. Performance characterization of a novel thin position-sensitive avalanche photodiode for 1 mm resolution positron emission tomography, *IEEE Trans. Nucl. Sci.*, 54, 415–421, 2007.
50. K.S. Shah et al. Positron sensitive avalanche photodiodes for gamma-ray imaging, *IEEE Trans. Nucl. Sci.*, 49, 1687–1692, 2002.
51. F.W. Lau et al. Analog signal multiplexing for PSAPD-based PET detectors: Simulation and experimental validation, *Phys. Med. Biol.*, 55, 7149–7174, 2010.
52. Y. Wu et al. A study of the timing properties of position-sensitive avalanche photodiodes, *Phys. Med. Biol.*, 54, 5155–5172, 2009.
53. G. Pratx and C.S. Levin. Bayesian reconstruction of photon interaction sequences for high-resolution PET detectors, *Phys. Med. Biol.*, 54, 5073–5094, 2009.
54. G. Chinn and C.S. Levin. A maximum NEC criterion for compton collimation to accurately identify true coincidences in PET, *IEEE Trans. Med. Imaging*, 30, 1341–1352, 2011.
55. C.S. Levin. Promising new photon detection concepts for high resolution clinical and preclinical positron emission tomography, *J. Nucl. Med.*, 53, 167–170, 2012.
56. A. Vandenbroucke et al. Performance characterization of a new high resolution PET scintillation detector. *Phys. Med. Biol.*, 55, 5895–5911, 2010.
57. Y. Gu et al. Effects of multiple-interaction photon events in a high resolution PET system that uses 3-D positioning detectors. *Med. Phys.*, 37, 5494–5508, 2010.
58. W.W. Moses et al. OpenPET: A flexible electronics system for radiotracer imaging, *IEEE Trans. Nucl. Sci.*, 57, 1–6, 2010.
59. M.E. Phelps. *PET Molecular Imaging and Its Biological Applications*. New York: Springer, 2004.
60. M. Rafecas et al. A monte-carlo study of high resolution PET with granulated dual layer detectors, *IEEE Trans. Nucl. Sci.*, 48, 1490–1495, 2001.
61. M. Rafecas et al. Inter crystal scatter in a dual layer, high resolution LSO-APD positron emission tomography, *Phys. Med. Biol.*, 48, 821–848, 2003.
62. Y. Yang et al. Depth of interaction resolution measurements for a high resolution PET detector using position sensitive avalanche photodiodes, *Phys. Med. Biol.*, 51, 2131–2142, 2006.

63. Y. Yang et al. A prototype PET scanner with DOI-encoding detectors, *J. Nucl. Med.*, 49, 1132–1140, 2008.
64. W.W. Moses, S.E. Derenzo, C.L. Melcher, and R.A. Manentet. A room temperature LSO/PIN photodiode PET detector module that measures depth of interaction, *IEEE. Trans. Nucl. Sci.*, 42, 1085–1089, 1995.
65. J.B. Mosset et al. Development of an optimized LSO/LuYAP phoswich detector head for the lausanne clearPET demonstrator, *IEEE Trans. Nucl. Sci.*, 53, 25–29, 2006.
66. J. Seidel, J.J. Vaquero, and M.V. Green. Resolution uniformity and sensitivity of the NIH ATLAS small animal PET scanner: Comparison to simulated LSO scanners without depth-of-interaction capability, *IEEE Trans. Nucl. Sci.*, 50, 1347–1350, 2003.
67. T. Ling, T.H. Burnett, T.K. Lewellen, and R.S. Miyaoka. Parametric positioning of a continuous crystal PET detector with depth of interaction decoding, *Phys. Med. Biol.*, 53, 1843–1863, 2008.
68. M.C. Maas et al. Experimental characterization of monolithic crystal small animal PET detectors read out by APD arrays, *IEEE Trans. Nucl. Sci.*, 53, 1071–1077, 2006.
69. D.R. Schaart et al. A novel, SiPM-array-based, monolithic scintillator detector for PET. *Phys. Med. Biol.*, 54, 3501–3512, 2009.
70. N. Zhang et al. Anode position and last dynode timing circuits for dual-layer BGO scintillator with PS-PMT based modular PET detectors, *IEEE Trans. Nucl. Sci.*, 49, 2203–2207, 2002.
71. M.M. Ter-Pogossian et al. Super PETTI: A positron emission tomography utilizing photon time-of-flight information, *IEEE Trans. Med. Imaging*, MI-1, 179–187, 1982.
72. W.H. Wong et al. Characteristics of small barium fluoride (BaF2) scintillator for high intrinsic resolution time-of-flight positron emission tomography, *IEEE Trans. Nucl. Sci.*, 31, 381–386, 1984.
73. N.A. Mullani et al. Dynamic imaging with high resolution time-of-flight PET camera TOFPET-I, *IEEE Trans. Nucl. Sci.*, 31, 609–613, 1984.
74. B. Mazoyer et al. Physical characteristics of TTV03, a new high spatial resolution time-of-flight positron tomography, *IEEE Trans. Nucl. Sci.*, 37, 778–782, 1990.
75. T.F. Budinger. Time-of-flight positron emission tomography: Status relative to conventional PET, *J. Nucl. Med.*, 24, 73–78, 1983.
76. W.W. Moses. Recent advances and future advances in time-of-flight PET, *Nucl. Instrum. Methods Phys. Res. A*, 580, 919–924, 2007.
77. J.S. Karp et al. Benefit of time-of-flight in PET: Experimental and clinical results, *J. Nucl. Med.*, 49, 462–470, 2008.
78. S. Surti and J.S. Karp. Experimental evaluation of a simple lesion detection task with time-of-flight PET, *J. Nucl. Med.*, 48, 471–480, 2007.
79. B.W. Jakoby, Y. Bercier, and M. Conti. Performance investigation of a time-of-flight PET/CT scanner, *IEEE Nucl. Sci. Symp. Conf. Rec.*, pp. 3738–3743, 2008.
80. B. Kemp et al. Clinical evaluation of a prototype time-of-flight PET/CT system, *J. Nucl. Med.*, 50, 1513, 2009.
81. Y. Liang et al. Feasibility study of sing cherenkov light crystals for time-of-flight PET Systems, *IEEE Nucl. Sci. Symp. Med. Imaging Conf. Rec.*, pp. M16–73, 2012.
82. W.W. Moses. Current trends in scintillator detectors and materials, *Nucl. Instr. Methods Phys. Res. A.*, 487, 123–128, 2002.

83. S. Surti, J.S. Karp, and G. Muehllehner. Image quality assessment of $LaBr_3$-based whole-body 3D PET scanners: A Monte Carlo evaluation, *Phys. Med. Biol.*, 49, 4593–4610, 2004.
84. M.E. Daube-Witherspoon et al. The imaging performance of a $LaBr_3$-based PET scanner, *Phys. Med. Biol.*, 55, 45–64, 2010.
85. S.I. Ziegler et al. Effects of scintillation light collection on the time resolution of a time-of-flight detector for annihilation quanta, *IEEE Trans. Nucl. Sci.*, 37, 574–579, 1990.
86. D.J. Herbert et al. The silicon photomultiplier for application to high-resolution positron emission tomography, *Nucl. Instrum. Methods Phys. Res.* A, 573, 84–87, 2007.
87. A. Nassalski et al. Multi pixel photon counters (MPPC) as an alternative to APD in PET applications, *IEEE Trans. Nucl. Sci.*, 57, 1008–1014, 2010.
88. D. Renker. New trends in photodetectors, *Nucl. Instrum. Methods Phys. Res. A*, 571, 1–6, 2007.
89. V.C. Spanoudaki and C.S. Levin. Investigating the temporal resolution limits of scintillation detection: Comparison between experiment and simulation, *Phys. Med. Biol.*, 56, 735–756, 2011.
90. V.C. Spanoudaki and C.S. Levin. Scintillation induced response in passively-quenched Si-based single photon counting avalanche diode arrays, *Opt. Express*, 19, 1665–1679, 2011.
91. V.C. Spanoudaki and C.S. Levin. Photo-detectors for time of flight positron emission tomography (ToF-PET), *Sensors*, 10, 10484–10505, 2010.
92. E. Roncali and S.R. Cherry. Application of silicon photomultipliers to positron emission tomography, *Ann. Biomed. Eng.*, 39, 1358–1377, 2011.
93. D.R. Schaart et al. First experiments with LaBr3:Ce crystals coupled directly to silicon photomultipliers for PET applications, *IEEE Nucl. Sci. Symp. Conf. Rec.*, pp. 3991–3994, 2008.
94. C. Kim, G.C. Wang, and S. Dolinsky. Multi-pixel photon counters for TOF PET detector and its challenges, *IEEE Trans. Nucl. Sci.*, 56, 2580–2585, 2009.
95. A. Stewart et al. Performance of 1-mm^2 silicon photomultiplier, *IEEE J. Quantum Electron.*, 44, 157–164, 2008.
96. M. Mazillo et al. Silicon photomultiplier technology at STMicroelectronics, *IEEE Trans. Nucl. Sci.*, 56, 2434–2442, 2009.
97. C. Degenhardt et al. The digital silicon photomultiplier–a novel sensor for the detection of scintillation light, *IEEE Nucl. Sci. Symp. Conf. Rec.*, pp. 2383–2386, 2009.
98. T. Frach et al. The digital silicon photomultiplier–principle of operation and intrinsic detector performance, *IEEE Nucl. Sci. Symp. Conf. Rec.*, pp. 1959–1965, 2009.
99. K. Yoon, C. Kim, B. Lee, and D. Lee. Single-chip CMOS image sensor for mobile applications, *IEEE J. Solid-State Circuits*, 37, 1839–1845, 2002.
100. F. Zappa, A. Lotito, and S. Tisa. Photon-counting chip for avalanche detectors, *IEEE Photonics Tech. Lett.*, 17, 184–186, 2005.
101. D. Renker and E. Lorenz. Advances in solid state photon detectors, *J. Instrum.*, 4, 04004, 2009.
102. R. Miyagawa and T. Kanade. CCD range-finding sensor, *IEEE Trans. Electron. Devices*, 44, 1648–1652, 1997.
103. R. Cubeddu et al. Time-resolved fluorescence imaging in biology and medicine, *J. Phys.*, 35, 61–76, 2002.

104. R.V. Krishnan et al. Development of a multiphoton fluorescence lifetime imaging microscopy system using a streak camera, *Rev. Sci. Instrum.*, 74, 2714–2721, 2003.
105. R.S. Goetzberger et al. Avalanche effects in silicon P-N junctions.2. Structurally perfect junctions, *J. Appl. Phys.*, 34, 1591, 1963.
106. PerkinElmer Optoelectronics (http://optoelectronics.perkinelmer.com) Technical notes: High speed APDs for Analytical and Biomedical Lowest Light Detection Applications.
107. M. Ghioni et al. Progress in silicon single-photon avalanche diodes, *IEEE J. Selected Topics Quantum Electron.*, 13, 852–862, 2007.
108. O. Elkhalili et al. A 4 × 64 pixel CMOS image sensor for 3-D measurement applications, *IEEE J. Solid State Circuits*, 39, 1208–1212, 2004.
109. L.P. Stoppa, M. et al. A CMOS 3-D imager based on single photon avalanche diode, *IEEE Trans. Circuits Syst. I*, 54, 8–12, 2007.
110. F.G. Tisa, A. Tosi, and F. Zappa. 100 kframe/s 8bit monolithic single-photon imagers, *Proceedings of the 38th European Solid-State Device Research Conference*, p. 4, 2008.
111. C. Niclass, C. Favi, T. Kluter, F. Monnier, and E. Charbon. Single-photon synchronous detection, *Proceedings of the 34th European Solid-State Circuits Conference*, pp. 114–117, 2008.
112. D. Palubiak et al. High-speed, single-photon avalanche-photodiode imager for biomedical applications, *IEEE Sens. J.*, 11, 2401–2412, 2011.
113. N. Faramarzpour, M.J. Deen, S. Shirani, and Q. Fang. Fully integrated single photon avalanche diode detector in standard CMOS 0.18-mu m technology, *IEEE Trans. Electron. Devices*, 55, pp. 760–767, March 2008.
114. M.A. Marwick and A.G. Andreou. Single photon avalanche photodetector with integrated quenching fabricated in TSMC 0.18 mu m 1.8 V CMOS process, *Electron Lett.*, 44, 643–644, 2008.
115. S.M. Sze. *Semiconductor Devices: Physics and Technology*, New York: Wiley, 1985.
116. M.A. Marwick and A.G. Andreou. Fabrication and testing of single photon avalanche detectors in the TSMC 0.18 μm CMOS technology, *41st Annual Conference on Information Sciences and Systems*, pp. 741–744, 2007.
117. M. Gronholm, J. Poikonen, and M. Laiho. A ring-oscillator-based active quenching and active recharge circuit for single photon avalanche diodes, *Circuit Theory and Design, European Conference on*, pp. 5–8, 2009.
118. F. Zappa et al. SPADA: Single-photon avalanche diode arrays, *IEEE Photonic Technol. Lett.*, 17, 657–659, 2005.
119. F. Zappa et al. Single-photon avalanche diode arrays for fast transients and adaptive optics, *IEEE Trans. Instrum. Meas.*, 55, 365–374, 2006.
120. I. Rech et al. Optical crosstalk in single photon avalanche diode arrays: A new complete model, *Opt. Express*, 16, 8381–8394, 2008.
121. C. Niclass et al. Design and characterization of a CMOS 3-D image sensor based on single photon avalanche diodes, *IEEE J. Solid-State Circuits*, 40, 1847–1854, 2005.
122. C. Niclass et al. A 128 × 128 single-photon image sensor with column-level 10-bit time-to-digital converter array, *IEEE J. Solid-State Circuits*, 43, 2977–2989, 2008.
123. D. Stoppa et al. Single-photon avalanche diode CMOS sensor for time-resolved fluorescence measurements, *Sens. J. IEEE*, 9, 1084–1090, 2009.

13

CT-SPECT/CT-PET

R. Glenn Wells

CONTENTS

13.1 Introduction

Nuclear medicine has been available as an imaging technology first in planar form in the 1950s and since the late 1960s as a tomographic technique. It is an extremely sensitive technique capable of measuring concentrations in the nanomolar to picomolar range. Using tracer technology, it provides functional information about a large number of different organs and systems within the human body. However, its poor spatial resolution and the often specific nature of the tracer uptake made the images difficult to understand for the untrained eye and earned nuclear medicine the moniker "unclear medicine" (von Schulthess 2004, Patel et al. 2009). The value of an anatomical reference had long been recognized and transmission/emission imaging had been proposed as early as 1966 (Kuhl et al. 1966). Many nuclear medicine studies like ventilation–perfusion scans and sentinel-node lymphoscintigraphy are frequently interpreted with side-by-side reference to an anatomical image, be it a chest x-ray or a crude body outline formed by the shadow of a flood transmission source.

The notion of acquiring both a nuclear medicine tomograph and a CT scan on the same camera, with a single patient bed, was first suggested in 1987 (Mirshanov 1987) but not developed. SPECT/CT imaging was further investigated by Hasegawa and colleagues at the University of California San Francisco starting in the late 1980s (Hasegawa et al. 1989, 1990), and produced the first modern configuration of a SPECT/CT camera in the mid-1990s (Blankespoor et al. 1996). They combined a 9800 Quick CT scanner with a single-head XR/T SPECT gamma camera (both devices from GE Healthcare) and integrated a patient bed that could be slid through gantries. They also demonstrated the feasibility of using the CT image to create an attenuation coefficient map that could be used for attenuation correction (Blankespoor et al. 1996, Hasegawa et al. 1999, Seo et al. 2008, Patton et al. 2009). This technology was taken up and first offered commercially as the Hawkeye CT system attached to a Millennium VG gamma camera (GE Healthcare) in 1999. The CT component was a single-slice, slow-rotation device, which led to long acquisition times and CT images that were not of diagnostic quality, but still sufficient for both attenuation correction and anatomical localization.

Meanwhile, Beyer et al. at the University of Pittsburgh also suggested the use of CT in combination with PET scanning (Beyer et al. 1994). Their work led to the first PET/CT prototype scanner in 1998 (Beyer et al. 2000, Patton et al. 2009). This unit combined both a diagnostic single-slice CT scanner (Somatom AR.SP, Siemens) and an ECAT-ART PET system (CTI PET Systems). At this time, CT technology was undergoing a giant boost in speed as multislice scanners were starting to be introduced. This made whole-body CT scanning considerably faster than the transmission-based options and reduced whole-body FDG study times by almost a factor of two. In addition, the localization aspect of overlaying the FDG oncology study on top of a diagnostic quality CT scan was a powerful draw for oncology. With reimbursement available for PET oncology in North America, this technology took off. Commercial PET/CT systems became available in 2001 and the additional benefit of the combined scanner became rapidly apparent. Within about 4 years, PET-only scanners were no longer being sold in North America and every PET system came combined with a CT scanner.

The success of PET-CT spurred further development in SPECT-CT. Though SPECT-only cameras have not been replaced by SPECT-CT to the same extent that PET-CT replaced PET, there are now a very large number of SPECT-CT cameras in use and a variety of configurations available. Traditional dual-head gamma cameras are available with everything from a Hawkeye CT scanner (upgraded to a 4-slice CT) to a high-speed 64-slice diagnostic quality CT (Patton et al. 2009). New developments in dedicated cardiac systems also have the option of a SPECT-CT configuration (NM 570c, GE Healthcare (Bocher et al. 2010); Cardius XPO, Digirad Inc. (Bai et al. 2010)).

Side-by-side use of anatomical x-ray/CT and function nuclear medicine images has been done for some time. However, the combined modality camera has many advantages for both the physician and the patient. This chapter addresses the advantages and disadvantages of combining nuclear medicine (NM) and CT and some of the methodology required to implement it.

13.2 Advantages of Localization

Nuclear medicine (NM) is a very sensitive technology allowing measurement of picomolar concentrations of tracer (Levin 2005, Seo et al. 2008). Accumulation of tracer at focal sites of cancer can allow us to detect very small lesions, but these lesions may appear as a bright focal source with little other background uptake in structures nearby. It is like a bright lightbulb in a dark room—easy to see, but difficult to tell exactly where it is. In the case of cancer detection, location can be extremely important. Accurate localization is essential for guiding biopsies, surgery, and radiation therapy. It provides information about the stage of the disease, indicating, for example, if the tumor has penetrated the adjacent bone or is contained in the surrounding soft tissue. Many studies have shown the benefits of having both nuclear medicine study and the CT study fused together as was highlighted in an excellent series of articles published in 2009 (Bockisch et al. 2009, Delbeke et al. 2009, Evan-Sapir 2009, Gnanasegaran et al. 2009, Kaufmann and Di Carli 2009).

In cancer imaging, the use of fused imaging in FDG-PET/CT improves the staging of cancer patients, and hence changes management compared to either modality alone or even both modalities evaluated side-by-side. The difference is 10%–15% or more and statistically significant (Czernin et al. 2007, Bockisch et al. 2009, Delbeke et al. 2009, Czernin et al. 2010). It has been shown for a variety of cancers including those of the head and neck, breast, lung, gastrointestinal tract, thyroid, and lymphoma (Delbeke et al. 2009). Combination of PET and CT also leads to improved lesion detection, improved diagnostic accuracy, and confirmation of small or subtle lesions.

Hybrid imaging improves the distinction between normal uptake and that representing true disease by accurately placing the activity within the correct anatomical structures. The combination has proven useful in guiding radiation therapy planning (Delbeke et al. 2009, Evan-Sapir 2009, Roach et al. 2010). It allows for a better definition of tumor boundaries and metastatic spread, which can help guide the definition of planned radiation fields during therapy and consequently leads to a sparing of normal tissues. Fused imaging is also useful in the follow-up of disease, particularly in the case of residual CT abnormalities. Whereas evaluation with CT alone is based on changes in tumor size, successful treatment may lead to tumor necrosis or halting of disease progression without immediate loss in tumor size. Functional imaging can more rapidly assess response to treatment by displaying a loss in metabolic activity in the treatment bed and distinguishing it from sites of activity outside the treated area (Evan-Sapir 2009). Finally, fused images can help guide biopsies by clarifying which potential tumors are most metabolically active and even within tumors, distinguishing areas that are less aggressive or even necrotic (Evan-Sapir 2009).

For SPECT/CT, the story is similar. One of the more popular uses for SPECT is a bone scan to evaluate for metastases based on Tc99m-labeled methylene diphosphonate (MDP). While sensitivity is very good for this test, it is complicated by tracer uptake due to a number of other processes such as degenerative joint diseases and osteomyelitis. The addition of CT can help to clarify diagnosis and improve the specificity of this test (Bockisch et al. 2009, Gnanasegaran et al. 2009, Mohan et al. 2010). Like FDG-PET, I-131, somatostatin receptor SPECT imaging, Ga67-white blood cell infection imaging and many other SPECT studies also benefit from localization (Delbeke et al. 2009).

In cardiac imaging, we know where the heart is so the benefit of CT is not to locate the organ, but rather to associate CT anatomy with function significance. A number of studies (Flotats et al. 2011) have demonstrated that there is a mismatch between the degree of stenosis and its functional significance. While <50% lesions are known to not degrade the function of the heart, and >90% lesions are known to definitely affect performance, in between there is a great deal more ambiguity. The DEFER trial (Pijls et al. 2007) showed that there was certainly a benefit to stenting lesions if they were functionally significant, as measured by a reduced fraction flow reserve by flow-wire. There was no benefit to stenting if the lesion was not functionally significant. CT-angiography (CTA) on its own shows that <50% of lesions with a 50% stenosis are functionally significant (Di Carli and Hachamovitch 2007). By registering the CTA with a perfusion study, it is possible to identify which lesions (Flotats et al. 2011), if any, have a functional significance and thus would benefit from intervention.

13.3 Attenuation with CT versus Transmission Scans

There are a number of factors that degrade the accuracy and quality of nuclear medicine images of which attenuation is the most important. The half-value thickness (HVT) for PET (511 keV) photons in water is 7.2 cm, but because a pair of photons is detected, the attenuation path length is the entire width of the patient, making the total attenuation much higher. For SPECT, the path lengths are shorter on average and the HVT is 4.5 cm, but the effects remain significant as is evidenced by the change in specificity of cardiac imaging with the application of attenuation correction (Garcia 2007).

Accurate correction for attenuation requires a patient-specific map of the linear attenuation coefficients of the tissues in the body. This was previously acquired by radioisotope transmission sources, but with the advent of PET/CT and SPECT/CT is now frequently obtained from a CT scan instead. There are a number of advantages to the use of CT for AC and these have bolstered its rise in popularity. There are also, though, a number of drawbacks. The advantages are a much shorter acquisition time, lower image noise, reduced staff exposure, improved consistency, and greater patient comfort. The disadvantages are increased patient exposure; larger, more complicated and more expensive equipment; a need for sequential acquisition of the emission and transmission images; and the speed of the CT acquisition.

13.4 CTAC: Advantages

Modern CT scanners rotate at speeds as high as 0.35 s/rotation. The slice thickness and spacing typically used for attenuation maps are on the order of 2.5 mm. The axial coverage of the CT scanner is 2–8 cm. The time required to obtain a transmission scan varies depending on the speed of the CT scanner and the field-of-view (FOV) that is to be acquired. It can range from a couple of seconds for a high-speed CT scan of a cardiac FOV to several minutes for a low-speed whole-body bone SPECT. In all cases though, the time required is a small fraction of the time needed for a radioisotope transmission scan. Radioisotope transmission images for PET required 3–5 min per bed position or axial FOV of the scanner. On a typical 6-bed whole-body study, the time for transmission imaging was 20–30 min. As emission imaging alone was generally 5 min per bed position, radioisotope transmission imaging almost doubled the total time required. Replacing the radioisotope scan with a high-speed CT reduces scan time by about 40% for whole-body studies (Even-Sapir et al. 2009).

The much greater intensity of the photon flux from an x-ray source compared to a radioisotope source also means that the CT images have much higher signal-to-noise. The noise in a reconstructed nuclear medicine image is a combination of the noise in the emission and transmission data (King et al. 1995, Kinahan et al. 1998). Compared to both radioisotope transmission scans and the emission scans themselves, the noise in the CT images is negligible and so using a CT scan for AC effectively eliminates the noise contribution from the transmission imaging (Figure 13.1).

The radioactive sources used for transmission imaging require periodic handling by the nuclear medicine staff to perform wipe tests to check for radioactive leakage. The sources also decay over time and so require replacement on a regular basis. For example, the half-lives of 153-Gd and 68-Ge, two sources typically used for transmission imaging, are 242 and 275 days respectively, about 2/3 of a year. Sources thus require replacement roughly every 2 years. During this time, the noise levels in the transmission scan and

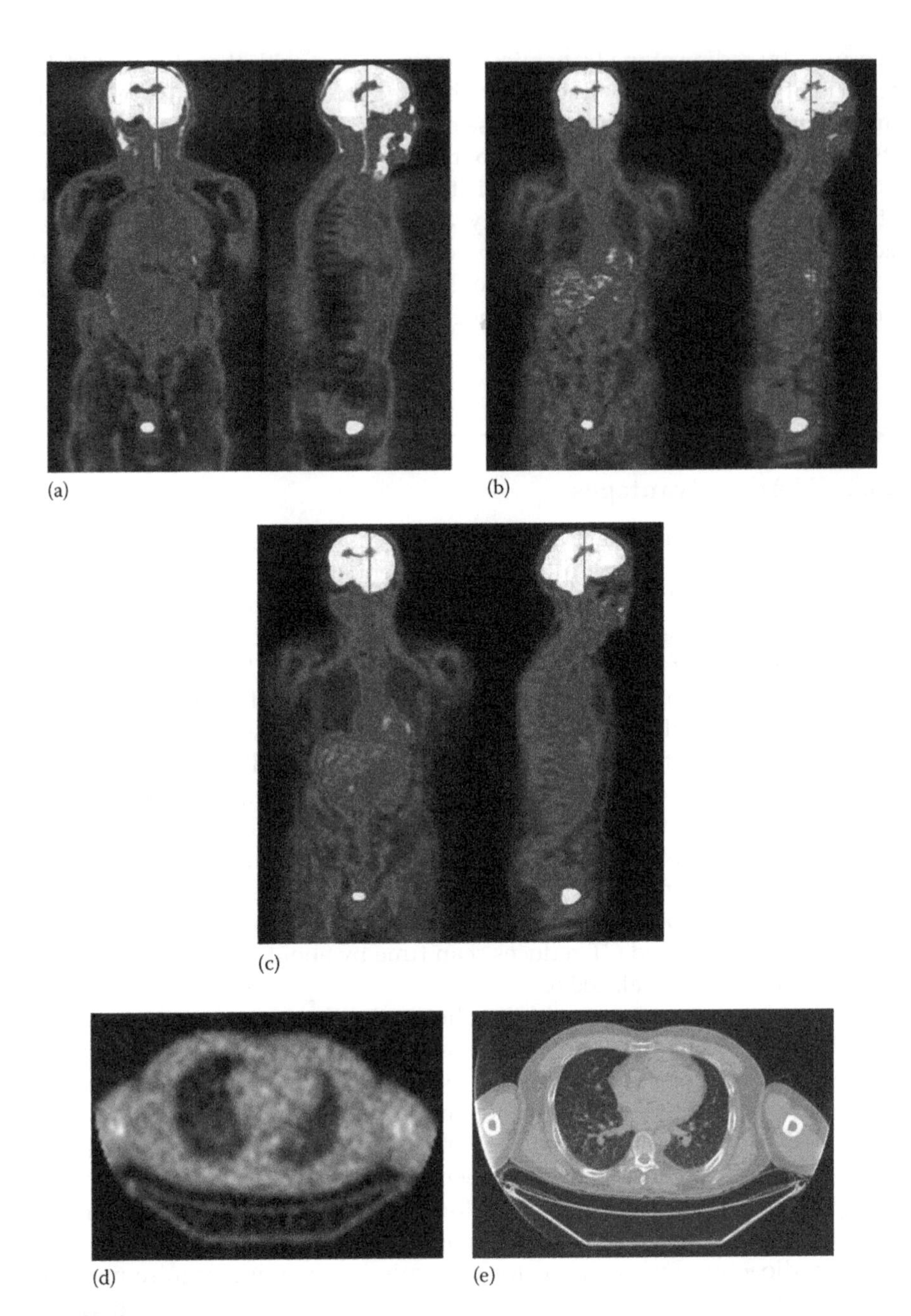

FIGURE 13.1
(See color insert.) Example images from an FDG PET/CT whole-body study. Shown is the same patient without attenuation correction (a), with correction using a Ge-68 attenuation map (b), with correction using a CT-based attenuation map (c), an example transverse slice through a Ge-68 transmission map (d), and the same slice through a CT map (e). For the PET images, white indicates highest amount of FDG activity.

the amount of interference between the transmission and the emission data will vary considerably and result in a continually changing accuracy in the transmission map. The handling of the radioactive sources results in additional exposure to the nuclear medicine staff that is eliminated with CT-AC.

Finally, the speed of the CT scan greatly reduces the total scan time and thus increases patient comfort and patient throughput. Increased patient comfort reduces the amount of patient motion and thereby the amount of motion-related artifacts in the images. The greater throughput increases the cost-effectiveness of the scanner and potentially reduces wait times as well.

13.5 CTAC: Disadvantages

The improvements in image quality of the transmission scan do not, unfortunately, come for free. The cost is an increase in the radiation exposure of the patient. While the exposure from a radioisotope transmission source was almost negligible next to the dose received from the injected emission tracer, the dose from the CT-scan can be a sizeable fraction of the total study dose (Patton et al. 2009). For example, a rest/stress cardiac SPECT scan with 99 mTc-tetrofosmin delivers a patient dose of 10 mSv. The exposure from transmission sources can be as low as 4 μSv (0.04%) (Perisinakis et al. 2002), but the exposure from a CT scan for attenuation correction would be about 0.8 mSv, about 200× higher. For a Rb82-PET rest/stress cardiac exam, the radiation dose is only 1.5 mSv and so the addition of the CT adds almost 50% to the total dose for the study. For whole body studies, transmission times are much longer due to the increased FOV, whereas the dose from the nuclear medicine test is constant resulting in an even greater percent contribution from the CT scan to the total study dose.

The addition of a diagnostic CT scanner to the PET or SPECT camera greatly increases its size, complexity, and cost. The increased cost is balanced by an improvement in diagnostic accuracy but can still be a barrier, particularly for small centers. The increase in complexity leads to more things that can go wrong and make the camera unavailable for use until it is repaired. The complexity also means a larger amount of time needs to be spent each morning on quality assurance tests by the technologists operating the scanner, further increasing operation costs. The increased bore length of the camera can be a problem for patients with issues like claustrophobia as the patient must be moved much deeper into the camera to obtain the images. Finally, the increased size, and need for x-ray shielding, may be a problem at some sites and require extensive renovations prior to being able to install the system.

A further disadvantage is that the CT and NM scans are acquired sequentially. The CT scanner typically uses a different set of detectors than the nuclear medicine camera, and consequently the transmission imaging is done either prior to or following the nuclear medicine acquisition. For a whole-body PET study, this can introduce a delay of 20 min or more between transmission and end of the emission scan. This delay is larger than would have occurred with radioisotope transmission imaging, wherein the transmission and emission data are interleaved at each bed position. This delay increases the likelihood of patient motion, leading to misregistration of the data sets.

13.6 Types of CT Used

There is a wide range of CT scanners used as part of an integrated PET/CT or SPECT/CT system. This is caused by both the rapidly changing environment of CT hardware and the desire by the manufacturers to offer cost-effective solutions, particularly for SPECT/CT. Early on in the development of PET/CT scanners, 1- and 2-slice CT scanners were offered commercially. Rapid improvements in CT technology have seen a leap from single-slice scanners to 320 slice scanners with dual-energy capabilities such that CT systems offered with early PET scanners are now considered obsolete. Presently, a 16-slice or better, high-speed diagnostic-quality CT is commonplace for PET/CT systems.

For SPECT/CT systems, there is a greater diversity of CT options available. The first unit introduced commercially also had a single-slice scanner (Seo et al. 2008, Patton et al. 2009). SPECT/CT systems with up to 64-slice CT scanners attached are now available, though lower-end units with only a few slices also remain on the market. The SPECT cameras also come with x-ray systems built on the SPECT gantry itself, rather than requiring a separate CT gantry. This reduces the cost of the system, but limits the speed at which the CT component can operate. These slow-rotation CTs spin at 2–5 rpm, but reduce the tube current and consequently the x-ray flux at which they operate down to 1–5 mA such that the radiation exposure to the patient is similar to that from a high-speed CT. The slow-rotation cameras are prone to CT artifacts caused by inconsistencies in the projection data introduced by patient motion such as breathing. The CT scans produced are thus not of diagnostic quality, but they are sufficient for tracer localization and attenuation correction.

Recent advances in CT acquisition protocols and data processing have allowed a reduction in the dose associated with CT scans, off-setting the increase in dose associated with using a CT scan for attenuation correction. These include approaches such as dose modulation as the tube rotates around the patient to alter the tube current in response to changing patient thickness, with the aim of maintaining similar signal to noise at all projection angles (Brisse et al. 2007). Prospective gating has been introduced for cardiac scans wherein the x-ray beam is targeted at specific phases of cardiac cycle through ECG gating. Finally, the introduction of iterative reconstruction for CT has led to an improvement in the signal to noise, or equivalently a 30%–50% reduction in patient radiation exposure for the same quality image (Marin et al. 2010, Fleischmann and Boas 2011, Park et al. 2012).

13.7 Creating an Attenuation Map from the CT Scan

To create an attenuation map from a CT scan requires several steps. The first step is the conversion from Hounsefield units into linear attenuation coefficients appropriate to the energy of the gamma ray(s) emitted by the nuclear medicine tracer. This is complicated by the fact that the x-ray beam is polyenergetic. X-rays are produced by accelerating electrons across a potential difference, typically 80–140 kV, and colliding them with a target. The x-rays are produced by Bremsstrahlung interactions in the target, generating photons with a range of energies up to a maximum equal to the accelerating potential. In the patient, the total mass attenuation coefficient varies with both the atomic number of the attenuating material (Z) and the energy of the incident photon (E). Thus, the relative attenuation of materials at different energies is different for different materials (Figure 13.2) and a single scaling factor

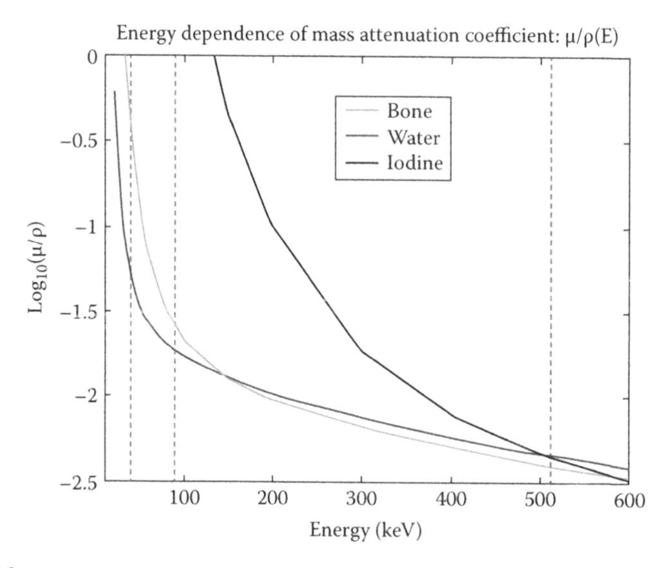

FIGURE 13.2
(See color insert.) Mass attenuation coefficients for water (similar to soft tissues), bone, and iodine. The range of energies for an x-ray beam (centered at 70 keV) is indicated by the dashed blue lines at left. The energy for PET imaging (511 keV) is shown with the dashed blue line at right.

cannot be used to convert a CT-map into an attenuation coefficient map. For example, at 70 keV, the mass attenuation of bone is greater than water, while at 511 keV the relationship is reversed.

Fortunately, the composition of most soft tissues is similar and it is reasonable to scale all of these in a similar manner. Bone, however, behaves differently and so a standard approach is to segment the CT-map into bone and soft tissues and scale each separately to the appropriate gamma-ray energy. The bone and soft-tissue maps are then recombined into a single attenuation map. The scaling factors are dependent on the spectral composition of the x-ray beam and so will be different for different accelerating potentials and may vary between scanners as well.

An additional complication is that the resolution of the CT scan is much higher than that of the nuclear medicine image. The difference in spatial resolution can lead to over- and undercorrection of the emission signal at sharp boundaries like the edges of organs (Meikle et al. 1993, King et al. 1995). To compensate for this, the CT image is filtered to remove high frequencies and match the resolution of the emission data. This blurred image is used for attenuation correction, but the full-resolution CT is still retained for the purpose of localization and the presentation of fused displays.

As the attenuation correction is based on the CT map, any artifacts in the CT can lead to artifacts in the nuclear medicine image. CT artifacts include beam-hardening, metal implants, CT contrast, truncation, and patient motion. Most CT scans will include at least a first-order beam-hardening correction that accounts for spectral changes through differing amounts of water. This corrects well for the soft-tissue component of beam-hardening, but does not correct for contributions from bone. The lack of bone correction leads to residual beam-hardening artifacts, particularly between boney structures such as inside of the skull or pelvic girdle. Beam hardening decreases the apparent attenuation and thus leads to an undercorrection of the nuclear medicine image. Advanced beam-hardening algorithms have been explored to reduce these effects, and new dual-energy CT scanners have the capability of generating mono-energy-equivalent CT images, which would remove this problem (Kyriakou et al. 2010, So and Lee 2011).

Truncation of the CT image can also be a problem. The field-of-view of the CT scanner is often smaller than that of the nuclear medicine device. For example, a typical CT field of view in a PET/CT system is 50 cm in diameter, but the bore size (and hence PET FOV) is 70–80 cm. This difference can lead to truncation of the CT image, particularly in the shoulders and arms of the patient. The truncation, and in particular bilateral truncation, can introduce severe streak artifacts into the CT and an increase in the Hounsfield unit (HU) of the CT image at the edge of the FOV. These inaccuracies are then propagated into the attenuation map and from there to the attenuation-corrected

NM image. To compensate for this, algorithms have been developed to extend the FOV of the CT in software by making basic assumptions about the nature of the truncated tissues and employing consistency conditions regarding the total attenuation seen by the CT at different projection angles (Mawlawi et al. 2006).

Metal objects can also be problematic for the CT image. They can completely attenuate an x-ray beam, but still be partially transparent to higher-energy gamma-rays. The CT image then contains streak artifacts and when converted to an attenuation map, the attenuation coefficients may be inappropriately large. Due to the very large attenuation, metal objects can often be identified automatically in the image and then software techniques can be used to compensate. A variety of approaches have been developed similar to the techniques used for higher-order beam-hardening correction such as separation and reprojection of the separate components, use of dual-energy CT, and interpolation of the projection data through the affected regions (Bamberg et al. 2011, Zhang et al. 2011, Joemai et al. 2012).

An additional problem is CT contrast. Typically iodinated, the attenuation from contrast scales very differently from either bone or soft tissues. At x-ray energies, it is very opaque, but it is almost completely transparent at higher energies like 511 keV. As contrast is frequently used for clinical CT, this complicates diagnostic CT procedures being performed in conjunction with a nuclear medicine exam, such as CT-angiography (CTA) combined with myocardial perfusion imaging. Fortunately, IV contrast disperses fairly rapidly, and studies have shown that it does not have a significant impact on the quality of PET images (Berthelsen et al. 2005). Oral contrasts, like barium, are more problematic as they can remain concentrated in the gastrointestinal tract for several days. An example is shown in Figure 13.3. The patient had undergone a CT procedure with oral contrast 3 days prior to the PET study. The inappropriate scaling of the attenuation map leads to artifactual uptake in the large intestine. The non-AC PET image clearly shows no such uptake present. The attenuation map can be corrected by proper segmentation and rescaling of the contrast within the CT and dual-energy CT may also be useful in mitigating this problem (Rehfeld et al. 2008). A good understanding of the source of the artifact and its appearance can also prevent misinterpretation (Groves et al. 2005).

Patient movement, both gross voluntary motion and physiological motion such as breathing or cardiac contractions, can lead to artifacts in the CT image. As the patient is moved through the CT scanner, one slice from the CT volume will be acquired with the patient in one position and another slice from a different position. For example, a mixture of slices acquired at expiration and inspiration can result in the appearance of a disjoint piece of liver apparently suspended in the lung. Another common respiratory artifact is a banana-shaped region of reduced apparent activity just above the diaphragm

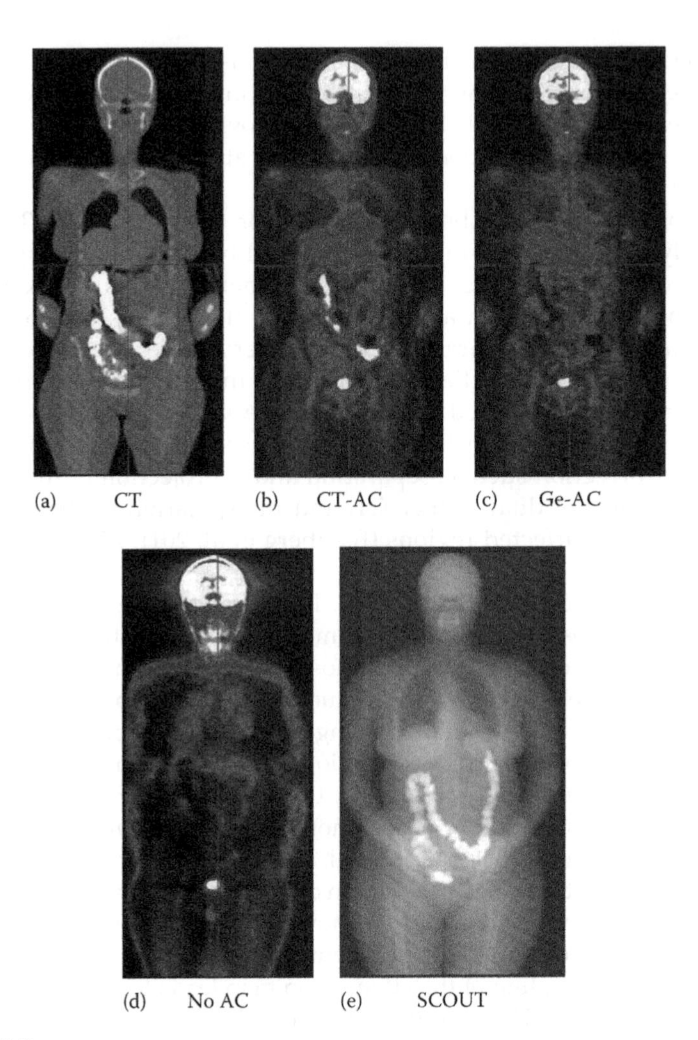

FIGURE 13.3
(See color insert.) Example of a whole-body FDG PET study in the presence of oral contrast. A coronal slice through the patient image volume is shown: CT (a), CT-based attenuation correction (b), Ge-based attenuation correction (c), and no attenuation correction (d). An x-ray through the patient is given in (e) clearly showing the oral contrast remaining in the large intestine. For the PET images, white indicates highest amount of FDG activity.

(Figure 13.4). In this case, the CT has been captured at a point of greater inspiration (and hence a lower diaphragm) than the mean position during PET acquisition resulting in an under correction of PET activity just above the diaphragm. Respiratory artifacts are avoided in diagnostic CT studies by having the patient hold their breath for the duration of the scan (usually only a few seconds with modern CT scanners). A breath-hold acquired with full inspiration is a poor representation of the attenuation experienced during a PET study with regular breathing averaged over several minutes.

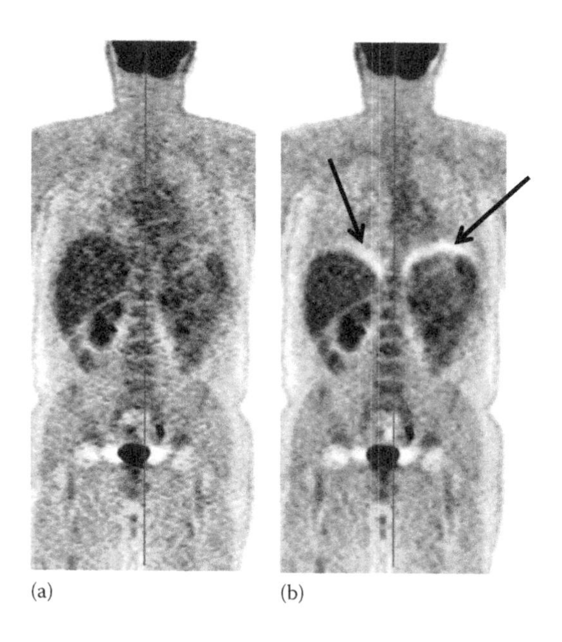

FIGURE 13.4
A common respiratory artifact in PET/CT. A coronal slice through the patient is shown for Ge-based attenuation correction (a) and CT-based attenuation correction (b). The area of reduced uptake indicated by the arrows in (b) is caused by an inconsistency in the position of the diaphragm between CT and PET imaging. For the PET images, black indicates high amounts of tracer activity.

Attempting to have the patient hold their breath at a representative moment is also unreliable. Respiratory motion has proven to be problematic, particularly for lung and cardiac studies (Pan et al. 2004, 2005, Gould et al. 2007, Bettinardi et al. 2010, Roach et al. 2010), and several different solutions have been proposed. The simplest solution is to capture a CT scan near end-expiration when patient movement is minimal—with a high-speed multi-slice scanner, it is possible to cover a small FOV such as is needed for cardiac studies in just a few seconds. This scan can then be aligned, using manual rigid-body registration tools, to the mean position of the NM scan (Khurshid et al. 2008, Wells et al. 2010). While successful in many cases, this approach cannot correct for deformations in the chest as the diaphragm moves and the lungs expand and contract (Gould et al. 2007, McQuaid and Hutton 2008). In this case, an average CT can be created that is more consistent with the time-averaged NM acquisition. This is done by acquiring a 4DCT—that is multiple CT volumes of the patient over the respiratory cycle—and then averaging them together (Pan et al. 2004). The acquisition of multiple CT volumes increases the patient radiation exposure, but the total exposure can be kept reasonably low, 1/4 mSv per bed position, by using low tube currents (Koepfli et al. 2004, Pan et al. 2006). Additionally, it is not necessary to acquire the entire whole-body CT as a 4DCT. Instead, one can focus on those portions of the patient that have the most motion, i.e., near the diaphragm, and patch in a time-average CT for that portion only (Pan et al. 2004). A final approach is a max-intensity CT (Alessio et al. 2007). This too uses a 4DCT, but instead of averaging the CT over time frames, it takes the maximum attenuation at each voxel in the volume. This increases the total attenuation in object, but has been shown to produce good results in clinical cardiac studies.

13.8 Additional Information from Diagnostic CT

One of the benefits of the combined scanner is that it provides a "one-stop shop" (Bockisch et al. 2009, Gnanasegaran et al. 2009, Czernin et al. 2010). CT is a powerful imaging modality in its own right; and if the CT on the hybrid scanner is of diagnostic quality, it is possible to perform the full suite of procedures of which CT is capable. Because the patient does not have to move from one scanner to another, there is exact concordance in the positioning and orientation of the patient. This makes image fusion and comparison of the two sets of data much easier. It also improves patient comfort and patient throughput as the time required to perform the tests is greatly reduced—there is only one patient visit.

CT provides better spatial resolution allowing detection of smaller abnormalities than might be found with PET. CT can also help to clarify sites of false-positive uptake on PET. The result is that PET/CT, compared to PET or CT alone, more accurately stages cancer and is superior for monitoring response to therapy for many different cancers such as in the lung, breast, colon, and lymphatic system (Czernin et al. 2010). CT can provide additional information in non-oncologic studies, for example, clarifying the cause of matched or mismatched defects in lung ventilation/perfusion studies (Roach et al. 2010).

With respect to cardiac evaluation, CT can add significantly to cardiac evaluation by PET or SPECT alone. Calcium scoring—a measure of the total amount of calcification in a patient's coronary arteries—and CT-angiography (CTA) have been shown to be independent indicators of patient prognosis (Schenker et al. 2008, Schwaiger et al. 2010, Chow et al. 2011, Hacker and Becker 2011), and having this information available leads to better assessment and risk stratification of patients. CTA also provides an excellent view of the coronary artery anatomy and coronary stenoses. CTA has very strong negative predictive value and fused with perfusion information can be used to help identify culprit lesions, those lesions that result in significant functional loss in the heart (Kaufman 2009, Santana 2009, Schwaiger et al. 2010).

The presence of an accurately coregistered high-resolution anatomical map also opens the door to using this information to enhance image quality in nuclear medicine. This can be done through the application of anatomical priors as part of a maximum a posteriori (MAP) reconstruction approach (Comtat et al. 2002, Lehovich et al. 2009, Vanhove et al. 2011, Cheng-Liao and Qi 2011). Radiotracers are expected to distribute fairly homogeneously through different compartments such as an organ and frequently are confined to those compartments. A sharp definition of the location of the anatomical boundaries restricts where the radiotracer may be within the body and thus provides information about the unknown tracer distribution that one is attempting to recover through image reconstruction.

13.9 Image Registration

Accurate localization, accurate attenuation correction, the use of anatomical priors, the fusing of anatomical CT information with functional NM images, indeed virtually all of the gains of a multimodal camera hinge on the accuracy of the coregistration of the images. Although hardware approaches minimize the source of some of the errors in registration, they cannot completely eliminate them (Gould et al. 2007, Goetze and Wahl 2007, Goetze et al. 2007) and further registration is often required. This final registration is often a rigid-body registration and is performed manually as a quality assurance procedure (Goetze et al. 2007). Manual intervention introduces uncertainty and variability into the accuracy of the final registration and can be a time-consuming process. The alternative is automated approaches to image registration. Semi-automated methods might take in user-designated landmarks to assist in registration. More automated approaches also exist that segment and align the organ of interest (Khurshid et al. 2008) or make use of consistency conditions to aid in registration (Alessio et al. 2010). For example, the total amount of activity seen in a patient should be the same, regardless of the angle of view, once you have compensated for attenuation. Thus, it is possible to adjust the position and orientation of the attenuation map until the difference between projections is minimized.

Even with hardware coregistered images from a hybrid camera, it is still necessary to perform some software registration following acquisition. One might then wonder about performing the entire registration by software and simply acquiring the images on separate machines. Indeed, some have argued to do just that. A number of sophisticated software registration packages have been developed and a recent review on the topic is given by Slomka and Baum (2009). The central advantage of software registration is flexibility. Images acquired previously, either at the same site or even among different sites, can be aligned. This reduces costs as it avoids repeating studies that have already been performed. It similarly saves time for the patient and avoids repeated radiation exposure in the case of tests like CT that use ionizing radiation. Use of software registration improves the efficiency of camera use. A CT scanner that takes a minute to acquire a whole-body attenuation map does not then sit idle during the 20 min PET acquisition or the 15 min SPECT study. A single CT scanner could potentially provide the attenuation map needs of multiple NM systems, greatly reducing the cost of the installation. Alternatively, it could be used for purely CT procedures, while the PET study is being performed. Likewise, particularly with SPECT/CT systems that use slow-rotation CT scanners, the transmission portion of the study can increase the time required for each exam, and thus reduce the number of SPECT studies that might be performed in a standard work day.

By allowing the alignment of any studies available, the physician is not restricted to the modalities available on a particular machine. Thus, one can combine SPECT and PET studies, incorporate MRI or ultrasound information, and use whatever images are available and relevant to the case at hand. This also allows for integration of images acquired at a later date based on the outcomes of the current test. Thus it is not necessary to decide upfront all the tests or procedures that need to be performed on a given patient. In addition, software registration allows for intramodality registration and thereby accurate serial evaluation of a patient during progression of disease or treatment. Software registration can also be used to accurately compare a patient's study with that of a population template. This can assist in identification of regions of interest and evaluation of the extent of disease.

A complication for software registration is the differences in the configuration of patient during the acquisition of the two images. Different days and different cameras mean rigid-body registration is not sufficient and so warping algorithms are often used to deform the volumes in a constrained manner so that they match. The problem is worse the longer the interval between scans, but even images acquired of the same patient on the same day on the same camera can be quite different unless great care is taken to ensure the patient is in exactly the same position. For example, differences are seen with cardiac SPECT rest/stress imaging during which two images of the patient are taken a few hours apart. Small movements of the arms between the two acquisitions can shift the distribution of soft tissues on the thorax and lead to variations in the amount of breast attenuation (Corbett and Ficaro 2000). Different protocols can also have different preferred patient positions such as arms above the head or arms resting at the patient's sides, which can be extremely difficult to correct for using software techniques. Different cameras have bed shapes that alter how a patient lies. An example is the flat table common for radiation therapy compared to the curved table that is typical for diagnostic imaging. For this reason, diagnostic images that are intending for integration into radiation therapy planning are often acquired on a flat table. A more extreme difference can occur with cardiac imaging; although supine imaging is most common, some cardiac cameras are now configured for upright imaging (e.g., Cardius XPO, Digirad). Finally, the longer the time delay between acquisitions, the greater the chance that the patient's morphology will change whether it is simple weight loss or gain or changes due to disease progression or treatment such as tumor shrinkage due to radiation therapy or surgery.

To correct for these changes, nonlinear registration algorithms are used. These "warping" algorithms modify the morphology of the patient to improve the alignment of one image with the other. A concern is that the warping could also distort the data of interest in the image and so great care must be taken to appropriately constrain the changes applied. Many different approaches have been developed from landmark registration, to mutual information, finite element models, and beyond. The interested reader is referred to (Slomka and Baum 2009) for a more complete introduction to this topic.

Another important application of registration algorithms has been in the area of motion compensation. Whether it is from voluntary motion or involuntary action such as breathing and cardiac contraction, patient motion leads to a blurring of the image data and a consequent change in the standard uptake value of tumors, variation in the size and intensity of lesions, and ambiguity in the position of the measured tracer uptake. Incorporated into pure list-mode reconstruction algorithms or working with data gated based on external motion measurement or physiologic signals, the registration algorithms can serve to reduce or remove the effects of motion. Efforts along these lines are being applied to both oncology and cardiology studies (Slomka et al. 2004, Gravier et al. 2006, Mair et al. 2006, Qiao et al. 2007, Dawood et al. 2008).

Many of the software registration algorithms are complicated and are not widely employed outside research centers. Registration is easiest when the two patient studies are acquired within a very short time of each other and with the patient in the same position. For this reason, most multimodal studies are presently acquired on multimodal cameras such as PET/CT and SPECT/CT systems. The large advantage of the combined camera is that it provides a registration that is approximately correct and that can be easily fine-tuned manually or with simple registration software. The great strength of software registration, though, is its flexibility and it will always remain a valuable tool in multimodal imaging.

13.10 MRI: PET/SPECT

Recently, there has been a surge in interest in combining nuclear medicine imaging with magnetic resonance imaging (MRI) (Cherry 2009). MRI is a powerful imaging technology that has many advantages over CT as a means of providing the anatomical information in a combined scanner. MRI uses radiofrequency radiation and is thus nonionizing, avoiding the radiation dose delivered by CT. This addresses one of the concerns of PET/SPECT-CT imaging, which is the increased dose from the CT component. MRI provides very high-resolution images with excellent soft-tissue contrast—an aspect poorly defined by CT. Different MRI acquisition pulse sequences allow probing of many different attributes within the body tissues such as differences in magnetic relaxation times (T1-weighted imaging), chemical composition (magnetic resonance spectroscopy), or diffusion properties of water (diffusion imaging). These properties can be quite complementary to the tracer-based imaging of nuclear medicine, and thus there can be seen great potential in the combination of these two techniques (Cherry 2009, Bouchelouche et al. 2010, Chen and Kinahan 2010, Czernin et al. 2010).

Unlike the typical PET-SPECT/CT configuration, the scanners combined with MRI are being designed for simultaneous acquisition of both the nuclear medicine and the MRI images. This addresses another issue with the current NM-CT systems, which is the delay between the CT and NM studies brought on by a serial acquisition strategy that can increase the risk of misregistration due to patient motion or produce other inconsistencies between the two data sets. Simultaneous acquisition is possible because of changes in the design of the PET detector elements. Photomultiplier tubes are very sensitive to the magnetic field environment and thus do not function well in the high magnetic field and rapidly changing gradient magnetic fields of the MRI scanner. Switching to solid-state amplification such as avalanche photodiodes and careful design of the detector electronics has made it possible to build PET-MR systems that do not interfere with each other (Cherry 2009). Though initially focused on small-animal and brain-only cameras, whole-body human systems have recently become commercially available (Biograph mMR, Siemens). Simultaneous acquisition provides exact coregistration and thus real-time evaluation of (and potentially correction for) patient voluntary and physiologic motion. It also ensures accurate localization of transient signals.

This technology is very new, however, and there remain several hurdles that must be overcome. The first of which is use of the MRI for attenuation correction of the PET/SPECT data set. Attenuation correction is essential to obtaining quantitatively accurate NM images. Absolute quantification is one of the strengths of PET imaging and needs to be maintained. The attenuation of photon signal from PET and SPECT is due to the interaction of photons with the patient soft tissues. The same interactions, albeit at different photon energies, generate the signal seen with a CT scanner and thus the CT directly measures the attenuation and can provide the map of attenuation coefficients needed for attenuation correction. The MRI signal strength is based on the relaxation properties of hydrogen and is not directly related to the effects causing attenuation in NM. The attenuation coefficient map must be inferred indirectly from the MRI. Different approaches have been explored to achieve this goal (Hofmann et al. 2009, 2011). One approach is to segment the image based on anatomical structures and then assign previously determined population attenuation coefficients. This works well in some areas, but is unable to capture the variable patient-specific attenuation in others such as the lungs. Another approach is nonlinear registration of a predetermined atlas (such as a CT scan), but this similarly does not address variable patient-specific attenuation. A final approach would be to register a previously acquired CT scan of the patient to a simultaneously acquired MRI image, though this reintroduces the radiation dose of the CT (Schreibmann et al. 2010). Some initial efforts at MRI-based attenuation correction have, nevertheless, had some success and work in this area is ongoing.

Additional problems for NM-MRI are those already associated with MRI. Metal implants are often a contraindication and might prevent use of this technology due to the high-magnetic fields. Metal can cause problems for CT, but algorithms have been developed that mitigate these effects and the residual impact is both small and local. Finally, cost of the equipment is also a consideration. MRI is an expensive technology and the cost-benefit analysis will need to be done, as it has been with PET-CT, to demonstrate its cost-effectiveness.

13.11 Summary

The use of SPECT/CT and PET/CT has seen tremendous growth over the past decade. The modalities are very complementary: SPECT and PET provide a very sensitive measure of the function of organs and systems in the body while CT provides structural anatomy with exquisite spatial resolution and excellent bone to soft tissue contrast. Integrating the two technologies is not without difficulty and can lead to artifacts. Nevertheless, the combination of anatomy and function has proven to have a greater diagnostic performance than either modality alone or even the two side by side. The success of the merger of these two imaging techniques has given strong support to the combination of other modalities as well, such as MRI which can provide superb soft-tissue contrast and a reduced radiation burden, making the future of multimodality imaging in medicine look very promising.

References

Alessio AM, Kinahan PE, Champley KM et al. Attenuation-emission alignment in cardiac PET/CT based on consistency conditions. *Med Phys*. 2010 Mar;37(3):1191–1200.

Alessio AM, Kohlmyer S, Branch K et al. Cine CT for attenuation correction in cardiac PET/CT. *J Nucl Med*. 2007 May;48(5):794–801.

Bai C, Conwell R, Kindem J et al. Phantom evaluation of a cardiac SPECT/VCT system that uses a common set of solid-state detectors for both emission and transmission scans. *J Nucl Cardiol.* 2010 Jun;17(3):459–469.
Bamberg F, Dierks A, Nikolaou K et al. Metal artifact reduction by dual energy computed tomography using monoenergetic extrapolation. *Eur Radiol.* 2011 Jul;21(7):1424–1429.
Berthelsen AK, Holm S, Loft A et al. PET/CT with intravenous contrast can be used for PET attenuation correction in cancer patients. *Eur J Nucl Med Mol Imaging.* 2005 Oct;32(10):1167–1175.
Bettinardi V, Picchio M, Di Muzio N et al. Detection and compensation of organ/lesion motion using 4D-PET/CT respiratory gated acquisition techniques. *Radiother Oncol.* 2010 Sep;96(3):311–316.
Beyer T, Kinahan PE, Townsend DW et al. The use of X-ray CT for attenuation correction of PET data. *Nucl Sci Symp Med Imaging Conf Rec.* 1994;4:1573–1577.
Beyer T, Townsend DW, Brun T et al. A combined PET/CT scanner for clinical oncology. *J Nucl Med.* 2000;41:1369–1379.
Blankespoor SC, Wu X, Kalki K et al. Attenuation correction of SPECT using X-ray CT on an emission-transmission CT system: Myocardial perfusion assessment. *IEEE Trans Nucl Sci.* 1996;41:2263–2274.
Bocher M, Blevis IM, Tsukerman L et al. A fast cardiac gamma camera with dynamic SPECT capabilities: Design, system validation and future potential. *Eur J Nucl Med Mol Imaging.* 2010 Oct;37(10):1887–1902.
Bockisch A, Freudenberg LS, Schmidt D et al. Hybrid imaging by SPECT/CT and PET/CT: Proven outcomes in cancer imaging. *Semin Nucl Med.* 2009 Jul;39(4):276–289.
Bouchelouche K, Turkbey B, Choyke P et al. Imaging prostate cancer: An update on positron emission tomography and magnetic resonance imaging. *Curr Urol Rep.* 2010 May;11(3):180–190.
Brisse HJ, Madec L, Gaboriaud G et al. Automatic exposure control in multichannel CT with tube current modulation to achieve a constant level of image noise: Experimental assessment on pediatric phantoms. *Med Phys.* 2007 Jul;34(7):3018–3033.
Chen DL, Kinahan PE. Multimodality molecular imaging of the lung. *J Magn Reson Imaging.* 2010 Dec;32(6):1409–1420.
Cheng-Liao J, Qi J. PET image reconstruction with anatomical edge guided level set prior. *Phys Med Biol.* 2011 Nov 7;56(21):68996918.
Cherry SR. Multimodality imaging: Beyond PET/CT and SPECT/CT. *Semin Nucl Med.* 2009 Sep;39(5):348–353.
Chow BJ, Small G, Yam Y et al. CONFIRM Investigators. Incremental prognostic value of cardiac computed tomography in coronary artery disease using CONFIRM: COroNary computed tomography angiography evaluation for clinical outcomes: An InteRnational Multicenter registry. *Circ Cardiovasc Imaging.* 2011 Sep;4(5):463–472.
Comtat C, Kinahan PE, Fessler JA et al. Clinically feasible reconstruction of 3D whole-body PET/CT data using blurred anatomical labels. *Phys Med Biol.* 2002;47:1–20.
Corbett JR, Ficaro EP. Attenuation corrected cardiac perfusion SPECT. *Curr Opin Cardiol.* 2000 Sep;15(5):330–306.
Czernin J, Allen-Auerbach M, Schelbert HR. Improvements in cancer staging with PET/CT: Literature-based evidence as of September 2006. *J Nucl Med.* 2007;48:78S–88S.

Czernin J, Benz MR, Allen-Auerbach MS. PET/CT imaging: The incremental value of assessing the glucose metabolic phenotype and the structure of cancers in a single examination. *Eur J Radiol*. 2010 Mar;73(3):470–480.

Dawood M, Buther F, Jiang X et al. Respiratory motion correction in 3-D PET data with advanced optical flow algorithms. *IEEE Trans Med Imaging*. 2008 Aug;27(8):1164–1175.

Delbeke D, Schöder H, Martin WH et al. Hybrid imaging (SPECT/CT and PET/CT): Improving therapeutic decisions. *Semin Nucl Med*. 2009 Sep;39(5):308–340.

Di Carli MF, Hachamovitch R. New technology for noninvasive evaluation of coronary artery disease. *Circulation*. 2007 Mar 20;115(11):1464–1480.

Even-Sapir E, Keidar Z, Bar-Shalom R. Hybrid imaging (SPECT/CT and PET/CT)—improving the diagnostic accuracy of functional/metabolic and anatomic imaging. *Semin Nucl Med*. 2009 Jul;39(4):264–275.

Fleischmann D, Boas FE. Computed tomography—old ideas and new technology. *Eur Radiol*. 2011 Mar;21(3):510–517.

Flotats A, Knuuti J, Gutberlet M et al. Cardiovascular Committee of the EANM, the ESCR and the ECNC. Hybrid cardiac imaging: SPECT/CT and PET/CT. A joint position statement by the European Association of Nuclear Medicine (EANM), the European Society of Cardiac Radiology (ESCR) and the European Council of Nuclear Cardiology (ECNC). *Eur J Nucl Med Mol Imaging*. 2011 Jan;38(1):201–212.

Garcia EV. SPECT attenuation correction: An essential tool to realize nuclear cardiology's manifest destiny. *J Nucl Cardiol*. 2007 Jan;14(1):16–24.

Gnanasegaran G, Barwick T, Adamson K et al. Multislice SPECT/CT in benign and malignant bone disease: When the ordinary turns into the extraordinary. *Semin Nucl Med*. 2009 Nov;39(6):431–442.

Goetze S, Brown TL, Lavely WC et al. Attenuation correction in myocardial perfusion SPECT/CT: Effects of misregistration and value of reregistration. *J Nucl Med*. 2007 Jul;48(7):1090–1095.

Goetze S, Wahl RL. Prevalence of misregistration between SPECT and CT for attenuation-corrected myocardial perfusion SPECT. *J Nucl Cardiol*. 2007 Apr;14(2):200–206.

Gould KL, Pan T, Loghin C et al. Frequent diagnostic errors in cardiac PET/CT due to misregistration of CT attenuation and emission PET images: A definitive analysis of causes, consequences, and corrections. *J Nucl Med*. 2007 Jul;48(7):1112–1121.

Gravier E, Yang Y, King MA et al. Fully 4D motion-compensated reconstruction of cardiac SPECT images. *Phys Med Biol*. 2006 Sep 21;51(18):4603–4619.

Groves AM, Kayani I, Dickson JC et al. Oral contrast medium in PET/CT: Should you or shouldn't you? *Eur J Nucl Med Mol Imaging*. 2005 Oct;32(10):1160–1166.

Hacker M, Becker C. The incremental value of coronary artery calcium scores to myocardial single photon emission computer tomography in risk assessment. *J Nucl Cardiol*. 2011 Aug;18(4):700–711.

Hasegawa BH, Gingold EL, Reilly SM et al. Description of a simultaneous emission-transmission CT system. *Proc SPIE*. 1990;1231:50–60.

Hasegawa BH, Reilly SM, Gingold EL et al. Design considerations for a simultaneous emission-transmission CT scanner. *Radiology*. 1989;173:414.

Hasegawa BH, Tang HR, Da Silva AJ et al. Implementation and applications of a combined CT/SPECT system. *IEEE Nucl Sci Symp Med Imaging Conf Rec*. 1999;3:1373–1377.

Hofmann M, Bezrukov I, Mantlik F et al. MRI-based attenuation correction for whole-body PET/MRI: Quantitative evaluation of segmentation- and atlas-based methods. *J Nucl Med*. 2011 Sep;52(9):1392–1399.

Hofmann M, Pichler B, Schölkopf B et al. Towards quantitative PET/MRI: A review of MR-based attenuation correction techniques. *Eur J Nucl Med Mol Imaging*. 2009 Mar;36 Suppl 1:S93–S104.

Joemai RM, de Bruin PW, Veldkamp WJ et al. Metal artifact reduction for CT: Development, implementation, and clinical comparison of a generic and a scanner-specific technique. *Med Phys*. 2012 Feb;39(2):1125–1132.

Kaufmann PA, Di Carli MF. Hybrid SPECT/CT and PET/CT imaging: The next step in noninvasive cardiac imaging. *Semin Nucl Med*. 2009 Sep;39(5):341–347.

Khurshid K, McGough RJ, Berger K. Automated cardiac motion compensation in PET/CT for accurate reconstruction of PET myocardial perfusion images. *Phys Med Biol*. 2008 Oct 21;53(20):5705–5718.

Kinahan PE, Townsend DW, Beyer T et al. Attenuation correction for a combined 3D PET/CT scanner. *Med Phys*. 1998 Oct;25(10):2046–2053.

King MA, Tsui BMW, Pan T. Attenuation compensation for cardiac single-photon emission computed tomographic imaging: Part 1. Impact of attenuation and methods of estimating attenuation maps. *J Nucl Cardiol*. 1995;2:513–524.

Koepfli P, Hany TF, Wyss CA et al. CT attenuation correction for myocardial perfusion quantification using a PET/CT hybrid scanner. *J Nucl Med*. 2004 Apr;45(4):537–542.

Kuhl DE, Hale J, Eaton WL. Transmission scanning: A useful adjunct to conventional emission scanning for accurately keying isotope deposition to radiographic anatomy. *Radiology*. 1966;87:278–284.

Kyriakou Y, Meyer E, Prell D et al. Empirical beam hardening correction (EBHC) for CT. *Med Phys*. 2010 Oct;37(10):5179–5187.

Lehovich A, Bruyant PP, Gifford HS et al. Impact on reader performance for lesion-detection/localization tasks of anatomical priors in SPECT reconstruction. *IEEE Trans Med Imaging*. 2009 Sep;28(9):1459–1467.

Levin CS. Primer on molecular imaging technology. *Eur J Nucl Med Mol Imaging*. 2005;32:S325–S345.

Mair BA, Gilland DR, Sun J. Estimation of images and nonrigid deformations in gated emission CT. *IEEE Trans Med Imaging*. 2006 Sep;25(9):1130–1144.

Marin D, Nelson RC, Schindera ST et al. Low-tube-voltage, high-tube-current multidetector abdominal CT: Improved image quality and decreased radiation dose with adaptive statistical iterative reconstruction algorithm—initial clinical experience. *Radiology*. 2010 Jan;254(1):145–153.

Mawlawi O, Erasmus JJ, Pan T et al. Truncation artifact on PET/CT: Impact on measurements of activity concentration and assessment of a correction algorithm. *AJR Am J Roentgenol*. 2006 May;186(5):1458–1467.

McQuaid SJ, Hutton BF. Sources of attenuation-correction artefacts in cardiac PET/CT and SPECT/CT. *Eur J Nucl Med Mol Imaging*. 2008 Jun;35(6):1117–1123.

Meikle SR, Dahlbom M, Cherry SR. Attenuation correction using count-limited transmission data in positron emission tomography. *J Nucl Med*. 1993 Jan;34(1):143–150.

Mirshanov DM. Transmission-Emission Computer Tomograph. Tashkent Branch. All-Union Research Surgery Center, USSR Academy of Medical Science, USSR, 1987.

Mohan HK, Gnanasegaran G, Vijayanathan S et al. SPECT/CT in imaging foot and ankle pathology-the demise of other coregistration techniques. *Semin Nucl Med.* 2010 Jan;40(1):41–51.

Pan T, Lee TY, Rietzel E et al. 4D-CT imaging of a volume influenced by respiratory motion on multi-slice CT. *Med Phys.* 2004 Feb;31(2):333–340.

Pan T, Mawlawi O, Luo D et al. Attenuation correction of PET cardiac data with low-dose average CT in PET/CT. *Med Phys.* 2006 Oct;33(10):3931–3938.

Pan T, Mawlawi O, Nehmeh SA et al. Attenuation correction of PET images with respiration-averaged CT images in PET/CT. *J Nucl Med.* 2005 Sep;46(9):1481–1487.

Park EA, Lee W, Kim KW et al. Iterative reconstruction of dual-source coronary CT angiography: Assessment of image quality and radiation dose. *Int J Cardiovasc Imaging.* 2012 Oct;28(7): 1775–1786.

Patel CN, Chowdhury FU, Scarsbrook AF. Hybrid SPECT/CT: The end of "unclear" medicine. *Postgrad Med J.* 2009 Nov;85(1009):606–613.

Patton JA, Townsend DW, Hutton BF. Hybrid imaging technology: From dreams and vision to clinical devices. *Semin Nucl Med.* 2009 Jul;39(4):247–263.

Perisinakis K, Theocharopoulos N, Karkavitsas N et al. Patient effective radiation dose and associated risk from transmission scans using 153Gd line sources in cardiac spect studies. *Health Phys.* 2002 Jul;83(1):66–74.

Pijls NH, van Schaardenburgh P, Manoharan G et al. Percutaneous coronary intervention of functionally nonsignificant stenosis: 5-year follow-up of the DEFER Study. *J Am Coll Cardiol.* 2007 May 29;49(21):2105–2111.

Qiao F, Pan T, Clark JW Jr et al. Joint model of motion and anatomy for PET image reconstruction. *Med Phys.* 2007 Dec;34(12):4626–4639.

Rehfeld NS, Heismann BJ, Kupferschläger J et al. Single and dual energy attenuation correction in PET/CT in the presence of iodine based contrast agents. *Med Phys.* 2008 May;35(5):1959–1969.

Roach PJ, Gradinscak DJ, Schembri GP et al. SPECT/CT in V/Q scanning. *Semin Nucl Med.* 2010 Nov;40(6):455–466.

Santana CA, Garcia EV, Faber TL et al. Diagnostic performance of fusion of myocardial perfusion imaging (MPI) and computed tomography coronary angiography. *J Nucl Cardiol.* 2009 Mar–Apr;16(2):201–211.

Schenker MP, Dorbala S, Hong EC et al. Interrelation of coronary calcification, myocardial ischemia, and outcomes in patients with intermediate likelihood of coronary artery disease: A combined positron emission tomography/computed tomography study. *Circulation.* 2008 Apr 1;117(13):1693–1700.

Schreibmann E, Nye JA, Schuster DM et al. MR-based attenuation correction for hybrid PET-MR brain imaging systems using deformable image registration. *Med Phys.* 2010 May;37(5):2101–2109.

Schwaiger M, Ziegler SI, Nekolla SG. PET/CT challenge for the non-invasive diagnosis of coronary artery disease. *Eur J Radiol.* 2010 Mar;73(3):494–503.

Seo T, Mari C, Hasegawa BH. Technological development and advances in single-photon emission computed tomography/computed tomography. *Semin Nucl Med.* 2008;38:177–198.

Slomka PJ and Baum RP. Multimodality image registration with software: State-of-the-art. *Eur J Nucl Med Mol Imaging.* 2009 Mar;36 Suppl 1:S44–S55.

Slomka PJ, Nishina H, Berman DS et al. "Motion-frozen" display and quantification of myocardial perfusion. *J Nucl Med.* 2004 Jul;45(7):1128–1134.

So A, Lee TY. Quantitative myocardial CT perfusion: A pictorial review and the current state of technology development. *J Cardiovasc Comput Tomogr*. 2011 Nov–Dec;5(6):467–481.

Vanhove C, Defrise M, Bossuyt A et al. Improved quantification in multiple-pinhole SPECT by anatomy-based reconstruction using microCT information. *Eur J Nucl Med Mol Imaging*. 2011 Jan;38(1):153–165.

von Schulthess GK. Positron emission tomography versus positron emission tomography/computed tomography: From "unclear" to "new-clear" medicine. *Mol Imaging Biol*. 2004 Jul–Aug;6(4):183–187.

Wells RG, Ruddy TD, DeKemp RA et al. Single-phase CT aligned to gated PET for respiratory motion correction in cardiac PET/CT. *J Nucl Med*. 2010 Aug;51(8):1182–1190.

Zhang X, Wang J, Xing L. Metal artifact reduction in X-ray computed tomography (CT) by constrained optimization. *Med Phys*. 2011 Feb;38(2):701–711.

14

Multimodality Imaging with MR/PET and MR/SPECT

Troy H. Farncombe

CONTENTS

14.1 Multimodality Imaging Systems

Since its inception, nuclear medicine imaging has always been known as a functional imaging technique rather than an anatomical imaging method. Three-dimensional images obtained from SPECT or PET devices depict the accumulation of radiopharmaceutical by specific cells, thus representing cellular function (see Figure 14.1a). As a result, these images lack the spatial information and instant recognition of other medical imaging techniques such as x-ray CT or MRI. The underlying information conveyed to an educated reader, however, is substantial as the distribution of radiopharmaceutical within the body provides physiological information about the body rather than strictly anatomical information. Nuclear medicine images often provide an early glimpse into disease progression as physiological changes occur prior to structural changes appearing. Even so, nuclear medicine

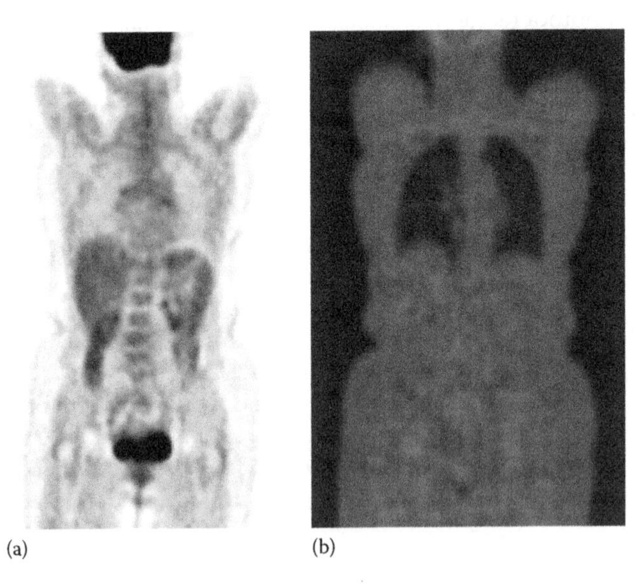

FIGURE 14.1
Typical whole-body PET scan showing the distribution of ^{18}FDG (a). A radionuclide transmission scan using Cs-137 (b). Note the absence of contrast between soft tissues.

has long been described as "unclear" medicine due to the fact that images typically lack any readily identifiable landmark information such as bones or lungs. In fact, an ideal nuclear medicine should contain no anatomical information whatsoever but rather depict the highly specific uptake of radiotracer in certain cell types. Admittedly, the lack of structural information may introduce some difficulties in localizing disease or framing the extent of disease. In order to provide some measure of anatomical structure and to correlate the radiopharmaceutical distribution with anatomy, numerous methods have been used.

14.1.1 Radionuclide Transmission Imaging

Among the first methods used to obtain anatomical information along with functional information was through the use of radionuclide transmission measurements [1–4]. This type of acquisition places a radioactive source opposite the detector and rotates it around the body with the radiation detector measuring the resultant transmitted radiation. By also measuring a blank transmission scan (i.e., no patient) and using the relation

$$I = I_0 e^{-\int \mu_i dl}$$

it is possible to reconstruct the 3D distribution of linear attenuation coefficients in the patient. Because the detectors operate in photon-counting mode rather than in current mode, images produced using transmission-based imaging suffer from poor spatial resolution and poor tissue contrast due to the low number of detected photons. In essence, images made using radionuclide transmission can distinguish soft tissue from air but little else (see Figure 14.1b). This is reasonably acceptable in order to perform attenuation compensation but generally is not of much value for accurately localizing disease.

The shortcomings of radionuclide transmission-based imaging have led to it being relegated to perform attenuation correction only. In fact, until the early 2000s, most PET scanners still utilized a Ge-68 or Cs-137 source for attenuation correction and even now, several SPECT machines offer radionuclide transmission sources (typically Gd-153) for attenuation correction.

14.1.2 CT-Based Co-Registration

Because of the limitations of radionuclide transmission imaging, many investigators turned to combining x-ray CT images with nuclear medicine PET or SPECT images [5,6]. Typically, PET or SPECT imaging would be performed on one system followed by CT imaging on another system, often in a different physical location. Care has to be taken to image the patient in the same orientation on both systems and often external fiducial markers are used in to aid in the co-registration process. Once acquired, PET or SPECT images would be matched to CT images by shifting the radionuclide images in three dimensions until they align with the CT. Image co-registration such as this often requires substantial operator interaction and a keen eye as image alignment is usually subjective.

In the late 1990s, the value of incorporating a high-resolution x-ray CT into a PET gantry was realized and the first PET/CT imaging system was developed [7,8]. By colocating the PET and CT in the same room and sharing a common patient bed, it became possible to obtain accurately co-registered PET and CT images in much reduced time. The benefits of PET/CT soon became apparent [9] and combined PET/CT was commercialized in the early 2000s. Sales of PET/CT have increased to the point where currently no major equipment manufacturer offers a PET-only device for routine clinical imaging. Similarly, SPECT/CT devices were soon developed and are now available commercially through all major vendors [10] are now available commercially through all major vendors.

14.1.3 Image Co-Registration

When performing radionuclide (PET or SPECT) and anatomical imaging (CT or MRI) with different physical systems, the accuracy of co-registration becomes paramount. In the simplest case, it is assumed that the patient can be represented as a rigid body and is free to move about only 6 degrees of freedom from one scan to the next. Thus, a given coordinate in one scan corresponds a transformed coordinate in the other scan via the transformation

$$\begin{bmatrix} x' \\ y' \\ z' \end{bmatrix} = \begin{bmatrix} \cos\theta\cos\varphi & \cos\xi\sin\varphi\sin\theta - \sin\xi\cos\theta & \cos\xi\sin\varphi\cos\theta + \sin\theta\sin\xi \\ \cos\varphi\sin\xi & \cos\xi\sin\varphi\sin\theta + \cos\xi\cos\theta & \sin\xi\sin\varphi\cos\theta - \sin\theta\cos\xi \\ -\sin\varphi & \sin\theta\cos\varphi & \cos\varphi\cos\theta \end{bmatrix} \times \begin{bmatrix} x \\ y \\ z \end{bmatrix} + \begin{bmatrix} \Delta x \\ \Delta y \\ \Delta z \end{bmatrix}$$

where θ, ϕ, ξ and Δx, Δy, Δz represent rotation and translations, respectively, about the three orthogonal axes.

Often in rigid-body transformations, it is up to the user to manipulate the transformation parameters in order to arrive at a suitably co-registered image. However, as this is a subjective assessment and dependent upon user interactions, it is often not reproducible from user to user. As a result, a number of methods [11] have been used to quantify the degree of co-registration, principal among them is the use of mutual information [12,13].

14.1.4 Image Display

Once anatomical (MR or CT) and functional (PET or SPECT) information has been gathered and co-registered, it must be presented to the interpreting physician in the most clear and concise manner possible. Often, even though images are co-registered, images are viewed separately in split windows. When the interpreter selects a location on either the anatomical or functional image, the viewer is usually shown three orthogonal views centered at the selected location.

As an alternative, it is common to present co-registered images simultaneously in a fused display using alpha blending [14]. In this type of display, each image is shown superimposed on the other but in a different color scale. For example, a PET and CT image can be displayed as a fused image by combining the red, green, and blue color channels of each image appropriately. Given the PET and CT images as 24 bit color images (8 bits in each channel), a fused image can be created with altered RGB color channels through the transformation

$$\begin{vmatrix} R \\ G \\ B \end{vmatrix} = \alpha \begin{vmatrix} R \\ G \\ B \end{vmatrix}_1 + (1-\alpha) \begin{vmatrix} R \\ G \\ B \end{vmatrix}_2$$

where α is given as an opacity value in the range 0–1 controlling to mixing of the two fused images (see Figure 14.2).

14.1.5 CT Attenuation Correction

Along with providing accurate spatial localization when overlaid with PET or SPECT images, co-registered x-ray CT images can also be used to

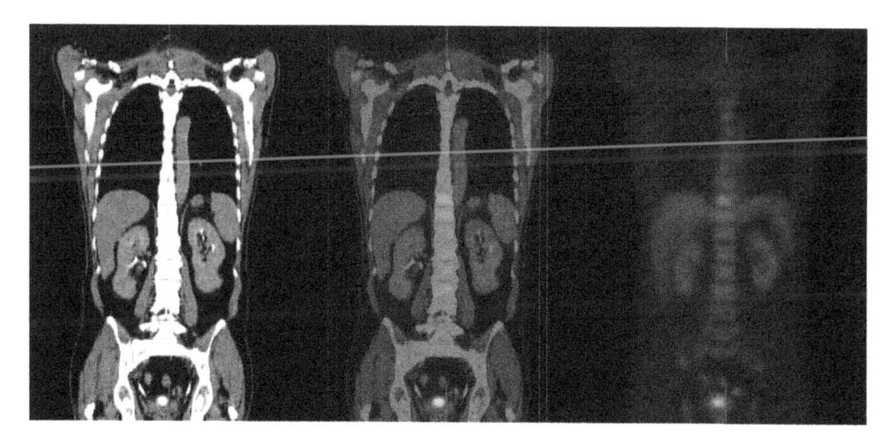

FIGURE 14.2
(See color insert.) Alpha blending of PET/CT images shown (from left to right): $\alpha = 0$ (CT only), $\alpha = 0.5$ (equal PET and CT), $\alpha = 1.0$ (PET only).

compensate for the attenuation of photons during the PET or SPECT data acquisition process [2,4,8,15]. As x-ray CT is acquired using a polyenergetic spectrum of x-rays, resultant images do not represent a linear attenuation coefficient but rather are represented in Hounsfield units (HU) where air is defined as −1000 HU and water as 0 HU with all other tissue types distributed accordingly along this scale. In order to perform accurate photon attenuation correction, CT images must first be converted into units of linear attenuation (i.e., cm^{-1}) specific to the PET or SPECT isotope energy. Various schemes exist for this [16–19], but in the simplest implementation, CT images are rescaled to linear attenuation coefficients via the expression

$$\mu^{PET} = \mu_{H_2O}^{PET}\left(\frac{HU}{1000}+1\right)$$

At low tissue densities, the conversion of CT number to linear attenuation coefficient follows this relation, but for CT numbers greater than soft tissue, this conversion is no longer linear with the same slope but rather becomes

$$\mu^{PET} = \mu_{H_2O}^{PET} + HU\left(\frac{\mu_{H_2O}^{CT}\left(\mu_{bone}^{PET}-\mu_{H_2O}^{PET}\right)}{1000\left(\mu_{bone}^{CT}-\mu_{H_2O}^{CT}\right)}\right)$$

In order to implement this type of conversion, calibrations must first be performed in order to evaluate μ_{bone}^{CT} and $\mu_{H_2O}^{CT}$ for each CT gantry and x-ray tube.

14.1.6 Clinical Applications of PET/CT and SPECT/CT

Probably the most common application of PET/CT imaging is ^{18}F–fluorodeoxyglucose (FDG) imaging for oncology [7]. With standard FDG-PET imaging, the distribution of FDG follows that of glucose metabolism (i.e., areas of high glucose metabolism result in large accumulations of FDG). As many malignant tumors have a high affinity for glucose, it follows that these tumors also would have a high affinity for FDG. Because FDG depicts cellular glucose metabolism, the resulting PET images lack any inherent anatomical information and typically have spatial resolutions on the order of 6–9 mm, thus making it difficult for interpreting physicians to pinpoint the exact location of a malignant tumor with certainty.

The introduction of PET/CT however has been able to change this. By acquiring a CT image of the same area of the body and overlaying the PET and CT on the same spatially co-registered image, it is possible to determine the location of FDG-avid tumors with a high degree of accuracy. In the case of head and neck cancers that typically spread to the neck lymph nodes, PET/CT has made it possible to correlate the uptake of FDG to specific lymph nodes, thus making surgical excision more precise [19,20].

Imaging ^{111}In-labeled white blood cells (WBCs) with SPECT has become an important procedure for determining sites of infection following orthopedic replacement surgery or in patients with fever of unknown origin. The immune system recruits leukocytes to fight bacteria at areas of infection and so labeling these cells permits physicians to localize injury. An ideal WBC SPECT image would lack any anatomical information as the distribution of WBCs would be specific *only* to the site of infection. Thus, the resultant image would depict a diffuse spot of activity with no anatomic context. As a means to provide some additional localization information, it is common to take advantage of the multi-isotope imaging capabilities of SPECT in order to perform simultaneous ^{99m}Tc-MDP bone imaging. The uptake of ^{99m}Tc-MDP is confined to the skeletal system and so provides a rough anatomical context to place the resultant WBC information. While dual-isotope imaging provides some measure of anatomical context, it fails to provide the spatial resolution needed to accurately localize infection. With the introduction of SPECT/CT, anatomical imaging is made much simpler and more precise as WBC distributions can now be correlated exactly with patient morphology [21–25].

Myocardial perfusion imaging with SPECT is one of the most commonly performed procedures in nuclear medicine and used to detect perfusion abnormalities in the heart. The distribution of ^{99m}Tc-labeled compounds such as sestamibi or tetrofosmin is related to the coronary arterial blood supply. As the blood supply to the heart is affected by the diameter of the coronary arteries, any reduction in perfusion is typically the result of a narrowing in the arteries, usually the result of either calcification of the lumen or buildup of atheromatous plaques (atherosclerosis) within the arteries. While useful for determining the extent of perfusion abnormalities, SPECT is incapable of determining the exact site of narrowing. However, with SPECT/CT imaging, it is possible to obtain anatomical information and correlate it with perfusion imaging in order to determine where the perfusion defects originate. As well, CT imaging is able to provide quantitative measures of calcification (calcium scoring) or 3D angiography (CTA) to depict the 3D coronary artery structure. By obtaining complementary information such as this in one single SPECT/CT exam, patients are able to be treated more effectively and in a more timely fashion [26].

14.1.7 Problems with CT-Based Co-Registration

While multimodality imaging with x-ray CT has proven itself valuable for a number of applications, it is not immune to problems. Because of the polyenergetic nature of x-ray production, a significant amount of low-energy x-rays exist in most spectra. Because of the energy dependence on photon attenuation, low-energy x-rays are preferentially attenuated as they pass through the body. The result is a progressive increase in the average x-ray energy, commonly referred to as beam hardening, the result of which produces, sometimes significant, streaking artifacts in resultant CT images (Figure 14.3).

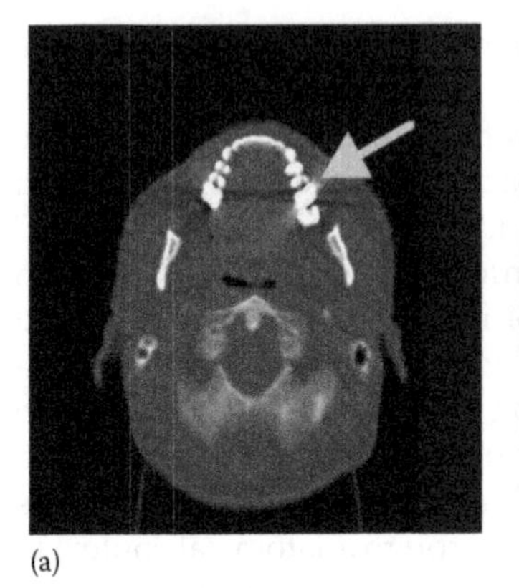

(a)

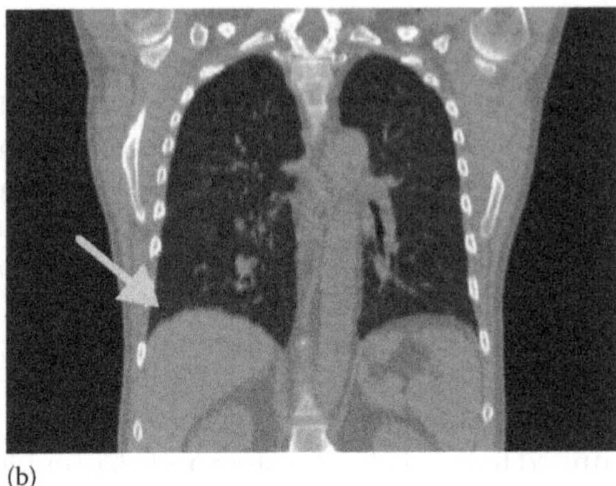

(b)

FIGURE 14.3
Examples of artifacts seen in x-ray CT scans. (a) Image showing streak artifacts resulting from beam hardening through dental fillings. (b) CT image of the thorax showing mild respiratory motion artifact. Note the stair step effect in the dome of the liver at the point of the arrow.

Materials such as Al or Cu are often used to reduce these low-energy x-rays as they pass from the x-ray tube, but usually this filtration does not remove all of this component. Images with beam hardening artifacts may lead to misinterpretation and incorrect attenuation compensation to the nuclear image data.

Because of the different time scales of data acquisition involved in nuclear medicine imaging and x-ray CT (minutes vs. seconds), it is possible that physiological processes occurring over short (or long) time scales can also interfere with the imaging method. For example, in the case of lung cancer imaging, FDG-PET scans may be acquired over the course of several minutes, thus motion artifacts as the result of normal tidal breathing essentially get "averaged out" over the course of imaging in order to obtain a single PET image. Contrast this to CT imaging where the entire lung volume may be imaged in a single breath-hold in just a few seconds. Since the PET image represents an average motion, CT and PET images are inherently misregistered [27,28], resulting in the possibility of incorrect localization.

While the addition of x-ray CT imaging has had a profound impact on nuclear medicine, there are some applications in which CT is not the optimal imaging modality. Because x-ray CT characterizes materials based on electron density (x-ray attenuation is primarily dependent upon photoelectric effect and Compton scattering), it follows that materials or tissues with similar electron densities may not produce much variation in x-ray attenuation. This is particularly true in imaging areas such as the brain where white and grey matter have very similar properties and cannot typically be distinguished with x-ray imaging.

Because of the shortcomings of x-ray CT imaging, there is interest in alternative approaches to perform anatomical/functional correlative imaging. Magnetic resonance imaging is one technique that is capable of high-resolution anatomical imaging that, at the same time, produces superior soft-tissue contrast compared to x-ray CT. As well, in contrast to x-ray CT that produces high-resolution images of anatomy, MRI is also capable of bridging the gap between anatomical and functional imaging by being capable of performing studies such as brain activation, chemical metabolism, or perfusion. As a result of these advantages, there is increasing interest in combining MR imaging with molecular imaging using PET and/or SPECT.

14.1.8 Requirements for PET/MR and SPECT/MR

MRI machines utilize high-strength magnetic fields and radio-frequency generators to manipulate the magnetic moments of certain atoms. Typical magnetic fields for clinical MRI machines range from 0.1 T (open field) to 4.7 T (fMRI) but with 3 T machines quickly becoming the norm. The high field used in MR requires that any material placed within the MRI field both have negligible effect on the field homogeneity and still operate normally under such a large magnetic field.

The design of most MRI systems also imposes a constraint on the size of the incorporated PET or SPECT device as high-field MRI systems use a cylindrical gantry with a fixed bore size. The ideal PET or SPECT imaging system that would be incorporated into an MRI device would need to have the following characteristics:

- Ability to operate in a high magnetic field with negligible effect on imaging performance
- Must be made of magnetically compatible materials and produce negligible effect on magnetic field homogeneity
- Small and relatively lightweight
- Produce image quality comparable to stand-alone systems

As can be expected, the logistics of incorporating PET or SPECT radiation detectors into a typical MR gantry are complex.

14.2 Basics of Pet and SPECT Imaging

In order to investigate the various approaches to combining MR with PET or SPECT, it is first important to describe the basics of radionuclide imaging and the imaging systems.

14.2.1 Scintillation Detectors

Overwhelmingly, scintillation detectors make up the majority of radiation detection methods used for PET or SPECT imaging. Briefly, these detectors usually consist of inorganic crystalline material that fluoresces in the presence of ionizing radiation. As many materials scintillate when exposed to ionizing radiation, the appropriate scintillator material must be selected for the chosen application. For example, in PET imaging, the requirement of the scintillator is to stop 511 keV photons and to distinguish two detections based on very short time scales. Thus, PET scintillators must have high stopping power (effectively a high density) and a very fast response time (i.e., short fluorescence time) [29]. Since most imaging used in SPECT utilizes lower-energy radionuclides, stopping power is less of a concern, but the ability to distinguish different photon energies is important. Thus, scintillators with high light output and good energy resolution are important. In the case of combined MR/PET or MR/SPECT, an appropriate scintillator must be chosen that meets the earlier mentioned criteria but is also compatible with the high magnetic field present in MR [30]. Table 14.1 presents some of the properties of the most common scintillators used in PET and SPECT imaging. As expected, the gadolinium-based scintillators, GSO and LGSO, have a very high magnetic susceptibility, thus making them unsuitable for PET/MR or SPECT/MR [30].

Regardless of the material used, any scintillator used for radionuclide imaging must be capable of absorbing the high-energy gamma rays originating from the radioactive decay and converting this energy into optical photons. When a high-energy gamma ray is incident upon a scintillator, three different interaction types are possible: photoelectric absorption, Compton scattering, or pair production. In typical nuclear medicine imaging, only the first two effects are relevant as pair production requires incoming gamma ray energies in excess of 1.022 MeV.

In the photoelectric effect, high-energy gamma rays interact with inner shell electrons of the scintillator. As the energy of the high-energy gamma ray

TABLE 14.1

Properties of Some Common Scintillators Used in Nuclear Medicine Imaging Equipment

Material	Density (g/cm³)	Effective Atomic Number (Z_{eff})	Emission Wavelength (nm)	Decay Time (μs)	Hygroscopic	Paramagnetic
NaI(Tl)	3.67	51	415	0.230	Yes	No
BGO	7.13	73	480	0.300	No	No
CsI(Tl)	4.51	54	540	0.68	Slight	No
CsI(Na)	4.51	54	420	0.63	Slight	No
LSO	7.40	65	420	0.040	No	No
LYSO	7.19	64	420	0.050	No	No
GSO	6.71	58	440	0.060	No	Yes
LGSO	6.5	59	415	0.065	Yes	Yes

Source: Lecomte, R., *Eur. J. Nucl. Med. Mol. Imaging*, 36, suppl. 1, S69, 2009.

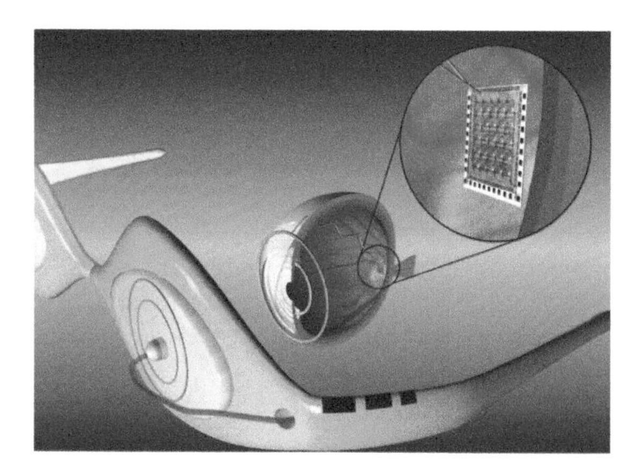

FIGURE 1.1
Example of the eyecam created and tested at the University of Southern California.

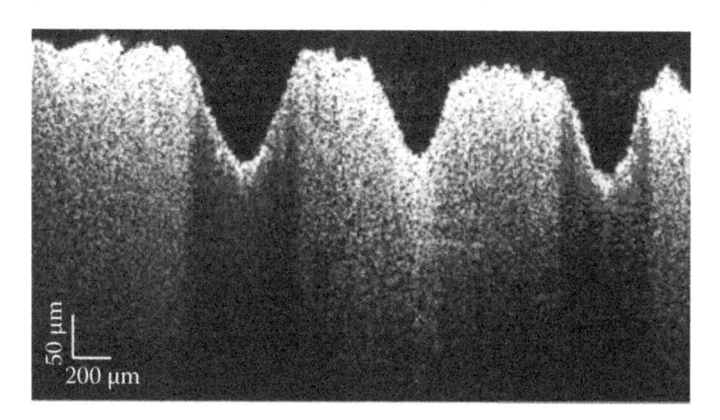

FIGURE 3.18
True color OCT image of a paper with a red, green, and blue line. (Taken from Robles, F.E., Light scattering and absorption spectroscopy in three dimensions using quantitative low coherence interferometry for biomedical applications, PhD thesis in Medical Physics, Duke University, Durham, NC, 2011.)

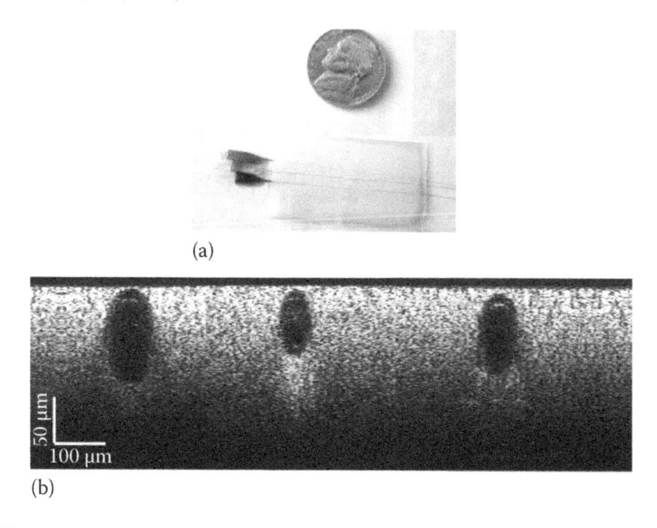

FIGURE 3.19
(a) Photograph and (b) METRiCS OCT image of tissue phantom consisting of intra-lipid and glass capillaries with food-color dyes. (Taken from Robles, F.E., Light scattering and absorption spectroscopy in three dimensions using quantitative low coherence interferometry for biomedical applications, PhD thesis in Medical Physics, Duke University, Durham, NC, 2011.)

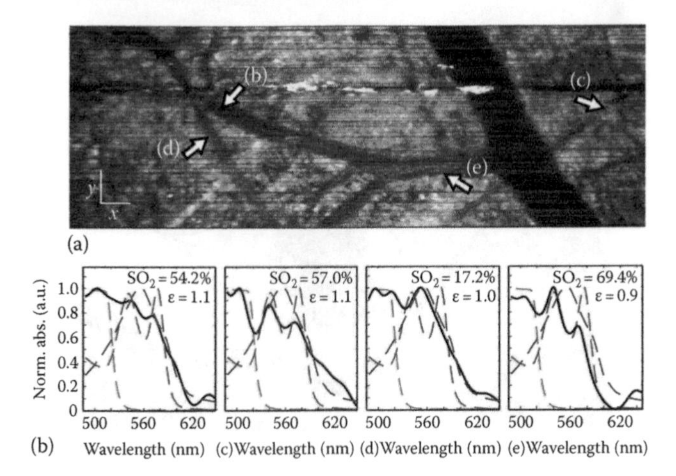

FIGURE 3.26
(a) *En face* (x–y) METRiCS OCT image of mouse dorsal skin flap window using exogenous contrast, and (b) spectral profiles from points indicated by arrows. The white x and y scale bars are 100 μm. The measured spectral profiles (black) are superposed with the theoretical oxy- (dotted red) and deoxy- (dotted blue) Hb normalized extinction coefficients, and normalized absorption of NaFS (dotted green). Also shown are the SO_2 levels and relative absorption of NaFS with respect to total Hb (ε ≡ NaFS/Hb). All spectra were selected from depths immediately beneath each corresponding vessel. (Taken from Robles, F.E. et al., *Nat. Photon.*, 5(12), 744, 2011.)

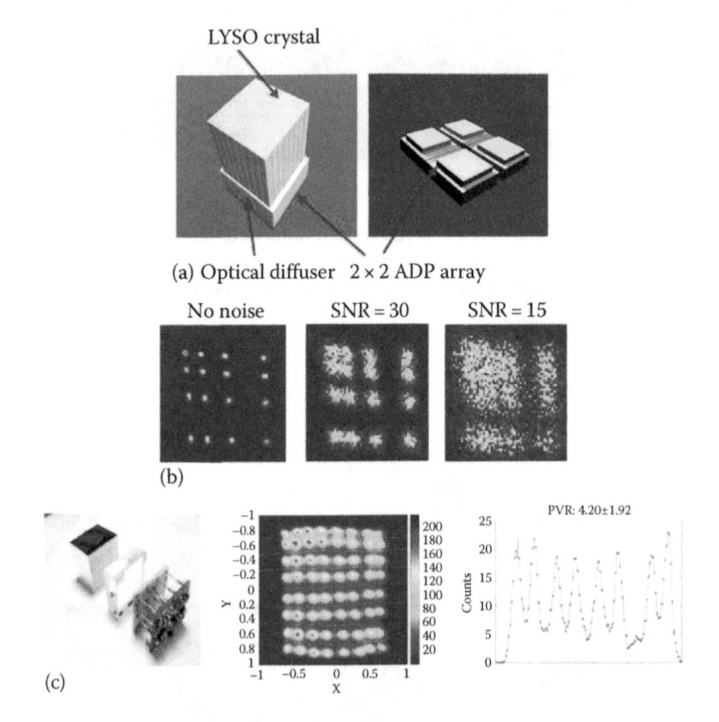

FIGURE 12.4
(a) Anger-logic PET block detector based on light multiplexing. A 2 × 2 array of large-area high-gain APDs was used. (b) Simulated flood maps under different SNR conditions of photodetectors (crystal dimensions: 2.5 × 2.5 × 20.0 mm³). Only one quadrant for the 8 × 8 array was simulated due to the symmetry. Significant degradation of the flood maps is observed as the SNR decreases. SNR refers to the ratio of the amplitude of 511 keV photopeak over the RMS noise of the detector when a single crystal is coupled directly to the photodetector. (c) The block detector prototype comprising the LYSO array (crystal dimensions: 2.75 × 3.0 × 20.0 mm³), the optical diffuser (9 mm thickness), the APD array, and the highly compact custom readout electronics. The crystal flood map and the 1-D profile for the PVR analysis are shown. (Adapted from Peng and Levin, *Phys. Med. Biol.*, 56, 3603, 2011.)

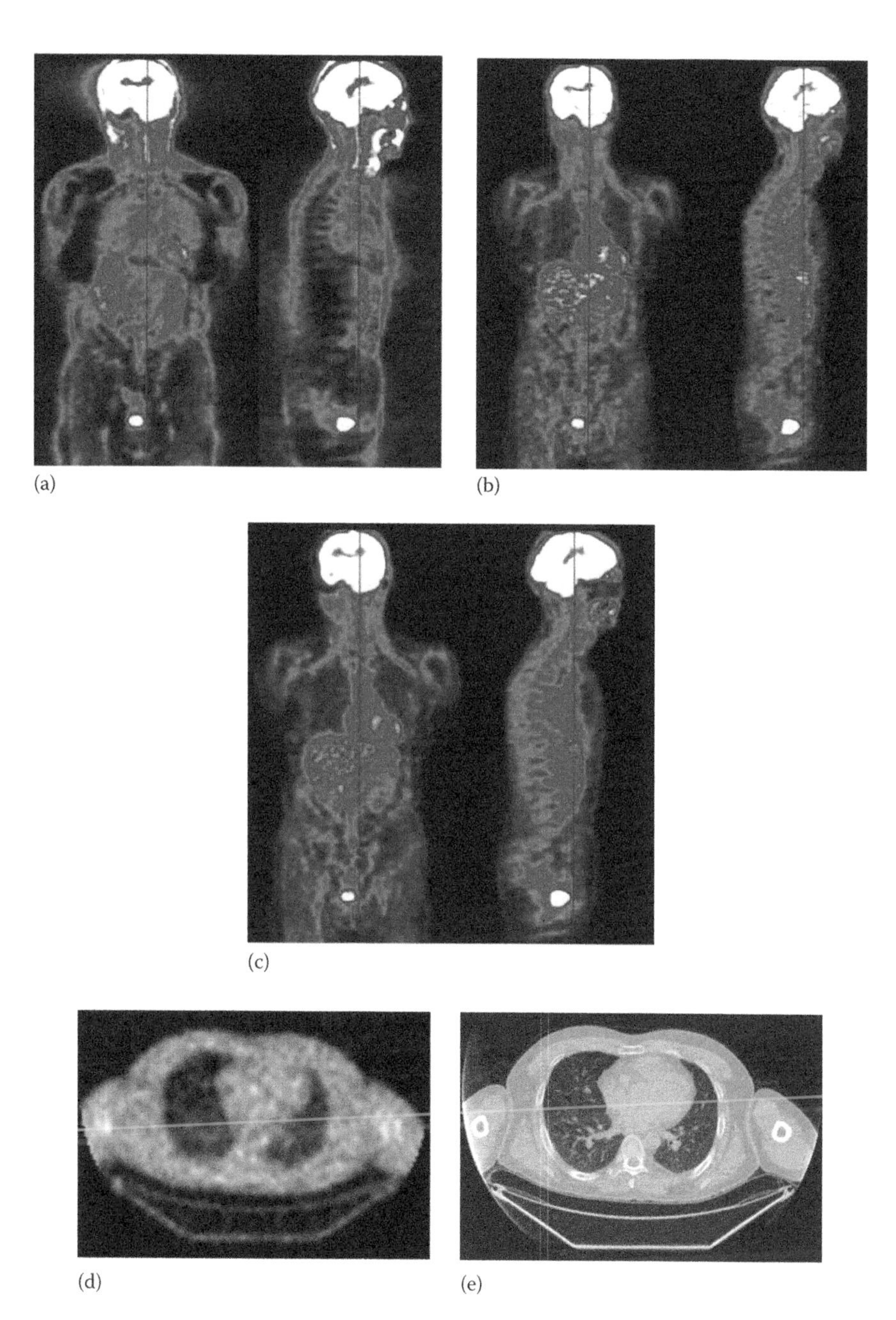

FIGURE 13.1
Example images from an FDG PET/CT whole-body study. Shown is the same patient without attenuation correction (a), with correction using a Ge-68 attenuation map (b), with correction using a CT-based attenuation map (c), an example transverse slice through a Ge-68 transmission map (d), and the same slice through a CT map (e). For the PET images, white indicates highest amount of FDG activity.

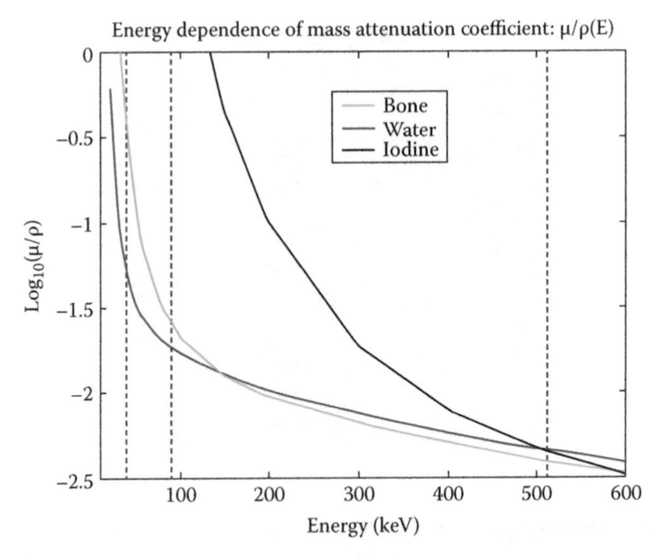

FIGURE 13.2
Mass attenuation coefficients for water (similar to soft tissues), bone, and iodine. The range of energies for an x-ray beam (centered at 70 keV) is indicated by the dashed blue lines at left. The energy for PET imaging (511 keV) is shown with the dashed blue line at right.

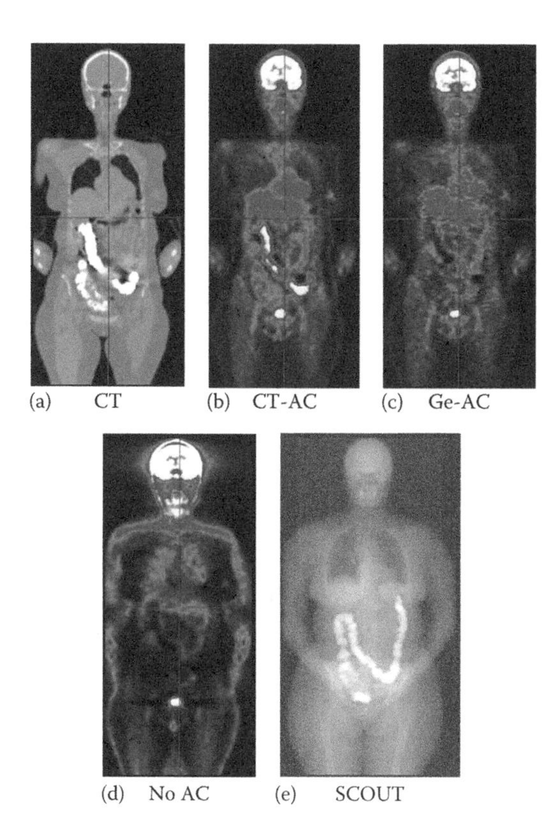

FIGURE 13.3
Example of a whole-body FDG PET study in the presence of oral contrast. A coronal slice through the patient image volume is shown: CT (a), CT-based attenuation correction (b), Ge-based attenuation correction (c), and no attenuation correction (d). An x-ray through the patient is given in (e) clearly showing the oral contrast remaining in the large intestine. For the PET images, white indicates highest amount of FDG activity.

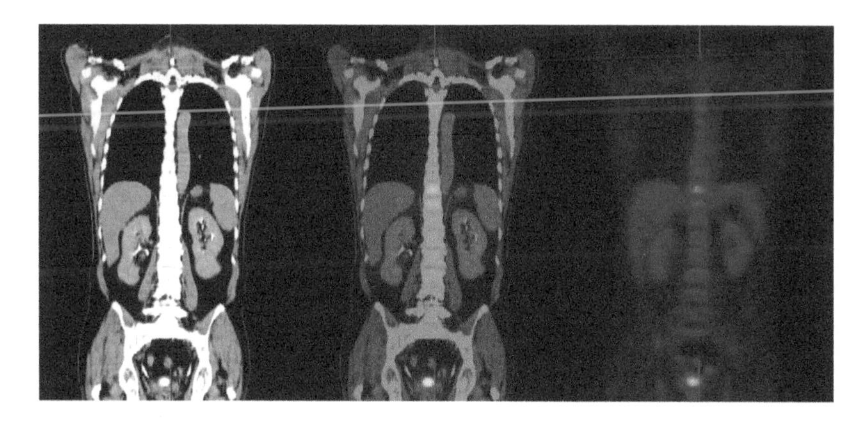

FIGURE 14.2
Alpha blending of PET/CT images shown (from left to right): $\alpha = 0$ (CT only), $\alpha = 0.5$ (equal PET and CT), $\alpha = 1.0$ (PET only).

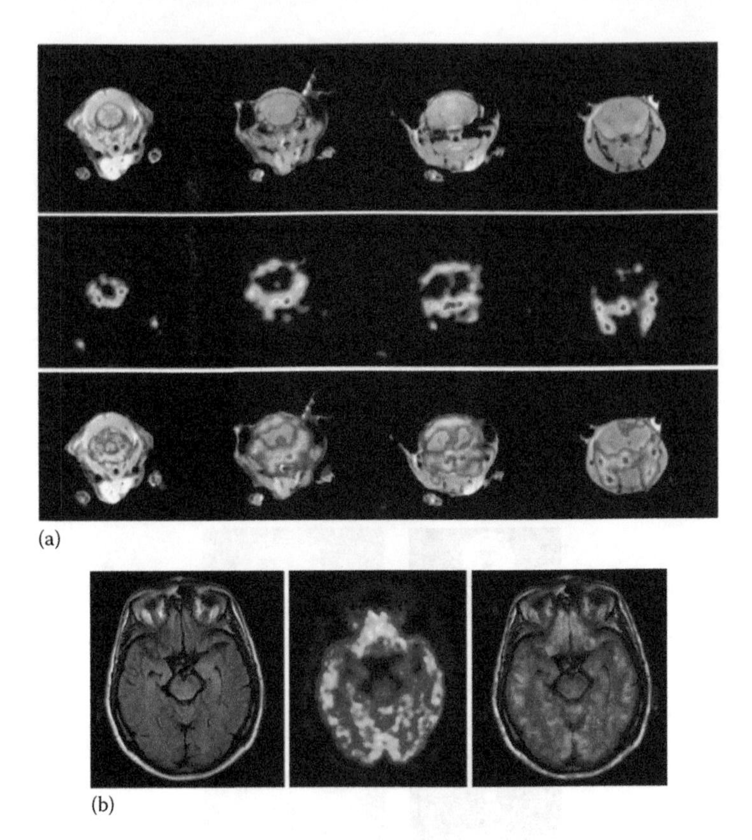

FIGURE 14.11
(a) First simultaneously acquired PET/MR images from the PSAPD PET detector ring developed at UC Davis. The PET image depicts [18]fluoride uptake within the skeletal system. (From Catana, C. et al., *J. Nucl. Med.*, 47, 1968, 2006.) (b) Simultaneous acquired PET/MR image depicting 18FDG uptake in the brain of a human subject. The MR image was acquired using a FLAIR sequence and the entire MR + PET procedure was acquired in 15 min. (From Schlemmer, H.P. et al., *Abdom. Imaging* 2008, http://www.SpringerLink.com)

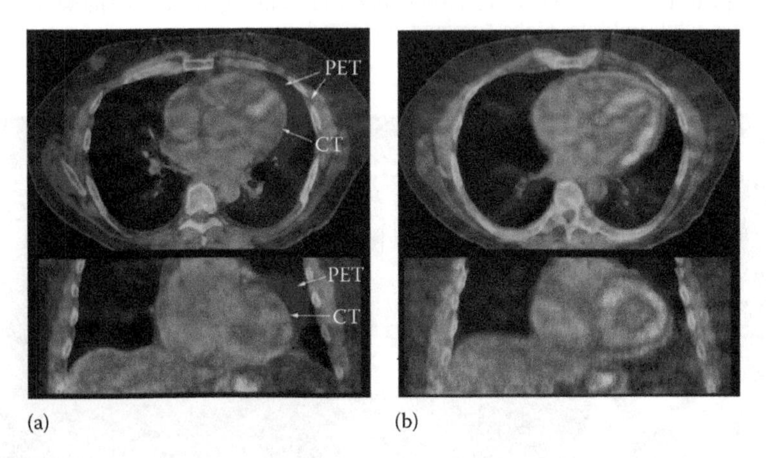

FIGURE 15.2
(a) Helical CT and PET fusion images in transaxial (top) and coronal (bottom) views showed marked misregistration. Arrows indicated the mismatch of heart borders between the helical CT and the PET images, and an artifactual defect in the myocardium. (b) For the same patient, cine average CT and PET fusion images showed no misregistration and artifactual defects. (Reprinted from Gould, K.L. et al., *J. Nucl. Med.*, 2007, 48(7), 1112. With permission.)

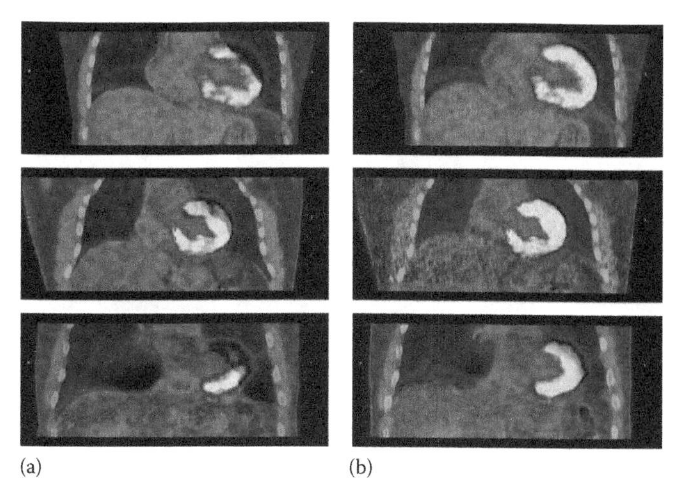

(a) (b)

FIGURE 15.6
Fused coronal images of the PET data with (a) helical CT and (b) CACT for two different patients. (Reprinted from Pan, T. et al., *Med. Phys.*, 2006, 33(10), 3931. With permission.)

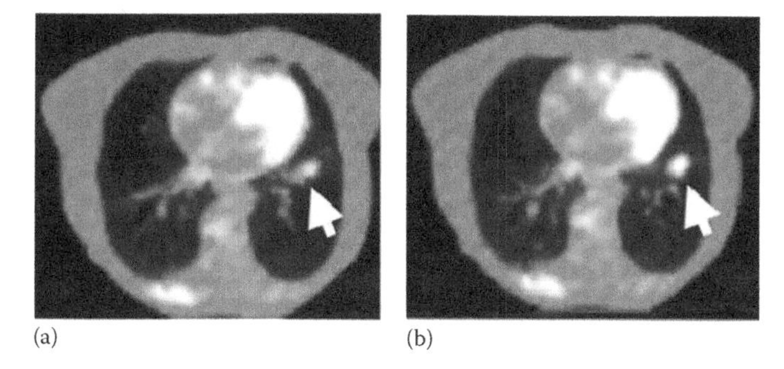

(a) (b)

FIGURE 15.9
Sampled transaxial fused PET/CT images demonstrated a thoracic lesion of a patient in (a) standard nongated PET acquisition, and (b) 4D PET/CT acquisition. (Reprinted from Nehmeh, S.A. et al., *Med. Phys.*, 2004, 31(12), 3179. With permission.)

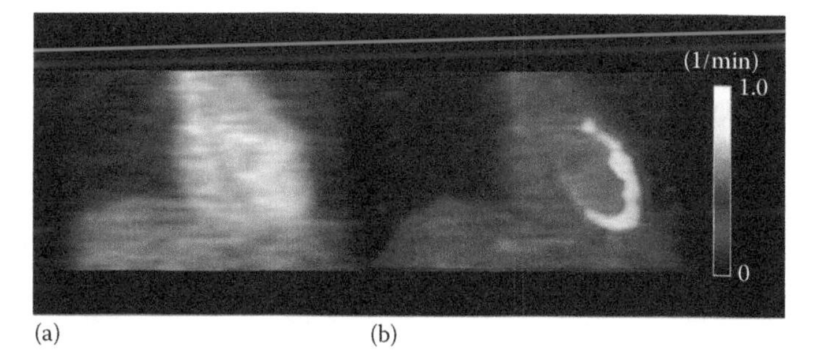

(a) (b)

FIGURE 17.8
Parametric image of K_1 generated from dynamic $H_2{}^{15}O$ PET using linear least squares method. (a) Static PET image. (b) K_1 parametric image superimposed on static image. (Modified from Figure 17.5 in Lee JS, et al., *J Nucl Med*. 2005, 46(10), 1687–1695 © by the Society of Nuclear Medicine, Inc.)

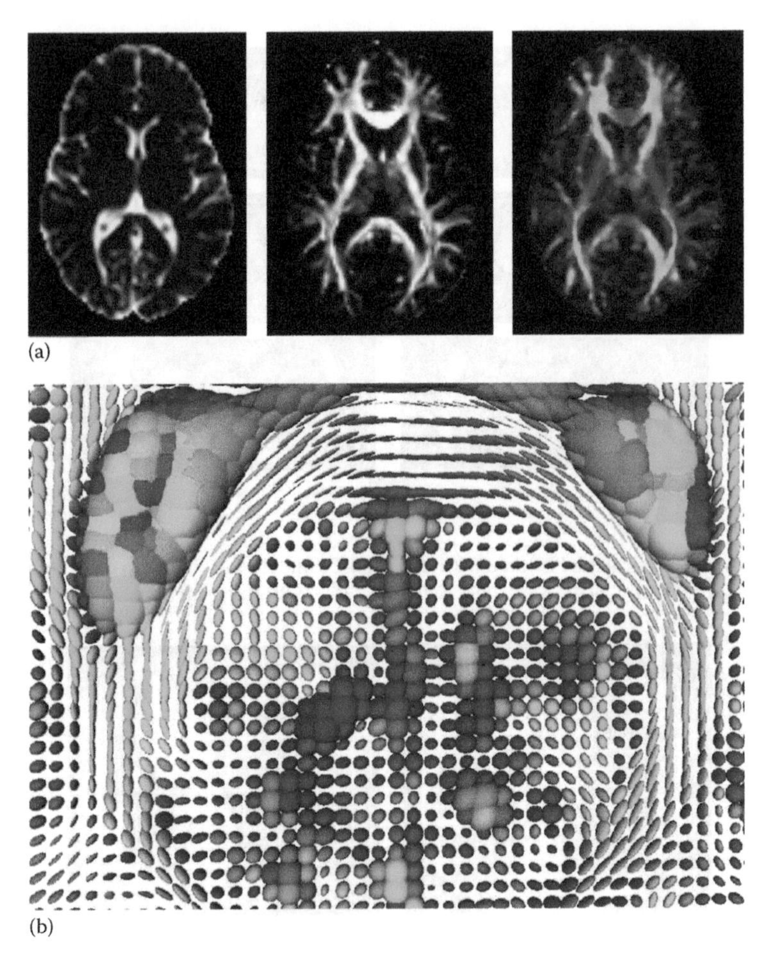

FIGURE 19.6
Various methods of visualizing the information contained in a diffusion tensor field. (a) MD (left), FA (center), and color-coded orientation map (right) and (b) ellipsoidal visualization. (Images generated using MedINRIA http://www-sop.inria.fr/asclepios/software/MedINRIA/ on data obtained from Mori, S., John Hopkins Medical Institute: Laboratory of Brain Anatomical MRI, in vivo human database, http://lbam.med.jhmi.edu/, accessed February 2009.)

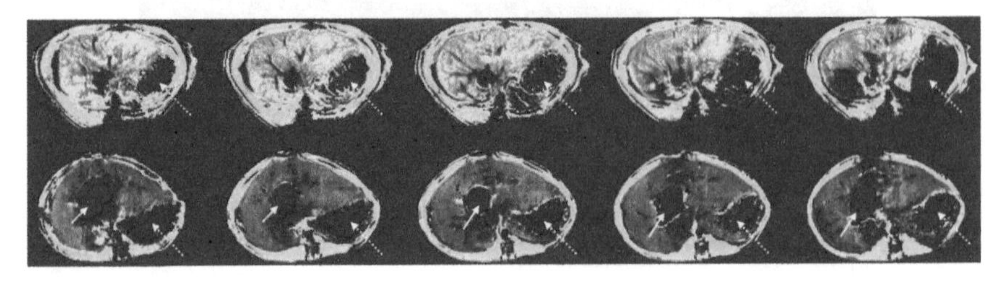

FIGURE 20.9
Color-coded T_{1rho} maps of a sham-operated rat liver (control, upper row) and a biliary duct ligation (BDL) rat liver 24 days postsurgery (lower row). BDL rat liver demonstrates higher T_{1rho} value (brighter) than the control rat liver (darker). Arrow: dilated biliary duct. Dotted arrow: gas in the stomach. (From Wang, Y.X. et al., *Radiology*, 259(3), 712–719, 2011.)

is higher than the binding energy of the electron, an electron will be ejected from the atom with a kinetic energy equal to the difference between the original gamma ray energy and the binding energy. In the Compton effect, the incoming gamma ray is initially absorbed by the scintillator atom, thus producing an electron ionization but also emitting a secondary photon. The energy of the ejected electron and the secondary photon is equal to the original gamma ray. The secondary photon is then free to interact via additional photoelectric or Compton interactions. In both cases, the ionization electrons that are produced give rise to the fluorescent properties of scintillators.

14.2.2 Photodetectors

In order for the radiation to be detected, the light produced in the scintillator material must be detected by a photodetector and converted to a measurable electrical signal. Various approaches exist for this conversion, but the most common is through the use of the photomultiplier tube, or PMT.

Essentially, a PMT consists of an evacuated glass housing with numerous components inside it. These components consist of (1) photocathode, (2) multiplying region, and (3) collector. The photocathode is on the front surface of the PM tube and is responsible for the conversion of incident light into electrons. This conversion takes place via the transfer of energy from the incoming photon to the material electrons. For most photocathode materials, this requires approximately 2–4 eV of energy for each liberation. Once liberated, these electrons must now transit the photocathode material, be emitted from the photocathode, and be accelerated to the first detection stage. In order to be emitted, the electrons require at least enough kinetic energy to escape the potential barrier holding the electrons to the material (i.e., the work function, typically a few eV) [31].

As a result of the energy required to liberate electrons (the work function), most photomultipliers are more sensitive to higher-energy light photons (i.e., shorter wavelengths) and in fact are optimal for scintillators like NaI(Tl) or Lutetium Oxyorthosilicate (LSO) that scintillate in the blue/green region of the visible light spectrum. After the electron is liberated, it must travel to the cathode surface in order to be ejected. Typically, electrons will only travel a very short distance (several nm) before being reabsorbed so it is imperative that the photocathode be kept as thin as possible. However, in keeping the cathode thin, it becomes less efficient in absorbing the photons from the scintillator in the first place. The efficiency in converting scintillation light to photoelectrons is deemed the quantum efficiency (Q.E.) and is typically in the range 20%–30% for modern PMTs [31].

At the heart of a PMT is the electron multiplication stage. An electron emitted from the surface of the photocathode will typically have very low kinetic energy (on the order of a few eV). If an electrode (the dynode) placed near the photocathode is held at a high positive potential relative to the photocathode (say, 100 V), then the emitted electron will be accelerated across the gap, thereby acquiring a kinetic energy of 100 eV en route. This electron will then collide with the dynode resulting in more electrons being liberated. As the creation of each electron requires sufficient energy to overcome the bandgap (2–3 eV), it is possible that the original 100 eV electron will ionize several electrons within the dynode material. However, not all of these electrons will have enough energy to traverse the dynode material and subsequently make it to the surface. As a result, only a small fraction of the electrons liberated will retain enough energy to be ejected from the dynode surface. When another dynode is placed in close proximity to the first dynode and again held at a positive potential relative to the first dynode, then the ejected electrons will be accelerated across this gap and will impact the next dynode, whereupon this process repeats itself. A typical PMT may have 6–12 such dynode stages, with each dynode stage increasing the number of electrons produced. At each dynode stage, the number of low-energy liberated electrons can be represented by δ. Thus, after N stages, the overall photomultiplier gain is given by

$$\text{Gain} = \alpha\delta^N$$

where α is the overall multiplier tube efficiency (i.e., the number of detected photoelectrons per number of emitted photoelectrons). It is not uncommon for modern PM tubes to achieve gains on the order of 10^6–10^7. It is also worth noting that the PMT gain is a function of the applied electrical potential. As a higher voltage is applied between the dynodes, more secondary electrons are produced when the incoming electron ionizes the dynode material and more of these secondary electrons are accelerated toward the next dynode. Additionally, each electron acquires more kinetic energy en route. So, while more electrons are incident upon the dynode, they penetrate the dynode material to a greater depth, thus making it more difficult for the secondary electrons to escape the material. Additionally, at typical room temperatures, there is a finite probability of an electron being spontaneously emitted from the photocathode even though a scintillation event did not occur. When operated with high operating bias, this single electron may result in a large photocurrent at the anode that may be mistaken for an event.

The final dynode of a photomultiplier is called the collector anode and has the highest potential difference between all the dynodes and the photocathode. At such, it represents the end of the line for secondary electron emissions. Between the anode and the high voltage, a load resistor is used to drain the photocurrent generated through the PMT.

As described, the successive PMT dynodes must maintain a positive potential difference between them in order that secondary electrons get accelerated from the current dynode to the next dynode. These voltages are typically configured in one of two ways:

1. *Positive bias*: In this case, the photocathode is held at zero potential and each successive dynode has a corresponding higher positive bias applied, with the full bias applied to the anode (see Figure 14.4).

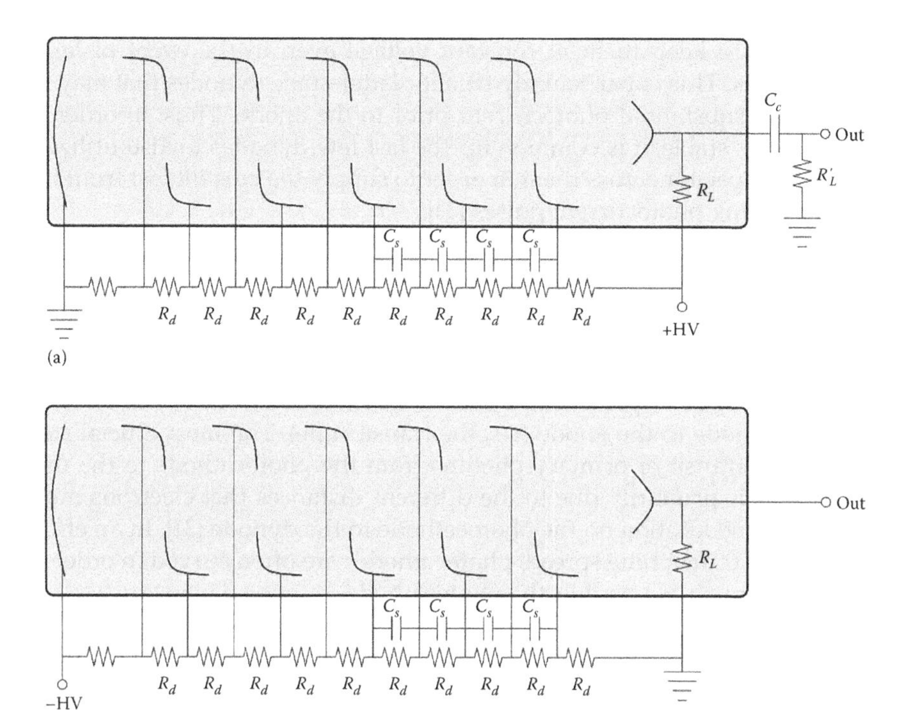

FIGURE 14.4
Schematic depicting the layout of a conventional PMT. The PMT is shown in the positive bias (a) and negative bias (b) configurations. The photocathode is shown on the left in both cases. (Adapted from Knoll, G.F.: *Radiation Detection and Measurement*. 3rd edn. 2000. Copyright Wiley-VCH Verlag GmbH & Co. KGaA.)

This is most usually applied via a resistive divider network that separates each dynode with a resistor. Because the anode voltage is high, the resultant photocurrent pulse ride will ride atop a large DC bias. To remove this bias, the anode output is then AC coupled by means of a capacitor placed in parallel with the load resistor. AC coupling ensures that only the photocurrent passes through the capacitor and not the underlying high-voltage bias.

2. *Negative bias*: In order to eliminate the need for AC coupling, it is instead possible to negatively bias the PMT (see Figure 14.4). In this configuration, instead of the anode being held at high positive bias, the photocathode is held at a high negative voltage. Again, each successive dynode is held at a higher positive voltage with the anode held at ground. In this way, the resulting photocurrent pulse is riding atop an inherent 0 V bias, thus negating any need for AC coupling.

With either bias configuration, it is important that the PMT current along the resistive network be kept large enough to supply the dynodes with enough bias to keep them at constant voltage even in the event of large photocurrents. This is particularly true for latter-stage dynodes that may be producing a substantial photocurrent prior to the anode. Thus, in order to keep this bias stable, it is common for the last few dynodes to also utilize a stabilizing capacitor across them in order to supply the current lost from the dynodes during photocurrent pulses [31].

When secondary electrons are liberated from each dynode, they typically have very low energy and are accelerated via the applied potential difference to the next dynode. This acceleration requires a finite amount of time for each electron as each electron will have a different amount of kinetic energy when ejected. Over the course of all electrons progressing through the dynodes, there will be a spread in the time it takes for electrons to transit from the photocathode to the anode (i.e., the transit time). The most crucial step is the initial transit of primary photons from the photocathode to the first dynode and is primarily due to the different distances that electrons must travel from one location on the photocathode to the dynode [31]. In an effort to reduce the transit time spread, photocathodes are often curved in order to equalize, as much as possible, the photocathode–dynode distance as well as incorporating focusing electrodes to guide the electrons to the dynode.

As PMTs utilize vacuum tubes and streams of low-energy accelerated electrons, they are inherently susceptible to magnetic fields. When placed in the vicinity of even low-strength magnetic fields, the electrons traversing from dynode to dynode are subject to magnetic forces that may affect the electron trajectories. Such effects have been seen even for small magnetic fields on the order of 10 Gauss [32], thus making PMTs impractical to utilize for photodetectors in conventional MRI devices. In an effort to reduce the effect of small magnetic fields, PMTs are usually wrapped in a thin magnetic shield such as Mu-metal [31].

14.2.3 Acquisition Electronics

The electrical signal output from a PMT, while amplified through the electron-multiplying process, is still rather small and must be further amplified and shaped [31]. A typical PMT output consists of a sharp drop in voltage (several nanoseconds) when the scintillator initially fluoresces followed by a longer tail, usually lasting several hundred nanoseconds. The amount of photocurrent produced by a photomultiplier is proportional to the initial photon energy and even though amplified through the PMT, the signal is still generally too small to be measured reliably. Thus, the output from the PMT is passed through a preamplifier (pre-amp) in order to yield an output signal that is proportional to, but greater than the input signal. This is usually accomplished through the use of either a voltage-sensitive (VSA) or charge-sensitive preamplifier (CSA). In order to overcome the capacitive loading of the connecting wires, it is usually located as close as possible to the photodetector.

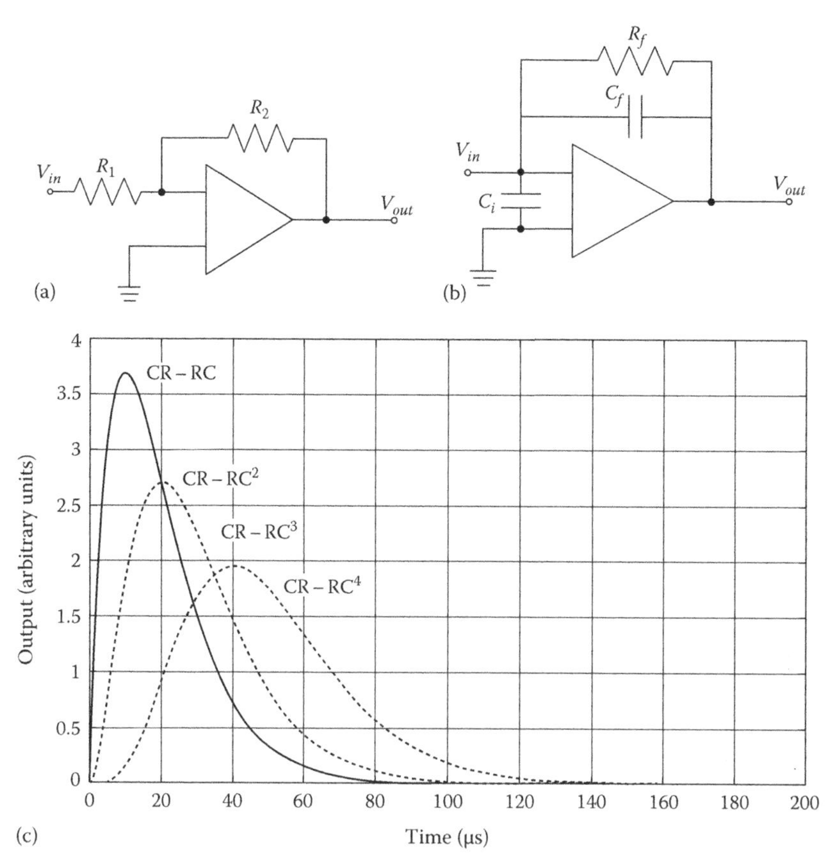

FIGURE 14.5
(a) Schematic of a VSA. The output V_{out} is proportional to the input signal V_{in} multiplied by ratio R_2/R_1 but is dependent upon the detector capacitance (not shown). (b) Schematic of a typical CSA. The output signal V_{out} is now independent of the detector capacitance and decays with time constant R_fC_f. (c) The effect of CR–RC^n pulse shaping on preamplifier signals. With more RC stages, the output pulse becomes more symmetric but with a longer rise and decay time.

A typical VSA is shown schematically in Figure 14.5a and consists of a basic inverting amplifier. Given an AC-biased input, the maximum voltage across the bias capacitor on the input side is given by

$$V_{\max} = \frac{Q}{C}$$

After passing through the amplifier, the input voltage is inverted and scaled by the ratio of the resistors such that

$$V_{out} = -\frac{R_2}{R_1} V_{in}$$

Such a preamplifier design works well for detectors whose capacitance does not change, such as PMTs, but for semiconductor detectors whose capacitance may change under different operating parameters, the ever-changing relationship between Q and V_{max} makes consistency difficult to achieve.

In contrast to the VSA, a CSA is shown schematically in Figure 14.5b. In this configuration, the output from the photodetector is first coupled through an input capacitor, C_i, before being passed through the amplifier. An input charge builds up on the feedback capacitor, C_f, and is dissipated through the feedback resistor, R_f. The subsequent output voltage through the CSA is thus

$$V_{out} = \frac{Q}{C_f}$$

In this scenario, the resultant output voltage is only a function of the integrated charge, Q, produced in the photodetector. Under realistic circumstances, the output pulse V_{out} will rise very quickly to a maximum that is proportional to the detected charge, followed by a slow decay described by the time constant R_fC_f. It is common for the initial rise time to be several ns long but the ensuing decay to proceed for several hundred microseconds.

As a result of these long tails, it is possible that in the presence of high count rates, each ensuing output pulse is added to some residual of the previous pulse. Thus, the detected absolute output voltage may be severely overestimated. If the count rate is high enough, it is possible that the output voltage will be above the saturation level of the amplifier thus making each detection appear with the same magnitude.

As it is only the height of the initial rise that is important, it is common to employ additional circuitry to provide pulse shaping to extract only this portion. Examples of such shapers include CR–RC circuits or Gaussian shapers. If the output pulse from a preamplifier (essentially a step function) is fed through a simple CR circuit, the output is a differentiated signal, equal to

$$V_{out} = V_{in}e^{-t/\tau}$$

where τ represents the time constant, RC. Similarly, if an input is directed through an RC circuit with the same time constant, τ, the output is an integration of the input equal to

$$V_{out} = V_{in}(1 - e^{-t/\tau})$$

If the preamplifier pulse is fed through the CR–RC circuit in series, the output signal is equal to

$$V_{out} = V_{in}\frac{t}{\tau}e^{-t/\tau}$$

By combining multiple RC stages, the net result is a CR–(RC)^n pulse shaper of which the output is described by

$$V_{out} = \frac{V_{in}}{n!}\left(\frac{t}{\tau}\right)^n e^{-t/\tau}$$

This type of shaper reduces the output tail further and achieves a Gaussian-shaped output function after the addition of only about 4 RC stages. While the output becomes more Gaussian in shape (as shown in Figure 14.5c) and the tails decay to baseline faster than with a standard CR–RC shaper, the initial rise time does become longer, thus limiting the peak counting rate.

In PET and SPECT imaging, in order to measure high count rates, it is common to clip the incoming PMT pulse while the scintillator is still scintillating. When clipped, the preamplifier will only integrate over a short time period and accumulate only a portion of the total charge produced. While the use of pulse clipping can improve the overall count rate capability of the detector, the trade-off is a larger noise component in the amplified signal due to less than complete charge being accumulated.

In current PET and SPECT designs, it is common for each shaped signal to then be digitized via an analog to digital converter (ADC). It is normally sufficient to use 8 or 10 bit flash ADCs for this task. In order to distinguish actual events from background noise, a summing and triggering circuit is used. Signals from all PMTs are passed through a summing circuit in order to determine the total signal produced during a scintillation event. The total signal is then compared to a preset threshold (situated well above background noise levels) and if the event signal is above this threshold, a trigger signal is sent to each PMT channel, thereby initiating the ADC conversion. Within each channel, a peak detect and hold circuit latches onto the peak of the shaped pulse and holds it until sampled by the ADC. Once sampled, the peak detect circuit is reset back to baseline to await the next pulse.

For each scintillation event, the location of the event must now be determined. SPECT imaging typically uses a large, monolithic scintillator coupled to an array of many photodetectors. For each scintillation event, multiple photodetectors will sense the light produced and by using all the signals, the (X,Y) location of the scintillation can be determined. This is typically performed using an Anger logic-type algorithm or some other variant whereby the scintillation location is determined via a weighted averaging in both the X and Y locations depending on the detector position, such as

$$X = \sum_{i=0}^{N} \alpha_i V_i \quad \text{and} \quad Y = \sum_{i=0}^{N} \beta_i V_i$$

where α and β are weighting factors for the ith digitized signal.

PET scanners typically consist of multiple modular detector blocks with each block comprised of a 2 × 2 array of PMTs coupled to a scintillator. For each scintillation, the light is shared among only 4 PMTs, thus making the localization somewhat simpler. Using the four digital PMT outputs, the (X,Y) location of the scintillation can be determined via the equations

$$X = \frac{(A+B)-(C+D)}{(A+B+C+D)} \quad \text{and} \quad Y = \frac{(A+C)-(B+D)}{(A+B+C+D)}$$

where A, B, C, and D correspond to the four PMT outputs, respectively.

The total gamma ray deposited energy (E) is simply determined by summing all the PMT signals. Once the (X,Y) locations are determined, it is then possible to display a histogram of detected positions on a computer monitor. In the case of SPECT, the detector then needs to rotate around the patient in order to acquire angular projections from which the 3D distribution of radioactivity can be determined. In the case of PET, an entire ring of detectors is positioned around the patient and acquires all possible projection angles simultaneously without any detector motion.

14.3 Imaging with MRI

14.3.1 Basics of MRI

MR imaging uses a high, static magnetic field ($\mathbf{B}_0$) in order to align the proton magnetization vector within the object. Typically, this magnetic field is on the order of 1.5–3 T and is produced using a superconducting magnet. This magnet is usually cylindrical with an open bore of approximately 70 cm diameter and 2 m long, although some low-field, open-gantry systems consist of a flat permanent magnet.

Within the bore is a set of three additional magnetic field-producing coils. These coils are the gradient coils $\mathbf{G}_x$, $\mathbf{G}_y$, and $\mathbf{G}_z$ and are used in conjunction with the main magnetic field to alter the local magnetic field within the bore. Also within the bore is a set of RF coils. These coils are used to transmit an RF pulse to the protons of the object, causing the magnetization vector to change as a result of the applied magnetic field. As the proton magnetization decays back to its normal state following RF excitation, a subsequent RF pulse is emitted from the protons, which is detected by the coil. By using a series of RF pulses in conjunction with altering the magnetic gradients within the bore, images can be obtained depicting localized properties of the magnetization vector. The RF coils may transmit and receive, transmit only, or receive only. In the case of transmit-only coils, a receiver set of coils must also be used to detect the emitted RF signal from the protons. It is important to note that RF coils should be placed as close as possible to the object being imaged in order to obtain the highest signal to noise ratio (SNR). This factor will become important when considering where PET or SPECT inserts can be placed within the bore.

MRI is able to distinguish tissues based on their proton density or magnetization properties such as their spin–spin (T2) or spin–lattice (T1) relaxation times. As these properties vary in the proximity of different tissues, MR images can depict subtle changes in tissues even though the tissue density does not vary much. Given the improvements in soft-tissue contrast in MRI compared to CT and the complementary functional information obtained, it seems natural to combine MR imaging with nuclear medicine imaging. In fact, combining PET or SPECT with MR images has been performed rather routinely using separate imaging systems and software co-registration methods. However, given the success of combined PET/CT and SPECT/CT, the combination of PET/MR or SPECT/MR in a single imaging gantry would seem to offer several advantages over separate imaging systems, not least of which is the improved diagnostic value of perfectly co-registered images. As we will see, however, the development of PET/MR or SPECT/MR is not necessarily as straightforward as the development of PET/CT or SPECT/CT.

The biggest hurdle with simply grafting a PET or SPECT machine onto an MR gantry is that the radionuclide imaging system must operate in the vicinity of a high magnetic field strength (typically 1.5–3 T but possibly upward of 9–14 T for animals). The high magnetic field creates problems for conventional nuclear instrumentation and so alternative techniques must be used. Some attempts have focused on using alternative MR configurations, while other attempts have investigated alternative nuclear detectors. We will investigate some of these approaches in the following section.

14.4 Alternative MR Configurations

In order to achieve higher spatial resolutions with higher SNR, there has been a push in MR imaging for higher and higher magnetic field strengths. In order to achieve these high magnetic fields, most MR machines utilize cryogenic superconducting magnets, making it impossible to cycle MR magnets on and off in order to provide sequential imaging with PET or SPECT. However, by using alternative approaches to generate the magnetic fields needed, it becomes possible to perform co-registered sequential nuclear/MR imaging.

14.4.1 Low-Field MR

As mentioned previously [32], even small magnetic fields can have a significant effect on the performance of PMTs. As such, conventional PMT-based radiation detectors cannot be placed in the vicinity of high-field MR scanners. Goertz et al. [33] however have taken the approach of using a non-superconducting, low-field MR (0.1 T) in conjunction with a single detector NaI(Tl) SPECT camera in order to conduct sequential SPECT/MR imaging in small animals. While resulting in lower spatial resolution and lower SNR compared to high-field MR, the advantage of using a low-field MR lies in the fact that the 5 Gauss line is situated only 15 cm from the edge of the main magnet, thereby making it possible to place a conventional PMT-based SPECT camera at this location. In this configuration, the animal is first imaged using the pinhole SPECT camera then manually moved the 15 cm or so to the MR system. Since the animal remains in the same imaging cell for both scans, any deviation in positioning can be remedied with a relatively simple rigid-body transformation. In fact, co-registration parameters between the MR and SPECT images were first determined using a three-tube geometric phantom and for all successive imaging experiments using animals, the SPECT scans were co-registered to the MR images using the same set of registration parameters.

While the low-field MR was not capable of providing high-resolution anatomical imaging or spectroscopy information, the spatial resolution and soft-tissue contrast was deemed of sufficient quality to be used for anatomical co-registration with SPECT. As well, it was pointed out that if the only requirement of the MR is for SPECT co-localization, then low-field MR is more than adequate for this purpose.

14.4.2 Field-Cycled MR

As another alternative to cryogenically maintained static magnetic fields, research has focused on using field-cycled MR [34–36]. In conventional MR, the main static magnetic field is used to both induce the magnetization within the object and create the magnetic environment for the transverse magnetization to precess. Thus, the field strength affects the amount of magnetization generated in the sample and must be kept as uniform as possible.

In contrast, the field-cycled MR consists of two nonsuperconducting magnets, one large field for generating the polarizing magnetization and a smaller homogeneous magnet for readout. The large field magnet is first applied creating a net magnetization in the object. After a short time (on the order of 1 s), the polarizing magnet is turned off, and the readout magnet is turned on, again for a short time. When the readout magnet is turned on, the net magnetization produced by the polarizing magnet will precess at the Larmor frequency while it decays. During this time, RF pulses and gradient fields are applied as in conventional MRI in order to gather the resonance signal.

Since the field-cycled MR does not utilize a static magnetic field, it is possible to incorporate conventional PMT-based radiation detectors into the gantry [36]. These detectors would not be operated simultaneously with the MR, but rather acquisitions would be interleaved between the MR pulse sequences when the main and readout magnetic fields are cycled off.

14.5 Alternative Radiation Detectors

Recent developments in radiation detection have made it possible to consider alternatives to the conventional scintillator/PMT arrangement.

14.5.1 Scintillation Detectors

In many respects, scintillation detectors still offer many advantages over other radiation detection methods. Reasonable cost, good energy resolution, and high detection efficiency are just some of the advantages scintillation detectors offer. As well, if appropriate scintillators are chosen, their relative insensitivity to high magnetic fields enables them to function properly in the presence of high magnetic fields. By the same token, the high electrical resistance of most scintillators reduces the effect of eddy currents within the detector material, thus reducing the influence of the scintillator on the magnetic field. For these reasons, scintillating detectors are still a good option for combined radionuclide/MR imaging. However, care must be taken to either place the photomultipliers at sufficient distance from the main magnet or to use alternative, magnetic-compatible photodetectors.

14.5.2 Fiber-Optic Coupling

The magnetic field at the center of most clinical MRI systems is typically between 1.5 and 3 T; however, this decreases rapidly with distance such that by ~3 m from the center, most magnetic fields are less than 10 mT [37]. Thus, if the sensitive PMTs and electronics of a PET or SPECT device can be placed in this lower-field area, then it would permit simultaneous imaging. This was the approach taken in the development of the first simultaneous PET/MR detectors [37–39].

The McPET I was a single, 38 mm diameter ring consisting of 48 – $2 \times 2 \times 10$ mm LSO scintillators with each crystal coupled to an individual pixel of a multichannel photomultiplier. This detector was integrated between the poles of a 0.2 T vertical field MRI. In order that the magnetic field exhibit negligible effect on the photomultipliers, the PMTs were placed ~3 m from the magnet and connected to the scintillators via a 4 m long and 2 mm diameter optical fiber. At this distance, the magnetic field strength was measured to be less than 0.1 mT.

As some light will inevitably be lost during transit through the fiber optic, the significantly greater light output of LSO compared to BGO lends itself well as the scintillator of choice even though the detection efficiency is slightly less. As the crystals were oriented with their long axis parallel to the axial direction, detection efficiency was somewhat reduced as only a 14% detection efficiency was realized, thus contributing to an energy resolution of 41%. Timing resolution for this system was also somewhat high at 20 ns. Nevertheless, this system was able to show that simultaneous PET and MR imaging was indeed possible using separated scintillators and photodetectors and that no noticeable distortion effects were seen in either the PET or MR images.

This system was further developed into the McPET II, which utilized 72 – 2 × 2 × 5 mm crystals oriented radially in a single slice and coupled to the same photodetection subsystem. By reorienting the crystals, an improvement in detection efficiency to 34% was obtained with similar energy resolution (45%) and timing resolution (26 ns) to the McPET I. This system was tested in magnetic fields up to 9.4 T without showing any noticeable effects on the PET images.

While the use of fiber-optic coupling showed the feasibility of simultaneous PET and MR imaging, a number of limitations are present with this approach. The attenuation of scintillator light output along the long optical fibers reduces the energy resolution and coincidence timing. As a result of the issues with using PMTs for photodetection, there has been increased interest in alternative types of photodetectors [40].

14.5.3 Photodiodes

Photodiodes are semiconductor photodetectors that convert incident light into electrical current proportional to the intensity of the incoming light. As such they have replaced PMTs in some applications [41,42]. Photodiodes consist either of PN-type (*p*- and *n*-type semiconductor layers) or PIN-type materials (*p*- and *n*-type layers with a depleted *i*-type region between them). Most photodiodes used in conjunction with scintillators are of the PIN type due to their superior sensitivity and responsiveness. Compared to PMTs, photodiodes offer higher intrinsic quantum efficiency, low power consumption, low operating bias, and insensitivity to magnetic fields.

As light photons enter the *p*-layer, electron/hole pairs are produced and collected on the boundary layers of the *i*-type region, driven there by an applied reverse bias across the PN contacts. Once at the anode, the electrons contribute to a small amount of current. Through the collection of many light photons, several electron/hole pairs are liberated and a small but detectable photocurrent is produced.

The inherent Q.E. of typical PIN photodiodes is 50%–70%, several times higher than PMTs [31]. However, photodiodes typically have a different response curve as a function of wavelength compared to photomultipliers.

Most PMTs have a peak Q.E. around 400 nm, thus making them suitable for use with common scintillators such as NaI(Tl), LSO, or BGO. However, photodiodes typically have responses peaked more toward the red end of the visible light spectrum, making them more efficient with scintillators such as CsI(Tl).

It is worth noting that for each visible light photon incident on the photodiode, a maximum of only one electron/hole pair will be generated (theoretical Q.E. of 100%) and so the conventional photodiode is a unity gain device. As most scintillation events only generate a few thousand scintillation photons, the resultant charge buildup is quite small for a single scintillation. Even so, with a low-noise CSA, it is often possible to detect the resultant charge buildup from a single event over the background detector noise.

With the small level of signal produced in photodiodes when operating in pulse mode, they are extremely sensitive to electronic noise. The two most important factors contributing to noise are the *photodiode capacitance* and the *leakage current*. Because of the PN junction, a photodiode has an inherent capacitance that increases with area but decreases with thickness. It is typical that photodiodes have capacitances on the order of 20–50 pF/cm^2 although some may be as high as 300 pF/cm^2. Because photodiodes are semiconductors, they will always show a small amount of conductivity as a result of thermal ionizations. This leakage current increases with temperature as well as with increasing thickness. Small fluctuations in the leakage current give rise to jitter that may obscure the small signal from a legitimate photo event. As the leakage current decreases with temperature, it is common for photodiodes to be actively cooled in order to reduce the dark current enough so as to be able to detect photo events.

In order to limit the device capacitance and dark current, photodiodes are typically only available in small sizes (typically less than 1 cm^2). Thus, it is tempting to consider using arrays of photodiodes coupled to large-area scintillators in order to replace conventional PMTs. The problem with this, as pointed out by Groom [31,43], is that as the light collection increases with photodiode area, so too does the inherent noise and as a result, very little gain in SNR is actually obtained. As a result, most applications of photodiodes as radiation detectors use individual scintillation crystals coupled to individual photodiodes [44]. With this design, most of the light produced in the scintillator interacts within a single photodiode, thus maximizing the SNR. Because of the relatively small size of photodiodes compared to the effective area of most imaging devices, designs incorporating photodiodes as photodetectors typically require the use of multichannel digital electronics and parallel processing. However, as a result, these devices do not suffer from the same dead-time considerations as conventional monolithic scintillators and Anger logic detectors.

The Digirad 2020Tc is the first commercially available photodiode-based gamma camera. This camera uses a 64×64 array of CsI(Tl) detectors (3×3 mm) with each detector pixel coupled to a PIN diode. Since there is no sharing of the scintillation light across multiple diodes, this camera does not suffer from excessive noise on each detector channel, although the system is also actively cooled to reduce background noise. It is noted that although this system uses PIN diodes, it has not been designed to operate in the vicinity of an MR.

14.5.3.1 Avalanche Photodiodes

As mentioned, photodiodes are unity gain devices and suffer from poor SNR when detecting the small signals produced in scintillation events. To mitigate this problem and achieve some level of signal amplification, the *avalanche photodiode* (APD) was developed. A typical APD device is shown in Figure 14.6a. In essence, an APD consists of a PN semiconductor, with a drift region between the p and n junctions. When a high reverse bias is applied across the junctions, it sets up a nonlinear electric field through the multiplying region. When exposed to light, electron/hole pairs are created in the drift region of which, because of higher electron mobility, results in the

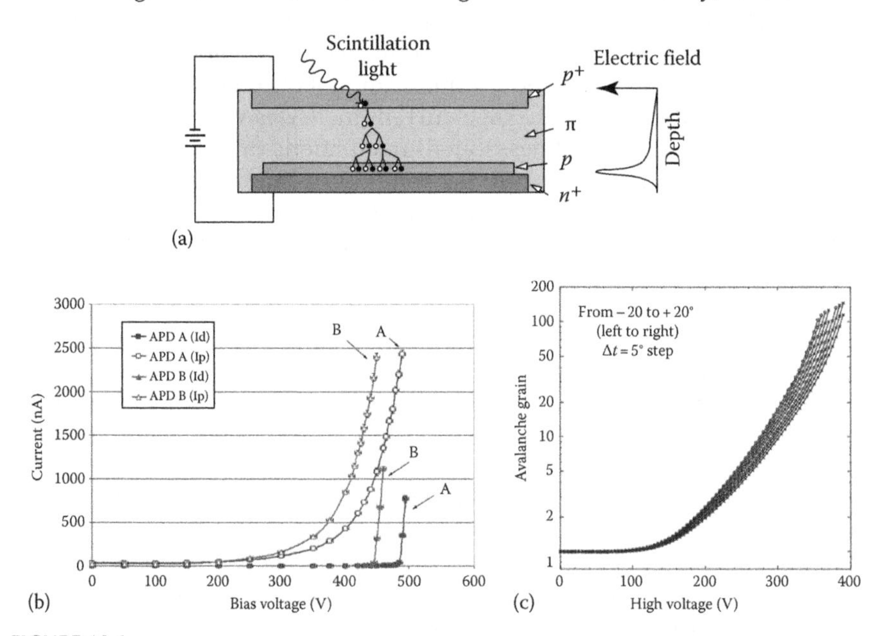

FIGURE 14.6
(a) Representation of a standard reach-through APD. Incoming light photons liberate electron/hole pairs in the semiconductor material that then are accelerated through the multiplying region, thus liberating more electron/hole pairs. The subsequent collection of electrons at the anode is amplified through the avalanche process. (b) Plot of gain vs. bias for two silicon avalanche photodiodes (Hamamatsu S8550–1010). Note the breakdown voltage is different for both diodes even though the model is the same. (c) Plot of gain vs. operating bias for a silicon APD (Hamamatsu S8550–55). In addition to an increase in gain at lower operating bias when cooled, the thermal noise content is significantly lowered when cooled, thus increasing the SNR. (From Kataoka, J. et al., *Nucl. Instr. Meth. Phys. Res. A*, 564, 300, 2006.)

electrons preferentially being pulled toward the anode. Because of the high electric field residing in this region, the electrons are accelerated and collide within the semiconductor matrix, thereby resulting in more ionizations and more electron/hole pairs. The secondary electrons are then free to be accelerated and undergo further ionizations, and so forth. Eventually, the freed electrons will reach the anode and result in an amplified electrical signal.

As the applied electrical potential increases, the number of impact ionizations in the multiplying region also increases, thereby resulting in increased signal gain. However, at high bias, the normally immobile holes will be attracted toward the cathode and produce impact ionizations en route. The additional electrons get pulled easily toward the anode, producing an additional, undesired, cascade effect. The point at which this occurs is the breakdown voltage (V_{br}) and is temperature dependent. When operated above breakdown voltage, the number of electron/hole pairs produced via the original interaction and subsequent multiplying process is no longer proportional to the intensity of incoming light. For this reason, APDs are generally operated at operating voltages somewhat below the breakdown voltage at a given operating temperature.

It is important to note that while the gain of an APD is greater than standard PIN diodes, the gain is dependent upon the operating voltage and the operating temperature [45]. Figure 14.6b depicts the system gain for a standard 10 × 10 mm APD exposed to light from a red LED over a range of operating voltages. As the bias is increased, the overall gain of the APD is also increased. However, in addition to increasing the APD signal, the noise also increases with bias. There comes a point prior to breakdown voltage where the noise is increasing faster than the useful signal, thus the SNR starts to decrease. It is at this point where the SNR is maximal that is usually optimal to operate the APD.

Figure 14.6c depicts the overall APD gain as a function of operating temperature. It is seen that by cooling the APD, at a given operating bias, the gain will be higher when at lower temperatures. As well, the breakdown voltage also decreases when cooled so that for a given gain, the APD can be operated with a lower operating bias [46,47]. Furthermore, the APD background noise also decreases when cooled thus improving the SNR. This temperature dependence on gain can become a problem, however, as it means the APD output varies as the device warms up or cools down. For most APDs, this variation in gain as a function of temperature is on the order of 2%–3%/°C, thereby making it rather challenging to operate an array of APDs without some form of active gain compensation [45].

One of the first APD-based radiation detectors for use with PET was the small animal PET developed by Lecomte et al., at the Universite de Sherbrooke [48–52]. This device utilized two rings of 256 – 3 × 5 × 20 mm BGO scintillators each coupled to an individual APD. This device has since been commercialized as the LabPET [53] and though the device has not been shown as of yet to operate in the vicinity of magnetic field, these devices were the first to show that a stable imaging system can be developed using APDs.

Recently, Pichler et al. [54–56] have replaced the PMT photodetectors on a clinical PET detector module with an array of APD detectors for use in simultaneous PET/MR. This detector uses a 3 × 3 array of 5 × 5 mm APDs coupled to LSO scintillators. The output of each APD is then fed into a CSA and associated electronics similar to PMT detectors.

In an effort to both improve the spatial resolution of large-area APDs and reduce the complexity of multichannel acquisition electronics when used with individual APD elements, position-sensitive APDs (PSAPDs) have been developed [57]. In essence, PSAPDs are large-area APDs (up to 14 × 14 mm) with a resistive layer on the back face upon which multiple contacts are placed (see Figure 14.7a). The usual configuration is to divide the back face into quadrants with separate outputs for each quadrant. Because the

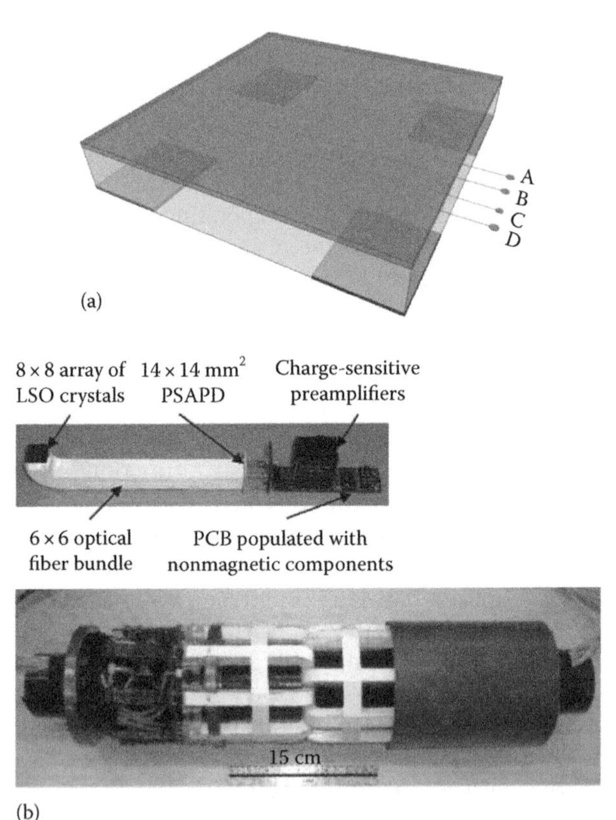

FIGURE 14.7
(a) Schematic of a PSAPD. Note the anode is divided into four quadrants, thus resulting in a charge division across the four anodes. The individual signals are combined in order to determine the (*X*,*Y*) location of the original scintillation event. (b) Preclinical PET/MR ring produced at UC Davis. The detector ring consists of a series of LSO detectors fiber-optically coupled to PSAPDs and CSAs. The entire ring and electronics is enclosed within a copper "can." (From Catana, C. et al., *J. Nucl. Med.*, 47, 1968, 2006.)

total charge produced in the APD is shared among four anodes, PSAPDs require high gain in order to overcome the inherent noise associated with each anode.

When coupled to an appropriate scintillator (usually pixelated CsI(Tl) or LSO), it is possible to acquire the four quadrant signals for each scintillation event and to subsequently determine the appropriate scintillator detector that gave rise to these signals via the same equations used for PET imaging (i.e., equations *X*, earlier). Given the relatively low SNR on each channel, PSAPDs also utilize a fifth detector channel on the top layer in order to obtain a summed signal over the entire array. This output is useful for obtaining low-noise information for subsequent energy discrimination. These devices have been investigated for use in preclinical imaging owing to the possibility of high resolution (300 μm [58]) both for PET [59,60] and SPECT [61].

Using these PSAPDs, Cherry et al. [62–65] have developed a preclinical PET insert for hybrid PET/MR imaging (Figure 14.7b). This device uses a ring of LSO detector modules fiber-optically coupled to PSAPDs. The entire ring consists of 16 detector modules with each detector module comprised of an 8 × 8 array of LSO scintillator crystals (1.43 × 1.43 × 6 mm). A short 6 × 6 fiber-optic bundle (1.95 × 1.95 mm per fiber) is used on each module to carry the light information from the scintillators to the 14 × 14 mm PSAPD. This bundle incorporated a 90° bend in order to minimize the radial extent of the insert. Each PSAPD utilized five outputs (four back layers for position information plus the top layer for timing and energy discrimination) with each channel passing through a low-noise CSA. As the APD and preamplifiers were placed inside the MR field, they were shielded with a cylinder of high-frequency laminate in order to minimize RF interference. Outputs from the preamplifiers were then fed to standard PET acquisition electronics located at a safe distance from the MR magnet. In order to reduce dark current, the PSAPDs were cooled to approximately −5°C by blowing cold nitrogen gas over them. Of note is the fact that a relatively large 40 ns coincidence timing window was used to discriminate coincidence events. This was explained by fact that the top layer PSAPD signal has a position-dependent delay as it was recognized that signals originating from the edge of the PSAPD would typically be measured sooner than signals originating from the center. Because the purpose of this system is preclinical imaging, it was noted that the large coincidence window would not represent a significant problem as anticipated random events will be quite low due to the relatively low radioactive doses administered to the animals.

As can be imagined, as the bias voltage is continually increased, additional impact ionizations occur within the APD at the same time some electrons are collected at the anode. However, there will come a point when more electrons are produced than can be collected at the anode. This is referred to as the breakdown voltage and at this point, the ensuing runaway production of electron/hole pairs leads to a large photocurrent that is no longer representative of the original light intensity. When operated in this way, the APD is said to be operating in Geiger mode as the detection of a single light photon in the APD results in a large photocurrent. Because the output is no longer proportional to the incident light intensity, APDs used as alternatives to PMTs for imaging are typically operated well below breakdown voltage. However, as the APD bias is increased above breakdown, the generation of electron/hole pairs increases exponentially. As the semiconductor material has some series resistance, more of the voltage is dropped across the resistance as the photocurrent grows, thus limiting the avalanche effect until at some point the voltage across the high-field region is reduced to breakdown when the generation of electrons is matched by the extraction of electrons at the anode. So, in the continued presence of applied bias, this large photocurrent will continue to flow indefinitely, thereby rendering the detector unresponsive to further photodetections. In order to further detect photons, the APD must be quenched.

Two means of quenching an APD are possible, passive quenching or active quenching. In the former case, the bias is simply removed from the APD once it is in the steady state. Once removed, the inherent resistance and capacitance results in a gradual decline in photocurrent until the avalanche process no longer occurs, thus resulting in dead time while the APD discharges. In active quenching, the APD bias is again removed after the avalanche process is initiated, but once the discharging starts, a quenching circuit steps in to shunt the APD and quickly discharge the device, thereby reducing the APD dead time. Once discharged, the bias can then be reapplied through the use of a switch. For simplicity, most applications of Geiger-mode APDs utilize a large series resistor, R_s, between the power supply and the APD, thus creating a virtual open circuit that continually charges the APD with time constant R_sC while allowing the APD to discharge with time constant RC.

For all diode-based photodetectors, every effort is made to reduce the number of high-energy gamma rays from interacting within the photodiode by using thick scintillators and making the diode as thin as possible. Even so, it is still possible that a gamma ray will penetrate the scintillator and interact within the diode material itself. When this occurs, electron/hole pairs are created the same as when low-energy light photons interact, thus giving rise to electrical signals. This is problematic as the signal from a direct ionization is several times larger than that which would be detected from the scintillator, thus making it possible that spurious events may have a large effect on subsequent processing.

14.5.3.2 Multi-Pixel Geiger-Mode Avalanche Photodiodes

Since the output signal of a single channel Geiger-mode APD is not proportional to the deposited gamma ray energy, APDs operated above V_{br} are not useful for imaging applications. However, as they are extremely sensitive, it is possible to measure single optical photons with these devices. If the APDs were miniaturized sufficiently and an array of such APDs configured,

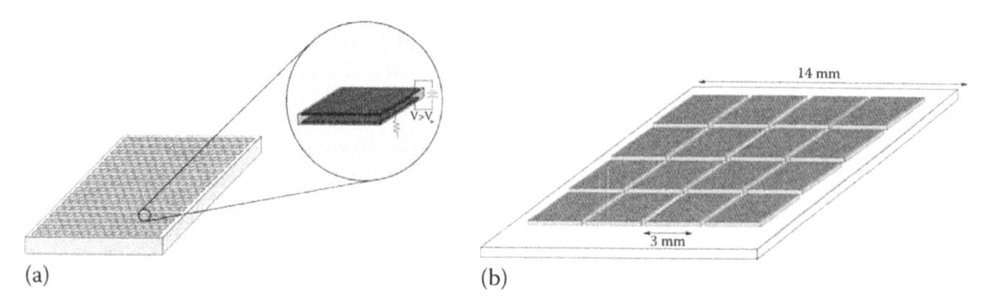

FIGURE 14.8
(a) Schematic of a multi-pixel Geiger-mode APD. Each APD element is typically between 25 and 50 μm in size, with the entire array usually about 1 mm × 1 mm. (b) Schematic representation of SiPM array using 4 × 4 SiPMs. The output of each SiPM can be independent (i.e., 16 channels per array) or summed together for a single output. When summed, the array behaves similar to a conventional PM tube and when operated in multichannel mode, the array becomes position sensitive for high-resolution applications.

then for a given scintillation event in a crystal, each APD element would be capable of detecting a single optical scintillation photon and outputting a large photocurrent. Since each Geiger-mode APD would output the same voltage when it detects a photon, simply summing up the total signal over all elements would give an indication of the total number of optical photons detected. This is the basis of the multi-pixel Geiger-mode APD, sometimes referred to as solid-state or silicon photomultipliers (SSPM or SiPM) [66–68], depicted schematically in Figure 14.8a.

Most SiPM designs use an array of several thousand APD detector cells all operating in Geiger mode. Each cell is typically between 20 and 50 μm square, thus the entire array is usually around 1–3 mm square. When a scintillation event occurs, the light photons will spread out through the scintillator and will be incident upon the SiPM that then initiates an avalanche process that creates a large current through the cell, thereby leading to an infinite gain if not for a quenching resistor integrated into each cell. As many light photons will be generated during a scintillation, many cells will undergo the avalanche cascade, thereby leading to an overall current proportional to the number of cells that undergo avalanche. As each cell can only react to a single light photon and output a given current, it is important that there are more cells than light photons expected and that the time required to reset each cell be as short as possible.

The overall gain of silicon photomultipliers approaches that of PMTs (around 10^6) [69] yet do not suffer from the same sensitivity to magnetic fields as PMTs. As well, they offer superior gain to either PIN diodes or APDs but with increased robustness, reduced noise, and the capability to operate at much reduced bias (~30 V). Because of these benefits, many research groups are currently investigating the use of SiPMs for PET or SPECT imaging [70–74]. Commercial SiPMs are available through suppliers such as Hamamatsu, SensL Ltd., and Zecotek among others. Further development of SiPMs has led to arrays of SiPMs such as the large-area SensL SPMArray. This device consists of a 4 × 4 array of 3 × 3 mm SiPMs with each SiPM being comprised of 3640 microcells (see schematic in Figure 14.8b). This array can be operated in single channel mode where all SiPMs in the array are summed together or in position-sensitive mode where each of the 16 elements outputs an independent signal. With this configuration, the SiPM array would normally be coupled to pixelated scintillators and the 16 outputs used to determine the site of photon interaction.

14.5.4 Solid-State Detectors

Over the past several years, there has been renewed interest in replacing the scintillator/photodetector combination with solid-state radiation detectors for PET and SPECT. The inherent robustness of semiconductor detectors in strong magnetic fields makes them an attractive alternative to scintillator/photodetectors; however, their relatively high cost due to low manufacturing yields has limited their use, until recently, to research only.

In essence, semiconductor radiation detectors behave similarly to PIN diodes as described earlier. When a high-energy gamma ray interacts within the semiconductor, a number of electron/hole pairs are produced. The migration of these electrons and holes across an applied electric field gives rise to a measurable electric charge at the electrode. The electron/hole cloud of charge that is created has been seen to remain relatively intact (less than 250 μm) as the charges migrate through the electric field, limiting the charge spread to less than 0.5 mm from the original interaction location [31], thus making them ideal for high-resolution PET or SPECT imaging.

14.5.4.1 Si(Li) and Ge(Li) Detectors

Lithium-doped silicon- and germanium-based semiconductors have been used extensively for spectroscopy applications due to their high energy resolution (2%–3% at 662 keV). While capable of extremely high energy resolution, Si(Li) and Ge(Li) semiconductors suffer from an inherently high thermal noise component, thus making it necessary to cryogenically cool the material. Additionally, the relatively low stopping power of silicon and germanium materials limits their usefulness to primarily spectroscopy applications although some use has been made in SPECT or PET applications [75].

14.5.4.2 CdTe and CdZnTe Detectors

The most commonly used semiconductor radiation detectors are cadmium telluride (CdTe) and cadmium zinc telluride (CdZnTe or CZT). While CdTe and CZT has a density sufficient to stop 511 keV photons for PET, most crystals of CZT can only be grown of sufficient thickness for low-energy applications (typically 3–5 mm thick). Thus, CZT arrays are primarily used for SPECT applications, although some groups have investigated them for use with PET [76].

CZT detectors are typically either pixelated into discrete detector elements or used as a monolithic block. In the former case, each pixel requires its own anode and cathode, thus resulting in an inherent improvement in spatial resolution as pixels are made smaller and smaller; however, this comes at the cost of increased readout complexity in the requirement for thousands of individual readout channels. In the case of large single blocks of CZT, there is a single common cathode for the entire block but individual anodes distributed in an array on the back side of the block. Because of the small amount of diffusion of the charge carrier cloud, there is minimal spread of the carriers as they traverse the detector block.

Recently, Gamma Medica-Ideas has produced a prototype CZT detector insert for preclinical SPECT/MR imaging [77,78], shown in Figure 14.9. This insert consists of 24 CZT detector modules arranged in three octagonal rings. Each module uses a single block of 25.4 × 25.4 × 5 mm CZT with a single cathode and with a 16 × 16 array of pixelated anodes. A −500 V potential bias is applied to the cathode, thus creating the large electric field required to draw electrons to the anodes to produce a measurable signal.

Each anode of the detector block is connected to a single channel of a multiple-channel application-specific integrated circuit (ASIC) [79]. In total, each ASIC has 128 input channels with each channel incorporating a CSA, pulse shaper, peak–hold circuit, and a threshold discriminator. Two ASICs are connected to each CZT module in order to acquire the entire 16 × 16 pixel array. The collimation system for this detector insert is comprised of an MR-compatible heavy-metal composite cylindrical sleeve, 35 mm in diameter. This sleeve

(a)

(b)

FIGURE 14.9
(a) CZT detector module used in the first MR-compatible SPECT detector. (b) SPECT/MR ring insert consisting of three rings of eight CZT detector modules. Each module uses a single pinhole collimator. (From Azman, S. et al., *IEEE Med. Imag. Conf. Rec.*, 2311, 2007.)

consists of a series of pinhole apertures, with each pinhole corresponding to one of the CZT detector modules.

In the past, CZT production has been plagued with low production yields and small crystal sizes. New production methods [80–82] offer the promise of delivering higher-yield materials with higher grade, thus making the transition from scintillators to solid-state detectors possible. At least two manufacturers now offer CZT-based clinical gamma cameras (GE and Spectrum Dynamics) for limited applications.

14.6 Other Considerations

14.6.1 Magnetic Compatibility

To reduce the influence of background radiation, PET and SPECT detectors are usually surrounded by a heavy-metal material that acts as radiation shielding. On stand-alone devices, this material is typically lead because of its high stopping power and relatively low cost. Because PET imaging utilizes coincidence detection, the required shielding is not very significant and in fact is usually used more to shield the photomultipliers from magnetic field effects. However, as SPECT detects a single photon at a time, any background radioactivity can have a significant effect on the imaging performance and so camera shielding is quite significant. Additionally, SPECT systems incorporate a heavy-metal collimator attached to the scintillator that limits the angle of acceptance of incoming gamma rays. Again, this collimator is typically constructed of lead or tungsten alloys.

In order to function in the high magnetic field of MR, it is necessary to consider the appropriateness of the radiation shielding and collimator materials used in PET and SPECT detectors [83]. Ideal PET and SPECT shielding materials would have the following properties:

1. High stopping power for radiation
2. Magnetically compatible (i.e., low magnetic susceptibility)
3. Nonconductive
4. Thin and lightweight

Properties (1) and (4) are very much related as, in general, for a material to have high stopping power, it must have a high effective atomic number, Z, and high density. Thus, there are trade-off considerations between weight and radiation stopping power that should be addressed for each application.

In order to measure magnetic compatibility, the use of magnetic susceptibility, χ, is most often used [84]. This property measures the extent of an induced magnetic field produced by the material when placed in a magnetic field and is defined as

$$\chi = \frac{M}{H}$$

where
M is the volume magnetization (magnetic moment per unit volume)
H is the magnetic field

Susceptibility is usually described in terms of parts per million (ppm) with the sign denoting the direction of the magnetization vector and the absolute value denoting the strength. For example, water has a low magnetic susceptibility of about -9×10^{-6}, while pure iron has a susceptibility of around 2×10^{5}. Depending on the application, it is common to also use relative permeability rather than susceptibility. In free space, the two quantities are related by the expression

$$\mu_r = \chi + 1$$

In selecting appropriate MRI-compatible materials, it is important to select materials both with low absolute susceptibility values and with similar susceptibilities to those materials that are being imaged [85]. In the former case, materials with high susceptibility will produce large magnetic field distortions and lead to severe image artifacts. Materials with widely different susceptibilities also produce image artifacts as local perturbations in the magnetic field will be created from the different materials. These perturbations may result in spin dephasing between voxels or geometric distortions by warping the imaging plane. In either case, so-called susceptibility artifacts are produced [86–88].

It is also important to consider the generation of eddy currents in any material introduced into the MR system. Eddy currents form when conductive materials are placed in the changing magnetic fields produced by the gradient coils or B_1 field. These induced fields may alter the magnetization vector flip angle, thereby resulting in image artifacts.

Along with artifacts produced in the MR images, it is also important to consider the effect of the magnetic fields and the radio frequencies on the performance of the radiation detection equipment. If used in conjunction with PET imaging, the magnetic field used in MR may actually have a beneficial effect on spatial resolution [89]. This is because the charged positrons produced as a result of the radioactive decay process will experience a Lorentz force on them as a result of moving through the strong magnetic field. This force will be directed perpendicular to the magnetic field so that the net effect is to produce a helical path. The radius of this path is given by

$$R = \frac{0.334}{B}\sqrt{(2m_pE_t) + E_t^2}$$

where

B is the magnetic field strength
m_p is the positron rest mass
E_t is the positron kinetic energy perpendicular to the magnetic field

While MRI systems have multiple magnetic field gradients, the main static field is the only magnet that appreciably affects the positron path. Given that this magnetic field is oriented axially, it is expected that spatial resolution improvements will only be seen along this direction. Nevertheless, improvements in spatial resolution are seen to be appreciable when the positron energy is significant, with little improvement with lower-energy positrons such as those emitted from F-18 [90]. Table 14.2 depicts some of the improvements seen for different nuclides in the presence of high B_0 fields.

Aside from the theoretical improvements in spatial resolution, most of the other consequences of the MR device will be detrimental to the operation of the radiation detectors. As already mentioned, the magnetic field will have a disastrous effect on PMT performance, thereby rendering them useless. Even if alternative detectors may be resilient to high magnetic fields, they may be susceptible to RF interference from the RF pulses used in MR imaging. This is particularly true of highly sensitive components such as APDs. The need to place additional electronics such as preamplifiers or ASICs in the vicinity of the magnetic field may also introduce problems with electrical interference. To overcome some of these effects, copper shielding may be used to reduce the amount of RF interference.

For SPECT imaging, the choice of collimator material introduces some problems of its own. The large lead collimator typically used for SPECT would

TABLE 14.2

Spatial Resolution Measurements Expected for Some of the Most Common PET Radionuclides in the Presence of a 10 T Magnetic Field

Radionuclide	Maximum Energy (MeV)	FWHM at 0 Tesla (mm)	FWHM at 10 Tesla (mm)
^{11}C	0.96	4.24 ± 0.07	3.73 ± 0.07
^{13}N	1.19	4.44 ± 0.06	3.80 ± 0.06
^{15}O	1.70	5.28 ± 0.10	3.80 ± 0.06
^{18}F	0.63	3.85 ± 0.06	3.70 ± 0.06
^{68}Ga	1.89	5.46 ± 0.10	3.86 ± 0.06
^{82}Rb	3.15	8.03 ± 0.15	3.91 ± 0.05

Source: Raylmann, R.R. et al., *IEEE Trans. Nucl. Sci.*, 43, 2406, 1996.

probably not be optimal for a SPECT-capable system as the electrical conductivity results in induced eddy currents when exposed to RF that may yield magnetic field inhomogeneities [83] or susceptibility artifacts in the MR images.

To produce 3D images using SPECT, a typical parallel-hole collimator must rotate around the object collecting views at multiple angles. This motion would most likely require the application of an electric motor, the exact placement of which, for obvious reasons, cannot be placed directly into the MR bore. Alternative approaches utilizing a ring of stationary cameras may alleviate some of the technical challenges involved in mechanical rotation. Designs that incorporate full detector rings [e.g., 91–93] with multiple pin-hole or coded aperture collimators would appear to be the most promising.

14.6.2 System Design

Given the space constraints within an MRI bore, it is important to consider the placement of the SPECT or PET system within the bore. In essence, three different designs for integrated nuclear/MR imaging are possible [94]: (1) separate systems, shared patient bed; (2) removable insert for PET or SPECT; and (3) fully integrated system (as shown in Figure 14.10). System (1) earlier would necessarily involve sequential imaging rather than simultaneous imaging and while the resultant images would be reasonably co-registered, it would still require at least twice the imaging time compared to simultaneous imaging. On the other hand, it has been shown that existing PMTs may be used when coupled

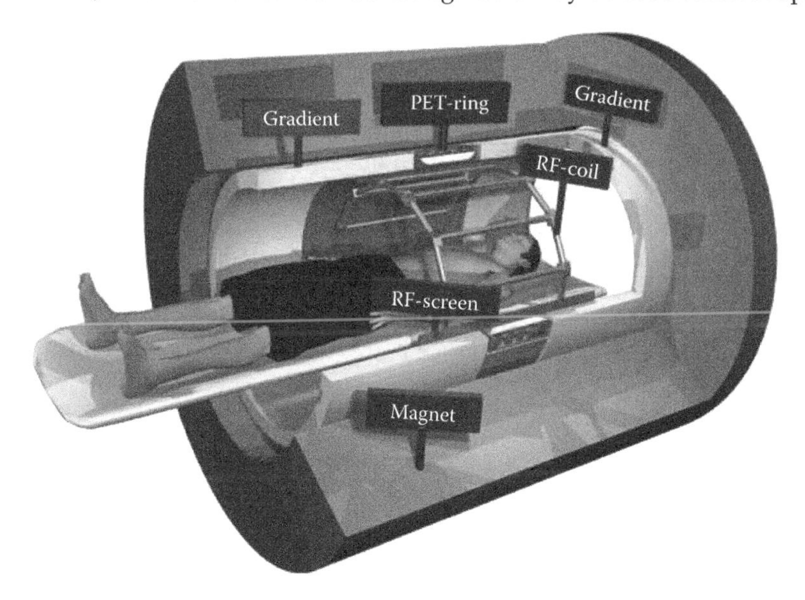

FIGURE 14.10
Theoretical integrated PET/MR system showing PET detector ring integrated into the gradient coil subsystem. This system would appear externally as a conventional MR system and be capable of whole-body PET/MR imaging. (Courtesy of Philips.com.)

to PET detectors via fiber optics and so system (1) may currently be clinically feasible. Systems (2) and (3) however would involve continued development of new radiation detection systems and in the case of (3), a complete redesign of the MRI and PET/SPECT system in order to integrate all components.

Additional, consideration must be made as to where the place the various needed for the respective imaging. As mentioned, in MRI imaging the receiver RF coils are usually placed as close as possible to the patient in order to maximize the SNR. In order to reduce costs and improve spatial resolution, PET or SPECT detectors must also be placed close to the patient being imaged thereby creating a conflict as to the exact placement of each detector system. It seems logical to place the RF coils closest to the patient, but the addition of these coils may introduce attenuation or scatter artifacts into the PET or SPECT data.

14.6.3 MR Attenuation Correction

As described previously, the conversion of CT numbers to linear attenuation coefficients must be performed in order to use the CT for attenuation correction [95]. While not necessarily trivial, this conversion is still rather straightforward as CT numbers do have a direct relation to photon attenuation (i.e., both related to electron density). The conversion of MR images to linear attenuation coefficient is not as straightforward, however, as MR images represent a measure of proton density and magnetic properties, not necessarily correlated to electron density. As a result, alternative approaches to derive linear attenuation coefficients must be devised.

The simplest approach is to simply segment MR images into air, soft tissue, and bone and to apply known attenuation coefficients to these regions. This technique has been successfully applied to neurological imaging [96] but does require some user intervention to aid in image segmentation. While working reasonably well for the brain, problems exist in extending this approach to the rest of the body that encompasses many more tissue types [97]. Alternative approaches have mapped MR images to standard attenuation coefficients [98] or have mapped MR to CT images in order to derive linear attenuation coefficients [99]. Such techniques require additional image co-registration in order to align the patient-specific MR with a generalized atlas. As PET/MR and SPECT/MR are still relatively early on in their development, ongoing research will no doubt improve the accuracy of MR-based attenuation correction schemes [100].

14.7 Conclusions

The development of MR-compatible PET and SPECT imaging devices has rapidly increased over the past several years with the development of new radiation detectors. The recent introduction of Geiger-mode silicon photomultipliers and improved production yields of high-quality CZT crystals has meant that there are now real alternatives to PMTs and scintillators for PET and SPECT. While the development of PET/MR and SPECT/MR may seem to offer many advantages over separate systems, with continued development of these systems come other issues relating to health-care costs [101] and patient safety [102] that will need to be addressed before the mainstream clinical adoption of these technologies. Given this, it is

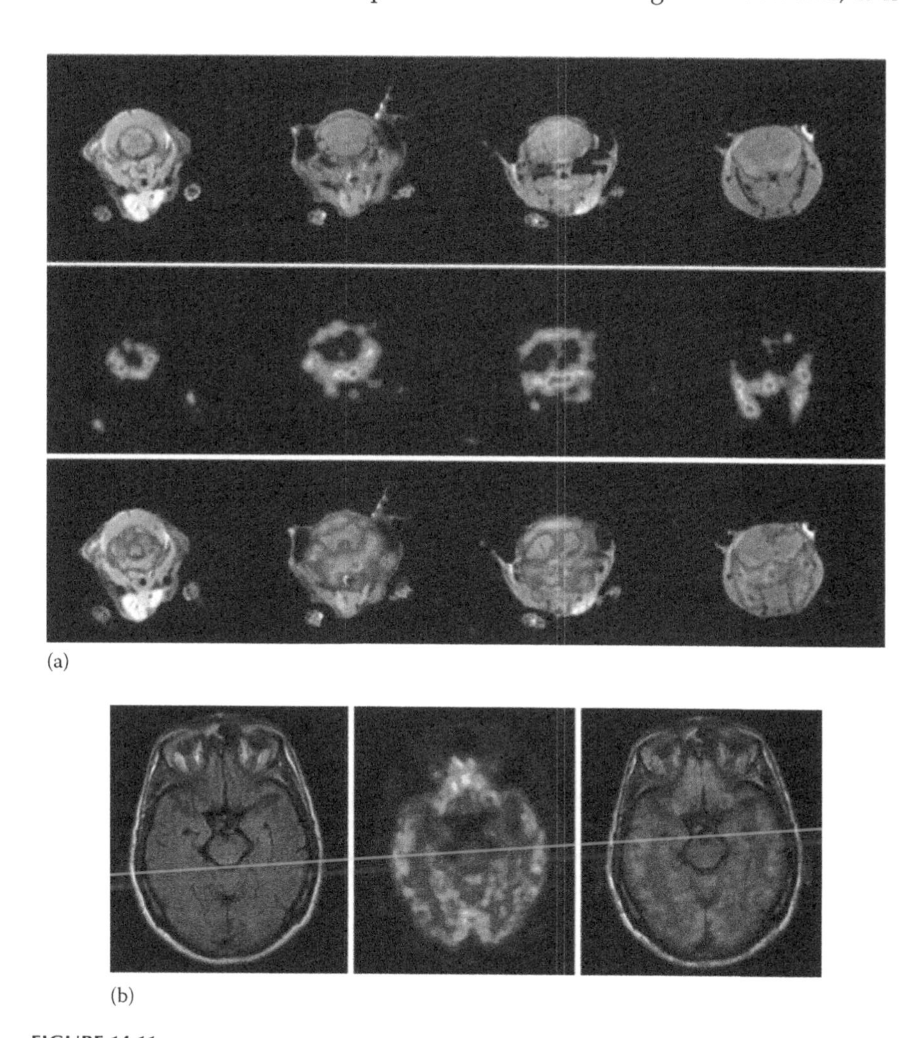

FIGURE 14.11
(See color insert.) (a) First simultaneously acquired PET/MR images from the PSAPD PET detector ring developed at UC Davis. The PET image depicts ^{18}fluoride uptake within the skeletal system. (From Catana, C. et al., *J. Nucl. Med.*, 47, 1968, 2006.) (b) Simultaneous acquired PET/MR image depicting 18FDG uptake in the brain of a human subject. The MR image was acquired using a FLAIR sequence and the entire MR + PET procedure was acquired in 15 min. (From Schlemmer, H.P. et al., *Abdom. Imaging*, 2008, http://www.SpringerLink.com)

not surprising that much of the early development on these devices has been applied to preclinical imaging [79,103,104]. Many soft-tissue tumor models cannot be adequately visualized with x-ray CT as soft-tissue contrast is not sufficient to delineate tumor from tissue. However, MRI, with its superior soft-tissue contrast, has proven itself very valuable for a lot of preclinical imaging studies. The ability to visualize the in vivo distribution of new radiotracers in the context of high-resolution, high-contrast MR imaging has provided valuable insights into drug metabolism and disease progression.

While many technical challenges still need to be overcome before PET/MR or SPECT/MR devices are commonplace, significant progress has been made to the point where prototype systems have been developed and initial imaging studies have been performed (see Figure 14.11a and b). Already, a number of potential clinical applications have been identified that will benefit from further SPECT/MR or PET/MR development. Applications such as oncology [105], neurology [106,107], cardiology [108], abdominal imaging [109], and cell trafficking [110,111] will keep pushing the need for these devices and we will no doubt continue to see great technological strides being made.

References

1. Bailey DL, Hutton BF, Walker PJ, Improved SPECT using simultaneous emission and transmission tomography, *J. Nucl. Med.*, 1987; 28 (5): 844–851.
2. Bailey DL, Transmission scanning in emission tomography, *Eur. J. Nucl. Med.*, 1998; 25 (7): 774–787.
3. deKemp RA, Nahmias C, Attenuation correction in PET using single photon transmission measurement, *Med. Phys.*, 1994; 21 (6): 771–778.
4. Celler A, Sitek A, Stoub E, Hawman P, Harrop R, Lyster D, Multiple line source array for SPECT transmission scans: Simulation, phantom and patient studies, *J. Nucl. Med.*, 1998; 39 (12): 2183–2189.
5. Evans AC, Marrett S, Torrescorzo J, Ku S, Collins L, MRI-PET correlation in three dimensions using a volume-of-interest (VOI) atlas, *J. Cereb. Blood Flow Metab.*, 1991; 11 (2): A69–A78.
6. Wahl RL, Quint LE, Cieslak RD, Aisen AM, Koeppe RA, Meyer CR, "Anatometabolic" tumor imaging: Fusion of FDG PET with CT or MRI to localize foci of increased activity, *J. Nucl. Med.*, 1993; 34 (7): 1190–1197.
7. Beyer T, Townsend DW, Brun T, Kinahan PE, Charron M, Roddy R, Jerin J, Young J, Byars L, Nutt R, A combined PET/CT scanner for clinical oncology, *J. Nucl. Med.*, 2000; 41: 1369–1379.
8. Kinahan PE, Townsend DW, Beyer T, Sashin D, Attenuation correction for a combined 3D PET/CT scanner, *Med. Phys.*, 1998; 25: 2046–2053.
9. Townsend DW, Beyer T, Blodget TM, PET/CT scanners: A hardware approach to image fusion, *Sem. Nucl. Med.*, 2003; 33 (3): 193–204.

10. Lang TF, Hasegawa BH, Liew SC, Brown JK, Blankespoor SC, Reilly SM, Gingld EL, Cann CE, Description of a prototype emission-transmission computed tomography imaging system, *J. Nucl. Med.*, 1992; 33: 1881–1887.
11. Hutton BF, Braun M, Software for image registration: Algorithms, accuracy, efficacy, *Sem. Nucl. Med.*, 2003; 33 (3): 180–192.
12. Pluim JPW, Maintz JBA, Viergever MA, Mutual information based registration of medical images: A survey, *IEEE Trans. Med. Imag.*, 2003; 22 (8): 986–1004.
13. Pluim JPW, Maintz JBA, Viergever MA, Image registration by maximization of combined mutual information and gradient information, *IEEE Trans. Med. Imag.*, 2000; 19 (8): 809–814.
14. Stokking R, Zubal GI, Viergever MA, Display of fused images: Methods, interpretation, and diagnostic improvements, *Sem. Nucl. Med.*, 2003; 31 (3): 219–227.
15. Kinahan PE, Hasegawa BH, Beyer T, X-ray-based attenuation correction for positron emission tomography/computed tomography scanners, *Sem. Nucl. Med.*, 2003; 33 (3): 166–179.
16. Carney JPJ, Townsend DW, Rappoport V, Bendriem B, Method for transforming CT images for attenuation correction in PET/CT imaging, *Med. Phys.*, 1996; 33 (4): 976–983.
17. Burger C, Goerres G, Schoenes S, Buck A, Lonn AHR, von Schulthess GK, PET attenuation coefficients from CT images: Experimental evaluation of the transformation of CT into PET 511-keV attenuation coefficients, *Eur. J. Nucl. Med.*, 2002; 29: 922–927.
18. LaCroix KJ, Tsui BMW, Hasegawa BH, Brown JK, Investigation of the use of x-ray CT images for attenuation compensation in SPECT, *IEEE Trans. Nucl. Sci.*, 1994; 41 (6): 2793–2799.
19. Pantvaidya GH, Agarwal JP, Deshpande MS, Rangarajan V, Singh V, Kakade A, D'Cruz AK, PET-CT in recurrent head neck cancers: A study to evaluate impact on patient management, *J. Surg. Oncol.*, 2009; 100 (5): 401–403.
20. Chowdhury FU, Bradley KM, Gleeson FV, The role of ^{18}F-FDG PET/CT in the evaluation of oesophageal carcinoma, *Clin. Radiol.*, 2008; 63 (12): 1297–1309.
21. Horger M, Eschmann SM, Pfannenberg C, Storek D, Vonthein R, Claussen CD, Bares R, Added value of SPECT/CT in patients suspected of having bone infection: Preliminary results, *Arch. Orthop. Trauma Surg.*, 2007; 127 (3): 211–221.
22. Nathan J, Crawford JA, Sodee DB, Bakale G, Fused SPECT/CT imaging of peri-iliopsoas infection using indium-111-labeled leukocytes, *Clin. Nucl. Med.*, 2006; 31 (12): 801–802.
23. Ingui CJ, Shah NP, Oates ME, Infection scintigraphy: Added value of single-photon emission computed tomography/computed tomography fusion compared with traditional analysis, *J. Comput. Assist. Tomogr.*, 2007; 31 (3): 375–380.
24. Bar-Shalom R, Yefremov N, Guralnik L, Keidar Z, Engel A, Nitecki S, Israel O, SPECT/CT using ^{67}Ga and ^{111}In-labeled leukocyte scintigraphy for diagnosis of infection, *J. Nucl. Med.*, 2006; 47 (4): 587–594.
25. Filippi L, Schillaci O, Usefulness of hybrid SPECT/CT in ^{99m}Tc-HMPAO-labeled leukocyte scintigraphy for bone and joint infections, *J. Nucl. Med.*, 2006; 47 (12): 1908–1913.
26. Schepis T, Gaemperli O, Koepfli P, Namdar M, Valenta I, Scheffel H, Leschka S et al., Added value of coronary artery calcium score as an adjunct to gated SPECT for the evaluation of coronary artery disease in an intermediate-risk population, *J. Nucl. Med.*, 2007; 48 (9): 1424–1430.

27. McQuaid SJ, Hutton BF, Sources of attenuation-correction artefacts in cardiac PET/CT and SPECT/CT, *Eur. J. Nucl. Med. Mol. Imag.*, 2008; 35 (6): 1117–1123.
28. Le Meunier L, Maass-Moreno R, Carrasquillo JA, Dieckmann W, Bacharach SL, PET/CT imaging: Effect of respiratory motion on apparent myocardial uptake, *J. Nucl. Cardiol.*, 2006; 13 (6): 821–830.
29. Saoudi A, Pepin CM, Lecomte R, Study of light collection in multi-crystal detectors, *IEEE Trans. Nucl. Sci.*, 2000; 47 (4); 1634–1639.
30. Yamamoto S, Kuroda K, Senda M, Scintillator selection for MR-compatible gamma detectors, *IEEE Trans. Nucl. Sci.*, 2003; 50 (5): 1683–1685.
31. Knoll GF, *Radiation Detection and Measurement*, 3rd edn., New York; Wiley, 2000.
32. Bieszk JA, Performance changes of an anger camera in magnetic fields up to 10 G, *J. Nucl. Med.*, 1986; 27: 1902–1907.
33. Goetz C, Breton E, Choquet P, Israel-Jost V, Constantinesco A, SPECT low-field MRI system for small-animal imaging, *J. Nucl. Med.*, 2008; 49 (1): 88–93.
34. Gilbert KM, Handler WB, Scholl TJ, Odegaard JW, Chronik BA, Design of field-cycled magnetic resonance systems for small animal imaging, *Phys. Med. Biol.*, 2006; 51: 2825–2841.
35. Handler WB, Gilbert KM, Peng H, Chronik BA, Simulation of scattering and attenuation of 511 keV photons in a combined PET/field-cycled MRI system, *Phys. Med. Biol.*, 2006; 51: 2479–2491.
36. Peng H, Investigation of new approaches to combined positron emission tomography and magnetic resonance imaging systems, PhD thesis, University of Western Ontario, Ottawa, Ontario, Canada, 2006.
37. Shao Y, Cherry SR, Farahani K, Slates R, Silverman RW, Meadors K, Bowery A, Siegel S, Development of a PET detector system compatible with MRI/NMR systems, *IEEE Trans. Nucl. Sci.*, 1997; 44 (3): 1167–1171.
38. Haak GM, Hammer BE, Christensen NL, Coupling scintillation light into optical fiber for use in a combined PET-MRI scanner, *Nucl. Instrum. Meth. A*, 1997; 399; 455–462.
39. Shao Y, Cherry SR, Farahani K, Meadors K, Siegel S, Silverman RW, Marsden PK, Simultaneous PET and MR imaging, *Phys. Med. Biol.*, 1997; 42: 1965–1970.
40. Lecomte R, Novel detector technology for clinical PET, *Eur. J. Nucl. Med. Mol. Imag.*, 2009; 36 (suppl 1): S69–S85.
41. Asghar M, Imrie DC, Silicon photodiodes as photomultiplier replacements in industrial scintillation counters, *J. Phys. E Sci. Instr.*, 1989; 22: 1012–1015.
42. Levin CS, Hoffman EJ, Investigation of a new readout scheme for high resolution scintillation crystal arrays using photodiodes, *IEEE Trans. Nucl. Sci.*, 1997; 44 (3): 1208–1213.
43, Groom DE, Silicon photodiode detection of bismuth Germanate scintillation light, *Nucl. Instr. Meth. A.*, 1984; 219: 141–148.
44. Gruber GJ, Moses WW, Derenzo SE, Wang NW, Beuville E, Ho MH, A discrete scintillation camera module using silicon photodiode readout of CsI(Tl) crystals for breast cancer imaging, *IEEE Tran. Nucl. Sci.*, 1998; 45: 1063–1068.
45. Kataoka J, Sato R, Ikagawa T, Kotoku J, Kuramoto Y, Tsubuku Y, Saito T et al., An active gain-control system for avalanche photo-diodes under moderate temperature variations, *Nucl. Instr. Meth. Phys. Res. A*, 2006; 564: 300–307.
46. Su YK, Change CY, Wu TS, Temperature dependent characteristics of a PIN avalanche photodiode (APD) in Ge, Si and GaAs, *Opt. Quant. Electron.*, 1979; 11: 109–117.

47. Hakim NZ, Saleh BEA, Teich MC, Generalized excess noise factor for avalanche photodiodes of arbitrary structure, *IEEE Trans. Electron Dev.*, 1990; 37 (3): 599–610.
48. Lighthouse AW, McIntyre RJ, Lecomte R, Schmitt D, A bismuth Germanate-avalanche photodiode module designed for use in high resolution positron emission tomography, *IEEE Trans. Nucl. Sci.*, 1986; 33 (1): 456–459.
49. Marriott CJ, Cadorette JE, Lecomte R, Scasnar V, Rousseau J, van Lier JE, High resolution PET imaging and quantitation of pharmaceutical biodistributions in a small animal using avalanche photodiode detectors, *J. Nucl. Med.*, 1994; 35: 1390–1397.
50. Lecomte R, Cadorette J, Rodrigue S, Lapointe D, Rouleau D, Bentourkia M, Yao R, Msaki P, Initial results from the Sherbrooke avalanche photodiode positron tomograph, *IEEE Trans. Nucl. Sci.*, 1996; 43 (3): 1952–1957.
51. Lecomte R, Cadorette J, Jouan A, Heon M, Rouleau D, Gauthier G, High resolution positron emission tomography with a prototype camera based on solid state scintillation detectors, *IEEE Trans. Nucl. Sci.*, 1990; 37 (2): 805–811.
52. Cadorette J, Rodrigue S, Lecomte R, Tuning of avalanche photodiode PET camera, *IEEE Trans. Nucl. Sci.*, 1993; 40 (4): 1062–1066.
53. Fontaine R, Belanger F, Viscogliosi N, Semmaoui H, Tetrault MA, Michaud JB, Pepin C, Cadorette J, Lecomte R, The hardware and signal processing architecture of LabPET™, a small animal APD-based digital PET scanner, *IEEE Trans. Nucl. Sci.*, 2009; 56 (1): 3–9.
54. Pichler BJ, Judenhofer MS, Catana C, Walton JH, Kneilling M, Nutt RE, Siegel SB, Claussen CD, Cherry SR, Performance test of an LSO-APD detector in a 7-T MRI scanner for simultaneous PET/MRI, *J. Nucl. Med.*, 2006; 47: 639–647.
55. Pichler BJ, Swann BK, Rochelle J, Nutt RE, Cherry SR, Siegel SB, Lutetium oxyorthosilicate block detector readout by avalanche photodiode arrays for high resolution animal PET, *Phys. Med. Biol.*, 2004; 49: 4305–4319.
56. Ziegler SI, Pichler BJ, Boening G, Rafecas M, Pimpl W, Lorenz E, Schmitz N, Schwaiger M, A prototype high-resolution animal positron tomograph with avalanche photodiode arrays and LSO crystals, *Eur. J. Nucl. Med. Mol. Imag.*, 2001; 28 (2): 136–143.
57. Shah KS, Farrell R, Grazioso R, Harmon ES, Karplus E, Position-sensitive avalanche photodiodes for gamma-ray imaging, *IEEE Trans. Nucl. Sci.*, 2002; 49 (4): 1687–1692.
58. Dokhale PA, Silverman RW, Shah KS, Grazioso R, Farrell R, Glodo J, McClish MA, Entine G, Tran V-H, Cherry SR, Performance measurements of a depth-encoding PET detector module based on position-sensitive avalanche photodiode read-out, *Phys. Med. Biol.*, 2004; 49: 4293–4303.
59. Levin CS, Foudray AMK, Olcott PD, Habte F, Investigation of position sensitive avalanche photodiodes for a new high-resolution PET detector design, *IEEE Trans. Nucl. Sci.*, 2004; 51 (3): 805–810.
60. Burr KC, Ivan A, LeBlanc J, Zelakiewicz S, McDaniel DL, Kim CL, Ganin A et al., Evaluation of a position sensitive avalanche photodiode for PET, *IEEE Trans. Nucl. Sci.*, 2003; 50 (4): 792–796.
61. Després P, Funk T, Shah KS, Hasegawa BH, Monte Carlo simulations of compact gamma cameras based on avalanche photodiodes, *Phys. Med. Biol.*, 2007; 52 (11): 3057–3074.
62. Catana C, Wu Y, Judenhofer MS, Qi J, Pichler BJ, Cherry SR, Simultaneous acquisition of multislice PET and MR images: Initial results with a MR-compatible PET scanner, *J. Nucl. Med.*, 2006; 47: 1968–1976.

63. Mackewn JE, Struhl D, Hallett WA, Halstead P, Page RA, Keevil SF, Williams SCR, Cherry SR, Design and development of an MR-compatible PET scanner for imaging small animals, *IEEE Trans. Nucl. Sci.*, 2005; 52 (5): 1376–1380.
64. Judenhofer MS, Catana C, Swann BK, Siegel SB, Jung W, Nutt RE, Cherry SR, Claussen CD, Pichler BJ, PET/MR images acquired with a compact MR-compatible PET detector in a 7-T magnet, *Radiology*, 2007; 244 (3): 807–814.
65. Catana C, Procissi D, Wu Y, Judenhofer MS, Qi J, Pichler BJ, Jacobs RE, Cherry SR, Simultaneous in vivo positron emission tomography and magnetic resonance imaging, *Proc. Natl. Acad. Sci. U S A*, 2008; 105 (10): 3705–3710.
66. Aull BF, Loomis AH, Young DJ, Heinrichs RM, Felton BJ, Daniels PJ, Landers DJ, Geiger-mode avalanche photodiodes for three-dimensional imaging, *Lincoln Lab. J.*, 2002; 13 (2): 335–350.
67. Pellion D, Borrel V, Esteve D, Therez F, Bony F, Bazer-Bachi AR, Gardou JP, APD detectors in the Geiger photon counter mode, *Nucl. Instr. Meth. Phys. Res. A*, 2006; 567: 41–44.
68. Renker D, Geiger-mode avalanche photodiodes, history, properties and problems, *Nucl. Instr. Meth. Phys. Res. A*, 2006; 567: 48–56.
69. Musienko Y, Reucroft S, Swain J, The gain, photon detection efficiency and excess noise factor of multi-pixel Geiger-mode avalanche photodiodes, *Nucl. Instr. Meth. Phys. Res. A*, 2006; 567: 57–61.
70. Nassalski A, Moszynski M, Syntfeld-Kazuch A, Szczesniak T, Swiderski L, Wolski D, Batsch T, Baszak J, Silicon photomultiplier as an alternative for APD in PET/MRI applications, *IEEE Med. Imag. Conf. Rec.*, Dresden, Germany, 2008; pp. 1620–1625.
71. Llosa G, Battiston R, Belcari N, Boscardin M, Collazuol G, Corsi F, Dalla Betta G-F et al., Novel silicon photomultipliers for PET applications, *IEEE Trans. Nucl. Sci.*, 2008; 55 (3): 877–881.
72. Herbert DJ, Saveliev V, Belcari N, D'Ascenzo N, Del Guerra A, Golovin A, First results of scintillator readout with silicon photomultiplier, *IEEE Trans. Nucl. Sci.*, 2006; 53 (1): 389–394.
73. Buzhan P, Dolgoshein B, Filatov L, Ilyin A, Kaplin V, Karakash A, Klemin S et al., Large area silicon photomultipliers: Performance and applications, *Nucl. Instr. Meth. Phys. Res. A*, 2006; 567: 78–82.
74. Otte AN, Barral J, Dolgoshein B, Hose J, Klemin S, Lorenz E, Mirzoyan R, Popova E, Teshima M, A test of silicon photomultipliers as readout for PET, *Nucl. Instr. Meth. Phys. Res. A*, 2005; 545: 705–715.
75. Di Domenico G, Zavattini G, Cesca N, Auricchio N, Andritschke R, Schopper F, Kanbach G, SiliPET: An ultra-high resolution design of a small animal PET scanner based on stacks of double-sided silicon strip detector, *Nucl. Instr. Meth. Phys. Res. A*, 2007; 571: 22–25.
76. Drezet A, Monnet O, Mathy F, Montemont G, Verger L, CdZnTe detectors for small field of view positron emission tomographic imaging, *Nucl. Instr. Meth. Phys. Res. A*, 2007; 571: 465–470.
77. Depres P, Izaguirre EW, Liu S, Cirignano LJ, Kim H, Wendland MF, Shah KS, Hasegawa BH, Evaluation of a MR-compatible CZT detector, *IEEE Med. Imag. Conf. Rec.*, Honolulu, HI, 2007; pp. 4324–4326.
78. Wagenaar DJ, Nalcioglu O, Muftuler LT, Szawlowski M, Kapusta M, Pavlov N, Meier D, Maehlum G, Patt B, Development of MRI-compatible nuclear medicine imaging detectors, *IEEE Med. Imag. Conf. Rec.*, San Diego, CA, 2006; pp. 1825–1828.

79. Azman S, Gjaerum J, Meier D, Muftuler LT, Maehlum G, Nalcioglu O, Patt BE et al., A nuclear radiation detector system with integrated readout for SPECT/MR small animal imaging, *IEEE Med. Imag. Conf. Rec.*, Honolulu, HI, 2007; pp. 2311–2317.
80. Szeles C, Cameron SE, Ndap JO, Chalmers WC, Advances in the crystal growth of semi-insulating CdZnTe for radiation detector applications, *IEEE Trans. Nucl. Sci.*, 2002; 49 (5): 2535–2540.
81. Chen H, Awadalla SA, Mackenzie J, Redden R, Bindley G, Bolotnikov AE, Camarda GS, Carini G, James RB, Characterization of traveling heater method (THM) grown $Cd_{0.9}Zn_{0.1}Te$ crystals, *IEEE Trans. Nucl. Sci.*, 2007; 54 (4): 811–816.
82. Awadalla SA, Chen H, Mackenzie J, Lu P, Iniewski K, Marthandam P, Redden R et al., Thickness scalability of large volume cadmium zinc telluride high resolution radiation detectors, *IEEE Med. Imag. Conf. Rec.*, Dresden, Germany, 2008; pp. 58–62.
83. Strul D, Cash D, Keevil SF, Williams SCR, Marsden PK, Gamma shielding materials for MR-compatible PET, *IEEE Trans. Nucl. Sci.*, 2003; 50 (1): 60–69.
84. Jiles D, *Introduction to Magnetism and Magnetic Materials*, CRC Press, Boca Raton, FL, 1998.
85. Schenck JF, The role of magnetic susceptibility in magnetic resonance imaging: MRI magnetic compatibility of the first and second kinds, *Med. Phys.*, 1996; 23 (6): 815–850.
86. Ludeke KM, Roschmann P, Tischler R, Susceptibility artefacts in NMR imaging, *Magn. Reson. Imaging*, 1985; 3: 329–343.
87. Bakker CJG, Bhagwandien R, Moerland MA, Fuderer M, Susceptibility artifacts in 2DFT spin-echo and gradient echo imaging—The cylinder model revisited, *Magn. Reson. Imaging*, 1993; 11: 539–548.
88. Camacho CR, Plewes DB, Henkelman RM, Nonsusceptibility artifacts due to nonmetallic objects in MR imaging, *J. Magn. Reson. Imaging*, 1995; 5: 75–88.
89. Wirrwar A, Vosberg H, Herzog H, Halling H, Weber S, Muller-Gartner H, 4.5 Tesla magnetic field reduces range of high-energy positrons—Potential implications for positron emission tomography, *IEEE Trans. Nucl. Sci.*, 1997; 44 (2): 184–189.
90. Raylmann RR, Hammer BE, Christensen NL, Combined MRI-PET scanner: A Monte Carlo evaluation of the improvements in PET resolution due to the effects of a static homogeneous magnetic field, *IEEE Trans. Nucl. Sci.*, 1996; 43: 2406–2412.
91. Rowe R, Aarsvold JN, Barrett HH, Chen JC, Klien WP, Moore BA, Pang IW, Patton DD, White TA, A stationary hemispherical SPECT imager for three-dimensional brain imaging, *J. Nucl. Med.*, 1993; 34: 474–480.
92. Goertzen AL, Jones DW, Seidel J, Li K, Green MV, First results from the high-resolution mouseSPECT annular scintillation camera, *IEEE Trans. Med. Imaging*, 2005; 24 (7): 863–867.
93. Genna S, Smith AP, The development of ASPECT, an annular single crystal brain camera for high efficiency SPECT, *IEEE Trans. Nucl. Sci.*, 1988; 35 (1): 654–658.
94. Delso G, Ziegler S, PET/MRI system design, *Eur. J. Nucl. Med. Mol. Imaging*, 2009; 36: S86–S92.
95. Zaidi H, Is MR-guided attenuation correction a viable option for dual-modality PET/MR imaging? *Radiology*, 2007; 244 (3): 639–642.
96. Zaidi H, Montandon M, Slosman DO, Magnetic resonance imaging-guided attenuation and scatter corrections in three-dimensional brain positron emission tomography, *Med. Phys.*, 2003; 30 (5): 937–948.

97. Beyer T, Weigert M, Quick HH, Pietrzyk U, Vogt F, Palm C, Antoch G, Muller SP, Bockisch A, MR-based attenuation correction for torso-PET/MR imaging: Pitfalls in mapping MR to CT data, *Eur. J. Nucl. Med. Mol. Imaging*, 2008; 35: 1142–1146.
98. Hofmann M, Pichler BJ, Scholkopf B, Beyer T, Towards quantitative PET/MRI: A review of MR-based attenuation correction techniques, *Eur. J. Nucl. Med. Mol. Imaging*, 2009; 36 (suppl 1): S93–104.
99. Hofman M, Steinke F, Scheel V, Charpiat G, Farquhar J, Aschoff P, Brady M, Scholkopf, Pichler BJ, MRI-based attenuation correction for PET/MRI: A novel approach combining pattern recognition and atlas registration, *J. Nucl. Med.*, 2008; 49: 1875–1883.
100. Kops ER, Herzog H, Alternative methods for attenuation correction for PET images in MR-PET scanners, *IEEE Med. Imag. Conf. Rec.*, Honolulu, HI, 2007; pp. 4327–4330.
101. Goyen M, Debatin JF, Healthcare costs for new technologies, *Eur. J. Nucl. Med. Mol. Imaging*, 2009; 36 (suppl 1): S139–S143.
102. Brix G, Nekolla EA, Nosske D, Griebel J, Risks and safety aspects related to PET/MR examinations, *Eur. J. Nucl. Med. Mol. Imaging*, 2009; 36 (suppl 1): S131–S138.
103. Wehrl HF, Judenhofer MS, Wiehr S, Pichler BJ, Pre-clinical PET/MR: Technological advances and new perspectives in biomedical research, *Eur. J. Nucl. Med. Mol. Imaging*, 2009; 36 (suppl 1): S56–S68.
104. Cherry SR, Louie AY, Jacobs RE, The integration of positron emission tomography with magnetic resonance imaging, *Proc. IEEE*, 2008; 96 (3): 416–438.
105. Antoch G, Bockisch A, Combined PET/MRI: A new dimension in whole-body oncology imaging? *Eur. J. Nucl. Med. Mol. Imaging*, 2009; 36 (suppl 1): S113–S120.
106. Heiss WD, The potential of PET/MR for brain imaging, *Eur. J. Nucl. Med. Mol. Imaging*, 2009; 36 (suppl 1): S105–S112.
107. Schlemmer H, Pichler B, Wienhard K, Schmand M, Nahmias C, Townsend D et al., Simultaneous MR/PET for brain imaging: First patient scans, *J. Nucl. Med.*, 2007; 48 (6): 45P.
108. Nekolla SG, Matinez-Moeller A, Saraste A, PET and MRI in cardiac imaging: From validation studies to integrated applications, *Eur. J. Nucl. Med. Mol. Imaging*, 2009; 36 (suppl 1): S121–S130.
109. Schlemmer HP, Pichler BJ, Krieg R, Heiss WD, An integrated MR/PET system: Prospective applications, *Abdom. Imaging*, 2008; http://www.SpringerLink.com
110. Qiao H, Zhang H, Zheng Y, Ponde DE, Shen D, Gao F, Bakken AB et al., Embryonic stem cell grafting in normal and infracted myocardium: Serial assessment with MR imaging and PET dual detection, *Radiology*, 2009; 250 (3): 821–829.
111. Lee HY, Li Z, Chen K, Hsu AR, Xu C, Xie J, Sun S, Chen X, PET/MRI dual-modality tumor imaging using arginine-glycine-aspartic (RGD)-conjugated radiolabeled iron oxide nanoparticles, *J. Nucl. Med.*, 2008; 49 (8): 1371–1379.

15

Reducing Respiratory Artifacts in Thoracic PET/CT

Greta S.P. Mok, Tao Sun, and Chi Liu

CONTENTS

15.1 Introduction

It has been more than a decade since the introduction of the combined positron emission tomography and computed tomography (PET/CT) [1,2], which has been well recognized as the most sensitive tool for detecting neoplasm. The PET/CT technology, including detectors, reconstruction algorithms, and image analysis, matures over the years while researchers are still striving for better image quality and quantitation with lower radiation dose. To achieve this goal, various imaging degradation factors need to be compensated and we mainly focus on attenuation and respiratory motion blurring in this chapter.

For medical radionuclide imaging, the emitted gamma rays are attenuated in the body mainly via photoelectric absorption and Compton scattering before reaching the detectors. The photons originated from the center of the object would be more likely to be attenuated as compared to those that arise from the peripheral. Thus, this effect is positively related to patient body sizes and would contribute to gradually, artifactual decrease of activity toward the center of the object. This effect is also more prominent for tracers with lower energy and for regions with higher attenuation coefficients, e.g., bone and metallic implants. Attenuation correction is critical for improving image quality and quantitation, and is the most important correction in PET.

Initial efforts of attenuation correction in emission tomography include conjugate-counting techniques especially using the geometric mean of the conjugate projections and the Chang's multiplicative method [3]. In the beginning, both methods assume a constant attenuation coefficient across the patient body, which is a reasonable estimate for brain and abdomen, to obtain the attenuation correction factor (ACF) by estimating the body thickness via the preliminary uncorrected images [4]. However, one may expect this assumption does not hold for most other regions like thorax and pelvis, where the presence of air and bone cannot be neglected. The urge of the attenuation coefficient for each voxel led to the requirement of an additional transmission scan.

Before combined PET/CT scanners were invented, attenuation correction on standalone PET was performed mostly with an external ^{68}Ge 511-keV pin source that rotated around the patient. The pre-injection transmission scan required patients to stay on the bed of the scanner for the tracer distribution before the emission acquisition, thus limiting the patients' throughput. Though simultaneous acquisition with the emission scan was feasible [5], the post-injection transmission scan was usually performed right after the emission scan, thus saving a lot of time and also reducing the chances of patient movement between the transmission and the emission scans. However, the photons emitted from the patients could interfere with the transmission scan and degrade the count rate performance of the detectors. Thus, one needed to track the location of the external source and its targeted detectors for the transmission scan. The limited count rate of this transmission scan may substantially affect the quantitative accuracy or cause artifacts in the emission data [6]. The anatomical information provided by the transmission scan was also limited, thus it was not suitable for diagnostic purpose.

Nowadays, most clinical PET/CT machines are no longer equipped with ^{68}Ge-based source. Instead, x-ray CT-based attenuation correction was commonly used in the combined PET/CT scanner [7]. This new design produced huge numbers of photons in the transmission scan even at low tube currents. Hence, it is much faster than the ^{68}Ge-based attenuation correction scan and has lower statistical noise level [8,9]. Also, CT-based attenuation correction ensures the significant energy difference between the gamma-rays (511 keV) and x-rays (~30–120 keV), so that the transmission scan would not be contaminated by the emitted gamma-rays from the patients. At the same time, it is necessary to transform the CT images acquired from an integrated energy spectrum of ~30–120 keV to attenuation maps of 511 keV for attenuation correction in the emission data. This conversion usually involves segmentation of bone and nonbone materials, and/or applying the bilinear scaling methods [8,10]. The use of CT-based attenuation maps leads to a more accurate activity concentration values and better uniformity [11], and provides more anatomical information for the correspondingly registered emission data.

On the other hand, CT-based attenuation correction for PET is hampered by artifacts that are usually not seen in the standalone PET images. Examples are artifacts due to respiratory motion [12–14], truncation [15], metallic implants [16,17], and CT contrast agents [18,19]. Among all these factors, respiratory motion is generally considered to be the main problem in CT-based attenuation correction. For standalone diagnostic CT, the optimized protocol is to obtain a 3D helical acquisition of the thoracic cavity over a single full-inspiration breath-hold CT scan. When applied with the emission exam, this technique captures a snapshot of the thoracic cavity at a distinct respiratory phase and does not represent the time-averaged position of the thoracic structures as PET acquisition does. This misalignment between transmission and emission scan is most noticeable at the left lung especially for two regions: (i) tissues around lower thorax and upper abdomen, where the transmission scan may not be presented in the emission scan (Figure 15.1); (ii) tissues around the lung and left-ventricle interface, where myocardial uptake is prominent overlying the left lung of the CT in the fused images (Figure 15.2). In fact, for thoracic structures, more than 40% of the studies have misalignments between the measured and the true position [20]. Erdi et al. examined PET/CT images of five patients with multiple lung carcinoma lesions, and showed that the spatial mismatch resulted in upto a 30% error in the standardized uptake value (SUV) of the lesions [21]. Also, phantom studies showed the effect of motion can result in as much as 75% underestimation of the maximum activity concentrations [22]. In a simulation study with over 1000 real patient respiratory traces, in addition to SUV underestimation caused by motion blurring, mismatched attenuation correction can cause SUV overestimation for lower lung tumors which are close to the liver dome, leading to complicated SUV errors [23]. These PET/CT mismatch artifacts may lead to inaccurate localization of tumors and hence potential misdiagnoses [13,24,25].

Besides the matching of PET and CT data, the respiratory motion during PET acquisition itself also causes substantial degradations of spatial resolution, contrast-to-noise ratio, size and shape of the lesions, and quantitative accuracy. For PET/CT-guided radiotherapy, it can cause the tumor size to be overestimated by up to 130% [23], thus increasing the planning target volume. Dawood et al. showed that the respiratory motion can cause the thoracic/abdominal tumors to move with an amplitude of ~6–23 mm [26],

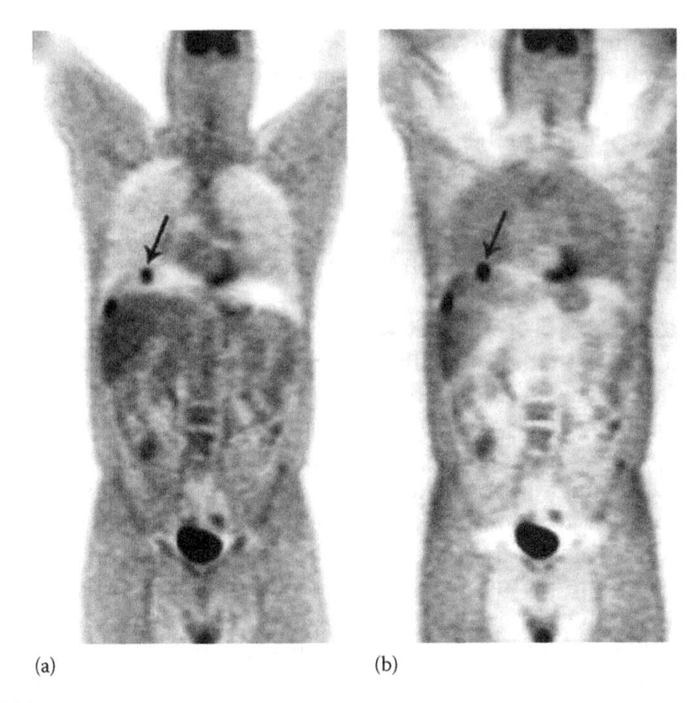

FIGURE 15.1
PET reconstructed images of a 58-year-old man with colon cancer. (a) A lesion at the dome of liver was mislocalized to right lung (arrow) because of the respiratory motion. (b) Images without attenuation correction showed that all lesions were confined to the liver. (Reprinted from Sureshbabu, W. and Mawlawi, O., *J. Nucl. Med. Technol.*, 2005, 33(3), 156. With permission.)

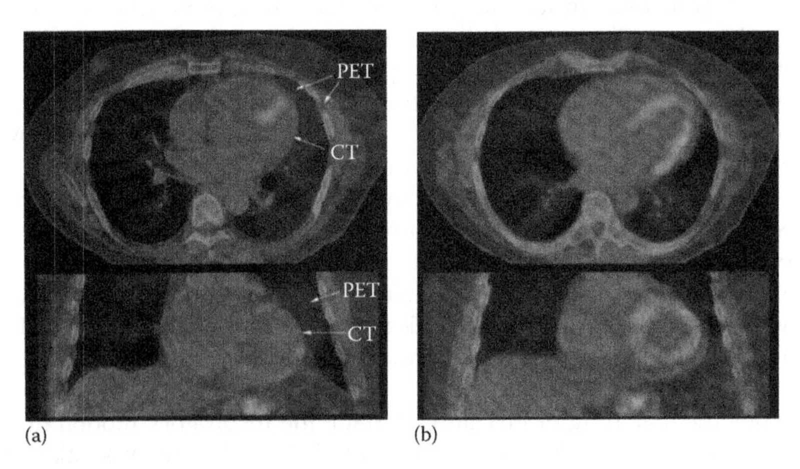

FIGURE 15.2
(See color insert.) (a) Helical CT and PET fusion images in transaxial (top) and coronal (bottom) views showed marked misregistration. Arrows indicated the mismatch of heart borders between the helical CT and the PET images, and an artifactual defect in the myocardium. (b) For the same patient, cine average CT and PET fusion images showed no misregistration and artifactual defects. (Reprinted from Gould, K.L. et al., *J. Nucl. Med.*, 2007, 48(7), 1112. With permission.)

and Erdi et al. demonstrated that the degree of blurring is proportional to the breathing magnitude [27]. Xu et al. showed that the blur extent was about 12.1 ± 3.7 mm in their blur reduction study [28].

Effective methods are warranted to reduce the PET/CT respiratory artifacts. Three main categories of these methods have been mostly investigated so far: breathing instruction, CT protocols, and gated four-dimensional (4D) PET/CT.

15.2 Breathing Instruction-Based Methods

15.2.1 Shallow Breathing

Instead of deep-inspiration breath-hold protocol in diagnostic CT, some studies acquire both CT and PET under continuous shallow breathing state, which has relatively smaller motion amplitudes. However, studies showed that this protocol resulted in an unsatisfactory evaluation of the lung parenchyma on CT images (Figure 15.3), where subcentimeter nodules can be missed to cause inaccurate comprehensive cancer staging [29].

Multidetector CT technology that employs six or more detector rows can reduce magnitude and frequency of respiratory artifacts in free breathing PET/CT [30]. This is because, with multidetector CT technology, the CT examination time of whole-body PET/CT scan is shortened to 20 s or less. Thus, patients' irregular breathing pattern will be significantly mitigated.

15.2.2 Normal End-Expiration Breath-Hold

Some researchers believe normal end-expiration breath-hold is the best option for transmission scan in thoracic PET/CT [31,32]. In a survey performed

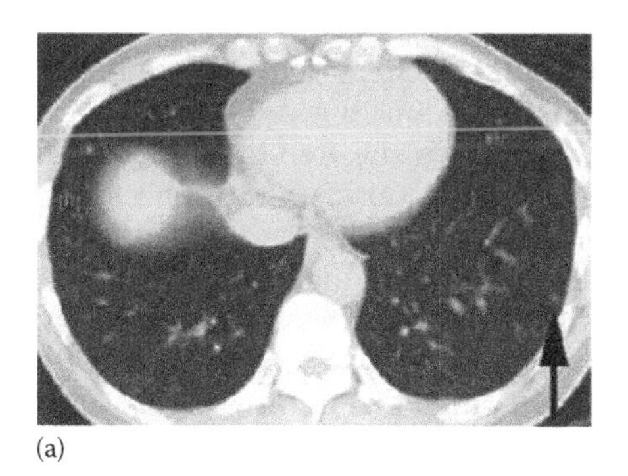
(a)

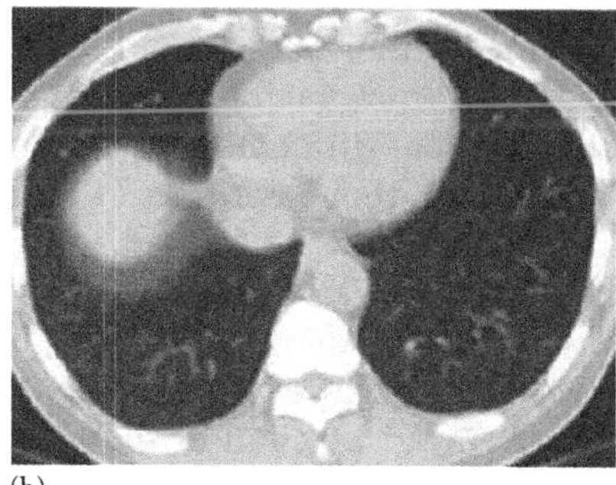
(b)

FIGURE 15.3
A 4-mm nodule in left lower lobe of lung (arrow) was seen on (a) deep-inspiration breath-hold CT, but was missed on (b) shallow-breathing CT for a patient with pancreatic cancer. (Reprinted from Allen-Auerbach, M. et al., *J. Nucl. Med.*, 2006, 47(2), 298. With permission.)

by Goerres et al. [10], CT scans were performed at four respiratory levels: free-breathing, maximal-inspiration, maximal-expiration, and normal-expiration. For different breathing protocols, multiple distances between the anatomical landmarks or a reference point in the CT and corresponding PET images were compared. Normal-expiration breath-hold showed the best results with better match and smaller range of measured distances as compared to other breathing patterns for the upper abdominal organs. Furthermore, this CT protocol reduced the occurrence and the severity of respiratory curvilinear artifacts on co-registered PET/CT images. Another study showed that reconstructed images from patients of normal-expiration breath-hold group had 28% less incidence rate of artifacts as compared with those from the free-breathing group [32]. One possible explanation is that a normal human spends most time in expiration in the whole respiratory cycle. Further comparison study showed that normal-expiration CT scan for attenuation correction gave good SUV recovery in the case of mobile tumors with size of 20 mm, even when the magnitude of breathing was large [33]. Attenuation correction using maximal-inspiration breath-hold protocol resulted in more serious underestimation of SUV, especially when breathing amplitude increased.

The clinical challenge of normal end-expiration breath-hold protocol is patient's compliance and comfort. The procedure needs to be rehearsed before transmission scan. Many patients are incapable of maintaining breath-hold for the duration of the whole-body CT examination, even for the most advanced multidetector CT scanner. It has been reported that 40%–60% patients with lung cancer are unable to consistently hold their breath for a long period [34]. This results in severe compensatory breathing artifacts at the midscan level, i.e., the lower thorax to the upper abdomen region, when the patient resumes breathing during the CT acquisition. These patients are suggested to use the shallow breathing protocol instead.

Even under the assumption of successful breath-hold, it is still doubtful if the normal end-expiration breath-hold CT will match with the PET data as it is difficult to instruct the patient to obtain a predefined position of the diaphragm. Furthermore, the shape of the diaphragm in normal end-expiration may differ from the free breathing in the PET acquisition, as breath-holding may generate different muscle tension [35] as in the free breathing state. Active expiration breath-hold originates from the muscles of the chest wall, and thus will push up the diaphragm and generate an abnormal shape of the diaphragm as compared to normal expiration without breath-holding. Also, normal end-expiration breath-hold CT still captures a snapshot of the thoracic cavity in a distinct respiratory phase and does not represent the time-averaged position of the thoracic structures as PET acquisition does. Hamill et al. [33] indicated good SUV recovery of normal-end expiration was actually due to canceling effects of underestimation due to motion blurring and overestimation due to attenuation. For small lung tumors near the diaphragm, their phantom study showed the inaccurate estimation of SUV after attenuation correction with normal-end expiration CT.

15.2.3 Deep-Inspiration Breath-Hold (DIBH) PET/CT

Nehmeh et al. proposed a deep-inspiration breath-hold protocol for both CT and PET (DIBH PET/CT) acquisition [36]. The patient was instructed to breathe deeply and then hold the breath, under the monitoring with an amplitude gating device, i.e., the real-time position management (RPM, Varian Medical Systems, Palo Alto, CA). Breath-hold CT data were acquired for about 16 s in the helical mode. PET scan was divided to nine 20 s frames, i.e., with a total acquisition of 3 min for one bed position. In the beginning of each PET frame, patient was instructed to breathe and hold the breath again as in DIBH CT. The inflation level was defined on the fly during DIBH CT session. The breath-hold signal exhibited a relaxation period of 1–2 s before it stabilized. Therefore, it was necessary to wait for 1–2 s before starting the acquisition. PET data were acquired for only one bed position. The whole process was shown in Figure 15.4. This method showed an increase in lesion

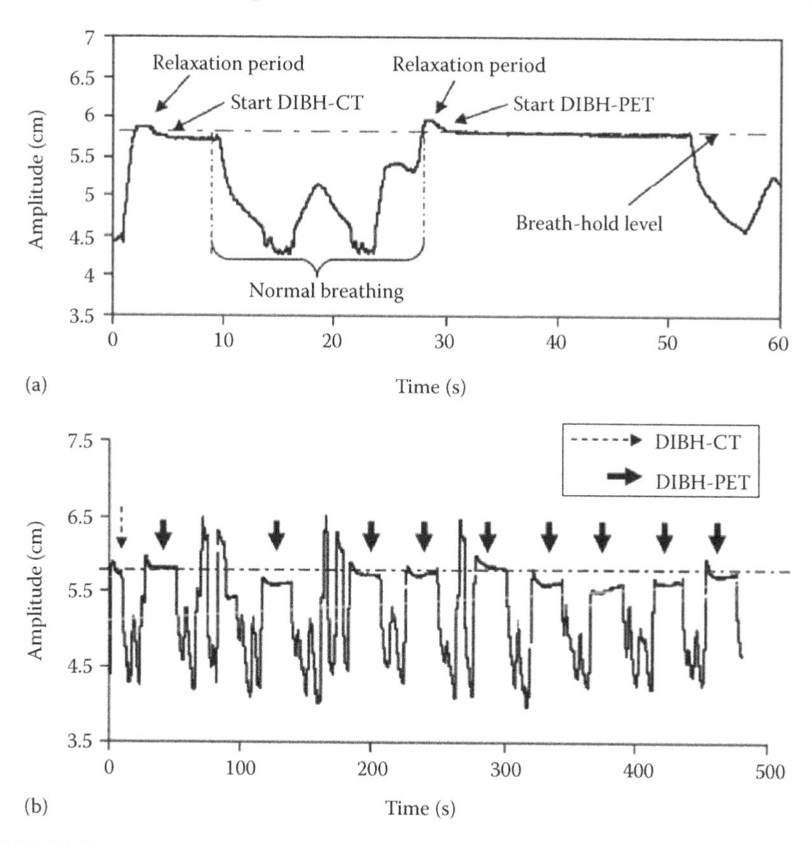

FIGURE 15.4
RPM breathing signals of a patient undergoing DIBH PET/CT acquisition mode. (a) The patient first did the breath-hold on CT and then started the first PET frame. Dashed line corresponded to the inflation level at which patient was instructed to hold the breath. (b) The whole DIBH PET/CT session. (Reprinted from Nehmeh, S.A. et al., *J. Nucl. Med.*, 2007, 48(1), 22. With permission.)

SUV of as much as 83% and a reduction in the distance between the centroids of PET and CT lesions as much as 49%, compared to the deep-inspiration breath-hold PET/CT. Similar study was proposed by Meirelles et al. [37] and Torizuka et al. [38]. They both showed increase in SUV and more precise localization and quantification of lesions when using the DIBH PET/CT technique. Better diagnostic accuracy can be achieved by using DIBH PET with a single 30 s, 45 s, or even 60 s scan in one frame [39,40].

By now, the DIBH PET/CT method was only explored in lung lesion studies. Additional external monitoring such as RPM and breathe-coaching are also needed to measure the respiratory cycle, ensuring the matching of the transmission and the emission respiratory phases. However, even a single breath-hold for 20 s may not be acceptable for senior patients or patients who have underlying lung diseases such as emphysema or pulmonary fibrosis [38]. Unsuccessful breath-hold due to irregular breathing pattern may cause highly variable results, which will affect the diagnostic accuracy.

15.3 CT Protocol–Based Methods

There are two main scan modes for CT acquisition, the axial (step-and-shoot) mode and the helical (spiral) mode. The step-and-shoot CT consists of two stages: (i) the patient remains stationary while the x-ray tube and gantry rotate around the patient to acquire a complete set of projections at a prescribed scanning location; (ii) the tube is off and the patient is translated to the next prescribed axial scanning location. The duty cycle (scan time/total time) of the step-and-shoot CT is ~50% at best. The helical CT [41,42] was then introduced to solve the interscan missing problem as in the step-and-shoot CT. The CT data are acquired while the patient is continuously transported through the gantry with a constant speed. The duty cycle of the helical CT is nearly 100%. As the scanning speed performance can be substantially improved, the scanner is able to image a given volume in a shorter time as compared to the step-and-shoot mode.

Specific CT protocols based on the axial mode or helical mode have been proposed for attenuation correction in PET images, with the same idea to acquire CT data over many respiratory cycles to ensure the CT attenuation map matches with the PET. Thus, breathing-induced artifacts such as PET/CT spatial mismatch and underestimation of the SUV would be reduced.

15.3.1 Slow CT

Lagerwaard et al. introduced slow CT under helical mode to average several respiratory cycles over the scan duration [43]. The tube rotation time of a single-slice CT was slowed down to 4 s/rot. Pitch was set to 1. Compared with the diagnostic deep-inspiration breath-hold CT, slow CT was closer to the average position of the thoracic cavity structures in PET.

In another study from Sorensen et al. [44], severe "gap" artifacts between successive PET reconstructed slices using slow CT-based attenuation correction were observed. It was possibly due to the inconsistencies in the projections from large respiratory motions combined with slow CT propagated into the attenuation-corrected PET images. Hence, this method may not be appropriate for attenuation correction in PET/CT for large respiratory motion amplitudes.

15.3.2 Low-Pitch CT

New generations of CT provide more configurations for helical mode. Nye et al. set the pitch as 0.562, which was the lowest value allowed by the scanner [45]. This pitch value increased the axial sampling during free-breathing mode without increasing the tube rotation time. A total of 16 s scan was used to cover the chest cavity. Final CT data were matched with the PET slice thickness for attenuation correction and to suppress respiratory motion artifacts. Compared to deep-inspiration breath-hold CT, this low-pitch CT reduced the number of problematic studies from 71% to 28%.

One concern for low-pitch CT is the radiation burden. Due to the potential longer exposure time, higher radiation dose is delivered to the patients for the low-pitch CT. The radiation dose was approximately 2.3 mSv as compared to a 1.8 mSv of the diagnostic breath-hold CT. The dose length product was 153 mGy, also higher than the conventional method of 117 mGy [45]. The relative high radiation dose hampers the low-pitch CT from common clinical applications.

15.3.3 Cine Average CT (CACT)

Pan et al. introduced 4D CT for PET/CT attenuation correction [46], see Chapter 7 in this book. 4D CT provided images of all phases of the breathing cycle for tumor staging and radiation therapy treatment planning [47]. Cine mode technique, modified from step-and-shoot mode, acquired repeated axial CT images at each table position for a certain time period. This method produced even thinner slice thickness than those of a low-pitch helical scan.

Pan et al. [46,48] and Cook et al. [49] averaged the images of 10 phases in 4D CT to form a respiratory cine average CT (CACT) at each table position (Figure 15.5) to cover the whole thorax region. The acquisition time for each table position is 5.9 s. The average CT of the thorax was then combined with the helical CT data of the regions outside the thorax, e.g., abdomen, to make up the integrative CT images. This method greatly improved the registration accuracy and tumor SUV as compared to attenuation correction using deep-inspiration or normal end-expiration breath-hold helical CTs (Figure 15.6). Similar results were achieved by Gould et al. [20].

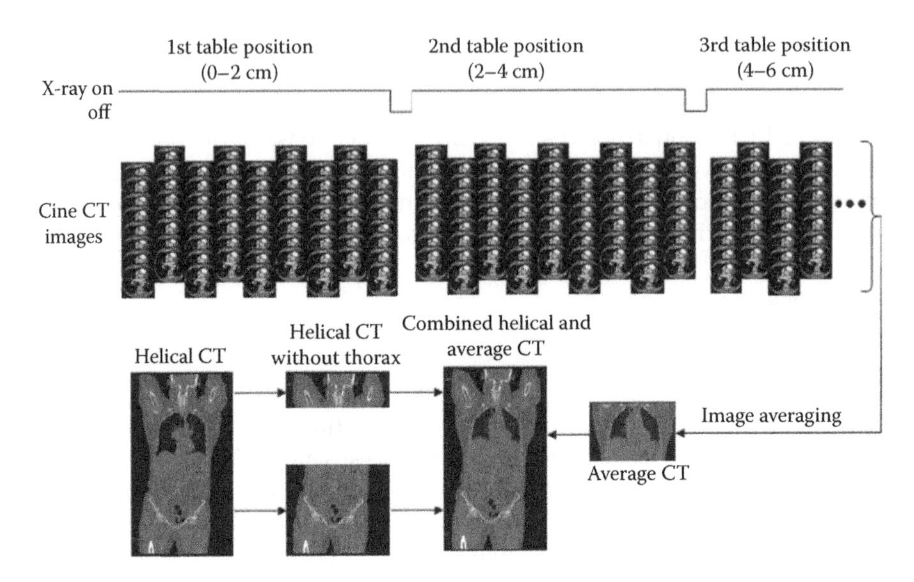

FIGURE 15.5
Top row: After cine scan at the first location, the table travels another 2 cm. After averaging the cine images, a full set of CT images can then be formed by combining the helical CT images above and below the thorax and the CACT images of the thorax for attenuation correction in the PET emission data. (Reprinted from Pan, T. et al., *Med. Phys.*, 2006, 33(10), 3931. With permission.)

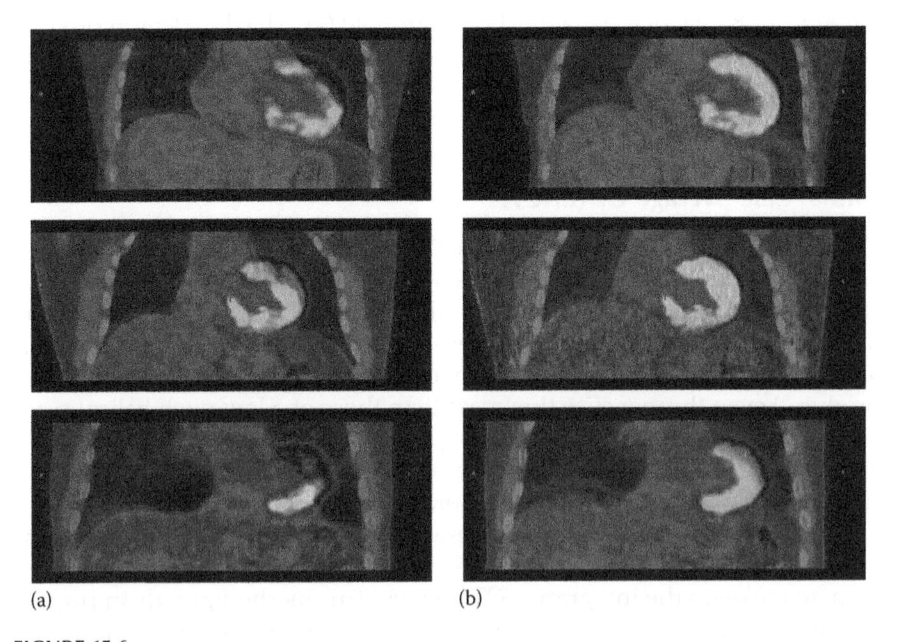

FIGURE 15.6
(See color insert.) Fused coronal images of the PET data with (a) helical CT and (b) CACT for two different patients. (Reprinted from Pan, T. et al., *Med. Phys.*, 2006, 33(10), 3931. With permission.)

Alessio et al. further evaluated both average and intensity maximum images of 4D CT for the PET/CT misalignment reductions [50]. Fewer misalignments were observed with intensity-maximum 4D CT. Moreover, compared to helical CT, 4D CT was more flexible to be retrospectively averaged or processed.

There are mainly two adverse problems for CACT: (i) increasing the CT acquisition time due to the longer period of each bed position; (ii) increasing the radiation dose. Lowering the tube current can potentially reduce the dose by a wide margin [48].

15.3.4 Interpolated Average CT (IACT)

An alternative method for CACT is interpolated average CT (IACT) to reduce radiation dose with similar image quality [51]. CT images of desired phases, e.g., end-inspiration and end-expiration phases from a respiratory cycle, are used to generate the velocity matrix using deformable image registrations such as optical flow method (OFM) [52] or B-splines. Interpolated phases are then obtained via linear interpolations or based on the organ movement functions [53]. The IACT is then calculated by averaging the original and interpolated phases. In a clinical study, Huang et al. obtained the IACTs using different numbers of desired phases from a cine CT [51]. The PET images were then reconstructed using different IACTs, CACT, and helical CT for attenuation correction. The maximum SUV difference between the use of IACT and CACT was about 3%. The radiation dose using IACT with two original phases, i.e., end-expiration and end-inspiration, could be potentially reduced by 85% as compared to the use of CACT.

A further simulation study showed that IACT was a robust method that worked for maximum respiratory motion amplitude of up to ~3 cm [54]. Another study investigated the clinical feasibility of IACT and its potential radiation dose reduction using an active breathing controller (ABC) [55]. If the CT scans were acquired during the voluntary breath-holds performed by the patients themselves, they could probably not represent the normal breathing state as in the PET acquisition. Thus, the ABC was introduced to capture the desired breathing phases. It is a noninvasive device that integrated a spirometer, an air mask, and a tube-valve system. The user can manually close the valve depending on the read-in breathing signal to suspend the patient's breathing for a desired period of time (<10 s). The results in a preliminary clinical study showed that IACT using ABC provided improved image quality as compared to conventional helical CT (Figure 15.7).

The aforementioned methods all aimed to correct the misalignments between PET and CT by modifying the CT images to match with the PET images. However, respiratory motion blur always exists in the static PET, and it cannot be compensated by only matching the CT with the static PET images. The following techniques aim to correct the motion blurring by further processing the PET raw data or reconstructed images.

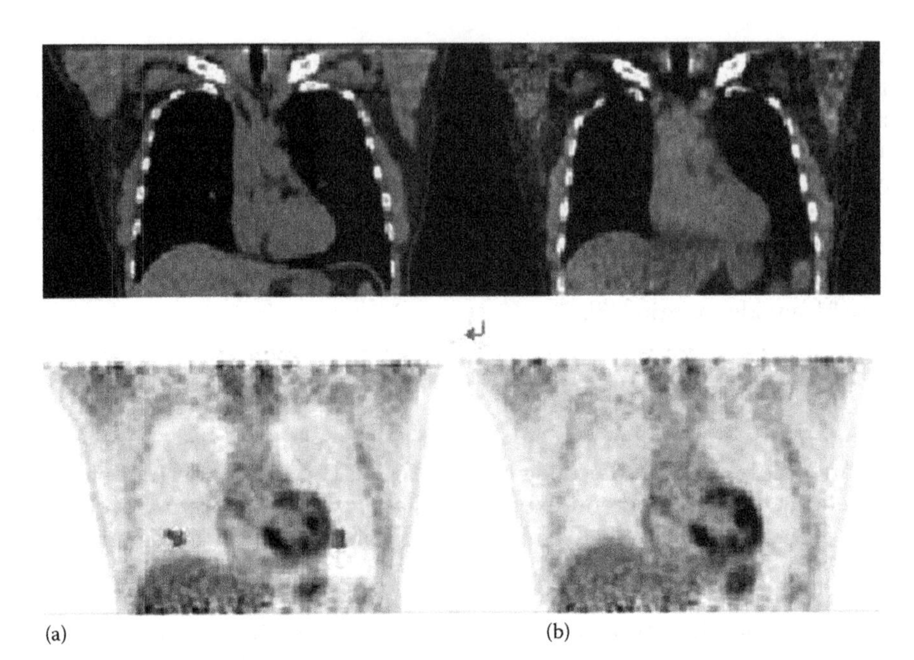

(a) (b)

FIGURE 15.7
(a) Sample images of helical CT (top) and its corresponding PET reconstructed images (bottom). (b) Sample images of IACT (top) and its corresponding PET reconstructed images (bottom). Red arrows: misalignment artifacts. (From Sun, T. et al. Low dose interpolated average CT for PET/CT attenuation correction using an Active Breathing Controller (ABC), In Conference Record of *IEEE Nuclear Science Symposium Medical Imaging Conference*, Anaheim, CA, October 28–November 3, 2012.)

15.4 Gated 4D PET/CT

Respiratory gated PET was firstly proposed for the standalone PET in brain scan to remove patient motions [56]. To reduce smearing due to the breathing motions and improve quantification of ^{18}F-FDG uptake in lung lesions, Nehmeh et al. proposed respiratory gated PET [57]. PET data were acquired into discrete bins in synchrony with the breathing cycle. In this study, 10 bins data were acquired for a FOV, with 300–500 ms time interval between each 1 min gated bin (Figure 15.8). The lesion motion was negligible due to the short acquisition interval and thereby approximately motion-free images were obtained. During the gated PET acquisition, different respiratory tracking systems, such as RPM, pressure-monitor belt, and spirometer, have been used to monitor the respiratory motion and generate a trigger for the PET scan at predefined amplitude or time phase. Studies using this technique showed an improvement in the target-to-background ratio and a more accurate measurement of the SUV [57,58].

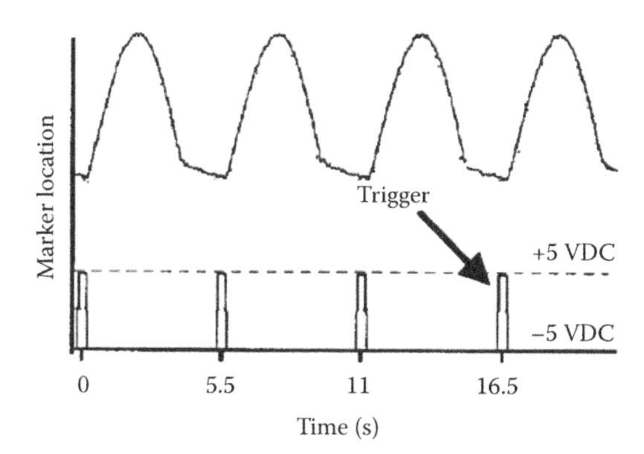

FIGURE 15.8
Respiratory gating signals indicated by RPM marker location and the corresponding electrical triggers for PET acquisition. (Reprinted from Nehmeh, S.A. et al., *Med. Phys.*, 2002, 29(3), 366. With permission.)

Each gated PET image is usually directly corrected by an unsynchronized breath-hold CT. Hence, the reconstructed image quality in some gates is even worse than those of the non-gated PET with attenuation correction by a conventional breath-hold CT due to worse mismatch. Instead of manipulating CT protocols to match the static PET, 4D PET/CT applied 4D CT on gated PET images to facilitate phase-matched CT-based attenuation correction.

In 4D PET/CT imaging, the previously described 4D CT and gated PET are combined [59]: the 4D CT images are spatially matched with the gated PET images. In synchrony with the externally monitored breathing signal, respiratory gated PET data are acquired into discrete bins. 4D CT data are acquired and sorted according to their phases to generate a respiratory gated CT data. To make the CT images in each phase coincide with the PET images, the gated CT images at each bin are then spatially rebinned and resliced in the axial direction. The gated PET data are then corrected for attenuation with the corresponding gated CT data.

In a clinical study, Nehmeh et al. measured distances of the lesion centroids between the gated PET and the phase matched 4D CT [59]. The result showed an improvement in lesion registration of PET and CT up to 41% as compared to the registration between gated PET and deep-inspiration breath-hold CT (Figure 15.9). Also, a reduction in PET derived tumor volumes of up to 42% and an increase in lesion SUV of up to 16% were also found. Similar studies [60–63] all showed that 4D PET/CT provided superior results as compared to gated/non-gated PET with a breath-hold CT for attenuation correction.

15.4.1 Registration-Based Methods

One may notice that in 4D PET/CT, each PET bin only contains a small fraction of detected PET events. The resultant reconstructed images, therefore, have higher noise level, thus a poor signal-to-noise ratio (SNR).

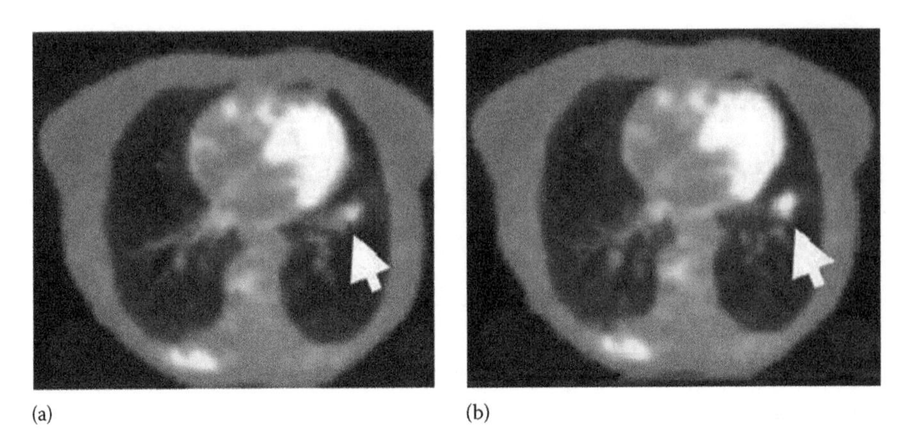

FIGURE 15.9
(See color insert.) Sampled transaxial fused PET/CT images demonstrated a thoracic lesion of a patient in (a) standard nongated PET acquisition, and (b) 4D PET/CT acquisition. (Reprinted from Nehmeh, S.A. et al., *Med. Phys.*, 2004, 31(12), 3179. With permission.)

In principle, it is better to utilize all PET information from the whole respiratory cycle instead of only one phase. One method to achieve this goal is to transform each respiratory PET bin to a referenced target bin that corresponds optimally to a matched CT phase from the gated CT data before attenuation correction (Figure 15.10). Several researchers used the motion vectors derived from gated CT and/or gated PET using different registration methods to guide the transformations, and there are two main categories of these applications.

1. Cardiac imaging

   The movement of the heart due to breathing is difficult to model. It can be approximated by a six-parameter rigid transformation

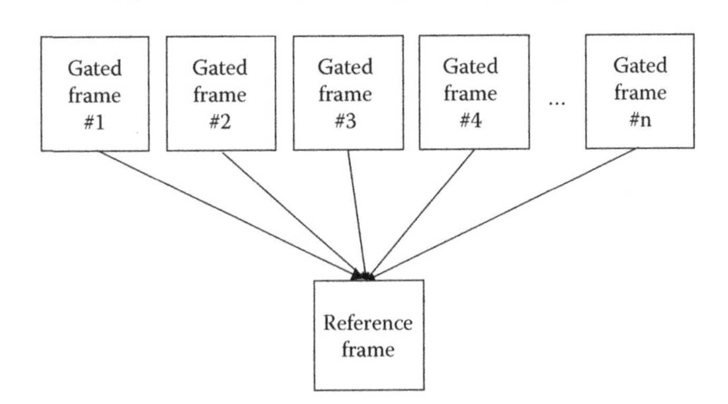

FIGURE 15.10
In 4D PET/CT gated schemes with registration, PET images are reconstructed using information from the complete dataset by registering all the frames to the reference frame.

with simple translational and rotational steps [64,65]. Livieratos et al. used this algorithm to reposition the heart to a referenced bin using the list-mode projection data before reconstruction [66]. The rigid motion was calculated from the simulated six PET respiratory gates of the digitized phantom. Substantial respiratory motion compensation in the myocardium was observed in the reconstructed images as compared with the ones without transformation in the list-mode data. Recovery of uptake contrast especially in the defect, apical, and basal myocardial areas were clearly demonstrated using the proposed method. McQuaid et al. proposed a realignment method to register a single CT with different gated PET frames, based on a rigid registration of the heart and a statistical shape model of the diaphragm [67]. Their patient studies showed that this technique led to PET reconstructed images that were closer to those with attenuation correction using a corresponding gated CT. The improved diaphragm matching between PET and CT images resulted in better quantitative accuracy. Klein et al. [68] indicated that an affine model may provide better registrations and motion correction than the technique restricted to the six-parameter rigid body assumption. Affine linear transform algorithm is a rigid motion transform for shear and compression, which is appropriate for deformation of the heart especially at the right coronary artery and left ventricle [65]. Although only small improvements were observed after performing affine model as compared to the rigid transformation model, further improvements are expected in next-generation PET scanners with better system resolution.

2. Lung imaging

Lung motion due to respiration is nonrigid in nature, and the deformation amplitude of the lungs during respiration is larger than that of the heart. Thus, CT attenuation maps are not always coregistered to PET images, especially at the lower thorax and diaphragm where large deformations often happen. This will lead to inaccurate PET reconstructed images. Therefore, for lung imaging, nonrigid motion correction algorithms would be superior to rigid motion correction algorithms.

Thorndyke et al. proposed a retrospective stacking method [69] for gated PET image processing. They retrospectively grouped all the gated PET images into several bins along measured amplitude of the respiratory cycle using RPM. These bins were then registered one-by-one to a referenced bin through B-spline deformable transformation, and then stacked and averaged to form a composite image. This method yielded reduced blurring and increased SNR in the reconstructed images as compared to conventional gated

4D PET/CT. Dawood et al. used the optical flow method (OFM) to calculate 3D motion vectors from the gated PET images [26]. For gated PET, OFM was further developed for estimating motion vectors between any bin and the target bin. The reconstructed image contained all PET information with minimal motion, leading to more accurate attenuation correction and quantification. The improvement of contrast slightly varied for lung lesion of different locations and sizes [33], and the smaller lesions (7 and 11 mm) suffered from more significant partial volume effect (PVE) as compared to the bigger lesions [70,71].

The 4D PET/CT transform algorithm is very computationally intensive. It also requires gating hardware to support the acquisition. In the meantime, this method may not be feasible for most clinics because of the high complexity of implementation. A practical alternate is quiescent period gating [72]. Studies showed that patients tend to spend more breathing time dwelling at the end-expiration quiescent period and also breathe to the same end-expiration location [23]. These results indicated that imaging at end-expiration could reduce motion blurring, while still retaining a large fraction of detected events in PET. Based on these findings, quiescent period gating techniques retrospectively extracted only the end-expiration quiescent portion of the patient breathing cycles from the PET list mode data according to external monitored motion signal to form a single PET image volume that had the optimal signal and noise trade-off. For attenuation correction, CT image acquired during end-expiration phase had the best match with end-expiration PET data and thus was used. This technique has been implemented in commercial PET/CT scanners.

15.4.2 Reconstruction-Based Methods

While registration-based 4D PET/CT methods process either the PET raw projection data or the reconstructed images by registering and averaging individual gated frames, motion compensation can also be done by incorporating deformations information among different frames into the iterative reconstruction process. The reconstruction-based method may lead to superior image quality than the registration-based method.

Lamare et al. [73] indicated the elastic motion correction can be integrated into the system matrix in the PET reconstruction process. If P is the system matrix that describes the PET system, the data acquisition in PET can be represented by the following equation:

$$m = Pf \tag{15.1}$$

where

m is the measured datasets

f is the radioactive distribution at the referenced respiratory location

The 3D spatial transformations for correcting the respiratory motion were obtained from the simulated 4D CT data. Deformation matrices were derived among all individual frames (acquired at time t) and the referenced frame of the full exhalation (acquired at time t_0)

$$m_t = P_t f \tag{15.2}$$

where

m_t is the measured datasets at variable time t

P_t is the system matrices at variable time t

f is the radioactive distribution at time t_0

This elastic-based respiratory motion correction can be integrated into the one-pass list-mode expectation-maximization (OPL-EM) algorithm

$$f^{k+1} = \frac{f^k}{S} \sum_N P_t^T \frac{1}{P_t f^k} \tag{15.3}$$

where

S is the sensitivity image used to correct for attenuation and normalization

N is the number of temporal gated frames

The results showed that incorporating the spatial transformation into reconstruction leads to a superior contrast of 20%–30% on average in the recovered lesion, in comparison to registering the reconstructed images of individual gated frames. Thus, motion-free PET images would be obtained after the attenuation correction with the end-expiration CT data. Similar results were also demonstrated in [74,75].

Grotus et al. [76] introduced the 4D joint-estimation algorithm to form a new 4D-OS (ordered subset)-EM reconstruction for gated PET images. They assumed the activity distribution at time t $f(i,t)$ of each voxel i can be written as a linear combination of a small number (N) of temporal basis functions

$$f(i,t) = \sum_{n=1}^{N} b_n(t) w_n(i) \tag{15.4}$$

Only the N basis functions $b_n(t)$ and the weights of each voxel $w_n(i)$ or these basis functions needed to be estimated. As the basis functions span the whole temporal range, basis functions and weight image functions were updated iteratively as in the OS-EM technique

$$w^{k+1} = \frac{w^k}{S} \sum_N Cb_t^T P_t^T \frac{1}{Cb_t P_t w^k} \tag{15.5}$$

$$b^{k+1} = \frac{b^k}{S} \sum_N Cw_t^T P_t^T \frac{1}{Cw_t P_t b^k} \tag{15.6}$$

where
- S is the sensitivity image used to correct for attenuation and normalization
- P_t is the system matrices at variable time t
- Cb_t and Cw_t are the matrices used to connect the basis function $b_n(t)$ and the weight function $w_n(i)$

The N basis functions $b_n(t)$ were initialized as sinusoids with all sinusoids having the same frequency but with different phases. The weight image functions $w_n(i)$ are initialized as uniform images. The weight image functions and the basis functions were both updated for every subset by running Equations 15.5 and 15.6 successively.

The advantage of this method was that it did not need any estimation of the deformation between gated PET images and image registrations. This method did not require 4D CT data, and radiation dose was significantly reduced. The 4D reconstruction incorporating motion compensation can yield a better trade-off between SNR and bias in estimating the activity for moving tumor than nongated and independent gating without registration methods [76].

The approaches mentioned earlier usually lead to low-noise image of a gated frame, which still contains intraframe motion, and thus are not completely motion-free. Liu et al. proposed to use the correlation between internal organ motion and external monitored motion to derive the complete rigid motion information of a tumor or an internal organ, such as heart, at high temporal resolution [61]. The derived internal motion information was then incorporated into list mode reconstruction to generate event-by-event motion corrected images without intraframe motion and noise amplification [77]. The referenced respiratory location that all list mode events were registered to was chosen to match with the acquired helical CT images [78]. Since this approach is designed to correct motion for a known target organ or tumor with assumed rigid motion, it is better suited for quantification applications of a known object rather than detecting unknown abnormalities.

15.5 Deconvolution PET/CT

Deconvolution techniques have been utilized successfully to remove the blurring of medical images, such as CT [79], MRI [80], and solve the PVE problem in PET [81,82]. This technique could reduce the inherent motion blur in PET images without any gating.

Naqa et al. suggested that the motion blur of the lung lesion can be modeled as an convolution between the true PET image and the local motion blurring kernel (MBK) [83]

$$I_{obs} = I_{true} \otimes \text{MBK}.$$

where

I_{obs} was the observed blurred PET image
I_{true} was the true PET image
$\otimes$ represented the convolution process

There were three steps for their method: (i) estimation of the breathing motion model from 4D CT data; (ii) conversion of the motion estimates into the MBK; (iii) deconvolution based on an EM iterative algorithm using the MBK. The clinical results showed this method was promising for either large or small tumors.

Chang et al. proposed a joint respiratory motion and PVE correction approach to improve the accuracy of PET image quantification [84]. In this regard, the observed PET image can be modeled as the convolution between PVE-blurred PET image and the MBK (Figure 15.11)

$$I_{obs} = (I_{true} \otimes \text{PSF} + n) \otimes \text{MBK}$$

where n is the additive noise. This method was implemented for both phantom and patient studies. The authors indicated that this technique has the potential to improve the accuracy of PET quantification in clinical environment as both PVE and respiratory motion have an effect of decreasing the tumor SUV based on their results.

There are three limitations for respiratory motion deblurring with deconvolution method: (i) Deconvolution works on the aggregated reconstructed images, but not on phase by phase basis. If the tumor moved in a nonrigid way, which sometimes happens, the over- or underestimation in tumor activity concentration can lead to inaccurate PET image quantification. (ii) This method is applied at a regional level, but not for the entire image. It can only correct the motion blur in a tumor-related region rather than the whole PET image, since most body structures move with a different pattern as compared to the tumor. (iii) Deconvolution process generally amplifies image noise, and thus may affect image quality and quantitative accuracy.

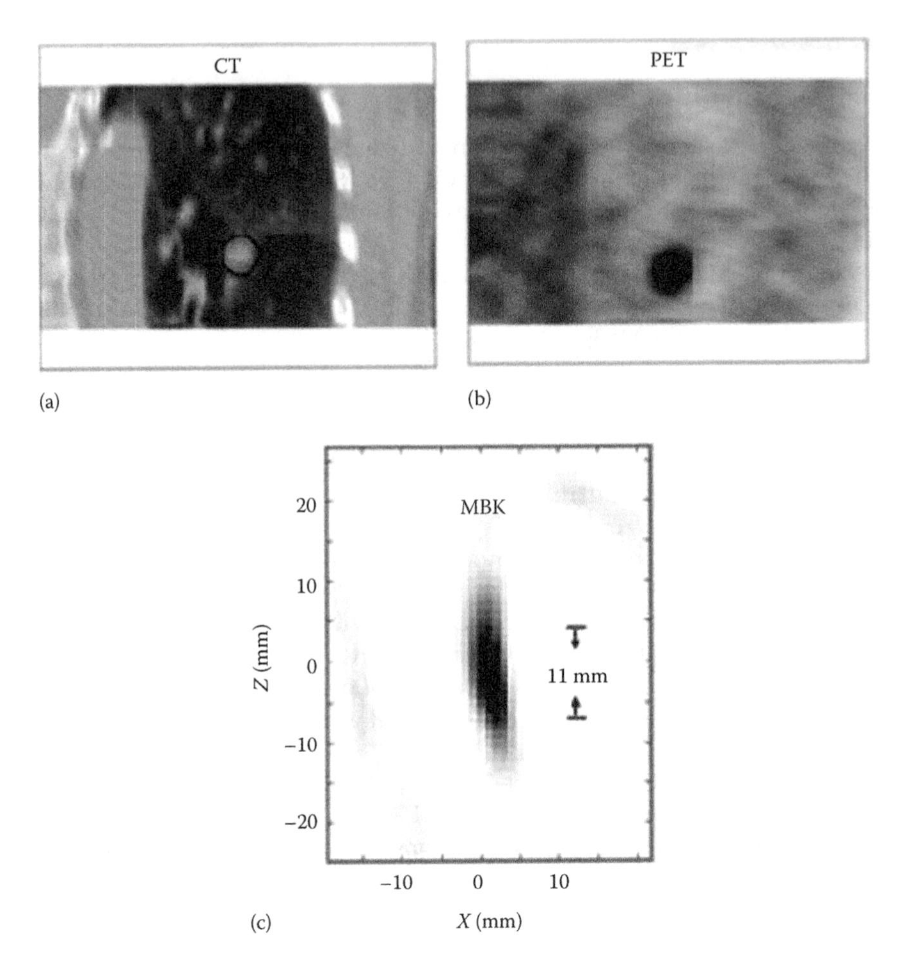

FIGURE 15.11
(a) The coronal view of the sampled 4D CT image, (b) the coronal view of the uncorrected PET image, and (c) the MBK estimated from 4D CT. (Reprinted from Chang, G. et al., *Med. Phys.*, 2010, 37(12), 6221. With permission.)

15.6 Conclusion

The CT-based method is better than the conventional ^{68}Ge-based transmission method for attenuation correction in PET in many aspects. However, it is hampered by respiratory artifacts. Several approaches to correct these artifacts in PET/CT images have been reviewed.

Attenuation correction with breath-hold CT is the easiest way to obtain the reconstructed PET image. Among different breathing patterns, normal end-expiration breath-hold CT shows the best result. However, patient compliance is critical to ensure the breath-hold duration. Some patients with heart disease or lung tumor usually do not have normal cardiopulmonary function, and it is hard to coach them to hold and release their breath consistently over many cycles during the examination.

With the computational capability of advanced computers, pre- and postprocessing techniques are preferred to achieve a better PET/CT registration. Further averaging the CT data over one or many respiratory cycles matches the PET data better. However, PET attenuation correction with slow CT results in severe inconsistent artifacts in the reconstructed image. Low-pitch CT and CACT overcome the inconsistent artifacts with the expense of higher radiation dose, which is a significant concern for patient safety. An alternative method using IACT for attenuation correction is introduced to further reduce radiation dose with similar image quality of CACT, but none of these methods can reduce the inherent motion blur in PET images.

Gated 4D PET/CT is an ongoing research technique to eliminate spatial blurring of the emission data. However, it takes huge amount of efforts for data acquisition and processing, and it may only be feasible for research institutions at this stage. Quiescent period gating is a practical approach to reduce motion blur with minimal noise increase and has been implemented in commercial scanners. To further improve image quality of reconstructed images, reconstruction-based method and deconvolution-based method are proposed. However, these methods need further evaluation on the clinical patients.

Each respiratory artifact reduction technique has its own advantages and disadvantages, and the optimum approach may probably be task- or patient-dependent. While improving image quality is important, the actual implementation of the respiratory artifact reduction technique highly depends on the robustness and complexity of the clinical setup.

Acknowledgments

This work was supported in part by the Multi-Year Research Grants (MYRG185(Y3-L3)-FST11-MSP & MYRG077(Y2-L2)-FST12-MSP) of University of Macau, Macau, an internal support from the Department of Diagnostic Radiology of Yale University, CTSA Grant (UL1 RR024139) from the National Institutes of Health (NIH), USA, and a research contract from Siemens Medical Solutions.

References

1. Beyer T, Townsend DW, Brun T, Kinahan PE, Charron M, Roddy R, et al. A combined PET/CT scanner for clinical oncology. *J Nucl Med*. 2000 Aug;41(8):1369–1379.
2. Townsend DW, Beyer T, Kinahan P, Meltzer CC, Brun T, Nutt R. The SMART scanner: A combined PET/CT tomograph for clinical oncology. *Radiology*. 1998 Nov;209P:169–170.
3. Chang LT. A method for attenuation correction in radionuclide computed tomography. *IEEE Trans Nucl Sci*. 1978;25(1):638–643.
4. Bergstrom M, Litton J, Eriksson L, Bohm C, Blomqvist G. Determination of object contour from projections for attenuation correction in cranial positron emission tomography. *J Comput Assist Tomogr*. 1982;6(2):365–372.
5. Meikle SR, Bailey DL, Hooper PK, Eberl S, Hutton BF, Jones WF. Simultaneous emission and transmission measurements for attenuation correction in whole-body *Pet. J Nucl Med*. 1995 Sep;36(9):1680–1688.
6. Meikle SR, Dahlbom M, Cherry SR. Attenuation correction using count-limited transmission data in positron emission tomography. *J Nucl Med*. 1993 Jan;34(1):143–144.
7. Kinahan PE, Hasegawa BH, Beyer T. X-ray-based attenuation correction for positron emission tomography/computed tomography scanners. *Semin Nucl Med*. 2003 Jul;33(3):166–179.
8. Kinahan P, Townsend DW, Beyer T, Sashin D. Attenuation correction for a combined 3D PETCT scanner. *Med Phys*. 1998;25(10):2046–2053.
9. Nakamoto Y, Osman M, Cohade C, Marshall LT, Links JM, Kohlmyer S, et al. PET/CT: Comparison of quantitative tracer uptake between germanium and CT transmission attenuation-corrected images. *J Nucl Med*. 2002 Sep 1;43(9):1137–1143.
10. Goerres GW, Kamel E, Heidelberg TN, Schwitter MR, Burger C, von Schulthess GK. PET-CT image co-registration in the thorax: Influence of respiration, *Eur J Nucl Med Mol Imaging*. 2012;29(3):351–360.
11. van Dalen JA, Visser EP, Vogel WV, Corstens FH, Oyen WJ. Impact of Ge-68/Ga-68-based versus CT-based attenuation correction on PET. *Med Phys*. 2007; 34(3):889–897.
12. Beyer T, Antoch G, Blodgett T, Freudenberg LF, Akhurst T, Mueller S. Dual-modality PET/CT imaging: The effect of respiratory motion on combined image quality in clinical oncology. *Eur J Nucl Med Mol Imaging*. 2003;30(4):588–596.
13. Osman MM, Cohade C, Nakamoto Y, Wahl RL. Respiratory motion artifacts on PET emission images obtained using CT attenuation correction on PET-CT. *Eur J Nucl Med Mol Imaging*. 2003 Apr;30(4):603–606.
14. Sureshbabu W, Mawlawi O. PET/CT imaging artifacts. *J Nucl Med Technol*. 2005 Sep 1;33(3):156–161.
15. Beyer T, Bockisch A, Kühl H, Martinez M-J. Whole-body 18F-FDG PET/CT in the presence of truncation artifacts. *J Nucl Med*. 2006 Jan;47(1):91–99.
16. Goerres GW, Hany F, Kamel E, von Schulthess GK, Buck A. Head and neck imaging with PET and PET/CT: Artefacts from dental metallic implants. *Eur J Nucl Med Mol Imaging*. 2002;29(3):367–370.

17. Kamel EM, Burger C, Buck A, von Schulthess GK, Goerres GW. Impact of metallic dental implants on CT-based attenuation correction in a combined PET/CT scanner. *Eur Radiol*. 2003 Apr;13(4):724–728.
18. Antoch G, Freudenberg LS, Egelhof T, Stattaus J, Jentzen W, Debatin JF, et al. Focal tracer uptake: A potential artifact in contrast-enhanced dual-modality PET/CT Scans. *J Nucl Med*. 2002 Oct 1;43(10):1339–1342.
19. Dizendorf E, Hany TF, Buck A, von Schulthess GK, Burger C. Cause and magnitude of the error induced by oral CT contrast agent in CT-based attenuation correction of PET emission studies. *J Nucl Med*. 2003 May;44(5):732–738.
20. Gould KL, Pan T, Loghin C, Johnson NP, Guha A, Sdringola S. Frequent diagnostic errors in cardiac PET/CT due to misregistration of CT attenuation and emission PET images: A definitive analysis of causes, consequences, and corrections. *J Nucl Med*. 2007;48(7):1112–1121.
21. Erdi YE, Nehmeh SA, Pan T, Pevsner A, Rosenzweig KE, Mageras G, et al. The CT motion quantitation of lung lesions and its impact on PET-measured SUVs. *J Nucl Med*. 2004;45(8):1287–1292.
22. Pevsner A, Nehmeh SA, Humm JL, Mageras GS, Erdi YE. Effect of motion on tracer activity determination in CT attenuation corrected PET images: A lung phantom study. *Med Phys*. 2005;32(7):2358.
23. Liu C, II LAP, Alessio AM, Kinahan PE. The impact of respiratory motion on tumor quantification and delineation in static PET/CT imaging. *Phys Med Biol*. 2009;54(24):7345–7362.
24. Cook G, Wegner E, Fogelman I. Pitfalls and artifacts in FDG PET and PET/CT oncologic imaging. *Semin Nucl Med*. 2004;34(2):122–133.
25. Nehmeh SA, Erdi YE. Respiratory motion in positron emission tomography/computed tomography: A review. *Semin Nucl Med*. 2008;38(3):167–176.
26. Dawood M, Buther F, Xiaoyi J, Schafers KP. Respiratory motion correction in 3-D PET data with advanced optical flow algorithms. *IEEE Trans Med Imaging*. 2008;27(8):1164–1175.
27. Erdi YE, Nehmeh SA, Pan T, Pevsner A, Rosenzweig KE, Mageras G, et al. The CT motion quantitation of lung lesions and its impact on PET-measured SUVs. *J Nucl Med*. 2004 Aug;45(8):1287–1292.
28. Xu QS, Yuan KH, Ye DT. Respiratory motion blur identification and reduction in ungated thoracic PET imaging. *Phys Med Biol*. 2011 Jul 21;56(14):4481–4498.
29. Allen-Auerbach M, Yeom K, Park J, Phelps M, Czernin J. Standard PET/CT of the chest during shallow breathing is inadequate for comprehensive staging of lung cancer. *J Nucl Med*. 2006;47(2):298–301.
30. Beyer T, Rosenbaum S, Veit P, Stattaus J, Muller SP, Difilippo FP, et al. Respiration artifacts in whole-body (18)F-FDG PET/CT studies with combined PET/CT tomographs employing spiral CT technology with 1 to 16 detector rows. *Eur J Nucl Med Mol Imaging*. 2005 Dec;32(12):1429–1439.
31. Goerres GW, Kamel E, Seifert B, Burger C, Buck A, Hany TF, et al. Accuracy of image coregistration of pulmonary lesions in patients with non-small cell lung cancer using an integrated PET/CT system. *J Nucl Med*. 2002;43(11):1469–1475.
32. de Juan R, Seifert B, Berthold T, von Schulthess GK, Goerres GW. Clinical evaluation of a breathing protocol for PET/CT. *Eur Radiol*. 2004;14(6):1118–1123.
33. Hamill JJ, Bosmans G, Dekker A. Respiratory-gated CT as a tool for the simulation of breathing artifacts in PET and PET/CT. *Med Phys*. 2008;35(2):576.

34. Senan S, De Ruysscher D, Giraud P, Mirimanoff R, Budach V. Literature-based recommendations for treatment planning and execution in high-dose radiotherapy for lung cancer. *Radiother Oncol*. 2004;71(2):139–146.
35. Vogel WV, van Dalen JA, Wiering B, Huisman H, Corstens FHM, Ruers TJM, et al. Evaluation of image registration in PET/CT of the liver and recommendations for optimized imaging. *J Nucl Med*. 2007;48(6):910–919.
36. Nehmeh SA, Erdi YE, Meirelles GSP, Squire O, Larson SM, Humm JL, et al. Deep-inspiration breath-hold PET/CT of the thorax. *J Nucl Med*. 2007 Jan 2007;48(1):22–26.
37. Meirelles GSP, Erdi YE, Nehmeh SA, Squire OD, Larson SM, Humm JL, et al. Deep-inspiration breath-hold PET/CT: Clinical findings with a new technique for detection and characterization of thoracic lesions. *J Nucl Med*. 2007 May;48(5):712–719.
38. Torizuka T, Tanizaki Y, Kanno T, Futatsubashi M, Yoshikawa E, Okada H, et al. Single 20-second acquisition of deep-inspiration breath-hold PET/CT: Clinical feasibility for lung cancer. *J Nucl Med*. 2009;50(10):1579–1584.
39. Yamaguchi T, Ueda O, Hara H, Sakai H, Kida T, Suzuki K, et al. Usefulness of a breath-holding acquisition method in PET/CT for pulmonary lesions. *Ann Nucl Med*. 2009;23(1):65–71.
40. Nagamachi S, Wakamatsu H, Kiyohara S, Fujita S, Futami S, Arita H, et al. The reproducibility of deep-inspiration breath-hold 18F-FDG PET/CT technique in diagnosing various cancers affected by respiratory motion. *Ann Nucl Med*. 2010;24(3):171–178.
41. Kalender WA, Seissler W, Klotz E, Vock P. Spiral volumetric CT with single-breath-hold technique, continuous transport, and continuous scanner rotation. *Radiology*. 1990;176(1):181–183.
42. Kalender WA. X-ray computed tomography. *Phys Med Biol*. 2006;51(13):R29.
43. Lagerwaard FJ, Van Sornsen de Koste JR, Nijssen-Visser MRJ, Schuchhard-Schipper RH, Oei SS, Munne A, et al. Multiple "slow" CT scans for incorporating lung tumor mobility in radiotheraphy planning. *Int J Radiat Oncol Biol Phys*. 2001;51(4):932–937.
44. van Sörnsen de Koste JR, Lagerwaard FJ, Schuchhard-Schipper RH, Nijssen-Visser MRJ, Voet PWJ, Oei SS, et al. Dosimetric consequences of tumor mobility in radiotherapy of stage I non-small cell lung cancer—an analysis of data generated using 'slow' CT scans. *Radiother Oncol*. 2001;61(1):93–99.
45. Nye JA, Esteves F, Votaw JR. Minimizing artifacts resulting from respiratory and cardiac motion by optimization of the transmission scan in cardiac PET/CT. *Med Phys*. 2007;34(6):1901.
46. Pan T, Mawlawi O, Nehmeh SA, Erdi YE, Luo D, Liu HH, et al. Attenuation Correction of PET images with respiration-averaged CT images in PET/CT. *J Nucl Med*. 2005 Sep 1;46(9):1481–1487.
47. Pan T, Lee T-Y, Rietzel E, Chen GTY. 4D-CT imaging of a volume influenced by respiratory motion on multi-slice CT. *Med Phys*. 2004;31(2):333.
48. Pan T, Mawlawi O, Luo D, Liu HH, Chi P-cM, Mar MV, et al. Attenuation correction of PET cardiac data with low-dose average CT in PET/CT. *Med Phys*. 2006;33(10):3931.
49. Cook RA, Carnes G, Lee TY, Wells RG. Respiration-averaged CT for attenuation correction in canine cardiac PET/CT. *J Nucl Med*. 2007 May;48(5):811–818.
50. Alessio AM, Kohlmyer S, Branch K, Chen G, Caldwell J, Kinahan P. Cine CT for attenuation correction in cardiac PET/CT. *J Nucl Med*. 2007;48(5):794–801.

51. Tzung-Chi H, Mok GSP, Wang S-J, Wu T-H, Zhang G. Attenuation correction of PET images with interpolated average CT for thoracic tumors. *Phys Med Biol.* 2011;56(8):2559.
52. Guerrero T, Zhang G, Huang T-C, Lin K-P. Intrathoracic tumour motion estimation from CT imaging using the 3D optical flow method. *Phys Med Biol.* 2004;49(17):4147–4161.
53. Lujan AE, Balter JM, Haken RKT. A method for incorporating organ motion due to breathing into 3D dose calculations in the liver: Sensitivity to variations in motion. *Med Phys*. 2003;30(10):2643–2649.
54. Mok G, Sun T, Huang T, Vai M, Interpolated average CT for attenuation correction in PET-A simulation study. *IEEE Trans Biomed Eng*. 2013.
55. Sun T, Wu T-H, Wu N-Y, Mok GSP. Low dose interpolated average CT for PET/CT attenuation correction using an Active Breathing Controller (ABC). In conference record of *IEEE Nuclear Science Symposium Medical Imaging Conference*, Anaheim, CA, Oct 28–Nov 3, 2012.
56. Picard Y, Thompson CJ. Motion correction of PET images using multiple acquisition frames. *IEEE Trans Med Imaging*. 1997;16(2):137–144.
57. Nehmeh SA, Erdi YE, Ling CC, Rosenzweig KE, Squire OD, Braban LE, et al. Effect of respiratory gating on reducing lung motion artifacts in PET imaging of lung cancer. *Med Phys*. 2002;29(3):366.
58. Boucher L, Rodrigue S, Lecomte R, Bénard F. Respiratory gating for 3-dimensional PET of the thorax: Feasibility and initial results. *J Nucl Med*. 2004 February 1;45(2):214–219.
59. Nehmeh SA, Erdi YE, Pan T, Pevsner A, Rosenzweig KE, Yorke E, et al. Four-dimensional (4D) PET/CT imaging of the thorax. *Med Phys*. 2004;31(12):3179.
60. Ponisch F, Richter C, Just U, Enghardt W. Attenuation correction of four dimensional (4D) PET using phase-correlated 4D-computed tomography. *Phys Med Biol*. 2008 Jul 7;53(13):N259–N268.
61. Wells RG, Ruddy TD, DeKemp RA, DaSilva JN, Beanlands RS. Single-phase CT aligned to gated PET for respiratory motion correction in cardiac PET/CT. *J Nucl Med*. 2010;51(8):1182–1190.
62. Nagel CCA, Bosmans G, Dekker ALAJ, Öllers MC, De Ruysscher DKM, Lambin P, et al. Phased attenuation correction in respiration correlated computed tomography/positron emitted tomography. *Med Phys*. 2006;33(6):1840.
63. Liu C, Alessio AM, Kinahan PE. Respiratory motion correction for quantitative PET/CT using all detected events with internal-external motion correlation. *Med Phys*. 2011;38(5):2715–2723.
64. McLeish K, Hill DLG, Atkinson D, Blackall JM, Razavi R. A study of the motion and deformation of the heart due to respiration. *IEEE Trans Med Imaging*. 2002;21(9):1142–1150.
65. Shechter G, Ozturk C, Resar JR, McVeigh ER. Respiratory motion of the heart from free breathing coronary angiograms. *IEEE Trans Med Imaging*. 2004;23(8):1046–1056.
66. Livieratos L, Stegger L, Bloomfield PM, Schafers K, Bailey DL, Camici PG. Rigid-body transformation of list-mode projection data for respiratory motion correction in cardiac PET. *Phys Med Biol*. 2005;50(14):3313.
67. McQuaid SJ, Lambrou T, Hutton BF. A novel method for incorporating respiratory-matched attenuation correction in the motion correction of cardiac PET-CT studies. *Physics in Medicine and Biology*. 2011 May 21;56(10):2903–2915.

68. Klein GJ, Reutter RW, Huesman RH. Four-dimensional affine registration models for respiratory-gated PET. *IEEE Trans Nucl Sci*. 2001;48(3):756–760.
69. Thorndyke B, Schreibmann E, Koong A, Xing L. Reducing respiratory motion artifacts in positron emission tomography through retrospective stacking. *Med Phys*. [10.1118/1.2207367]. 2006;33(7):2632.
70. Soret M, Bacharach SL, Buvat I. Partial-volume effect in PET tumor imaging. *J Nucl Med*. 2007 June;48(6):932–945.
71. Visvikis D, Lamare F, Bruyant P, Boussion N, Cheze Le Rest C. Respiratory motion in positron emission tomography for oncology applications: Problems and solutions. *Nucl Instrum Methods Phys Res A*. 2006;569(2):453–457.
72. Liu C, Alessio A, Pierce L, Thielemans K, Wollenweber S, Ganin A, et al. Quiescent period respiratory gating for PET/CT. *Med Phys*. 2010;37(9):5037–5043.
73. Lamare F, Carbayo MJL, Cresson T, Kontaxakis G, Santos A, Rest CCL, et al. List-mode-based reconstruction for respiratory motion correction in PET using non-rigid body transformations. *Phys Med Biol*. 2007;52(17):5187–5204.
74. Li T, Thorndyke B, Schreibmann E, Yang Y, Xing L. Model-based image reconstruction for four-dimensional PET. *Med Phys*. 2006;33(5):1288–1298.
75. Qiao F, Pan T, Clark JW Jr., Mawlawi OR. A motion-incorporated reconstruction method for gated PET studies. *Phys Med Biol*. 2006;51(15):3769–3783.
76. Grotus N, Reader AJ, Stute S, Rosenwald JC, Giraud P, Buvat I. Fully 4D list-mode reconstruction applied to respiratory-gated PET scans. *Phys Med Biol*. 2009;54(6):1705.
77. Chan C, Jin X, Fung EK, Mulnix T, Carson RE, CL. Event-by-event respiratory motion correction with 3-dimensional internal-external motion correlation. In conference record of *IEEE Nuclear Science Symposium and Medical Imaging Conference*, Anaheim, CA, Oct 28–Nov 3, 2012;2117–2122.
78. Alessio AM, Kinahan PE, Champley KM, Caldwell JH. Attenuation-emission alignment in cardiac PET/CT based on consistency conditions. *Med Phys*. 2010;37(3):1191–1200.
79. Ming J, Ge W, Skinner MW, Rubinstein JT, Vannier MW. Blind deblurring of spiral CT images. *IEEE Trans Med Imaging*. 2003;22(7):837–845.
80. Sourbron S, Luypaert R, Van Schuerbeek P, Dujardin M, Stadnik T, Osteaux M. Deconvolution of dynamic contrast-enhanced MRI data by linear inversion: Choice of the regularization parameter. *Magn Reson Med*. 2004;52(1):209–213.
81. Kirov AS, Piao JZ, Schmidtlein CR. Partial volume effect correction in PET using regularized iterative deconvolution with variance control based on local topology. *Phys Med Biol*. 2008;53(10):2577.
82. Faber TL, Raghunath N, Tudorascu D, Votaw JR. Motion correction of PET brain images through deconvolution: I. Theoretical development and analysis in software simulations. *Phys Med Biol*. 2009;54(3):797.
83. Naqa IE, Low DA, Bradley JD, Vicic M, Deasy JO. Deblurring of breathing motion artifacts in thoracic PET images by deconvolution methods. *Med Phys*. 2006;33(10):3587–3600.
84. Chang G, Chang T, Pan T, John W, Clark J, Mawlawi OR. Joint correction of respiratory motion artifact and partial volume effect in lung/thoracic PET/CT imaging. *Med Phys*. 2010;37(12):6221–6232.

16

Image Reconstruction for 3D PET

Jinyi Qi

CONTENTS

16.1 Introduction

PET scanners use multiple rings of detectors that surround the patient to record coincidence photon pairs produced by positron annihilation. Early PET scanners placed septa between rings to collect coincidence events only within each detector ring. This is commonly referred to as 2D PET since the data consist of a stack of 2D sinograms and can be reconstructed using a

2D filtered back projection algorithm slice by slice.* With the developments of fast scintillators and electronics, modern PET scanners no longer require septa and can accept all coincidence events. This results in a factor of 4–7 increase in sensitivity and hence increases the signal-to-noise ratio (SNR). PET without septa is commonly referred to as fully 3D PET. The higher sensitivity of fully 3D PET can be used to reduce injection dose or scan time or to improve image quality. The data, however, are overcomplete and require new reconstruction algorithms.

In this chapter, we will first analyze the overcompleteness of the fully 3D PET data and present rebinning methods that can reduce the fully 3D data to a stack of 2D sinograms while preserving the data quality. We then focus our attention to model-based iterative reconstruction methods, which can be applied to either fully 3D PET data or rebinned 2D data. Finally we present recent developments in direct reconstruction of parametric images from dynamic PET data.

16.2 Fourier Analysis of Fully 3D PET Data

To illustrate the overcompleteness of the fully 3D PET data, we resort to the Fourier transform of parallel projections. Without loss of generality, we model the PET scanner as a cylinder of radius R and length L with its axis along the z-axis. Following the notation in [LDM+99], we specify each line of response (LOR) by four parameters: (s, φ, z, δ) where s and φ are the radial offset and azimuthal angle in the transaxial planes, respectively, z is the axial coordinate of the point midway between the two detectors, and $\delta \equiv \tan\theta$ with θ being the co-polar angle between the LOR and the transaxial planes. Let us also denote the source distribution by $f(x, y, z)$ and its 3D Fourier transform by $F(\nu_x, \nu_y, \nu_z)$.

The LOR data can be represented by

$$p(s,\phi,z,\delta) = \int_{-\infty}^{\infty} f(s\cos\phi - t\sin\phi, s\sin\phi + t\cos\phi, z + t\delta)dt \tag{16.1}$$

* This picture is rather simplified since 2D systems do allow detection of events between *adjacent* rings. These are used to reconstruct additional transaxial images, so that the thickness of each plane is half of the axial extent of a single detector ring and the number of reconstructed planes in a 2D scanner is usually $2P - 1$ where P is the number of detector rings.

Taking 2D Fourier transform of $p(s, \varphi, z, \delta)$ with respect to (s, z),* we get

$$P(\nu_s, \phi, \nu_z, \delta) = \int\int\int_{-\infty}^{\infty} f(\underbrace{s\cos\phi - t\sin\phi}_{x'}, \underbrace{s\sin\phi + t\cos\phi}_{y'}, \underbrace{z + t\delta}_{z'}) e^{-j2\pi(s\nu_s + z\nu_z)}\, dt\, ds\, dz \tag{16.2}$$

$$= \int\int\int_{-\infty}^{\infty} f(x', y', z') e^{-j2\pi[(x'\cos\phi + y'\sin\phi)\nu_s + z'\nu_z + (x'\sin\phi - y'\cos\phi)\delta\nu_z]}\, dx'\, dy'\, dz' \tag{16.3}$$

$$= \int\int\int_{-\infty}^{\infty} f(x', y', z') e^{-j2\pi[x'(\nu_s\cos\phi + \delta\nu_z\sin\phi) + y'(\nu_s\sin\phi - \delta\nu_z\cos\phi) + z'\nu_z]}\, dx'\, dy'\, dz' \tag{16.4}$$

$$= F(\nu_x, \nu_y, \nu_z)\Big|_{\substack{\nu_x = \nu_s\cos\phi + \delta\nu_z\sin\phi = \nu'\cos(\phi - \sigma) \\ \nu_y = \nu_s\sin\phi - \delta\nu_z\cos\phi = \nu'\sin(\phi - \sigma)}} \tag{16.5}$$

where

$$\sigma = \arctan\left(\frac{\nu_z\delta}{\nu_s}\right) \tag{16.6}$$

$$\nu' = \nu_s\sqrt{1 + \left(\frac{\nu_z\delta}{\nu_s}\right)^2} \tag{16.7}$$

Note that the 2D Fourier transform of $p(s, \varphi, z, \delta)$ lines up with the 3D Fourier transform of the image along ν_z-axis. This is due to the cylindrical geometry of PET scanners that maintains a fixed axial sampling distance for projections of all co-polar angles. However, in $\nu_x - \nu_y$ plane, the Fourier transform samples of $P(\nu_s, \varphi, \nu_z, \delta)$ and $F(\nu_x, \nu_y, \nu_z)$ do not line up nicely. As illustrated in Figure 16.1, $P(\nu_s, \varphi, \nu_z, \delta)$ for $0 \leq \varphi < 2\pi$ form a set of rotating lines with angle φ and radial distance of $\nu_z\delta$ in the $\nu_x - \nu_y$ plane. When $\nu_z\delta = 0$, $P(\nu_s, \varphi, \nu_z, \delta)$ form the polar coordinates of the corresponding $\nu_x - \nu_y$ plane. When $\nu_z\delta \neq 0$, there is a circular region in the $\nu_x - \nu_y$ plane that is not sampled by $P(\nu_s, \varphi, \nu_z, \delta)$. The radius of the missing region is equal to $|\nu_z\delta|$.

In 3D frequency space, the missing region for the projections of a given co-polar angle θ is a cone along the ν_z-axis with the cone angle equal to $\theta \equiv \arctan(\delta)$ (see Figure 16.2 for an illustration). Therefore, only the projections with $\delta = 0$ completely cover the 3D frequency space; oblique projections

* Here we assume there is no truncation of PET data along the axial direction. We will return to the issue of axial truncation later.

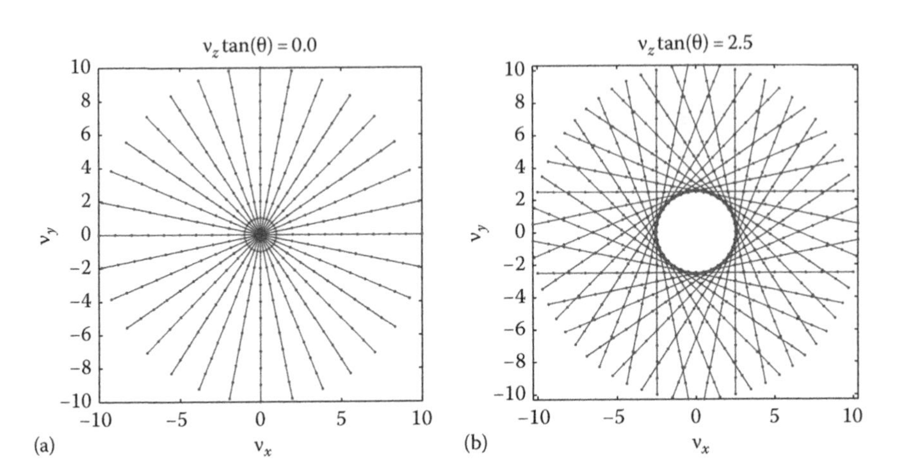

FIGURE 16.1
The sampling pattern of $P(\nu_s, \varphi, \nu_z, \delta)$, $0 \leq \nu_s < \infty$, $0 \leq \varphi < 2\pi$ in the $\nu_x - \nu_y$ plane for the cases of (a) $\nu_z\delta = 0$ and (b) $\nu_z\delta = 2.5$.

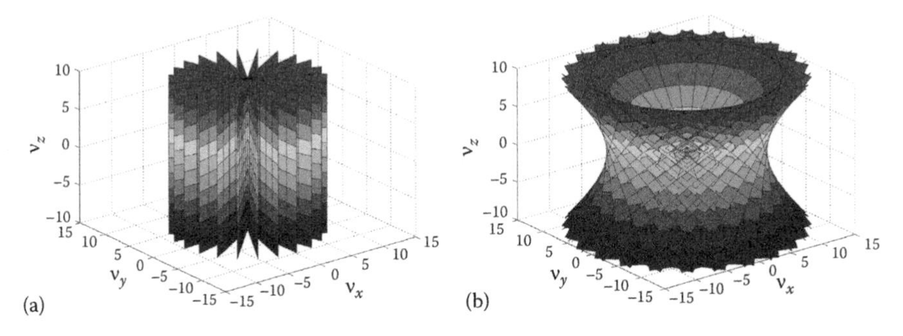

FIGURE 16.2
The distribution of $P(\nu_s, \varphi, \nu_z, \delta)$ in the 3D frequency space for (a) $\delta = 0$ and (b) $\delta = 1.2$.

($\delta \neq 0$) are insufficient by themselves to recover the object being imaged, but they provide redundant measurements for the frequency components outside the missing cone region.

16.3 Fourier Rebinning

While the oblique projections are incomplete by themselves, they can be used to average out noise and hence improve the SNR of the reconstructed images. There are two ways to perform the averaging operation. The first approach is to backproject all projections (both direct and oblique ones) into the image space and then apply a 3D filter to compensate the variable sampling density in the frequency space. Exchanging the order of the filtering operation and back projection results in the 3D filtered back projection algorithm [Col80].

However, the limited axial extent of the scanner results in missing data in the oblique sinograms, which have to be estimated before the filtering operation can be applied [KR89]. The second approach is to perform the averaging in the frequency domain, which is commonly referred to as Fourier rebinning. Fourier rebinning converts fully 3D PET data into a stack of 2D sinograms, which can be reconstructed using either a 2D FBP or an iterative method. Computationally Fourier rebinning is more efficient than 3D back projection and hence will be the focus here.

16.3.1 Exact Rebinning Methods: FOREX and FORE-J

The essential step of Fourier rebinning is to interpolate Fourier samples of oblique projections, $P(\nu_s, \phi, \nu_z, \delta \neq 0)$, onto the Fourier samples of the direct projections, $P(\nu_s, \phi, \nu_z, \delta = 0)$. Because Fourier samples of oblique projections all line up along ν_z-axis, the interpolation is only required inside each $\nu_x - \nu_y$ plane. Considering the periodic nature of $P(\nu_s, \phi, \nu_z, \delta)$ in ϕ direction, the interpolation along ϕ can be easily performed in the Fourier space. Taking the 1D Fourier transform of $P(\nu_s, \phi, \nu_z, \delta)$ with respect to ϕ, we get

$$\hat{P}(\nu_s, k, \nu_z, \delta) = \int_0^{2\pi} e^{-2\pi jk\phi} P(\nu_s, \phi, \nu_z, \delta) d\phi \tag{16.8}$$

$$= \int_0^{2\pi} e^{-2\pi jk\phi} F(\nu' \cos(\phi - \sigma), \nu' \sin(\phi - \sigma), \nu_z) d\phi \tag{16.9}$$

$$= e^{-2\pi jk\sigma} \int_0^{2\pi} e^{-2\pi jk\phi} F(\nu' \cos\phi, \nu' \sin\phi, \nu_z) d\phi \tag{16.10}$$

Let $\delta = 0$, then $\sigma = 0$, $\nu' = \nu_s$, and Equation 16.10 reduces to

$$\hat{P}(\nu_s, k, \nu_z, 0) = \int_0^{2\pi} e^{-2\pi jk\phi} F(\nu_s \cos\phi, \nu_s \sin\phi, \nu_z) d\phi \tag{16.11}$$

Substituting Equation 16.11 back into Equation 16.10, we get

$$\hat{P}(\nu_s, k, \nu_z, \delta) = e^{-2\pi jk\sigma} \hat{P}(\nu', k, \nu_z, \delta = 0) \tag{16.12}$$

or equivalently

$$\hat{P}(\nu_s, k, \nu_z, 0) = e^{2\pi jk\sigma} \hat{P}(\nu', k, \nu_z, \delta) \tag{16.13}$$

Equations 16.12 and 16.13 provide the direct relationship between $\hat{P}(\nu_s, k, \nu_z, 0)$ and $\hat{P}(\nu_s, k, \nu_z, \delta)$. Specifically, Equation 16.12 can be used to estimate oblique projections ($\delta \neq 0$) from direct sinograms ($\delta = 0$), which is commonly referred to as inverse Fourier rebinning [LDM+99]; Equation 16.13 can be used to rebin oblique sinogram onto direct sinograms, that is, Fourier rebinning. The interpolation along ϕ is achieved by a linear phase shift in the Fourier space, and any 1D interpolation method can be used along ν_s (linear interpolation is the most common choice).

One difficulty in direct application of Equation 16.13 to fully 3D PET data is that the PET data $p(s, \phi, z, \delta \neq 0)$ are known only for a limited z range $|z| < (L/2 - |R\delta|)$, which are not sufficient for computing the axial Fourier transform. One solution is to estimate the missing oblique projections using Equation 16.12, which gives rise to the following FOREX algorithm [LDM+99]:

1. Initialize a stack of Fourier transform of 2D sinograms $\hat{P}_{2D}(\nu', k, \nu_z) = 0$.
2. Compute the 3D Fourier transform of the direct sinograms $\hat{P}(\nu', k, \nu_z, \delta = 0)$.
3. For each sampled δ perform the following:
   a. Calculate $P'(\nu_s, k, z, \delta)$, the 2D Fourier transform of the data with respect to s and ϕ, in the measured axial range $|z| < (L/2 - |R\delta|)$.
   b. Estimate $\hat{P}(\nu_s, k, \nu_z, \delta)$ from $\hat{P}(\nu_s, k, \nu_z, \delta = 0)$ using Equation 16.12 and take the inverse 1D Fourier transform of $\hat{P}(\nu_s, k, z, \delta)$ with respect to ν_z to estimate $P'(\nu_s, k, z, \delta)$ in the unmeasured range $|z| \geq (L/2 - |R\delta|)$.
   c. Merge the results of steps 3(a) and 3(b) to get the complete $P'(\nu_s, k, z, \delta)$, and take the 1D Fourier transform with respect to z to get $\hat{P}(\nu_s, k, \nu_z, \delta)$.
   d. Rebin $\hat{P}(\nu_s, k, \nu_z, \delta)$ into $\hat{P}_{2D}(\nu', k, \nu_z)$ using Equation 16.13 with 1D linear interpolation in ν_s.
4. Normalize the rebinned sinogram $\hat{P}_{2D}(\nu', k, \nu_z)$ to compensate the variable number of contributions to each frequency component.
5. Take the 3D inverse Fourier transform of $\hat{P}_{2D}(\nu', k, \nu_z)$ to get a stack of 2D sinograms $p_{2D}(s, \phi, z)$. These sinograms can be reconstructed with any 2D reconstruction algorithm.

The drawback of the FOREX algorithm is that the estimation of missing projections can be time consuming. To avoid the estimation of missing data, Defrise and Liu [DL99] exploited the local consistency conditions of PET data using John's equation. They showed that any consistent PET data must satisfy the following second-order PDE (John's equation)

$$\frac{\partial^2 p(s,\phi,z,\delta)}{\partial z\,\partial\phi} + \frac{\partial^2 p(s,\phi,z,\delta)}{\partial\delta\,\partial s} = -s\delta\frac{\partial^2 p(s,\phi,z,\delta)}{\partial z^2} \tag{16.14}$$

In Fourier domain, John's equation becomes

$$k\frac{\partial P'(\nu_s,k,z,\delta)}{\partial z}+\nu_s\frac{\partial P'(\nu_s,k,z,\delta)}{\partial \delta}=-\delta\frac{\partial^3 P'(\nu_s,k,z,\delta)}{\partial \nu_s\,\partial z^2} \tag{16.15}$$

For any slice z_0 and for any $\nu_s \neq 0$, Equation 16.15 can be rewritten as

$$\frac{d}{d\delta}P'\left(\nu_s,k,z_0+\left(\frac{k}{\nu_s}\right)\delta,\delta\right)=-\frac{\delta}{\nu_s}\frac{\partial^3 P'(\nu_s,k,z_0+(k/\nu_s)\delta,\delta)}{\partial \nu_s\,\partial z^2} \tag{16.16}$$

Integrating Equation 16.16 on both sides results in a new rebinning formula:

$$P'_{\text{reb}}(\nu_s,k,z_0)=P'\left(\nu_s,k,z_0+\left(\frac{k}{\nu_s}\right)\delta,\delta\right)+\int_0^{\delta}d\delta'\frac{\delta'}{\nu_s}\frac{\partial^3 P'(\nu_s,k,z_0+(k/\nu_s)\delta',\delta')}{\partial \nu_s\,\partial z^2} \tag{16.17}$$

The new rebinning method is commonly referred to as FORE-J. It operates directly on the 2D Fourier transform of PET sinograms and thus does not require any missing projections. However, the high-order partial derivatives make it more susceptible to noise than the FOREX algorithm.

16.3.2 Approximate Rebinning: FORE

For PET system with a relatively small acceptance angle δ_{max}, a fast approximate rebinning algorithm can be obtained by using small-angle approximation in Equation 16.13. Using Taylor series, we have

$$\sigma=\arctan(0)+\frac{\nu_z\delta}{\nu_s}+\cdots \tag{16.18}$$

$$\nu'=\nu_s\left(1+\frac{1}{2}\left(\frac{\nu_z\delta}{\nu_s}\right)^2+\cdots\right) \tag{16.19}$$

Keeping only the first-order terms, we get $\sigma\approx\nu_z\delta/\nu_s$ and $\nu'\approx\nu_s$. Then Equation 16.13 reduces to

$$\hat{P}(\nu_s,k,\nu_z,0)=e^{2\pi jk(\nu_z\delta/\nu_s)}\hat{P}(\nu_s,k,\nu_z,\delta) \tag{16.20}$$

Taking the 1D inverse Fourier transform with respect to ν_z results in

$$P'(\nu_s,k,z,\delta)=P'\left(\nu_s,k,z-\frac{k\delta}{\nu_s},\delta=0\right) \tag{16.21}$$

The rebinning formula (21) is commonly referred to as FORE [DKT+97]. It can also be derived by dropping the second term on the right-hand side of the FORE-J equation (16.17). Compared with FOREX and FORE-J, FORE can be substantially faster. However, the approximation is valid only when $\delta/\nu_s \ll 1$. At low radial frequency, the accuracy of Equation 16.21 breaks down. Therefore, low-frequency components are rebinned using only direct sinograms ($\delta = 0$). Details can be found in [DKT+97].

16.4 Model-Based Reconstruction

16.4.1 Data Model

Model-based reconstruction methods can adopt arbitrarily accurate system models and easily handle missing data caused by axial truncation and/or gaps between detectors. In addition, model-based reconstruction methods allow explicit modeling of statistical distribution of measurement noise. The combination of improved modeling of the detection process and improved handling of statistical noise offers the possibility for enhanced performance of PET with both high count data (where model accuracy limits resolution) and low count data (where statistical noise limits resolution).

In the following we represent the lexicographically ordered elements of the image by $\mathbf{f} = \{f_j, j = 1 \ldots N\}$ and the elements of the measured sinograms by $\mathbf{y} = \{y_i, i = 1, \ldots M\}$. The basic model for the expectation of the data is

$$\bar{y} = E[y] = \mathbf{Pf} + \mathbf{r} + \mathbf{s} \tag{16.22}$$

where the elements of the projection matrix $\mathbf{P} \in IR^{M \times N}$, p_{ij}, contain the probabilities of detecting an emission from voxel site j at detector pair i, $\mathbf{r}$ and $\mathbf{s}$ represent the expectations of random and scattered events, respectively.

Scattered events refer to coincidence detection after one or both of the photons have undergone Compton scattering. Given the distribution of the source image and an image of the linear attenuation coefficient, an accurate scatter profile can be computed using the Klein–Nishina formula for Compton scatter [OJB92, WNC95, MLC96, Oll96, Wat00] or Monte Carlo simulation [LDH95]. Since the scatter profiles are smooth, it is possible to compute them with reasonable computational load from a low-resolution, preliminary reconstruction of the emission source. Once this is estimated, the scatter contribution can be viewed as a known offset in the mean of the data in (16.22) rather than as an explicit function of the data that must be recomputed with each new estimate of the image.

Random coincidences (henceforth called "randoms") are caused by the detection of two independent photons within the coincidence timing window.

The random contribution to the data is a function of the length of this timing window and of the source activity. By simply delaying the timing window by a fixed amount, one can obtain data that consist of purely randoms and with the same mean number of counts as for the non-delayed window. Assuming that all detectors function independently of each other and the event rate remains constant, the expected randoms between a pair of detectors can be computed from their single-event rates (single rates):

$$r_i = 2\tau s_{i_1} s_{i_2} \tag{16.23}$$

where

s_{i_1} and s_{i_2} are the single rates of the two detectors forming the ith LOR
τ is the coincidence timing window width

For systems that measure single rates of individual detectors, Equation 16.23 is often used to estimate the expectation of random events. For systems without single rate measurements, one can use Equation 16.23 to obtain a smooth random estimate by fitting the delayed window measurements to the model [CH86, MLC96]. In the following we will assume that the scatter and random components in the data have been estimated.

16.4.2 Maximum-Likelihood and Penalized-Likelihood Methods

PET data $\boldsymbol{y}$ are well modeled as independent Poisson random variables with distribution [YF02]

$$p(\boldsymbol{y} \mid \mathbf{f}) = \prod_{i=1}^{M} \frac{\overline{y}_i^{\,y_i} e^{-\overline{y}_i}}{y_i!} \tag{16.24}$$

The corresponding log likelihood, after dropping constants, is

$$\mathcal{L}(\boldsymbol{y} \mid \mathbf{f}) = \sum_{i=1}^{M} y_i \log \overline{y}_i - \overline{y}_i \tag{16.25}$$

The mean $\overline{\boldsymbol{y}}$ is related to the image through the affine transform (16.22).

Maximum likelihood (ML) estimates of the image can be obtained by maximizing the log-likelihood function in (16.25). However, the resulting images are often very noisy due to ill conditioning. Some form of regularization is required to produce acceptable images. Often this is accomplished simply by starting with a smooth initial estimate and terminating an ML search before convergence. The drawback of the iteration-based regularization is that the image properties are hard to control explicitly. Alternatively regularization can be introduced into the objective function through the penalized-likelihood (PL) (or equivalently maximum a posteriori (MAP)) formulation.

PL reconstruction estimates the unknown image by maximizing a PL function

$$\hat{f} = \arg\max_{f\geq 0} \Phi(f), \quad \Phi(f) = \mathcal{L}(y \mid f) - \beta U(f) \tag{16.26}$$

where $U(f)$ is a regularization function. The most commonly used regularization is the image roughness penalty that is defined as

$$U(f) = \frac{1}{2}\sum_{j=1}^{N}\sum_{k\in\mathcal{N}_j} \kappa_{jk}\psi(x_j - x_k) \tag{16.27}$$

where
- $\psi(t)$ is a potential function
- κ_{jk} is the weighting factor related to the distance between pixel j and pixel k in the neighborhood N_j

For a 3D problem, the neighbors of an internal voxel would be the nearest 6 voxels for a first-order model, or the nearest 26 voxels for a second-order model (with appropriate modifications for the boundaries of the lattice). The regularization parameter β controls the trade-off between data fidelity and spatial smoothness. When β goes to zero, the reconstructed image approaches the ML estimate.

The potential functions $\psi_{jk}(f_j - f_k)$ are chosen to encourage local smoothness of PET images. All have the basic property of being monotonic nondecreasing functions of the absolute intensity difference $|(f_j - f_k)|$. A wide range of functions have been studied in the literature that attempt to produce local smoothing while not removing or blurring true boundaries or edges in the image. One common choice of $\psi(t)$ in PET image reconstruction is the quadratic function

$$\psi(t) = \frac{1}{2}t^2 \tag{16.28}$$

The disadvantage of the quadratic regularization is that it may over-smooth edges and small objects when a large β is used. One potential function that can preserve edges is the absolute value function

$$\psi(t) = |t| \tag{16.29}$$

but it is not differentiable at $t = 0$. A similar function but with continuous second-order derivatives is the Lange function [Lan90]

$$\psi(t) = \delta\left(\frac{|t|}{\delta} - \log\left(1 + \frac{|t|}{\delta}\right)\right) \tag{16.30}$$

which approximates the quadratic function when $|t| \ll \delta$ and approaches the absolute function for $|t| \gg \delta$. Other examples of nonquadratic convex potential functions include the hyperbola function $\sqrt{t^2+\delta^2}$, the Huber function [Hub81], and the l_p penalty ($p \geq 1$) [BS93]. Non-convex potential functions have also been proposed to form even sharper edges in reconstructed images (e.g., [GM85]). In addition, the regularization function can be modified to include anatomical information to preserve sharp transitions at organ boundaries, for example, [LY91, GLRZ93, SC97, WCL04, SPR+11, CLQ11].

16.5 Optimization Algorithms

The necessary and sufficient conditions for an image f^* to be a maximizer of $\Phi(\mathbf{f})$ subject to $f \geq 0$ are given by the Kuhn–Tucker conditions:

$$\left.\frac{\partial}{\partial f_j}\Phi(\mathbf{f})\right|_{f=f^*} \begin{cases} =0, & f_j^* > 0 \\ \geq 0, & f_j^* = 0 \end{cases} \quad j=1,\ldots,N \tag{16.31}$$

Finding an ML or MAP image estimate is equivalent to solving the set of coupled nonlinear equations given by the Kuhn–Tucker conditions. Since closed-form solutions do not generally exist, iterative estimation algorithms are used. The underlying principle of iterative optimization algorithms is as follows: starting from an initial estimate, f^0, find a sequence of images, f^n, $n = 1, 2,\ldots$, that converges to f^*, as $n \to \infty$. All algorithms can be written in the following update form:

$$\boldsymbol{f}^{n+1} = \boldsymbol{f}^n - \alpha^n \boldsymbol{a}^n \tag{16.32}$$

where $\mathbf{a}^n$ is a vector of the same dimension as the image and represents a search direction and α^n is a scalar step size. Different approaches to finding $\mathbf{a}^n$ and α^n give rise to different algorithms [QL06].

16.5.1 Gradient-Based Algorithms

Standard gradient-based algorithms have been used to solve emission tomography problems since the early 1970s [Goi72, BG74]. As the name implies, the search direction is calculated from the gradient of the objective function at the current iterate. The simplest method is that of steepest ascent, which sets $\mathbf{a}^n$ equal to the gradient and chooses the step size α^n using a 1D line search algorithm, such as the Newton–Raphson method or the Armijo rule [Lue84]. Convergence of steepest ascent is slow but can be enhanced using positive definite preconditioning matrices, which are used to modify the search direction [LM86, Kau87, RRK92]. Preconditioners alone have limited effectiveness, but when combined with the conjugate gradient algorithm, rather than steepest ascent, they produce substantial gains in convergence rates [Kau93, MLCZ94, MLC96, QLC+98, QLH+98]. A general form of the preconditioned conjugate gradient (PCG) algorithm is

$$\begin{pmatrix} \mathbf{f}^{n+1} & = & \mathbf{f}^n + \alpha^n \mathbf{a}^n : \alpha \text{ from line search} \\ \mathbf{a}^n & = & \mathbf{d}^n + \beta^{n-1} \mathbf{a}^{n-1} \\ \mathbf{d}^n & = & \mathbf{C}^n \mathbf{g}^n \\ \beta^{n-1} & = & \dfrac{(\mathbf{g}^n - \mathbf{g}^{n-1})' \mathbf{d}^n}{\mathbf{g}^{n-1'} \mathbf{d}^{n-1}} \end{pmatrix} \tag{16.33}$$

where $\mathbf{g}^n$ is the gradient vector of the objective function at $f = f^n$, $\mathbf{C}^n$ is the preconditioning matrix, and α^n is the step size found using a line search. For nonquadratic functions, $\mathbf{a}^n$ is not guaranteed to be an ascent direction so that it is necessary to check whether this is the case at each iteration, that is, $\mathbf{g}^n$, $\mathbf{a}^n > 0$, and reset $\mathbf{a}^n$ to $\mathbf{d}^n$ if the condition is violated [Lue84].

Ideally the preconditioner would be the inverse of the Hessian of the objective function. Exact computation of the inverse of the Hessian is impractical and approximations are used instead. A simple but effective diagonal preconditioner, inspired by the form of the ML expectation maximization (EM) algorithm described later, is [Kau93]

$$\mathbf{C}^n = \text{diag}\left\{ \frac{f_j^n + \delta}{\sum_i p(i,j)} \right\} \tag{16.34}$$

where δ is a small positive number to ensure that $\mathbf{C}^n$ is positive definite. An alternative diagonal preconditioner uses the inverse of the diagonal elements of the Hessian matrix [JSS00]. More complicated approaches to approximating the inverse of the Hessian include Fourier-based methods [CP+93, FB99] and matrix factorization [CH97].

One of the major challenges in the use of gradient-based methods in emission tomography is inclusion of the non-negativity constraint. The simplest approach is to restrict the step size in (16.32) so that all f_j^{n+1} are nonnegative, but this restriction slows down or even prevents convergence [Kau87]. The bent line search overcomes the problem by allowing a larger step size so that voxel values can take negative values; a truncation operator is then applied to remove negative values before moving to the next iteration. A refinement of the bent line search uses a second line search in the direction formed as the difference between the truncated update and the image from the previous iteration [Kau93]. A similar method is the active set approach in which an unconstrained search is performed over the subset of voxels identified as not belonging to the "active set" of zero-valued voxels. The active set is then updated based on the gradient at the current estimate [MLC96]. Penalty functions can be used to convert the constrained optimization to an unconstrained one [Lue84], but care must be taken to modify the preconditioner to account for the effect of active penalties on the Hessian [MLCZ94]. Primal–dual interior point methods can also be used to handle the positivity constraint [JSS00].

16.5.2 Coordinate Accent (CA) Algorithms

Seeking easier means to deal with the non-negativity constraint, researchers have also turned to coordinate-wise algorithms. While there are a number of variations on this basic theme [Fes94, BS96, SB93, ZSSB00], the essence of these methods is to update each voxel in turn so as to maximize the objective function with respect to that voxel. Given the current estimate $\mathbf{f}^n$, the update for voxel $j = 1 + (n - 1) \bmod N$ is

$$f_j^{n+1} = \arg\max_{f \geq 0} \Phi(f_1^n, f_2^n, \cdots, f_{j-1}^n, f, f_{j+1}^n, \cdots, f_N^n)$$

$$f_l^{n+1} = f_l^n, \quad l \neq j \tag{16.35}$$

The sequential update changes the multidimensional optimization problem into a series of 1D optimizations and makes the imposition of non-negativity constraints straightforward. Although many algorithms can be used to solve the 1D maximization problem, exact solution of (16.35) requires repeated computation of the derivative of the objective function, which in turn involves forward and back projection and would greatly increase the per iteration computational cost. To solve this problem, Gaussian approximations of the log-likelihood function have been used [SB93, Fesr94]. [BS96] approximate the log-likelihood function by a quadratic function at each iteration using a Newton–Raphson approach. However, the Newton–Raphson update does not guarantee monotonicity and convergence. A method to restore these desired properties is to use functional substitution [ZSSB00].

Coordinate ascent algorithms can achieve fast convergence rates if given a good initial image (e.g., a filtered back projection reconstruction). However, when starting from a uniform image, initial convergence can be very slow. This is because in CA algorithms high-frequency components converge much faster than low-frequency components [SB93, Fes94]. In addition, if a CA algorithm updates voxels in a raster-scan fashion, then the algorithm will exhibit a faster convergence rate in the scan direction than in the orthogonal direction. Therefore, it is preferable to update the image voxels using either four different raster-scan orderings or a random ordering. We should note that besides simultaneous update and sequential update algorithms, there exist algorithms that update a group of voxels at a time [FFCL97, ZSSB00]. These methods can potentially reduce reconstruction time by grouping the voxels properly, but are more difficult to implement.

16.5.3 Optimization Transfer Algorithms

Instead of maximizing the original objective function directly as in the gradient and CA algorithms, optimization transfer methods replace the original cost function at each step with a surrogate function, which when maximized is guaranteed to increase the value of the original function. By choosing the surrogate functions carefully, reductions in computation time and speedup in convergence can be realized. In PET reconstruction, commonly used surrogated functions are either quadratic or separable, or both.

Two conditions are generally required for the surrogate function $\phi(f; f^n)$:

$$\Phi(f) - \Phi(f^n) \geq \phi(f; f^n) - \phi(f^n; f^n) \tag{16.36}$$

$$\nabla\Phi(f)\,|_{f=f^n} = \nabla_f \phi(f; f^n)\,|_{f=f^n} \tag{16.37}$$

Condition (16.36) guarantees that any increase in the surrogate function $\phi(f; f^n)$ will result in at least the same amount of increase in the original objective function $\Phi(f)$; condition (16.37) guarantees that f^n will not be a maximizer of the surrogate function $\phi(f; f^n)$ if it is not also a maximizer of the original objective function $\Phi(f)$. These two conditions guarantee that the original objective function can be monotonically increased by maximizing the surrogate function at each iteration:

$$f^{n+1} = \arg\max_f \phi(f; f^n) \tag{16.38}$$

Figure 16.3 shows a 1D example of optimization transfer where a series of quadratic functions are used as surrogate functions to maximize a nonquadratic function $\Phi(f)$. We start at $f^{(1)} = 12$. Maximizing the first surrogate function $\phi(f; f^{(1)} = 12)$ results in $f^{(2)} = 8.06$. Then the second surrogate function $\phi(f; f^{(2)} = 8.06)$ is constructed and its maximizer provides $f^{(3)} = 6.67$. While the maximizer of each surrogate function $\phi(f; f^n)$ does not coincide with that of the original function $\Phi(f)$, maximizing $\phi(f; f^n)$ guarantees an increase in $\Phi(f)$. Applying this procedure repeatedly, we obtain a sequence of $\{f^n\}$ that eventually converges to the maximizer of $\Phi(f)$. A proof of convergence of optimization transfer algorithms is provided in [FH95]. A comprehensive review can be found in [LHY00].

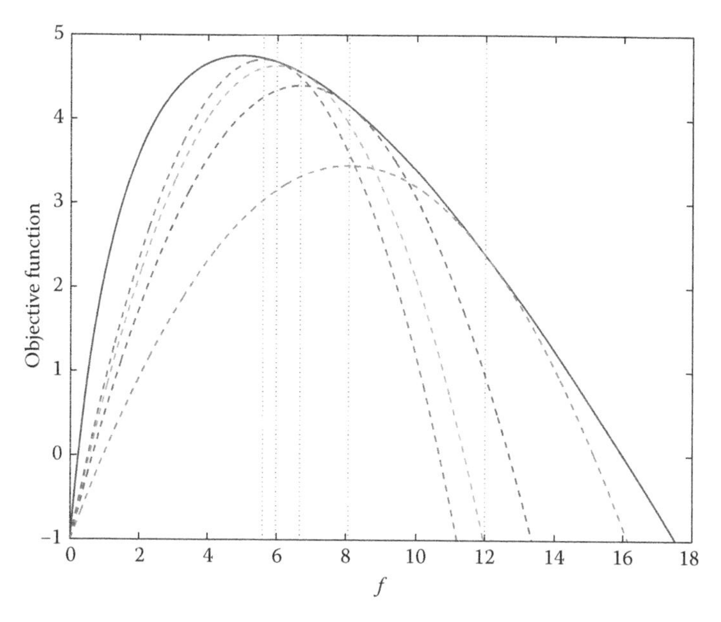

FIGURE 16.3
A 1D example of optimization transfer where a nonquadratic function $\Phi(f) = 6 \log (f + 1) - (f + 1)$ (solid line) is substituted by a series of quadratic functions $\phi(f; f^n)$ (dashed lines) starting at $f^{(1)} = 12$. The sequence obtained by maximizing the quadratic surrogate functions is $\{f^{(1)} = 12, f^{(2)} = 8.06, f^{(3)} = 6.67, f^{(4)} = 6.00, f^{(5)} = 5.59, \ldots\}$ (marked by vertical dotted lines).

16.5.3.1 EM Algorithm

The EM algorithm of [DLR77], which was first applied to maximum-likelihood PET reconstruction by [SV82] and [LC84], is the best known example of an optimization transfer algorithm in emission tomography. The EM algorithm is a general framework for computing maximum-likelihood estimates through introduction of a "complete" but unobserved data set. Each iteration consists of an E-step, computing the conditional expectation of the log likelihood of the complete data, and an M-step, maximization of this conditional expectation with respect to the image. In the optimization transfer framework, the E-step in the EM algorithm can be simply viewed as computing a surrogate function. Here we derive the surrogate function $\phi(\mathbf{f}; \mathbf{f}^n)$ using the concavity of the log-likelihood function and Jensen's inequality, without involving the complete data concept [Dep93].

For any function in the form of $\varphi\left(\sum_j w_j x_j\right)$, where $\phi(\cdot)$ is a concave function, Jensen's inequality guarantees the following [LHY00]:

$$\varphi\left(\sum_j w_j x_j\right) \geq \sum_j \frac{w_j b_j}{\sum_l w_l b_l} \varphi\left(\frac{\sum_l w_l b_l}{b_j} x_j\right) \tag{16.39}$$

when all components of the vectors $\{w_j\}$ and $\{b_j\}$ are positive. The equal sign is satisfied when $b_j = x_j$ for all j's.

Applying Jensen's inequality to the $\log\left(\sum_j p_{ij} f_j + r_i + s_i\right)$ term in (16.25) and using $\{w_j = p_{ij}, j = 1,\ldots,N, w_{N+1} = r_i + s_i\}$, $\{x_j = f_j, j = 1,\ldots,N, x_{N+1} = 1\}$, $\{b_j = f_j^n, j = 1,\ldots,N, b_{N+1} = 1\}$, we get

$$\begin{aligned}
\mathcal{L}(y\,|\,\mathbf{f}) &= \sum_{i=1}^{M} y_i \log\left(\sum_j p_{ij} f_j + r_i + s_i\right) - \sum_j p_{ij} f_j \\
&\geq \sum_{i=1}^{M} y_i \sum_j \frac{p_{ij} f_j^n}{\hat{y}_i^n} \log\left(\frac{\hat{y}_i^n}{f_j^n} f_j\right) + y_i \frac{r_i + s_i}{\hat{y}_i^n} \log \hat{y}_i^n - \sum_j p_{ij} f_j \\
&= \sum_{i=1}^{M} \left[y_i \sum_j \frac{p_{ij} f_j^n}{\hat{y}_i^n} \log(f_j) - \sum_j p_{ij} f_j \right] + c_1^n
\end{aligned} \tag{16.40}$$

where $\hat{y}_i^n = \sum_j p_{ij} f_j^n + r_i + s_i$ is the forward projection of the current estimate $\mathbf{f}^n$ and

$$c_1^n = \sum_{i=1}^{M} y_i \sum_j \frac{p_{ij} f_j^n}{\hat{y}_i^n} \log\left(\frac{\hat{y}_i^n}{f_j^n}\right) + y_i \frac{r_i + s_i}{\hat{y}_i^n} \log \hat{y}_i^n$$

is a constant independent of $\mathbf{f}$ and hence will be omitted in the following optimization.

Therefore, we obtain the following surrogate function for the Poisson log-likelihood function $\mathcal{L}(y|f)$:

$$\phi^{\mathcal{L}}(f; f^n) = \sum_j \left(\sum_{i=1}^{M} \frac{p_{ij} y_i}{\hat{y}_i^n} \right) f_j^n \log(f_j) - f_j \sum_{i=1}^{M} p_{ij} \tag{16.41}$$

It is easy to prove that $\phi_1(f; f^n)$ satisfies the two conditions in (16.36) and (16.37). One virtue of $\phi_1(f; f^n)$ is that the maximization can be carried out for each pixel independently:

$$f_j^{n+1} = \arg\max_{f_j} \left(\sum_{i=1}^{M} \frac{p_{ij} y_i}{\hat{y}_i^n} \right) f_j^n \log(f_j) - f_j \sum_{i=1}^{M} p_{ij} \tag{16.42}$$

Taking the derivative with respect to f_j and setting it to zero, we obtain the ML EM algorithm:

$$f_j^{n+1} = \frac{f_j^n}{\sum_i p_{ij}} \sum_i \frac{p_{ij} y_i}{\sum_l p_{il} f_l^n + r_i + s_i} \tag{16.43}$$

The EM algorithm (16.43) converges monotonically to a global maximizer of the likelihood function [VSK85] and guarantees non-negativity. The simple update equation together with these properties makes the EM algorithm an attractive approach. Unfortunately its converge rate is very slow, an issue we return to later.

The EM algorithm has been extended to MAP reconstruction directly where it is generally referred to as generalized EM (GEM). In the case of spatially independent priors, such as the gamma [LBL87, WG97] or independent Gaussian prior [LH87, HKM+00], the objective function in the M-step remains separable and can be maximized directly. However, for the coupled penalty function in (16.27), the objective function is not separable and direct solution is generally not possible. Instead one or more iterations of gradient or coordinate ascent are performed during the M-step [HL89].

An alternative approach is to construct a separable surrogate function for the penalty term $U(f)$. One form of the surrogate function is given by De Pierro [Pie95, LHy00] using the following inequality of a convex function $\psi(\cdot)$:

$$\psi\left(\sum_l c_l x_l\right) \leq \sum_l \alpha_l \psi\left(\frac{c_l}{\alpha_l}(x_l - b_l) + \sum_l c_l b_l\right) \tag{16.44}$$

where

$\alpha_l \geq 0$

$\sum_l \alpha_l = 1$

$\alpha_l > 0$ whenever $c_l \neq 0$

Applying inequality (16.44) to the pair-wise penalty $U(f)$ in (16.27) with $\{c_j = 1, c_k = -1, c_l = 0, l \neq j, l \neq k\}$ and $\{\alpha_j = \alpha_k = 1/2, \alpha_l = 0, l \neq j, l \neq k\}$, we get a separable surrogate function for convex penalty functions:

$$\begin{aligned}\phi^{\mathcal{U}}\left(f;f^n\right) &= \frac{1}{2}\sum_{j=1}^{N}\sum_{k\in\mathcal{N}_j}\frac{\kappa_{jk}}{2}\left\{\psi\left(2(f_j - f_j^n) + f_j^n - f_k^n\right) + \psi\left(-2(f_k - f_k^n) + f_j^n - f_k^n\right)\right\}\\ &= \frac{1}{2}\sum_{j=1}^{N}\sum_{k\in\mathcal{N}_j}\kappa_{jk}\psi\left(2f_j - f_j^n - f_l^n\right)\end{aligned} \tag{16.45}$$

Combining $\phi^{\mathcal{L}}$ and $\phi^{\mathcal{U}}$ together, we have the surrogate function for the PL function $\Phi(f)$ in (16.26):

$$\begin{aligned}\phi(f;f^n) &= \phi^{\mathcal{L}}(f;f^n) - \beta\phi^{\mathcal{U}}(f;f^n)\\ &= \sum_j\left\{\left(\sum_{i=1}^{M}\frac{p_{ij}y_i}{\hat{y}_i^n}\right)f_j^n\log(f_j) - f_j\sum_{i=1}^{M}p_{ij} - \frac{\beta}{2}\sum_{k\in\mathcal{N}_j}\kappa_{jk}\psi\left(2f_j - f_j^n - f_l^n\right)\right\}\end{aligned} \tag{16.46}$$

The surrogate function in (16.46) is separable in f and can be maximized pixel by pixel. In particular, when $\psi(t) = \frac{1}{2}t^2$, we have a closed-form solution for f_j^{n+1} [LH87]:

$$f_j^{n+1} = \frac{2C}{B + \sqrt{B^2 + 4AC}} \tag{16.47}$$

where

$$A = 2\beta \sum_{k \in \mathcal{N}_j} \kappa_{jk} \tag{16.48}$$

$$B = \sum_i p_{ij} - \beta \sum_{k \in \mathcal{N}_j} \kappa_{jk} \left(f_j^n + f_k^n \right) \tag{16.49}$$

$$C = f_j^n \sum_i \frac{p_{ij} y_i}{\hat{y}_i^n} \tag{16.50}$$

When $\beta = 0$, Equation 16.47 reduces to the ML EM update Equation 16.43; when β is large, f_j^{n+1} approaches $\sum_{k \in \mathcal{N}_j} \kappa_{jk} \left(f_j^n + f_k^n \right) / \left(2 \sum_{k \in \mathcal{N}_j} \kappa_{jk} \right)$, a weighted average of its current value and neighboring pixels.

16.5.3.2 Quadratic Surrogate and Coordinate Ascent (QSCA)

Another type of surrogate function that can be optimized relatively easily is quadratic functions. The most popular quadratic surrogate function (also known as paraboloidal surrogate function) for PET was proposed by Fessler and Erdogan in [FE98]. We start the derivation by rewriting the Poisson log-likelihood function in the following form:

$$\mathcal{L}(y \mid f) = \sum_i h_i(l_i) \tag{16.51}$$

where $l_i = \sum_j p_{ij} f_j$ and

$$h_i(l) = y_i \log(l + r_i + s_i) - (l + r_i + s_i) \tag{16.52}$$

One property of $h_i(l)$ is that its second derivative $\ddot{h}_i(l)$ increases monotonically and approaches zero as $l \to \infty$. Consequently, when $r_i + s_i > 0$, $|\ddot{h}_i(l)|$ is bounded for $l \geq 0$ with the maximum achieved at $l = 0$, and hence a quadratic surrogate function exists [LHY00].

Let $\dot{h}_i(l)$ denote the first derivative of $h_i(l)$. The quadratic surrogate function with the optimal curvature is given by [FE98]

$$\phi_{QS}^{\mathcal{L}}\left(f \mid f^n\right) = \sum_i \left[h_i\left(l_i^n\right) + \dot{h}_i\left(l_i^n\right)\left(l_i - l_i^n\right) - \frac{1}{2}\eta_i\left(l_i^n\right)\left(l_i - l_i^n\right)^2 \right] \tag{16.53}$$

where $l_i^n = \sum_j p_{ij} f_j^n$ and the optimal curvature is

$$\eta_i(l) = \begin{cases} \dfrac{2}{l^2}[h_i(l) - h_i(0) - l\dot{h}_i(l)], & l > 0 \\ -\ddot{h}_i(l), & l = 0 \end{cases} \tag{16.54}$$

Note that the quadratic surrogate function is only valid for $r_i + s_i > 0$ and $l_i \geq 0$. There are 1D examples of this quadratic surrogate function that are shown in Figure 16.3.

A quadratic surrogate function $\phi_{QS}^{\mathcal{U}}\left(\kappa \mid \kappa^n\right)$ can also be computed for non-quadratic penalty functions when the second derivative of the penalty function, $\ddot{\psi}(t)$, is a nonincreasing function of $|t|$. In this case, a proper surrogate function is given by [EF99a]

$$\phi_{QS}^{\mathcal{U}}\left(f \mid f^n\right) = \frac{1}{2}\sum_{j=1}^{N}\sum_{k\in\mathcal{N}_j} \kappa_{jk}\left[\psi\left(u_{jk}^n\right) + \dot{\psi}\left(u_{jk}^n\right)\left(f_j - f_k - u_{jk}^n\right) + \frac{1}{2}\rho_{jk}^n\left(f_j - f_k - u_{jk}^n\right)^2 \right] \tag{16.55}$$

where $u_{jk}^n = f_j^n - f_k^n$ and the curvature ρ_{jk}^n is

$$\rho_{jk}^n = \begin{cases} \dfrac{\dot{\psi}(u_{jk}^n)}{u_{jk}^n}, & u_{jk}^n \neq 0 \\ \ddot{\psi}(0), & u_{jk}^n = 0 \end{cases} \tag{16.56}$$

By combining the surrogate functions for the likelihood and penalty function, the non-separable quadratic surrogate (SQS) function for the PL function is

$$\begin{aligned} \phi_{QS}\left(f \mid f^n\right) &= \phi_{QS}^{\mathcal{L}}\left(f \mid f^n\right) - \beta\phi_{QS}^{\mathcal{U}}\left(f \mid f^n\right) \\ &= \Phi\left(f^n\right) + \left(g^n\right)^T \Delta f - \frac{1}{2}\Delta f^T \mathbf{W}^n \Delta f \end{aligned} \tag{16.57}$$

where $\Delta f = f - f^n$. The gradient vector g^n and the curvature matrix W^n are calculated by

$$g^n = P^T\left(\frac{y}{\hat{y}^n} - 1\right) - \beta v \tag{16.58}$$

$$W^n = P^T \text{diag}[\eta^n] P + \beta R \tag{16.59}$$

where diag [·] denotes a diagonal matrix and the elements of v and R are $v_j = \sum_{k \in \mathcal{N}_j} \kappa_{jk} \dot{\psi}(u_{jk}^n)$ and

$$R_{jk} = \begin{cases} -\kappa_{jk}\rho_{jk}^n, & j \neq k \\ \sum_{k \in \mathcal{N}_j} \kappa_{jk}\rho_{jk}^n, & j = k \end{cases} \tag{16.60}$$

A coordinate ascent algorithm can be used to maximize the non-SQS function in (16.57). The procedure can be written as

$$f_j^{n+1} = \arg\max_{f_j \geq 0} \phi_j(f_j) \tag{16.61}$$

$$\phi_j(f_j) = \hat{g}_j^n (f_j - f_j^n) - \frac{1}{2} W_{jj}^n (f_j - f_j^n)^2 \tag{16.62}$$

where the gradient $\hat{g}_j^n$ with respect to x_j is

$$\hat{g}_j^n = g_j^n + \sum_{l<j} W_{jl}^n (f_l^{n+1} - f_l^n) \tag{16.63}$$

Setting the derivative of $\phi_j(f_j)$ to zero and solving the resulting equation, we get the update equation

$$f_j^{n+1} = \left[f_j^n + \frac{\hat{g}_j^n}{W_{jj}^n} \right]_+ \tag{16.64}$$

where $[x]_+ = \max(0, x)$.

16.5.3.3 Separable Quadratic Surrogate (SQS) Algorithm

The coordinate ascent algorithm requires column access of the system matrix *P*, which is difficult to parallelize and incompatible with ordered-subset (OS) method (see Section 16.5.4). Using the concave property of $\phi_{QS}(f|f^n)$ in (16.57), we can construct a separable surrogate $\phi_{SQS}(f|f^n)$ [EF99b, AF03]:

$$\phi_{SQS}\left(f \mid f^{n}\right)=\Phi\left(f^{n}\right)+\left(g^{n}\right)^{T} \Delta f-\frac{1}{2} \Delta f^{T} \operatorname{diag}[\widehat{w}^{n}] \Delta f \tag{16.65}$$

where the elements of $\hat{w}^n$ are given by

$$\hat{w}_j^n = \sum_i |W_{ij}^n| \tag{16.66}$$

Readers can verify that $\operatorname{diag}[\hat{w}^n] - W^n$ is a positive semi-definite matrix. Thus $\phi_{SQS}(f|f^n)$ always lies below $\phi_{QS}(f|f^n)$ with matched function value and gradient at $f = f^n$.

The new surrogate can be maximized for all the pixels simultaneously. Maximizing $\phi_{SQS}(f|f^n)$ with respect to f subject to $f \geq 0$ results in

$$f_j^{n+1} = \left[f_j^n + \frac{g_j^n}{\hat{w}_j^n} \right]_+ \tag{16.67}$$

Equations 16.67 and 16.64 share a common form, but with different gradient and curvature values. Because $\hat{w}_j^n \geq W_{jj}^n$, the update in (16.67) is less than that in (16.64), and hence the SQS algorithm is generally slower in convergence than the QSCA algorithm. However, the simultaneous update of the SQS algorithm is easily parallelizable and can also be accelerated by OS methods.

16.5.4 Ordered-Subset Algorithms

All of the methods described so far update the image at each iteration using the entire data set. The computation cost can be substantially reduced by dividing the data into subsets and using one subset at each iteration. The best example is the ordered-subset EM (OSEM) algorithm, which is now widely used in clinical applications. While the original OSEM algorithm [HL94] and earlier methods employing subsets of the data [HLS90, HSLR90] are not convergent, convergent forms have since been developed as we describe later.

OS methods divide the objective function into a sum of sub-objective functions as

$$\Phi(f) = \sum_{q=1}^{N_S} \Phi_q(f) \tag{16.68}$$

where each sub-objective function $\Phi_q(f)$ is a function of a subset of the data. We use $\{S_q\} \subseteq \{1, 2, \ldots, M\}$ to denote the collection of the indices of the projections in the qth subset of the data. The subsets are defined so that $\{1,2,\ldots,M\} = \bigcup_{q=1}^{N_s} S_i$, where N_s is the number of subsets and $S_i \cap S_j = \emptyset$ for $i \neq j$. At each iteration, only one $\Phi_q(f)$ is involved in the calculation. Let n denote the index for a complete cycle and q the index for a sub-iteration, and define $f^{(n,0)} = f^{n-1}, f^{(n,N_s)} = f^n$.

The OSEM algorithm applies the EM algorithm to each sub-objective function $\Phi_q(f)$ in turn. The update equation is given by

$$f_j^{(n,q)} = \frac{f_j^{(n,q-1)}}{\sum_{i \in S_q} p_{ij}} \sum_{i \in S_q} \frac{p_{ij} y_i}{\sum_l p_{il} f_l^{(n,q-1)} + r_i + s_i} \tag{16.69}$$

$$\text{for} \quad j = 1,\cdots,N, \quad q = 1,\cdots,N_s$$

To avoid directional artifacts, the projections are usually chosen to have maximum separation in angle in each subset. [HL94] further recommend that the subsets are chosen so that an emission from each pixel has equal probability of being detected in each of the subsets (subset balance), that is, $\sum_{i \in S_q} p_{ij}$ is independent on q. In practice this can be difficult to achieve due to spatially varying attenuation and detector sensitivities.

OSEM produces remarkable improvements in convergence rates in the early iterations. For a reasonable number of subsets (e.g., 32 in [HL94]), almost linear speedup with respect to the number of subsets can be achieved in the early iterations. For this reason OSEM is by far the most widely used iterative ECT algorithm in clinical and research PET and SPECT studies. In the case of 3D PET, the data can be Fourier rebinned into stacked 2D sinograms before application of a 2D OSEM algorithm [CK+98, LKF+04].

The major drawback of the OSEM algorithm is that it generally does not converge if the number of subsets remains greater than one. Incremental OS methods achieve convergence through the introduction of an augmented cost function [Ahn04]. Consider a surrogate $\phi_q(f; \bar{f})$ for the sub-objective function $\Phi_q(f)$ that satisfies

$$\phi_q(f; f) = \Phi_q(f), \ \forall f \tag{16.70}$$

$$\phi_q(f; \bar{f}) \leq \Phi_q(f), \ \forall f, \bar{f} \tag{16.71}$$

Then for all $\bar{f}$, $\max_{\bar{f}} \phi_q(f; \bar{f}) = \phi_q(f; f) = \Phi_q(f)$. It follows that if we construct an augmented cost function as

$$F\left(f; \bar{f}_1, \ldots, \bar{f}_{N_s}\right) = \sum_{q=1}^{N_s} \phi_q\left(f; \bar{f}_q\right) \tag{16.72}$$

then the maximizer of $\Phi(f)$ can be found as

$$\hat{f} = \arg\max_{f} \Phi(f) = \arg\max_{f} \max_{\bar{f}_1,\ldots,\bar{f}_{N_s}} F\left(f;\bar{f}_1,\ldots,\bar{f}_{N_s}\right) \tag{16.73}$$

Incremental OS methods alternately update f and one $\bar{f}_q$ at each step, that is,

$$f^{(n+1,q)} = \arg\max_{f} F\left(f;\bar{f}_1^{n+1},\ldots,\bar{f}_{q-1}^{n+1},\bar{f}_q^{n},\ldots,\bar{f}_{N_s}^{n}\right) \tag{16.74}$$

$$\bar{f}_q^{n+1} = f^{(n+1,q)} \qquad \text{for } q = 1,\cdots,N_s \tag{16.75}$$

Using the EM surrogate function in (16.40) for each subset of the data, that is,

$$\phi_q\left(f;\bar{f}\right) = \sum_j \left(\bar{f}_j \sum_{i\in S_q} \frac{p_{ij}y_i}{\sum_l p_{il}\bar{f}_l + r_i + s_i} \log\left(\frac{\hat{y}_i^n}{\bar{f}_j^n} f_j\right) + \sum_{i\in S_q} y_i \frac{r_i + s_i}{\hat{y}_i^n} \log \hat{y}_i^n - f_j \sum_{i\in S_q} p_{ij} \right) \tag{16.76}$$

we obtain the update equation of the convergent OSEM (COSEM) algorithm [HRKG04, KHRG04]:

$$f_j^{(n+1,q)} = \frac{1}{\sum_i p_{ij}} \sum_{q'=1}^{N_s} e_{q',j} \tag{16.77}$$

where

$$e_{q',j} = \begin{cases} f_j^{(n+1,q')} \sum_{i\in S_q} \dfrac{p_{ij}y_i}{\sum_l p_{il} f_l^{(n+1,q')} + r_i + s_i}, & q' < q \\ f_j^{(n,q')} \sum_{i\in S_q} \dfrac{p_{ij}y_i}{\sum_l p_{il} f_l^{(n,q')} + r_i + s_i}, & q' \geq q \end{cases} \tag{16.78}$$

COSEM increases the augmented objective function $F\left(f;\bar{f}_1,\ldots,\bar{f}_{N_s}\right)$ at every iteration, but does not guarantee a monotonic increase of the original objective function $\Phi(f)$. In emission tomography, the dominant cost of forward and back projection is associated with the update of the surrogate functions $\phi_q(f;\bar{f}_q)$ as a result of changes in $\bar{f}_q$. In COSEM this corresponds to the evaluation of $e_{q',j}$; however, this quantity remains unchanged at sub-iteration q for all $q' \neq q$. Therefore, by updating only one sub-objective function $\phi_q(f; \bar{f}_q)$ at a time, incremental OS methods reduce the computation cost by a factor similar to conventional OS methods. Unfortunately, incremental OS methods, while faster than their non-OS counterparts, are slower than the conventional OS methods at early iterations. Faster convergence can therefore be achieved by starting with a conventional OS algorithm, but switching to a convergent incremental OS algorithm at later iterations [HRKG04, Ahn04]. This can be taken a step further by noting that, asymptotically, PCG methods converge faster than the OS methods, and consequently one could switch from incremental OS to PCG at later iterations [LAL05].

16.6 Direct Dynamic PET Reconstruction

16.6.1 Dynamic PET Data Model

Because PET acquires all projection data simultaneously, PET can monitor the temporal change of radiotracer distribution over the entire scan duration. The concentration of the radiotracer at a given pixel at time t can be described as the following tracer kinetic model [ST02]:

$$C_T(t;\boldsymbol{\kappa}, f_v) = (1 - f_v)q(t;\boldsymbol{\kappa}) * C_P(t) + f_v C_{wb}(t) \qquad (16.79)$$

where

$\boldsymbol{\kappa}$ is a vector that contains all the kinetic parameters that determines the tracer uptake in the tissue
f_v is the fractional volume of blood in the tissue
$q(t; \boldsymbol{\kappa})$ is the impulse response function
$C_P(t)$ is the tracer concentration in plasma
$C_{wb}(t)$ is the tracer concentration in whole blood
"$*$" denotes the convolution operator

The model in (16.79) is general and includes compartmental models and simplified linear models as special cases. For example, the impulse response $q(t)$ of the commonly used three-compartment model is given by

$$q(t;\boldsymbol{\kappa}) = \frac{K_1}{\Delta\alpha}\left(k_3 + k_4 - \alpha_1 \quad \alpha_2 - k_3 - k_4\right)e^{-\alpha(t)} \qquad (16.80)$$

where $\boldsymbol{\kappa} = \{K_1, k_2, k_3, k_4\}$ are the rate constants between compartments, $\Delta\alpha = \alpha_2 - \alpha_1$ with $\alpha_{1,2} = \frac{1}{2}(k_2 + k_3 + k_4) \mp \frac{1}{2}\left[(k_2 + k_3 + k_4)^2 - 4k_2k_4\right]^{1/2}$, and $e^{-\alpha(t)} = \left(e^{-\alpha_1 t}, e^{-\alpha_2 t}\right)^T$. For simplicity, we will use $\boldsymbol{\theta}_j$ to denote the collection of kinetic parameters and fractional volume $\{\boldsymbol{\kappa}, f_v\}$ for pixel j and $\boldsymbol{\theta} \equiv \{\boldsymbol{\theta}_j\}$ to denote the parametric images.

A dynamic PET scan is often divided into multiple consecutive time frames, with each frame containing coincidence events recorded from the start of the frame till the end of the frame. The image intensity at pixel j in time frame m, $f_m(\boldsymbol{\theta}_j)$, is then given by

$$f_m(\theta_j) = \sum_{t_{m,s}}^{t_{m,e}} C_T(\tau, \theta_j) e^{-\lambda\tau} d\tau \tag{16.81}$$

where

$t_{m,s}$ and $t_{m,e}$ denote the start and end times of frame m, respectively
λ is the decay constant of the radiotracer

Substituting $f_m(\boldsymbol{\theta})$ into Equation 16.22, the expected PET data in time frame m are

$$\bar{y}_m(\theta) = \boldsymbol{P} f_m(\theta) + \boldsymbol{r} + \boldsymbol{s} \tag{16.82}$$

The goal of dynamic PET is to estimate $\boldsymbol{\theta}$ for each pixel from dynamic PET measurements. Conventional methods usually reconstruct a sequence of emission images from measured projection data first and then fit the time activity curve (TAC) in each voxel to a kinetic model. To obtain an accurate estimate, the resolution and noise distribution of reconstructed emission images should be modeled in the kinetic modeling. However, exact modeling of noise distribution in emission images reconstructed by iterative methods is extremely difficult because the noise is space variant and object dependent. Often the space-varying noise variance and correlations between pixels are simply ignored in the kinetic modeling step, which can lead to suboptimal results. Direct reconstruction of parametric images from raw projection data solves this problem by combining kinetic modeling and emission image reconstruction into a single formula. It allows accurate modeling of noise statistics in data and hence can be statistically more efficient [CL85, KBMS05, WFQ08].

In theory, one can directly apply the algorithms described in Section 16.5, such as the EM and PCG, to the optimization problems of direct parametric image reconstruction [CL85, MBPC97, KBMS05, TTT08, WFQ08]. However, it has been found that the coupling between the spatial image reconstruction and the temporal model can greatly affect the convergence rate of these algorithms, especially when the temporal basis functions are highly correlated, such as those used in the Patlak model and spectral analysis [TTT07]. An alternative approach is to decouple the kinetic modeling from the emission image reconstruction using the optimization transfer principle [LHY00] as described later.

16.6.2 Linear Kinetic Models

Under a linear kinetic model, a TAC is represented as a linear combination of a set of basis functions:

$$f_{jm}(\theta_j) = \sum_{k=1}^{n_k} b_{mk}\theta_{jk} \tag{16.83}$$

where

b_{mk} is the (m, k)th element of a basis matrix $\boldsymbol{B} \in IR^{n_m \times n_k}$ of which each column denotes a temporal basis function

θ_{jk} is the coefficient (kinetic parameter) of the kth basis function at pixel j

n_k is the total number of basis functions

Examples of the basis functions include B-splines [RGH00, RGH02, NQAL02, LAAL07], spectral bases [MMC+98, MBPC97, RSC+06, RCS+07], and the Patlak model [WFQ08, TTT08].

By representing dynamic PET data and parametric images as long vectors,* the expectation of dynamic PET data can be written in the following matrix-vector formula:

$$\bar{\boldsymbol{y}} = (\boldsymbol{B} \otimes \boldsymbol{P})\theta + \boldsymbol{r} + \boldsymbol{s} \tag{16.84}$$

where $\otimes$ denotes the Kronecker product.

The ML EM algorithm used in static PET can be directly applied to the reconstruction of linear parametric images by considering $B \otimes P$ as a single system matrix, which results in the following update equation:

$$\theta_{jk}^{n+1} = \frac{\theta_{jk}^{n}}{\left(\sum_m b_{mk}\right)\left(\sum_i p_{ij}\right)} \sum_i \sum_m p_{ij} b_{mk} \frac{y_{im}}{\hat{y}_{im}^{n}} \tag{16.85}$$

where

$$\hat{\boldsymbol{y}}^{n} = (\boldsymbol{B} \otimes \boldsymbol{P})\theta^{n} + \boldsymbol{r} + \boldsymbol{s}$$

Unfortunately, the EM algorithm in (16.85) can be very slow in convergence for dynamic PET reconstruction, when the temporal basis functions are highly correlated [TTT07]. To address this problem, Wang and Qi proposed a

* For example, $\bar{y} = [\bar{y}_{11}, \ldots, \bar{y}_{n_i 1}, \bar{y}_{12}, \ldots, \bar{y}_{n_i 2}, \ldots, \bar{y}_{1 n_m}, \ldots, \bar{y}_{n_i n_m}]^T$ where n_i is the total number of sinogram elements in each frame and n_m is the total number of frames.

nested EM algorithm [WQ10], in which the image reconstruction and kinetic modeling are decoupled in each iteration by the introduction of a surrogate function in the image space. The surrogate function $S_{\mathcal{L}}(\boldsymbol{\theta}; \boldsymbol{\theta}^n)$ for the log-likelihood function is borrowed directly from the ML EM algorithm [CL85]:

$$\mathcal{S}_{\mathcal{L}}\left(\theta;\theta^{n}\right)=\sum_{j}\left(\sum_{i}p_{ij}\right)\left(\sum_{m}\hat{f}_{jm}^{n+1}\log f_{jm}\left(\theta_{j}\right)-f_{jm}\left(\theta_{j}\right)\right) \tag{16.86}$$

where $\hat{f}_{jm}^{n+1}$ is defined as

$$\hat{f}_{jm}^{n+1}=\frac{f_{jm}(\theta_{j}^{n})}{\sum_{i}p_{ij}}\sum_{i}p_{ij}\frac{y_{im}}{\hat{y}_{im}^{n}} \tag{16.87}$$

The new kinetic parameter estimate at iteration $n + 1$ can be obtained by maximizing $\mathcal{S}_{\mathcal{L}}(\theta;\hat{\theta}^{n})$ with respect to $\boldsymbol{\theta}$. Since $S_{\mathcal{L}}(\boldsymbol{\theta}; \boldsymbol{\theta}^n)$ is separable in voxels, the maximization can be carried out voxel by voxel, that is,

$$\theta_{j}^{n+1}=\arg\max_{\theta_{j}}\sum_{m}\hat{f}_{jm}^{n+1}\log f_{jm}\left(\theta_{j}\right)-f_{jm}\left(\theta_{j}\right) \tag{16.88}$$

which can be solved by the following EM update equation:

$$\theta_{jk}^{n,l+1}=\frac{\theta_{jk}^{n,l}}{\sum_{m}b_{mk}}\sum_{m}b_{mk}\frac{\hat{f}_{jm}^{n+1}}{f_{jm}(\theta_{j}^{n,l})},\quad l=1,\ldots,n_{l} \tag{16.89}$$

where l is the sub-iteration number and $\hat{\theta}_{jk}^{n+1}\equiv\hat{\theta}_{jk}^{n,n_l+1}$.

The algorithm is referred to as the "nested EM" algorithm because an EM iteration is nested in the M-step. Each full iteration of the nested EM algorithm consists of one iteration of (16.87) and multiple iterations of (16.89). The first step (16.87) resembles an emission image reconstruction and the second step (16.89) is an update of the linear kinetic parameters based on the intermediate image $\left\{\hat{f}_{jm}^{n+1}\right\}$. It is interesting to note that the traditional EM algorithm in (16.85) is just a special case of the nested EM algorithm with $n_l = 1$. Because the size of matrix $\boldsymbol{B}$ is much smaller than that of the system matrix $\boldsymbol{P}$, the computational cost of (16.89) is much less than that of (16.87). Therefore, the nested EM algorithm can accelerate the convergence rate of the direct reconstruction by running multiple iterations of (16.89) without affecting the overall computational time. A similar concept to improve convergence speed was also used in the coordinate descent optimization for nonlinear kinetic models, where linear parameters were updated more often than nonlinear parameters [KBMS05].

16.6.3 Nonlinear Kinetic Models

The optimization transfer principle can also be applied to direct reconstruction of nonlinear kinetic parameters. Both the non-separable and SQS functions can be used to transfer direct parametric image reconstruction into a pixel-wise least squares fitting problem [WQ09b]. Here we only describe the separable case.

Replacing Δf in (16.65) with $f_m(\boldsymbol{\theta}) - f_m(\boldsymbol{\theta}^n)$ and summing over all frames, we can obtain a separable surrogate $\phi_{SQS}(\boldsymbol{\theta}|\boldsymbol{\theta}^n)$ for parametric image reconstruction:

$$\phi_{SQS}(\boldsymbol{\theta} \mid \boldsymbol{\theta}^n) = \sum_m (g_m^n)^T \left[f_m(\boldsymbol{\theta}) - f_m(\boldsymbol{\theta}^n) \right] - \frac{1}{2} \left[f_m(\boldsymbol{\theta}) - f_m(\boldsymbol{\theta}^n) \right]^T \operatorname{diag}\left[\widehat{w}_m^n \right] \left[f_m(\boldsymbol{\theta}) - f_m(\boldsymbol{\theta}^n) \right] \tag{16.90}$$

where g_m^n and $\widehat{w}_m^n$ are calculated based on $f^n = f_m(\boldsymbol{\theta}^n)$ using (16.58) and (16.66), respectively. We omit a constant that is independent of $\boldsymbol{\theta}$ in (16.90) for simplicity.

Maximizing $\phi_{SQS}(\boldsymbol{\theta}|\boldsymbol{\theta}^n)$ with respect to $\boldsymbol{\theta}$ can be written into the following least squares formulation

$$\theta_j^{n+1} = \arg\min_{\theta_j} \frac{1}{2} \sum_m \hat{w}_{jm}^n \left(f_m(\theta_j) - \hat{f}_{jm}^{n+1} \right)^2 \tag{16.91}$$

where

$$\hat{f}_{jm}^{n+1} = f_m(\theta_j^n) + \frac{g_{jm}^n}{\hat{w}_{jm}^n} \tag{16.92}$$

Equation 16.92 has the same form as the SQS algorithm for static image reconstruction, and (16.91) can be solved by an existing nonlinear least squares (NLS) algorithm for kinetic modeling. By decoupling the image update and kinetic modeling at each iteration, the direct reconstruction algorithm can take advantage of well-developed fitting algorithms for kinetic modeling, such as the Levenberg–Marquardt algorithm for general compartment models [Mar63] and the basis function algorithm for one-tissue compartment models [GLHC97, WJK+05, BKR+05]. Therefore, the method can be adapted to different kinetic models as long as an NLS fitting algorithm exists for that model. Furthermore, it also allows using different kinetic models for different organs in an image as long as the information about the organ boundaries and the corresponding kinetic models is available before the reconstruction. For regions that do not conform to a kinetic model, one can simply use frame-based TACs. An extension of the method to the simplified reference tissue model can also be found in [WQ09a].

16.7 Summary

We have described Fourier rebinning and iterative image reconstruction methods for 3D PET. While we focused on penalized maximum-likelihood method with Poisson likelihood function and pair-wise penalties, the optimization algorithms can be applied to maximization of other objective functions. Finally we should point out that besides the penalized maximum-likelihood formulation, there exist other Bayesian estimators that can be applied to PET image reconstruction, such as those that compute the posterior mean using Markov chain Monte Carlo methods [Gre96, Wei97, HBJ+97, Sit11].

Acknowledgments

The author would like to acknowledge grant support ROIEB000194 and RC4EB012836 from the National Institute of Biomedical Imaging and Bioengineering.

References

[AF03] S. Ahn and J. A. Fessler. Globally convergent image reconstruction for emission tomography using relaxed ordered subsets algorithms. *IEEE Transactions on Medical Imaging*, 22(5):613–626, 2003.

[Ahn04] S. Ahn. Convergent algorithms for statistical image reconstruction in emission tomography. PhD thesis, The University of Michigan, St Ann Arbor, MI, 2004.

[BG74] T. F. Budinger and G. T. Gullberg. Three-dimensional reconstruction in nuclear medicine emission imaging. *IEEE Transactions on Nuclear Science*, NS-21:2–20, June 1974.

[BKR+05] R. Boellaard, P. Knaapen, A. Rijbroek, G. J. J. Luurtsema, and A. A. Lammertsma. Evaluation of basis function and linear least squares methods for generating parametric blood flow images using o-15-water and positron emission tomography. *Molecular Imaging and Biology*, 7(4):273–285, 2005.

[BS93] C. Bouman and K. Sauer. A generalized Gaussian image model for edge-preserving MAP estimation. *IEEE Transactions on Image Processing*, 2(3):296–310, July 1993.

[BS96] C. Bouman and K. Sauer. A unified approach to statistical tomography using coordinate descent optimization. *IEEE Transactions on Image Processing*, 5(3):480–492, March 1996.

[CH86] M. E. Casey and E. J. Hoffman. Quantitation in positron emission computed tomography: 7. A technique to reduce noise in accidental coincidence measurements and coincidence efficiency calibration. *Journal of Computer Assisted Tomography*, 10(5):845–850, September/October 1986.

[CH97] G. Chinn and S. C. Huang. A general class of preconditioners for statistical iterative reconstruction of emission computed tomography. *IEEE Transactions on Medical Imaging*, 16(1):1–10, 1997.

[CK+98] C. Comtat, P. Kinahan et al. Fast reconstruction of 3D pet data with accurate statistical modeling. *IEEE Transactions on Nuclear Science*, 45:1083–1089, 1998.

[CL85] R. E. Carson and K. Lange. The EM parametric image reconstruction algorithm. *Journal of America Statistics Association*, 80(389):20–22, 1985.

[CLQ11] J. Cheng-Liao and J. Qi. PET image reconstruction with anatomical edge guided level set prior. *Physics in Medicine and Biology*, 56(21):6899–6918, 2011.

[Col80] J. G. Colsher. Fully three-dimensional positron emission tomography. *Physics in Medicine and Biology*, 25(1):103–115, 1980.

[CP+93] N. H. Clinthorne, T. S. Pan et al. Preconditioning methods for improved convergence rates in iterative reconstruction. *IEEE Transactions on Medical Imaging*, 12(1):78–83, March 1993.

[Dep93] A. R. Depierro. On the relation between the isra and the em algorithm for positron emission tomography. *IEEE Transactions on Medical Imaging*, 12(2):328–333, 1993.

[DKT+97] M. Defrise, P. E. Kinahan, D. W. Townsend, C. Michel, M. Sibomana, and D. F. Newport. Exact and approximate rebinning algorithms for 3-D PET data. *IEEE Transactions on Medical Imaging*, 16(2):145–158, April 1997.

[DL99] M. Defrise and X. A. Liu. A fast rebinning algorithm for 3D positron emission tomography using John's equation. *Inverse Problems*, 15(4):1047–1065, 1999.

[DLR77] A. P. Dempster, N. M. Laird, and D. B. Rubin. Maximum likelihood from incomplete data via the EM algorithm. *Journal of Royal Statistical Society, Series B*, 39(1):1–38, 1977.

[EF99a] H. Erdogan and J. A. Fessler. Monotonic algorithms for transmission tomography. *IEEE Transactions on Medical Imaging*, 18(9):801–814, 1999.

[EF99b] H. Erdogan and J. A. Fessler. Ordered subsets algorithms for transmission tomography. *Physics in Medicine and Biology*, 44(11):2835–2851, 1999.

[FB99] J. A. Fessler and S. D. Booth. Conjugate-gradient preconditioning methods for shift-variant image reconstruction. *IEEE Transactions on Image Processing*, 8(5):688–699, 1999.

[FE98] J. A. Fessler and H. Erdogan. A paraboloidal surrogates algorithm for convergent penalized-likelihood emission image reconstruction. *1998 IEEE Nuclear Science Symposium and Medical Imaging Conference*, 2:1132–1135, 1998.

[Fes94] J. A. Fessler. Penalized weighted least-squares image reconstruction for PET. *IEEE Transactions on Medical Imaging*, 13:290–300, June 1994.

[FFCL97] A. Jeffrey. E. P. Fessler, N. Ficaro, H. Clinthorne, and L. Kenneth. Grouped-coordinate ascent algorithms for penalized-likelihood transmission image reconstruction. *IEEE Transactions on Medical Imaging*, 16(2):166–175, April 1997.

[FH95] J. A. Fessler and A. O. Hero. Penalized maximum-likelihood image reconstruction using space-alternating generalized EM algorithms. *IEEE Transactions on Image Processing*, 4:1417–1429, October 1995.

[GLHC97] R. N. Gunn, A. A. Lammertsma, S. P. Hume, and V. J. Cunningham. Parametric imaging of ligand-receptor binding in PET using a simplified reference region model. *Neuroimage*, 6(4):279–287, 1997.

[GLRZ93] G. Gindi, M. Lee, A. Rangarajan, and I. George Zubal. Bayesian reconstruction of functional images using anatomical information as priors. *IEEE Transactions on Medical Imaging*, 12(4):670–680, December 1993.

[GM85] S. Geman and D. E. McClure. Bayesian image analysis: An application to single photon emission tomography. *Proceedings of Statistical Computing Section of the American Statistical Association*, Statistical computing section, pp. 12–18, 1985.

[Goi72] M. Goitein. Three-dimensional density reconstruction from a series of two-dimensional projections. *Nuclear Instruments and Methods*, 101(3):509–518, 1972.

[Gre96] P. J. Green. MCMC in image analysis. In W.R. Gilks, S. Richardson, and D.J. Spiegelhalter, eds., *Markov Chain Monte Carlo in Practice*, Chapter 21, pp. 381–399. Chapman & Hall, London, U.K., 1996.

[HBJ+97] D. M. Higdon, J. E. Bowsher, V. E. Johnson, T. G. Turkington, D. R. Gilland, and R. J. Jaszczak. Fully Bayesian estimation of Gibbs hyperparameters for emission computed tomography data. *IEEE Transactions on Medical Imaging*, 16:516–526, 1997.

[HKM+00] R. H. Huesman, G. J. Klein, W. W. Moses, J. Qi, B. W. Reutter, and P. R. G. Virador. List mode maximum likelihood reconstruction applied to positron emission mammography with irregular sampling. *IEEE Transactions on Medical Imaging*, 19:532–537, 2000.

[HL89] T. Hebert and R. Leahy. A generalized EM algorithm for 3-D Bayesian reconstruction from Poisson data using Gibbs priors. *IEEE Transactions on Medical Imaging*, 8(2):194–202, June 1989.

[HL94] H. Malcolm Hudson and Richard S. Larkin. Accelerated image reconstruction using ordered subsets of projection data. *IEEE Transactions on Medical Imaging*, 13(4):601–609, December 1994.

[HLS90] T. Hebert, R. Leahy, and M. Singh. 3D ML reconstruction for a prototype SPECT system. *Journal of the Optical Society of America. Part A, Optics and Image Science*, 7(7):1305–1313, July 1990.

[HRKG04] I.-T. Hsiao, A. Rangarajan, P. Khurd, and G. Gindi. An accelerated convergent ordered subsets algorithm for emission tomography. *Physics in Medicine and Biology*, 49(11):2145–2156, 2004.

[HSLR90] S. Holte, P. Schmidlin, A. Linden, and G. Rosenqvist. Iterative image reconstruction for positron emission tomography: A study of convergence and quantitation problems. *IEEE Transactions on Nuclear Science*, 37(2):629–635, 1990.

[Hub81] P. J. Huber. *Robust Statistics*. John Wiley & Sons, New York, 1981.

[JSS00] C. A. Johnson, J. Seidel, and A. Sofer. Interior-point methodology for 3-D PET reconstruction. *IEEE Transactions on Medical Imaging*, 19(4):271–285, 2000.

[Kau87] L. Kaufman. Implementing and accelerating the EM algorithm for positron emission tomography. *IEEE Transactions on Medical Imaging*, 6(1):37–51, March 1987.

[Kau93] L. Kaufman. Maximum likelihood, least squares, and penalized least squares for PET. *IEEE Transactions on Medical Imaging*, 12(2):200–214, June 1993.

[KBMS05] M. E. Kamasak, C. A. Bouman, E. D. Morris, and K. Sauer. Direct reconstruction of kinetic parameter images from dynamic PET data. *IEEE Transactions on Medical Imaging*, 24(5):636–650, 2005.

[KHRG04] P. Khurd, I. T. Hsiao, A. Rangarajan, and G. Gindi. A globally convergent regularized ordered-subset EM algorithm for list-mode reconstruction. *IEEE Transactions on Nuclear Science*, 51(3):719–725, 2004.

[KR89] P. E. Kinahan and J. G. Rogers. Analytic 3D image reconstruction using all detected events. *IEEE Transactions on Nuclear Science*, 36:964–968, 1989.

[LAAL07] Q. Z. Li, E. Asma, S. Ahn, and R. M. Leahy. A fast fully 4-D incremental gradient reconstruction algorithm for list mode pet data. *IEEE Transactions on Medical Imaging*, 26(1):58–67, 2007.

[LAL05] Q. Li, S. Ahn, and R. Leahy. Fast hybrid algorithms for PET image reconstruction. In *IEEE Nuclear Science Symposium Conference Record*, pp. 1933–1937, Puerto Rico, 2005.

[Lan90] K. Lange. Convergence of EM image reconstruction algorithms with Gibbs smoothing. *IEEE Transactions on Medical Imaging*, 9(4):439–446, December 1990. Correction: 10(2):228, June 1991.

[LBL87] K. Lange, M. Bahn, and R. Little. A theoretical study of some maximum likelihood algorithms for emission and transmission tomography. *IEEE Transactions on Medical Imaging*, 6(2):106–114, June 1987.

[LC84] K. Lange and R. Carson. EM reconstruction algorithms for emission and transmission tomography. *Journal of Computer Assisted Tomography*, 8(2):306–316, April 1984.

[LDH95] C. S. Levin, M. Dahlbom, and E. J. Hoffman. A Monte Carlo correction for the effect of Compton scattering in 3-D PET brain imaging. *IEEE Transactions on Nuclear Science*, 42(4):1181–1185, 1995.

[LDM+99] X. Liu, M. Defrise, C. Michel, M. Sibomana, C. Comtat, P. Kinahan, and D. Townsend. Exact rebinning methods for three-dimensional PET. *IEEE Transactions on Medical Imaging*, 18(8):657–664, August 1999.

[LH87] E. Levitan and G. T. Herman. A maximum a posteriori probability expectation maximization algorithm for image reconstruction in emission tomography. *IEEE Transactions on Medical Imaging*, 6(3):185–192, September 1987.

[LHY00a] K. Lange, D. R. Hunter, and I. Yang. Optimization transfer using surrogate objective functions. *Journal of Computational and Graphical Statistics*, 9(1):1–20, 2000.

[LKF+04] K. Lee, P. E. Kinahan, J. A. Fessler, R. S. Miyaoka, M. Janes, and T. K. Lewellen. Pragmatic fully 3D image reconstruction for the MiCES mouse imaging PET scanner. *Physics in Medicine and Biology*, 49:4563–4578, 2004.

[LM86] R. M. Lewitt and G. Muehllehner. Accelerated iterative reconstruction for positron emission tomography based on the EM algorithm for maximum likelihood estimation. *IEEE Transactions on Medical Imaging*, 5(1):16–22, March 1986.

[Lue84] D. G. Luenberger. *Linear and Nonlinear Programming*, 2nd edn. Addison-Wesley Publishing Company, Reading, MA, 1984.

[LY91] R. Leahy and X. H. Yan. Incorporation of anatomical MR data for improved functional imaging with PET. In A. Colchester and D. Hawkes, eds., *Information Processing in Medical Imaging*, pp. 105–120. LNCS, Vol 511, Springer, Heidelberg, Berlin, 1991.

[Mar63] D. W. Marquardt. An algorithm for least-squares estimation of nonlinear parameters. *Journal of the Society for Industrial and Applied Mathematics*, 11(2):431–441, 1963.

[MBPC97] J. Matthews, D. Bailey, P. Price, and V. Cunningham. The direct calculation of parametric images from dynamic PET data using maximum-likelihood iterative reconstruction. *Physics in Medicine and Biology*, 42(6):1155–1173, 1997.

[MLC96] E. Mumcuoglu, R. Leahy, and S. Cherry. Bayesian reconstruction of PET images: Methodology and performance analysis. *Physics in Medicine and Biology*, 41:1777–1807, 1996.

[MLCZ94] E. U. Mumcuoglu, R. Leahy, S. R. Cherry, and Z. Zhou. Fast gradient-based methods for Bayesian reconstruction of transmission and emission PET images. *IEEE Transactions on Medical Imaging*, 13(4):687–701, December 1994.

[MMC+98] S. R. Meikle, J. C. Matthews, V. J. Cunningham, D. L. Bailey, L. Livieratos, T. Jones, and P. Price. Parametric image reconstruction using spectral analysis of PET projection data. *Physics in Medicine and Biology*, 43(3):651–666, 1998.

[NQAL02] T. E. Nichols, J. Qi, E. Asma, and R. M. Leahy. Spatiotemporal reconstruction of list-mode PET data. *IEEE Transactions on Medical Imaging*, 21(4):396–404, 2002.

[OJB92] J. M. Ollinger, G. C. Johns, and M. T. Burney. Model-based scatter correction in three dimensions. In *IEEE Nuclear Science Symposium Conference Record*, Vol. 2, pp. 1249–1251, Orlando, FL, 1992.

[Oll96] J. M. Ollinger. Model-based scatter correction for fully 3D PET. *Physics in Medicine and Biology*, 41:153–176, 1996.

[Pie95] A. R. De Pierro. A modified expectation maximization algorithm for penalized likelihood estimation in emission tomography. *IEEE Transactions on Medical Imaging*, 14(1):132–137, March 1995.

[QL06] J. Qi and R. M. Leahy. Iterative reconstruction techniques in emission computed tomography. *Physics in Medicine and Biology*, 51:R541–R578, 2006.

[QLC+98] J. Qi, R. M. Leahy, S. R. Cherry, A. Chatziioannou, and T. H. Farquhar. High resolution 3D Bayesian image reconstruction using the microPET small animal scanner. *Physics in Medicine and Biology*, 43(4):1001–1013, 1998.

[QLH+98] J. Qi, R. M. Leahy, C. Hsu, T. H. Farquhar, and S. R. Cherry. Fully 3D Bayesian image reconstruction for ECAT EXACT HR+. *IEEE Transactions on Nuclear Science*, 45(3):1096–1103, June 1998.

[RCS+07] A. J. Reader, J. C. Matthews, F. C. Sureau, C. Comtat, R. Trebossen, and I. Buvat. Fully 4D image reconstruction by estimation of an input function and spectral coefficients. *2006 IEEE Nuclear Science Symposium and Medical Imaging Conference*, 5:3260–3267, 2007.

[RGH00] B. W. Reutter, G. T. Gullberg, and R. H. Huesman. Direct least-squares estimation of spatiotemporal distributions from dynamic SPECT projections using a spatial segmentation and temporal b-splines. *IEEE Transactions on Medical Imaging*, 19(5):434–450, 2000.

[RGH02] B. W. Reutter, G. T. Gullberg, and R. H. Huesman. Effects of temporal modelling on the statistical uncertainty of spatiotemporal distributions estimated directly from dynamic SPECT projections. *Physics in Medicine and Biology*, 47(15):2673–2683, 2002.

[RRK92] N. Rajeevan, K. Rajgopal, and G. Krishna. Vector-extrapolated fast maximum likelihood estimation algorithms for emission tomography. *IEEE Transactions on Medical Imaging*, 11(1):9–20, March 1992.

[RSC+06] A. J. Reader, F. C. Sureau, C. Comtat, R. Trebossen, and I. Buvat. Joint estimation of dynamic PET images and temporal basis functions using fully 4D ML-EM. *Physics in Medicine and Biology*, 51(21):5455–5474, 2006.

[SB93] K. Sauer and C. Bouman. A local update strategy for iterative reconstruction from projections. *IEEE Transactions on Signal Processing*, 41(2):534–548, February 1993.

[SC97] S. Sastry and R. E. Carson. Multimodality Bayesian algorithm for image reconstruction in positron emission tomography: A tissue composition model. *IEEE Transactions on Medical Imaging*, 16(6):750–761, 1997.

[Sit11] A. Sitek. Reconstruction of emission tomography data using origin ensembles. *IEEE Transactions on Medical Imaging*, 30(4):946–956, 2011.

[SPR+11] S. Somayajula, C. Panagiotou, A. Rangarajan, Q. Li, S. R. Arridge, and R. M. Leahy. PET image reconstruction using information theoretic anatomical priors. *IEEE Transactions on Medical Imaging*, 30(3):537–549, 2011.

[ST02] K. C. Schmidt and F. E. Turkheimer. Kinetic modeling in positron emission tomography. *Quarterly Journal of Nuclear Medicine*, 46(1):70–85, 2002.

[SV82] L. A. Shepp and Y. Vardi. Maximum likelihood reconstruction for emission tomography. *IEEE Transactions on Medical Imaging*, 1(2):113–122, October 1982.

[TTT07] C. Tsoumpas, F. Turkheimer, and K. Thielemans. Convergence properties of algorithms for direct parametric estimation of linear models in dynamic PET. *Nuclear Science Symposium Conference Record, 2007*, 4:3034–3037, 2007.

[TTT08] C. Tsoumpas, F. E. Turkheimer, and K. Thielemans. Study of direct and indirect parametric estimation methods of linear models in dynamic positron emission tomography. *Medical Physics*, 35(4):1299–1309, 2008.

[VSK85] Y. Vardi, L. A. Shepp, and L. Kaufman. A statistical model for positron emission tomography. *Journal of the American Statistical Association*, 80(389):8–37, March 1985.

[Wat00] C. C. Watson. New, faster, image-based scatter correction for 3D PET. *IEEE Transactions on Nuclear Science*, 47(4):1587–1594, 2000.

[WCL04] C. H. Wang, J. C. Chen, and R. S. Liu. Development and evaluation of MRI based Bayesian image reconstruction methods for pet. *Computerized Medical Imaging and Graphics*, 28(4):177–184, 2004.

[Wei97] I. S. Weir. Fully Bayesian reconstructions from single-photon emission computed tomography data. *Journal of the American Statistical Association*, 92(437):49–60, 1997.

[WFQ08] G. B. Wang, L. Fu, and J. Y. Qi. Maximum a posteriori reconstruction of the patlak parametric image from sinograms in dynamic PET. *Physics in Medicine and Biology*, 53(3):593–604, 2008.

[WG97] W. Wang and G. Gindi. Noise analysis of MAP-EM algorithms for emission tomography. *Physics in Medicine and Biology*, 42:2215–2232, 1997.

[WJK+05] H. Watabe, H. Jino, N. Kawachi, N. Teramoto, T. Hayashi, Y. Ohta, and H. Iida. Parametric imaging of myocardial blood flow with O-15-water and PET using the basis function method. *Journal of Nuclear Medicine*, 46(7):1219–1224, 2005.

[WNC95] C. C. Watson, D. Newport, and M. E. Casey. A single scatter simulation technique for scatter correction in 3D PET. In *International Meeting on Fully Three Dimensional Image Reconstruction in Radiology and Nuclear Medicine*, pp. 255–268, Aix-les-Bain, France, 1995.

[WQ09a] G. Wang and J. Qi. Direct reconstruction of PET receptor binding parametric images using a simplified reference tissue model. *Proceedings of SPIE*, 7258:72580V1–72580V8, 2009.

[WQ09b] G. Wang and J. Qi. Generalized algorithms for direct reconstruction of parametric images from dynamic PET data. *IEEE Transactions on Medical Imaging*, 28(11):1717–1726, 2009.

[WQ10] G. Wang and J. Qi. Acceleration of the direct reconstruction of linear parametric images using nested algorithms. *Physics in Medicine and Biology*, 55:1505–1517, 2010.

[YF02] D. F. Yu and J. A. Fessler. Mean and variance of coincidence counting with deadtime. *Nuclear Instruments and Methods in Physics Research Section A: Accelerators, Spectrometers, Detectors and Associated Equipment*, 488(12):362–374, 2002.

[ZSSB00] J. Zheng, S. S. Saquib, K. Sauer, and C. A. Bouman. Parallelizable Bayesian tomography algorithms with rapid, guaranteed convergence. *IEEE Transactions on Image Processing*, 9:1745–1759, 2000.

17

Tracer Kinetic Analysis for PET and SPECT

Jae Sung Lee and Dong Soo Lee

CONTENTS

17.1 Introduction

17.1.1 Purpose

Positron emission tomography (PET) and single photon emission computed tomography (SPECT) are major nuclear medicine imaging modalities, which enables the noninvasive (and tomographic) measurement of radioactive tracer (radiotracer) concentration in a living biological system, with the spatial resolution <0.1 g of tissue and temporal resolution <1 min. In particular, PET is most widely used in the tracer kinetic studies because PET has superior spatial resolution, sensitivity, and quantitative accuracy in the measurement of radiotracer concentration.

Spatial distribution of a radiotracer in the body is time varying, and it depends on a number of physiological and biological components, such as blood flow, substrate transport, and biochemical reactions. Tracer kinetics is the mathematical description of the movement of radiotracer within a system, and the rate of radiotracer movement (or the change of radiotracer concentration) often provides direct information on the rate of a biological process. Therefore, we can trace and understand the physiological and biological dynamic processes through the tracer kinetic analysis.

17.1.2 Procedures

In the typical tracer kinetic studies, we first define the biological parameters to be determined. A tracer, which follows a substrate physiology without disturbing the system, is then introduced into the body (mostly through venous injection). After the tracer injection, radiation (gamma ray) emitted from the tracer is measured, using a PET or gamma camera, to produce a picture of the body showing where the tracer has moved. Dynamic scan protocol with multiple frames should be used to estimate the time-varying distribution of radiotracer. During the scan, the arterial (or venous) blood samples are collected, continuously or intermittently, to measure the radioactive concentration in blood, which is used as the input function in tracer kinetic analysis. Sometimes, the blood activity concentration is obtained noninvasively by drawing a region of interest (ROI) on the left ventricle (or atrium) of the heart or large arteries (i.e., abdominal aorta) in images. Mathematical model that suitably describes the movement of radiotracer is then applied to estimate the parameters.

Applying the volume of interest (VOI), drawn through multiple adjacent slices, is a useful way to reduce noise in the time-varying concentration of radiotracer (so-called time–activity curve). Because the manual drawing ROI (or VOI) is time- and labor-consuming, automatically defined ROI (or VOI) is sometimes used.

17.2 Principles of Tracer Kinetic Modeling

17.2.1 Standard Kinetic Parameters

Standard kinetic parameters can be estimated based on the time dependence of radioactivity concentrations in specific organs and tissues. Those parameters include peak concentration, plateau concentration, time to peak concentration, area under the time–activity curve, elimination rate (biological half-life), and mean residence time.[1,2] If the concentration of a radiopharmaceutical plateaus after certain time by any mechanism, the plateau concentration can be estimated by a single static image acquisition without dynamic acquisition. The standard uptake value (SUV) is commonly used as a minimum standard measure of such a plateau concentration of ^{18}F-flurodeoxyglucose (^{18}F-FDG) and other radiopharmaceuticals, and is calculated from concentration, injected dose, and body weight (or body surface area).[3]

17.2.2 Tracer Kinetic Modeling

Mathematical modeling of radiopharmaceutical kinetics can provide more specific information (rate constants) relating to uptake, washout, metabolism, and the distribution of radiopharmaceuticals.[2] Two classes of models are commonly used for kinetics of radiotracers: noncompartmental and compartmental models.

17.2.2.1 Compartmental Modeling

Compartmental model is a model consisting of a finite number of interconnected compartments, each of which represents the effective amount of radiotracer and is assumed to behave as if it is well-mixed and kinetically homogeneous.[4]

Compartments are interconnected by pathways, which represent fluxes of materials and/or biochemical conversions (Figure 17.1, Figures 17.3 through 17.5).

17.2.2.1.1 Blood Flow

^{15}O labeled water ($H_2{}^{15}O$) is an inert and diffusible tracer, which is used to measure blood flow by PET. To estimate blood flow by $H_2{}^{15}O$ PET, a single-tissue compartment model based on Fick's principle is used (Figrue 17.1).[5,6]

The rate of change in concentration of $H_2{}^{15}O$ in tissue is equal to the rate of $H_2{}^{15}O$ entering the tissue minus the rate of tracer leaving the tissue per unit volume. Thus,

$$\frac{dC_t(t)}{dt} = f \cdot C_a(t) - f \cdot C_v(t) \tag{17.1}$$

where

t is a time (min)
$C_t(t)$ is the tissue concentration of $H_2{}^{15}O$ (μCi/g)
$C_a(t)$ is the arterial concentration of $H_2{}^{15}O$ (μCi/mL)
$C_v(t)$ is the venous concentration of $H_2{}^{15}O$ (μCi/mL)
f is the blood flow per unit weight of tissue (mL/min/g)

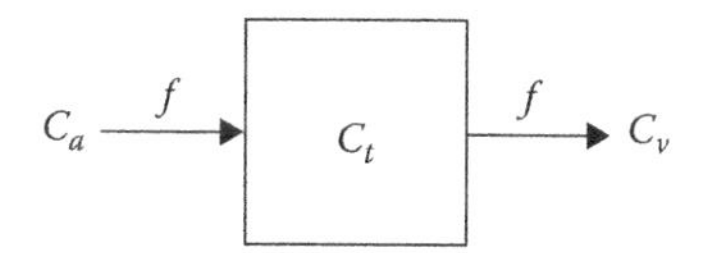

FIGURE 17.1
Single-tissue compartment model for $H_2{}^{15}O$. C_t is the tissue concentration of $H_2{}^{15}O$ (μCi/g), C_a the arterial concentration of $H_2{}^{15}O$ (μCi/mL), C_v the venous concentration of $H_2{}^{15}O$ (μCi/mL), and f the blood flow per unit weight of tissue (mL/min/g).

Assuming that water is freely diffusible in the tissue space, and that the ^{15}O distribution reaches equilibrium instantaneously, venous concentration is related to the tissue concentration by

$$C_v(t) = \frac{1}{p} C_t(t) \tag{17.2}$$

where p is tissue/blood partition coefficient (mL/g), which is a measurement of the ratio of the water content in the tissue to that in the blood at equilibrium. Inserting Equation 17.2 into Equation 17.1 gives

$$\frac{dC_t(t)}{dt} = f \cdot C_a(t) - \frac{f}{p} C_t(t) \tag{17.3}$$

Solving differential Equation 17.3 for $C_t(t)$ gives

$$C_t(t) = f \int_0^t C_a(\tau) \cdot e^{-\frac{f}{p}(t-\tau)} d\tau = f \cdot C_a(t) \otimes e^{-\frac{f}{p} \times t} \tag{17.4}$$

where $C_t(t = 0) = 0$ is assumed, and $\otimes$ denotes the convolution integral. Figure 17.2 illustrates tissue time-activity curves of various blood flow values, being generated using Equation 17.4.

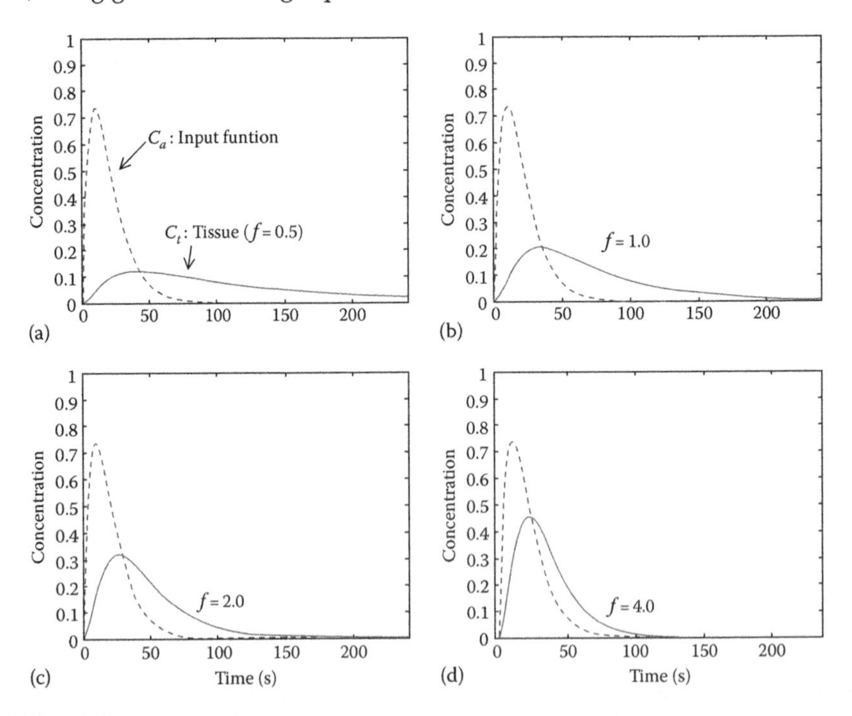

FIGURE 17.2
Tissue time–activity curves with different values of blood flow (f) as described by Equation 17.4 (partition coefficient p is fixed at 0.9). (a) f = 0.5, (b) f = 1.0, (c) f = 2.0, and (d) f = 4.0.

In terms of the measurement of cerebral blood flow, two unknown parameters, blood flow (f) and partition coefficient (p), are commonly estimated.

In the measurement of myocardial blood flow, the measured tissue time–activity curves by PET can be related to the ideal one by incorporating the partial volume and spillover effects[7,8]:

$$\tilde{C}_t(t) = \alpha \cdot C_t(t) + V_a \cdot C_a(t) = \alpha \cdot f \cdot C_a(t) \otimes e^{-\frac{f}{p} \times t} + V_a \times C_a(t) \tag{17.5}$$

where

α is the perfusible tissue fraction (PTF, g/mL) or recovery coefficient for the partial volume correction of the tissue time–activity curve

V_a is the arterial blood volume fraction (mL/mL) of the ROI. By fixing p (= 0.91 mL/g), three unknown parameters (α, f, V_a) can be estimated

17.2.2.1.2 Glucose Metabolism

^{18}F-FDG, an ^{18}F-labeled analogue of glucose, is a tracer used to study glucose metabolism by PET. FDG enters the tissue and is phosphorylated into FDG-6-phosphate (FDG-6-PO_4), which cannot be further metabolized by the glycolytic pathway. Therefore, a three-compartment model (a two-tissue compartment model) is commonly used for ^{18}F-FDG (Figure 17.3).[9]

This system can be described by the following set of differential equations:

$$\frac{dC_e(t)}{dt} = K_1 C_p(t) - k_2 C_e(t) - k_3 C_e(t) + k_4 C_m(t) \tag{17.6}$$

$$\frac{dC_m(t)}{dt} = k_3 C_e(t) - k_4 C_m(t) \tag{17.7}$$

where

$C_p(t)$ is the plasma concentration of FDG (μCi/mL)

$C_e(t)$ is the concentration of exchangeable FDG (μCi/g)

$C_m(t)$ is the concentration of FDG-6-PO_4 (μCi/g)

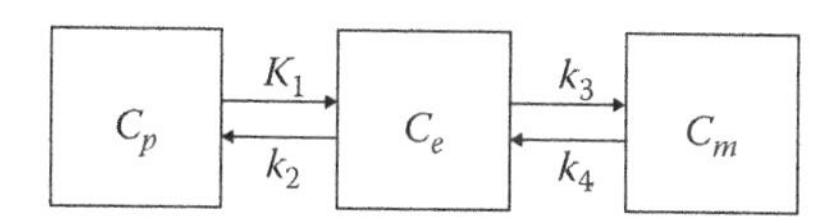

FIGURE 17.3
Three-compartment (two-tissue compartment) model for ^{18}F-FDG. C_p is the plasma concentration of FDG (μCi/mL), C_e the concentration of exchangeable FDG (μCi/g), and C_m the concentration of FDG-6-PO_4 (μCi/g). The rate constants K_1, k_2, k_3, and k_4 are defined as delivery (mL/min/g), washout (min^{-1}), phosphorylation of FDG (min^{-1}), and dephosphorylation of FDG-6-P (min^{-1}), respectively.

The rate constants K_1, k_2, k_3, and k_4 are defined as delivery (mL/min/g), washout (min^{-1}), phosphorylation of FDG (min^{-1}), and dephosphorylation of FDG-6-P (min^{-1}), respectively.

The measured tissue time–activity curve by PET is the sum of the solution of the aforementioned two equations[10]:

$$C_t(t) = C_e(t) + C_m(t) = K_1 C_p(t) \otimes \left[\left(\frac{k_3 + k_4 - q_1}{q_2 - q_1} \right) e^{-q_1 t} - \left(\frac{k_3 + k_4 - q_2}{q_2 - q_1} \right) e^{-q_2 t} \right]$$

$$q_1, q_2 = \frac{(k_2 + k_3 + k_4) \mp \sqrt{(k_2 + k_3 + k_4)^2 - 4k_2 k_4}}{2} \quad (17.8)$$

Making the assumption that k_4 is zero (the dephosphorylation of FDG-6-PO_4 is relatively slow), the analytic solution of the tissue time–activity curve is described by

$$C_t(t) = \frac{K_1}{k_2 + k_3} C_p(t) \otimes \left[k_3 + k_2 e^{-(k_2 + k_3)t} \right] \quad (17.9)$$

Individual rate constants can be estimated using linear or nonlinear regression analysis to fit the aforementioned equation to the measured tissue time–activity curve. The net transport rate (K_{in}) of FDG, $K_1 k_3/(k_2+k_3)$, can be calculated using individual rate constants determined by the regression analysis of Equation 17.8 or 17.9, or by Gjedde-Patlak graphical analysis.[11–13]

Finally, the glucose utilization rate is obtained using the following equation:

$$\frac{1}{LC} \frac{K_1 k_3}{k_2 + k_3} C_p \quad (17.10)$$

where

LC is a lumped constant, which is used to calibrate the differences in transport and phosphorylation between FDG and glucose

C_p represents the glucose concentration in blood[9,14]

17.2.2.1.3 Receptor-Ligand Binding

The concentration of receptors (B_{max}) and the equilibrium dissociation constant (K_d) for the specific binding of ligands to receptors, or their ratios (binding potentials) can be estimated using the compartment modeling method with a plasma compartment and three tissue compartments (free, nonspecific binding, and specific binding, respectively).[14,15] If the exchange between the free and nonspecific compartments is rapid enough and/or the nonspecific binding is small relative to the free compartment, the model can

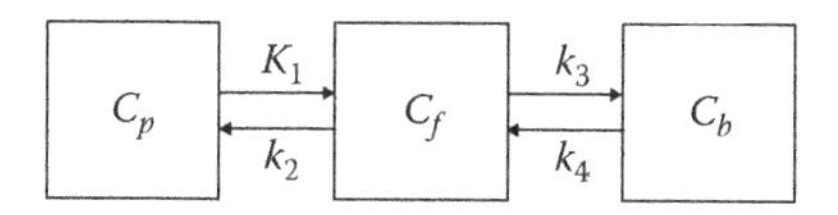

FIGURE 17.4
Three-compartment (two-tissue compartment) model for a receptor-rich region. C_p is the concentration of radioligand in plasma (μCi/mL), C_f the free or nonspecifically bound radioligand (μCi/g), and C_b the radioligand specifically bound by receptors (μCi/g). The rate constants K_1, k_2, k_3, and k_4 are defined as the delivery (mL/min/g), washout (min^{-1}), forward receptor–ligand reaction (min^{-1}), and reverse receptor–ligand reaction (min^{-1}), respectively.

be reduced to a two-tissue compartment model.[16] Each compartment in this model represents the concentration of radioligand in plasma (C_p, μCi/mL), free or nonspecifically bound radioligand (C_f, μCi/g), and radioligand specifically bound by receptors (C_b, μCi/g) (Figure 17.4).

Changes in C_f and C_b are described by

$$\frac{dC_f(t)}{dt} = K_1C_p(t) - k_2C_f(t) - k_3(t)C_f(t) + k_4C_b(t) \tag{17.11}$$

$$\frac{dC_b(t)}{dt} = k_3(t)C_f(t) - k_4C_b(t)$$

$$k_3(t) = k_{on}\left(B_{\max} - \frac{C_b(t)}{SA}\right),\ k_4 = k_{off} \tag{17.12}$$

where the rate constants K_1, k_2, k_3, and k_4 are defined as the delivery (mL/min/g), washout (min^{-1}), forward receptor-ligand reaction (min^{-1}), and reverse receptor-ligand reaction (min^{-1}), respectively. k_{on} is the bimolecular association rate constant (g/pmol/min), k_{off} the unimolecular dissociation rate constant (min^{-1}), B_{max} the density of receptor sites available for radioligand binding (apparent B_{max}, pmol/g), and SA the specific activity (radio activity per mole of a labeled compound, μCi/pmol).[17,18]

If the SA is high, then $C_b(t)$/SA (occupancy of receptors by the labeled compound) is negligibly small, $k_3(t)$ (= $k_{on}(B_{max} - C_b(t)/SA)$) can then be approximated by $k_{on}B_{max}$, and the above Equations 17.11 and 17.12 can be written as a first-order equation without the ($C_b(t)C_f(t)$) multiplication term and an analytical solution can be derived (the same form as Equation 17.8). Otherwise, the analytic solution of this equation cannot be simply obtained. Numerical solutions for C_f and C_b are, therefore, required for time–activity curves with low SA. Binding potential (*BP*) is obtained from $B_{max}/K_d = k_{on}B_{max}/k_{off}$, and is equal to k_3/k_4 for high SA data.

In imaging studies, volume of distribution is the ratio of the concentration of radiotracer in a region of tissue to that in plasma at equilibrium. At equilibrium, also the Equations 17.11 and 17.12 will be zero. Thus, the volume of distribution of each compartment in Figure 17.4 and their sum can be related with the rate constants as follows:

$$V_f = \frac{C_f}{C_p} = \frac{K_1}{k_2} \tag{17.13}$$

$$V_b = \frac{C_b}{C_p} = \frac{K_1 k_3}{k_2 k_4} \tag{17.14}$$

$$V_t = V_f + V_b = \frac{K_1}{k_2}\left(1 + \frac{k_3}{k_4}\right) \tag{17.15}$$

Consequently, the *BP* can be related with the volume of distribution or their ratio (distribution volume ratio: *DVR*) as follows:

$$BP = \frac{V_b}{V_f} = \frac{V_t - V_f}{V_f} = \frac{V_t}{V_f} - 1 = DVR - 1 \tag{17.16}$$

For some radioligands, reference regions that do not have high-affinity receptors are available. The single-tissue compartment model is used for such reference regions (Figure 17.5). If we assume that the distribution volumes of free and nonspecifically bound radioligand in an ROI and in a reference region are the same ($K_1 / k_2 = K_1' / k_2'$) for some radioligands, biological constraint can be applied to improve the stability of rate constant estimates or to replace the arterial input function by the time–activity curve for a reference region. The ratio of K_1' and k_2', which are obtained by fitting the time–activity curve from the reference region, can be fixed during rate constant estimation for ROIs.[19,20]

The reference tissue model is a method in which this constraint is used to derive the relationship between the time–activity curves of an ROI and of reference tissue; moreover, the analytical solution of the time–activity curves for an ROI can be fitted without an arterial input function.[21] Lammertsma and Hume derived the following equation for the tracer (i.e., [^{11}C]raclopride), the kinetics of which cannot be distinguished in free (+ nonspecific) and

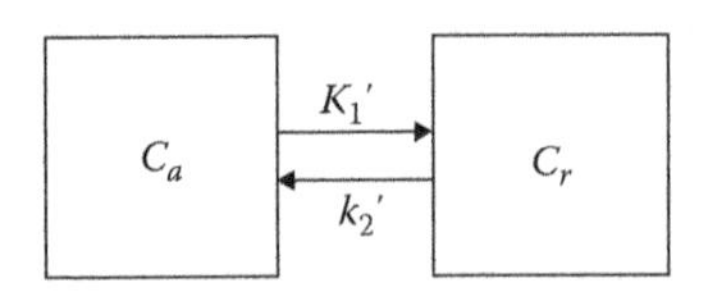

FIGURE 17.5
Two-compartment (single-tissue compartment) for a reference region: C_r is the concentration of the free or nonspecifically bound radioligand in the reference region (μCi/g).

specific compartments, since the exchange of tracer between those two compartments is rapid compared to that between plasma and free compartments (simplified reference tissue model).[22]

$$C_t(t) = R_1 C_r(t) + \left\{ k_2 - \frac{R_1 k_2}{(1+BP)} \right\} C_r(t) \otimes e^{\frac{-k_2 t}{(1+BP)}} \tag{17.17}$$

where $R_1 = k_1/k_2$ and BP is k_3/k_4.

17.2.2.2 Noncompartmental Modeling

In the noncompartmental approach, specific kinetic parameters are estimated without the detailed compartmentalization of the system.[4] The complexity of the systems is usually lumped into some macro parameters. Graphical analysis is an example of this approach, in which the differential equations of compartmental analysis are converted into linear plots.[11–13,23–25] The slope and intercept of these plots are then used to determine the kinetic parameters. Spectral analysis is another method used for PET/SPECT data, in which the model is not strictly constrained by the number of compartments and the relationship between compartments.[26,27]

17.2.2.2.1 Graphical Analysis

Although the graphical methods are independent of particular model structures, if a model structure describes the transfer of tracer, the slope of the linear equation can be related to combinations of the model parameters.[23] Since linear regression is employed in graphical analysis, it requires only the computation of the analytic solution of the slopes. The Gjedde-Patlak plot is the most widely used method for irreversibly binding tracers (Figure 17.6a).[11–13] If the transport of a tracer to the third compartment is irreversible ($k_4 = 0$), differential equations (i.e., Equations 17.6 and 17.7) for the three-compartment model can be rearranged into the following equation:

$$\frac{C_t(t)}{C_p(t)} = \frac{K_1 k_3}{k_2 + k_3} \frac{\int_0^t C_p(\tau) d\tau}{C_p(t)} + V_e, \tag{17.18}$$

for $t > t'$ when the distribution volume of the reversible part (V_e) is constant.

The slope of the aforementioned equation equals to the net transfer rate of tracer from plasma to the irreversible compartment.

For the quantification of reversible tracers ($k_4 \neq 0$), Logan graphical methods are used to determine the distribution volumes of reference region and region with specific binding and/or their ratio with or without the arterial

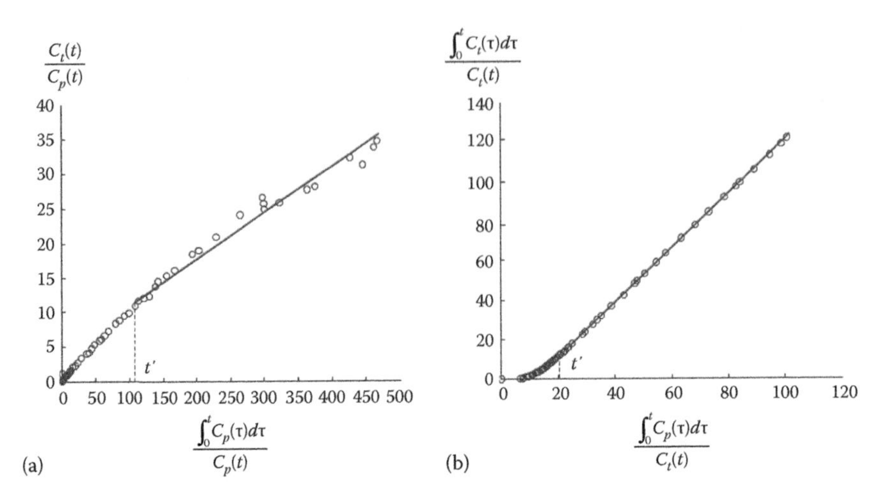

FIGURE 17.6
Graphical analysis. (a) Gjedde-Patlak plot. (b) Logan plot.

input function (Figure 17.6b).[24,25] The Logan equation for specific binding region shown in Figure 17.4 is

$$\frac{\int_0^t C_t(\tau)d\tau}{C_t(t)} = \frac{K_1}{k_2}\left(1+\frac{k_3}{k_4}\right)\frac{\int_0^t C_p(\tau)d\tau}{C_t(t)} + b = V_t\frac{\int_0^t C_p(\tau)d\tau}{C_t(t)} + b, \tag{17.19}$$

for $t > t'$ when the intercept b becomes constant.

For reference region with no specific binding (Figure 17.5), the Logan plot can be represented as follows:

$$\frac{\int_0^t C_r(\tau)d\tau}{C_r(t)} = \frac{K_1'}{k_2'}\frac{\int_0^t C_p(\tau)d\tau}{C_r(t)} + b \tag{17.20}$$

By substituting the integral term of input function that can be obtained from Equation 17.19 into Equation 17.20 with the assumption of $K_1/k_2 = K_1'/k_2'$, we can also derive the following equation in which the reference tissue time–activity curve is regarded as input function:

$$\frac{\int_0^t C_t(t)dt}{C_t(t)} = DVR\frac{\int_0^t C_r(t)dt}{C_t(t)} + b \tag{17.21}$$

where the slope, DVR, is the distribution volume ratio.

17.2.2.2.2 Spectral Analysis

In the case of a new tracer of unknown kinetics, noncompartmental tracer kinetic modeling techniques are necessary initially to identify the kinetic components present in the measured data and to determine the pharmacokinetic variables with relatively few model assumptions. Data-driven approaches like spectral analysis can be used for this purpose.[26,27]

In spectral analysis, the observed time course of a labeled drug is fitted using a linear combination of possible tissue response curves, each of which is a single exponential in time convolved with the input function C_p:

$$C_t(t) = \frac{1}{t_{i+1} - t_i} \int_{t_i}^{t_{i+1}} \left(\sum_{j=1}^{N} C_p(\tau) \otimes \alpha_j e^{-\beta_j t} \right) dt \quad \left(\alpha_j \geq 0, \quad \lambda \leq \beta_j \leq 1 \right) \tag{17.22}$$

where

N is the maximal number of basis functions allowed in the model
λ is the decay constant of the radioisotope
the t_i ($i = 1, \ldots, n$) values are the frame times

Values of α are determined, which best fit the measured data at predefined β values on the interval [λ, 1(fastest measurable dynamic)]. The fitted values of α, together with the corresponding chosen values of β, define the tissue unit impulse-response function:

$$h(t) = \sum_{j=1}^{N} \alpha_j e^{-(\beta_j - \lambda)t} \tag{17.23}$$

from which the pharmacokinetic parameters can be calculated as follows:

$$K_1 = h(0) = \sum_{j=1}^{N} \alpha_j, \quad VD = \int_0^{\infty} h(\tau) d\tau = \sum_{j=1}^{N} \frac{\alpha_j}{\beta_j - \lambda}, \quad MRT = \frac{VD}{K_1} \tag{17.24}$$

where

K_1 is the rate constant for the delivery of tracer from the plasma to tissue
VD is the fractional volume of the distribution
MRT is the mean residence time of the tracer in the sampled tissue[27]

17.3 Parameter Estimation and Parametric Images

To estimate unknown parameters (i.e., rate constants) from the tissue time–activity curves and input function, these parameters are conventionally fitted by nonlinear least squares (NLS) analysis. However, the result of the parameter estimation using NLS is dependent on the initial guessing of parameters. Thus, a poor initial guess results in an incorrect result at local minima of the cost function and slow convergence. In addition, an appropriate convergence threshold and constraints on the parameters should be determined by experience. The NLS method also requires considerable computation time to estimate parameters.[28] This method is, therefore, impractical for a pixel-by-pixel analysis designed to generate parametric images of these parameters, which is important for clinical and research purposes, since parametric images provide us with anatomically oriented information about the kinetic parameters, without the spatial resolution loss associated with the ROI method. Thus, the linear analysis methods, such as graphical analysis, linear least squares (LLS), and linear weighted least squares (LWLS) are commonly used for generating parametric images.

In weighted least squares methods, the structure of the error variance can be concerned by weighting the measurement error with the relative accuracy of the measurement.[28] However, direct estimation of the error variance of every pixel is hard to achieve practically. An approximate weighting formula is, therefore, commonly used.[29,30] Generalized least squares (GLS) and generalized weighted least squares (GWLS) are the generalized formula of LLS and LWLS, respectively. Since the dependency between the measurement error at each sampling point in the LLS and LWLS methods leads to bias in the estimated parameters, GLS and GWLS were suggested as alternatives to eliminate this bias.[29,30]

17.3.1 Nonlinear Least Squares (NLS)

Let us begin explain these methods using an example of the single-tissue compartment model with two unknown parameters (K_1 and k_2). Recall that the model equation for the single-tissue compartment model with two parameters is described by Equation 17.3. Replacing f and f/p with K_1 and k_2, respectively,

$$\frac{dC_t(t)}{dt} = K_1 C_a(t) - k_2 C_t(t) \tag{17.25}$$

The solution of Equation 17.25 is

$$C_t(t) = K_1 C_a(t) \otimes e^{-k_2 t} \tag{17.26}$$

Parameter estimation is performed by defining an optimization cost function and then by choosing those parameter values that minimize the function. The most common cost function is the residual sum of squares, which is the sum of the squared deviations between the measured tissue

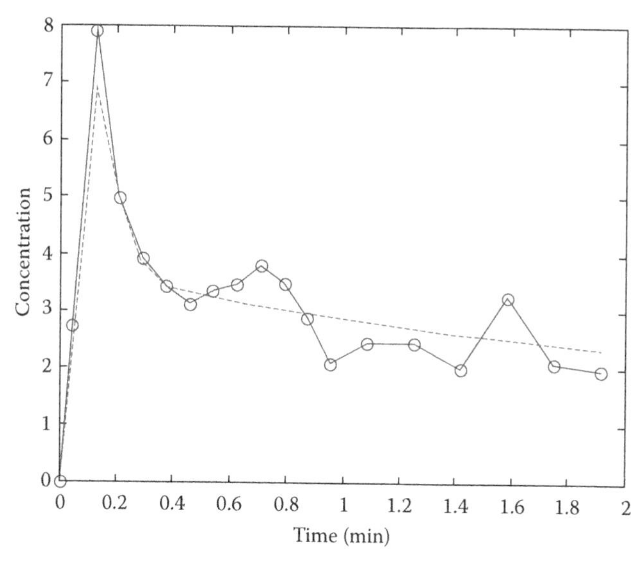

FIGURE 17.7
Parameter estimation by curve fitting. Circles on solid line represent the measured data and dashed line is the fitted (or predicted) tissue time–activity curve.

time–activity curve and predicted tissue time–activity curve obtained using estimated parameters (Figure 17.7). When the NLS method is used, the parameters are varied iteratively using a Gaussian–Newton or a Levenberg–Marquardt algorithm until the following residual sum of squares reaches a minimum.[31,32]

$$\sum_{i=1}^{n} w_i \left[C_t(t_i) - K_1 C_a(t) \otimes e^{-k_2 t} \right]^2 \tag{17.27}$$

where w_i is the weight associated with the ith measurement, and reflects the relative accuracy of the measurement. The aforementioned equation is general formulation, which describes the structure of the error variance, and the technique is called nonlinear weighted least squares. If we set all w_i values to be equal, the method is simply called nonlinear least squares.

Under the assumptions of a zero mean, an uncorrelated measurement error, and errorless independent variables, the optimal choice of weights is[28]

$$w_i = Var(e_i)^{-1} \tag{17.28}$$

where e_i is the error in ith measurement.

17.3.2 Linear Least Squares (LLS)

17.3.2.1 Single-Tissue Compartment Models

Integrating Equation 17.25 from time 0 to t_i, where $t_i = t_1, t_2, \ldots, t_n$ are the sampling times of the tissue measurements, gives

$$C_t(t_1) = K_1 \int_0^{t_1} C_a(\tau)d\tau - k_2 \int_0^{t_1} C_t(\tau)d\tau + \varepsilon_1$$
$$C_t(t_2) = K_1 \int_0^{t_2} C_a(\tau)d\tau - k_2 \int_0^{t_2} C_t(\tau)d\tau + \varepsilon_2$$
$$\vdots$$
$$C_t(t_n) = K_1 \int_0^{t_n} C_a(\tau)d\tau - k_2 \int_0^{t_n} C_t(\tau)d\tau + \varepsilon_n \quad (17.29)$$

where $\varepsilon_1, \varepsilon_2, \ldots, \varepsilon_n$ are error terms.

These equations can be considered a set of linear equations in which the time integration of the input function and of the tissue time–activity curve are independent variables, the instantaneous tissue time–activity curve value is a dependent variable, and K_1 and k_2 are its coefficients.

Rearranging these equations into a matrix form gives

$$\mathrm{y} = \mathrm{X}\theta + \varepsilon \quad (17.30)$$

where

$$\mathrm{y} \equiv \begin{bmatrix} C_t(t_1) & C_t(t_2) & \cdots & C_t(t_n) \end{bmatrix}^T \quad (17.31)$$

$$\mathrm{X} \equiv \begin{bmatrix} \int_0^{t_1} C_a(\tau)d\tau & \int_0^{t_1} C_t(\tau)d\tau \\ \int_0^{t_2} C_a(\tau)d\tau & \int_0^{t_2} C_t(\tau)d\tau \\ \vdots & \vdots \\ \int_0^{t_n} C_a(\tau)d\tau & \int_0^{t_n} C_t(\tau)d\tau \end{bmatrix} \quad (17.32)$$

$$\varepsilon \equiv \begin{bmatrix} \varepsilon_1 & \varepsilon_2 & \cdots & \varepsilon_n \end{bmatrix}^T \quad (17.33)$$

$$\theta \equiv \begin{bmatrix} K_1 & -k_2 \end{bmatrix}^T \quad (17.34)$$

If we assume that the error terms are independent of each other, an estimate of θ can be obtained by using the following equation:

$$\hat{\theta}_{\text{LLS}} = (X^T X)^{-1} X^T y \tag{17.35}$$

Suppose that the variance is not constant, the estimate θ then is

$$\hat{\theta}_{\text{LWLS}} = (X^T W X)^{-1} X^T W y \tag{17.36}$$

where W is an n by n diagonal matrix consisting of the weights w_i. This later method is called the linear weighted least squares (LWLS) method.

Although error terms in Equation 17.29 are assumed to be mutually independent, the assumption is incorrect since the integration periods overlap so that the later error terms contain previous errors. This dependency between error terms leads to a bias in the estimated parameters when the LLS method is used. The GLS method was suggested to overcome this bias problem.[29,30] GLS iteratively upgrades the estimates, which are initially obtained from the LLS, and the final estimates are unbiased. The GLS estimate is given as

$$\hat{\theta}_{\text{GLS}} = (Z^T Z)^{-1} Z^T r \tag{17.37}$$

where

$$r \equiv \begin{bmatrix} C_t(t_1) - \hat{k}_2 C_t(t_1) \otimes e^{-\hat{k}_2 t_1} \\ C_t(t_2) - \hat{k}_2 C_t(t_2) \otimes e^{-\hat{k}_2 t_2} \\ \vdots \\ C_t(t_n) - \hat{k}_2 C_t(t_n) \otimes e^{-\hat{k}_2 t_n} \end{bmatrix} \tag{17.38}$$

$$Z \equiv \begin{bmatrix} C_a(t_1) \otimes e^{-\hat{k}_2 t_1} & C_t(t_1) \otimes e^{-\hat{k}_2 t_1} \\ C_a(t_2) \otimes e^{-\hat{k}_2 t_2} & C_t(t_2) \otimes e^{-\hat{k}_2 t_2} \\ \vdots & \vdots \\ C_a(t_n) \otimes e^{-\hat{k}_2 t_n} & C_t(t_n) \otimes e^{-\hat{k}_2 t_n} \end{bmatrix} \tag{17.39}$$

Equation 17.37 is used repeatedly with $\hat{k}_2$ updated for each iteration.

As for the LWLS method, the GWLS estimate is

$$\hat{\theta}_{\text{GWLS}} = (Z^T W Z)^{-1} Z^T W r \tag{17.40}$$

where $\boldsymbol{r}$ and $\boldsymbol{Z}$ have the forms shown in Equations 17.38 and 17.39.

This LLS and related estimation methods can be also applied to the extended model for incorporating the partial volume and spillover effects.[33]

From Equation 17.5, the ideal time–activity curves can be expressed with the measured time–activity curves according to the following equations:

$$C_a(t) = \tilde{C}_a(t) \tag{17.41}$$

$$C_t(t) = \frac{\tilde{C}_t(t) - V_a \cdot \tilde{C}_a(t)}{\alpha} \tag{17.42}$$

Substituting the aforementioned two equations into Equation 17.25, and multiplying both sides by α,

$$\frac{d}{dt}\left\{\tilde{C}_t(t) - V_a \cdot \tilde{C}_a(t)\right\} = K_1 \cdot \tilde{C}_a(t) - k_2\left\{\tilde{C}_t(t) - V_a \cdot \tilde{C}_a(t)\right\} \tag{17.43}$$

where K_1 is the product of α and f.

Rearranging this equation and integrating both sides from time 0 to each PET sampling point yields the following linear equation:

$$\tilde{C}_t(t) = P_1 \cdot \tilde{C}_a(t) + P_2 \int_0^t \tilde{C}_a(\tau)\,d\tau + P_3 \int_0^t \tilde{C}_t(\tau)\,d\tau \tag{17.44}$$

$$\begin{aligned} P_1 &\equiv V_a \\ P_2 &\equiv K_1 + k_2 \cdot V_a \\ P_3 &\equiv -k_2 \end{aligned} \tag{17.45}$$

Figure 17.8 shows the parametric image of K_1 generated from the dynamic myocardial $H_2{}^{15}O$ PET image using the method described earlier.[33]

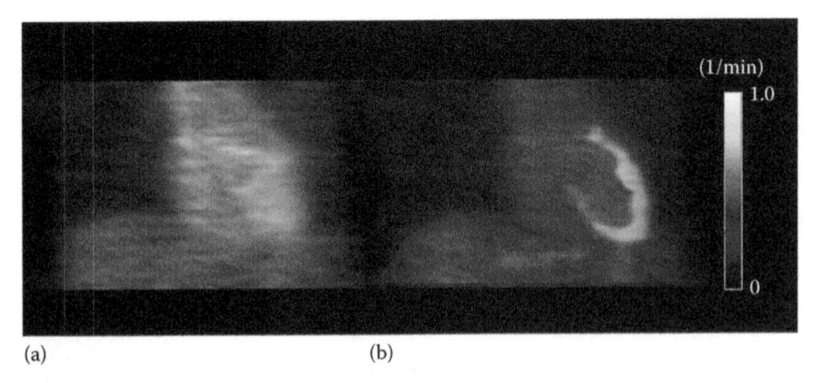

FIGURE 17.8
(See color insert.) Parametric image of K_1 generated from dynamic $H_2{}^{15}O$ PET using linear least squares method. (a) Static PET image. (b) K_1 parametric image superimposed on static image. (Modified from Figure 17.5 in Lee JS, et al., *J Nucl Med*. 2005, 46(10), 1687–1695 © by the Society of Nuclear Medicine, Inc.)

17.3.2.2 Two-Tissue Compartment Models

17.3.2.2.1 Irreversible Binding

From differential equations of the two-tissue compartment model with irreversible binding, two multiple linear regression model equations can be derived.[34] The functional equation of the first formula (multiple linear analysis for irreversible radiotracer 1: MLAIR1) takes the form:

$$C_t(t) = P_1 C_p(t) + P_2 \int_0^t C_p(\tau)\,d\tau + P_3 \int_0^t C_t(\tau)\,d\tau + P_4 \int_0^t \int_0^\tau C_p(s)\,ds\,d\tau \tag{17.46}$$

where the macro parameters $P_1 \sim P_4$ are given by

$$\begin{aligned} P_1 &= V_a \\ P_2 &= K_1 + k_2 V_a + k_3 V_a \\ P_3 &= -(k_2 + k_3) \\ P_4 &= K_1 k_3 \end{aligned} \tag{17.47}$$

The net accumulation rate K_{in} can then be acquired using the following equation:

$$K_{in} = \frac{K_1 k_3}{k_2 + k_3} = -\frac{P_4}{P_3} \tag{17.48}$$

The MLAIR1 is a useful alternative to the Gjedde-Patlak plot because the determination of a linear interval is not necessary. However, the error propagation associated with the division calculation on the macro parameters (Equation 17.48), to obtain K_{in}, is a possible limitation of MLAIR1 for the voxel-wise estimations of K_{in} for the generation of parametric images because of the high noise level in the individual time–activity curves of each voxel. On the contrary, MLAIR2 is a formula in which K_{in} could be directly estimated from the coefficient of an independent variable as follows:

$$\int_0^t C_t(\tau)\,d\tau = P_1' \int_0^t \int_0^\tau C_p(s)\,ds\,d\tau + P_2' \int_0^t C_p(\tau)\,d\tau + P_3' C_p(t) + P_4' C_t(t) \tag{17.49}$$

where the macro parameters are given by the following equations and can also be obtained by LLS estimation:

$$\begin{aligned} P_1' &= \frac{K_1 k_3}{k_2 + k_3} = K_{in} \\ P_2' &= \frac{K_1}{k_2 + k_3} + V_a \\ P_3' &= \frac{V_a}{k_2 + k_3} \\ P_4' &= -\frac{1}{k_2 + k_3} \end{aligned} \tag{17.50}$$

In Figure 17.9, the parametric images of K_{in} generated using Gjedd-Patlak plot (a) and MLAIR2 (b) are compared.[34]

17.3.2.2.2 Reversible Binding

Rearrangement of Equation 17.19 gives (multilinear analysis 1: MA1)[35]:

$$C_t(t) = -\frac{V_t}{b}\int_0^t C_p(\tau)d\tau + \frac{1}{b}\int_0^t C_t(\tau)d\tau \tag{17.51}$$

Multiple linear regression on the aforementioned equation yields smaller bias in V_t estimation than the simple linear regression on the original form (Equation 17.19) because noisy term $C_t(t)$ appears only in the dependent variable in Equation 17.51.

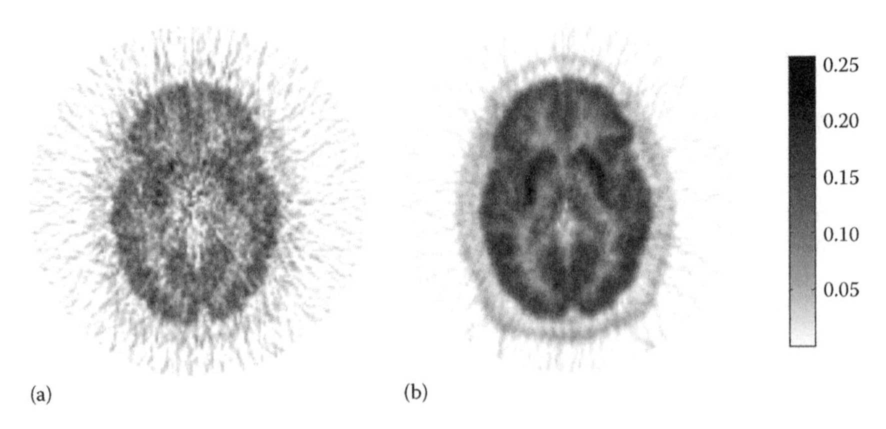

FIGURE 17.9
Parametric images of net accumulation rate K_{in} of [^{11}C]MeNTI using Gjedd-Patlak plot (a) and MLAIR2 (b). (Modified from Figure 17.6 in reference Kim, S.J. et al., *J. Cereb. Blood Flow Metab.* 2008, 28(12), 1965.)

Another multiple linear regression formula for tracers with reversible biding is MA2:

$$C_t(t) = P_1\int_0^t\int_0^\tau C_p(s)\,ds\,d\tau + P_2\int_0^t\int_0^\tau C_t(s)\,ds\,d\tau + P_3\int_0^t C_t(\tau)\,d\tau + P_4\int_0^t C_p(\tau)\,d\tau \tag{17.52}$$

where the macro parameters $P_1 \sim P_4$ are given by

$$\begin{aligned} P_1 &= k_2k_4V_t \\ P_2 &= -k_2k_4 \\ P_3 &= -(k_2+k_3+k_4) \\ P_4 &= K_1 \end{aligned} \tag{17.53}$$

The V_t can then be acquired using the following equation:

$$V_t = -\frac{P_1}{P_2} \tag{17.54}$$

References

1. Aboagye EO, Price PM, Jones T. *In vivo* pharmacokinetics and pharmacodynamics in drug development using positron-emission tomography. *Drug Discov Today*. 2001;6(6):293–302.
2. Fischman AJ, Alpert NM, Rubin RH. Pharmacokinetic imaging: a noninvasive method for determining drug distribution and action. *Clin Pharmacokinet*. 2002;41(8):581–602.
3. Lowe VJ, Hoffman JM, DeLong DM, Patz EF, Coleman RE. Semiquantitative and visual analysis of FDG-PET images in pulmonary abnormalities. *J Nucl Med*. 1994;35(11):1771–1776.
4. Cobelli C, Foster D, Toffolo G. *Tracer Kinetics in Biomedical Research: From Data to Model*. New York: Kluwer Academic/Plenum Publishers; 2000.
5. Kety SS. The theory and application of the exchange inert gas at the lung and tissues. *Pharmacol Rev*. 1951;3:1–41.
6. Kety SS. Measurement of local blood flow by the exchange of an inert, diffusible substance. *Methods Med Res*. 1960;8:228–236.
7. Herrero P, Markham J, Myears DW, Weiheimer CJ, Bergmann SR. Measurement of myocardial blood flow with positron emission tomography: correction for count spillover and partial volume effects. *Math Comput Model*. 1988;11:807–812.
8. Iida H, Rhodes CG, de Silva R, Yamamoto Y, Araujo LI, Maseri A, et al. Myocardial tissue fraction-correction for partial volume effects and measure of tissue viability. *J Nucl Med*. 1991;32:2169–2175.

9. Phelps ME, Huang SC, Hoffman EJ, Selin C, Sokoloff L, Kuhl DE. Tomographic measurement of local cerebral glucose metabolic rate in humans with (F-18)2-fluoro-2-deoxy-D-glucose: validation of method. *Ann Neurol*. 1979;6(5):371–388.
10. Frost JJ, Douglass KH, Mayberg HS, Dannals RF, Links JM, Wilson AA, et al. Multicompartmental analysis of [^{11}C]-carfentanil binding to opiate receptors in humans measured by positron emission tomography. *J Cereb Blood Flow Metab*. 1989;9(3):398–409.
11. Gjedde A. Calculation of cerebral glucose phosphorylation from brain uptake of glucose analogs in vivo: a re-examination. *Brain Res*. 1982;257(2):237–274.
12. Patlak CS, Blasberg RG. Graphical evaluation of blood-to-brain transfer constants from multiple-time uptake data. Generalizations. *J Cereb Blood Flow Metab*. 1985;5(4):584–590.
13. Patlak CS, Blasberg RG, Fenstermacher JD. Graphical evaluation of blood-to-brain transfer constants from multiple-time uptake data. *J Cereb Blood Flow Metab*. 1983;3(1):1–7.
14. Schmidt KC, Turkheimer FE. Kinetic modeling in positron emission tomography. Q *J Nucl Med*. 2002;46(1):70–85.
15. Ichise M, Meyer JH, Yonekura Y. An introduction to PET and SPECT neuroreceptor quantification models. *J Nucl Med*. 2001;42(5):755–763.
16. Meyer JH, Ichise M. Modeling of receptor ligand data in PET and SPECT imaging: a review of major approaches. *J Neuroimaging*. 2001;11(1):30–39.
17. Wong DF, Gjedde A, Wagner HN Jr. Quantification of neuroreceptors in the living human brain. I. Irreversible binding of ligands. *J Cereb Blood Flow Metab*. 1986;6(2):137–146.
18. Huang SC, Barrio JR, Phelps ME. Neuroreceptor assay with positron emission tomography: equilibrium versus dynamic approaches. *J Cereb Blood Flow Metab*. 1986;6(5):515–521.
19. Kuwabara H, Cumming P, Reith J, Leger G, Diksic M, Evans AC, et al. Human striatal L-dopa decarboxylase activity estimated in vivo using 6-[^{18}F]fluorodopa and positron emission tomography: error analysis and application to normal subjects. *J Cereb Blood Flow Metab*. 1993;13(1):43–56.
20. Wong DF, Yung B, Dannals RF, Shaya EK, Ravert HT, Chen CA, et al. *In vivo* imaging of baboon and human dopamine transporters by positron emission tomography using [^{11}C]WIN 35,428. *Synapse*. 1993;15(2):130–142.
21. Hume SP, Myers R, Bloomfield PM, Opacka-Juffry J, Cremer JE, Ahier RG, et al. Quantitation of carbon-11-labeled raclopride in rat striatum using positron emission tomography. *Synapse*. 1992;12(1):47–54.
22. Lammertsma AA, Hume SP. Simplified reference tissue model for PET receptor studies. *Neuroimage*. 1996;4(3 Pt 1):153–158.
23. Logan J. Graphical analysis of PET data applied to reversible and irreversible tracers. *Nucl Med Biol*. 2000;27(7):661–670.
24. Logan J, Fowler JS, Volkow ND, Wolf AP, Dewey SL, Schlyer DJ, et al. Graphical analysis of reversible radioligand binding from time-activity measurements applied to [N-^{11}C-methyl]-(-)-cocaine PET studies in human subjects. *J Cereb Blood Flow Metab*. 1990;10(5):740–747.
25. Logan J, Fowler JS, Volkow ND, Wang GJ, Ding YS, Alexoff DL. Distribution volume ratios without blood sampling from graphical analysis of PET data. *J Cereb Blood Flow Metab*. 1996;16(5):834–840.

26. Cunningham VJ, Jones T. Spectral analysis of dynamic PET studies. *J Cereb Blood Flow Metab*. 1993;13(1):15–23.
27. Meikle SR, Matthews JC, Brock CS, Wells P, Harte RJ, Cunningham VJ, et al. Pharmacokinetic assessment of novel anti-cancer drugs using spectral analysis and positron emission tomography: a feasibility study. *Cancer Chemother Pharmacol*. 1998;42(3):183–193.
28. Phelps ME, Mazziotta J, Schelbert H. *Positron Emission Tomography and Autoradiography: Principle and Application for the Brain and Heart*. New York: Raven Press; 1986.
29. Feng D, Huang S-C, Wang Z, Ho D. An unbiased parametric imaging algorithm for nonuniformly sampled biomedical system parameter estimation. *IEEE Trans Med Imaging*. 1996;15:512–518.
30. Feng D, Wang Z, Huang S-C. A study on statistically reliable and computationally efficient algorithms for generating local cerebral blood flow parametric images with positron emission tomography. *IEEE Trans Med Imaging*. 1993;12:182–188.
31. Bard Y. *Nonlinear Parameter Estimation*. New York: Academic Press; 1974.
32. Marquardt DW. An algorithm for least squares estimations of nonlinear parameters. *J Soc Ind Appl Math*. 1963;11:431–441.
33. Lee JS, Lee DS, Ahn JY, Yeo JS, Cheon GJ, Kim SK, Park KS, Chung JK, Lee MC. Generation of parametric image of regional myocardial blood flow using $H_2{}^{15}O$ dynamic PET and a linear least-squares method. *J Nucl Med*. 2005;46(10):1687–1695.
34. Kim SJ, Lee JS, Kim YK, Frost J, Wand G, McCaul ME, Lee DS. Multiple linear analysis methods for the quantification of irreversibly binding radiotracers. *J Cereb Blood Flow Metab*. 2008;28(12):1965–1977.
35. Ichise M, Toyama H, Innis RB, Carson RE. Strategies to improve neuroreceptor parameter estimation by linear regression analysis. *J Cereb Blood Flow Metab*. 2002;22(10):1271–1281.

18

Multicoil Parallel MRI

Angshul Majumdar and Rabab Ward

CONTENTS

18.1 Introduction

Magnetic resonance imaging (MRI) is a safe medical imaging modality that can produce high-quality images. Other popular imaging modalities do not satisfy both these desirable properties—ultrasound is safe but generally yields images of poorer quality, whereas x-ray computer tomography (CT) can yield good-quality images but at the cost of potential health hazard. The main shortcoming of MRI (compared to ultrasound and CT) is its prolonged data acquisition time. The large scan time results in (1) inconvenience to patients as they have to spend a considerable amount of time in the scanner and (2) degradation in the quality of static MR images by introducing motion artifacts. In the past two decades, a considerable amount of effort has been expended in devising ways to reduce the scan time in MRI.

Broadly, there are two approaches to reduce the MRI scan time: hardware (i.e., physics) and software (i.e., signal processing) based approaches.

In the hardware-based approach, to expedite the scan process, the internal mechanisms of the scanner are changed. The introduction of multichannel receiver coils (parallel MRI) over a single channel is a classic example of hardware-based acceleration. Software-based acceleration does not interfere with the mechanism of the scanner. The basic idea behind reducing the scan time is to acquire less samples than is conventionally required and then use a smart reconstruction technique to obtain the image.

This chapter discusses recent developments in software-based acceleration techniques for both single-channel and multichannel MRI scanners. These developments are based on the theory of compressed sensing (CS) [1–3]. CS addresses the problem of recovering a sparse signal from its undersampled measurements. The MRI is a classic application of CS techniques. CS has recently paved the path for accelerating MRI scans even from single-channel scanners!

Before discussing how CS has improved MRI acceleration, in Section 18.2, we will briefly discuss the theory of CS in Section 18.2. In Section 18.3, we will review recent developments in CS-based MRI acceleration for single-channel coils. Section 18.4 will discuss the application of CS in multichannel parallel MRI acquisition. Finally, in Section 18.5, the conclusions will be discussed.

18.2 Theory of Compressed Sensing

CS [1–3] studies the problem of solving an underdetermined system of linear equations whose solution is known to be a sparse vector, i.e., estimation of the vector x from the following:

$$y_{m\times 1} = \Phi_{m\times n} x_{n\times 1} + \eta_{n\times 1}, \quad m < n \tag{18.1}$$

where

$y_{m\times 1}$ is the observation

$\Phi_{m\times n}$ is the measurement operator

$\eta_{n\times 1}$ is the noise

In general, the inverse problem (18.3) has an infinite number of solutions. But when the solution $x_{n\times 1}$ is s-sparse (s nonzeroes), the solution is typically unique [4]. Given enough processing time and computational power, one can find a solution to (18.3) by a brute force combinatorial search.

Mathematically, the combinatorial search can be represented as follows:

$$\min_x \|x\|_0 \text{ subject to } \|y - \Phi x\|_2^2 \le \sigma \tag{18.2}$$

Here, we drop the dimensionality of the matrices and vectors to simplify the notation.

Please note that in (18.2), $\|.\|_0$ is not a norm, it only counts the number of nonzeroes in the vector, and σ is proportional to the variance of the noise.

Solving the inverse problem (18.1) via (18.2) requires the number of measurements m to be only $\geq 2s$. This can be understood intuitively. According to the assumption that the solution x is s-sparse, i.e., it has $2s$ degrees of freedom, s positions and the corresponding s values imply that it is possible to recover the image when the number of samples is more than $2s$.

Unfortunately, (18.2) is an non-deterministic polynomial (NP) hard optimization problem; the solution x can only be recovered from $2s$ samples through an exhaustive search. It is possible to solve (18.2) approximately using a greedy algorithm [5–7]. To guarantee a solution, such approximate algorithms pose strict constraints on the nature of the inverse problem. In many practical situations, these constraints are not met. Fortunately, theoretical studies in CS [1–4] show that it is possible to solve the inverse problem (18.1) by allowing a convex relaxation of the NP hard objective function in (18.2) as follows:

$$\min_x \|x\|_1 \text{ subject to } \|y - \Phi x\|_2^2 \leq \sigma \tag{18.3}$$

where $\|.\|_1$ is the sum of the absolute values of the vector.

The l_1-norm is the tightest convex envelope for NP hard l_0-norm. Convex optimization (18.3) guarantees a solution under far weaker constraints than required by approximate greedy algorithms. This is a strong result that guarantees the solution of an NP hard problem (18.2) by a convex relaxation that can be solved by quadratic programming (18.3). However, compared to the $2s$ samples required by the NP hard problem (18.2), the convex surrogate (18.3) requires considerably more samples, given by

$$m > Ck \log(n) \tag{18.4}$$

For all practical problems, we are interested in decreasing the number of measurement samples as much as possible, but at the same time, we should also be able to recover the solution by a tractable polynomial time algorithm. Intuitively, if the value of p in the l_p-norm used in (18.5) is a fraction between zero and one, we hope to recover the solution by using less number of samples than required by (18.3). This intuition is theoretically proven in [8–11].

When the value of p lies between zero and one, the optimization problem becomes nonconvex:

$$\min_x \|x\|_p^p \text{ subject to } \|y - \Phi x\|_2^2 \leq \sigma \; 0 < p < 1 \tag{18.5}$$

The number of equations (samples) needed to solve the inverse problem (18.1) via nonconvex optimization (18.5) is [11]

$$n \geq C_1 k + pk \log(n) \tag{18.6}$$

As the value of p decreases, the second term almost vanishes, and the number of equations only grows linearly with the number of nonzeroes in the solution (same order as NP hard solution). Notice that the number of equations required by the nonconvex program (18.5) lies between those of the NP hard (18.2) and the convex (18.3) program.

18.3 Accelerating Single-Channel MRI Scans

The data acquisition model for MRI is expressed as

$$y = Fx + \eta \tag{18.7}$$

where
- x represents the image to be reconstructed
- y represents the k-space data
- F is the Fourier transform that maps the k-space to the image domain
- η is the noise assumed to be normally distributed

The term k-space is actually the Fourier frequency domain. In MRI literature, "k-space" is a more popular term.

For accelerating the MRI scans, the number of k-space samples collected (length of y) should be less than the number of pixels in the image. Therefore, the problem of reconstructing the image x, from its undersampled measurements, is an underdetermined inverse problem. In recent years, CS-based techniques have become popular in reconstructing MR images by exploiting their sparsity in a transform domain. We will discuss the CS-based reconstruction techniques in Section 18.3.1. But CS is not the only solution to this problem. We have showed that it is possible to reconstruct these images by exploiting their rank deficiency. We will discuss it in Section 18.3.2.

18.3.1 Sparse Modeling of MR Images

The theory of CS was developed for sparse signals. Medical images are not sparse in the pixel/spatial domain. But they have an approximately sparse (compressible) representation in the wavelet domain. The wavelet analysis and synthesis equations are

$$\begin{aligned} &\text{Analysis}: \ \alpha = Wx \\ &\text{Synthesis}: \ x = W^T\alpha \end{aligned} \tag{18.8}$$

where
x is the image
W is the wavelet transform
α is the approximately sparse wavelet transform coefficient

We can write the inverse problem (18.7), in terms of the wavelet transform coefficients, as

$$y = FW^T\alpha + \eta \tag{18.9}$$

Following the concepts of CS in Section 18.2, the wavelet coefficients can be recovered by solving

$$\min_{\alpha} \|\alpha\|_1 \text{ subject to } \|y - FW^T\alpha\|_2^2 \leq \sigma \tag{18.10}$$

Once the wavelet coefficient vector α is obtained, the MR image can be reconstructed by applying the synthesis equation. Even though the theory behind recovery of approximately sparse signals has been only developed for the convex case of l_1-norm minimization, it has been shown experimentally that l_p-norm minimization [11,12] can stably recover the wavelet coefficients by solving

$$\min_{\alpha} \|\alpha\|_p^p \text{ subject to } \|y - FW^T\alpha\|_2^2 \leq \sigma \tag{18.11}$$

In [12], MR image reconstruction experiments were carried out on simulated Shepp–Logan phantoms instead of real images. Experiments on some real MR images were carried out in [11]. The reconstruction accuracy obtained from nonconvex l_p-norm minimization showed minor improvements over l_1-norm minimization.

The earlier direct application of CS to the MR image reconstruction problem results in the sparse synthesis prior optimization (18.10) or (18.11). The synthesis prior is only applicable when the image can be modeled as a sparse set of coefficients in a transform domain that satisfies the synthesis and analysis in Equation 18.8. There are powerful signal models such as the piecewise constant model for images that can be expressed in synthesis–analysis equations. A piecewise constant image is sparse under finite differencing, and the corresponding sparsity promoting optimization is called the total variation (TV) minimization:

$$\min_x TV(x) \text{ subject to } \|y - Fx\|_2^2 \leq \sigma \tag{18.12a}$$

where $TV(x) = \sum \sqrt{D_h(x)^2 + D_v(x)^2}$ and D_h and D_v are the horizontal and vertical differencing operators.

TV minimization has been used previously in MR image reconstruction [13,14].

TV falls under the general category of co-sparse analysis prior. Co-sparsity is not a well-researched technique in the CS literature but is rapidly gaining interest [15,16] since it allows for more powerful signal modeling than sparse synthesis prior.

However, the TV prior is not a very appropriate choice for MR reconstruction; in a comparative study [17] between convex synthesis prior (l_1-norm minimization) and TV minimization, it was found that the former yields slightly better reconstruction results.

To improve the individual shortcomings of wavelets and TV, [18,19] proposed a mixed prior combining the two, i.e., solving the following:

$$\min_\alpha \|\alpha\|_1 + \gamma TV(W^T\alpha) \text{ subject to } \|y - FW^T\alpha\|_2^2 \leq \sigma \tag{18.12b}$$

Here, the TV prior is defined in the image domain. The factor γ balances the relative importance of the TV prior and the l_1-norm prior. It is a free parameter and needs to be decided by the user. There is no analytical basis to fix it.

In [18,19], it is claimed that the mixed prior (18.12) yields superior results compared to the synthesis prior (18.9) optimization. A more recent work [20] proposed a nonconvex mixed prior by combining the l_p-norm with TV:

$$\min_\alpha \|\alpha\|_p^p + \gamma TV(W^T\alpha) \text{ subject to } \|y - FW^T\alpha\|_2^2 \leq \sigma \tag{18.13}$$

Solving the nonconvex mixed prior (18.13) gives slight improvements in reconstruction accuracy over the convex mixed prior (18.12) and the nonconvex synthesis prior (18.10).

As mentioned earlier, the TV prior is an analysis prior. Similarly, one can use wavelet-based analysis prior in the following fashion:

$$\min_x \|Wx\|_p^p \text{ subject to } \|y - Fx\|_2^2 \leq \sigma \tag{18.14}$$

For $p = 1$, this is a convex problem. When orthogonal* wavelets are used, the analysis and synthesis prior formulations (18.14) and (18.10) are the same theoretically. But they are different for redundant tight frames† (curvelets, complex dual-tree wavelets, etc.). This follows directly from the definition of orthogonal and tight-framed transforms. A comparative study on analysis and synthesis priors on orthogonal and redundant wavelets was carried out in [21]. The empirical conclusions from the work are as follows:

1. For orthogonal wavelets, the analysis and the synthesis prior yields the same results. This is to be expected since it follows theoretically (see appendix for mathematical explanation).
2. Synthesis prior on redundant wavelets yields the worst results.
3. Analysis prior on redundant wavelets yields the best possible results.

There is no theoretical study that explains the results in [21]. The analysis prior formulation was introduced into MR image reconstruction for the first time in [22]. It is not a popular choice for MR image reconstruction. However, in that work, we showed that with proper choice of wavelet, the analysis prior formulation can yield the best reconstruction results.

In summary, we see that in general there are three CS priors that have been used for MR image reconstruction—synthesis prior (18.11), analysis prior (18.14), and mixed prior (l_p-norm and TV) (18.12):

$$\text{Synthesis prior: } \min \|\alpha\|_p^p \text{ subject to } \|y - FW^T\alpha\|_2^2 \leq \sigma$$

$$\text{Analysis prior: } \min \|Wx\|_p^p \text{ subject to } \|y - Fx\|_2^2 \leq \sigma$$

$$\text{Mixed prior: } \min \|\alpha\|_p^p + \gamma TV(W^T\alpha) \text{ subject to } \|y - FW^T\alpha\|_2^2 \leq \sigma$$

Most previous works in MR image reconstruction used either the synthesis prior or the mixed prior. In a previous work, we proposed the analysis prior optimization for MR image reconstruction [21]. The problem with such disparate studies is that during practical MR image reconstruction, one does not know which reconstruction algorithm will produce the best possible result. Thus, there is a need to evaluate all these algorithms on a uniform dataset. In this chapter, we compare the three different priors on real MR images in order to ascertain which prior yields the best possible results.

---

* $W^TW = I = WW^T$.

† $W^TW = I \neq WW^T$.

18.3.1.1 Experimental Evaluation

The results in this section have been reproduced from [23], where we performed an experimental study on three different MR images (Figure 18.1). The ground-truth data are collected by fully sampling the k-space on a uniform Cartesian grid. All the images are of size 256 × 256 pixels. The T2-weighted brain and the phantom data have been obtained from [18]. These data were collected by a 1.5 T scanner with echo time of 85 ms. The T2-weighted rat's spinal cord was acquired by a 7 T scanner at UBC MRI Lab with echo time of 13 ms. For all the images, spin echo sequence was used. We simulated Cartesian undersampling (where lines in the vertical direction randomly are omitted [16]).

In our experiments, the number of vertical lines for Cartesian sampling was fixed at 96. This corresponds to a sampling ratio of 37.5% (acceleration factor of 2.7) for Cartesian sampling. For Cartesian sampling, 33% of the samples were collected from the center of the k-space, while the rest was randomly collected from the remaining k-space. We ran the experiments on different values of p between zero and one for the three types of prior and found that on an average, the best results are obtained for $p = 0.8$.

The Haar wavelets have been used as the sparsifying transform. The synthesis prior and the analysis prior algorithm yielded the same answer for orthogonal wavelets. For redundant wavelet transform, the synthesis prior yielded worse results than the orthogonal wavelet transform, but the analysis prior with redundant wavelet transform gave better results. The same phenomenon had been observed in previous studies [21]. Therefore, for the synthesis prior and the mixed prior (which is a combination of synthesis and TV prior), the orthogonal Haar transform was used only; for the analysis prior, the redundant Haar transform is used.

For quantitative evaluation, the normalized mean squared error (NMSE) was used for comparing the results in Table 18.1.

NMSE values do not speak about the image quality. Therefore, for qualitative evaluation, we show the reconstructed images and the difference

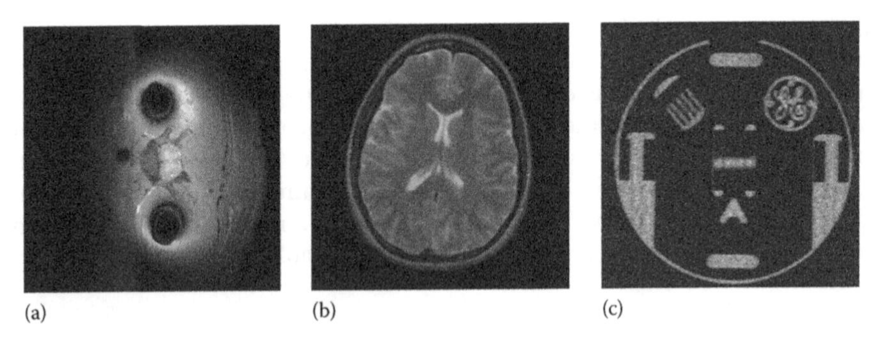

(a) (b) (c)

FIGURE 18.1
(a) Rat (spine), (b) brain, and (c) phantom.

TABLE 18.1

Reconstruction Accuracy for Cartesian Undersampling ($p = 0.8$)

Image Name	Synthesis Prior Orthogonal Haar	Mixed Prior	Analysis Prior Redundant Haar
Rat	0.1244	0.0955	0.0675
Brain	0.1830	0.1165	0.0709
Phantom	0.3409	0.2321	0.1847

images in Figure 18.2. The difference images have been magnified five times for visual clarity.

From Figure 18.2, it is easily discernible that the analysis prior optimization yields visually superior images compared to the others. In all the images, the reconstruction artifacts are blocky; this is because the Haar wavelets have been used as the sparsifying transform.

A more detailed comparative study on this topic can be found in [23].

18.3.2 Modeling the MR Image as a Low-Rank Matrix

From the discussion in Section 18.2, we see that it is possible to solve for a vector of size n from its lower-dimensional projections (m) provided the number of nonzero entries in the vector s is sufficiently small (18.4). A n dimensional vector with s nonzero elements has only $2s$ (s positions and corresponding s values) degrees of freedom. CS is able to find a solution when the number of measurements is sufficiently larger than the degrees of freedom in the solution. The same idea applies to low-rank matrix recovery from undersampled projections. A matrix of size N x N but of rank r has only $r(2n - r)$ degrees of freedom. As long as the number of lower-dimensional projections is sufficiently larger than the number of degrees of freedom, it is possible to recover the matrix.

In this section, we show that transform domain sparsity is not the only way the information content of the image is compactly captured. In many cases, singular value decomposition (SVD) can also efficiently represent the information content of the MR image. Therefore, instead of using CS for recovering the matrix from undersampled k-space measurements, it is possible to exploit the low-rank property of the images to recover them.

Owing to the spatial redundancy, MR images do not have full rank, i.e., for an image of size $N \times N$, its rank is $r < N$. For an image of rank "r," the number of degrees of freedom is $r(2N - r)$. Thus, when $r \ll N$, the number of degrees of freedom is much less than the total number of pixels N^2. This idea has been successfully used in recent studies to propose SVD-based compression schemes for images [24,25]. Just as CS exploits sparsity in the transform domain to reconstruct the MR image from subsampled k-space data, this approach will exploit the low-rank property of the medical images for their reconstruction.

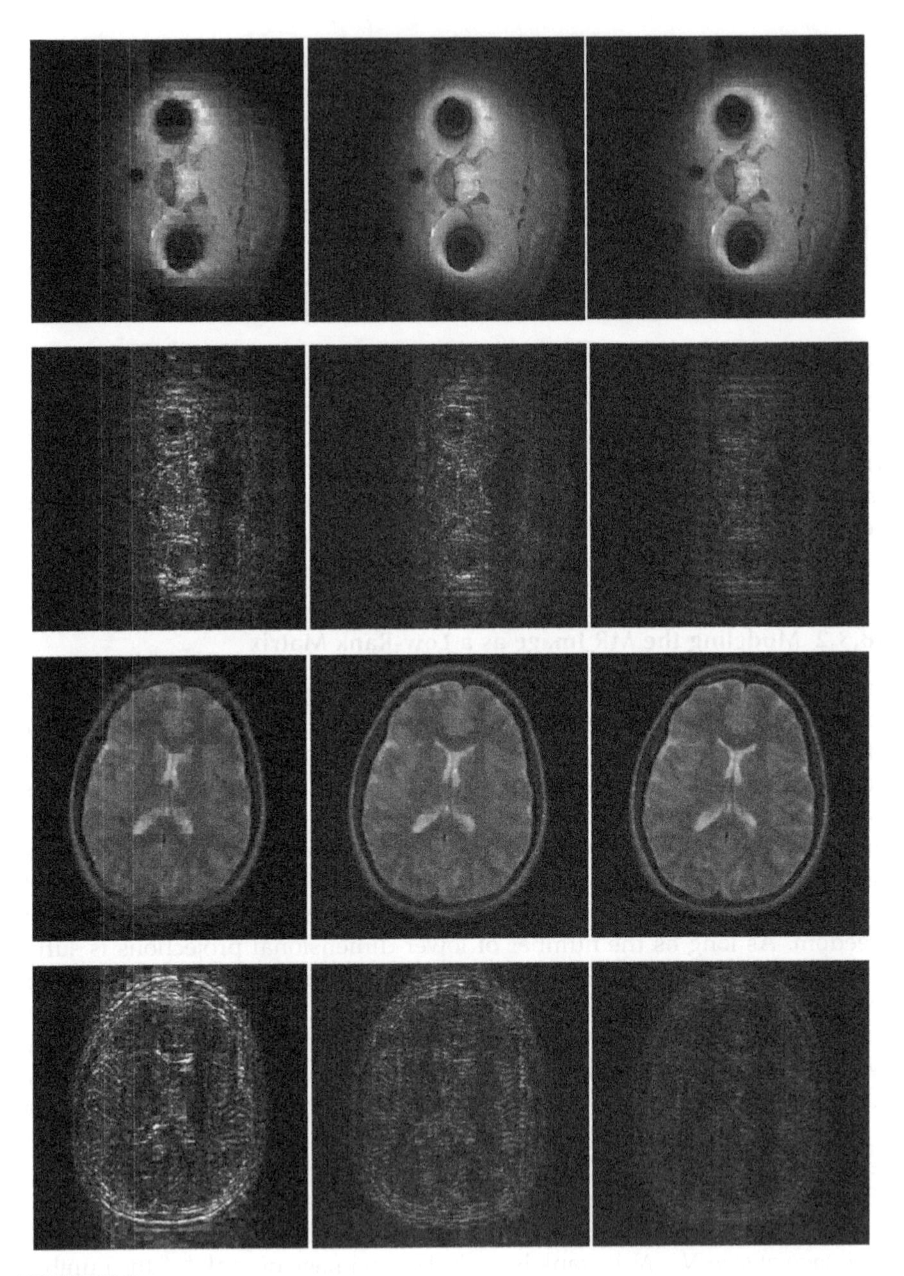

FIGURE 18.2
Reconstructed and difference images for Cartesian undersampling. For all the rows: left image—synthesis prior, middle image—mixed prior, and right image—analysis prior. Odd rows represent the reconstructed images, and the even rows represent difference images.

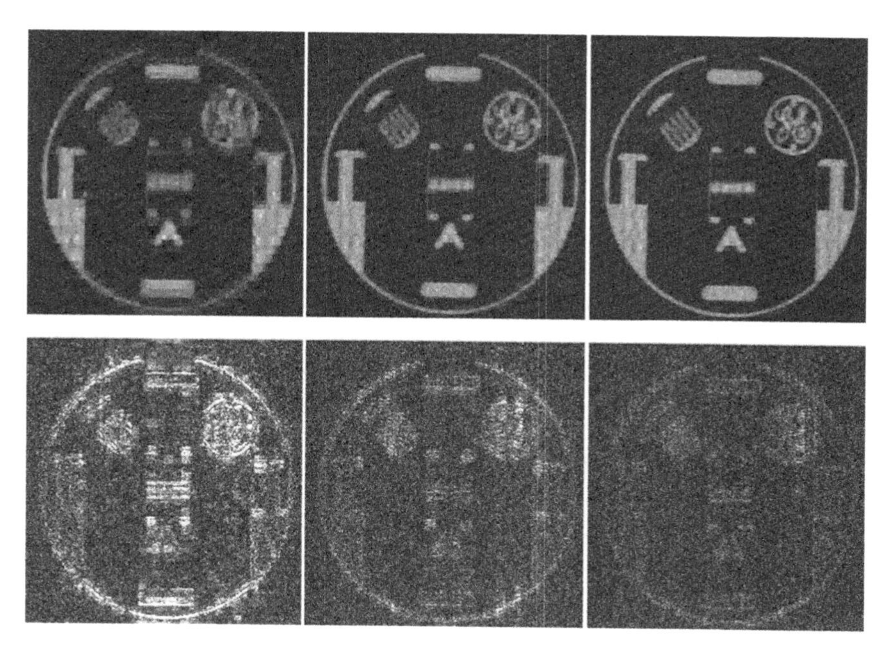

FIGURE 18.2 (continued)
Reconstructed and difference images for Cartesian undersampling. For all the rows: left image—synthesis prior, middle image—mixed prior, and right image—analysis prior. Odd rows represent the reconstructed images, and the even rows represent difference images.

The image can be recovered from its subsampled k-space measurements by exploiting its rank deficiency in the following manner:

$$\min \; rank(X) \text{ subject to } \|y - Fx\|_2^2 \leq \sigma \tag{18.15}$$

where
X is the image
x is the vectorized version of the image

As in l_0-norm minimization, the rank minimization problem (18.15) is NP hard. In CS, the NP hard problem is bypassed by replacing the l_0-norm by its nearest convex surrogate—the l_1-norm. Similarly for (18.15), it is intuitive to replace the NP hard objective function of rank minimization by its tightest convex relaxation—the nuclear norm. Therefore, in [30], we have proposed to solve the following problem instead:

$$\min \; \|X\|_* \text{ subject to } \|y - Fx\|_2^2 \leq \sigma \tag{18.16}$$

where $\|X\|_*$ is the nuclear norm of the image X, which is defined as the sum of its singular values.

The fact that the rank of the matrix can be replaced by its nuclear norm for optimization purpose is justified theoretically as well. Recent theoretical studies by Recht et al. [26–28] and others [29] show the equivalence of (18.15) and (18.16). We refrain from theoretical discussion on this subject and ask the interested reader to peruse References [25–29].

The work [30] is motivated by the efficacy of l_p-norm minimization in nonconvex CS. The actual problem in CS is to solve the l_0-norm minimization problem (18.2). This problem being NP hard is replaced by its closest convex surrogate the l_1-norm (18.3). The l_1-norm minimization problem is more popular in CS because its envelope is convex and therefore easier to theoretically analyze. However, the l_p-norm better approximates the original NP hard problem. Consequently, it gives better results [12,20,22,23] both theoretically and practically.

In the aforesaid work [30], our main target was to solve the NP hard rank minimization problem (18.15). But following previous studies in CS, we propose to replace the NP hard rank minimization problem by its closest nonconvex surrogate the Schatten-p-norm instead of the convex nuclear norm. The Schatten-p-norm is defined as

$$\|X\|_{p*} = \sum_i \sigma_i^p, \text{ where } \sigma_i\text{'s are singular values of } X \tag{18.17}$$

In [30], we proposed to solve the MR image reconstruction problem by minimizing its Schatten-p norm directly. Mathematically, the optimization problem is as follows:

$$\min \|X\|_{p*} \text{ subject to } \|y - Fx\|_2^2 \leq \sigma \tag{18.18}$$

Before our work [30], there was no existing algorithm to solve the Schatten-p-norm minimization problem. For the first time, we proposed an efficient algorithm to solve this problem (18.18) in [30]. The interested reader should refer to the aforementioned reference for the algorithm.

18.3.2.1 Experimental Evaluation

We carried out the experiments on three slices from the BrainWeb (two slices) and National Institute of Health (NIH) (one slice) databases (see Figure 18.3). Radial sampling was employed to collect the k-space data. Our proposed reconstruction technique is compared against the standard CS-based reconstruction method called the CSMRI [18]. One of the fastest known l_1-norm minimization algorithms called the spectral projected gradient L1 (SPGL1) [31] was used as the solver from CSMRI. The Haar wavelet

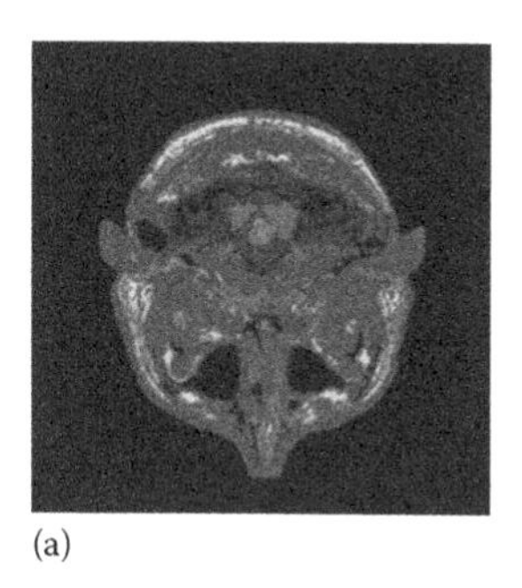
(a)

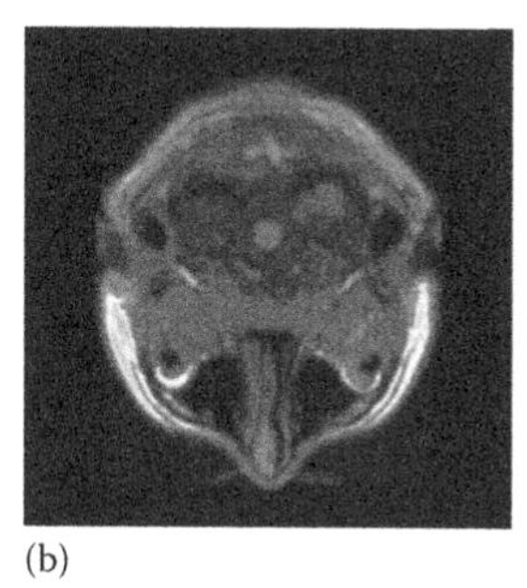
(b)

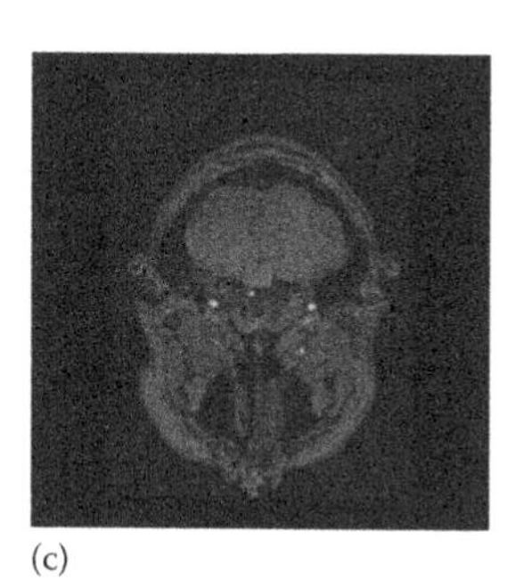
(c)

FIGURE 18.3
Ground-truth images: (a) BrainWeb new, (b) BrainWeb old, and (c) NIH.

is used as the sparsifying transform for CSMRI. Our proposed reconstruction approach, based on Schatten-p-norm minimization, gave the best results when the value of p is 0.9. Therefore, for all the experiments, we kept the fixed p at the said value.

We will provide both quantitative and qualitative reconstruction results. As in the previous section for quantitative comparison, the NMSE between the reconstructed image and the ground truth is computed. However, we also provide the reconstructed images for qualitative assessment.

In the first set of experiments, it is assumed that the data are noise free. The number of radial sampling lines has been varied from 50 to 110 in steps of 20. In Table 18.1, the NMSE values are shown, and in Table 18.2, the reconstruction times are tabulated.

Table 18.2 shows that for the BrainWeb new and the NIH data, the reconstruction error from our proposed method is slightly higher (1%–2%) and for the BrainWeb old data, the error is slightly lower (around 1%) than CSMRI. However, these slight variations are practically indistinguishable in the reconstructed images. The reconstructed images from the proposed approach and CSMRI are shown in Figure 18.4 for visual quality assessment.

In the following table, we compare the reconstruction times required for our proposed method and by the CS-based method.

TABLE 18.2
Comparison of NMSE for Proposed and CS Reconstruction

Slice Name	Technique	50 Lines	70 Lines	90 Lines	110 Lines
BrainWeb new	CSMRI	0.19387	0.14855	0.11762	0.09765
	Proposed	0.19573	0.15926	0.13452	0.11390
BrainWeb old	CSMRI	0.13199	0.09500	0.07082	0.05461
	Proposed	0.12066	0.08042	0.06012	0.04407
NIH	CSMRI	0.23270	0.18807	0.15649	0.12988
	Proposed	0.23475	0.19456	0.16738	0.14236

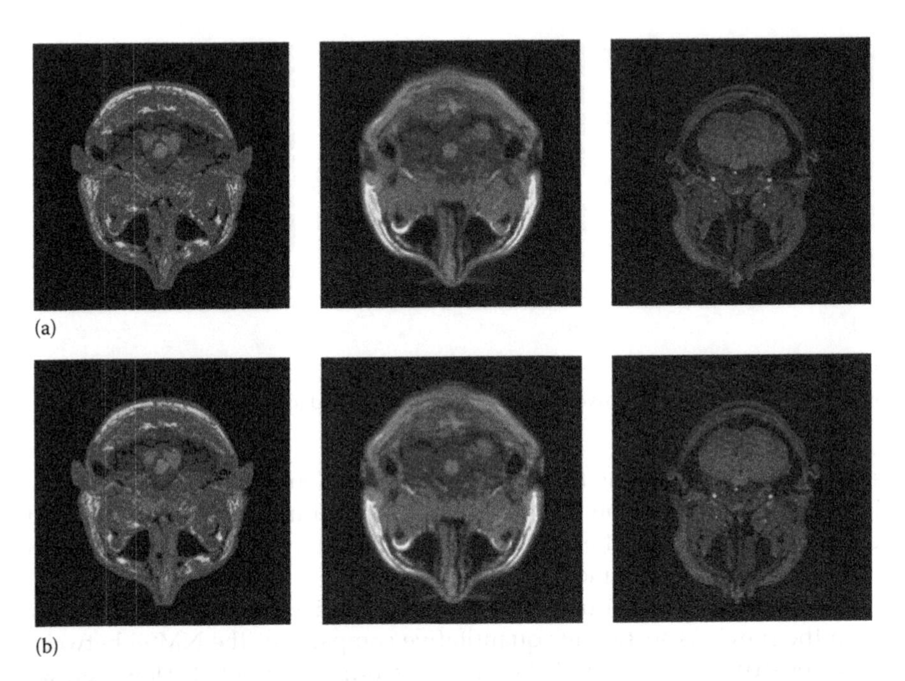

FIGURE 18.4
Reconstruction results (220 lines). (a) CSMRI reconstruction and (b) reconstruction from proposed approach. Left to right—BrainWeb new, BrainWeb old, and NIH.

Table 18.3 shows that our proposed approach is about 5–10 times faster than CSMRI. It must be remembered that the SPGL1 is one of the most sophisticated and fastest l_1-norm minimization algorithms where our algorithm has been implemented naively. The core computational demand of our algorithm is in computing the SVD in each iteration. Computing the SVD naively is time consuming since we already know that the matrix is rank deficient. Our current algorithm can be made even faster by employing PROPACK [32] for computing the SVD.

TABLE 18.3
Comparison of Reconstruction Times

Slice Name	Technique	50 Lines	70 Lines	90 Lines	110 Lines
Brain web new	CSMRI	83.4761	120.084	202.632	61.5545
	Proposed	9.04923	10.0613	32.3338	8.55673
Brain web old	CSMRI	82.7161	82.3672	130.733	85.9018
	Proposed	8.25629	10.9292	22.4166	10.0542
NIH	CSMRI	92.767	92.0046	206.72	99.7556
	Proposed	13.3627	12.7485	45.1203	11.7448

CS-based MR image reconstruction techniques have been successful in reconstructing image from subsampled k-space measurements. Our proposed low-rank model of the MR image aims at the same goal but from an entirely new perspective. The reconstruction results show that the proposed approach and the CS-based techniques provide virtually indistinguishable results. However, the low-rank model has a distinct advantage in terms of speed. It is about 5–10 times faster than one of the fastest CS-based reconstruction algorithms.

18.4 Accelerating Multichannel Scans

The objective of multicoil parallel MRI is to reduce the data acquisition time. Instead of scanning the whole k-space by a single receiver coil, parallel MRI uses multiple coils. Depending on the physical position of the receiver coils in the scanner, they have different field of views and thereby different sensitivity profiles/maps. Acceleration is achieved since each of the receiver coils partially samples the k-space. The sampling mask used to undersample the k-space is the same for all the coils. The ratio of the total number of possible k-space samples to the number of samples actually collected by each coil is the "acceleration factor." In parallel MRIs, the problem is to reconstruct the underlying MR image given the partial k-space samples acquired by each coil.

Mathematically, the multicoil data acquisition process can be formulated as

$$y_i = RFS_i x + \eta_i, \; i = 1 \ldots C \tag{18.19}$$

where

y_i is the k-space data collected by the ith coil
F is the Fourier transform operator
R is the undersampling mask
S_i is the sensitivity map of the ith coil
x is the image to be reconstructed (vectorized image consisting of q pixels)
η is the noise (a complex vector of length p)

In Equation 18.19, the image to be reconstructed x and the sensitivity maps S_i's are the unknowns. The basic parallel MR image reconstruction techniques—sensitivity encoding (SENSE) [33] and simultaneous acquisition of spatial harmonics (SMASH) [34]—explicitly require the sensitivity maps to be available. More recent techniques like generalized autocalibrating partially parallel acquisitions (GRAPPA) [35] estimate the sensitivity-dependent interpolation weights from the autocalibration signal (ACS) lines.

If the sensitivity maps are available, then SENSE is physically the most optimal reconstruction technique in terms of mean squared error. It is based on solving the following inverse problem:

$$y = RFSx + \eta \tag{18.20}$$

where

$$y = \begin{bmatrix} y_1 \\ \dots \\ y_C \end{bmatrix}$$

$$S = \begin{bmatrix} S_1 \\ \dots \\ S_1 \end{bmatrix}$$

$$\eta = \begin{bmatrix} \eta_1 \\ \dots \\ \eta_C \end{bmatrix}$$

R and F carry their predefined meanings.

Estimating the sensitivity maps is not trivial. The accuracy of image reconstruction is sensitive to the accuracy with which the sensitivity maps have been computed. But even with this shortcoming, SENSE is by far the most widely used parallel MR image reconstruction technique. All commercial scanners use modifications of the basic SENSE method—Philips uses SENSE, Siemens uses mSENSE, General Electric uses ASSET, and Toshiba uses SPEEDER [36].

In the section, we will discuss the basic SENSE method and its advancements in recent times based on CS techniques. In recent times, some robust SENSE algorithms have been developed that are less sensitive to the estimated sensitivity maps compared to the previous ones. In the next section, we discuss a technique that enjoys the same mathematical rigor as SENSE but does not require sensitivity map estimation either implicitly or explicitly. This is by far the most robust multicoil reconstruction algorithm that can achieve better reconstruction results than commercial and state-of-the-art algorithms in parallel MRI.

18.4.1 SENSE-Based Reconstruction

The physical data acquisition model for the multicoil parallel MRI is expressed by Equation 18.20. In SENSE, it is assumed that the coil sensitivities are known, i.e., S is given. The first study that proposed the SENSE framework [33] solves the reconstruction problem by the least squares optimization:

$$\min_x \|y - RFSx\|_2 \tag{18.21}$$

However, in practical scenarios, a regularized version of the least squares problem is solved [37–39]:

$$\min_x \|y - RFSx\|_2 + \lambda\Phi(x) \tag{18.22}$$

where
$\Phi(x)$ is the regularization term
λ is the regularization parameter

The regularization function Φ can be the l_1-norm of the wavelet transform of the image; it can also be an isotropic or anisotropic TV of the image.

Recently, CS-based techniques have been incorporated into the SENSE framework [40]. The reconstruction problem is formulated as

$$\arg\min_x \|\Psi x\|_1 \text{ subject to } \|y - RFSx\|_2 \leq \varepsilon \tag{18.23}$$

This formulation (18.23) is named SparSENSE. Instead of incorporating the sparsity of the image in the transform domain, the information regarding the rank deficiency of the image matrix can also be used in the SENSE framework. This leads to the following nuclear norm-regularized SENSE reconstruction (NNSENSE) problem [41]:

$$\arg\min_x \|X\|_* \text{ subject to } \|y - RFSx\|_2 \leq \varepsilon \tag{18.24}$$

Here, $\|X\|_*$ is the nuclear norm of the image matrix. It is the sum of the singular values of X. This is a convex metric and is a special case of the Schatten-p-norm where p equals 1.

All the aforementioned studies assume that the sensitivity maps are known. Thus, the accuracy of the reconstruction technique depends on the provided sensitivity maps. However, the estimates of the sensitivity maps are not always accurate. In order to get a robust reconstruction, a joint estimation of the image and sensitivity map has been proposed in JSENSE [42]. The sensitivity map for each coil is parameterized as a polynomial of degree N whose coefficient is a. The SENSE equation (18.20) is expressed as

$$y = RFS(a)x + \eta \tag{18.25}$$

where the sensitivity for the ith coil at position (m,n) is given by $s_i(m,n) = \sum_{p=0}^{N}\sum_{q=0}^{N} a_{i,p,q}x^p y^q$, where (m,n) denotes the position and N and a denote the degree of the polynomial and its coefficients, respectively.

In order to jointly estimate the coil sensitivities (given the polynomial formulation) and the image, the following problem has to be solved:

$$\arg\min_{a,x} \|y - RFS(a)x\|_2 \quad (18.26)$$

However, it has been argued in [42] that solving for all the variables jointly is an intractable nonconvex problem; hence, the image and the polynomial coefficients were solved iteratively, as follows:

$$\arg\min_{x} \|y - RFS(a)x\|_2 \quad (18.27)$$

In (18.27), it is assumed that the coil sensitivity parameters (a) are given, and the problem is to solve for the image x. In the next step, it is assumed that the image x is given, and based on that, the coil sensitivity parameters a are estimated:

$$\arg\min_{a} \|y - RFS(a)x\|_2 \quad (18.28)$$

Problem (18.26) is nonconvex, and there is no guarantee that solving it by iterating between (18.27) and (18.28) will converge. However, with reasonably good initial estimates of the sensitivity maps, JSENSE provides considerable improvement over SENSE reconstruction.

The JSENSE was regularized using the CS framework [43]. It was assumed that the image is sparse in a finite difference domain and the coil sensitivity is sparse in Chebyshev polynomial basis or in Fourier domain. Therefore, it is proposed in [43] to solve the problem by alternately solving the following two optimization problems:

$$\hat{x} = \min_{x} \|y - RFSx\|_2 + \lambda_x TV(x) \quad (18.29)$$

$$\hat{S} = \min_{x} \|y - RFXs\|_2 + \lambda_s \Psi \|s\|_1 \quad (18.30)$$

where Ψ is the Chebyshev polynomial basis.

In (18.29), the image is solved by TV regularization assuming the sensitivity map is known. In (18.30), the sensitivity map is estimated assuming the image is known. Here, Ψ is the sparsifying basis (Chebyshev polynomial basis or Fourier basis) for the sensitivity maps.

The problem with this formulation is that the basis used for sparsifying the sensitivity maps and the measurement basis is not incoherent. This violates the basic premises of CS-based reconstruction [44] and thus does not yield good reconstruction results.

To overcome the problem of modeling the sensitivity maps as a sparse set of Fourier coefficients, we propose an alternate model in this chapter. We propose to model the sensitivity maps as rank-deficient matrices. This is a new assumption. In order to corroborate it empirically, we show plots of singular value decay of the two sensitivity maps (one and three) for an 8-channel MRI scanner in Figure 18.5a. It shows that the singular values of the sensitivity maps decay fast and thus the sensitivity maps are approximately rank deficient.

The MR image can be either modeled as a sparse set of transform coefficients or modeled as a rank-deficient matrix [30,41]. In this chapter, we propose modeling the sensitivity maps as rank-deficient matrices. Assuming

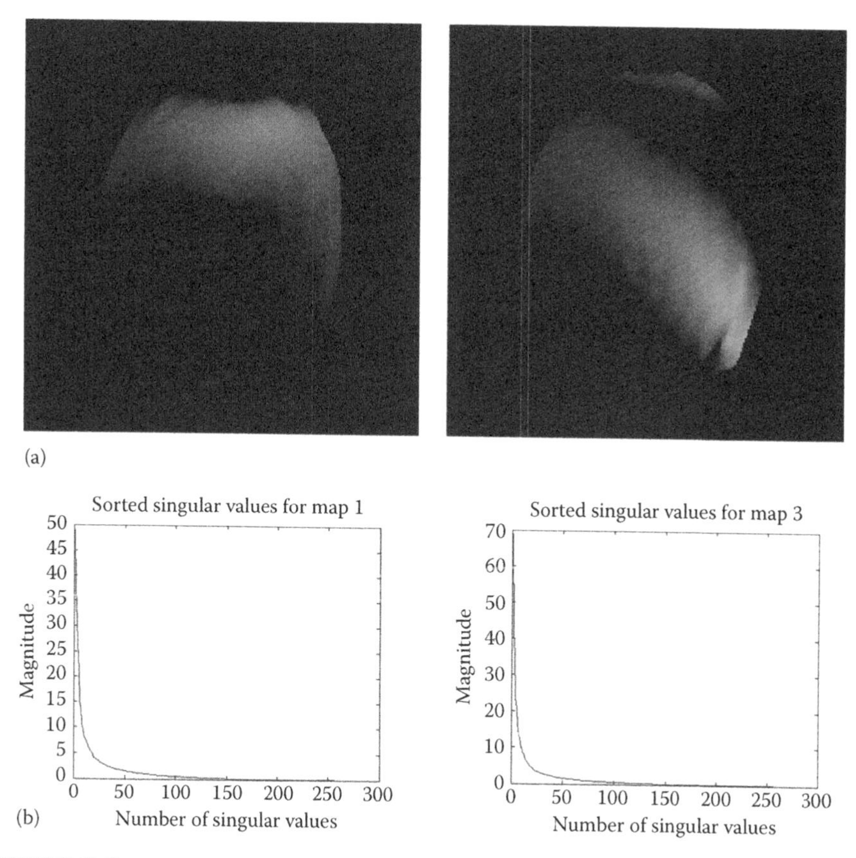

FIGURE 18.5
(a) Sensitivity maps and (b) decay of singular values.

that the image is sparse in a transform domain, the parallel MRI problem can be formulated as follows:

$$\min_{x,S} \|y - RFSx\|_2 \text{ subject to}$$
$$\|Wx\|_1 + \gamma TV(x) \le \tau_1 \qquad (18.31)$$
$$\|S_i\|_* \le \tau_2, \forall i$$

where
 τ_1 and τ_2 are the constraints on the transform domain sparsity of the image and on the rank deficiencies of the sensitivity maps, respectively
 γ is the term balancing the relative importance of the transform domain sparsity and TV

Alternately, when the image is modeled as a rank-deficient matrix, one would ideally solve the following problem:

$$\min_{x,S} \|y - RFSx\|_2 \text{ subject to}$$
$$\|X\|_* \le \tau_1 \qquad (18.32)$$
$$\|S_i\|_* \le \tau_2, \forall i$$

where τ_1 represents the constraint on the rank deficiency of the MR image to be reconstructed.

Unlike [43], we exploit the rank deficiency of the sensitivity maps in order to estimate them. The incoherency criterion required by CS-based techniques does not arise in our formulation. The proposed method is called *i*SENSE.

Both (18.31) and (18.32) are nonconvex problems similar to the JSENSE formulation. Solving for the MR image and the sensitivity maps jointly is an intractable problem. Following the approach in JSENSE, we propose to iteratively solve the optimization problems by the following algorithm.

Algorithm 1

Initialize: Use an initial estimate of the sensitivity maps.
Iteratively solve steps 1 and 2 until convergence.

Step 1: Using the initial or the estimated sensitivity maps obtained in step 2, solve for the image either by solving a CS problem

$$\min_{x} \|y - RFSx\|_2 \text{ subject to } \|Wx\|_1 + \gamma TV(x) \le \tau_1$$

(continued)

or by solving nuclear norm minimization

$$\min_x \|y - RFSx\|_2 \text{ subject to } \|X\|_* \leq \tau_1$$

Step 2: Using the estimate of the image from Step 1, solve for the sensitivity maps by the nuclear norm minimization:

$$\min_S \|y - RFSx\|_2 \text{ subject to } \|S_i\|_* \leq \tau_2, \forall i$$

There are two issues with Algorithm 1—(i) There are no efficient solvers for the stated CS or nuclear norm minimization problems; (ii) it is not an easy task to specify values for τ_1 and τ_2. This is because obtaining estimates on the sparsity of the image or the rank deficiencies of image and the sensitivity maps is very difficult. Both these issues can be resolved by solving the optimization problems in an equivalent alternative form where the constraint is on the data fidelity term instead of on the sparsity or rank deficiency. This leads to the following algorithm.

Algorithm 2

Initialize: An initial estimate of the sensitivity maps.
Iteratively solve steps 1 and 2 until convergence.

Step 1: Using the initial or the estimated sensitivity maps obtained in step 2, solve for the image either by solving a CS problem

$$\min_x \|Wx\|_1 + \gamma TV(x) \text{ subject to } \|y - RFSx\|_2 \leq \varepsilon_1$$

or by solving nuclear norm minimization

$$\min_x \|X\|_* \text{ subject to } \|y - RFSx\|_2 \leq \varepsilon_1$$

Step 2: Using the estimate of the image from Step 1, solve for the sensitivity maps by nuclear norm minimization:

$$\min_S \|S_i\|_* \text{ subject to } \|y - RFSx\|_2 \leq \varepsilon_2, \forall i$$

Algorithm 2 requires solving CS and nuclear norm minimization problems. Efficient algorithms are available for both of them. It should be noted that

Step 1 (which assumes that the sensitivity maps are given) is the same as SparSENSE or NNSENSE as the case may be.

The problems (18.31) and (18.32) are nonconvex. Hence, their solution is not guaranteed to reach a unique minimum. Therefore, it is not possible to theoretically analyze convergence of Algorithm 2. For similar reasons, previous studies [42,43] did not analyze the convergence of iterative SENSE algorithms. However, in this work, we found that practically the said algorithm converges in three iterations.

18.4.1.1 Experimental Evaluation

All the experiments were performed on a 64 bit AMD Athlon processor with 4 GB of RAM running Windows 7. The simulations were carried out on MATLAB® 2009.

There are five sets of ground-truth data used for the experimental evaluation (Figure 18.6). The brain data have been used previously in [19]. The brain data come from a fully sampled T1-weighted scan of a healthy volunteer. The volunteer was scanned using spoiled gradient echo sequence with the following parameters—echo time = 8 ms, repetition time = 17.6 ms, and flip angle = 20°. The scan was performed on a GE Signa EXCITE 1.5 T scanner, using an eight-channel receiver coil. The eight-channel data for Shepp–Logan phantom were simulated using the B1 simulator (http://maki.bme.ntu.edu.tw/tool_b1.html) using the default settings for eight-channel receiver coil. The UBC MRI Lab prepared three datasets. The first one is a phantom created by doping saline with Gd-DTPA. The second and third ones are the ex vivo and in vivo images of a rat's spinal cord. A four-coil Bruker scanner was used for acquiring the data using a FLASH sequence. The ground truth has been formed by sum-of-squares reconstruction of the multichannel images. All the images had a native resolution of 256 × 256 pixels.

This method only requires an initial estimate of the sensitivity maps. For this reason, we have used the simple sum-of-squares method for sensitivity estimation [33,42]. The initial sensitivity maps were computed in the following manner:

- For each coil, the center of the k-space of the image was fully sampled (ACS lines). One-third of all the lines were used as ACS lines. The central k-space region was apodized by a Kaiser window of parameter 4 [42].

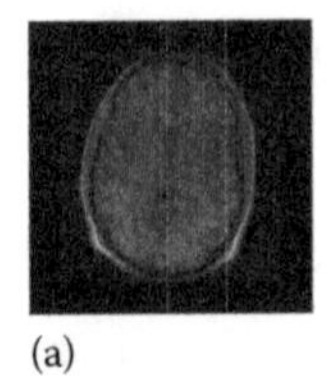
(a)

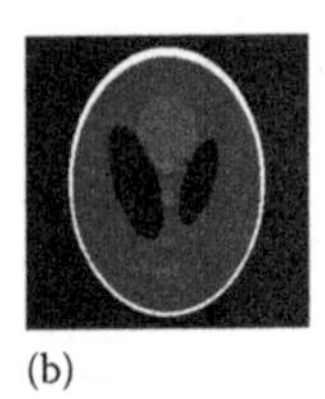
(b)

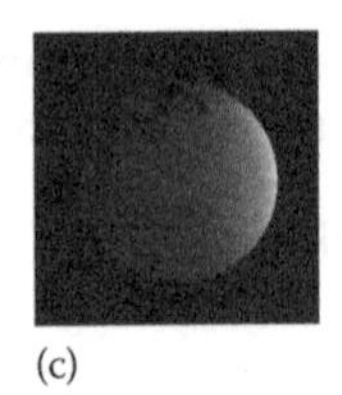
(c)

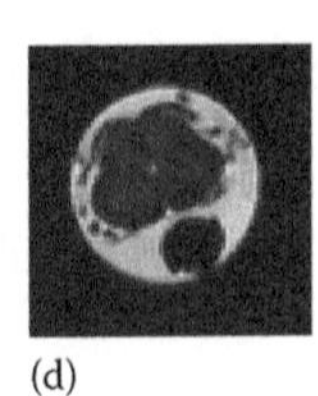
(d)

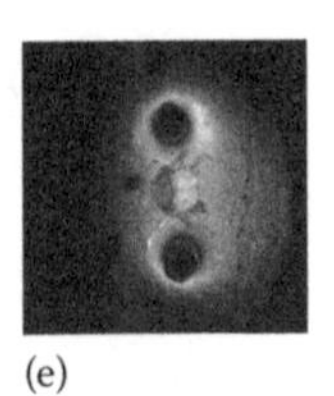
(e)

FIGURE 18.6
Ground-truth images: (a) brain, (b) Shepp–Logan phantom, (c) UBC phantom, (d) rat ex vivo, and (e) rat in vivo.

- The corresponding low-frequency images were computed by using the 2D inverse Fourier transform.
- The low-frequency images for each coil were normalized by dividing them by the sum-of-squares image. These were used as the initial estimates of sensitivity maps.

Our method needs only two parameters (ε_1 and ε_2 in Algorithm 2) to be specified; the values of ε_1 and ε_2 had been fixed at 1 and 10, respectively. The choice behind the particular parametric values can be understood intuitively. From previous studies, it is known that by using CS-based and nuclear norm regularization-based techniques, it is possible to estimate the image to a high degree of accuracy. Therefore, while estimating the image, the data fidelity term (ε_1) is fixed at a small value. But estimating the sensitivity map to a high degree of accuracy is difficult. For this reason, the data fidelity term (ε_2) during sensitivity map estimation is fixed at a larger value (compared to ε_1).

In [45], it is argued that the sensitivity maps should be normalized by their combined sum-of-squares map. The proposed *i*SENSE method models the sensitivity maps as rank-deficient matrices; thus, the estimated sensitivity maps are not normalized. During actual implementation, once the sensitivity maps are estimated, they are normalized by their sum-of-squares map before being used for image estimation.

We have compared our proposed *i*SENSE method with state-of-the-art reconstruction techniques like SparSENSE [41] and the NNSENSE [40] and a well-known traditional reconstruction technique GRAPPA [35]. All these are single-pass methods and assume either implicitly (GRAPPA) or explicitly (SparSENSE and NNSENSE) the knowledge of sensitivity maps. We also compared our work against the recently developed technique of JSENSE [42]. JSENSE alternately estimates the MR image and the sensitivity maps and is similar in spirit to our proposed *i*SENSE.

It must be noted that SparSENSE and NNSENSE are actually Step 1 of our proposed algorithm. We found that our proposed *i*SENSE converges in three iterations; after which, there is no significant improvement in results.

In this work, we simulate a variable density subsampling scheme where the center of the k-space is fully sampled, while the rest of the k-space is sampled sparsely. The k-space is fully sampled along the frequency encoding (x-axis) direction but is undersampled along the phase encoding direction. For the eight-channel data (brain and Shepp–Logan phantom), the acceleration factor is 4, while for the four-channel data (UBC MRI slices 1 and 7), the acceleration factor is 2. The dense sampling of the center of k-space is required for estimating the sensitivity maps.

First, we show the quantitative experimental results. The measurement for reconstruction accuracy is NMSE between the ground-truth image formed by sum-of-squares reconstruction from the fully sampled k-space data and the magnitude image obtained by reconstruction (by various techniques)

TABLE 18.4

*i*SENSE Reconstruction Accuracy after 1st, 2nd, and 3rd Iterations

Name of Image	Technique	1st Iteration	2nd Iteration	3rd Iteration
Brain	*i*SENSE CS	0.17	0.09	0.05
	*i*SENSE NN	0.17	0.09	0.05
Shepp–Logan phantom	*i*SENSE CS	0.11	0.07	0.05
	*i*SENSE NN	0.09	0.07	0.06
UBC MRI phantom	*i*SENSE CS	0.14	0.10	0.07
	*i*SENSE NN	0.18	0.11	0.08
Rat ex vivo	*i*SENSE CS	0.10	0.06	0.04
	*i*SENSE NN	0.11	0.07	0.04
Rat in vivo	*i*SENSE CS	0.19	0.14	0.10
	*i*SENSE NN	0.19	0.15	0.11

from partially sampled k-space data. For our proposed method *i*SENSE, *i*SENSE CS denotes the case where the image reconstructed by SparSENSE and *i*SENSE NN denotes the case where image is reconstructed by NNSENSE.

The reconstruction accuracy from our proposed method *i*SENSE is shown in Table 18.4. The results are shown for all three iterations.

Next, we compare our proposed technique *i*SENSE against the methods mentioned earlier. It has been reported in [42] that the JSENSE method converges in three to four iterations. We ran the JSENSE algorithm for four iterations; the reconstruction accuracy after the final iteration is reported. It must be noted that after the first iteration of *i*SENSE, the obtained result corresponds to SparSENSE (for the CS-based technique) and NNSENSE (for the nuclear norm regularization-based technique).

The numerical results indicate that our proposed *i*SENSE method is better than the techniques compared against, at least for the datasets used in this work. However, the NMSE is not always the best metric for reconstruction accuracy. Table 18.5 shows that in terms of NMSE, JSENSE is almost as accurate as our proposed *i*SENSE CS and *i*SENSE NN for all the datasets and GRAPPA has similar reconstruction accuracy as our proposed *i*SENSE

TABLE 18.5

Comparison of Reconstruction Accuracy from Various Techniques

Name of Image	*i*SENSE CS	*i*SENSE NN	SparSENSE	NNSENSE	JSENSE	GRAPPA
Brain	0.05	0.05	0.17	0.17	0.10	0.08
Shepp–Logan phantom	0.05	0.06	0.11	0.09	0.09	0.13
UBC MRI phantom	0.07	0.08	0.14	0.18	0.12	0.10
Rat ex vivo	0.04	0.04	0.10	0.11	0.06	0.09
Rat in vivo	0.10	0.11	0.19	0.19	0.13	0.17

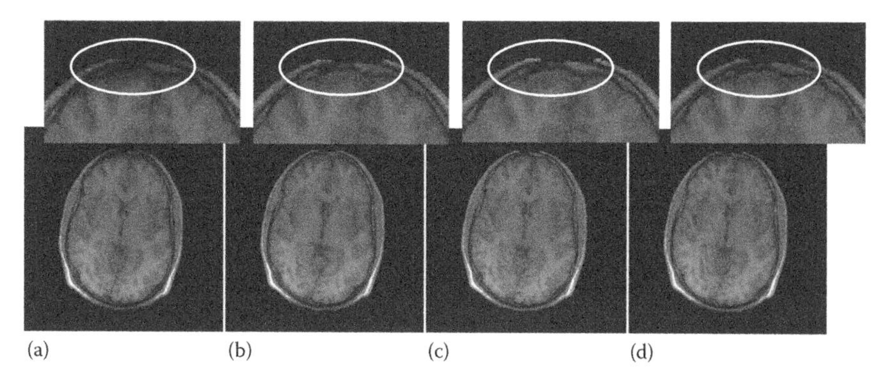

FIGURE 18.7
Brain images reconstructed by *i*SENSE CS: (a) after 1st iteration (SparSENSE), (b) after 2nd iteration, (c) after 3rd (final) iteration, and (d) ground truth.

methods for the brain, Shepp–Logan phantom, and the UBC MRI phantom. We will show in Figures 18.9 through 18.12 that in terms of qualitative reconstruction, our proposed method gives considerably better results than JSENSE and GRAPPA.

Before comparing *i*SENSE with JSENSE and GRAPPA, we will show the benefit of *i*SENSE reconstruction. Especially for clinical MRI, one is interested in preserving minute anatomical details. This is evident from Figures 18.7 and 18.8. Here, we show, for the brain image, how the fine anatomical details are recovered after each iteration of our proposed *i*SENSE technique.

The first image (a) shows the reconstructed image after Step 1. This image (a) corresponds to SparSENSE (Figure 18.7 and NNSENSE (Figure 18.8). The second image (b) shows the reconstructed image after two iterations, i.e., Step 1 → Step 2 → Step 1. The third image (c) shows the reconstructed image

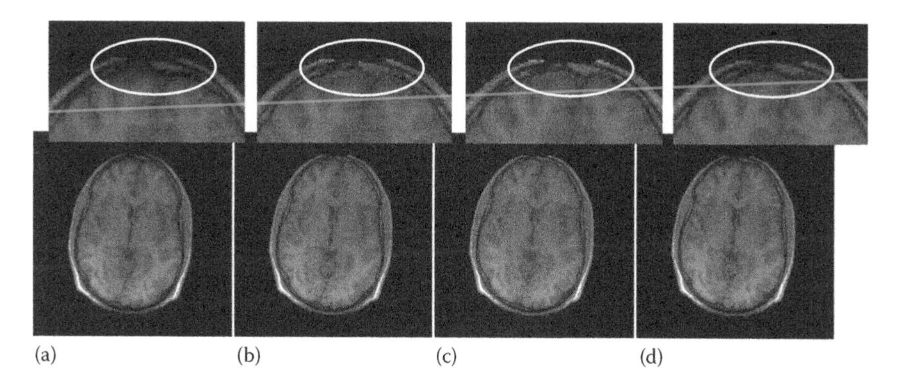

FIGURE 18.8
Brain images reconstructed by *i*SENSE NN: (a) after 1st iteration (NNSENSE), (b) after 2nd iteration, (c) after 3rd (final) iteration, and (d) ground truth.

after three iterations, i.e., Step 1 → Step 2 → Step 1 → Step 2 → Step 1. There is no significant improvement after the third iteration. The fourth image (d) is the ground-truth image.

The top portion of each image has been magnified for better clarity. It is easy to see from the magnified portions that in the SparSENSE and NNSENSE reconstruction, some fine anatomical details are missing (encircled area). After the second iteration, some of the details are captured. The third iteration shows slight improvement over the second iteration. By the third iteration, the anatomical detail captured is as good as the ground truth.

Quantitative results in Table 18.5 do not give a clear idea regarding the qualitative aspects of reconstruction. In Figures 18.9 through 18.12, it is shown how the *i*SENSE method fairs over GRAPPA and JSENSE. We show the reconstructed and difference images for the real datasets (brain, UBC MRI phantom, rat ex vivo, rat in vivo). The difference images are computed by taking the absolute difference between the ground-truth and the reconstructed magnitude image. The difference images have been magnified 10 times for visual clarity.* The reconstruction results from *i*SENSE CS and *i*SENSE NN are almost similar. They do not have any quantitative or discernible differences. In this work, we therefore show the results from *i*SENSE

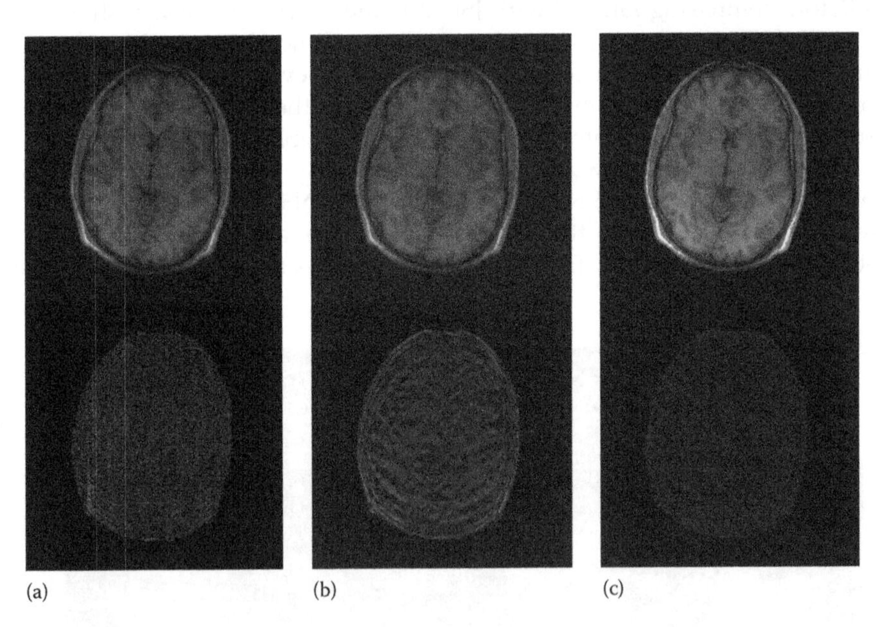

FIGURE 18.9
Reconstructed (1st row) and difference (2nd row) images of brain. (a) GRAPPA, (b) JSENSE, and (c) *i*SENSE.

* The images, especially the difference images, should be viewed on a computer screen. The contrast resolution required for observing the fine variations is not fulfilled by most printers.

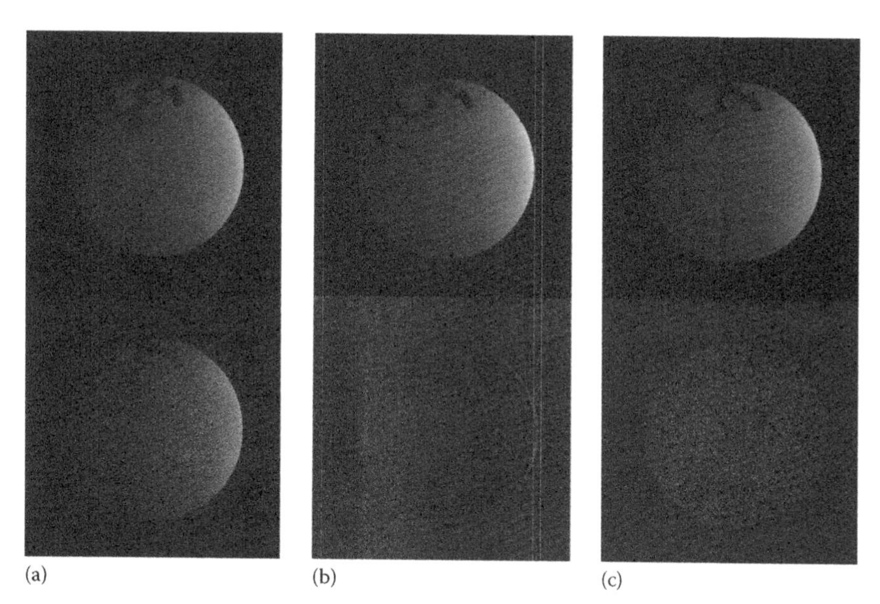

FIGURE 18.10
Reconstructed (1st row) and difference (2nd row) images of UBC MRI phantom 1. (a) GRAPPA, (b) JSENSE, and (c) *i*SENSE.

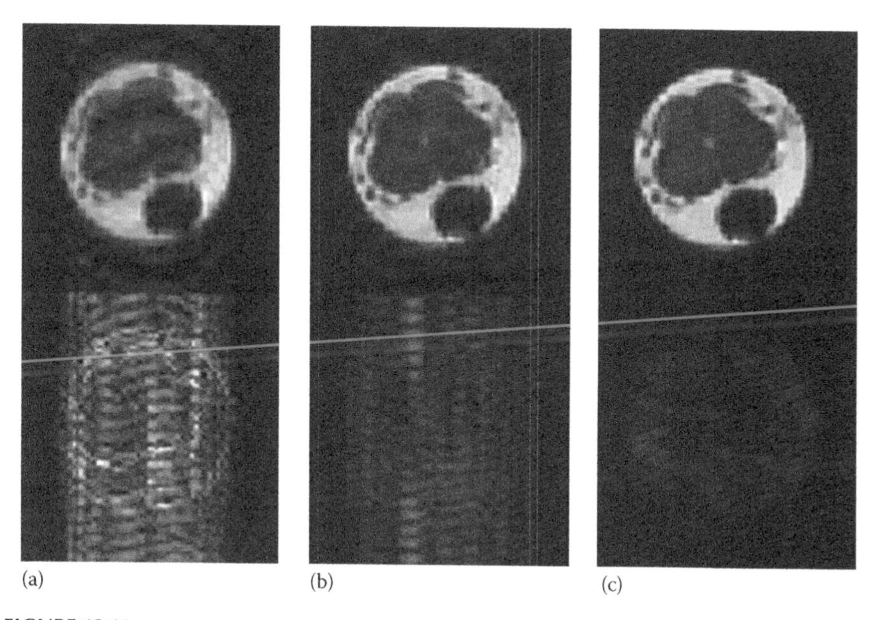

FIGURE 18.11
Reconstructed (1st row) and difference (2nd row) images of rat ex vivo 1. (a) GRAPPA, (b) JSENSE, and (c) *i*SENSE.

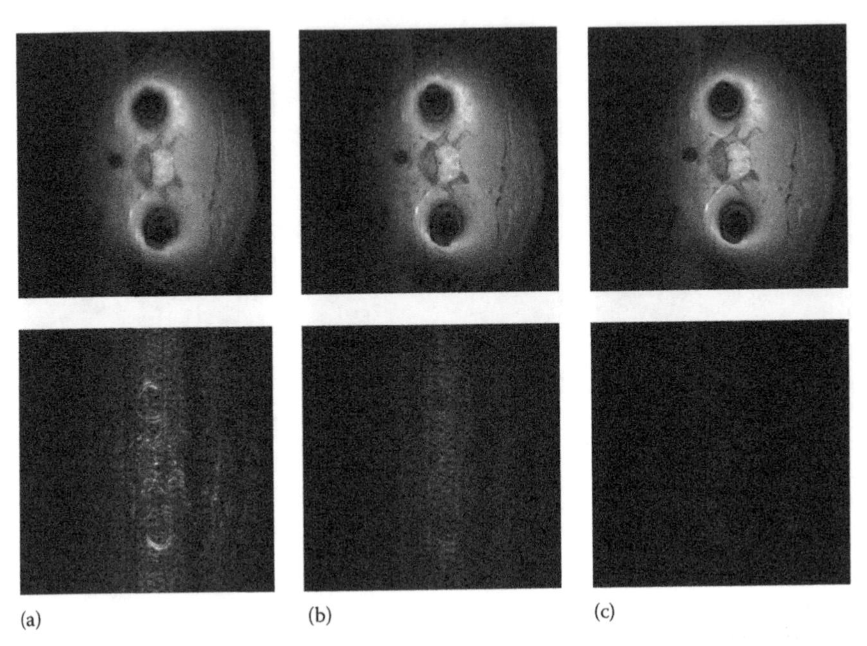

FIGURE 18.12
Reconstructed (1st row) and difference (2nd row) images of rat in vivo 1. (a) GRAPPA, (b) JSENSE, and (c) *i*SENSE.

CS only. We show the results for *i*SENSE CS, JSENSE, and GRAPPA, since the reconstruction results for SparSENSE and NNSENSE have already been shown in Figures 18.7 and 18.8.

Both GRAPPA and JSENSE reconstruction methods show visible aliasing artifacts for all the images. This is especially discernible from the difference images. Our proposed *i*SENSE yields the best reconstruction results for all the images with very little reconstruction artifacts (the difference image is almost dark in all cases).

In most cases, the reconstruction times for static MRI reconstruction are not at a premium. However, it is worthwhile to compare the speed of the reconstruction algorithms especially when the *i*SENSE NN gives about an order of improvement in speed over the *i*SENSE CS-based method. The reconstruction times are shown in Table 18.6. The results reported here are at par with results from previous studies [30,41], where it was shown that

TABLE 18.6
Reconstruction Times (Rounded in Seconds) for *i*SENSE CS and *i*SENSE NN

Method Used	Brain	Shepp–Logan Phantom	UBC MRI Phantom	Rat Ex Vivo	Rat In Vivo
*i*SENSE CS	415	230	240	242	279
*i*SENSE NN	35	23	23	24	28

NN regularization yields about an order of magnitude faster reconstruction compared to CS-based techniques.

Both the methods *i*SENSE CS and *i*SENSE NN recover images with high accuracy in terms of quantitative and qualitative evaluation. However, given the present challenges of using low-rank techniques in 3D reconstruction, the CS-based reconstruction technique is more practical.

18.4.2 Calibration-Free Parallel MRI Reconstruction

All previous multicoil parallel MR image reconstruction algorithms require some form of parameter estimation pertaining to the sensitivity profiles. The image domain methods require the estimate of the sensitivity profiles from the calibration data, while the frequency domain methods require computing the linear interpolation weights from the calibration data. In both cases, the assumption is that the estimates based on the calibration data should hold for the unknown image (to be reconstructed) as well. For image domain methods, there are different ways to compute the sensitivity profiles, and the reconstruction results are dependent on the technique used to compute them. For frequency domain methods, besides the calibration data, the interpolation weights also depend on the calibration kernel. In short, all known methods for multicoil parallel MRI are dependent on their respective parameter estimation stage. In the section on Experimental Evaluation, we will show that the well-known image domain and frequency domain methods are indeed sensitive to the results of the calibration stage.

In order to alleviate the sensitivity issues associated with calibration, we propose a solution that does not require calibration in any form. Our method is named CaLM (Calibration-Less Multicoil) MRI [55].

We rewrite the physical data acquisition model (18.19) for multicoil MRI:

$$y_i = RFS_i x + \eta_i, \quad i = 1 \ldots C$$

This can be expressed in a slightly different fashion as

$$y_i = RFx_{S_i} + \eta_i, \quad i = 1 \ldots C \tag{18.33}$$

where $x_{S_i} = S_i x$ are the sensitivity-encoded images for each coil.

Incorporating the wavelet transform into (18.33), we get

$$y_i = RFW^T \alpha_{S_i} + \eta_i, \quad i = 1 \ldots C \tag{18.34}$$

In compact matrix–vector notation, (18.34) can be expressed as

$$y = E_W \alpha_S + \eta \tag{18.35}$$

where

$$y = \begin{bmatrix} y_1 \\ \dots \\ y_C \end{bmatrix}, \quad E_W = \begin{bmatrix} RFW^T & 0 & 0 \\ 0 & \dots & 0 \\ 0 & 0 & RFW^T \end{bmatrix}, \quad \alpha = \begin{bmatrix} \alpha_{S_1} \\ \dots \\ \alpha_{S_C} \end{bmatrix} \text{ and } \eta = \begin{bmatrix} \eta_1 \\ \dots \\ \eta_C \end{bmatrix}$$

The problem is to solve the inverse problem (18.35). This is an underdetermined system of equations (the length of α_{S_i} is the same as the total number of pixels in the image, but the length of y_i is less than the number of pixels as the k-space is undersampled) unlike SENSE or CS SENSE where the system is overdetermined. Thus, solving (18.35) requires the conditions of CS recovery to hold. These conditions are as follows:

1. The solution α_S is sparse in some sense.
2. The matrix E_W should follow the restricted isometric property (RIP) [23].

It is easy to argue that the α_S (wavelet coefficients) will be sparse. But we will argue that α_S will also be group sparse. In a group-sparse vector, all the coefficients are segregated into groups. Group sparsity means that in that vector, only a few groups will have nonzero coefficients, and the rest will be all zeros. Theoretically, group sparsity leads to stronger reconstruction accuracy compared to ordinary sparsity [46,47]. It is not possible to ensure that E_W will satisfy RIP. However, the sampling pattern can be designed to yield good results and will be discussed in the next subsection.

18.4.2.1 Group-Sparse Reconstruction

The wavelet transform encodes the discontinuities (singularities) in the images. The wavelet transform coefficients corresponding to the discontinuities are of high value, while those related to smooth areas are of low values (zero or near about zero). MR images are mostly smooth with a small number of discontinuities, and thus the wavelet coefficients of MR images are sparse.

In all previous parallel MRI studies, the underlying assumption was that the sensitivity profile is smooth. This is the reason why it can be computed from the low-frequency components of the k-space or by polynomial curve fitting. Since the sensitivity profile is smooth, it does not introduce jump discontinuities when it represents the underlying image; neither does it get rid of any existing discontinuities. Thus, the discontinuities that existed in the original MR image x also exist in the sensitivity-encoded images x_{S_i}.

This does not imply that the levels of discontinuities in the sensitivity-encoded image would be the same as the original image, but the locations of the discontinuities will be the same. This in turn implies that the wavelet coefficients of the original image will be of similar values to those of the sensitivity-encoded images at corresponding positions; that is, if for a particular index the wavelet coefficient is large in the original MR image (x), then the wavelet coefficients of the sensitivity-encoded images (x_{S_i}) will be high at the corresponding index. Similarly, for low-valued wavelet coefficients, the correspondence holds as well.

In summary, the wavelet coefficients corresponding to different sensitivity-encoded images will have similar values at corresponding positions/indices.

Now, the full wavelet coefficient vector α_S in (18.35) is usually expressed as

$$\left\{\underbrace{\alpha_{S_1}(1),\alpha_{S_1}(2),\ldots,\alpha_{S_1}(N)}_{\alpha_{S_1}}\ \underbrace{\alpha_{S_2}(1),\alpha_{S_2}(2),\ldots,\alpha_{S_2}(N)}_{\alpha_{S_2}}\ \cdots\ \underbrace{\alpha_{S_C}(1),\alpha_{S_C}(2),\ldots,\alpha_{S_C}(N)}_{\alpha_{S_C}}\right\}^T \quad (18.36)$$

where N is the total number of wavelet coefficients for each image. This wavelet coefficient vector α_S can also be grouped according to the position/index of the wavelet coefficients as follows:

$$\left\{\underbrace{\alpha_{S_1}(1),\alpha_{S_2}(1),\ldots,\alpha_{S_C}(1)}_{\alpha(1)}\ \underbrace{\alpha_{S_1}(2),\alpha_{S_2}(2),\ldots,\alpha_{S_C}(2)}_{\alpha(2)}\ \cdots\ \underbrace{\alpha_{S_1}(N),\alpha_{S_2}(N),\ldots,\alpha_{S_C}(N)}_{\alpha(N)}\right\}^T \quad (18.37)$$

It is concluded that in (18.37), the coefficients in each of the groups ($\alpha(i)$) will have similar values, i.e., the coefficients in any group will all be of high values or all be of very small values (zeroes or close to zero). In other words, α_S will be group sparse.

18.4.2.1.1 Synthesis Prior Formulation

The inverse problem (18.35) should be solved by taking into account the fact that the solution is group sparse. A group-sparse solution $\hat{\alpha}_S$ can be obtained by solving the following $l_{2,1}$ mixed-norm optimization problem [47–49]:

$$\hat{\alpha}_S = \min_{\alpha_S} \|\alpha_S\|_{2,1} \text{ such that } \|y - E_W\alpha_S\|_2^2 \leq \sigma \quad (18.38)$$

where the $l_{2,1}$ mixed-norm is defined as $\|v\|_{2,1} = \sum_{j=1}^{N}\left(\sum_{k=1}^{C} v_j^2(k)\right)^{1/2}$ and there are N groups and C coefficients in each group.

The l_2 norm $\left(\sum_{k=1}^{C} v_j^2(k)\right)^{1/2}$ over the group of coefficients enforces the selection of the entire group of coefficients (j), whereas the summation over the l_2-norm enforces group sparsity, i.e., the selection of only a few groups.

18.4.2.1.2 Analysis Prior Formulation

The formulation (18.38) solves for the wavelet transform coefficients of the image and is called the synthesis prior formulation. Following our previous study in single-channel MRI reconstruction [22], we propose to solve (18.35) with analysis prior constraints, since analysis prior is known to give better results as seen in Section 18.3.1 and in [21–23].

For framing the analysis prior group sparsity problem, we require expressing (18.34) in the following form:

$$y = F_D x_S + \eta \tag{18.39}$$

where

$$y = \begin{bmatrix} y_1 \\ \cdots \\ y_C \end{bmatrix}, \quad F_D = \begin{bmatrix} RF & 0 & 0 \\ 0 & \cdots & 0 \\ 0 & 0 & RF \end{bmatrix}, \quad x = \begin{bmatrix} x_{S_1} \\ \cdots \\ x_{S_C} \end{bmatrix} \quad \text{and} \quad \eta = \begin{bmatrix} \eta_1 \\ \cdots \\ \eta_C \end{bmatrix}$$

The group-sparse analysis prior reconstruction proceeds by solving the following optimization problem:

$$\hat{x}_S = \min_{x_S} \left\| W_D x_S \right\|_{2,1} \text{ such that } \left\| y - F_D x_S \right\|_2^2 \le \sigma \tag{18.40}$$

where

$$W_D = \begin{bmatrix} W & 0 & 0 \\ 0 & \cdots & 0 \\ 0 & 0 & W \end{bmatrix}$$

Analysis prior is not widely used in CS applications. Hence, there is no out-of-the-box algorithm to solve (18.40). We propose solving the analysis prior group-sparse problem using the majorization–minimization approach. The algorithm for solving the analysis prior group-sparse optimization problem has been derived in [51].

18.4.2.2 Experimental Evaluation

The reconstruction accuracy from the method described earlier (CaLM MRI) is dependent on the sampling trajectory used. Unlike SENSE-based approaches, the proposed formulation requires solving an underdetermined system of equations. The accuracy of the reconstruction is heavily dependent

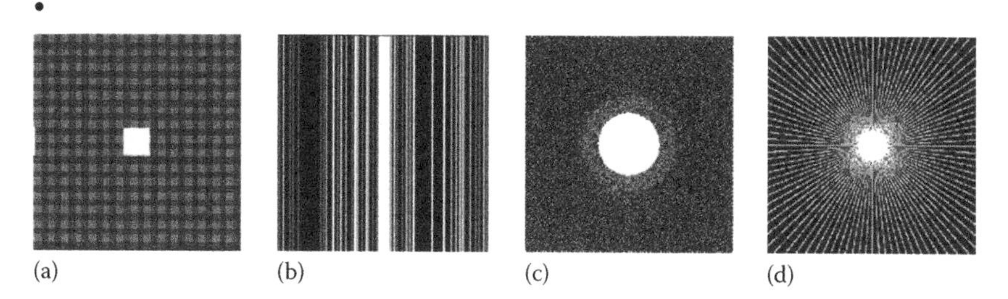

FIGURE 18.13
(a) Uniform sampling (26%), (b) random sampling (25%), (c) Gaussian sampling (20%), and (d) radial sampling (23%).

on the RIP concept in CS [3]. Thus, in this chapter, we experiment with four different sampling schemes (Figure 18.13):

1. Periodic undersampling of k-space
2. Random omission of k-space lines in vertical direction
3. Gaussian random sampling of k-space
4. Radial sampling

Periodic undersampling is traditionally the most often used parallel MRI acceleration method. However, it is unsuitable for CS reconstruction. As we will see in the results section, it yields the worst results for CaLM MRI.

Randomly skipping k-space lines in the vertical direction has been used previously for CS-based MRI [18]. This gives better results than periodic undersampling.

Gaussian sampling yields extremely good results. In general, such sampling points cannot be efficiently acquired. However, when undersampling both phase-encoded directions in 3D scans, they can be efficiently acquired in two dimensions. Such sampling has been used in the past [52,53].

Radial sampling is by far the best suited practical and efficient sampling scheme for CS-based MR image reconstruction techniques. It is very fast and could be used for real-time MR data acquisition and is robust to object motion and ghosting artifacts.

There are four sets of ground-truth data used for our experimental evaluation (Figure 18.14). The brain data and the Shepp–Logan phantom have been used previously in [54]. The brain data come from a fully sampled T1-weighted scan of a healthy volunteer. The volunteer was scanned using spoiled gradient echo sequence with the following parameters—echo time = 8 ms, repetition time = 17.6 ms, and flip angle = 20°. The scan was performed by a GE Signa EXCITE 1.5 T scanner, using an eight-channel receiver coil. The eight-channel data for Shepp–Logan phantom were simulated. The UBC MRI Lab prepared a phantom by doping saline with Gd-DTPA. An MRI scanner with four-channel RF coil was used for acquiring the data using

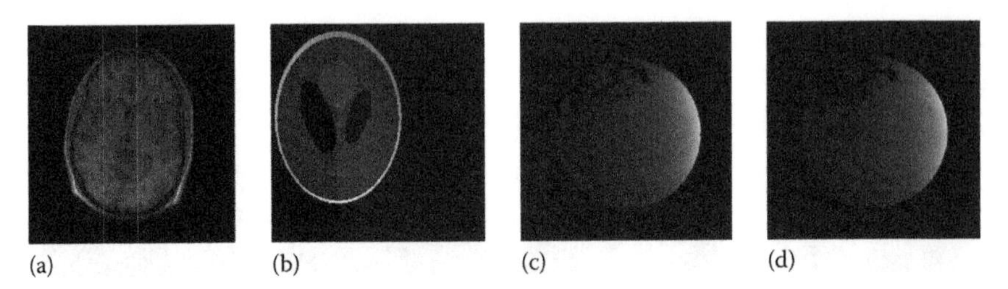

FIGURE 18.14
Ground-truth images: (a) brain, (b) Shepp–Logan phantom, (c) UBC MRI phantom slice 1, and (d) UBC MRI phantom slice 7.

a FLASH sequence. In this work, we have selected slices one and seven for experimentation. The ground truth is formed by sum-of-squares reconstruction of the multichannel images.

We compare the proposed CaLM MRI with three state-of-the-art autocalibrated methods—GRAPPA, l1SPIRiT, and CS SENSE. For non-Cartesian sampling, the GRAPPA operator gridding (GROG) method is used instead of GRAPPA. For CS SENSE and CaLM MRI, the mapping from non-Cartesian k-space to the Cartesian image space is the nonuniform fast Fourier transform (NUFFT).

For the CS SENSE and the proposed CaLM MRI, we tried orthogonal and redundant versions of Haar, Daubechies wavelets, and orthogonal fractional spline. But we found the best results were obtained for complex dual-tree wavelets. This has a small redundancy (factor of two) and has proved to yield good results on MR images in the past [22].

For CS SENSE, the sensitivity profiles are estimated in the fashion shown in [13]. A window at the center of the k-space is densely sampled, from which a low-resolution image for each coil is obtained. These images are combined by sum of squares. The sensitivity map is computed by dividing the low-resolution image of the corresponding coil by the combined sum-of-squares image.

The two objectives in this work are as follows:

1. To show that the reconstruction from existing parallel MRI techniques (CS SENSE, GRAPPA, and SPIRiT) is sensitive to the calibration stage. CaLM MRI however is not dependent on the calibration stage.
2. To show that our proposed CaLM MRI yields results that are at par with the best results obtained from state-of-the-art parallel MRI techniques—CS SENSE, GRAPPA, and SPIRiT.

For CS SENSE, the reconstruction accuracy is sensitive to the size of the calibration window, i.e., the size of the window used for densely sampling the center of the k-space. GRAPPA and SPIRiT are sensitive to the size of the calibration kernel and the calibration window. Here, the sizes of the calibration window and the calibration kernel are varied to show that the existing parallel MRI reconstruction methods are sensitive to the calibration stage.

In the following, we first show the quantitative reconstruction results in Tables 18.7 through 18.10. The measure for reconstruction accuracy is the NMSE. Tables 18.7 through 18.9 show how the reconstruction accuracies from GRAPPA, l1SPIRiT, and CS SENSE are sensitive to their corresponding calibration stages. Table 18.10 tabulates results for CaLM MRI.

Tables 18.7 and 18.8 show the reconstruction results by the frequency domain methods—GRAPPA and l1SPIRiT. The results are plotted for three different calibration configurations and four different sampling schemes.

The first configuration is the default configuration as suggested in [54]. The calibration uses the maximum possible samples from the center of the k-space, and the kernel size is 5 × 5. The second configuration has a window size of 16 × 16 and a kernel size of 3 × 3, and the third configuration has a window size of 32 × 32 and a kernel size of 9 × 9. In the following tables, the configuration is written as window size and kernel size, i.e., 16, 3 denote a window size of 16 and kernel size of 3.

In Table 18.9, the reconstruction results from CS SENSE are tabulated. For CS SENSE, the sensitivity profile needs to be estimated/calibrated. In SENSE, the size of the calibration window determines the sensitivity map. Here, we show the results for three window sizes—16 × 16, 32 × 32, and 64 × 64.

TABLE 18.7

NMSE for GRAPPA Reconstruction

	Brain			Phantom			Slice 1			Slice 7		
Sampling Scheme	**Max, 5**	**16, 3**	**32, 9**	**Max, 5**	**16, 3**	**32, 9**	**Max, 5**	**16, 3**	**32, 9**	**Max, 5**	**16, 3**	**32, 9**
Uniform	0.06	0.04	0.06	0.02	0.06	0.04	0.09	0.12	0.07	0.09	0.13	0.09
Random	0.04	0.06	0.07	0.23	0.32	0.32	0.10	0.12	0.10	0.11	0.14	0.11
Gaussian	0.14	0.11	0.15	0.05	0.03	0.11	0.05	0.09	0.06	0.05	0.09	0.05
Radial	0.14	0.11	0.25	0.10	0.05	0.15	0.08	0.11	0.07	0.08	0.12	0.07

TABLE 18.8

NMSE for l1SPIRiT Reconstruction

	Brain			Phantom			Slice 1			Slice 7		
Sampling Scheme	**Max, 5**	**16, 3**	**32, 9**	**Max, 5**	**16, 3**	**32, 9**	**Max, 5**	**16, 3**	**32, 9**	**Max, 5**	**16, 3**	**32, 9**
Uniform	0.06	0.06	0.08	0.03	0.11	0.15	0.07	0.11	0.07	0.06	0.11	0.08
Random	0.13	0.15	0.16	0.21	0.41	0.32	0.15	0.19	0.13	0.17	0.20	0.13
Gaussian	0.07	0.05	0.07	0.06	0.06	0.07	0.03	0.06	0.05	0.03	0.07	0.05
Radial	0.07	0.09	0.07	0.09	0.08	0.11	0.03	0.06	0.05	0.04	0.07	0.06

TABLE 18.9

NMSE for CS SENSE Reconstruction

Sampling Scheme	Brain			Phantom			Slice 1			Slice 7		
	16 × 16	32 × 32	64 × 64	16 × 16	32 × 32	64 × 64	16 × 16	32 × 32	64 × 64	16 × 16	32 × 32	64 × 64
Uniform	0.09	0.14	0.18	0.04	0.06	0.07	0.14	0.10	0.18	0.13	0.11	0.20
Random	0.40	0.43	0.63	0.44	0.19	0.39	0.39	0.32	0.43	0.39	0.30	0.44
Gaussian	0.16	0.19	0.21	0.04	0.06	0.07	0.11	0.08	0.15	0.11	0.08	0.15
Radial	0.28	0.19	0.25	0.04	0.07	0.07	0.11	0.08	0.14	0.10	0.08	0.13

TABLE 18.10

NMSE for CaLM MRI Reconstruction

Sampling Scheme	Synthesis Prior				Analysis Prior			
	Brain	**Phantom**	**Slice1**	**Slice7**	**Brain**	**Phantom**	**Slice1**	**Slice7**
Uniform	0.15	0.32	0.21	0.20	0.12	0.22	0.18	0.15
Random	0.18	0.21	0.11	0.12	0.13	0.21	0.12	0.12
Gaussian	0.08	0.02	0.06	0.06	0.07	0.02	0.04	0.05
Radial	0.10	0.02	0.05	0.05	0.07	0.01	0.05	0.06

For GRAPPA and l1SPIRiT, the average variation of reconstruction error is about 15% of the mean. For CS SENSE, the variation in reconstruction error is larger; it is about 25% of the mean. Moreover, there is not a single configuration that yields the best reconstruction results consistently.

Our proposed method CaLM MRI does not require any calibration. In the following table, the results from CaLM MRI method are shown. We show the results for both the synthesis and the analysis prior.

Even though we have shown the results for the analysis and the synthesis priors, we had already argued that the results from the analysis prior are expected to yield better results. This is in line with our previous work [22]. As mentioned before, CaLM MRI does not show good results for uniform undersampling. For the other Gaussian and radial sampling schemes, CaLM MRI method is at par with or even better than the best reconstruction results from GRAPPA, l1SPIRiT, and CS SENSE.

For visual qualitative evaluation, we provide the reconstructed brain images. Radial and Gaussian samplings show good results for all the reconstruction methods. For GRAPPA, l1SPIRiT, and CS SENSE, the images corresponding to the best configuration are shown. Since the UBC MRI slices are similar in nature and have very close reconstruction accuracies, we only show results for slice 1 during qualitative evaluation in Figures 18.15 and 18.16.

l1SPIRiT yields by far the best reconstruction results in terms of visual assessment. Both GRAPPA and CS SENSE show reconstruction artifacts. The reconstructed images are grainy (much like taking a picture with a digital camera in low light with high film speed). CaLM MRI (both analysis and synthesis prior) does not produce grainy reconstruction artifacts. For the synthetic Shepp–Logan phantom and the UBC MRI slices, our method is as good as l1SPIRiT. But for the anatomical image of the brain, CaLM MRI method produces a slightly smooth reconstruction.

The results from CaLM MRI are at par with the best results (at least for the data available with us) obtained from state-of-the-art multicoil parallel MR image reconstruction. The main advantage of our CaLM MRI method is that it does not require any calibration, unlike the other methods. This is a significant advantage since it is known that the parametric reconstruction methods are sensitive to the calibration stage.

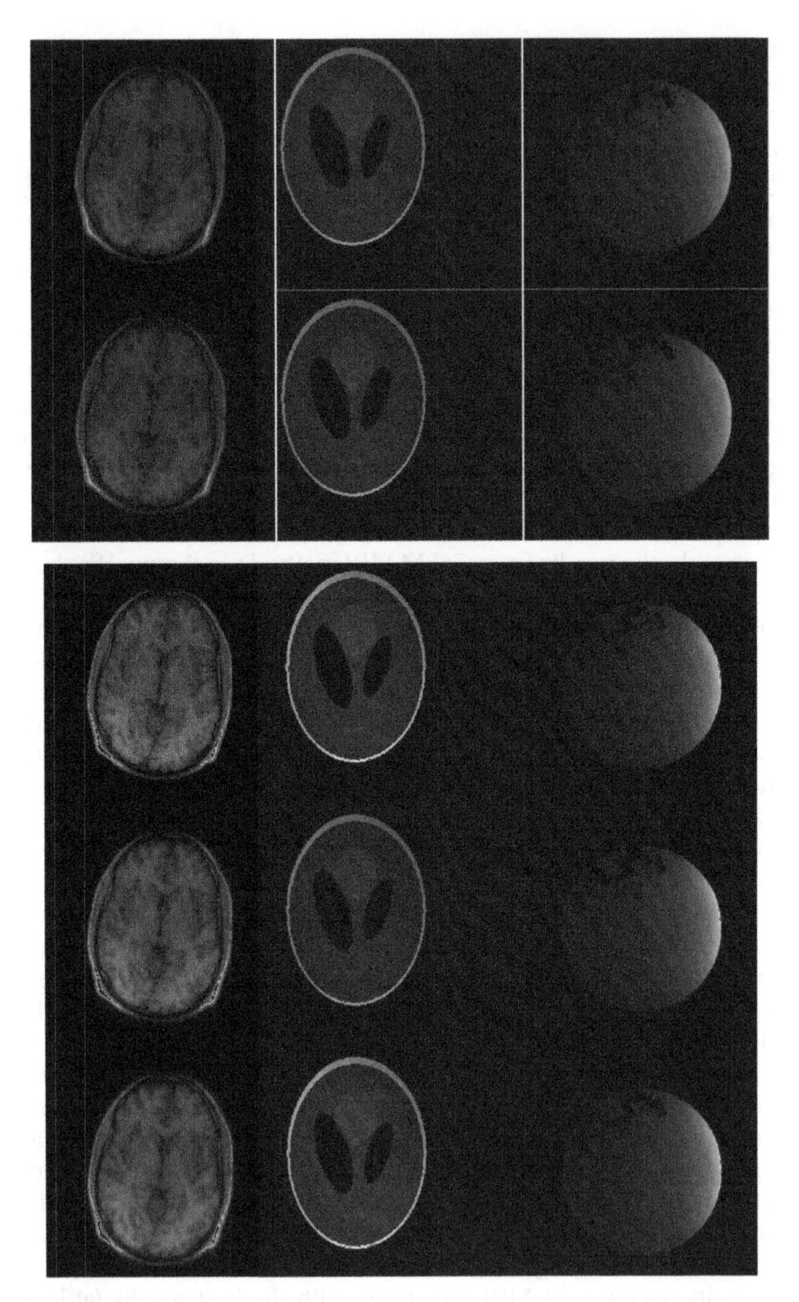

FIGURE 18.15
Reconstructed images from radial sampling. Left to right: brain, Shepp–Logan phantom, and UBC MRI slice 1; top to bottom: l1SPIRiT, GRAPPA, CS SENSE, CaLM MRI synthesis, and CaLM MRI analysis.

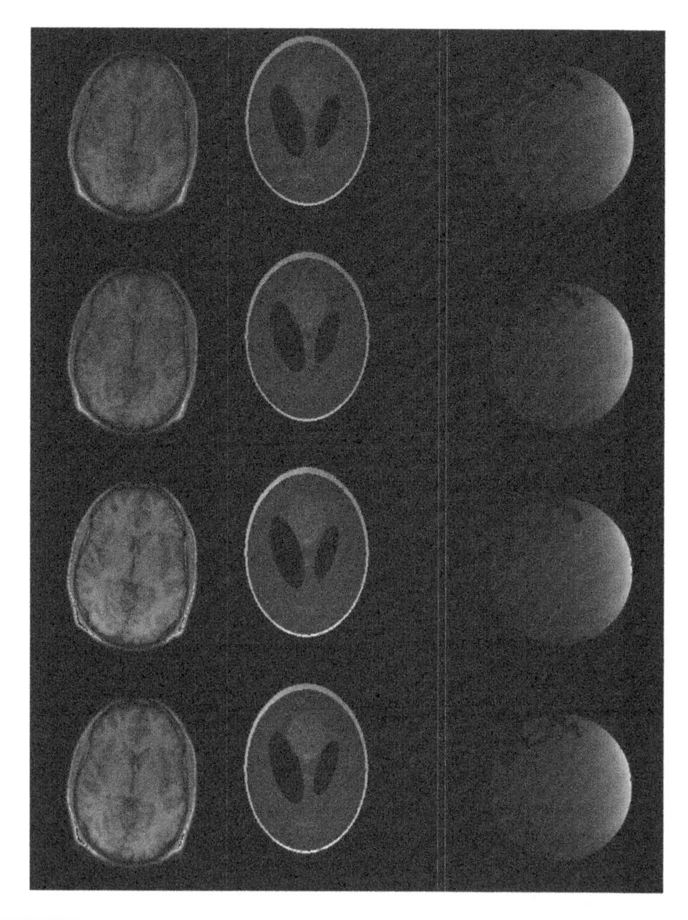

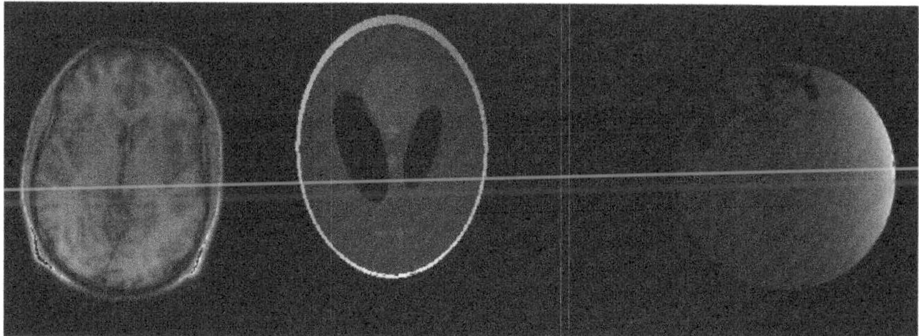

FIGURE 18.16
Reconstructed images from Gaussian sampling. Left to right: brain, Shepp–Logan phantom, and UBC MRI slice 1; top to bottom: l1SPIRiT, GRAPPA, CS SENSE, CaLM MRI synthesis, and CaLM MRI analysis.

18.5 Conclusion

This chapter discusses some recent techniques in accelerating single-channel and multichannel MRI scans. Most of these techniques are based on the development of a newly developed branch of signal processing called CS. The main idea is to collect less k-space samples than is traditionally acquired and to use a smart reconstruction method to obtain good-quality images. CS-based techniques exploit the spatial redundancy of the images in order to reconstruct them from subsampled k-space measurements. The fundamentals of CS are briefly discussed in Section 18.2.

In Section 18.3.1, we discuss the problem of single-channel MRI reconstruction. There are different CS formulations that can be used for image recovery. The problem is that different researchers claim the superiority of their own method over others. This becomes a source of confusion to the practitioner who does not know which method to use. We review the different CS-based techniques that are used for MR image recovery and evaluate them via thorough experimentation. We find that the analysis prior formulation yields the best reconstruction results.

CS-based methods exploit the spatial redundancy of an MR image in terms of its transform domain sparsity. In Section 18.3.2, we showed that instead of exploiting the sparsity (in a transform domain), it is possible to recover the image by exploiting its rank deficiency. We showed that this technique yields the same reconstruction accuracy as CS-based methods but has about an order of magnitude faster reconstruction time.

Section 18.4 discusses the problem of multichannel parallel MRI reconstruction. In the parallel MRI problem, both the image and the sensitivity maps are unknown. However, joint solution of the two leads to an intractable problem. All parallel MRI reconstruction algorithms assume that the sensitivity maps either are provided or can be estimated from the data itself. Traditionally, the SENSE is an optimal method for parallel MRI reconstruction when the sensitivity maps are known. The sensitivity maps are either estimated from calibration scans or estimated from the actual data itself. Both of these estimation techniques are error prone, and the error introduced during the sensitivity map estimation unfortunately affects the final reconstructed image.

In Section 18.4.1, we discuss the variations of the basic SENSE technique. After the advent of CS, SENSE reconstruction is regularized by introducing sparsity promoting terms into the optimization problem. This, however, does not lead to any major improvement—the problem introduced from the sensitivity map estimation still remains. In recent studies, it has been shown how the SENSE reconstruction can be made robust against initial errors introduced during the sensitivity map estimation process. These methods iteratively reconstruct the maps and the MR image. They start with the initial estimates of the maps. Given the maps, they reconstruct the image.

In the next iteration, they estimate the maps given the image. This procedure continues till convergence. There are two such studies—JSENSE and *i*SENSE. Both of them are more robust than the SENSE method. The *i*SENSE yields better reconstruction than the JSENSE.

The *i*SENSE and JSENSE though robust are not insensitive to the map estimation errors. In Section 18.4.2, we discuss our proposed CaLM MRI method. This technique does not require any explicit or implicit estimation of the sensitivity maps. It performs better than commercial and state-of-the-art techniques in MRI reconstruction.

References

1. D. Donoho, Compressed sensing, *IEEE Transactions on Information Theory*, 52(4), 1289–1306, 2006.
2. E. Candès, Compressive sampling, *International Congress of Mathematicians*, 3, 1433–1452, 2006.
3. E. Candès, J. Romberg, and T. Tao, Stable signal recovery from incomplete and inaccurate measurements, *Communications on Pure and Applied Mathematics*, 59(8), 1207–1223, 2006.
4. D. L. Donoho, For most large underdetermined systems of linear equations the minimal l1-norm solution is also the sparsest solution, *Communications on Pure and Applied Mathematics*, 59(6), 797–829, 2006.
5. J. Tropp and A. Gilbert, Signal recovery from random measurements via orthogonal matching pursuit, *IEEE Transactions on Information Theory*, 53(12), 4655–4666, 2007.
6. D. Needell and R. Vershynin, Uniform uncertainty principle and signal recovery via regularized orthogonal matching pursuit, *Foundations of Computational Mathematics*, 9(3), 317–334, 2009.
7. T. Blumensath and M. E. Davies, Stagewise weak gradient pursuits, *IEEE Transactions on Signal Processing*, 57(11), 4333–4346, 2009.
8. R. Saab, R. Chartrand, and Ö. Yilmaz, Stable sparse approximation via nonconvex optimization, *IEEE International Conference on Acoustics, Speech, and Signal Processing*, Las Vegas, NV, pp. 3885–3888, 2008.
9. R. Saab and Ö. Yılmaz, Sparse recovery by non-convex optimization—Instance optimality, *Applied and Computational Harmonic Analysis*, 29(1), 30–48, 2010.
10. R. Chartrand and V. Staneva, Restricted isometry properties and nonconvex compressive sensing, *Inverse Problems*, 24(035020), 1–14, 2008.
11. R. Chartrand, Exact reconstruction of sparse signals via nonconvex minimization, *IEEE Signal Processing Letters*, 14, 707–710, 2007.
12. J. Trzasko and A. Manduca, Highly undersampled magnetic resonance image reconstruction via homotopic l0-minimization, *IEEE Transactions on Medical Imaging*, 28(1), 106–121, 2009.
13. K. T. Block, M. Uecker, and J. Frahm, Iterative image reconstruction using a total variation constraint, *Magnetic Resonance in Medicine*, 57(6), 1086–1098, 2007.

14. B. Liu, L. Ying, M. Steckner, J. Xie, and J. Sheng, Regularized sense reorthogonalization, using iteratively refined total variation method, *IEEE International Symposium on Biomedical Imaging*, Washington, DC, pp. 121–124, 2007.
15. S. Nam, M. E. Davies, M. Elad, and R. Gribonval, The co-sparse analysis model and algorithms, *Applied and Computational Harmonic Analysis*, 34(1), 30–56, 2013.
16. M. Elad, P. Milanfar, and R. Rubinstein, Analysis versus synthesis in signal priors, *Inverse Problems*, 23, 947–968, 2007.
17. M. Guerquin-Kern, D. Van De Ville1, C. Vonesch, J.-C. Baritaux, K. P. Pruessmann, and M. Unser, Wavelet-regularized reconstruction for rapid MRI, *IEEE International Symposium on Biomedical Imaging*, Boston, MA, pp. 193–196, 2009.
18. M. Lustig, D. L Donoho, and J. M. Pauly, Sparse MRI: The application of compressed sensing for rapid MR imaging, *Magnetic Resonance in Medicine*, 58(6), 1182–1195, 2007.
19. S. Ma, W. Yin, Y. Zhang, and A. Chakraborty, An efficient algorithm for compressed MR imaging using total variation and wavelets, *IEEE International Conference on Computer Vision and Pattern Recognition*, Anchorage, AK, 2008.
20. R. Chartrand, Fast algorithms for nonconvex compressive sensing: MRI reconstruction from very few data, *IEEE International Symposium on Biomedical Imaging*, Boston, MA, pp. 262–265, 2009.
21. I. W. Selesnick and M. A. T. Figueiredo, Signal restoration with overcomplete wavelet transforms: Comparison of analysis and synthesis priors, *Proceedings of SPIE*, 7446 (Wavelets XIII), 2009.
22. A. Majumdar and R. K. Ward, Under-determined non-Cartesian MR reconstruction, *Medical Image Computing and Computer Assisted Intervention*, Beijing, China, pp. 513–520, 2010.
23. A. Majumdar and R. K. Ward, On the choice of compressed sensing priors: An experimental study, *Signal Processing: Image Communication*, 27(9), 1035–1048, 2012.
24. I. Baeza, J. A. Verdoy, R. J. Villanueva, and J. V. Oller, SVD lossy adaptive encoding of 3D digital images for ROI progressive transmission, *Image and Vision Computing*, 28(3), 449–457, 2010.
25. A. Al-Fayadh, A. J. Hussain, P. Lisboa, and D. Al-Jumeily, An adaptive hybrid image compression method and its application to medical images, *IEEE International Symposium on Biomedical Imaging*, Paris, France, pp. 237–240, 2008.
26. B. Recht, W. Xu, and B. Hassibi, Necessary and sufficient conditions for success of the nuclear norm heuristic for rank minimization, *47th IEEE Conference on Decision and Control 2008*, Cancun, Mexico, pp. 3065–3070, 2008.
27. B. Recht, W. Xu, and B. Hassibi, Null space conditions and thresholds for rank minimization, *Mathematical Programming*, Cary, NC, 2010.
28. B. Recht, M. Fazel, and P. A. Parrilo, Guaranteed minimum rank solutions to linear matrix equations via nuclear norm minimization, *SIAM Review*, 52, 471–501, 2010.
29. K. Dvijotham and M. Fazel, A nullspace analysis of the nuclear norm heuristic for rank minimization, *IEEE International Conference on Acoustics, Speech, and Signal Processing*, Dallas, TX, March 2010.
30. A. Majumdar and R. K. Ward, An algorithm for sparse MRI reconstruction by Schatten p-norm minimization, *Magnetic Resonance Imaging*, 29(3), 408–417, 2011.
31. E. van den Berg and M. P. Friedlander, Probing the Pareto frontier for basis pursuit solutions, *SIAM Journal on Scientific Computing*, 31(2), 890–912, 2008.

32. R. M. Larsen, Lanczos bidiagonalization with partial reorthogonalization, Department of Computer Science, Aarhus University, Technical Report, DAIMI PB-357, September 1998. http://soi.stanford.edu/~rmunk/PROPACK/
33. K. P. Pruessmann, M. Weiger, M. B. Scheidegger, and P. Boesiger, SENSE: Sensitivity encoding for fast MRI, *Magnetic Resonance in Medicine*, 42, 952–962, 1999.
34. D. K. Sodickson and W. J. Manning, Simultaneous acquisition of spatial harmonics (SMASH): Fast imaging with radiofrequency coil arrays, *Magnetic Resonance in Medicine*, 38(4), 591–603, 1997.
35. M. A. Griswold, P. M. Jakob, R. M. Heidemann, M. Nittka, V. Jellus, J. Wang, B. Kiefer, and A. Haase, Generalized autocalibrating partially parallel acquisitions (GRAPPA), *Magnetic Resonance in Medicine*, 47(6), 1202–1210, 2002.
36. M. Blaimer, F. Breuer, M. Mueller, R. M. Heidemann, M. A. Griswold, and P. M. Jakob, SMASH, SENSE, PILS, GRAPPA: How to choose the optimal method, *Topics in Magnetic Resonance Imaging*, 15(4), 223–236, 2004.
37. F. H. Lin, T. Y. Huang, N. K. Chen, F. N. Wang, S. M. Stufflebeam, J. W. Belliveau, L. L. Wald, and K. K. Kwong, Functional MRI using regularized parallel imaging acquisition, *Magnetic Resonance in Medicine*, 54(2), 343–353, 2005.
38. S. Ramani and J. A. Fessler, Parallel MR image reconstruction using augmented Lagrangian methods, *IEEE Transactions on Medical Imaging*, 30(3), 694–706, 2011.
39. B. Liu, K. King, M. Steckner, J. Xie, J. Sheng, and L. Ying, Regularized sensitivity encoding (SENSE) reconstruction using Bregman iterations, *Magnetic Resonance in Medicine*, 61, 145–152, 2009.
40. D. Liang, B. Liu, J. Wang, and L. Ying, Accelerating SENSE using compressed sensing, *Magnetic Resonance in Medicine*, 62(6), 1574–1584, 2009.
41. A. Majumdar and R. K. Ward, Nuclear norm regularized SENSE reconstruction, *Magnetic Resonance Imaging*, 30(2), 213–221, 2012.
42. L. Ying and J. Sheng, Joint image reconstruction and sensitivity estimation in SENSE (JSENSE), *Magnetic Resonance in Medicine*, 57, 1196–1202, 2007.
43. C. Fernández-Granda and J. Sénégas, L1-norm regularization of coil sensitivities in non-linear parallel imaging reconstruction, *International Society for Magnetic Resonance in Medicine 2010*, Stockholm, Sweden, 2010.
44. E. J. Candès and J. Romberg, Sparsity and incoherence in compressive sampling, *Inverse Problems*, 23(3), 969–985, 2007.
45. D. K. Sodickson and C. A. McKenzie, A generalized approach to parallel magnetic resonance imaging, *Medical Physics*, 28(8), 1629–1643, 2001.
46. R. G. Baraniuk, V. Cevher, M. F. Duarte, and C. Hegde, Model-based compressive sensing, arXiv:0808.35 72v5.
47. J. Huang and T. Zhang, The benefit of group sparsity, *Annals of Statistics*, 38(4), 1978–2004, 2010.
48. E. van den Berg and M. P. Friedlander, Theoretical and empirical results for recovery from multiple measurements, *IEEE Transactions on Information Theory*, 56(5), 2516–2527, 2010.
49. A. Majumdar and R. K. Ward, Non-convex group sparsity: Application to color imaging, *IEEE International Conference on Acoustics, Speech, and Signal Processing*, Dallas, TX, pp. 469–472, 2010.
50. A. Majumdar and R. K. Ward, Accelerating multi-echo T2 weighted MR imaging: Analysis prior group sparse optimization, *Journal of Magnetic Resonance*, 210(1), 90–97, 2011.

51. A. Majumdar and R. K. Ward, Compressive color imaging with group sparsity on analysis prior, *IEEE International Conference on Image Processing*, Hong Kong, China, pp. 1337–1340, 2010.
52. R. Chartrand, Nonconvex compressive sensing and re construction of gradient-sparse images: Random vs. tomographic Fourier sampling, *IEEE International Conference on Image Processing (ICIP)*, San Diego, CA, 2008.
53. M. Lustig, D. L. Donoho, and J. M. Pauly, Rapid MR imaging with compressed sensing and randomly under-sampled 3DFT trajectories, *Proceedings of the 14th Annual Meeting of ISMRM*, Seattle, WA, 2006.
54. M. Lustig and J. M. Pauly, SPIRiT: Iterative self- consistent parallel imaging reconstruction from arbitrary k-space, *Magnetic Resonance in Medicine*, 64, 457–471, 2010.
55. A. Majumdar and R. K. Ward, Calibration-less multi-coil MR image reconstruction, *Magnetic Resonance Imaging*, 30(7), 1032–1045, 2012.

19

Brain Connectivity Mapping and Analysis Using Diffusion MRI

Brian G. Booth and Ghassan Hamarneh

CONTENTS

19.1 Diffusion-Weighted Image Acquisition

Diffusion magnetic resonance imaging (dMRI) is a powerful imaging protocol that allows for the assessment of the organization and integrity of fibrous tissue. The imaging works by measuring the diffusion of water molecules within the body. This diffusion is restricted by cell membranes and as such, rates of diffusion are far less in directions perpendicular to fibrous tissue than parallel to the fibers. With enough diffusion measurements along different directions in 3D, we can noninvasively obtain a profile of the diffusion at various points within the imaged subject.

The diffusion profiles obtained from dMRI have had a significant impact on the analysis of neural connectivity within the white matter of the brain. Neural pathways, dubbed fiber tracts, can be traced out using the directional information from the diffusion profiles. This process is known as tractography and, due to noise, motion artifacts, and partial voluming effects, is a computationally difficult problem.

We present here an examination of the current state of tractography and dMRI. In particular, we look at the computational challenges inherent in this area and the open problems that remain.

19.1.1 Biological Basis for Diffusion MRI

The biological basis for dMRI dates back to 1828 when botanist Robert Brown noticed the continuous and random motion of pollen grains suspended in water [29]. What Brown had discovered was later determined to be the motion of water molecules due to thermal agitation [58]. This motion, now known as Brownian motion or diffusion, was later characterized by A. Einstein [41], resulting in Einstein's equation:

$$r^2 = 6dt \tag{19.1}$$

What Einstein's equation characterized was that the square of the average displacement of molecules (r) with a given diffusion rate (d) is proportional to the observation time (t). If we can measure this molecular displacement over a fixed time, we can obtain the diffusion rate of different substances under different conditions.

As the majority of the human body is water [46], the diffusion phenomenon occurs within us as well. While the diffusion process is random, our cell structures can restrict or hinder the motion of water molecules [19]. As such, the diffusion of water molecules in our body depends on the microstructure of our tissues. Fast molecular diffusion occurs within and around a cell as there are few microstructures to inhibit motion. Diffusion through the cell however is slower as the cell membrane and other structures (e.g., myelin sheaths in the brain white matter) restrict molecular motion.

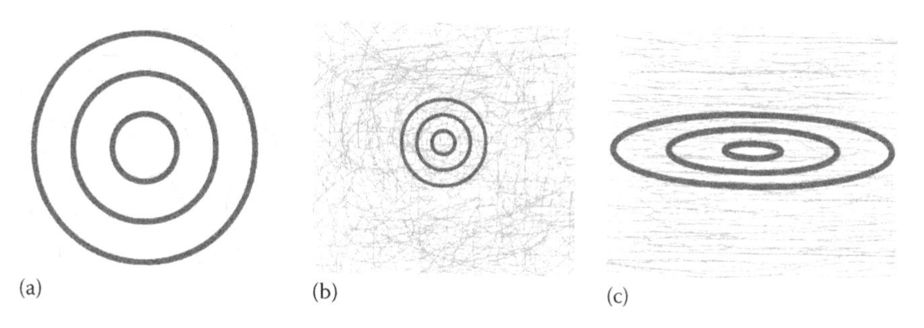

FIGURE 19.1
Synthetic examples of the diffusion seen in CSF (a), gray matter (b), and white matter (c) within the brain. The diffusion rates for various directions are shown in bold. (Adapted from Alexander, D.C., *Visualization and Processing of Tensor Fields*, Chapter 5: an introduction to computational diffusion MRI: the diffusion tensor and beyond, Springer Berlin Heidelberg, Heidelberg, Germany, pp. 83–106, 2006.)

Since the diffusion of water within the body is dependent on local cell structure, we can discuss how different organizations of these structures affect diffusion rates. Consider, for example, the human brain where functional regions (gray matter) are connected by a collection of neural pathways (white matter). Figure 19.1 presents diffusion measures for the brain's corticospinal fluid (CSF), gray matter, and white matter, respectively. When the cell structure is minimal as in CSF, we see fast isotropic diffusion. More complex cell structure that is not consistently organized, such as gray matter, shows slower, but still isotropic, diffusion. Yet if the local cell structure is organized in a consistent orientation, as it is in white matter, the diffusion rates become anisotropic, that is, they vary with regard to direction [19].

These diffusion differences within the brain are potentially useful cues in analyzing brain structure and function. For example, measuring the average diffusion rate or the anisotropy of a tissue can give us significant information about the tissue's organization and integrity [76]. Diffusion measurements would be most informative in white matter regions where the orientation of the microstructure can be inferred from the diffusion. This microstructural orientation within the brain's white matter is in turn known to describe the direction of neural pathways in the brain [20]. As a result, by measuring the diffusion using Einstein's equation, we could infer the orientational structure of the brain's white matter and in turn map out the brain's neural pathways. This is precisely what dMRI is used to accomplish.

19.1.2 Diffusion-Weighted Image Acquisition

To understand how diffusion can be measured through MRI, we must first address the basic concepts on nuclear magnetic resonance. All elementary

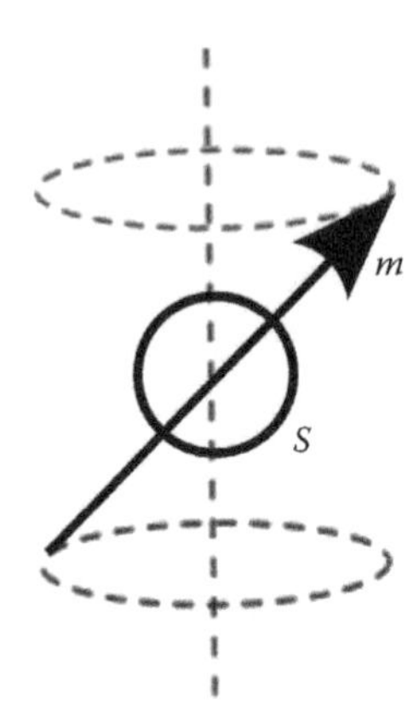

FIGURE 19.2
Nuclear spin s generating a magnetic moment m. The particle spins around a rotational axis shown here in gray. (Adapted from Lenglet, C. et al., Diffusion tensor magnetic resonance imaging: Brain connectivity mapping. Technical Report 4983, INRIA, October 2003.)

particles possess a physical property of spin (s) as seen in Figure 19.2. This spin property rotates the particle around its nucleus, thereby giving the particle a magnetic moment (m). This magnetic moment can then be manipulated using nuclear magnetic resonance. As the body is mostly water, the spins of hydrogen atoms within water molecules become a good candidate for MRI.

19.1.2.1 Magnetic Resonance Imaging

MRI is comprised of three principal steps: precession, resonance, and relaxation. We consider each in turn.

Precession: A static magnetic field $\mathbf{B}_0$ is applied to the body. This magnetic field aligns the rotational axis of each spin with its field direction. These spins now rotate, or precess, around the same magnetic axis. Note that roughly an equal number of spins will be aligned with the positive direction of the magnetic axis as with the negative direction and the overall signal generated during precession will be minimal.

Resonance: With the magnetic field $\mathbf{B}_0$ in place, a second, weaker, magnetic pulse is applied to the body in the direction $\mathbf{g}$. This second field results in the magnetic moment m of each spin aligning with the pulse direction $\mathbf{g}$. The spin's axis of rotation remains aligned with $\mathbf{B}_0$. The result of the resonance phase is to cause the net magnetism of the spin to veer away from the main magnetic field $\mathbf{B}_0$. This change in the magnetic field induces a small current within the subject.

Relaxation: The second magnetic pulse is removed and the magnetic moments of the hydrogen atoms realign with $\mathbf{B}_0$. As this realignment occurs, the changing magnetic field generated by the realignment of the spins induces a current in the coil of the MRI scanner. From this current, two common measurements can be taken:

1. Spin–spin relaxation time (T_2): The amount of time it takes for the magnetism in the direction of **g** to reduce to 37% of its original value
2. Spin–lattice relaxation time (T_1): The amount of time it takes for the magnetism in the direction of $\mathbf{B}_0$ to recover 63% of the magnetism it lost when the second gradient was applied in the direction **g**

These relaxation times can be visualized at multiple locations in the brain, resulting in what are known as T_1 and T_2 weighted images.

19.1.2.2 Diffusion-Weighted Imaging

As magnetic resonance imaging depends on the magnetic moments of hydrogen atoms, Stejskal and Tanner were able to develop a sequence of precession, resonance, and relaxation periods that allow MRI to measure the movement of hydrogen atoms over time and in turn the water molecules of which they are a part [95]. Le Bihan and Breton later adapted this MR image sequence to the scanning of the human body [24]. This imaging sequence is summarized in Figure 19.3 for a given angular direction **g**.

The sequence in Figure 19.3 assumes that the magnetic field $\mathbf{B}_0$ has been applied and that the spins are precessing around $\mathbf{B}_0$. In this state, a magnetic pulse is applied at an angle of 90° from the direction of $\mathbf{B}_0$. This pulse aligns the spins that were separately aligned to either the positive or negative $\mathbf{B}_0$ axis. Once the spins are aligned, the 90° pulse is removed and a second pulse, known as a gradient pulse, is applied in the direction **g**. This gradient pulse senses the induced current to a specific angular direction.

A third magnetic pulse in the direction 180° from $\mathbf{B}_0$ follows the gradient pulse, then the gradient pulse is reapplied. The 180° pulse plays a key role in that it flips the spin direction of the atoms to the opposite of what they were during the precession phase. As a result of this flip, the current induced by stationary atoms during the application of the second gradient pulse will cancel out the current induced by the same atoms during the first gradient pulse [67]. Therefore, the resulting signal measured after all gradient pulses have been applied relates solely to the molecules experiencing motion in the direction **g**.

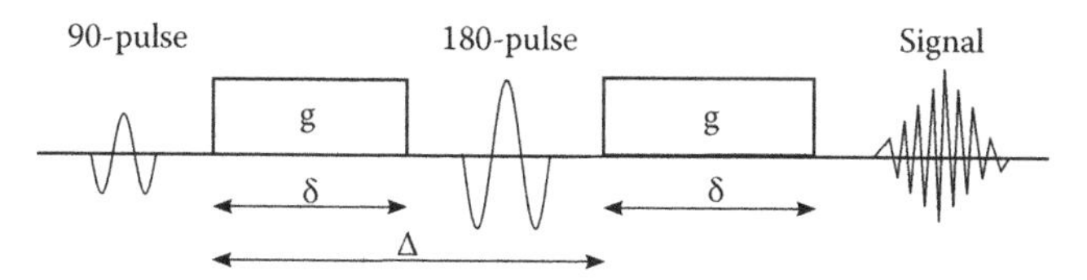

FIGURE 19.3
The Stejskal–Tanner diffusion-weighted imaging sequence. (Adapted from Westin, C.-F. et al., *Med. Image Anal.*, 6, 93, 2002.)

The T_2 relaxation time is then measured from this final signal for multiple locations in the brain and visualized in what are known as diffusion-weighted images (DWIs). Figure 19.4 displays a conventional T_2 image next to sample DWIs for various gradient directions **g**. Note here that rapid diffusion results in fast T_2 relaxation times, resulting in a low intensity in the DWI.

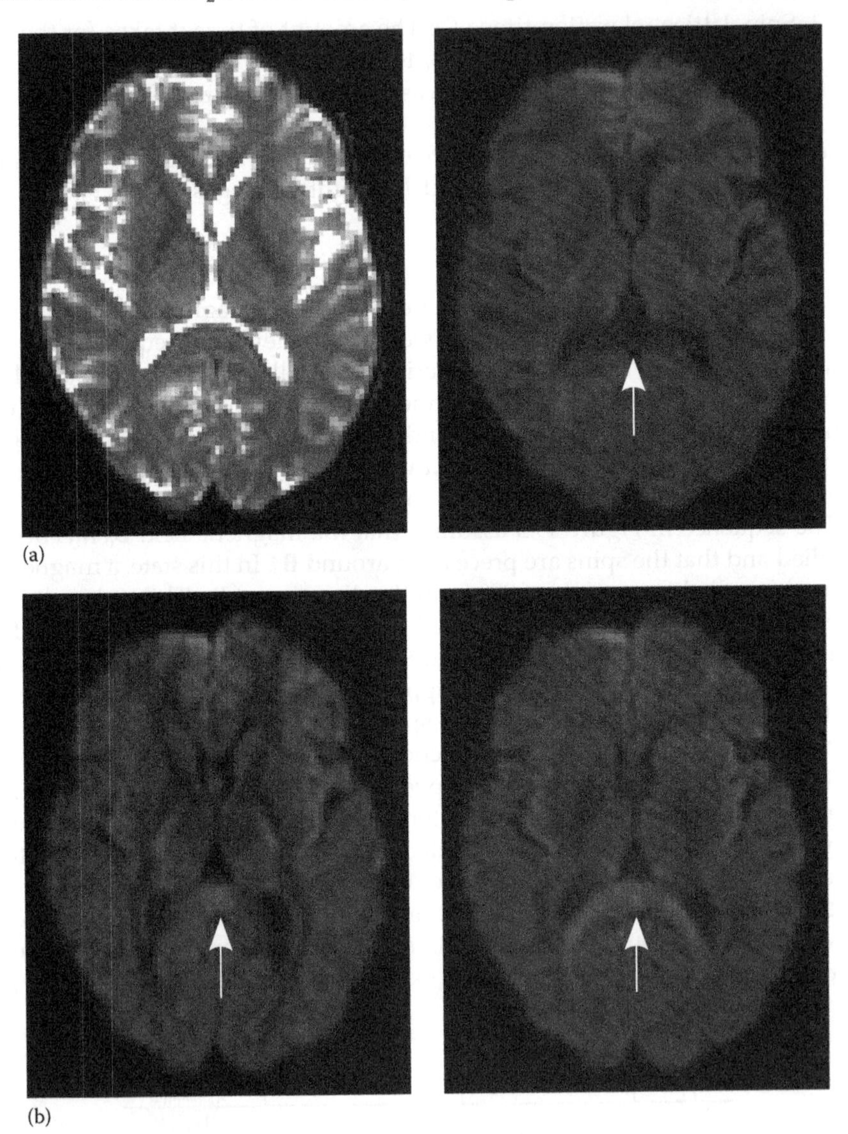

FIGURE 19.4
Axial slices of (left to right) (a) standard T_2 image and (b) its corresponding DWIs from gradient pulses in the horizontal, vertical, and out-of-plane directions. Note the differences in measured diffusion in the splenium due to gradient direction (highlighted by the white arrows). (Adapted from Jones, D.K., *Cortex*, 44(8), 936, September 2008.)

Further note the different rates of diffusion for different directions within the brain's white matter as pointed out by the white arrows in Figure 19.4.

From the DWI for gradient direction **g**, the diffusion rate (d) can be computed using the Stejskal–Tanner equation:

$$S = S_0 \exp(-bd) \tag{19.2}$$

where

S is the DWI intensity
S_0 is the standard
T_2 is the image intensity
b is the diffusion weighting [95]

The diffusion weighting b is in turn proportional to the strength and duration of the gradient pulse. The T_2 image used in (19.2) is typically referred to in this context as a B_0 image as it is acquired without the application of the gradient pulses (i.e., $b = 0$). The scalar d is commonly referred to as the apparent diffusion coefficient (*ADC*).

19.1.3 Correction of Image Artifacts

To obtain a full understanding of dMRI, we must acknowledge how the quality of the DWIs is affected or limited by the image acquisition process. All further analysis is going to depend on the accuracy of these diffusion measurements, and as such, we must address the presence of noise and imaging artifacts within these DWIs.

dMRI is susceptible to various artifacts, the three most common being eddy currents, subject motion, and Rician noise [13]. Let us consider each in turn.

19.1.3.1 Eddy Currents

As seen in the diffusion imaging sequence in Figure 19.3, multiple magnetic gradient pulses are applied in rapid succession. Switching between these gradients can result in fluctuations in the scanner's magnetic field. These fluctuations induce what are known as eddy currents in the coil of the MRI scanner. The eddy currents interfere with the currents induced by the scanned subject, thereby distorting the resulting DWIs [13].

Much is known of eddy currents, namely, that they are dependent on the magnitude of the gradient pulse, independent of the subject being scanned, and that they result in related geometric and intensity distortions in DWIs [51]. The geometric distortion produced from eddy currents has been shown to consist of a translation, scale, and shear of the resulting image and is commonly rectified using affine registration [27,51,70]. The DWIs are registered to a T_2-weighted image with the mutual information similarity measure showing the best results [70]. As the T_2 image is acquired without gradient pulses that produce eddy currents, it is assumed to be free of geometric distortion, thereby making it an appropriate template to which we can register the DWIs. Intensity corrections are then calculated directly from the magnitudes of the shear, scaling, and translations of the affine warp [51,70].

One benefit of eddy currents being independent of the subject scanned is that the affine warp used in the correction can be obtained by imaging a physical phantom with known ground truth [34]. This warp can then be applied to later subject scans.

19.1.3.2 Subject Motion

Depending on the number of DWIs being acquired, the length of a dMRI scan can range from a couple of minutes [74] to a few hours [103]. During that time, the subject may move both voluntarily involuntarily (e.g., breathing). As a result, the same voxel location in two DWIs is not guaranteed to correspond to the same anatomical location in the subject.

While correcting for subject motion in a single image has been well studied (see [98] for a survey), the problem of correcting motion between separate DWIs has yet to receive a strong theoretical treatment [13]. Even so, two main approaches have been proposed to correct for subject motion between DWIs, both involving image registration. First, we can, as with eddy current correction, align the DWIs to a T_2-weighted image with the mutual information similarity measure [70,90]. The alternative approach is to model the diffusion at each voxel (as discussed further in Section 19.2) and to align images so as to minimize the residual of the model fit [7,10]. A recent quantitative comparison of these approaches suggests that both methods are equally capable of correcting for subject motion [91].

Note however that if the rotational motion of the subject is large, the directions of the applied gradient pulses would need to be corrected as well [66,90].

19.1.3.3 Rician Noise

Any environment is going to contain a certain amount of background noise. In the case of dMRI, this noise has been well modeled using a Rician distribution [16] given as

$$p(x \mid \mu, \sigma) = \frac{x}{\sigma^2} \exp\left(-\frac{x^2 + \mu^2}{2\sigma^2}\right) I_0\left(\frac{x\mu}{\sigma^2}\right) \tag{19.3}$$

where

x is the observed image intensity
μ is the noise-free signal
σ is the standard deviation of the noise
I_0 is the zeroth-order Bessel function of the first kind

At high signal to noise ratios, the Rician distribution is occasionally approximated using a Gaussian distribution [43]. This additive noise can have an adverse effect on the diffusion rates calculated using the Stejskal–Tanner equation, particularly for images taken at a high diffusion weighting [60].

Historically, variational methods have been applied to remove this Rician noise, with anisotropic filtering [82] and total variation regularization [16,43], both showing success. Weighted-mean filtering approaches have also been used [97,107]. The main conceptual difference in noise removal in DWIs is whether to denoise one image at a time (the scalar approach) or all images at once (the vector approach) [43]. Recent results suggest that the vector-based algorithms improve signal to noise to a greater extent [97].

19.2 Modeling Local Diffusion Patterns

Since the introduction of dMRI, two key advancements have propelled the field to where it is today: first, the introduction of the diffusion tensor by Basser et al. [14] and second, the introduction of higher angular resolution diffusion imaging (HARDI) [101]. The introduction of the diffusion tensor brought forth the concept of modeling the diffusion rates from the DWIs as a 3D function within each voxel. HARDI, on the other hand, allowed us to increase the complexity of these models to better represent the local diffusion properties. This section will show how these two contributions underlie the ability to perform brain connectivity analysis.

19.2.1 Diffusion Tensor Model

While we have shown that the Stejskal–Tanner equation (19.2) relates diffusion rates to the DWI intensities, we can consider a more general formulation of the diffusion properties at a voxel. Since water molecules undergo random Brownian motion, we can consider a probability density function (PDF) $p_t(\mathbf{x})$ describing the probability of a water molecule experiencing a displacement $\mathbf{x}$ over the observation time t. It has been shown that the distribution p_t is related to the DWI intensities via the Fourier transform [5].

$$\frac{S(\mathbf{g})}{S_0} = \int_{\mathbf{x}\in\mathbb{R}^3} p_t(\mathbf{x})\exp(-ib\mathbf{g}\mathbf{x})\,d\mathbf{x} \tag{19.4}$$

As, $S(\mathbf{g})$ represents the diffusion-weighted signal for the gradient direction $\mathbf{g}$, S_0 is the unweighted B_0 image signal, and b is the diffusion weighting. With enough DWIs $S(\mathbf{g})$, the Fourier transform can be inverted to obtain p_t. This is known as *q-space* imaging [103]. In practice however, the number of DWIs required to accurately perform the inversion leads to scanning times on the order of hours [54] that is generally not available in a clinical setting. As a result, it has become common to assume a model for p_t, the simplest model being a zero-mean Gaussian:

$$p_t(\mathbf{x}) = \frac{1}{\sqrt{(2\pi)^3 \, |2t\mathbf{D}|}} \exp\left(\frac{-\mathbf{x}^T \mathbf{D}^{-1} \mathbf{x}}{4t} \right) \tag{19.5}$$

where the covariance is $2t\mathbf{D}$.

Plugging the Fourier transform of (19.5) into (19.4) results in a more general case of the Stejskal–Tanner equation:

$$S(\mathbf{g}) = S_0 \exp\left(-b\mathbf{g}^T \mathbf{D}\mathbf{g}\right) \tag{19.6}$$

The 3×3 second-order positive-definite symmetric matrix $\mathbf{D}$ is referred to as the diffusion tensor [14]. It contains six unique elements and therefore six DWIs are required, along with the B_0 image, to estimate the tensor. The DWIs are obtained from uniform, noncollinear gradient directions so as to not favor a given direction in the tensor fitting process. These seven images can be obtained with an MRI scan on the order of 1–2 min [74], thereby making it a clinically feasible imaging protocol.

Many factors affect the quality of the diffusion tensors. As mentioned earlier, noise, motion, and distortions in the DWIs will result in poor tensor estimates. Aside from post-processing the DWIs, it is also common to obtain DWIs from more than six gradient directions in order to overfit the tensor, thereby reducing the effect of having some corrupted DWI signals [60]. The fitting procedure also affects the quality of the resulting tensors. The simplest approach is to take the logarithm of (19.6) and fit the tensor using least squares [14]. This approach, however, does not ensure that the resulting tensor be positive definite (i.e., have positive eigenvalues). Nonlinear fitting allows for this constraint and generally results in a less noisy tensor field [60], especially if spatial regularization is incorporated into the fitting procedure [102]. Additional approaches include using a weighted least squares fitting of the log-transformed equation (19.6) that is used to detect and remove outlier DWI signals prior to the final tensor fit [33].

19.2.2 Tensor Image Visualization

The power of the diffusion tensor lies in its ability to measure and visualize more detailed properties of the diffusion than the scalar DWIs provide. For example, we can look at how the rates of diffusion vary with direction or calculate the average diffusion rate at each voxel. In fact, significant diagnostic information can be obtained from the diffusion tensor by analyzing its principal components obtained through the tensor's eigendecomposition [50]. Given a diffusion tensor **D**, we can obtain the eigendecomposition:

$$\mathbf{D} = \begin{bmatrix} \mathbf{e}_1 & \mathbf{e}_2 & \mathbf{e}_3 \end{bmatrix} \begin{bmatrix} \lambda_1 & & \\ & \lambda_2 & \\ & & \lambda_3 \end{bmatrix} \begin{bmatrix} \mathbf{e}_1 & \mathbf{e}_2 & \mathbf{e}_3 \end{bmatrix}^T \tag{19.7}$$

where the eigenvalues are positive and sorted in descending order (i.e., $\lambda_1 \geq \lambda_2 \geq \lambda_3$). The eigenvectors $\mathbf{e}_1$, $\mathbf{e}_2$, $\mathbf{e}_3$ are considered the main axes of diffusion, while the eigenvalues encode the rates of diffusion along each corresponding axis. Given this interpretation, we can visualize the diffusion tensor as an ellipsoid as shown in Figure 19.5. The axes of the ellipsoid are the eigenvectors of the tensor, while the tensor's eigenvalues describe the ellipsoid's stretch along each axis. Another interpretation of the ellipsoid is as an isoprobability surface of the Gaussian diffusion model given in (19.5).

The tensor eigendecomposition allows for the computation of two key diffusion properties: the mean diffusivity (MD) and the fractional anisotropy (FA) [58,105]. These two measures respectively capture the mean and variance of the diffusion rate with respect to direction. They are computed from the tensor's eigenvalues as

$$MD = \frac{1}{3}\left(\lambda_1 + \lambda_2 + \lambda_3\right) \tag{19.8}$$

$$FA = \sqrt{\frac{3}{2}}\,\frac{\sqrt{(\lambda_1 - MD)^2 + (\lambda_2 - MD)^2 + (\lambda_3 - MD)^2}}{\sqrt{\lambda_1^2 + \lambda_2^2 + \lambda_3^2}} \tag{19.9}$$

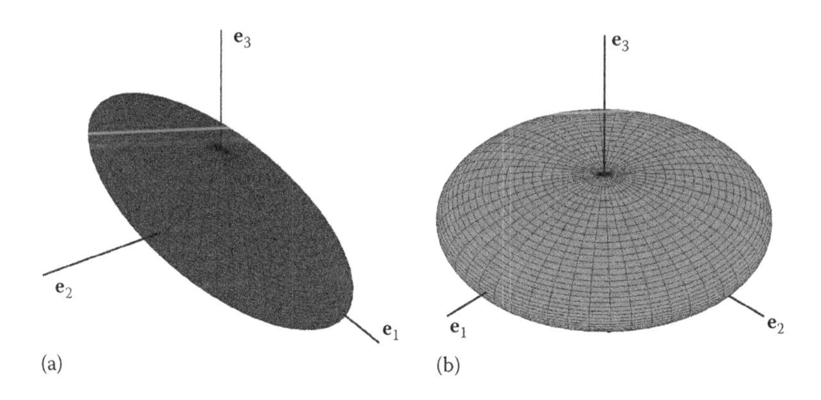

FIGURE 19.5
Examples of the ellipsoidal representation of prolate (a) and oblate (b) diffusion tensors. (Adapted from Jones, D.K., *Cortex*, 44(8), 936, September 2008.)

Examples of MD and FA on a slice of the brain are shown in Figure 19.6. From these images, we observe the brain microstructure described in Section 19.1.1. Note that the MD is higher in the ventricles than the rest of the brain due to the lack of tissue structure. Conversely, the FA is highest in the white matter of the brain due to coherent orientation of

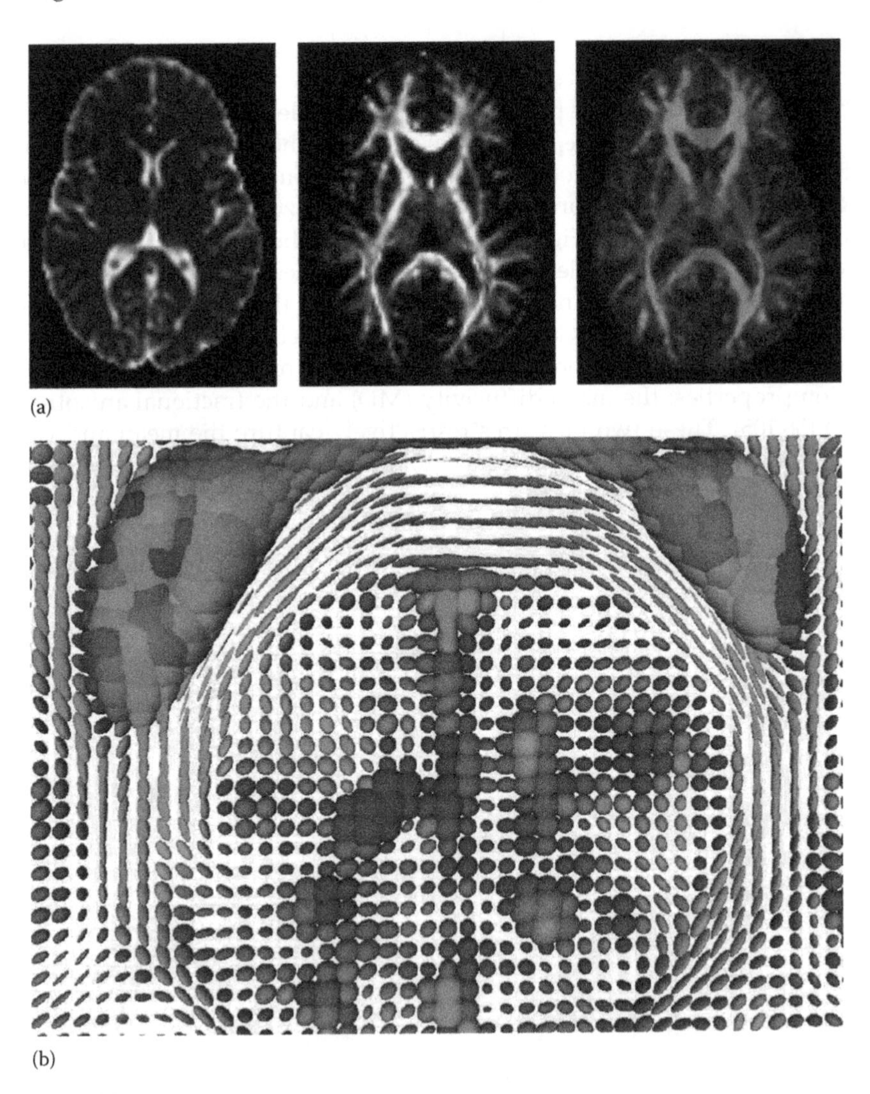

FIGURE 19.6
(See color insert.) Various methods of visualizing the information contained in a diffusion tensor field. (a) MD (left), FA (center), and color-coded orientation map (right) and (b) ellipsoidal visualization. (Images generated using MedINRIA http://www-sop.inria.fr/asclepios/software/MedINRIA/ on data obtained from Mori, S., John Hopkins Medical Institute: Laboratory of Brain Anatomical MRI, in vivo human database, http://lbam.med.jhmi.edu/, accessed February 2009.)

the tissue microstructure. We can further estimate the orientation of this microstructure as being equivalent to $\mathbf{e}_1$. Of course, the quality of this estimate will depend in part on the FA. Low FA would imply a less coherent orientation in the tissue microstructure, making the estimation of this orientation not well founded.

Other scalar measures have been generated to characterize both the shape and anisotropy of diffusion tensors, but FA and MD are most commonly used in practice. A review of other scalar tensor measures can be found in [105,114].

Aside from visualizing the scalar FA and MD maps, approaches have been developed to display the tensor's orientation information as well. The two most common approaches are shown in Figure 19.6. First, the primary diffusion direction (PDD) $\mathbf{e}_1$ can be visualized as a color image, where the RGB values are $R = FA|\mathbf{e}_1 \cdot [1,0,0]|$, $G = FA|\mathbf{e}_1 \cdot [0,1,0]|$, and $B = FA|\mathbf{e}_1 \cdot [0,0,1]|$ [80]. Such a scheme allows for an intuitive visualization of orientation weighted by the orientation strength, yet color assignments are not unique. For example, the color yellow would be assigned to voxels with $\mathbf{e}_1 = [1, 1, 0]$ and $\mathbf{e}_1 = [-1, 1, 0]$, leading to ambiguity of the underlying diffusion direction. As a result, it is occasionally necessary to visualize the tensor ellipsoids themselves as seen in Figure 19.6b. Generating a less ambiguous color representation of tensor data remains an area of open research [115].

Regardless of visualization strategy, the value of the orientation information in dMRI is significant as it allows us to infer the orientation of neural pathways in the white matter of the brain. If we take, for example, the FA in Figure 19.6, we can observe four major neural pathways. The forceps minor can be seen in the upper portion of the image arching upward in a U shape. A similar looking pathway, the forceps major, can be seen in the bottom half of the image as an inverted U shape. Flanking the forceps major on either side are the optic radiations. These pathways, as seen in dMRI, agree with histological studies [31], thereby making dMRI a powerful tool for mapping out these neural pathways noninvasively.

19.2.3 High Angular Resolution Diffusion Models

While the diffusion tensor model provides a powerful tool for visualizing and assessing the microstructure of brain tissue, it suffers from a significant limitation: the assumption that diffusion follows a Gaussian model. While this model may hold for simple examples such as those in Figure 19.1, there exist many situations where the local diffusion is more complex.

Take, for example, this situation shown in Figure 19.7a. This example shows a mixture of fibrous tissues oriented along the positive and negative diagonal directions, resulting in a diffusion profile in bold. Ideally, we would like to model this diffusion as shown in Figure 19.7b. Unfortunately, the diffusion tensor model assumes ellipsoidal Gaussian diffusion. As a result, we would obtain for this example the tensor shown in Figure 19.7c. This tensor would misleadingly suggest that diffusion is equal for all directions in the

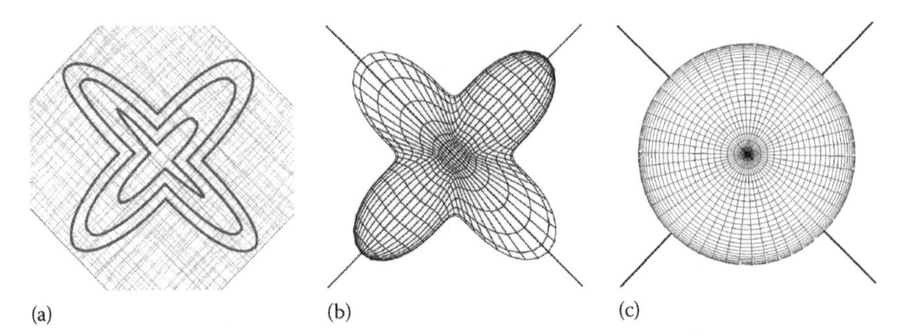

FIGURE 19.7
Example of crossing fibers and how they are modeled using dMRI. (a) Crossing fibers, (b) corresponding diffusion ODF, and (c) corresponding diffusion tensor. (Adapted from Alexander, D.C., *Visualization and Processing of Tensor Fields*, Chapter 5: an introduction to computational diffusion MRI: The diffusion tensor and beyond, Springer, Berlin Heidelberg, Germany, pp. 83–106, 2006; Descoteaux, M. et al. *IEEE Trans. Med. Imaging*, 28(2), 269, 2009; Westin, C.-F. et al., *Med. Image Anal.*, 6, 93, 2002, respectively.)

plane of the crossing. Further, the primary eigenvector of the tensor is not guaranteed to align with either fiber direction.

Such an example is common in the white matter of the brain. The neural pathways are made up of aligned tissue fibers whose diameter is on the order of microns [3]. In contrast, the resolution of DWIs is typically on the order of millimeters cubed. As a result, this type of averaging of diffusion from multiple pathways is unavoidable. In fact, it has been estimated that at least one-third [21] to two-thirds [39] of voxels in the brain may exhibit this crossing fiber property. As a result, attempts have been made to come up with better diffusion models.

Tuch et al. first proposed the use of more descriptive diffusion models by showing that there are regions in the brain where fibers cross [101]. They noted that in order to detect these crossing fibers, DWIs from a greater number of diffusion directions, and at a higher gradient weighting, were required. Thus was born the concept of higher angular resolution diffusion imaging (HARDI).

Initial attempts to model more complicated diffusion profiles revolved around fitting multiple tensors to the DWI signals [101]. A mixture of Gaussian's model was assumed and the Stejskal–Tanner equation was updated to incorporate the mixture:

$$\frac{S(\mathbf{g})}{S_0} = \sum_i f_i \exp\left(-b\mathbf{g}^T \mathbf{D}_i \mathbf{g}\right) \tag{19.10}$$

Each tensor $\mathbf{D}_i$ has a corresponding volume fraction f_i representing the fraction of the local diffusion the tensor represents. Later work using the CHARMED [8,9] and FORECAST [6] methods assumed a particular shape for each fitted tensor. The former approach attempts to model intra-fiber and extra-fiber diffusion using prolate (cigar-like) and spherical tensors, respectively. The latter approach models prolate tensors with an equal and known MD. More recent work has instead assumed a mixture of Wishart (MOW) distributions—effectively a distribution over tensors—as the choice of diffusion model [56,57].

While these mixture model approaches allow for the same intuitive representation as the single tensor model, they also have their limitations. These include the following:

- The number of tensors being fitted to the DWI signal has to be specified ahead of time. While there has been work on estimating this number from the data [5,101], there is no ground truth specification for the number of tensors to fit at a voxel.
- There is no guarantee that the assumed shape of the fitted tensors is appropriate for the underlying diffusion. If the shape assumption is poor, the volume fractions can be poorly estimated [6].
- The mixture model, and not the underlying mixture components, is fit to the DWIs. While the peaks of the mixture model will align with the directions of maximal diffusion, there is no guarantee that the peaks of the underlying distributions will align with these directions as well [101].

Some of these limitations have been addressed in recent work. For example, instead of fitting a fixed number of tensors to the data, volume fractions can be calculated for a set of basis tensors [56,57,69]. Those tensors with a volume fraction above a given threshold are maintained to model the diffusion. Also, instead of using the mixture components for further analysis, the mixture model itself is used to analyze the diffusion [6,56,57].

On the other end of the spectrum, model-free approaches have also been proposed to capture local diffusion properties. Again, Tuch instituted this approach through the introduction of *q-ball* imaging [100]. Based on the earlier *q-space* approach described by the Fourier transform in (19.4), Tuch noticed that the directional dependence of the diffusion rate is the most commonly used information for dMRI analysis and that the radial distance component of the diffusion does not play a significant role. As a result, instead of modeling the diffusion as a PDF $p(\mathbf{x})$, where $\mathbf{x}$ is a vector of any length, *q-ball* imaging models the diffusion orientation distribution function (ODF) $\psi(\theta, \varphi)$, where θ, φ are spherical angles. As such, the ODF captures the probability of diffusion along different angular directions (θ, φ) but without a radial distance parameter. An estimation of the ODF can be more efficiently obtained through the use of the Funk–Radon transform [100]. By ignoring the radial component, the *q-ball* ODF can be estimated with fewer DWI samples than the original PDF from *q-space* imaging, leading to more reasonable scanning times.

Two other model-free approaches have also gained traction in the dMRI community. First, the diffusion orientation transform (DOT) shares similarities with *q-ball* imaging as both are based on the earlier *q-space* approach. In contrast, DOT assumes diffusion decays exponentially along the radial direction and uses this assumption to perform the Fourier transform in (19.4) using fewer DWI samples [79]. The DOT diffusion ODF is then obtained by analytically integrating the resulting PDF along the radial direction.

An alternative model-free approach is Jansons and Alexander's persistent angular structure (PAS) approach [54]. The goal behind PAS is to find a diffusion PDF $p(\mathbf{x})$ from (19.4) that is smooth yet captures the key angular structure of the diffusion. This goal is achieved through optimization by finding a PDF $p(\mathbf{x})$ that maximizes entropy while minimizing the error in fitting to the DWI samples. A Lagrange multiplier is used to weight the two competing terms [54].

One of the key limitations of the model-free HARDI approaches is precisely that a model is not assumed. In areas of low anisotropy, both PAS and *q-ball* imaging can overestimate the directional dependence of the diffusion as a result of image noise [54,100]. This overestimation can result in spurious maxima in the diffusion ODFs. While work has been done in reviewing and comparing different HARDI approaches [1,3,4,87], there is generally no consensus as to which HARDI model is best suited to represent diffusion MR characteristics.

As with the diffusion tensor, the quality of the diffusion ODF is going to depend in part on the algorithm used to fit the ODF to the diffusion data. Recently, Jian and Vemuri pointed out that many of these diffusion model fitting algorithms can be unified using a spherical deconvolution framework [55]. Given DWI signals $S(\mathbf{g})$ and the B_0 image S_0, the diffusion ODF ψ can be considered as a deconvolved version of the signal:

$$\frac{S(\mathbf{g})}{S_0} = \int_{\mathcal{M}} R(\mathbf{g}, \theta, \phi)\psi(\theta, \phi) \tag{19.11}$$

where

R is the convolution kernel

$\mathcal{M}$ is a manifold, typically $\mathbb{R}^3$ in the case of a PDF and $\mathbb{S}^2$ for the ODF

With a discretization of $\mathcal{M}$, (19.11) can be converted into a linear least squares problem. However, the kernel matrix representing R is typically ill-conditioned and highly sensitive to noise in the sampled $S(\mathbf{g})$ [55]. As a result, nonlinear fitting approaches are recommended [2,55].

19.2.4 HARDI Representation and Visualization

While a diffusion PDF, such as the Gaussian PDF used with the diffusion tensor, provides information on the radial aspect of the measured diffusion, the concept of using a diffusion ODF has become commonplace with HARDI data [55]. As such, an efficient representation of the ODF would be useful for further computation. Since the diffusion ODF is a spherical function, the most popular choice for its representation is a real spherical harmonic expansion [6,40,55]. The diffusion ODF ψ can be represented as

$$\psi(\theta,\phi) = \sum_{\ell=0}^{K} \sum_{m=-\ell}^{\ell} F_\ell^m Y_\ell^m(\theta,\phi) \tag{19.12}$$

where integers ℓ and m are the degree and order of the harmonics, respectively.

The basis harmonics Y_ℓ^m are given as

$$Y_\ell^m = \begin{cases} \sqrt{\dfrac{2\ell+1}{4\pi}} P_\ell^0(\cos(\phi)), & \text{if } m = 0 \\ \sqrt{2}\sqrt{\dfrac{2\ell+1}{4\pi}\dfrac{(\ell-m)!}{(\ell+m)!}} P_\ell^m(\cos(\phi))\cos(m\theta), & \text{if } m > 0 \\ \sqrt{2}\sqrt{\dfrac{2\ell+1}{4\pi}\dfrac{(\ell-m)!}{(\ell+m)!}} P_\ell^{-m}(\cos(\phi))\sin(m\theta), & \text{if } m < 0 \end{cases} \tag{19.13}$$

where P_ℓ^m is the associated Legendre function of degree ℓ and order m.

As the ODF is antipodally symmetric, only the even-degree basis harmonics are used [40]. Typically, the expansion is limited to degree $\ell \leq 8$ to suppress noise artifacts in the resulting ODF [39]. Other ODF representations have also seen limited use, including von Mises–Fisher distributions [88] and fourth-order tensors [11].

The notions of MD and FA have also been extended to HARDI diffusion ODFs, with the latter being referred to in this context as generalized anisotropy (GA). As before, the two measures correspond to the mean and variance of the diffusion ODF ψ:

$$MD = \frac{1}{4\pi}\int \psi(\theta,\phi)dS \qquad GA = \frac{1}{4\pi}\int (\psi(\theta,\phi) - MD)^2 dS \tag{19.14}$$

Unlike the diffusion tensor model, MD and GA generally do not have an elegant solution. Analytical solutions have been proposed for both measures [11,78] but involve ad hoc scaling and normalization weights. Examples of MD and GA images are shown in Figure 19.8.

Finally, we note that the visualization of the orientation information in the diffusion ODF typically involves visualizing the spherical ODFs themselves as seen in Figure 19.8.

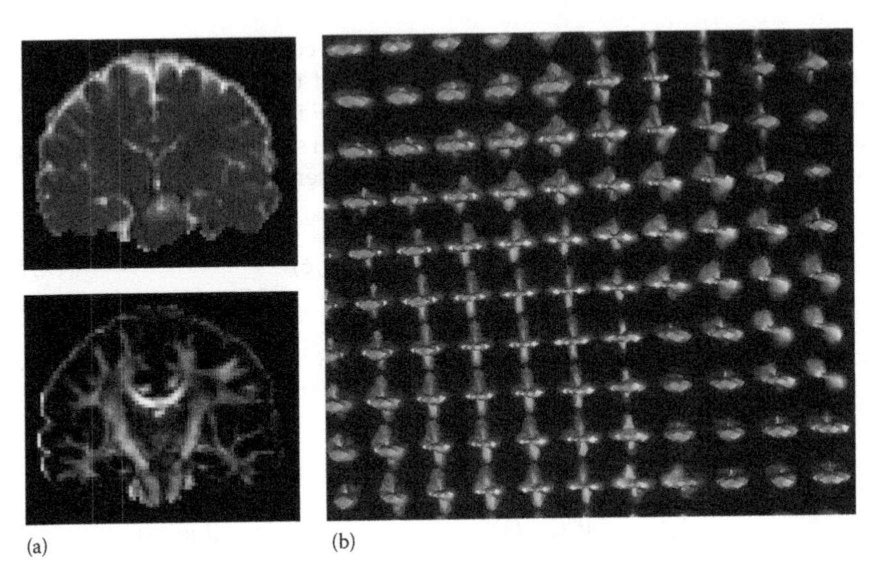

(a) (b)

FIGURE 19.8
Sample visualization techniques for diffusion ODFs obtained from HARDI. (a) MD (top) and GA (bottom) and (b) *q-ball* diffusion ODFs.

19.2.5 HARDI versus the Diffusion Tensor

Despite the presence of these HARDI models that better represent the underlying diffusion properties, the use of the diffusion tensor model still persists in a clinical setting [74–76]. There are various reasons for the use of what is perceived to be an inferior model and these reasons highlight some of the limitations of HARDI:

- The number of gradient directions, and in turn DWIs, required for the reconstruction of HARDI models is still significantly larger than for diffusion tensor imaging. With scanning time as a bottleneck, the opportunity to obtain enough DWIs for a HARDI reconstruction remains in many cases a luxury.
- To observe non-Gaussian diffusion, the strength of the magnetic gradients used in the scan is increased [101]. Increasing the gradient strength increases diffusion rates, which in turn are inversely proportional to relaxation time and DWI intensity. If we increase the gradient strength enough, the DWI intensities can fall below the noise floor, an effect seen with HARDI imaging settings [60].
- Recent research suggests that limitations of the tensor model with regard to crossing fibers can be overcome by taking into account neighborhood information [12,92]. With such advancements, it remains unclear if HARDI can provide information that cannot be recovered from a diffusion tensor image.

Due to these aforementioned reasons and the wealth of diffusion tensor medical research [76], the tensor model cannot be ignored.

19.2.6 Diffusion MR Image Regularization

While modeling local diffusion through tensors or ODFs provides useful information with regard to orientation dependence and anisotropy, we must not lose sight of the fact that the resulting diffusion profiles are estimates of the underlying diffusion and are open to error. While noise reduction can be performed on DWIs as seen in Section 19.1.3, any remaining noise can be amplified due to a poor model fit. Further, the choice of model, particularly the tensor model, can result in a poor fit to the diffusion measurements. These potential errors will affect any further analysis of the diffusion data and so it is important that they be addressed.

Methods for addressing model error generally fall into two categories: correcting for noise and correcting for model choice. We look at these in turn.

19.2.6.1 Noise Reduction

Total variational regularization has been a popular choice for noise removal in the diffusion tensor images [35,37,43,85,99,108]. Generally, total variation approaches involve the minimization of an energy functional:

$$E_{TV}(\mathcal{I}) = E_{data}(I_{orig}, \mathcal{I}) + E_{diff}(\mathcal{I}) \tag{19.15}$$

where E_{data} is a term that measures the distance from the noise reduced image $\mathcal{I}$ from the original version of the image I_{orig}. The term E_{diff} is the Perona and Malik anisotropic diffusion term that controls the smoothing in the image [104]. The variations between the proposed methods relate to what is regularized and how they maintain the constraints of the diffusion tensor. In some cases, only the principal diffusion direction (i.e., the tensor's primary eigenvector) is regularized [108]. In other cases, the tensor's eigenvectors and eigenvalues are regularized separately [37,99]. Still other approaches regularize all tensor elements at once, using the Cholesky factorization [35] or tensor distance metrics [43,85] to ensure positive definiteness of the diffusion tensors.

Other regularization approaches include graph-based anisotropic diffusion filtering of the tensor elements [110], Markov random field minimization of the principal diffusion directions [86], and bilateral [49] and nonlocal means [107] filtering using tensor distance metrics.

19.2.6.2 Model Correction

While noise in the diffusion measurements is one source of inaccuracy, the chosen diffusion model may also be inaccurate. As seen earlier, the diffusion tensor model is unable to accurately represent diffusion profiles in areas of crossing fibers. Recent work has looked at determining crossing fiber locations from neighborhood information in tensor images. Barmpoutis et al. estimate an ODF called a *tractosema* by performing a neighborhood kernel integration with a specialized kernel [12]. Meanwhile, Savadjiev et al. produce a similar ODF by measuring the probability of diffusion along 3D curves through the surrounding neighborhood [92]. These approaches are able to resolve certain crossing fiber situations, yet it remains unclear whether all crossing fibers can be discovered using some similar post-processing on diffusion tensor images.

Despite attempts to fully address noise and inconsistencies within the data, it is unlikely that we will ever be sure that all imperfections will be removed. As such, recent work has looked at quantifying this error using statistical methods [52,106].

19.3 Brain Connectivity Mapping from dMRI

The orientation information in dMRI is incredibly valuable in mapping out structure in the white matter of the brain. As diffusion is strongest along the fiber tracts that make up neural pathways, the directions of maximal diffusion at each voxel location can be used to help reconstruct the fiber tracts, thereby mapping out connectivity in the brain. The problem of mapping out these connections is known as tractography and is complicated by many factors. We have already mentioned two: poor diffusion model fitting and noisy diffusion measurements. Here, we look at how these complications, and others, affect existing algorithms as well as how the tractography results are used.

19.3.1 Streamline Tractography

The earliest approaches to the tractography problem surrounded tracing out 3D curves that followed the direction of strongest diffusion [15,72]. These 3D curves, known as streamlines, evolve using the following Euler equation:

$$\mathbf{r}(s_1) = \mathbf{r}(s_0) + \alpha\varepsilon_1(\mathbf{r}(s_0)) \tag{19.16}$$

where $\mathbf{r}$ is the streamline curve parametrized by its length from a given seed point and s_i are points along the curve. ε_1 is the PDD at the given location on the curve and the choice of notation comes from the use of the diffusion tensor's primary eigenvector as the PDD. The PDD acts as the tangent to the streamline as it evolves with stepsize α, where α is sufficiently smaller than the voxel size to limit discretization effects on the evolving curve. These aspects are further shown in Figure 19.9. While the aforementioned Euler equation most easily describes the streamline evolution, a higher-order Runge–Kutta method is commonly used to improve numerical stability in

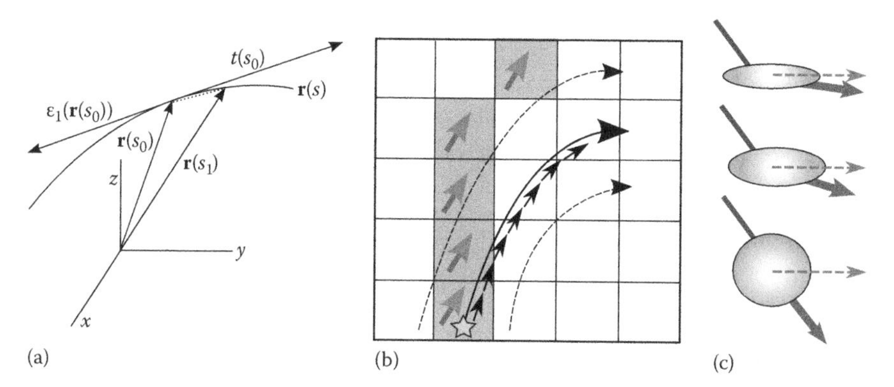

FIGURE 19.9
Examples of streamline evolution. Streamlines evolve in the direction tangent to the local PDD. Stepsizes in the evolution equation are chosen sufficiently small so as to avoid poor tracking due to discretization. (a) Streamline tangent to PDD, (b) effect of stepsize on streamline tractography, and (c) tensor deflection. (Adapted from Basser, P.J. et al., *Magn. Reson. Med.*, 44, 625, 2000; Lazar, M. et al., *Hum. Brain Map.*, 18, 306, 2003; Mori, S. et al., *Ann. Neurol.*, 45(2), 265, 1999, respectively.)

the streamline evolution [15]. The streamline evolution continues until the PDD becomes unreliable. Typically, the reliability of the PDD is captured through either FA [15] or via neighborhood PDD coherence [72]. This initial tractography approach is referred to in the literature as the *fiber assignment by continuous tracking* (FACT) method.

One of the concerns with the FACT approach is that it follows the PDD regardless of whether the PDD at a voxel is an accurate estimation of fiber tract orientation. In areas of lower anisotropy (but still above the termination threshold of FACT), the PDD may become more unreliable. In these situations, we may wish to regulate the effect of the local PDD on the direction of the evolving streamline. This is the idea behind the tensor deflection (TEND) approach [65]. In this algorithm, the local diffusion tensor **D** is used to deflect the incoming streamline curve as given by the evolution equation:

$$\mathbf{r}(s_1) = \mathbf{r}(s_0) + \alpha \mathbf{D}(\mathbf{r}(s_0)) \cdot \mathbf{r}(s_0) \tag{19.17}$$

The greater the anisotropy of the tensor, the more reliable the PDD and therefore the stronger the deflection of the streamline fiber. An example of this evolution is shown in Figure 19.9c.

By deflecting the incoming streamline curve, TEND implicitly creates a curvature constraint on the evolving streamline. The streamline can only bend as much as a diffusion tensor will allow. In some situations, this curvature constraint can cause the TEND algorithm to deviate from a high-curvature fiber tract, thereby generating a poor result [38]. To compensate for this effect, the tensorline approach was proposed [65] that evolves the streamline curve based on a weighted combination of (19.16) and (19.17). As a result, the curvature can be turned on and off based on local anisotropy or prior knowledge. Streamline tractography has also been extended to multitensor models [23] and HARDI using extracted ODF maxima [26]. In these cases, streamlines follow the PDD of minimal angle with the incoming curve. A further review can be found in [112].

From a computational standpoint, these streamline approaches have many limitations, namely:

- Streamline approaches only follow one tract at a time. The algorithm cannot naturally handle situations where tracts branch or cross. One approach to address this concern is to perform a brute force implementation of the algorithm where every point of the brain is, in turn, used as a seed. The tracts that are kept are ones that flow through one or more regions of interest [73]. Even so, this brute force approach doesn't guarantee that crossing or kissing (i.e., barely touching) fibers are appropriately handled.
- These algorithms, particularly the FACT algorithm, assume that the principal diffusion direction is an accurate and error-free estimate of the fiber direction. Any error in the PDD measurement will result in accumulation of error with each step taken. Error in the PDD, while small at each step, can accumulate to the point where the streamline can "jump" into a neighboring tract, thereby giving a false display of anatomical connectivity [73].
- Despite the aforementioned issues, these tractography algorithms present a binary result: a 3D space curve. There is no representation of the confidence or accuracy of the resulting streamline tract. Attempts are being made to quantify that confidence from streamlines [116], but addressing this problem is still in the early stages.

Even so, streamline tractography has been successful in detecting major fiber tracts like the forceps major shown in Figure 19.10a.

19.3.2 Probabilistic Tractography

One of the major concerns with tractography is the amount of confidence we can have in the accuracy of the generated tracts. As such, significant work has gone into performing tractography from a probabilistic point of view. Given points A and B in a diffusion MR image $\mathcal{I}$, we can define the probability of a tract connecting A and B as

$$p(A \to B \mid \mathcal{I}) = \sum_{n=1}^{\infty} \int_{\Omega_{AB}^{n}} p(n) p(\mathbf{v}_{1:n} \mid \mathcal{I}) d\Omega_{AB}^{n} \tag{19.18}$$

where n is the length of the tract, $\mathbf{v}_{1:n}$ is a random path of length n, and Ω_{AB}^{n} is the space of fiber tracts of length n that connect A to B [42]. Given the exponential number of paths in the space Ω_{AB}^{n}, this integration cannot be done

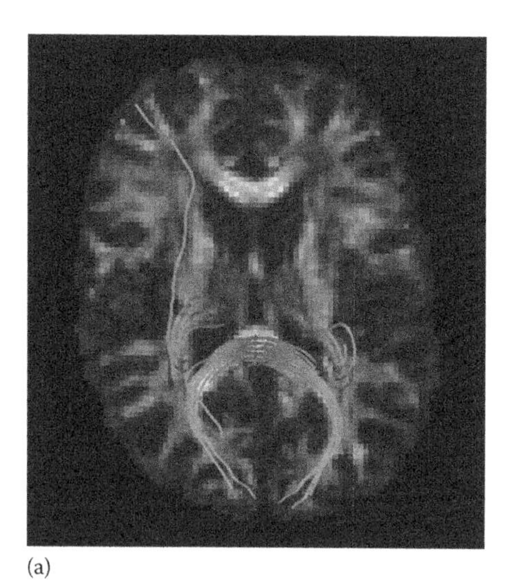
(a)

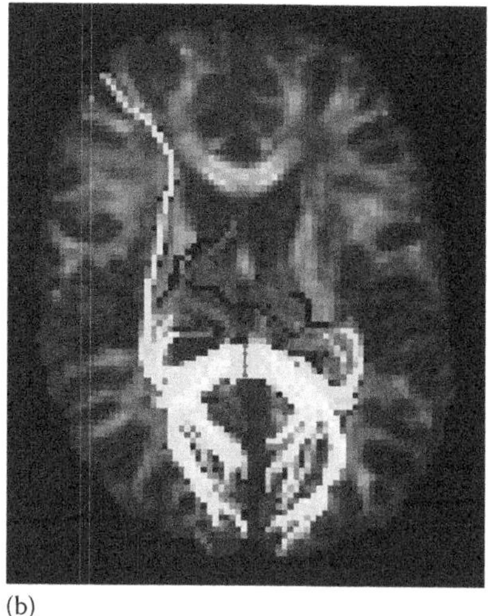
(b)

FIGURE 19.10
Examples of streamline and probabilistic tractography applied to a seed region in the splenium of the corpus callosum. Note that since probabilistic tractography uses streamline tractography as an underlying mechanism, the results are similar. (a) Streamline tractography and (b) probabilistic tractography.

analytically. Instead, the probability $p(A \rightarrow B \mid \mathcal{I})$ is sampled through the use of Markov chain Monte Carlo (MCMC) [22,25,42,63].

Conceptually, MCMC-based probabilistic tractography shares many similarities with streamline tractography algorithms. Both trace out 3D streamlines by following a local tangent vector. The differences with probabilistic tractography approaches are that instead of exclusively using the principal diffusion direction as the local tangent to the curve, we sample each tangent vector $\mathbf{v}_i$ from a given distribution $p_i(\mathbf{v}_i \mid \mathbf{v}_{i-1}, \mathcal{I})$. Also, we repeat the streamline tractography many times from the same seed A. Each resulting streamline is considered a sample of $p(A \rightarrow B \mid \mathcal{I})$. With enough of these samples (K), we can obtain a reasonable approximation of the probability that regions A and B are connected:

$$p(A \rightarrow B \mid \mathcal{I}) = \sum_{n=1}^{\infty} \sum_{k=1}^{K} p(n) \frac{\vartheta(\mathbf{v}_{1:n}^{k})}{K} \tag{19.19}$$

The function $\vartheta(\mathbf{v}_{1:n}^{k})$ is equal to one if path k connects regions A and B and zero otherwise. The prior probability $p(n)$ is usually taken to be uniform, thereby being unbiased to path length. Effectively, the probability that A and B are connected is equal to the fraction of random paths that connect A and B [25]. Probability maps containing the values from (19.19) can then be displayed and analyzed. An example is shown in Figure 19.10b.

While many probabilistic tractography algorithms have been proposed, their key differences seem to lie in how the distribution $p_i(\mathbf{v}_i|\mathbf{v}_{i-1},\mathcal{I})$ models the local tangent vectors $\mathbf{v}_i$ that make up each random path. In the succeeding text are the most popular approaches:

Diffusion profile: We noted earlier that the diffusion tensor describes a Gaussian model of diffusion. As such, this Gaussian model is commonly used to describe the distribution $p_i(\mathbf{v}_i|\mathbf{D}_i)$ [48,63]. The incoming tract direction $\mathbf{v}_{i-1}$ is modeled as being independent of the local diffusion tensor $\mathbf{D}_i$, thereby splitting the tangent vector distribution into two terms:

$$p_i(\mathbf{v}_i \mid \mathbf{v}_{i-1},\mathcal{I}) = p_i(\mathbf{v}_i,\mathcal{I})p(\mathbf{v}_i \mid \mathbf{v}_{i-1}) \tag{19.20}$$

where the diffusion data is the local diffusion tensor $\mathcal{I} = \mathbf{D}_i$. The conditional probability $p(\mathbf{v}_i|\mathbf{v}_{i-1})$ is typically chosen as a binary distribution with $p(\mathbf{v}_i|\mathbf{v}_{i-1}) = 1$ if the angle between $\mathbf{v}_i$ and $\mathbf{v}_{i-1}$ is less than ninety degrees (and zero otherwise).

These Gaussian approaches typically replace the diffusion tensor $\mathbf{D}$ with a scaled version $\mathbf{D}^\alpha$. As expected, the diffusion data contain diffuse information and α is used to decrease probabilities perpendicular to the PDD. Sample values for α range from 2 [48] to 7 [63].

HARDI versions have also followed a similar approach with a sharpened version of the diffusion ODF used for $p_i(\mathbf{v}_i, \mathcal{I})$ [39,62].

Heuristic approaches: Parker et al. proposed the probabilistic index of connectivity (PICo) approach where the tangent vector distribution also takes the form in (19.20). The key difference is how the diffusion data are used. The distribution $p_i(\mathbf{v}_i, \mathcal{I})$ is not Gaussian in this case and instead is replaced by a heuristic distribution based on local anisotropy [81]. The PDD is taken as the mean tangent direction with a cone of uncertainty whose apex angle is a function of a local anisotropy measure.

Bayesian formulations: The tangent vector distribution can also be described using Bayes' rule with respect to the diffusion data:

$$p_i(\mathbf{v}_i \mid \mathbf{v}_{i-1},\mathcal{I}) = \frac{p(\mathcal{I} \mid \mathbf{v}_i,\mathbf{v}_{i-1})p(\mathbf{v}_i \mid \mathbf{v}_{i-1})}{p(\mathcal{I})} \tag{19.21}$$

The posterior distribution $p(\mathcal{I}\,|\mathbf{v}_i, \mathbf{v}_{i-1})$ captures how well the diffusion model fits the DWI samples and is typically approximated using a Gaussian distribution on the model's residual fit [21,22,36,42]. Friman et al. update the curvature prior $p(\mathbf{v}_i|\mathbf{v}_{i-1})$ to be the dot product between the two tangent vectors [42].

Statistical bootstrap: Instead of assuming some distribution for the tangent vector, some have used bootstrap techniques to approximate the distribution from multiple samples [59,64,106]. During the image acquisition process, multiple DWIs can be obtained for each gradient direction. When it comes to fitting a diffusion model, we can do so by fitting to a randomly selected subset of the DWIs. This fitting process can be repeated for many image subsets, thereby generating multiple diffusion MR images. The distribution of the PDDs generated from this set of dMRIs can then be used as a model-free approximation to $p_i(\mathbf{v}_i, \mathcal{I})$.

In the absence of multiple DWI acquisitions, wild bootstrap can be performed [59,106]. In this situation, noise is added to the DWIs by using random perturbations of the residual of the model fit. The fitting is then reperformed for each set of noise-simulated DWIs to obtain multiple diffusion MR images from which the distribution $p_i(\mathbf{v}_i, \mathcal{I})$ can be generated.

Probabilistic tractography approaches have the advantage of characterizing uncertainty in the tractography algorithm. Even so, these methods too have their limitations:

- As each step taken along a tract contains some uncertainty, the connection probabilities we obtain using this approach are inevitably linked to the length of the tract. As such, we cannot interpret these probabilities as a measure of tract quality since they are not invariant to length [58].
- As with streamline tractography, noise can still cause the maxima of the tangent vector distribution to be off. There exists no mechanism in the tractography algorithm to correct for this error.
- The number of path samples required to approximate (19.18) is commonly on the order of thousands [22,42,59]. This results in significant computational cost and running times on the order of an hour or more for a given seed point [111]. Some recent work has tried to address this issue through, for example, the use of particle filters [111].

19.3.3 Front Propagation Tractography

A third set of tractography algorithms can be described as front propagation approaches where some form of information propagates outward from a given seed region at a speed proportional to the amount of fiber tract evidence. The information propagated by the front can then be used to reconstruct the fiber tracts. These algorithms can be divided into three main groups based on their computational aspects.

19.3.3.1 Fast Marching Tractography

Conceptually, the fast marching tractography approaches are distinguished by the calculation of a time of arrival of the propagating front for each voxel [83]. This arrival time T is related to the speed of the front F via the Eikonal equation

$$|\nabla T|F = 1 \quad \text{or} \quad T(\mathbf{r}_i) = T(\mathbf{r}_{i-1}) + \frac{|\mathbf{r}_i - \mathbf{r}_{i-1}|}{F(\mathbf{r}_i)} \tag{19.22}$$

where $\mathbf{r}_i$ and $\mathbf{r}_{i-1}$ are neighboring voxels on opposite sides of the propagating front.

With the arrival times calculated for all voxels, fiber tracts can be delineated by performing gradient descent on the arrival time map. By generating tracts in this fashion, situations of branching and merging fibers are handled naturally through the propagation of the front.

The speed F of the front is set based on the presence or absence of a fiber tract. A common choice is to use the diffusion profile as the speed function [67,68,77,93], thereby ensuring faster speed along directions of faster diffusion. Another choice is neighborhood PDD coherence [83]. By making the front speed an indicator of tract presence, we can characterize the tract's "quality" as some function of the speed. One approach is to characterize the confidence of a tract γ by its weakest link τ [83]:

$$\zeta(\gamma) = \min_{\tau} F(\gamma(\tau)) = \min_{\tau} \frac{1}{|\nabla T(\gamma(\tau))|} \tag{19.23}$$

While measuring tract quality in this fashion is a heuristic approach, it does provide us with a measure of confidence that is invariant to path length.

19.3.3.2 Tractography via Flow Simulation

Instead of using an arrival time map for tract reconstruction, we can interleave the two operations, thereby recovering the tract as we propagate the front. Noting that diffusion is fastest along a fiber tract, some researchers [17,18,32,45,61,77,96,113,117] have proposed that we simply simulate the diffusion and reconstruct candidate tracts through the analysis of the diffusion front.

The diffusion is simulated using Fick's second law [77], given as

$$\frac{\partial u}{\partial t} = \nabla\cdot(\psi \nabla u) \tag{19.24}$$

where

u is the local molecular concentration

ψ is the diffusion function (either the tensor $\boldsymbol{D}$ [18,45,61,77,113] or a diffusion ODF in the case of HARDI [32])

Given a seed point, the diffusion process is simulated for a fixed time t. The resulting concentration map u is then thresholded to obtain the hard diffusion front shown in Figure 19.11a. The voxels along the diffusion front are then scored based on a set of criteria to determine the likelihood that they are on a fiber tract [32,61,113]. Sample criteria include distance from the seed point, FA, and path curvature [61]. The diffusion is then simulated at each candidate point and the process repeats itself.

While this approach has generally gone out of favor due to the ad hoc criteria used to select fiber tract points, the ideas generated by this tractography approach have been applied elsewhere. One example is the work of O'Donnell

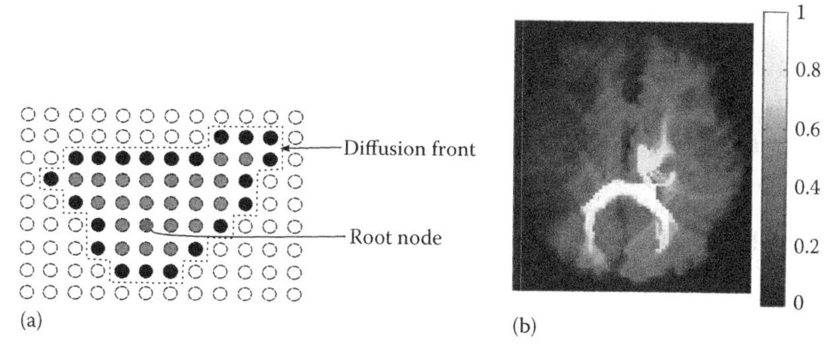

FIGURE 19.11
Examples of front propagation tractography, specifically the representation of the diffusion front and the connection strengths generated using minimal path tractography on a seed in the splenium of the corpus callosum. (a) An example diffusion front and (b) example of minimum path tractography. (Adapted from Kang, N. et al., *IEEE Transactions on Medical Imaging*, 24(9), 1127, September 2005, Booth, B.G. and Hamarneh, G. Exact integration of diffusion orientation distribution functions for graph-based diffusion MRI analysis, *Proceedings of ISBI*, pp. 935–938, 2011; respectively.)

et al. where the steady state flux (i.e., $\partial u/\partial t = 0$) is solved for and tracts are generated that maximize the resulting flux [77]. A similar approach is used by Hageman et al. where instead of modeling diffusion, they model fluid flow using the Navier–Stokes equation [47]. By using the fluid flow model, Hageman et al. are capable of modeling further concepts of the flow (e.g., viscosity).

19.3.3.3 Minimal Path Tractography Algorithms

A third set of tractography algorithms also display this concept of front propagation: graph-based minimal path algorithms [53,94,109]. These tractography algorithms discretize the image space into a graph formulation and use Dijkstra's algorithm to obtain the path of strongest diffusion. In this case, the front being propagated is the boundary between the visited and unvisited nodes.

To ensure the shortest path is the path of strongest diffusion, the edge weights in the graph are set to $w(e_{ij}) = -log(P_{diff}(i,j))$, where the pseudo-probability P_{diff} is given as

$$P_{diff}(i,j) = \frac{1}{Z}\left(\int_{(\theta,\phi)\in\beta_i} \psi_i(\theta,\phi)dS + \int_{(\theta,\phi)\in\beta_j} \psi_j(\theta,\phi)dS \right) \tag{19.25}$$

where

ψ_i is the diffusion ODF at voxel i

Z is a normalizing constant

β_i is the solid angle around the graph edge between i and j [53]

As with the fast marching algorithm, we can consider a "weakest link" connection strength here as well by selecting the largest edge weight along the tract.

One must be concerned when using this method to ensure that the angular discretization provided by the edge connectivity is fine enough to avoid diverging effects similar to those in Figure 19.9. An example of this form of tractography is shown in Figure 19.11b.

19.4 Conclusions

dMRI provides us with the ability to analyze brain connectivity noninvasively. By measuring the diffusion of water molecules along various directions in 3D and knowing that cell structure restricts molecular diffusion, we are able to infer the directional organization and integrity of fibrous tissue. Further modeling of these diffusion measurements allows us to assess characteristics such as bulk diffusivity and anisotropy. The directional dependence of the diffusion can also be used to trace out the imaged axonal fibers.

Various computational aspects of dMRI have been presented in this chapter, from acquiring diffusion-weighted MRI to modeling the diffusion through the use of diffusion tensors or diffusion ODFs to uncovering neural pathways with tractography algorithms. These analysis techniques have become even more established with their incorporation into software packages like FSL, MedINRIA, Camino, and TrackVis. These aspects of image analysis merely scratch the surface of what may be possible with this relatively new imaging technique. Already, there is work being done in the areas of segmentation [118], registration [119], and statistics of dMRI data [84]. It is hoped that continued work in dMRI will culminate in the ability to generate a human connectome: a detailed connectivity map of the human brain [30].

References

1. D. Alexander. A comparison of q-ball and PASMRI on sparse diffusion MRI data. In *Proceedings of International Society of Magnetic Resonance in Medicine (ISMRM)*, Kyoto, Japan, p. 90, 2004.
2. D.C. Alexander. Maximum entropy spherical deconvolution for diffusion MRI. In *Proceedings of Information Processing for Medical Imaging—IPMI*, Glenwood Springs, Co, Vol. 19, pp. 76–87, 2005.
3. D.C. Alexander. Multiple-fiber reconstruction algorithms for diffusion MRI. *Annals of the New York Academy of Sciences*, 1046:113–133, 2005.
4. D.C. Alexander. *Visualization and Processing of Tensor Fields*, Chapter 5: An introduction to computational diffusion MRI: The diffusion tensor and beyond, pp. 83–106. Springer Berlin Heidelberg, Heidelberg, Germany, 2006.

5. D.C. Alexander, G.J. Barker, and S.R. Arridge. Detection and modeling of non-Gaussian apparent diffusion coefficient profiles in human brain data. *Magnetic Resonance in Medicine*, 48:331–340, 2002.
6. A.W. Anderson. Measurement of fiber orientation distributions using high angular resolution diffusion imaging. *Magnetic Resonance in Medicine*, 54:1194–1206, 2005.
7. J.L.R. Andersson and S. Skare. A model-based method for retrospective correction of geometric distortions in diffusion-weighted EPI. *NeuroImage*, 16:177–199, 2002.
8. Y. Assaf and P. Basser. Composite hindered and restricted model of diffusion (CHARMED) MR imaging of the human brain. *NeuroImage*, 27(1):48–58, 2005.
9. Y. Assaf, R.Z. Freidlin, G.K. Rohde, and P.J. Basser. New modeling and experimental framework to characterize hindered and restricted water diffusion in brain white matter. *Magnetic Resonance in Medicine*, 52:965–978, 2004.
10. Y. Bai and D.C. Alexander. Model-based registration to correct for motion between acquisitions in diffusion MR imaging. In *Proceedings of International Symposium on Biomedical Imaging—ISBI*, Paris, France, pp. 947–950, 2008.
11. A. Barmpoutis, M.S. Hwang, D. Howland, J.R. Forder, and B.C. Vemuri. Regularized positive-definite fourth order tensor field estimation from DW-MRI. *NeuroImage*, 45:S153–S162, 2009.
12. A. Barmpoutis, B.C. Vemuri, D. Howland, and J.R. Forder. Extracting tractosemas from a displacement probability field for tractography for DW-MRI. In D. Metaxas, L. Axel, G. Fichtinger, and G. Székely, eds., *Medical Image Computing and Computer-Assisted Intervention—MICCAI 2008*, Vol. 5241 of *LNCS*, pp. 9–16. Springer, Heidelberg, Germany, 2008.
13. P.J. Basser and D.K. Jones. Diffusion-tensor MRI: Theory, experimental design and data analysis—A technical overview. *NMR in Biomedicine*, 15:456–467, 2002.
14. P.J. Basser, J. Mattiello, and D. LeBihan. MR diffusion tensor spectroscopy and imaging. *Biophysical Journal*, 66:259–267, 1994.
15. P.J. Basser, S. Pajevic, C. Pierpaoli, J. Duda, and A. Aldroubi. In vivo fiber tractography using DT-MRI data. *Magnetic Resonance in Medicine*, 44:625–632, 2000.
16. S. Basu, T. Fletcher, and R. Whitaker. Rician noise removal in diffusion tensor MRI. In *Medical Image Computing and Computer-Assisted Intervention—MICCAI*, Vol. 4190 of *LNCS*, pp. 117–125. Springer, Heidelberg, Germany, 2006.
17. P.G. Batchelor, D.L.G. Hill, D. Atkinson, F. Calamanten, and A. Connelly. Fibre-tracking by solving the diffusion-convection equation. In *Proceedings of International Society of Magnetic Resonance in Medicine (ISMRM)*, Honolulu, Vol. 10, 2002.
18. P.G. Batchelor, D.L.G. Hill, F. Calamante, and D. Atkinson. Study of connectivity in the brain using the full diffusion tensor from MRI. In M.F. Insana and R.M. Leahy, eds., *Proceedings of Information Processing in Medical Imaging: IMPI 2001*, Davis, CA, pp. 121–133, 2001.
19. C. Beaulieu. The basis of anisotropic water diffusion in the nervous system—A technical review. *NMR in Biomedicine*, 15:435–455, 2002.
20. C. Beaulieu. The biological basis of diffusion tractography. In *Proceedings of International Symposium on Biomedical Imaging (ISBI)*, Arlington, VA, pp. 347–350, 2006.
21. T.E.J. Behrens, H.J. Berg, S. Jbabdi, M.F.S. Rushworth, and M.W. Woolrich. Probabilistic diffusion tractography with multiple fibre orientations: What can we gain? *NeuroImage*, 34:144–155, 2007.

22. T.E.J. Behrens, M.W. Woolrich, M. Jenkinson, H. Johansen-Berg, R.G. Nunes, S. Clare, P.M. Matthews, J.M. Brady, and S.M. Smith. Characterization and propagation of uncertainty in diffusion-weighted MR imaging. *Magnetic Resonance in Medicine*, 50:1077–1088, 2003.
23. O. Bergmann, G. Kindlmann, S. Peled, and C.-F. Westin. Two-tensor fiber tractography. In *Proceedings of International Symposium on Biomedical Imaging (ISBI)*, Washington, DC, pp. 796–799, 2007.
24. D.L. Bihan and E. Breton. Imagerie de diffusion in vivo par résonance magnétique nucléaire. *Compte Rendus de l'Académie de Sciences Paris*, 301:1109–1112, 1985.
25. M. Björnemo, A. Brun, R. Kikinis, and C.-F. Westin. Regularized stochastic white matter tractography using diffusion tensor MRI. In T. Dohi and R. Kikinis, eds., *Medical Image Computing and Computer-Assisted Intervention—MICCAI*, Tokoyo, Japan, Vol. 2488, pp. 435–442, 2002.
26. L. Bloy and R. Verma. On computing the underlying fiber directions from the diffusion orientation distribution function. In D. Metaxas, L. Axel, G. Fichtinger, and G. Székely, eds., *Medical Image Computing and Computer-Assisted Intervention—MICCAI 2008*, Vol. 5241 of *LNCS*, pp. 1–8. Springer, Heidelberg, Germany, 2008.
27. N. Bodammer, J. Kaufmann, M. Kanowski, and C. Tempelmann. Eddy current correction in diffusion-weighted imaging using pairs of images acquired with opposite diffusion gradient polarity. *Magnetic Resonance in Medicine*, 51:188–193, 2004.
28. B.G. Booth and G. Hamarneh. Exact integration of diffusion orientation distribution functions for graph-based diffusion MRI analysis. In *Proceedings of ISBI*, Chicago, pp. 935–938, 2011.
29. R. Brown. A brief account of microscopical observations made in the months of June, July and August 1827 on the particles contained in the pollen of plants; and on the general existence of active molecules in organic and inorganic bodies. *Philosophical Magazine*, 4:161, 1828.
30. E. Bullmore and O. Sporns. Complex brain networks: Graph theoretical analysis of structural and functional systems. *Nature Review Neuroscience*, 10(3):186–198, March 2009.
31. U. Bürgel, K. Amunts, L. Hoemke, H. Mohlberg, J.M. Gilsbach, and K. Zilles. White matter fiber tracts of the human brain: Three-dimensional mapping at microscopic resolution, topography and intersubject variability. *NeuroImage*, 29:1092–1105, 2006.
32. J.S.W. Campbell, K. Siddiqi, V.V. Rymar, A.F. Sadikot, and G.B. Pike. Flow-based fiber tracking with diffusion tensor and q-ball data: Validation and comparison to principal diffusion direction techniques. *NeuroImage*, 27:725–736, 2005.
33. L.-C. Chang, D.K. Jones, and C. Pierpaoli. RESTORE: Robust estimation of tensors by outlier rejection. *Magnetic Resonance in Medicine*, 53:1088–1095, 2005.
34. B. Chen, H. Guo, and A.W. Song. Correction for direction-dependent distortions in diffusion tensor imaging using matched magnetic field maps. *NeuroImage*, 30:121–129, 2006.
35. O. Christiansen, T.-M. Lee, J. Lie, U. Sinha, and T.F. Chan. Total variation regularization of matrix-valued images. *International Journal of Biomedical Imaging*, 2007(27432):11, 2007.
36. P.A. Cook, H. Zhang, S.P. Awate, and J.C. Gee. Atlas-guided probabilistic diffusion-tensor fiber tractography. In *Proceeding of International Symposium on Biomedical Imaging (ISBI)*, Paris, France, pp. 951–954, 2008.

37. O. Coulon, D.C. Alexander, and S. Arridge. Diffusion tensor magnetic image regularization. *Medical Image Analysis*, 8:47–67, 2004.
38. S. Crettenand, S.D. Meredith, M.J. Hoptman, and R.B. Reilly. Quantitative analysis and comparison of diffusion tensor imaging tractography algorithms. In *Proceedings of Irish Signals and Systems Conference—ISSC*, Dublin, Ireland, pp. 105–110, 2006.
39. M. Descoteaux, R. Deriche, T.R. Knösche, and A. Anwander. Deterministic and probabilistic tractography based on complex fibre orientation distributions. *IEEE Transaction Medical Imaging*, 28(2):269–286, 2009.
40. M. Descoteaux, E. Angelino, S. Fitzgibbons, and R. Deriche. Regularized, fast, and robust analytical q-ball imaging. *Magnetic Resonance in Medicine*, 58:497–510, 2007.
41. A. Einstein. Über die von der molekularkinetischen theorie der Wärme geforderte bewegung von in ruhenden flussigkeiten suspendierten teilchen. *Annalen der Physik*, 4:549–590, 1905.
42. O. Friman, G. Farneback, and C.-F. Westin. A Bayesian approach for stochastic white matter tractography. *IEEE Transactions on Medical Imaging*, 25(8):965–978, 2006.
43. C. Frindel, M. Robini, P. Croisille, and Y.-M. Zhu. Comparison of regularization methods for human cardiac diffusion tensor MRI. *Medical Image Analysis*, 13:405–418, 2009.
44. J. Gee and D. Alexander. *Welckert and Hagen: Visualization and Image Processing of Tensor Fields*, Chapter 20—Diffusion tensor image registration. Springer, Berlin, Germany, 2005.
45. D. Gembris, H. Schumacher, and D. Suter. Solving the diffusion equation for fiber tracking in the living human brain. In *Proceedings of International Society of Magnetic Resonance in Medicine (ISMRM)*, Glasgow, UK, p. 1529, 2001.
46. A.C. Guyton. *Textbook of Medical Physiology*, 8th edn. W.B. Saunders, Philadelphia, PA, 1991.
47. N.S. Hageman, A.W. Toga, K. Narr, and D.W. Shattuck. A diffusion tensor imaging tractography algorithm based on navier-stokes fluid mechanics. *IEEE Transactions on Medical Imaging*, 28:348–360, 2009.
48. P. Hagmann, J.-P. Thiran, L. Jonasson, P. Vandergheynst, S. Clarke, P. Maeder, and R. Meuli. DTI mapping of human brain connectivity: Statistical fibre tracking and virtual dissection. *NeuroImage*, 19:545–554, 2003.
49. G. Hamarneh and J. Hradsky. Bilateral filtering of diffusion tensor magnetic resonance images. *IEEE Transactions on Image Processing*, 16(10):2463–2475, October 2007.
50. K.M. Hasan, P.J. Basser, D.L. Parker, and A.L. Alexander. Analytical computation of the eigenvalues and eigenvectors in DT-MRI. *Journal of Magnetic Resonance*, 152:41–47, 2001.
51. J.C. Haselgrove and J.R. Moore. Correction for distortion of echo-planar images used to calculate the apparent diffusion coefficient. *Magnetic Resonance in Medicine*, 36:960–964, 1996.
52. M.O. Irfanoglu, C.G. Koay, S. Pajevic, R. Machiraju, and P.J. Basser. Diffusion tensor field registration in the presence of uncertainty. In *Medical Image Computing and Computer-Assisted Intervention—MICCAI*, Vol. 5761 of *LNCS*, pp. 181–189. Springer, Heidelberg, Germany, 2009.

53. Y. Iturria-Medina, E.J. Canales-Rodríguez, L. Melie-García, P.A. Valdés-Hernández, E. Martínez-Montes, Y. Alemán-Gómez, and J.M. Sánchez-Bornot. Characterizing brain anatomical connections using diffusion weighted MRI and graph theory. *NeuroImage*, 36:645–660, 2007.
54. K.M. Jansons and D.C. Alexander. Persistant angular structure: New insights from diffusion magnetic resonance imaging data. *Inverse Problems in Physics*, 19:1031–1046, 2003.
55. B. Jian and B.C. Vemuri. A unified computational framework for deconvolution to reconstruct multiple fibers from diffusion weighted MRI. *IEEE Transaction Medical Imaging*, 26(11):1464–1471, 2007.
56. B. Jian and B.C. Vemuri. Multi-fiber reconstruction from diffusion MRI using mixture of Wisharts and sparse deconvolution. In N. Karssemeijer and B. Lelieveldt, eds., *Proceedings of Image Processing for Medical Imaging (IPMI)*, Vol. 4584 of *LNCS*, pp. 384–395. Springer, Heidelberg, Germany, 2007.
57. B. Jian, B.C. Vemuri, E. Özarslan, P.R. Carney, and T.H. Mareci. A novel tensor distribution model for the diffusion-weighted MR signal. *NeuroImage*, 37:164–176, 2007.
58. D.K. Jones. Studying connections in the living human brain with diffusion MRI. *Cortex*, 44(8):936–952, September 2008.
59. D.K. Jones. Tractography gone wild: Probabilistic fiber tracking using the wild bootstrap with diffusion MRI. *IEEE Transactions on Medical Imaging*, 27(9):1268–1274, September 2008.
60. D.K. Jones and P.J. Basser. Squashing peanuts and smashing pumpkins: How noise distorts diffusion-weighted MR data. *Magnetic Resonance in Medicine*, 52:979–993, 2004.
61. N. Kang, J. Zhang, E.S. Carlson, and D. Gembris. White matter fiber tractography via anisotropic diffusion simulation in the human brain. *IEEE Transactions on Medical Imaging*, 24(9):1127–1137, September 2005.
62. I. Kezele, M. Descoteaux, C. Poupon, F. Poupon, and J.-F. Mangin. Spherical wavelet transform for ODF sharpening. *Medical Image Analysis*, 14:332–342, 2010.
63. M.A. Koch, D.G. Norris, and M. Hund-Georgiadis. An investigation of functional and anatomical connectivity using magnetic resonance imaging. *NeuroImage*, 16:241–250, 2002.
64. M. Lazar and A.L. Alexander. Bootstrap white matter tractography (BOOT-TRAC). *NeuroImage*, 24:524–532, 2005.
65. M. Lazar, D.M. Weinstein, J.S. Tsuruda, K.M. Hasan, K. Arfanakis, M.E. Meyerand, B. Badie, H.A. Rowley, V. Haughton, A. Field, and A.L. Alexander. White matter tractography using diffusion tensor deflection. *Human Brain Mapping*, 18:306–321, 2003.
66. A. Leemans and D.K. Jones. The B-matrix must be rotated when correcting for subject motion in DTI data. *Magnetic Resonance in Medicine*, 61:1336–1349, 2009.
67. C. Lenglet, R. Deriche, and O. Faugeras. Diffusion tensor magnetic resonance imaging: Brain connectivity mapping. Technical Report 4983, INRIA, Sophia Antipolis, France, October 2003.
68. C. Lenglet, R. Deriche, and O. Faugeras. Inferring white matter geometry from diffusion tensor MRI: Application to connectivity mapping. In *Proceedings of European Conference on Computer Vision (ECCV)*, Prague, CZE, pp. 127–140, 2004.
69. A.D. Leow, S. Zhu, K. McMahon, G.I. de Zubicaray, M. Meredith, M. Wright, and P.M. Thompson. The tensor distribution model. In *Proceedings of the International Symposium on Biomedical Imaging (ISBI)*, Paris, France, pp. 863–866, May 2008.

70. J.-F. Mangin, C. Poupon, C. Clark, D. Le Bihan, and I. Bloch. Distortion correction and robust tensor estimation for MR diffusion imaging. *Medical Image Analysis*, 6:191–198, 2002.
71. S. Mori. John Hopkins Medical Institute: Laboratory of Brain Anatomical MRI, in vivo human database. http://lbam.med.jhmi.edu/ (accessed February 2009).
72. S. Mori, Barbara J. Crain, V.P. Chacko, and Peter C.M. van Zijl. Three-dimensional tracking of axonal projections in the brain by magnetic resonance imaging. *Annals of Neurology*, 45(2):265–269, 1999.
73. S. Mori and P.C.M. van Zijl. Fiber tracking: Principles and strategies—A technical review. *Nuclear Magnetic Resonance in Biomedicine*, 15:468–480, 2002.
74. P. Mukherjee, J.I. Berman, S.W. Chung, C.P. Hess, and R.G. Henry. Diffusion tensor MR imaging and fiber tractography: Theoretic underpinnings. *American Journal of Neuroradiology*, 29:632–641, April 2008.
75. P. Mukherjee, S.W. Chung, J.I. Berman, C.P. Hess, and R.G. Henry. Diffusion tensor MR imaging and fiber tractography: Technical considerations. *American Journal of Neuroradiology*, 29:843–852, May 2008.
76. P.G.P. Nucifora, R. Verma, S.-K. Lee, and E.R. Melhem. Diffusion-tensor MR imaging and tractography: Exploring brain microstructure and connectivity. *Radiology*, 245(2):367–384, November 2007.
77. L. O'Donnell, S. Haker, and C.-F. Westin. New approaches to estimation of white matter connectivity in diffusion tensor MRI: Elliptic PDEs and geodesics in a tensor-warped space. In *Medical Image Computing and Computer-Assisted Intervention—MICCAI 2002*, pp. 459–466, Tokyo, Japan, 2002.
78. E. Ozarslan, B.C. Vemuri, and T.H. Mareci. Generalized scalar measures for diffusion MRI using trace, variance, and entropy. *Magnetic Resonance in Medicine*, 53:866–876, 2005.
79. E. Özerslan, T. Shepherd, B. Vemuri, S. Blackband, and T. Mareci. Resolution of complex tissue microarchitecture using the diffusion orientation transform (DOT). *NeuroImage*, 31(3):1086–1103, 2006.
80. S. Pajevic and C. Pierpaoli. Color schemes to represent the orientation of anisotropic tissues from diffusion tensor data: Application to white matter fiber tract mapping in the human brain. *Magnetic Resonance in Medicine*, 42:526–540, 1999.
81. G.J.M. Parker, H.A. Haroon, and C.A.M. Wheeler-Kingshott. A framework for streamline-based probabilistic index of connectivity (PICo) using a structural interpretation of MRI diffusion measurements. *Journal of Magnetic Resonance Imaging*, 18:242–254, 2003.
82. G.J.M. Parker, J.A. Schnabel, M.R. Symms, D.J. Werring, and G.J. Barker. Nonlinear smoothing for reduction of systematic and random errors in diffusion tensor imaging. *Journal of Magnetic Resonance Imaging*, 11:702–710, 2000.
83. G.J.M. Parker, C.A.M. Wheeler-Kingshott, and G.J. Barker. Estimating distributed anatomical connectivity using fast marching methods and diffusion tensor imaging. *IEEE Transactions on Medical Imaging*, 21(5):505–512, 2002.
84. O. Pasternak, N. Sochen, and P.J. Basser. The effect of metric selection on the analysis of diffusion tensor MRI data. *NeuroImage*, 49:2190–2204, 2010.
85. X. Pennec, P. Fillard, and N. Ayache. A Riemannian framework for tensor computing. *International Journal of Computer Vision*, 66(1):41–66, 2006.
86. C. Poupon, C.A. Clark, V. Frouin, J. Régis, I. Bloch, D. Le Bihan, and J.-F. Mangin. Regularization of diffusion-based direction maps for the tracking of brain white matter fascicles. *NeuroImage*, 12:184–195, 2000.

87. A. Ramirez-Manzanares, P.A. Cook, and J.C. Gee. A comparison of methods for recovering intra-voxel white matter fiber architecture from clinical diffusion imaging scans. In *Medical Image Computing and Computer-Assisted Intervention—MICCAI*, Vol. 5241 of *LNCS*, pp. 305–312. Springer, Heidelberg, Germany, 2008.
88. Y. Rathi, O. Michailovich, M.E. Shenton, and S. Bouix. Directional functions for orientation distribution estimation. *Medical Image Analysis*, 13:432–444, 2009.
89. C.A.-L. Rodrigo de Luis-Garcia and C.-F. Westin. *Tensors in Image Processing and Computer Vision*, Chapter Segmentation of Tensor Fields: Recent Advances and Perspectives. Springer, London, U.K., 2009.
90. G.K. Rohde, A.S. Barnett, P.J. Basser, S. Marenco, and C. Pierpaoli. Comprehensive approach for correction of motion and distortion in diffusion-weighted MRI. *Magnetic Resonance in Medicine*, 51:103–114, 2004.
91. K.E. Sakaie and M.J. Lowe. Quantitative assessment of motion correction for high angular resolution diffusion imaging. *Magnetic Resonance Imaging*, 28:290–296, 2010.
92. P. Savadjiev, J.S.W. Campbell, G.B. Pike, and K. Siddiqi. 3D curve inference for diffusion MRI regularization and fibre tractography. *Medical Image Analysis*, 10:799–813, 2006.
93. G. Sebastiani, F. de Pasquale, and P. Barone. Quantifying human brain connectivity from diffusion tensor MRI. *Journal of Mathematical Imaging and Vision*, 25:227–244, 2006.
94. S.N. Sotiropoulos, L. Bai, P.S. Morgan, C.S. Constantinescu, and C.R. Tench. Brain tractography using Q-ball imaging and graph theory: Improved connectivities through fibre crossings via a model-based approach. *NeuroImage*, 49:2444–2456, 2010.
95. E.O. Stejskal and J.E. Tanner. Spin diffusion measurements: Spin echoes in the presence of a time-dependent field gradient. *The Journal of Chemical Physics*, 42(1):288–292, January 1965.
96. J.-D. Tournier, F. Calamante, D.G. Gadian, and A. Connelly. Diffusion-weighted magnetic resonance imaging fibre tracking using a front evolution algorithm. *NeuroImage*, 20:276–288, 2003.
97. A. Tristan-Vega and S. Aja-Fernandez. DWI filtering using joint information for DTI and HARDI. *Medical Image Analysis*, 14:205–218, 2010.
98. T.P. Trouard, Y. Sabharwal, M.I. Altbach, and A.F. Gmitro. Analysis and comparison of motion correction techniques in diffusion-weighted imaging. *Journal of Magnetic Resonance Imaging*, 6(6):925–935, 1996.
99. D. Tschumperle and R. Deriche. Diffusion tensor regularization with constraints preservation. In *Proceedings of Computer Vision and Pattern Recognition—CVPR*, Kawai, HI, Vol. 1, pp. 948–953, 2001.
100. D.S. Tuch. Q-ball imaging. *Magnetic Resonance in Medicine*, 52:1358–1372, 2004.
101. D.S. Tuch, T.G. Reese, M.R. Wiegell, N. Makris, J.W. Belliveau, and V.J. Wedeen. High angular resolution diffusion imaging reveals intravoxel white matter fiber heterogeneity. *Magnetic Resonance in Medicine*, 48:577–582, 2002.
102. Z. Wang, B.C. Vemuri, Y. Chen, and T.H. Mareci. A constrained variational principle for direct estimation and smoothing of the diffusion tensor field from complex DWI. *IEEE Transactions on Medical Imaging*, 23(8):930–939, 2004.
103. V. Wedeen, T. Reese, D. Tuch, M. Wiegel, J.-G. Dou, R. Weiskoff, and D. Chessler. Mapping fiber orientation spectra in cerebral white matter with fourier-transform diffusion MRI. In *Proceedings of the International Society of Magnetic Resonance in Medicine*, Denver, CO, p. 82, 2000.

104. J. Weickert and T. Brox. Diffusion and regularization of vector- and matrix-valued images. Technical Report 58, Universität des Saarlandes, Saarbrücken, 2002.
105. C.-F. Westin, S.E. Maier, H. Mamata, A. Nabavi, F.A. Jolesz, and R. Kikinis. Processing and visualization for diffusion tensor MRI. *Medical Image Analysis*, 6:93–108, 2002.
106. B. Whitcher, D.S. Tuch, J.J. Wisco, A.G. Sorensen, and L. Wang. Using the wild bootstrap to quantify uncertainty in diffusion tensor imaging. *Human Brain Mapping*, 29:346–362, 2008.
107. N. Wiest-Daessle, S. Prima, P. Coupe, S.P. Morrissey, and C. Barillot. Non-local means variants for denoising of diffusion-weighted and diffusion tensor MRI. In *Medical Image Computing and Computer-Assisted Intervention—MICCAI*, Vol. 4792 of *LNCS*, pp. 344–351. Springer, Heidelberg, Germany, 2007.
108. E. Yoruk and B. Acar. Structure preserving regularization of DT-MRI vector fields by nonlinear anisotropic diffusion filtering. In *Proceedings of European Signal Processing Conference (EUSIPCO)*, Antalya, Turkey, p. 4, 2005.
109. A. Zalesky. DT-MRI fiber tracking: A shortest paths approach. *IEEE Transaction Medical Imaging*, 27(10):1458–1571, 2008.
110. Fan Zhang and E.R. Hancock. Tensor MRI regularization via graph diffusion. In *Proceedings of British Machine Vision Conference (BMVC)*, Edinburgh, UK, pp. 578–589, 2006.
111. F. Zhang, E.R. Handcock, C. Goodlett, and G. Gerig. Probabilistic white matter fiber tracking using particle filtering and von Mises-Fisher sampling. *Medical Image Analysis*, 13:5–18, 2009.
112. J. Zhang, H. Ji, N. Kang, and N. Cao. Fiber tractography in diffusion tensor magnetic resonance imaging: A survey and beyond. Technical Report 437–05, University of Kentucky, Lexington, KY, April 2005.
113. J. Zhang, N. Kang, and S.E. Rose. Approximating anatomical brain connectivity with diffusion tensor MRI using kernel-based diffusion simulations. In *Proceedings of Information Processing for Medical Imaging—IPMI*, Vol. 3565 of *LNCS*, pp. 64–75. Springer, Heidelberg, Germany, 2005.
114. K. Nand, R. Abugharbieh, B. Booth, and G. Hamarneh. Detecting structure in diffusion tensor MR images. In Lecture Notes in Computer Science, *Medical Image Computing and Computer-Assisted Intervention (MICCAI)*, Toronto, CA, 6892:90–97, 2011.
115. G. Hamarneh, C. McIntosh, and M. Drew. Perception-based visualization of manifold-valued medical images using distance-preserving dimensionality reduction. *IEEE Transactions on Medical Imaging (IEEE TMI)*, 30(7):1314–1327, 2011.
116. C. Brown, B.G. Booth, and G. Hamarneh. K-Confidence: Assessing uncertainty in tractography using k optimal paths. In *IEEE International Symposium on Biomedical Imaging (IEEE ISBI)*, San Francisco, 250–253, 2013.
117. B.G. Booth and G. Hamarneh. Global multi-region competitive tractography. In *IEEE workshop on Mathematical Methods for Biomedical Image Analysis (IEEE MMBIA)*, Breckenridge, CO, 73–78, 2012.
118. B.G. Booth and G. Hamarneh. A cross-sectional piecewise constant model for segmenting highly curved fiber tracts in diffusion MR images. In Lecture Notes in Computer Science, *Medical Image Computing and Computer-Assisted Intervention (MICCAI)*, Nagoya, Japan, to appear, 2013.
119. B.G. Booth and G. Hamarneh. Consistent information content estimation for diffusion tensor MR images. *In IEEE Conference on Healthcare Informatics, Imaging and Systems Biology (IEEE HISB)*, San Jose, 166–173, 2011.

20

T_{1rho} MR Imaging: Principle, Technology, and Application

Jing Yuan and Yi-Xiang J. Wang

CONTENTS

20.1 Principle of T_{1rho} Relaxation Time and T_{1rho} MRI

20.1.1 From T_1, T_2 to T_{1rho} Relaxation Time

Magnetic resonance imaging (MRI) is a powerful and versatile medical imaging modality and has been extensively applied to routine clinical practice. In addition to its nonionizing radiation nature, a distinct advantage of MRI compared to other imaging modalities like computed tomography (CT), x-ray, and nuclear medicine is that MRI is able to provide superior and versatile soft tissue contrasts based on the intrinsic properties of tissues. Proton density–weighted contrast, T_1-weighted contrast, and T_2-weighted contrast are three common contrast mechanisms routinely used for clinical applications.

The process of the radiofrequency (RF) excited spins (or protons) returning to the equilibrium, realigned with the original longitudinal direction of the static main magnetic field B_0, is called relaxation. Spin–lattice relaxation rate R_1 characterizes the rate at which the longitudinal magnetization component recovers exponentially toward its thermodynamic equilibrium along B_0 direction. The inverse of R_1 is called spin–lattice relaxation time T_1. Spin–spin relaxation rate R_2 and its inverse T_2 is used to characterize the procedure that the transverse magnetization component decays exponentially toward zero in the plane perpendicular to B_0. T_1 relaxation involves the interaction of spins with the surround environment (lattice), including atoms and molecules, of varying sizes and shapes, in spontaneous motions at different frequencies. The interaction efficiency between spin and lattice maximizes at the Larmor frequency. If a tissue contains more atoms and molecules at the motion frequencies around Larmor frequency, the spin–lattice interaction and energy exchange will be more active and lead to the fast energy dissipation of spins and hence the shorter T_1 relaxation time. Different from T_1, T_2 relaxation primarily involves the dephasing procedure in the transverse plane due to the slight difference of resonant frequency and processing rate for each spin. After a short while, the phases of spins are no longer synchronized due to the different processing rates. This incoherent phase distribution is reflected by the decay of the bulk transverse magnetization macroscopically. The faster dephasing procedure is associated with shorter T_2 relaxation time and vice versa. Based on the differences of T_1 and T_2 for different tissues, MR images with T_1-weighted contrast and T_2-weighted contrast can be generated by manipulating MRI pulse sequences and imaging parameters.

T_{1rho} (or $T_{1\rho}$) relaxation time is fully named as spin–lattice relaxation time in the rotating frame and presents a relaxation procedure different from T_1 and T_2. T_{1rho}-weighted contrast also provides extra information of tissues beyond T_1- and T_2-weighted contrasts. T_{1rho} relaxation time was first described long ago but applied for MRI only from the late 1980s. As mentioned earlier, magnetization undergoes T_2 relaxation in the transverse plane and decays exponentially due to the dephasing process. However, if an external RF pulse is applied aligned with the spins with the same processing frequency as spins in the transverse plane, the spins are considered stationary relative to this external RF field in the rotating frame. In the presence of this external RF field, spins undergo with a slower relaxation rate than the normal R_2, which is named as spin–lattice relaxation rate in the rotating frame, or R_{1rho}. Intuitively, the inverse of R_{1rho} is named as T_{1rho} relaxation time. This external RF pulse to slow down the decay rate of transverse magnetization is called spin-lock RF pulse. The reason for the longer T_{1rho} than T_2 could be explained by the effect of the spin-lock pulse field strength on spins. Similar to the alignment of spins along the main magnetic field B_0, some spins in the transverse plane would be aligned with the spin-lock pulse field strength B_{SL}, rotating at a rate determined by the Larmor equation, FSL = γB_{SL}, where γ is the gyromagnetic ratio of hydrogen atom of 42.58 MHz/T and FSL is called spin-lock frequency. Note that although FSL has a unit of Hz as frequency, it is proportional to the spin-lock field strength and used to evaluate the field strength of a spin-lock pulse. Here, we could neglect the precession rate of spins with B_0 in the rotating frame because the spin-lock pulse is assumed on-resonance. The impose of a spin-lock pulse to some degree forces spins to precess along the spin-lock field direction and alters the tendency of the spins to precess at their own individual frequencies in the transverse plane. As such, the phases of the processing spins with spin-lock field are more coherent compared to the absence of the spin-lock field. Consequently, the dephasing procedure of spins in the transverse plane is slowed down, which leads to a slower relaxation rate of R_{1rho} than R_2 and hence a longer relaxation time T_{1rho} than T_2. It is intuitive to imagine that the higher spin-lock field strength is able to more strongly "lock" the transverse spins along the spin-lock field direction and results in a longer T_{1rho} relaxation time. It is analogous to the fact that T_1 relaxation time increases with the main magnetic field strength B_0. On the other hand, with the reduction of spin-lock field strength, T_{1rho} reduces and finally equals to T_2 if the spin-lock field strength reduces to zero.

In a short summary, T_{1rho} relaxation could be considered as a slowed-down T_2 relaxation in the presence of the spin-lock field strength. T_{1rho} relaxation behave also like T_1 relaxation in the spin-lock magnetic field strength, although much smaller than B_0. Meanwhile, the spin-lock field is often applied in the transverse plane rather than the longitudinal direction of B_0. The value of T_{1rho} is usually larger than T_2 while smaller than T_1.

As mentioned earlier, T_1 relaxation is sensitive to the spin–lattice motional processes that are at or around the Larmor frequency. For modern high field MRI scanners, the corresponding proton Larmor frequency for the main magnetic field is very high, for instance, about 64 MHz at 1.5T and 128 MHz at 3T. For biological tissues, we may be more interested in low-frequency motional processes (100 Hz to few KHz) in the spin–lattice interaction as these processes may provide useful information on composition of macromolecules such as protein and proton exchange between water and macromolecules. All such information could be closely associated with physiological and pathological processes and eventually benefit clinical practice. However, it is difficult to directly detect and characterize these low-frequency motional processes through T_1 relaxation time at very low B_0 due to both the unavailability of such extremely low field scanner and the poor signal-to-noise ratio (SNR) at low field. T_{1rho} imaging provides us a viable approach to study low-frequency motional processes in biological tissues that is not feasible by using T_1 and T_2 imaging. The spin-lock frequency can be tailored for studies of biological processes at different frequencies. In addition, T_{1rho} imaging can also be performed on high field MRI scanner to take the advantage of excellent image SNR.

20.1.2 Spin-Lock Radiofrequency Pulse and T_{1rho} Imaging

In Section 20.1.1, we only introduce how to make spins precess around the spin-lock field direction in the rotating frame by applying a spin-lock RF pulse. This procedure refers to T_{1rho}-weighted magnetization preparation. Like any other magnetization preparations such as saturation, inversion, and magnetization transfer, T_{1rho}-weighted magnetization preparation itself cannot generate T_{1rho}-weighted MR images unless it is combined with the subsequent pulse sequences to acquire the T_{1rho} relaxation–prepared MR signal. This section will briefly introduce the basic implementation of T_{1rho} imaging by the combination of spin-lock RF pulse and acquisition pulse sequence.

The diagram of a normal spin-lock pulse cluster is illustrated in Figure 20.1. The initial hard pulse P_1 is applied along X direction and flips magnetization into the transverse plane. A spin-lock pulse P_{SL} is applied subsequently along Y direction with a duration of TSL and an RF field strength of B_{SL}. Spin-lock frequency FSL is determined by $FSL = \gamma B_{SL}$, and the flip angle of P_{SL} is calculated by $\Phi_Y = 2\pi \cdot FSL \cdot TSL$ (in rad). A final hard pulse P_2 is applied opposite to P_1 subsequent to P_{SL} to return the T_{1rho}-relaxed transverse magnetization to the longitudinal axis. The evolution of magnetization under the application of this spin-lock pulse in the rotating frame is also depicted in Figure 20.1. Note the status of magnetization at different time points (t_0–t_3) and the decay of the transverse magnetization along B_{SL}. Since it is in the rotating frame, the on-resonance transverse magnetization is always along Y direction.

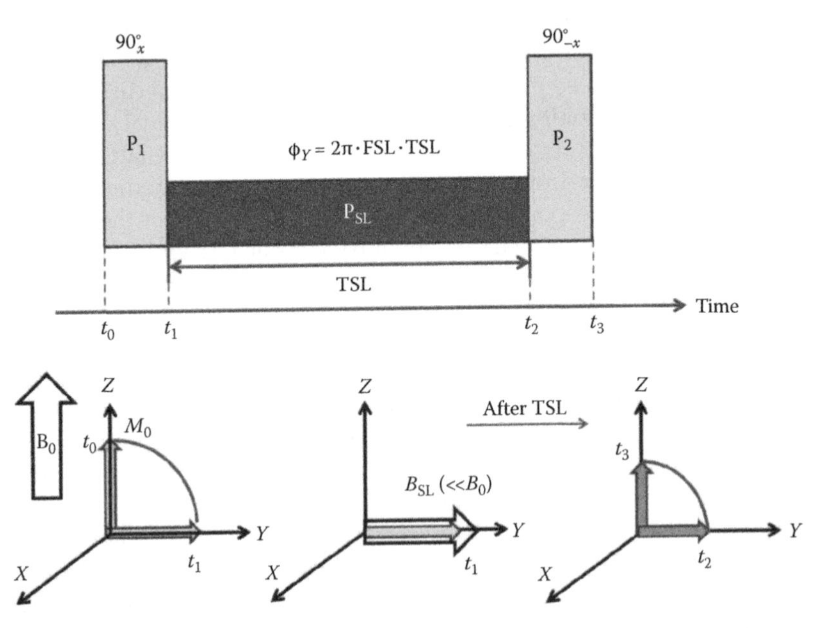

FIGURE 20.1
The diagram of a normal spin-lock pulse cluster and the magnetization revolution during the application of this spin-lock pulse.

Figure 20.2 illustrates the block diagram for a typical T_{1rho} imaging experiment. First, a spin-lock pulse is applied for the preparation of T_{1rho}-weighted magnetization. After the T_{1rho}-weighted magnetization is flipped back to the longitudinal plane, the residual transverse magnetization is destroyed by a strong gradient crusher. Then a normal acquisition pulse sequence, like spin echo (SE)–type or gradient echo (GRE)–type sequences, either in 2D or 3D, is applied for image acquisition. Finally, a sufficiently long (a couple of times of T_1) wait time is often applied after image acquisition to allow magnetization to fully restore in the longitudinal direction for the next T_{1rho} preparation. In practice, other module blocks can be integrated into the T_{1rho} imaging diagram as shown in Figure 20.3 according to the requirements on specific clinical applications. For example, other magnetization modules like fat suppression or fluid suppression can also be applied before or after the spin-lock pulse if necessary. Meanwhile, the combination of spin-lock

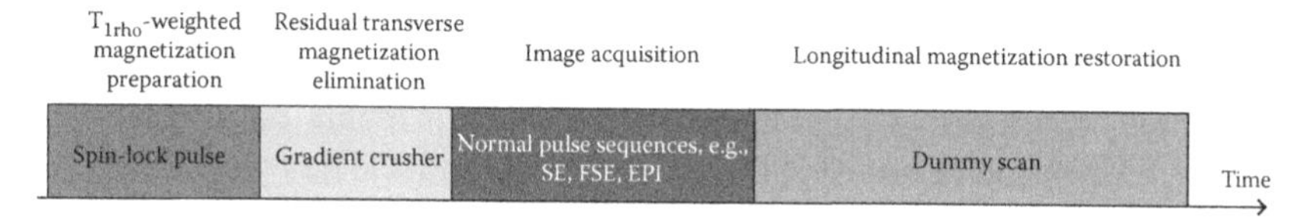

FIGURE 20.2
Block diagram for a typical T_{1rho} imaging experiment. The purpose of each module is described above the block.

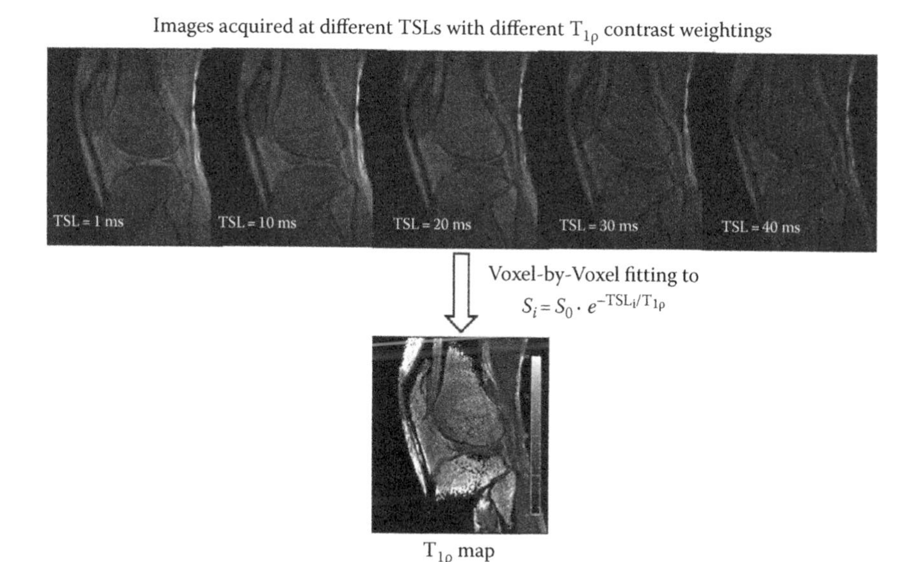

FIGURE 20.3
The illustration of T_{1rho} mapping from a series of images with different T_{1rho} contrast weightings, acquired with different spin-lock times (TSLs). Note that some voxels are missing in the T_{1rho} map due to the poor goodness of fit to the mono-exponential decay model.

pulse and acquisition pulse sequence can be flexible. A T_{1rho} preparation by spin-lock pulse may be applied for each *k*-space line acquisition. More often, spin-lock pulse is played out once and then followed by acquisitions of multiple *k*-space lines to save the total scan time if the T_{1rho}-weighted contrast does not reduce much. However, it is worth noting that the acquired image in T_{1rho} imaging actually does not show purely T_{1rho}-weighted contrast. Instead, the image usually shows a composite contrast because other contrasts associated with the applied pulse sequence and imaging parameters could be imposed on the T_{1rho}-weighted contrast during image acquisition.

20.1.3 T_{1rho}-Weighted Contrast, T_{1rho} Mapping, and T_{1rho} Dispersion

T_{1rho} MRI is normally utilized in three forms. They are T_{1rho}-weighted contrast imaging, T_{1rho} mapping, and T_{1rho} dispersion imaging. T_{1rho}-weighted contrast imaging is the most basic form of T_{1rho} imaging. It applies spin-lock technique to generate particular T_{1rho}-weighted image contrast, hopefully offering additional information for diagnosis beyond images with traditional T_1- or T_2-weighted contrasts. T_{1rho}-weighted contrast has been demonstrated to improve delineation of breast lesions [1]. Because muscle has faster relaxation rates and greater magnetization transfer (MT) than blood, T_{1rho}-weighted contrast has been used to suppress the background myocardium to enhance myocardium-blood contrast in cardiac imaging [2]. T_{1rho}-weighted contrast has also been applied for differentiation of hepatic hemangiomas from metastatic focal liver lesions [3] and characterization of focal liver lesions [4]. T_{1rho}-weighted image was also applied for head and neck for recognition of benign and malignant tumors [5]. T_{1rho}-weighted contrast was also proposed and used for functional MRI (fMRI) to obtain better contrast than the normal T_{2*}-weighted blood oxygen level-dependent (BOLD) contrast [6].

T_{1rho}-weighted contrast imaging involves only a single spin-lock time at a certain spin-lock frequency to generate an appropriate T_{1rho}-weighted contrast level. As comparison, T_{1rho} mapping involves at least two, usually multiple, spin-lock times at a certain spin-lock frequency to obtain a series of images with different levels of T_{1rho}-weighted contrast. Voxel-wise image intensities with different TSLs are then fitted to a mono-exponential decay mathematical model to calculate the voxel-wise T_{1rho} values, which is called T_{1rho} map (Figure 20.3). As T_{1rho} mapping involves multiple TSLs, the scan time of T_{1rho} mapping is longer than T_{1rho}-weighted contrast imaging. However, T_{1rho} mapping gives voxel-wise T_{1rho} values that are independent of acquisition sequence as quantitative biomarkers rather than the qualitative T_{1rho}-weighted contrast for the better characterization of tissue properties. T_{1rho} mapping is the most common form of T_{1rho} imaging and has been most intensively used for various applications, which would be introduced in detail in the section of "applications of T_{1rho} MRI."

T_{1rho} dispersion is the dependence of T_{1rho} relaxation time with spin-lock field strength, or spin-lock frequency. Therefore, T_{1rho} (or R_{1rho}) dispersion imaging usually involves T_{1rho} (or R_{1rho}) quantification or mapping for at least two spin-lock frequencies. Because T_{1rho} is sensitive to the low-frequency motional processes around or at spin-lock frequency, T_{1rho} dispersion imaging becomes a powerful tool to study the low-frequency spin–lattice interaction in biological subjects and provides useful information like proton exchange between water and macromolecules.

The simplest T_{1rho} dispersion imaging is performed with two-FSL T_{1rho} dispersion factor (or coefficient, effect) defined as the ratio of T_{1rho} value (or T_{1rho}-weighted image intensity) at low and high FSL [5,7,8]. An example of this two-FSL dispersion imaging is the quantification of cerebral $H_2{}^{17}O$ concentration for ^{17}O-labeled blood contrast agent to study the oxygenation of brain and myocardium [9–12].

More comprehensive T_{1rho} dispersion imaging performs T_{1rho} quantification at multiple FSLs and plots the curve of T_{1rho} variation on the dependence of FSL to obtain the T_{1rho} dispersion spectrum [12–19]. Chemical exchange (CE), usually represented by proton exchange, is thought to play a crucial role in T_{1rho} dispersion [20–22]. In biological subjects, proton exchange occurs between the large free water pool and some macromolecules that contain labile protons and resonate at offset frequencies from water. When proton exchange rate is much smaller than or comparable to the offset frequency of labile protons, it is often referred to slow or intermediate exchange. If proton exchange rate is much larger than the offset frequency, it is referred to fast exchange. Bloch–McConnell equations [23] quantitatively describe the time evolution of the magnetization of a multi-pool model under proton exchange between different pools and form the theoretical basis of CE investigation through T_{1rho} dispersion. Bloch–McConnell equations show that T_{1rho} is a complicated function of many parameters, including T_1 and T_2, the offset frequency of the labile proton pool, the pool population ratio of free water and labile protons, proton exchange rates and spin-lock field strength. In a two-pool model and under the assumption of fast exchange, the analytical expression of R_{1rho} has been given [24]. Trott et al. later proposed a more general theoretical R_{1rho} relaxation model beyond fast-exchange limit for a two-pool system that is applicable for the investigation of proton exchange at different rates [25]. The theoretical model of R_{1rho} relaxation could be also extended to a multiple-pool CE system [26,27].

It is worth noting that CE has been recently exploited as a novel and powerful contrast mechanism for MRI, usually performed in the forms of either T_{1rho} dispersion imaging or chemical exchange saturation transfer (CEST) imaging [28,29]. In CEST imaging, different irradiation frequencies of a longer but weaker saturation RF pulse than spin-lock pulse for T_{1rho} imaging are applied to obtain the signal intensity ratios compared to the unsaturated signal intensity (defined as magnetization transfer ratio (MTR)) as a function of frequency offset with respect to water, often referred as the Z-spectrum. To eliminate direct water saturation effect, asymmetric MTR, the subtraction of the MTR at the opposite offset frequency of labile proton pool and that at the offset frequency is proposed. Voxel-wise map of asymmetric MTR is used to visualize the CE-based contrast. CEST imaging has been extensively applied for slow to intermediate CE study of important molecules to physiology and pathology [30–34] and many exogenous contrast agents have been developed to enhance the CEST contrast [35–38]. Despite the different acquisition and interpretation form by CEST and T_{1rho} dispersion imaging, both offer CE information of physiological subjects, although CEST is traditionally described to be sensitive to slow to intermediate proton exchange, while T_{1rho} dispersion is described more sensitive to fast proton exchange. Theoretical analysis is able to reveal the underlying correlations between the models of CEST and T_{1rho} [39]. In practice, T_{1rho} and CEST imaging could be used alternatively or combined together to study the CE processes [39–42].

20.2 Techniques of T_{1rho} MRI

20.2.1 Design of Spin-Lock Radiofrequency Pulse

The most basic spin-lock pulse has been introduced in Section 20.1.2. Theoretically, the magnetization along the spin-lock pulse field direction decays mono-exponentially with regard to spin-lock time TSL at the rate of R_{1rho}, and the magnetization perpendicular to the spin-lock pulse field direction decays also mono-exponentially with TSL but with the rate of R_{2rho}. However, in practice, the actual spin-lock pulse field direction and strength could be obscured by the presence of B_0 and B_1 field inhomogeneities due either to the imperfection of MRI hardware or to the susceptibilities and heterogeneities of the imaged subject. In the presence of B_0 and B_1 field inhomogeneities, the effective spin-lock field direction and strength may be considerably deviated from the nominal spin-lock direction and strength, leading to the poor alignment of the magnetization along the nominal spin-lock direction and the complicated magnetization evolution rather than the ideal mono-exponential decay along the nominal spin-lock direction. An illustration of the nominal and true spin-lock pulse strength, as well as the effective spin-lock strength and direction in the rotating frame, is shown in Figure 20.4. B_1 inhomogeneities lead to the deviation of true spin-lock frequency (FSL_t) from the nominal spin-lock frequency (FSL_n). In the presence of both B_0 and B_1 inhomogeneities, the magnetization nutates about the effective spin-lock field direction indicated by z' in Figure 20.4, deviated from Z direction with an angle $\theta = \tan^{-1}(\omega_1/\Delta\omega_0) = \tan^{-1}(FSL/\gamma\Delta B_0)$. The effective spin-lock field strength is calculated as $\omega_{eff} = (\Delta\omega_0^{\,2} + \omega_1^{\,2})^{1/2}$. This poor alignment and the complicated magnetization evolution are finally presented in the form of banding-like artifacts on T_{1rho}-weighted images and also result in the errors of T_{1rho} quantification using the traditional mono-exponential decay model. As seen from Figure 20.4, a simple solution to reduce the influence of B_0 inhomogeneity is to use a spin-lock pulse with strong spin-lock field strength

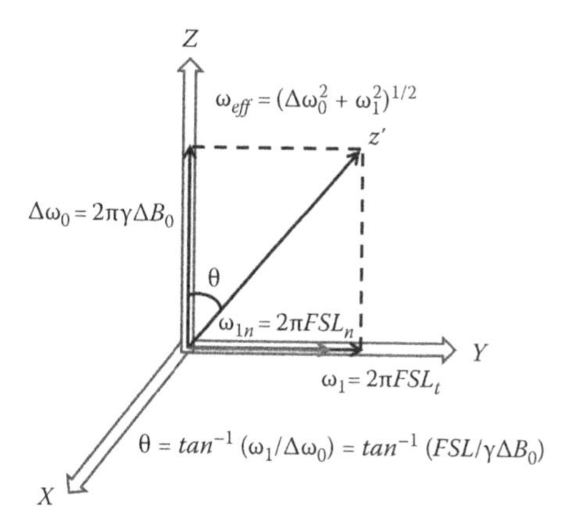

FIGURE 20.4
Illustration of the nominal and true spin-lock pulse strength, as well as the effective spin-lock strength and direction in the rotating frame.

and large FSL so that θ approaches 90° and $\omega_{eff} \approx \omega_1$. However, this is restricted by the scanner hardware performance as well as the regulation of specific absorption rate (SAR) on human scan. B_0 shimming and B_1 calibration are also helpful but often not sufficient to alleviate the banding artifact. Therefore, the improved design of spin-lock RF pulse is vital to the T_{1rho}-weighted image quality and the accurate quantification of T_{1rho} relaxation time.

To precisely trace the magnetization evolution at each time point during an SL pulse, full Bloch analysis is necessary but too complicated. To simplify the analysis while maintaining the acceptable accuracy, the transient effect to the steady state is usually neglected. As such, the magnetization evolution can be conveniently traced by the multiplication of a series of rotation and relaxation matrices to present each RF component in the cluster of a composite spin-lock pulse. An instantaneous RF pulse is represented in the form of matrix notation $R_\varphi(\phi)$, where R denotes a rotation matrix, φ is the pulse field orientation, and φ is the pulse flip angle. The magnetization evolution under an RF pulse could be generally expressed by $M_{t1} = R_\varphi(\phi)M_{t0}$, where M_{t0} and M_{t1} denote the magnetization before and after the pulse excitation, respectively. The basic rotation matrices that rotate magnetization about the *x*-, *y*-, or *z*-axis by an angle φ, in three dimensions, are expressed, respectively, as

$$R_x(\varphi) = \begin{bmatrix} 1 & 0 & 0 \\ 0 & \cos\varphi & \sin\varphi \\ 0 & -\sin\varphi & \cos\varphi \end{bmatrix};\; R_y(\varphi) = \begin{bmatrix} \cos\varphi & 0 & -\sin\varphi \\ 0 & 1 & 0 \\ \sin\varphi & 0 & \cos\varphi \end{bmatrix};$$

$$R_z(\varphi) = \begin{bmatrix} \cos\varphi & \sin\varphi & 0 \\ -\sin\varphi & \cos\varphi & 0 \\ 0 & 0 & 1 \end{bmatrix} \tag{20.1}$$

Note that each of these basic magnetization vector rotations typically appears clockwise when the axis about which they occur points toward the observer, and the coordinate system is right-handed. The spin-lock relaxation matrix under a spin-lock pulse with duration T is given by Equation 20.2, assuming the original magnetization is along z direction:

$$E_\rho(T) = \begin{bmatrix} e^{-T/T_{2\rho}} & 0 & 0 \\ 0 & e^{-T/T_{2\rho}} & 0 \\ 0 & 0 & e^{-T/2T_{1\rho}} \end{bmatrix} = \begin{bmatrix} E_{2\rho}(T) & 0 & 0 \\ 0 & E_{2\rho}(T) & 0 \\ 0 & 0 & E_{1\rho}(T) \end{bmatrix} \quad (20.2)$$

20.2.1.1 Normal Spin-Lock Pulse

Assuming the flip angle of β for the tip-down pulse P_1 and the tip-up pulse P_2 and α for the spin-lock component SL, $\alpha = 2\pi \cdot \text{FSL} \cdot \text{TSL}$, for a given spin-lock frequency FSL, the magnetization evolution is expressed in Equation 20.3:

$$M(\text{TSL}) = R_{-x}(\beta) \cdot R_y(\alpha) \cdot E_\rho(TSL) \cdot R_x(\beta) \cdot M(t_0), \quad \text{where } M(t_0) = [0 \quad 0 \quad M_0]^T \quad (20.3)$$

Since P_1 and P_2 usually have much shorter duration than SL, the relaxation during P_1 and P_2 is negligible. By substituting Equations 20.1 and 20.2 into Equation 20.3, the final longitudinal magnetization M_z is easily derived as given by [43]

$$M_z = \left(\sin^2\beta \cdot e^{-TSL/T_{1\rho}} + \cos^2\beta \cdot \cos\alpha \cdot e^{-TSL/T_{2\rho}}\right) M_0 \quad (20.4)$$

According to Equation 20.4, M_z follows an exponential decay as long as β is equal to $\pi/2$ or 90°. Otherwise, in the presence of B_1 inhomogeneities and β no longer is equal to $\pi/2$, M_z shows a composite function to many factors of T_{1rho}, T_{2rho}, and α, reflected in the image as banding artifact. Because α is dependent on both TSL and FSL, the location distribution of artifacts varies with TSL and FSL as well.

20.2.1.2 Rotary Echo Spin-Lock Pulse

To compensate for artifacts caused by B_1 inhomogeneity, a rotary echo spin-lock pulse (Figure 20.5) was proposed [43]. This rotary echo SL pulse is divided into two segments (SL1 and SL2 in Figure 20.5) with the equal duration but the opposed phase and effectively refocuses the rotation spin-lock phase accumulation.

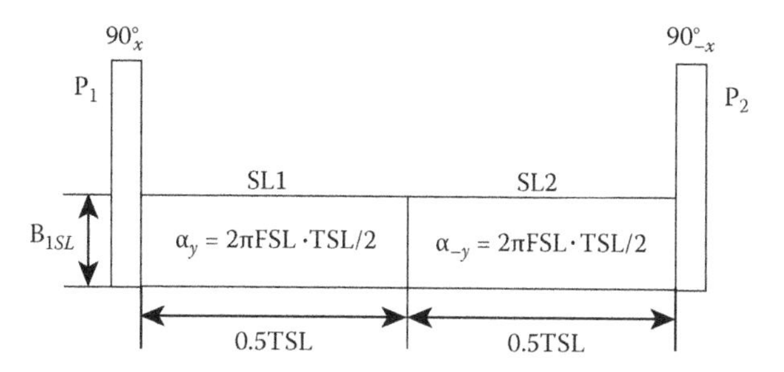

FIGURE 20.5
Structure of a rotary echo spin-lock pulse. Spin-lock section is divided into two segments (SL1 and SL2) with the equal duration but the opposed phase.

The magnetization evolution under the application of a rotary echo spin-lock pulse can be expressed by

$$M(\text{TSL}) = R_{-x}(\beta)\cdot R_{-y}(\alpha)\cdot E_\rho\left(\frac{\text{TSL}}{2}\right)\cdot R_y(\alpha)\cdot E_\rho\left(\frac{\text{TSL}}{2}\right)\cdot R_x(\beta)\cdot M(t_0) \quad (20.5)$$

Note that here, $\alpha = 2\pi \cdot \text{FSL} \cdot (\text{TSL}/2)$. The final longitudinal magnetization M_z is derived as

$$M_z = \left(\sin^2\beta \cdot e^{-\text{TSL}/T_{1\rho}} + \cos^2\beta \cdot e^{-\text{TSL}/T_{2\rho}}\right)M_0 \quad (20.6)$$

Comparing Equations 20.4 and 20.6, the term of $\cos\alpha$ has been completely eliminated in Equation 20.6, so M_z no longer shows TSL- or FSL-dependent artifact distribution. However, it is worth pointing out that rotary echo pulse could still show some artifacts when β not equal to $\pi/2$. Meanwhile, M_z does not show pure T_{1rho}-weighted contrast and may be contaminated by T_{2rho}-weighted contrast in the presence of B_1 inhomogeneities.

In the presence of B_0 inhomogeneities, the magnetization evolution could also be traced quantitatively by [44]

$$\begin{aligned} M(\text{TSL}) &= R_{-x}(\beta)\cdot R_{z''}(\alpha)E_\rho\left(\frac{\text{TSL}}{2}\right)\cdot R_{z'}(\alpha)E_\rho\left(\frac{\text{TSL}}{2}\right)\cdot R_x(\beta)\cdot M(t_0) \\ &= R_{-x}(\beta)\cdot R_x(-\theta)R_z(\alpha)E_\rho\left(\frac{\text{TSL}}{2}\right)R_x(\theta)\cdot R_x(\theta)R_z(\alpha)E_\rho\left(\frac{\text{TSL}}{2}\right) \\ &\quad R_x(-\theta)\cdot R_x(\beta)\cdot M(t_0) \end{aligned} \quad (20.7)$$

where θ is the direction of the effective spin-lock field as shown in Figure 20.3. The final longitudinal magnetization M_z is given by

$$M_z = \left(e^{-TSL/T_{2\rho}}\left(\cos^2\beta - \cos^2\theta\right)\left(\sin^2\alpha - \cos^2\alpha\cos 2\theta\right) + e^{-TSL/2T_{1\rho}}e^{-TSL/2T_{2\rho}}\cos\alpha\sin^2 2\theta + e^{-TSL/T_{1\rho}}\left(\cos^2\beta - \sin^2\theta\right)\cos 2\theta\right)\cdot M_0 \tag{20.8}$$

Equation 20.8 shows that the resultant longitudinal magnetization by a rotary echo SL pulse in the presence of B_0 inhomogeneities does not show a pure T_{1rho} contrast, but a very complicated composite contrast dependent on many factors, including T_{1rho}, $T2_{rho}$, B_0 inhomogeneity, TSL, and FSL. Although complicated, Equation 20.8 provides a possible means to simultaneously quantify T_{1rho}, T_{2rho}, and B_0 inhomogeneity even in the presence of banding artifacts, particularly at low spin-lock frequencies [44].

20.2.1.3 B_1 and B_0 Insensitive Composite Spin-Lock Pulse

Rotary echo pulse is still susceptible to B_0 inhomogeneities. Witschey et al. [45] proposed new B_1 and B_0 insensitive composite spin-lock pulses to further reduce the sensitivity of rotary echo spin-lock pulses to B_0 inhomogeneities by inserting an 180° refocusing pulse between the two rotary echo spin-lock pulse segments of SL1 and SL2 (Figure 20.6). This 180° refocusing pulse works just similar to the refocusing pulse used in the normal spin-echo acquisition pulse sequences. The spins nutating about the effective spin-lock field direction at the first spin-echo pulse section SL1 reversed and then return to the original position after processing about the effective spin-lock field direction at the second spin-lock section. The analytical expression of the magnetization under this B_1 and B_0 insensitive composite spin-lock pulse could be derived as

$$M_z = \Big[e^{-TSL/T_{2\rho}}\left(\cos^2\beta - \cos^2\theta\right)\left(\sin^2\alpha\cos 2\beta - \cos^2\alpha\cos^2\theta + \cos 2\beta\cos^2\alpha\sin^2\theta\right) + 0.5\cdot e^{-TSL/2T_{1\rho}}e^{-TSL/2T_{2\rho}}(1+\cos 2\beta) \left(\cos\alpha\sin^2 2\theta + 2\sin\alpha\cos\theta - 2\cos 2\beta\sin\alpha\cos\theta\right) + e^{-TSL/T_{1\rho}}\left(\cos^2\beta - \sin^2\theta\right)\left(\cos^2\theta\cos 2\beta - \sin^2\theta\right)\Big]\cdot M_0 \tag{20.9}$$

The detail derivation is neglected here. In Equation 20.9, the flip angle of the refocusing pulse is written as 2β. If $\beta = 90°$, Equation 20.9 is simplified exactly as Equation 20.6.

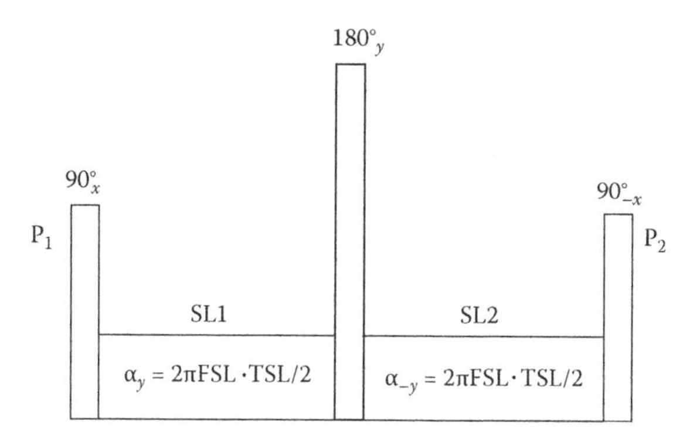

FIGURE 20.6
Structure of a composite spin-lock pulse insensitive to B_0 and B_1 inhomogeneities. A 180° refocusing pulse is inserted between the two rotary echo spin-lock pulse segments of SL1 and SL2.

Another revision of this B_1 and B_0 insensitive composite spin-lock pulse is to change the direction of the tip-up pulse P_2 from $-x$ to x [45]. The purpose of this alternation is to compensate for the imperfect flip angle β and make it more robust to B_1 inhomogeneities. The analytical magnetization expression for this revision could also be obtained using the theory frame as and readers are suggested to derive by themselves.

20.2.1.4 Dixon's Composite Spin-Lock Pulse and Phase Cycling

In 1996, Dixon et al. proposed a composite spin-lock pulse in the form of 90(y)–135(x)–360(x,SL)–135(x)–90($-y$) for myocardial suppression. This composite pulse is tolerant to shimming and frequency errors for the cases of small off-resonance. Meanwhile, the strength and duration of each section in this composite pulse are required to meet some particular quantitative relationship. Actually, the magnetization evolution under Dixon's composite spin-lock pulse could also be calculated through the approach of multiplication of a series of rotation and relaxation matrices as used earlier. The result suggests that the effectiveness and robustness may be restricted in practice if the B_0 or B_1 inhomogeneities are relatively large and the strength and duration requirement for each pulse elements are not well fulfilled.

The idea of phase cycling has also been proposed to reduce the spin-lock artifacts by using two T_{1rho}-weighted images. The first T_{1rho}-weighted image is acquired using a normal spin-lock pulse and the second image is acquired using the same spin-lock pulse but with the flip angle of P_2 inversed. The final image is the subtraction of the acquired two images [46]. Using the theoretical analysis approach given earlier, it can be shown that the longitudinal magnetization after this process is

$$M_z = 2\sin^2\beta \cdot e^{-\mathrm{TSL}/T_{1\rho}} M_0 \tag{20.10}$$

As seen from Equation 20.10, phase cycling is robust to B_1 inhomogeneities. Even the tip pulse flip angle β is not exactly 90°, the resulting magnetization still follows the mono-exponential decay and suffers no contamination from T_{2rho} relaxation and the spin-lock flip angle α relating to FSL and TSL. On the other hand, the disadvantage of phase cycling is also obvious. The acquisition of two images doubles the scan time.

Recently, Chen et al. [47] combined the Dixon's composite spin-lock pulse and phase cycling together, named phase-cycled composition spin-lock (PCC-SL), for B_0 and B_1 field inhomogeneity compensation, and compared the performance of PCC-SL with individual Dixon's approach, phase cycling, and other composite spin-lock pulses. The results showed that PCC-SL outperformed other approaches in artifact reduction and T_{1rho} quantification.

In summary, many approaches have been proposed to design composite spin-lock pulses for T_{1rho} imaging to reduce spin-lock artifact and hence achieve more accurate T_{1rho} quantification. Although significantly outperform compared to the traditional spin-lock pulse, the availability of these composite spin-lock pulses on clinical scanners is still limited. Meanwhile, effectiveness and robustness of these composite spin-lock pulses may still be compromised in practice and have to be further validated, in particular for the cases of long spin-lock duration, low spin-lock frequency, and pronounced susceptibilities.

20.2.2 Toward Fast T_{1rho} Imaging at High Field with Low Specific Absorption Rate

To apply T_{1rho} imaging for clinical applications, RF energy deposition by the spin-lock pulses has to be concerned and the associating SAR has to meet the requirement of relevant safety regulations. SAR is normally defined as the amount of RF energy per unit mass per unit time deposited into the imaged subject during an imaging experiment. For example, the U.S. Food and Drug Administration (FDA) has established guidelines to regulate the allowable maximum SAR for clinical imaging, that is, 4 W/kg averaged over the whole body for any 15 min period, 3 W/kg averaged over the head for any 10 min period, or 8 W/kg in any gram of tissue in the extremities for any period of 5 min.

According to the SAR model proposed by Collins et al. in 1998 [48], SAR is proportional to the square of the RF pulse flip angle and the square of static field strength B_0. In general, T_{1rho} imaging is SAR intensive due to the long spin-lock time and hence large flip angle. Furthermore, spin-lock pulses with strong field strength and spin-lock frequency are often preferable for the purpose of artifact reduction, which makes the SAR problem more prominent in practice, particularly for high field MRI scanners. In order to maintain the SAR within the allowable limits, acquisition repetition time (TR) for T_{1rho} imaging is often significantly lengthened, resulting in a proportional increase of the total scan time and hence the reduction of patient comfort as well as the increased proneness to motion artifact.

The most straightforward and convenient method for the reduction of SAR is to reduce the spin-lock field strength or spin-lock frequency. Unfortunately, the fact is that this method is rarely used because the low spin-lock frequency also brings the severe problem of spin-lock artifact in the presence of B_0 and B_1 homogeneities and hinders the accurate quantification of T_{1rho}. In this aspect, the use of various dedicated composite spin-lock pulse to reduce spin-lock artifact is beneficial for SAR reduction. Another approach is to quantify T_{1rho} using the model different from the mono-exponential one even in the presence of artifact, if the magnetization evolution and its relationship with imaging parameters could be analytically derived [44]. This approach does not require the complicated design of composite spin-lock pulses, but the analytical T_{1rho} model has to be derived individually for each kind of spin-lock pulse. Sensitivity, accuracy, and precision of T_{1rho} fitting according to these complicated models with regard to imaging parameters have to be concerned and investigated carefully to justify the fitting results.

For T_{1rho} mapping studies in which multiple TSLs are usually applied, the reduction of the TSL number is also an effective means to reduce the total spin-lock RF energy deposition and the total scan time, although the averaged SAR does not reduce. Theoretically, two TSLs are sufficient to calculate T_{1rho} based on the mono-exponential model because T_{1rho} is the only unknown parameter to be fitted. However, the calculated T_{1rho} may be susceptible to noise, and the accuracy of T_{1rho} calculation is much dependent on the TSL applied, true T_{1rho} value, and SNR. Yuan et al. [49] studied the accuracy of T_{1rho} mapping by using two TSLs through theoretical analysis and Monte Carlo simulation. Segmented T_{1rho} acquisition could also be applied to reduce SAR for each TSL. In this approach, spin-lock pulse is not applied for each k-space line acquisition. Instead, every spin-lock pulse for T_{1rho} preparation is followed by acquisitions of multiple *k*-space lines. The extreme case of segmented T_{1rho} acquisition is the single-shot T_{1rho} acquisition in which all *k*-space lines are acquired after the sole T_{1rho} preparation.

Spin-lock RF pulse is an important source of RF energy deposition but not the only one because it has to be combined with normal acquisition pulse sequences. Therefore, other general approaches to reduce SAR could also be employed. For example, the use of partial Fourier acquisition and parallel imaging [50–52] is helpful to reduce the phase-encoding numbers and hence, the number of spin-lock pulses applied. Compressed sensing reconstruction [53,54] is also beneficial and some preliminary results have been presented [55]. It is well known that MR image contrast is majorly determined by the central *k*-space lines, while the peripheral *k*-space lines determine the fine edge detail. Wheaton et al. [56] proposed a partial *k*-space acquisition approach in which a full power spin-lock pulse is applied to only the central phase-encode lines of *k*-space, while the remainder of *k*-space receives a low-power spin-lock pulse. SAR was reduced by 40%, while the error of T_{1rho} mapping was only 2%, as demonstrated on human brain T_{1rho} imaging at 1.5T. For some specific applications in which only a fraction of the whole image may be interested, such as spine and cartilage T_{1rho} imaging, reduced field-of-view (rFOV) method [57,58] could be applied to reduce the phase-encoding numbers and hence SAR and total scan time. For the application of T_{1rho} imaging at ultrahigh field strength B_0 higher than 3T, the design of spin-lock pulse based on parallel excitation [59] should be promising to reduce SAR.

20.3 Applications of T_{1rho} MRI

During the recent years, T_{1rho}-weighted contrast, T_{1rho} relaxation time mapping, and T_{1rho} dispersion have been extensively studied, covering various organs and a wide range of disease processes, particularly in the areas of intervertebral discs, knee cartilage, and spine. T_{1rho} MRI is a relative time-consuming and SAR-intensive technique; imaging of these areas benefits from that there is very limited physiological motion. Till now, all applications with T_{1rho} MRI remain research phase.

20.3.1 Applications in Intervertebral Disc and Articular Cartilage

Early signs of disc degeneration are manifested by biochemical changes, including a loss of proteoglycans (PGs), a loss of osmotic pressure, and hydration. In the later stages of disc degeneration, morphological changes occur, including a loss of disc height, disc herniation, annular tears, and radial bulging. While lumbar spinal fusion is used for surgical treatment of low back pain with advanced degeneration, earlier stages of disc degeneration may be amenable to emerging alternative treatments (e.g., cell therapy, growth factor therapy) that may preclude the morbidity associated with fusion. Noninvasive quantitative assessments for these early degenerative changes are needed. T_{1rho} relaxation time measurement that reflects the intrinsic material properties of disc tissues may have the potential to detect subtle differences in tissue composition that may not be apparent with T_2-weighted image-based assessment; therefore, it is likely to be more useful for early disc changes.

In an in vivo study at 1.5T by Auerbach et al., T_{1rho}-weighted image was found to have a great dynamic range than T_2-weighted images. In addition, T_{1rho} relaxation time correlated significantly with disc degeneration ($r = -0.51$, $P < 0.01$) assessed from conventional T_2-weighted images according to the Pfirrmann classification system. T_{1rho} imaging allows for spatial measurements on a continuous rather than an integer-based grading, minimizing the potential for observer bias. In one recent study of lumbar MRI at 3T [60], T_{1rho} was measured using a rotary echo spin-lock pulse (spin-lock frequency of 500 Hz) embedded in a 3D balanced fast field-echo sequence and compared to T_2 values in regions-of-interests (ROIs) including nucleus pulposus (NP) and annulus fibrosus (AF) (Figure 20.7). Eight- and five-level

FIGURE 20.7
An example of placement of ROIs over NP (#), anterior AF (*), and posterior AF (∧) in one disc of T_2-weighted image (T2WI), T_{1rho} map, and T_2 map.

disc degeneration semiquantitative grading was performed [60–62]. It was found for NP, T_{1rho} and T_2 decrease quadratically with disc degeneration grades and have no significant trend difference (Figure 20.8). For AF, T_{1rho} decreases linearly as the disc degenerated and has a slope of −3.02 and −4.56 for eight- and five-level gradings, respectively, while the slopes for T_2 values are −1.43 and −1.84, respectively, being significantly flatter than those of T_{1rho}

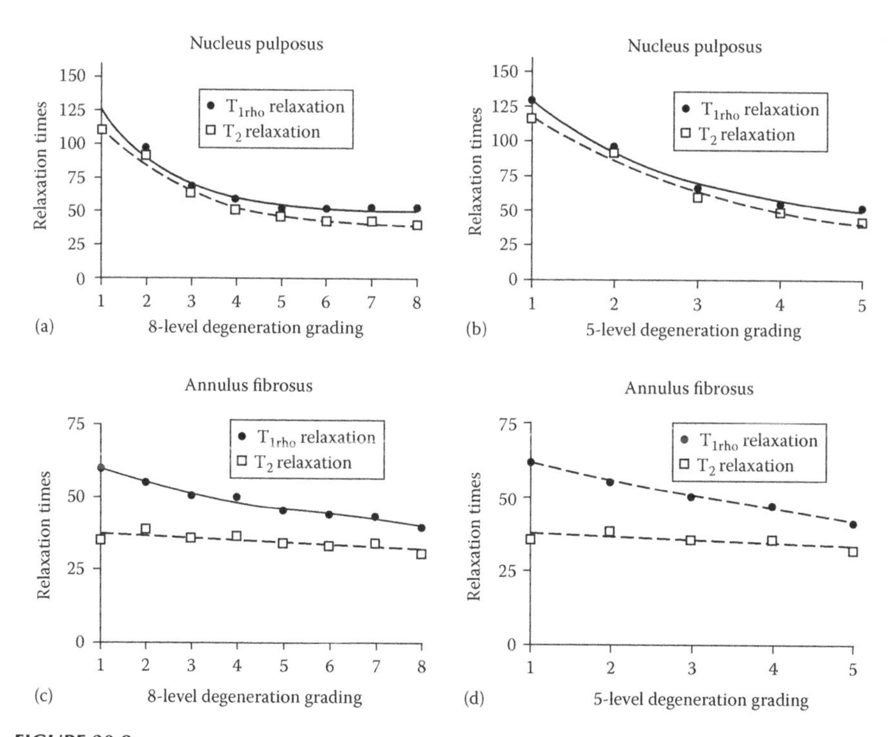

FIGURE 20.8
T_{1rho} and T_2 relaxation times decrease in (a, b) nucleus pulposus and (c, d) annulus fibrosus as disc degeneration grade progresses.

($P < 0.001$, Figure 20.2). Therefore, T_{1rho} is a more sensitive biomarker for early disc AF degeneration. The T_{1rho} reduction in AF during disc degeneration may represent some macromolecular changes that occur during the process. It is also noted in this study that, for both NP and AF, discs of grade 5/8–8/8 degenerations had similar T_{1rho} and T_2 relaxation times without significant statistical difference. The eight-level grading system offers information on disc space narrowing, that is, <30% reduction in disc height for grade 6/8, 30%–60% reduction in disc height for grade 7/8, and >60% reduction in disc height for grade 8/8 [62]. As disc space narrowing is more likely to cause low back pain, these results suggest that even when T_{1rho} and T_2 relaxation times are routinely measured for disc degeneration assessment in the future, the eight-level grading system will still be useful for clinical evaluation, particularly for those with severe disc degeneration and disc space narrowing.

Osteoarthritis (OA) is a multifactorial disease characterized primarily by the progressive loss of articular cartilage. Standard MRI is useful to visualize cartilage directly yet morphologic imaging shows damage at a stage when cartilage is already irreversibly lost. With the improvement in cartilage resurfacing procedures and development of disease-modifying drugs for OA, there is a need to develop a noninvasive method to monitor early cartilage degeneration or restoration. Standard MR is inconclusive in quantifying early degenerative changes of the cartilage matrix, especially biochemical changes such as PG loss. Early events in the development of cartilage matrix breakdown include the loss of PGs, changes in water content, and molecular level changes in collagen. The extracellular matrix in the articular cartilage provides a motion-restricted environment to water molecules. Increased T_2 values were reported previously in degenerated cartilage in both animal models and in human subjects. In an effort to correlate the T_2 relaxation times with biochemical changes in cartilage, previous in vitro studies have reported that T_2 correlated poorly with PG content [63,64] and PG cleavage did not affect T_2 values significantly [65]. Instead, T_2 can be affected mainly by collagen content and orientation and/or water content [66,67]. It has been observed that loss of PG is an initiating event in early OA, while neither the content nor the type of collagen is altered in early OA [68]. Therefore, lack of specificity to quantify PG loss may make T_2 relaxation time less appealing for early detection of cartilage degeneration.

Changes to the extracellular matrix, such as PG loss, therefore may be reflected in measurements of T_{1rho}. PG depletion–induced changes in T_{1rho} relaxation and dispersion were first found ex vivo in healthy bovine specimens [64,69]. Initial studies in human subjects showed elevated T_{1rho} values in patients with OA [46]. Li et al. [70] studied 16 healthy volunteers without clinical or radiological evidence of OA and 10 patients with OA using a 3T scanner. They found that the average T_{1rho} and T_2 values were significantly increased in OA patients compared with controls (52.04 ± 2.97 ms vs. 45.53 ± 3.28 ms for T_{1rho}) and 39.63 ± 2.69 ms vs. 34.74 ± 2.48 ms for T_2. Increased T_{1rho} and T_2 values were correlated with increased severity in radiographic and MR grading of OA. T_{1rho} had a larger range and higher effect size than T_2. It was concluded that T_{1rho} relaxation time may be a more sensitive indicator for early cartilage degeneration than T_2 relaxation time. In a later study, Li et al. also studied the spatial variation of T_{1rho} and T_2 values in patellar cartilage was studied in different cartilage layers [71]. Increased T_{1rho} and T_2 values are found more heterogeneous in osteoarthritic cartilage. T_{1rho} and T_2 values show different spatial distributions that may provide complementary information regarding cartilage degeneration in OA.

The role of specific biochemical changes in the altered MR signal intensity is still not well understood in the reported literature. Makela et al. [21] and Duvvuri et al. [15] suggested that proton exchange between chemically shifted NH and OH groups of PG and the tissue water could be an important relaxation mechanism contributing to T_{1rho} relaxation. In the articular cartilage study, the loss of PGs results in an increase in T_{1rho} relaxation time [69]. R_{1rho} relaxation rate has been shown to decrease linearly with decreasing PG content in ex vivo bovine patellae and has been proposed as a more specific indicator of PG content than T_2 relaxation in trypsinized cartilage and in human cartilage specimens obtained from patients with severe OA who underwent total knee replacement [15]. On the other hand, T_{1rho} is reported to increase with sulfated glycosaminoglycan content in degenerative discs [72]. While the reduced water content is a contributing factor of the reduction of T_{1rho} and T_2 with degenerated discs [72], T_{1rho} and T_2 relaxations represent different contrast mechanisms and may provide complementary information about the integrity of the disc, and the differences in relaxation times may be linked to the varying tissue compositions and material properties.

20.3.2 Neurological Applications

T_{1rho} variation has been reported to be associated with a number of neurodegenerative diseases and also has potentials to be an early biomarker for cerebral ischemia. Alzheimer's disease (AD) is a common neurodegenerative dementia of the elderly and the identification of cognitive decline in its earliest manifestations and the heterogeneity for AD has been a research focus of neuroimaging for long. AD-induced cerebral atrophy as revealed by conventional MRI is nonspecific and is hard to be differentiated with changes due to normal aging. Parkinson's disease (PD) is another neurodegenerative disease in which dementia is common. PD is associated with the loss of dopaminergic neurons in the substantia nigra. Dementia affects approximately 40% of PD patients and the risk for the development of dementia in PD is six times higher than in non-PD age-matched population [73]. Early detection of AD and PD by noninvasive imaging means is crucial for the patient's treatment.

Borthakur et al. [74] measured the T_{1rho} of plaque burden in a mouse model of AD. A significant decrease in T_{1rho} was observed in the cortex and hippocampus of 12- and 18-month-old animals compared to the age-matched controls. A correlation between changes in T_{1rho} and the age of the animals was also found. The results suggest that T_{1rho} may be a sensitive method for noninvasively determining AD-related pathology. In a further study, T_{1rho} was measured on patients with clinically diagnosed AD, mild-cognitive impairment (MCI), and in age-matched cognitively normal control subjects in order to compare T_{1rho} values with changes in brain volume in the same regions of the brain [75]. A statistically significant ($P < 0.01$) increase in both the gray matter (GM) and white matter (WM) in the medial temporal lobe (MTL) was seen in AD patients over age-matched controls. However, AD and MCI cohorts only displayed significant difference ($P < 0.01$) in WM. In some following studies, increased T_{1rho} was similarly found on AD and MCI cohorts compared to normal control subjects [76–78].

Nestrasil et al. [79] reported that compared to controls, the substantia nigra of PD subjects also had increased T_{1rho}. Haris et al. [77] measured T_{1rho} in the hippocampus in the brain of control, AD, PD, and PD patients with dementia (PDD). They found that the serial measurement of T_{1rho} in both AD and PD provides information on disease progression and contributes to early diagnosis.

Gröhn et al. showed that T_{1rho} could serve as a sensitive MRI indicator of cerebral ischemia [14,80]. In a rat study, T_{1rho} prolongs within minutes after a drop in the cerebral blood flow. T_{1rho} dispersion increases by approximately 20% on middle cerebral artery (MCA) occlusion. The T_{1rho} dispersion change dynamically increases to be 38% ± 10% by the first 60 min of ischemia in the brain region destined to develop infarction. Following reperfusion after 45 min of MCA occlusion, the tissue with elevated T_{1rho} dispersion (yet normal diffusion) develops severe histologically verified neuronal damage; thus, the former parameter unveils an irreversible condition earlier than currently available MRI methods. Makela et al. [81] quantitatively measured T_{1rho} and magnetization transfer (MT) of acute cerebral ischemia in rat. T_{1rho} changes in the acute phase of ischemia were found to coincide with both elevated maximal MT and amount of MT. These changes occurred independent of the overall MT rate and in the absence of net water gain to the tissue. The extensive prolongation of T_{1rho} was thought to be associated with water accumulation. There are other animal studies of cerebral ischemia by using T_{1rho} imaging to investigate the variation of water fraction [82] and to estimate the onset time [83]. In another recent pilot study, an increased T_{1rho} was found in aging rats [84] and rats with spontaneous hypertension [85]. However, probably due to medical emergency nature of stroke patients, to the best of our knowledge, there is few on T_{1rho} measurement on human subjects so far.

20.3.3 Liver Imaging, Cancer Imaging, and Other Miscellaneous Applications

With biliary duct ligation (BDL) and carbon tetrachloride intoxication–induced rat liver fibrosis models, recent studies have showed that MR T_{1rho} imaging is able to detect liver fibrosis, and the degree of fibrosis is correlated with the degree of elevation of the T_{1rho} measurements [86,87]. On the other hand, liver acute edema only leads to minimal T_{1rho} increase [86]. In carbon tetrachloride intoxication–induced rat liver fibrosis models, T_{1rho} MRI has also been confirmed to assess liver fibrosis regression [86]. T_{1rho} MRI has been applied for human liver and consistent liver $T_{1\rho}$ measurement has been achieved for healthy volunteers [88]. Pilot clinical study proved T_{1rho} is increased in patients with liver cirrhosis [89] (20.9, 20.10).

Early pilot studies have been carried out to explore the value of T_{1rho} and T_{1rho} dispersion in breast tumors, head and neck tumors, and skeletal muscle diseases [5,7,8,90,91]. T_{1rho} measurement has been tested to assess early

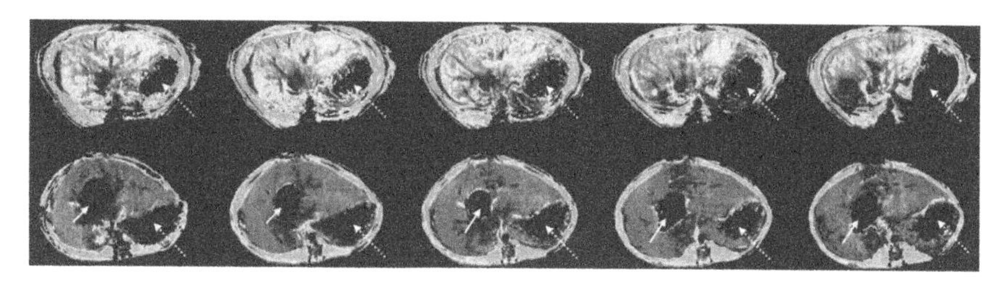

FIGURE 20.9
(See color insert.) Color-coded T_{1rho} maps of a sham-operated rat liver (control, upper row) and a biliary duct ligation (BDL) rat liver 24 days postsurgery (lower row). BDL rat liver demonstrates higher T_{1rho} value (brighter) than the control rat liver (darker). Arrow: dilated biliary duct. Dotted arrow: gas in the stomach (From Wang, Y.X. et al., *Radiology*, 259(3), 712–719, 2011).

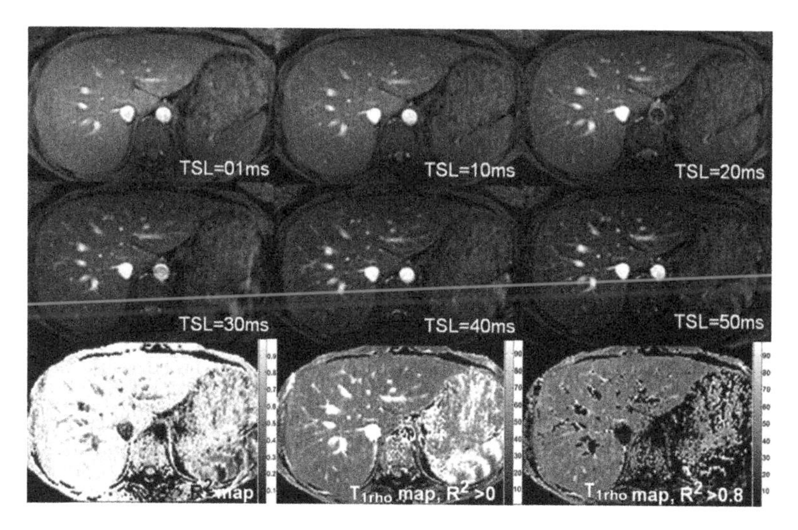

FIGURE 20.10
The upper and middle rows show liver T_{1rho}-weighted images acquired with spin-lock times (TSL) ranged from 1 to 50 ms. Note that vessels demonstrate high signal on T_{1rho}-weighted images. The lower row shows coefficient of determination (R^2) map (left), T_{1rho} map without R^2 evaluation (middle), and T_{1rho} map with $R^2 > 0.8$ evaluation (right).

therapeutic response of tumors. Duvvuri et al. [92] compared T_{1rho} and T_2 measurements on a murine radiation–induced fibrosarcoma model before and after cyclophosphamide treatment. It was shown that the number of pixels exhibiting T_{1rho} values longer than controls in viable regions of the tumor increased significantly as early as 18 h after drug administration and remained elevated up to 36 h after treatment. Although a trend of increasing T_2 relaxation time relative to controls was noted in viable regions of the tumor 36 h after treatment, the changes were not statistically significant. Histological examination indicated a decrease in mitotic index that paralleled the changes in T_{1rho}. Kettunen et al. [93] assessed the effectiveness of T_{1rho} dispersion and the low FSL T_{1rho} in noninvasive monitoring of gene therapy response in BT4C glioma in rats and concluded that T_{1rho} with low FSL is an excellent MRI marker of early gene therapy response in gliomas. Using the same BT4C glioma animal model, T_{1rho} and T_{2rho} MRI were demonstrated to be useful tools to quantify early changes in water dynamics reflecting treatment response during gene therapy [94].

T_{1rho} imaging has also been applied for cardiac MRI. Muthupillai et al. [95] evaluated acute myocardial infarction (MI) using a contrast-enhanced T_{1rho}-weighted cine turbo field-echo MR sequence and a delayed-enhancement sequence. With T_{1rho} weighting, percentage enhancement of irreversibly injured myocardium was 68% ± 41%, compared with 23% ± 24% without T_{1rho} weighting ($P < 0.006$). It was suggested that the addition of T_{1rho} weighting to contrast-enhanced cine turbo field-echo MR sequences offers a new contrast enhancement mechanism for characterization of acutely infarcted myocardium. The influence of contrast dose and time of imaging after contrast administration on the tissue characterization of MI using T_{1rho} imaging was studied by Huber et al. [96]. Recently, Witschey et al. [97] used T_{1rho} imaging for chronic MI characterization on a Yorkshire swine model. Results showed that T_{1rho}-weighted imaging using long TSL enables high discrimination between infarct and myocardium and suggested that T_{1rho} may be useful to visualizing MI without the need for exogenous contrast agents for clinical cardiac applications.

20.4 Conclusion

T_{1rho} relaxation is the relaxation process under the application of a spin-lock pulse in the rotating frame. T_{1rho} relaxation time usually has the value between T_1 and T_2 and provides a new contrast mechanism that is different from T_1- and T_2-weighted contrast. T_{1rho} imaging can be performed in the forms of T_{1rho}-weighted image, T_{1rho} mapping, and T_{1rho} dispersion. T_{1rho} imaging is useful to study low-frequency motional processes and CE in biological tissues. T_{1rho} imaging, particularly at low spin-lock frequency, is sensitive to B_0 and B_1 inhomogeneities. Various composite spin-lock pulses have been proposed to alleviate the influence of field inhomogeneities so as to reduce the banding-like spin-lock artifacts. T_{1rho} imaging could be SAR intensive and time consuming, which should be well addressed for clinical applications. T_{1rho} imaging has been proposed and applied for many applications, covering various tissues and a wide range of disease processes, and is promising for routine clinical use in the future.

References

1. Santyr GE, Fairbanks EJ, Kelcz F, Sorenson JA. Off-resonance spin locking for MR imaging. *Magn Reson Med* 1994;32(1):43–51.
2. Dixon WT, Oshinski JN, Trudeau JD, Arnold BC, Pettigrew RI. Myocardial suppression in vivo by spin locking with composite pulses. *Magn Reson Med* 1996;36(1):90–94.
3. Halavaara JT, Lamminen AE, Bondestam S, Sepponen RE, Tanttu JI. Spin lock magnetic resonance imaging in the differentiation of hepatic haemangiomas and metastases. *Br J Radiol* 1995;68(815):1198–1203.
4. Halavaara JT, Sepponen RE, Lamminen AE, Vehmas T, Bondestam S. Spin lock and magnetization transfer MR imaging of local liver lesions. *Magn Reson Imaging* 1998;16(4):359–364.
5. Markkola AT, Aronen HJ, Paavonen T et al. Spin lock and magnetization transfer imaging of head and neck tumors. *Radiology* 1996;200(2):369–375.
6. Hulvershorn J, Borthakur A, Bloy L et al. T1rho contrast in functional magnetic resonance imaging. *Magn Reson Med* 2005;54(5):1155–1162.
7. Lamminen AE, Tanttu JI, Sepponen RE, Pihko H, Korhola OA. T1rho dispersion imaging of diseased muscle tissue. *Br J Radiol* 1993;66(789):783–787.
8. Markkola AT, Aronen HJ, Paavonen T et al. T1rho dispersion imaging of head and neck tumors: a comparison to spin lock and magnetization transfer techniques. *J Magn Reson Imaging* 1997;7(5):873–879.
9. McCommis KS, He X, Abendschein DR, Gupte PM, Gropler RJ, Zheng J. Cardiac 17O MRI: toward direct quantification of myocardial oxygen consumption. *Magn Reson Med* 2010;63(6):1442–1447.
10. Tailor DR, Baumgardner JE, Regatte RR, Leigh JS, Reddy R. Proton MRI of metabolically produced H2 17O using an efficient 17O2 delivery system. *Neuroimage* 2004;22(2):611–618.
11. Tailor DR, Roy A, Regatte RR et al. Indirect 17(O)-magnetic resonance imaging of cerebral blood flow in the rat. *Magn Reson Med* 2003;49(3):479–487.
12. Reddy R, Stolpen AH, Leigh JS. Detection of 17O by proton T1rho dispersion imaging. *J Magn Reson B* 1995;108(3):276–279.
13. Rommel E, Kimmich R. T1 rho dispersion imaging and volume-selective T1rho dispersion weighted NMR spectroscopy. *Magn Reson Med* 1989;12(3):390–399.
14. Gröhn OHJ, Kettunen MI, Makela HI et al. Early detection of irreversible cerebral ischemia in the rat using dispersion of the magnetic resonance imaging relaxation time, T1rho. *J Cereb Blood Flow Metab* 2000;20(10):1457–1466.

15. Duvvuri U, Goldberg AD, Kranz JK et al. Water magnetic relaxation dispersion in biological systems: the contribution of proton exchange and implications for the noninvasive detection of cartilage degradation. *Proc Natl Acad Sci U S A* 2001;98(22):12479–12484.
16. Borthakur A, Wheaton AJ, Gougoutas AJ et al. In vivo measurement of T1rho dispersion in the human brain at 1.5 tesla. *J Magn Reson Imaging* 2004;19(4):403–409.
17. Koskinen SK, Niemi PT, Kajander SA, Komu ME. T1rho Dispersion profile of rat tissues in vitro at very low locking fields. *Magn Reson Imaging* 2006;24(3):295–299.
18. Virta A, Komu M, Kormano M. T1rho of protein solutions at very low fields: dependence on molecular weight, concentration, and structure. *Magn Reson Med* 1997;37(1):53–57.
19. Virta A, Komu M, Lundbom N et al. Low field T1rho imaging of myositis. *Magn Reson Imaging* 1998;16(4):385–391.
20. Cobb JG, Xie J, Gore JC. Contributions of chemical exchange to T1rho dispersion in a tissue model. *Magn Reson Med* 2011;66(6):1563–1571.
21. Makela HI, Grohn OH, Kettunen MI, Kauppinen RA. Proton exchange as a relaxation mechanism for T1 in the rotating frame in native and immobilized protein solutions. *Biochem Biophys Res Commun* 2001;289(4):813–818.
22. Desvaux H, Berthault P. Study of dynamic processes in liquids using off-resonance rf irradiation. *Prog Nucl Magn Reson Spectrose* 1999;35(4):295–340.
23. McConnell HM. Reaction rates by nuclear magnetic resonance. *J Chem Phys* 1958;28:430–431.
24. Davis DG, Perlman ME, London RE. Direct measurements of the dissociation-rate constant for inhibitor-enzyme complexes via the T1 rho and T2 (CPMG) methods. *J Magn Reson B* 1994;104(3):266–275.
25. Trott O, Palmer AG, 3rd. R1rho relaxation outside of the fast-exchange limit. *J Magn Reson* 2002;154(1):157–160.
26. Palmer AG, 3rd, Massi F. Characterization of the dynamics of biomacromolecules using rotating-frame spin relaxation NMR spectroscopy. *Chem Rev* 2006;106(5):1700–1719.
27. Trott O, Palmer AG, 3rd. Theoretical study of R(1rho) rotating-frame and R2 free-precession relaxation in the presence of n-site chemical exchange. *J Magn Reson* 2004;170(1):104–112.
28. van Zijl PC, Yadav NN. Chemical exchange saturation transfer (CEST): what is in a name and what isn't? *Magn Reson Med* 2011;65(4):927–948.
29. Ward KM, Aletras AH, Balaban RS. A new class of contrast agents for MRI based on proton chemical exchange dependent saturation transfer (CEST). *J Magn Reson* 2000;143(1):79–87.
30. Cai K, Haris M, Singh A et al. Magnetic resonance imaging of glutamate. *Nat Med* 2012;18(2):302–306.
31. Ling W, Regatte RR, Navon G, Jerschow A. Assessment of glycosaminoglycan concentration in vivo by chemical exchange-dependent saturation transfer (gagCEST). *Proc Natl Acad Sci U S A* 2008;105(7):2266–2270.
32. van Zijl PC, Jones CK, Ren J, Malloy CR, Sherry AD. MRI detection of glycogen in vivo by using chemical exchange saturation transfer imaging (glycoCEST). *Proc Natl Acad Sci U S A* 2007;104(11):4359–4364.
33. Zhou JY, Payen JF, Wilson DA, Traystman RJ, van Zijl PCM. Using the amide proton signals of intracellular proteins and peptides to detect pH effects in MRI. *Nat Med* 2003;9(8):1085–1090.

34. Zhou J, Lal B, Wilson DA, Laterra J, van Zijl PC. Amide proton transfer (APT) contrast for imaging of brain tumors. *Magn Reson Med* 2003;50(6):1120–1126.
35. Sherry AD, Woods M. Chemical exchange saturation transfer contrast agents for magnetic resonance imaging. *Ann Rev Biomed Eng* 2008;10:391–411.
36. McMahon MT, Gilad AA, DeLiso MA, Berman SM, Bulte JW, van Zijl PC. New "multicolor" polypeptide diamagnetic chemical exchange saturation transfer (DIACEST) contrast agents for MRI. *Magn Reson Med* 2008;60(4):803–812.
37. Aime S, Delli Castelli D, Terreno E. Highly sensitive MRI chemical exchange saturation transfer agents using liposomes. *Angew Chem Int Ed Engl* 2005;44(34):5513–5515.
38. Zhang S, Merritt M, Woessner DE, Lenkinski RE, Sherry AD. PARACEST agents: modulating MRI contrast via water proton exchange. *Acc Chem Res* 2003;36(10):783–790.
39. Jin T, Autio J, Obata T, Kim SG. Spin-locking versus chemical exchange saturation transfer MRI for investigating chemical exchange process between water and labile metabolite protons. *Magn Reson Med* 2011;65(5):1448–1460.
40. Jin T, Wang P, Zong X, Kim SG. Magnetic resonance imaging of the Amine-Proton EXchange (APEX) dependent contrast. *Neuroimage* 2012;59(2):1218–1227.
41. Kogan F, Singh A, Cai K, Haris M, Hariharan H, Reddy R. Investigation of chemical exchange at intermediate exchange rates using a combination of chemical exchange saturation transfer (CEST) and spin-locking methods (CESTrho). *Magn Reson Med* 2012;68(1):107–119.
42. Cobb JG, Xie J, Li K, Gochberg DF, Gore JC. Exchange-mediated contrast agents for spin-lock imaging. *Magn Reson Med* 2012;67(5):1427–1433.
43. Charagundla SR, Borthakur A, Leigh JS, Reddy R. Artifacts in T(1rho)-weighted imaging: correction with a self-compensating spin-locking pulse. *J Magn Reson* 2003;162(1):113–121.
44. Yuan J, Li Y, Zhao F, Chan Q, Ahuja AT, Wang YX. Quantification of T1rho relaxation by using rotary echo spin-lock pulses in the presence of B0 inhomogeneity. *Phys Med Biol* 2012;57(15):5003–5016.
45. Witschey WR, 2nd, Borthakur A, Elliott MA et al. Artifacts in T1 rho-weighted imaging: compensation for B(1) and B(0) field imperfections. *J Magn Reson* 2007;186(1):75–85.
46. Li X, Han ET, Ma CB, Link TM, Newitt DC, Majumdar S. in vivo 3T spiral imaging based multi-slice T(1rho) mapping of knee cartilage in osteoarthritis. *Magn Reson Med* 2005;54(4):929–936.
47. Chen W, Takahashi A, Han E. Quantitative T(1)(rho) imaging using phase cycling for B(0) and B(1) field inhomogeneity compensation. *Magn Reson Imaging* 2011;29(5):608–619.
48. Collins CM, Li S, Smith MB. SAR and B1 field distributions in a heterogeneous human head model within a birdcage coil. Specific energy absorption rate. *Magn Reson Med* 1998;40(6):847–856.
49. Yuan J, Zhao F, Griffith JF, Chan Q, Wang YX. Optimized efficient liver T1rho mapping using limited spin lock times. *Phys Med Biol* 2012;57(6):1631–1640.
50. Griswold MA, Jakob PM, Heidemann RM et al. Generalized autocalibrating partially parallel acquisitions (GRAPPA). *Magn Reson Med* 2002;47(6):1202–1210.
51. Pruessmann KP, Weiger M, Scheidegger MB, Boesiger P. SENSE: sensitivity encoding for fast MRI. *Magn Reson Med* 1999;42(5):952–962.

52. Sodickson DK, Manning WJ. Simultaneous acquisition of spatial harmonics (SMASH): fast imaging with radiofrequency coil arrays. *Magn Reson Med* 1997;38(4):591–603.
53. Liang D, Dibella EV, Chen RR, Ying L. k-t ISD: dynamic cardiac MR imaging using compressed sensing with iterative support detection. *Magn Reson Med* 2012;68(1):41–53.
54. Lustig M, Donoho D, Pauly JM. Sparse MRI: the application of compressed sensing for rapid MR imaging. *Magn Reson Med* 2007;58(6):1182–1195.
55. Yuan J, Liang D, Zhao F, Li Y, Wang Y-X, Ying L. k-T ISD Compressed sensing reconstruction for T1ρ mapping: a study in rat brains at 3T. *Proceedings 20th Scientific Meeting, International Society for Magnetic Resonance in Medicine*, Melbourne, Australia, 2012: p. 4197.
56. Wheaton AJ, Borthakur A, Corbo M, Charagundla SR, Reddy R. Method for reduced SAR T1rho-weighted MRI. *Magn Reson Med* 2004;51(6):1096–1102.
57. Yuan J, Zhao TC, Tang Y, Panych LP. Reduced field-of-view single-shot fast spin echo imaging using two-dimensional spatially selective radiofrequency pulses. *J Magn Reson Imaging* 2010;32(1):242–248.
58. Rieseberg S, Frahm J, Finsterbusch J. Two-dimensional spatially-selective RF excitation pulses in echo-planar imaging. *Magn Reson Med* 2002;47(6):1186–1193.
59. Zhu Y. Parallel excitation with an array of transmit coils. *Magn Reson Med* 2004;51(4):775–784.
60. Wang Y-X, Zhao F, Griffith JF et al. T1rho and T2 relaxation times for lumbar disc degeneration: an in vivo comparative study at 3.0-Tesla MRI. *Eur Radiol* 2012;23(1):228–234.
61. Pfirrmann CW, Metzdorf A, Zanetti M, Hodler J, Boos N. Magnetic resonance classification of lumbar intervertebral disc degeneration. *Spine (Phila Pa 1976)* 2001;26(17):1873–1878.
62. Griffith JF, Wang YX, Antonio GE et al. Modified Pfirrmann grading system for lumbar intervertebral disc degeneration. *Spine (Phila Pa 1976)* 2007;32(24):E708–E712.
63. Toffanin R, Mlynarik V, Russo S, Szomolanyi P, Piras A, Vittur F. Proteoglycan depletion and magnetic resonance parameters of articular cartilage. *Arch Biochem Biophys* 2001;390(2):235–242.
64. Regatte RR, Akella SV, Borthakur A, Kneeland JB, Reddy R. Proteoglycan depletion-induced changes in transverse relaxation maps of cartilage: comparison of T2 and T1rho. *Acad Radiol* 2002;9(12):1388–1394.
65. Nieminen MT, Toyras J, Rieppo J et al. Quantitative MR microscopy of enzymatically degraded articular cartilage. *Magn Reson Med* 2000;43(5):676–681.
66. Duvvuri U, Reddy R, Patel SD, Kaufman JH, Kneeland JB, Leigh JS. T1rho-relaxation in articular cartilage: effects of enzymatic degradation. *Magn Reson Med* 1997;38(6):863–867.
67. Gray ML, Burstein D, Xia Y. Biochemical (and functional) imaging of articular cartilage. *Semin Musculoskelet Radiol* 2001;5(4):329–343.
68. Gray ML, Eckstein F, Peterfy C, Dahlberg L, Kim YJ, Sorensen AG. Toward imaging biomarkers for osteoarthritis. *Clin Orthop Relat Res* 2004;427(Suppl):S175–S181.
69. Akella SV, Regatte RR, Gougoutas AJ et al. Proteoglycan-induced changes in T1rho-relaxation of articular cartilage at 4T. *Magn Reson Med* 2001;46(3):419–423.

70. Li X, Benjamin Ma C, Link TM et al. In vivo T(1rho) and T(2) mapping of articular cartilage in osteoarthritis of the knee using 3 T MRI. *Osteoarthritis Cartilage* 2007;15(7):789–797.
71. Li X, Pai A, Blumenkrantz G et al. Spatial distribution and relationship of T1rho and T2 relaxation times in knee cartilage with osteoarthritis. *Magn Reson Med* 2009;61(6):1310–1318.
72. Johannessen W, Auerbach JD, Wheaton AJ et al. Assessment of human disc degeneration and proteoglycan content using T1rho-weighted magnetic resonance imaging. *Spine (Phila Pa 1976)* 2006;31(11):1253–1257.
73. Padovani A, Costanzi C, Gilberti N, Borroni B. Parkinson's disease and dementia. *Neurol Sci* 2006;27 Suppl 1:S40–S43.
74. Borthakur A, Gur T, Wheaton AJ et al. In vivo measurement of plaque burden in a mouse model of Alzheimer's disease. *J Magn Reson Imaging* 2006;24(5):1011–1017.
75. Borthakur A, Sochor M, Davatzikos C, Trojanowski JQ, Clark CM. T1rho MRI of Alzheimer's disease. *Neuroimage* 2008;41(4):1199–1205.
76. Haris M, McArdle E, Fenty M et al. Early marker for Alzheimer's disease: hippocampus T1rho (T(1rho)) estimation. *J Magn Reson Imaging* 2009;29(5):1008–1012.
77. Haris M, Singh A, Cai K et al. T1rho (T1rho) MR imaging in Alzheimer' disease and Parkinson's disease with and without dementia. *J Neurol* 2011;258(3):380–385.
78. Haris M, Singh A, Cai K et al. T(1rho) MRI in Alzheimer's disease: detection of pathological changes in medial temporal lobe. *J Neuroimaging* 2011;21(2):e86–e90.
79. Nestrasil I, Michaeli S, Liimatainen T et al. T1rho and T2rho MRI in the evaluation of Parkinson's disease. *J Neurol* 2010;257(6):964–968.
80. Grohn OH, Lukkarinen JA, Silvennoinen MJ, Pitkanen A, van Zijl PC, Kauppinen RA. Quantitative magnetic resonance imaging assessment of cerebral ischemia in rat using on-resonance T(1) in the rotating frame. *Magn Reson Med* 1999;42(2):268–276.
81. Makela HI, Kettunen MI, Grohn OH, Kauppinen RA. Quantitative T(1rho) and magnetization transfer magnetic resonance imaging of acute cerebral ischemia in the rat. *J Cereb Blood Flow Metab* 2002;22(5):547–558.
82. Jokivarsi KT, Niskanen JP, Michaeli S et al. Quantitative assessment of water pools by T 1 rho and T 2 rho MRI in acute cerebral ischemia of the rat. *J Cereb Blood Flow Metab* 2009;29(1):206–216.
83. Jokivarsi KT, Hiltunen Y, Grohn H, Tuunanen P, Grohn OH, Kauppinen RA. Estimation of the onset time of cerebral ischemia using T1{rho} and T2 MRI in rats. *Stroke* 2010;41(10):2335–2340.
84. Zhao F, Yuan J, Jiu T et al. A longitudinal study on age-related changes of T1rho relaxation in rat brain. *Proceedings 20th Scientific Meeting, International Society for Magnetic Resonance in Medicine*, Melbourne, Australia, 2012: p. 3758.
85. Zhao F, Zhang L-H, Yuan J, Chan Q, Yew D, Wang Y-X. A comparative study of brain regional T1rho values of spontaneously hypertensive rat and Wistar Kyoto rat. *Proceedings 20th Scientific Meeting, International Society for Magnetic Resonance in Medicine*, Melbourne, Australia, 2012: p. 3068.
86. Zhao F, Wang YX, Yuan J et al. MR T1rho as an imaging biomarker for monitoring liver injury progression and regression: an experimental study in rats with carbon tetrachloride intoxication. *Eur Radiol* 2012;22(8):1709–1716.

87. Wang YX, Yuan J, Chu ES et al. T1rho MR imaging is sensitive to evaluate liver fibrosis: an experimental study in a rat biliary duct ligation model. *Radiology* 2011;259(3):712–719.
88. Deng M, Zhao F, Yuan J, Ahuja AT, Wang YX. Liver T1rho MRI measurement in healthy human subjects at 3 T: a preliminary study with a two-dimensional fast-field echo sequence. *Br J Radiol* 2012;85(1017):e590–e595.
89. Wang Y-X, Zhao F, Wong WS, Yuan J, Kwong KM, Chan LY. Liver MR T1rho measurement in liver cirrhosis patients: a preliminary study with a 2D fast field echo sequence at 3T. *Proceedings 20th Scientific Meeting, International Society for Magnetic Resonance in Medicine*, Melbourne, Australia, 2012: p. 1289.
90. Santyr GE, Henkelman RM, Bronskill MJ. Spin locking for magnetic resonance imaging with application to human breast. *Magn Reson Med* 1989;12(1):25–37.
91. Virta A, Komu M, Lundbom N, Kormano M. T1 rho MR imaging characteristics of human anterior tibial and gastrocnemius muscles. *Acad Radiol* 1998;5(2):104–110.
92. Duvvuri U, Poptani H, Feldman M et al. Quantitative T1rho magnetic resonance imaging of RIF-1 tumors in vivo: detection of early response to cyclophosphamide therapy. *Cancer Res* 2001;61(21):7747–7753.
93. Kettunen MI, Sierra A, Narvainen MJ et al. Low spin-lock field T1 relaxation in the rotating frame as a sensitive MR imaging marker for gene therapy treatment response in rat glioma. *Radiology* 2007;243(3):796–803.
94. Sierra A, Michaeli S, Niskanen JP et al. Water spin dynamics during apoptotic cell death in glioma gene therapy probed by T1rho and T2rho. *Magn Reson Med* 2008;59(6):1311–1319.
95. Muthupillai R, Flamm SD, Wilson JM, Pettigrew RI, Dixon WT. Acute myocardial infarction: tissue characterization with T1rho-weighted MR imaging—initial experience. *Radiology* 2004;232(2):606–610.
96. Huber S, Muthupillai R, Lambert B, Pereyra M, Napoli A, Flamm SD. Tissue characterization of myocardial infarction using T1rho: influence of contrast dose and time of imaging after contrast administration. *J Magn Reson Imaging* 2006;24(5):1040–1046.
97. Witschey WR, Zsido GA, Koomalsingh K et al. In vivo chronic myocardial infarction characterization by spin locked cardiovascular magnetic resonance. *J Cardiovasc Magn Reson* 2012;14(1):37.

21

Brain Connectivity Assessed with Functional MRI

Aiping Liu, Junning Li, Martin J. McKeown, and Z. Jane Wang

CONTENTS

21.1 Introduction

In recent years, modern neuroimaging technologies have become increasingly important for studying brain function in vivo (Figure 21.1). In particular, functional magnetic resonance imaging (fMRI) enables the brain to be noninvasively assessed at excellent spatial resolution and relatively good temporal resolution.

21.1.1 fMRI and Brain Activity Analysis

FMRI utilizing blood oxygen level dependent (BOLD) contrast is based on the differing paramagnetic properties of oxy- and deoxy-hemoglobin [1]. BOLD is an indirect marker of neural activity, as it is based on focal blood flow modulated by local brain metabolism. The majority of fMRI analyses to date have examined changes in BOLD *amplitude* as a result of performing cognitive, motor, visual, sensory, taste, or even osmic tasks. In contrast, fMRI connectivity explores modulation of widespread, simultaneous BOLD signal fluctuations generated during a task, or, in the case of "resting state" studies, without any obvious external stimulus or task.

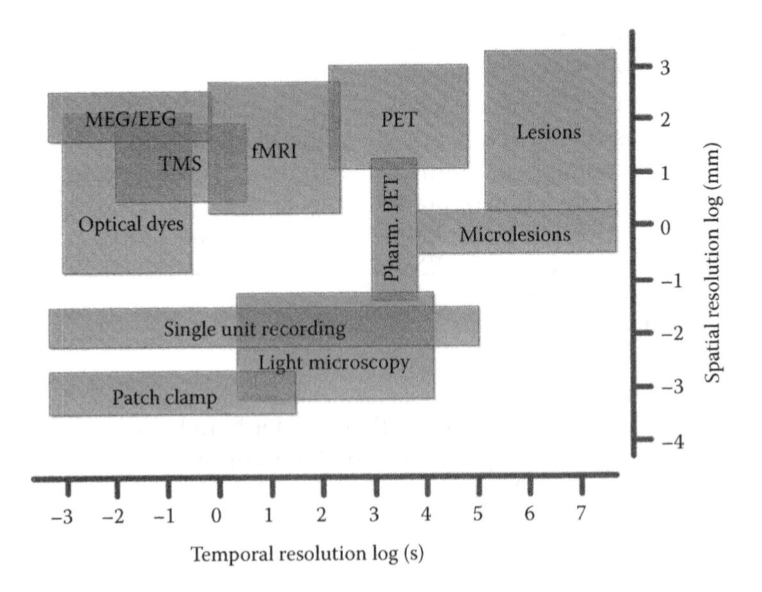

FIGURE 21.1
Temporal and spatial resolution of current neuroimaging technology. Note the noninvasive technologies suitable for assessment in human populations: TMS, transcranial magnetic stimulation; MEG, magnetoencephalography; EEG, electroencephalography; PET, positron emission tomography; and Pharm. PET, pharmacological PET. (Adapted from Churchland, P. and Terrence, S., *The Computational Brain*, MIT Press, Cambridge, MA, 1992.)

Proper biological interpretation of fMRI connectivity patterns is difficult, not least because the exact origin of all BOLD signal fluctuations is unclear, and it is known that non-neural signals, such as cardiac and respiration pulsations, induce fluctuations in the BOLD signal [2–4], hence preprocessing methods to reduce these influences are crucial for accurate fMRI connectivity analysis [2,3]. However, recent work looking at the relation between resting state BOLD fluctuations and anatomical connectivity assessed by diffusion tensor imaging (DTI) [5] strongly supports the notion that functional and anatomical connectivity are closely linked, implying that neural activity is a key component of spontaneous BOLD signal fluctuations [6]. DTI is a relatively new imaging method, used to measure structural brain connectivity by measuring the diffusion profile of water molecules in the brain, allowing, under fairly strong assumptions, the reconstruction of white matter tracts [7].

21.1.2 Behavioral Paradigms in fMRI

Because the exact amplitude of the BOLD signals is not directly comparable across subjects, it is standard to determine the *relative* differences in BOLD signal amplitude across two tasks (e.g., finger tapping vs. rest). Traditionally, this has been done in a block design, where the subjects may, for example, tap their fingers for 20–30s followed by 20–30s of rest, and the cycle is repeated. This alternating of experimental and control tasks will tend to make analysis methods more robust against any erroneous interpretations based on nonneuronal slow drifts in the signal and/or fatigue effects that would bias interpretations if the experimental task was done only at the beginning or end of the run. Block-related designs generally assume that any hemodynamic response to neuronal activity is saturated by rapidly and repeatedly performing the same task within a block. An alternate approach is to assess the BOLD response to a single stimulus, such as a simple motor movement. This has the advantage of comparing stimuli that might have the same loci of activation but different amplitudes of neuronal (and subsequent hemodynamic) response, but has the disadvantage of significantly prolonging scanning time, as the hemodynamic response must decay sufficiently before the next stimulus can be presented.

A more recent type of paradigm is the so-called resting-state study, whereby individuals are instructed to simply rest quietly with their eyes closed and remain awake. In this condition, spatially widespread, unprompted activity not attributable to specific external stimuli can be observed. Statistical analyses on the spontaneous fluctuations in the BOLD signal can then be performed to determine which regions covary and/or influence one another.

21.1.3 Ways to Assess Task-Related BOLD Effects

Task-related changes in fMRI can be complex, with changes in BOLD amplitude, the spatial extent of activation, and changes in connectivity. The vast majority of ways to infer task-related activation have utilized hypothesis-driven methods to examine changes in BOLD amplitude, like the standard general linear model (GLM). For example, the box-car design alluded earlier can be convolved with an estimate of the temporal profile of hemodynamic response resulting from an abrupt change in neural activity to specify the expected task-related changes seen in a voxel. This temporal profile, as well as other factors thought to influence the time course of the observed signals at each voxel can be considered as regressor variables, $\boldsymbol{G}$, in a linear regression equation, giving

$$\boldsymbol{X} = \boldsymbol{G}\beta + \varepsilon \tag{21.1}$$

where $\boldsymbol{X}$ is an n by v row mean-zero data matrix with n being the number of time points in the experiment and v being the total number of voxels in all slices, $\boldsymbol{G}$ is a specified n by p design matrix containing the time courses of all p factors hypothesized to modulate the BOLD signal, β is a p by v matrix of p spatial patterns to be estimated, and ε is a matrix of noise or residual errors typically assumed to be independent, zero-mean, and Gaussian distributed, that is, $N(0, \sigma^2)$. Once $\boldsymbol{G}$ is specified, standard regression techniques can be used to provide a least square estimate for the parameters in β. The statistical significance of these parameters can be considered to constitute spatial maps [8], one for each row in β, which correspond to the time courses specified in the columns of the design matrix.

A key component of the GLM is that the residuals are assumed to be Gaussian distributed. As the rows of X are not independent (as they constitute a nontemporally white time series), a noise model may be used to make the residuals temporally uncorrelated. This is typically an autoregressive (AR) model of order 1–2.

While less frequently studied, the spatial pattern of the BOLD signals can be altered by task-related activity. Spatial or "3D texture" descriptors such as 3D moment invariants that are invariant to the exact coordinate system, can be used to examine the effects of task-related changes in fMRI [9].

However, another critical way in which task performance can affect BOLD signals, and the main focus here, are changes in connectivity. Macroscopic brain connectivity has been increasingly recognized as being vital for understanding normal brain function and the pathophysiology of many neuropsychiatric diseases [10]. Assessing the functional connectivity between two brain loci can either occur at the individual voxel or the region of interest (ROI) level. The term "functional connectivity" is sometimes used quite loosely. It may refer to different ROIs covarying over time or it may refer to direct functional connections between ROIs. As is well-known in statistics, correlation does not imply causation, and thus the terms "functional connectivity" and "effective connectivity" are sometimes used to distinguish between correlative and causative relations between regions, respectively.

In this chapter, we are mainly interested in the methods of brain connectivity modeling, with particular emphasis on Bayesian network (BN) modeling methods for estimating the interactions between brain regions.

21.2 Brain Connectivity Modeling

With the increasing recognition of the importance of brain connectivity, there has been rapid development of many proposed mathematical models. In this section, we briefly introduce broad categories of fMRI connectivity modeling methods.

21.2.1 Brief Review on Brain Connectivity Modeling Methods

Brain connectivity patterns have been inferred based on bivariate analyses such as CT [11], frequency-based coherence analysis [12,13], mutual information [14], Granger causality derived from bivariate AR models [4], and so on. Multivariate models, including multivariate autoregression models (MAR) [15], structural equation models (SEM) [16], and dynamic causal models (DCM) [17] have also been proposed to assess brain connectivity. Other commonly used approaches include linear decomposition methods such as independent component analysis (ICA) [18], Least absolute selection and shrinkage operator (LASSO)-based methods [19,20], and BN models [21–23]. There are different (but not mutually exclusive) ways in which all these proposed brain connectivity modeling approaches can be categorized: exploratory vs. confirmatory, linear vs. nonlinear, static vs. dynamic, directional connectivity vs. bidirectional, and the subject-level vs. group-level modeling.

The most straightforward way to assess brain connectivity is the seed-based ROI method. This method starts by extracting the mean time course of the BOLD signal from an anatomically or functional activation-defined ROI, and then this time course is correlated with all other brain regions. The outcome of such a procedure is a functional connectivity map (fcMAP) representing how strongly different ROIs are functionally connected with the given region [24,25]. The main benefit of the seed-based method is that it results in an unambiguous functional map with a straightforward interpretation, while the main disadvantage is the dependence of resulting fcMAPs may depend critically on the choice of seed ROI [26,27].

The CT method [11] estimates how strongly two ROIs interact with each other by calculating the correlation coefficient between their activities, and computing this for all possible ROI pairs. If the correlation coefficient is deemed statistically significant, the two regions are considered associated with each other. The CT is rigorously defined statistically and is able to control the type I error rate, and also to estimate the strength of interactions.

However, standard pair-wise correlation cannot distinguish whether two components interact directly or indirectly through a third component [28].

21.2.1.1 Conditional-Dependence Measures

Instead of simply referring to different brain regions that are covarying, partial correlation can be employed to estimate if one brain region has direct influence over another [29], as it measures the normalized correlation with the effect of all other variables removed. The application of partial correlation in inferring the relationship between two variables is based on the conditional-independence test. The definition of conditional independence is as follows: X and Y are conditional-independent given Z if and only if $P(XY|Z) = P(X|Z)$ $P(Y|Z)$. It is similar to the pair-wise independence definition $P(XY) = P(X)$ $P(Y)$, but conditional on a third variable Z. Note that pair-wise independence does not imply conditional independence, and *vice versa* (Figure 21.2). For instance, the activities of two brain regions A and B are commonly driven by that of a third region C, then the activities of A and B maybe correlated in pair-wise fashion, but if the influence from C is removed, then their activities will become independent, as shown in Figure 2.12b. On the other hand, if the activities of two brain regions are correlated even after all possible influence from other regions are removed (as shown in Figure 21.2c), then very likely there is a direct connection between them (i.e., the two regions are conditionally dependent). Therefore, the conditional dependence implies that two brain regions are directly connected. More importantly, conditional independence is a key concept in multivariate analyses such as graphical modeling, where two nodes are connected if and only if the corresponding variables are not conditionally independent, which we will discussed in the next section.

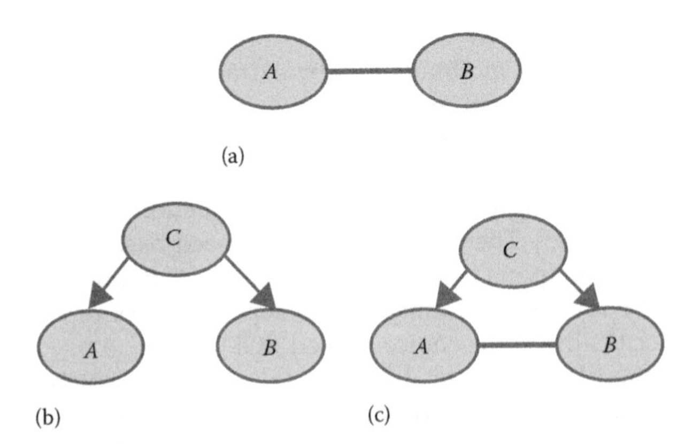

FIGURE 21.2
Types of statistical (in)dependence. (a) Unconditional (pair-wise) dependence. (b) Unconditional (pair-wise) dependence but not conditional dependence. (c) Unconditional (pair-wise) dependence but also conditional dependence.

It must be noted that conditional independence alone without temporal information is not enough to determine causal relationships, that is, the direction of connectivity. To infer the direction of the connections, one approach is to use Granger causality (GC) [4]. GC is based on the statistical hypothesis testing for determining whether one time series can be used to at least partially predict another. If knowing the past of A can predict B significantly more accurately than knowing the past of B alone, then A is said to cause B [4]. Another way to estimate the direction is using Patel's conditional-dependence measure [30]. It estimates the connectivity between A and B by measuring the imbalance between conditional probability $P(A|B)$ and $P(B|A)$. Both Granger causality and Patel's conditional-dependence measure are usually considered confirmatory methods as they generally need prior knowledge on the network model.

21.2.1.2 Linear Decomposition Methods

Different from the pair-wise analysis between two variables discussed before, linear decomposition methods, such as principal component analysis (PCA) or independent component analysis (ICA) [18] can be used to assess which voxels have a tendency to covary. For instance, ICA decomposes BOLD patterns into spatially independent maps and their associated time courses, which was a significant shift from the traditional hypothesis-based approach for fMRI analysis when first proposed [31]. Because no time courses of activation need to be specified *a priori*, it is ideally suited to assess resting-state fMRI data [32,33] or in situations where the anticipated activation patterns may deviate from normal. Thus, ICA analysis of fMRI has been widely used to study clinical populations, for example, Alzheimer's disease [34], depression [35], schizophrenia [36], mild cognitive impairment [37], and noncommunicative brain-damaged patients [38].

21.2.1.3 Clustering Methods

Similar to the linear decomposition methods, clustering techniques, such as the self-organizing map (SOM) [39,40], k-means clustering [41], hierarchical clustering [42], and graph clustering [43], are also data-driven approaches to explore unknown interactions between brain regions. They are based on the assumption that if the time courses of voxels and/or ROIs tend to cluster together, they likely have interactions between them. Clustering is usually implemented with fast and heuristic algorithms and thus is suitable for large-scale problems where it is difficult to perform rigorous statistical analysis. However, the data-driven feature also brings disadvantages, as certain algorithms may either fall in local optimal solutions or their convergence cannot be proved. Statistical criteria such as specificity and sensitivity generally cannot be theoretically analyzed for clustering methods.

21.2.1.4 Multivariate Statistical Methods

MAR model [15], SEM [16], and DCM [17] are popular multivariate regression models proposed to assess brain connectivity. The MAR model focuses on the lagged interactions and incorporates the covariance as well as temporal information across the samples. It represents one sample of a time series as the weighted sum of its previous samples,

$$y_t = \sum_{p=1}^{P} A_p y_{(t-p)} + e_t \tag{21.2}$$

where
- y_t is the K-dimensional vector denoting the values of K ROIs BOLD signal values at time t
- A_p is the AR coefficient matrix at time lag p
- e_t is the noise term

Different from the MAR model that infers the lagged information, SEM estimates the simultaneous interactions between brain ROIs. It can be formulated as follows:

$$y_t = My_t + e_t \tag{21.3}$$

where M represents the connection strengths matrix. DCM is distinguished from them by accommodating the nonlinear and dynamic activities between brain regions. It models neural activities as hidden variables as follows:

$$\dot{y}_t = \left(A + \sum_{j=1}^{J} u_t(j)B^j \right) y_t + Cu_t \tag{21.4}$$

where u_t is the exogenous input. Multivariate regression models formulated in these ways are statistically rigorous, flexible, and supported by many well-developed algorithms for various kinds of purposes, such as spectrum analysis, model selection, and statistical inference [44–49]. However, a major drawback of these models is that the computational cost grows exponentially with the number of ROIs. This typically restricts their use to confirmatory studies examining a few ROIs.

21.2.1.5 Sparse Methods

Since fMRI studies have relatively few time points (typically less than a couple of hundred) and the number of ROIs may be large (>20, say), modeling the brain connectivity is a difficult statistical inference problem. LASSO-based models combine computational efficiency with the ability to deal with high dimensionality, and hence such methods including the robust LASSO and the inverse covariance estimation have been proposed [19,20,50]. For instance, by fitting a model to all the variables, the graphical LASSO estimates a sparse *graphical model*, whereby ROIs are represented as vertices and variable-wise relationships are represented as edges [50].

21.2.1.6 Graphical Models

Graphical models are suitable for modeling the brain connectivity as their graphical nature assists in the biological interpretation of the connectivity patterns. The linear, non-Gaussian, acyclic model (LiNGAM) is a causal graphical model [51] that assumes that the variables have non-Gaussian distributions of nonzero variances, and identify the brain connectivity structure as a directed acyclic graph. The BN models, however, are the most popular graphical models proposed for studying the interactions between brain regions, and thus in the third section, we will particularly focus on the BN modeling methods.

21.2.2 Assessment of Network Modeling Methods and Directionality of Brain Connectivity

In a recent review paper, different connectivity estimation methods for use on fMRI time series were explored, using simulated fMRI data generated from a DCM model [52]. The results suggest that, in general, partial correlation, regularized inverse covariance estimation and several BN modeling methods give better performance for determining if a connection exists between ROIs. However, accurately determining the connection directionality is much more difficult to obtain, and several proposed directionality estimation methods, including Granger causality and Patel's conditional-dependence measure, may be unsuccessful. Nevertheless, a recently proposed combination method which first applies a structure-learning algorithm to estimate the existence of the connections, and then employs an orientation algorithm to detect the directionality of the fixed connection structure can provide accurate estimates, particularly at the group level [53]. An interesting ancillary of these simulations is that the exact spatial delineation of ROIs appears extremely important for the estimation of brain connectivity networks [52,54].

21.2.3 Group Modeling of Brain Connectivity

As biomedical research typically studies a group of subjects to make inferences about a population, rather than focusing on an individual subject, group analysis plays an important role in statistical modeling of brain connectivity.

However, group-level methods for modeling brain connectivity need to handle not only the variances and correlations across subjects, but also the fact that the exact structure of brain connectivity may vary across individuals.

Group studies are often closely linked to exploratory analysis. In contrast to confirmatory studies that usually involve verification of just a few preselected models, exploratory studies must search through a huge number of possible models to find one or a few models that are best supported by data. Thus, efficiency of the search strategy becomes important, especially since the number of possible network structures increases super-exponentially to the number of brain ROIs involved. For example, with just seven ROIs, there are more than a billion possible network structures.

Besides computational efficiency, accuracy is another important criterion for an exploratory method. In biological scenarios, the goal is not just adequately modeling the overall multivariate time series derived from multiple ROIs, but also that the structure of the model, from which biological interpretations are made, is accurately depicted. Colloquially, accuracy can be inferred from answering the questions: "How many of the connections in the model inferred from data are actually true?" "how many true connections can be detected by the model?" and "how many nonexisting connections in the model are falsely reported?" Therefore, error control is a crucial point in the design of reliable methods for discovering connectivity.

Several graphical methods have been proposed to infer group connectivity in neuroimaging. Bayesian model selection handles intersubject variability and error control well; however, its current proposed implementation does not scale well, making it more suitable for confirmatory, rather than an exploratory research [55]. A data-driven method for estimating large-scale brain connectivity using Gaussian modeling, and dealing with the variability between subjects by using optimal regularization schemes has been proposed [20]. Ramsey et al. describe and evaluate a combination of a multi-subject search algorithm and the orientation algorithm [53]. The group iterative multiple model estimation (GIMME) method is designed to model the connectivity for multi-subjects that can capture the shared structure across the subjects as well as obtaining the individual specific connectivity networks [56]. All these studies have provided initial steps for accurate group modeling; however, these group algorithms do not jointly take into account the intersubject variability, efficiency, and error rate control.

21.3 Bayesian Network (BN) Modeling for Brain Connectivity

BN approaches may provide a suitable framework to examining ways to derive accurate, computationally efficient group-level models of brain connectivity. The popularity of BN modeling for discovering neural interactions is likely

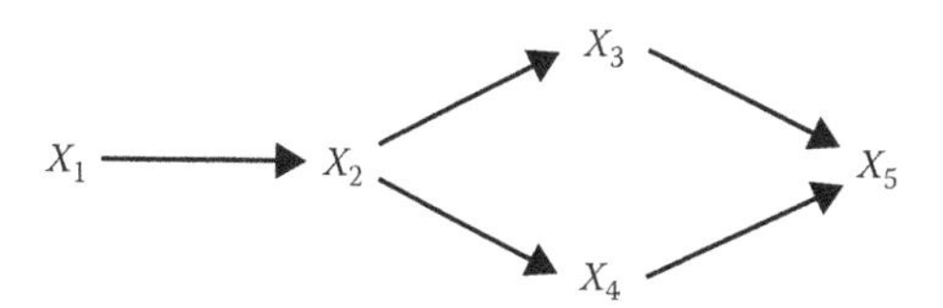

FIGURE 21.3
The illustration of a BN. The joint probability distribution can be factorized as $P(X) = P(X_1)P(X_2|X_1)P(X_3|X_2)P(X_4|X_2)P(X_5|X_3,X_4)$.

related to its graphical nature, making for intuitive interpretation of the results, and the rigorous statistical theory underlying the models [21–23,53,57–59]. The edges of a BN are directional, which is suitable for modeling the long-range interactions between brain regions. Furthermore, BNs are modular and flexible: given a network structure, different statistical methods can be used to describe the dependence relationships between nodes and their parent nodes. As demonstrated in Figure 21.3, activities between two brain regions may covary because the inference from one region is relayed to the other via intermediate region(s). Here we describe how BN modeling methods can be used to infer direct influence from fMRI signals and thus here we do not interpret "connectivity" simply as covariance between different regions, but rather that the direct influence of one brain region over another.

21.3.1 Bayesian Networks (BNs)

"D-separation" [60, p. 36] is the key concept for BNs to encode conditional-independence relationships. Suppose a and b are two vertices in a directed acyclic graph (DAG) G, C is a subset of vertices in graph G excluding a and b. a and b are d-separated by C if and only if for any undirected acyclic path M on G connecting a and b, there exists an intermediate vertex $m \in M$ other than a or b such that at least one of the following holds:

1. The two edges connected by $\boldsymbol{m}$ in path $\boldsymbol{M}$ do not meet head-to-head in graph $\boldsymbol{G}$ and $\boldsymbol{m} \in \boldsymbol{C}$.
2. The two edges connected by $\boldsymbol{m}$ in path $\boldsymbol{M}$ meet head-to-head in graph $\boldsymbol{G}$, and $\boldsymbol{m} \notin \boldsymbol{C}$.

The concept of d-separation between two vertices can be extended to that between two vertices sets. Vertex sets A and B are d-separated by vertex set C, where A, B, and C are disjoint, if and only if for any vertex pair $a \in A$ and $b \in B$ is d-separated by C.

With the concept of d-separation, a DAG encodes a set of conditional-independence relationships according to the directed Markov property which is defined as follows: if A, B, and C are three disjoint vertex sets, and A is d-separated by C from B, then X_A and X_B are conditionally independent given X_C, that is, $P(X_A,X_B|X_C) = P(X_A|X_C)P(X_B|X_C)$ [61, pp. 46–53].

Moreover, a probability distribution P is faithful to a DAG G [60, pp. 13, 81] if all and only the conditional-independence relationships derived from P are encoded by G. In general, a probability distribution may possess other independence relationships besides those encoded by a DAG. The faithfulness assumption requires that all the conditional-independence relationships in P are encoded by the d-separations derived from the graph structure.

A useful implication of the directed Markov property and faithfulness assumption is that if two vertices a and b are directly connected, then they are not d-separated by any combination of other vertices, and the variables that they denote are consequently conditional-dependent given any combination of other variable. When analyzing brain connectivity with BNs, we use vertices to denote brain ROIs, and edges to denote connections between ROIs. With the directed property, d-separated ROIs are conditional-independent given the activities of certain brain regions. ROIs that are not d-separated are potentially conditionally dependent given any combination of other ROIs. If the faithfulness assumption is also employed, then these non-d-separated ROIs are directly connected.

The joint probability distribution of a BN over random variables $X = [X_1, X_2, \ldots\ldots, X_n]$ can be factorized as

$$P(X) = \prod_{i=1}^{n} P(X_i \mid pa(X_i)) \tag{21.5}$$

where $pa(X_i)$ is the parents set of X_i, that is the vertex set directly pointing to X_i in the network. For instance, in Figure 21.3, the joint probability distribution can be factorized as follows:

$$P(X) = \prod_{i=1}^{5} P(X_i \mid pa(X_i)) = P(X_1)P(X_2 \mid X_1)P(X_3 \mid X_2)P(X_4 \mid X_2)P(X_5 \mid X_3, X_4) \tag{21.6}$$

Inferring network structure from data is a fundamental and challenging problem in the application of BNs. Because the number of possible network structures increases exponentially as the number of vertices increases, it is impossible to exhaustively search for the structure that best fits the data. A wide range of the structure-learning methods can be classified as score-based searches [62]. The goodness-of-fit of a network structure is typically measured by the Akaike information criterion (AIC), the Bayesian information criterion (BIC), or the Bayesian Dirichlet likelihood equivalent metric (BDE). The search strategies are typically greedy, such as hill-climbing, or stochastic, such as simulated annealing or Markov chain Monte Carlo [22]. Though goodness-of-fit scores can differentiate between different networks, they do not directly reflect the error rate of edges. However, the error rate of the network structure itself, rather than the fitting score, is more often concerned in studying brain connectivity.

21.3.2 PC_{fdr} Algorithm

In this part, we discuss the PC_{fdr} algorithm, a BN structure-learning method that is able to control the error rate of the network structure under user-specified levels.

The PC algorithm, a well-known network-learning method based on conditional-independence tests, is computationally efficient and asymptotically reliable. The pseudo-code of the original PC algorithm is given in Step 6(a) of Table 21.1, where lower-case letters a and b denote vertices, the upper-case letter C denotes a vertex set, $X_{\blacksquare}$ denotes variables associated with a vertex or a vertex set, and Adj(a,G) denotes vertices adjacent to a in graph G.

The PC algorithm gains its efficiency by incrementally shrinking the size of Adj(a,G) to avoid exhaustively testing all conditional-independence relationships. Its search depth d represents the number of conditional variables. The searching is initialized with the completely connected graph G and the

TABLE 21.1

The PC Algorithm in [60], and the PC_{fdr} Algorithm in [21]

Step	
1.	Form the complete undirected graph G, and set search depth $d = 0$.
2.	*Repeat*
3.	*Repeat*
4.	Select an ordered pair of variable vertices a and b such that they are adjacent in G and Adj(a,G) b has at least d vertices in it.
5.	*For* vertex subset $C \subseteq \text{Adj}(a, G)\backslash b$, and $\lvert C \rvert = d$ *do*
6.	(a) Test hypothesis $X_a \perp X_b \mid X_C$ and calculate the p-value for edge $a \sim b$ at the *subject level*. Control the *Type I error rate* to decide which edges should be removed from G, and then remove them, *or*
	(b) Test hypothesis $X_a \perp X_b \mid X_C$ and calculate the p-value for edge $a \sim b$ at the *subject level*. Control the *FDR* to decide which edges should be removed from G, and then remove them.
7.	Update G.
8.	*If* the edge between a and b is removed, *do*
9.	Break loop at Step 5.
10.	*End if*
11.	*End for*
12.	*Until* all existing edges have been tested.
13.	$d = d + 1$.
14.	*Until* cannot find a triple (a, b, C) for Step 5.

In Step 6, either (a) or (b) is executed, according to the targeted error type. (a) is the PC algorithm in [60], for controlling the type I error rate [60]; (b) is the original PC_{fdr} algorithm in [21], for controlling the FDR.

search depth $d = 0$. At each loop, for all the subsets of conditional variables C with cardinality equal to d, testing the hypothesis of conditional independence of any possible neighbor vertices a and b given subset C. Once C is found to disconnect a and b, then the connection between them is removed and the neighbors of vertices are updated accordingly. At the end of each loop, the search depth d is increased by one. In this way, the PC algorithm can efficiently recover the graph structure.

In contrast to the original PC algorithm which controls the type I error rate individually for each connection during the conditional-independence testing, the PC_{fdr} algorithm developed by Li and Wang (Step 6(b) of Table 21.1) is capable of asymptotically controlling the false discovery rate (FDR) under prespecified levels [21]. The FDR is defined as the expected ratio of falsely discovered positive hypotheses to all those discovered, which is one of the important error control criteria for multiple testing (see Table 21.2). Compared with the traditional type I and type II error rates, FDR is more reasonable in some applications such as bioinformatics and neuroimaging since it is directly related with the uncertainty of the discovered positive results (Table 21.3).

The PC_{fdr} algorithm is proved to be able to control the FDR under a user-specified level $q(q > 0)$ at the limit of large sample sizes if the following assumptions are satisfied:

(A1) The probability distribution P is faithful to a DAG G_{true}.

(A2) The number of vertices is fixed.

TABLE 21.2

Error Control Criteria for Multiple Testing

Types of Error Rate Control	Symbol	Definition
False discovery rate	FDR	$E\left(\frac{FP}{N2}\right)$
Type I error rate (false positive rate)	α	$E\left(\frac{FP}{T1}\right)$
Type II error rate (false negative rate)	β	$E\left(\frac{FN}{T2}\right)$
Specificity (true negative rate)	$1-\alpha$	$E\left(\frac{TN}{T1}\right)$
Sensitivity (true positive rate)	$1-\beta$	$E\left(\frac{TP}{T2}\right)$
Family wise error rate	FWER	$P(FP \geq 1)$

Related notations for the results of multiple testing are recorded in Table 21.3.

TABLE 21.3

Notations for the Results of Multiple Testing

	Testing Results		
Truth	**Negative**	**Positive**	**Total**
Negative	True negative (TN)	False positive (FP)	T1
Positive	False negative (FN)	True positive (TP)	T2
Total	N1	N2	

(A3) Given a fixed significant level of testing conditional-independence relationships, the power of detecting conditional-dependence relationships with statistical tests approaches 1 at the limit of large sample sizes.

Assumption (A1) is generally assumed when graphical models are applied, although it restricts the probability distribution P to a certain class. Assumption (A2) is usually implicitly stated, but here we explicitly emphasize it because it simplifies the proof. Assumption (A3) may seem demanding, but actually it can be easily satisfied by standard statistical tests if the data are identically and independently sampled. The detection power and the FDR of the PC_{fdr} algorithm at the limit of large sample sizes are elucidated in Theorems 21.1 and 21.2. The detailed proofs can refer to [21].

Theorem 21.1: Assuming (A1), (A2), and (A3), PC_{fdr} algorithm is able to recover all the true connections with probability 1 as the sample size approaches infinity:

$$\lim_{m \to \infty} P(E_{\text{true}} \subseteq E) = 1$$

where E_{true} denotes the set of the undirected edges derived from the true DAG G_{true}, E denotes the set of the undirected edges recovered with the algorithms, and m denotes the sample size.

Theorem 21.2: Assuming (A1), (A2), and (A3), the FDR of the undirected edges recovered with the PC_{fdr} algorithm approaches a value not larger than the user-specified level q as the sample size m approaches infinity:

$$\lim_{m \to \infty} supFDR(E, E_{\text{true}}) \leq q$$

where $FDR(E,E_{true})$ is defined as

$$\begin{cases} FDR(E,E_{\text{true}}) = E\left[\dfrac{|E \setminus E_{\text{true}}|}{|E|}\right] \\ \text{Define } \dfrac{|E \setminus E_{\text{true}}|}{|E|} = 0, \quad \text{if} \quad |E| = 0 \end{cases}$$

21.3.3 Extensions to the PC$_{fdr}$ Algorithm

The PC$_{fdr}$ algorithm was developed to provide a computationally efficient means to control the FDR of computed edges asymptotically. Nevertheless, the PC$_{fdr}$ algorithm is unable to accommodate *a priori* information about connectivity, and was designed to infer connectivity from a single subject rather than a group of subjects, potentially making its application to typical fMRI scenarios limited. Here we introduce two extensions to the original PC$_{fdr}$ algorithm and propose a multi-subject brain connectivity modeling approach by combining the two extensions, allowing it to incorporate prior knowledge and extending it to robustly assess the dominant brain connectivity in a group of subjects. We name the extension as the gPC$^{+}_{fdr}$ algorithm and the pseudo-code of the gPC$^{+}_{fdr}$ algorithm is given in Table 21.4.

The first extension is an adaptation of *a priori* knowledge, allowing users to specify which edges must appear in the network, which cannot, and which are to be learned from the data. Many applications require imposing prior knowledge into network learning. For example, analyzing causal relationship in time series may forbid backward connections from time $t + 1$ to t, such as in dynamic BNs. In some situations, researchers may want to exclude some impossible connections based on anatomical knowledge. Incorporating *a priori* knowledge into PC$_{fdr}$ algorithm allows for more flexibility in using the method and potentially leads to greater sensitivity in accurately discovering the true brain connectivity.

It handles prior knowledge with two inputs: E_{must}, the set of edges assumed to appear in the true graph, and E_{test}, the set of edges to be tested from the data. The original PC$_{fdr}$ algorithm can thus be regarded as a special case of the extended algorithm, by setting $E_{must} = \varnothing$ and E_{must} = {all possible edges}.

The second extension to PC$_{fdr}$ algorithm is a combination of the PC$_{fdr}$ algorithm and a mixed-effect model to robustly deal with intersubject variability. Suppose we have m subjects in a group. Then for subject i, the conditional independence between the activities of two brain regions a and b given other regions C can be measured by the partial correlation coefficient between $X_a(i)$ and $X_b(i)$ given $X_C(i)$, denoted as $r_{ab|C}(i)$. Here the index i indicates that these variables are for subject i. To focus on the group–subject relationship, in the following discussion we omit the subscript "$ab|C$," and simply use index "i" to emphasize that a variable is associated with subject i.

TABLE 21.4

The $\text{gPC}^{+}_{\text{fdr}}$ Algorithm

Input: Data X, the undirected edges E_{must} that are assumed to exist in the true undirected graph G_{true} according to prior knowledge, the undirected edges $E_{\text{test}}(E_{\text{must}} \cap E_{\text{test}} = \varnothing)$ to be tested from the data X, and the FDR level q for making inference about E_{test}

Output: An undirected graph G_{stop}, i.e., the value of G when the algorithm stops, or equivalently, E_{stop}, the edges in G_{stop}

1. Form an undirected graph G from $E_{\text{must}} \cup E_{\text{test}}$, and set search depth $d = 0$.
2. *Repeat*
3. *Repeat*
4. Select an ordered pair of variable vertices a and b such that they are adjacent in G and Adj(a,G) b has at least d vertices in it.
5. *For* vertex subset $C \subseteq \text{Adj}(a, G)b$, and $|C| = d$ *do*
6. Test hypothesis $X_a \perp X_b \mid X_C$ and calculate the p-value for edge $a \sim b$ at the *group level*. Control the *FDR* to decide which edges should be removed from G, and then remove them.
7. Update G.
8. *If* the edge between a and b is removed, *do*
9. Break loop at Step 5.
10. *End if*
11. *End for*
12. *Until* all existing edges have been tested.
13. $d = d + 1$.
14. Until $|\text{adj}(a, G)\backslash b| < d$ for any other of for every ordered pair of vertices a and b that a~b is in $E \cap E_{\text{test}}$.

To study the group-level conditional-independence relationships, a group-level model should be introduced for r_i. Since partial correlation coefficients are bounded and their sample distributions are not Gaussian, we apply Fisher's z-transformation to convert (estimated) partial correlation coefficients r to a Gaussian-like distributed z-statistic z, which is defined as

$$z = Z(r) = \frac{1}{2}\ln\left(\frac{1+r}{1-r}\right) \tag{21.7}$$

where

r is a (estimated) partial correlation coefficient

z is its z-statistic

The group model we employ is

$$z_i = z_g + e_i \tag{21.8}$$

where e_i follows a Gaussian distribution $N\left(0, \sigma_g^2\right)$ with zero mean and σ_g^2 variance. Consequently, the group-level testing of conditional independence is to verify whether the null hypothesis $z_g = 0$ is true or not.

Because z_i is unknown and can only be estimated, the inference of z_g should be conducted with $\hat{z}_i = Z(\hat{r}_i)$. If $X_a(i)$, $X_b(i)$, and $X_C(i)$ jointly follow a multivariate Gaussian distribution, then $\hat{z}_i$ asymptotically follows a Gaussian distribution $N\left(z_{i,}\sigma_i^2\right)$ with $\sigma_i^2 = 1/(N_i - p - 3)$, where N_i is the sample size of subject *i*'s data and *p* represents the number of variables in $X_c(i)$. Therefore, from Equation 21.8, we have

$$\hat{z}_i = z_g + e_i + \varepsilon_i \tag{21.9}$$

where ε_i follows $N\left(z_{i,}\sigma_i^2\right)$, and e_i follows $N\left(z_{i,}\sigma_g^2\right)$.

This is a mixed-effect model where ε_i denotes the within-subject randomness and e_i denotes the intersubject variability. At the group level, $\hat{z}_i$ follows a Gaussian distribution $N\left(z_{i,}\sigma_i^2,\sigma_g^2\right)$. Note that unlike regular mixed-effect models, the within-subject variance σ_i^2 in this model is known, because N_i and *p* are known given the data $X(i)$ and *C*. In general, $\sigma_i^2 = 1/(N_i - p - 3)$ is not necessarily equal to σ_j^2 for $i \neq j$, and the inference of z_g should be conducted in the manner of mixed models, such as estimating σ_g^2 with the restricted maximum likelihood (ReML) approach. However, if the sample size of each subject's data is the same, then σ_i^2 equals σ_j^2. For this balanced case, which is typically true in fMRI applications and as well the case in this paper, we can simply apply a t-test to $\hat{z}_i$'s to test the null hypothesis $z_g = 0$. We employed this t-test in our simulations and the analysis on the fMRI data presented in this paper. Replacing Step 6(b) in Table 21.1 of the single-subject PC_{fdr} algorithm (i.e., the within-subject hypothesis test) with the test of $z_g = 0$, we can extend the single-subject version of the algorithm to its group-level version. Such a testing approach significantly simplifies the estimation process and our simulation results presented later demonstrates that this method could still control the FDR at a user-specified error rate level.

21.3.4 Case Study in Parkinson's Disease

Here, in order to assess application performance using the BN modeling, we apply the extension of the PC_{fdr} algorithm for inferring group brain connectivity network to fMRI data collected from 20 subjects, 10 people with Parkinson's disease (PD), and 10 age-matched controls. All experiments were approved by the University of British Columbia Ethics Committee.

During the fMRI experiment, each subject was instructed to squeeze a bulb in their right hand to control an "inflatable" ring so that it smoothly passed through a vertically scrolling tunnel. The normal controls performed only one trial, while the Parkinson's subjects performed twice, once before L-dopa medication, and the other approximately an hour later, after taking medication.

TABLE 21.5

Brain Regions of Interest (ROIs)

Full Name of Brain Region	Symbol
Left/right lateral cerebellar hemispheres	lCER, rCER
Left/right globus pallidus	lGLP, rGLP
Left/right putamen	lPUT, rPUT
Left/right supplementary motor cortex	lSMA, rSMA
Left/right thalamus	lTHA, rTHA
Left/right primary motor cortex	lM1, rM1

"l" or "r" in the abbreviations stand for "left" or "right," respectively.

Three groups were categorized: group N for the normal controls, group P_{pre} for the PD patients before medication, and group P_{post} for the PD patients after taking L-dopa medication. For each subject, 100 observations were used in the network modeling. For details of the data acquisition and preprocessing, please refer to [63]. Twelve anatomically defined regions of interest (ROIs) were chosen based on prior knowledge of the brain regions associated with motor performance (Table 21.5).

We utilized the two extensions of the PC_{fdr} algorithm and learned the structures of first-order group dynamic BN from fMRI data. Because the fMRI BOLD signal can be considered as the convolution of underlying neural activity with a hemodynamic response function, we assumed that there must be a connection from each region at time t to its mirror at time $t + 1$.

We also assumed that there must be a connection between each region and its homologous region in the contralateral hemisphere as shown in Figure 21.4a. The TR-interval (i.e., sampling period) was a relatively long 1.985 s; we restricted ourselves to learn only connections between ROIs without time lags. In total, there are $12 + 6 = 18$ predefined connections, and $12 \times (12 - 1)/2 - 6 = 60$ candidate connections to be tested. The brain connectivity networks (with the target FDR of 5%) learned for the normal (group N) and PD groups before (group P_{pre}) and after (group P_{post}) medication are compared in Figure 21.4. Note the connection between the cerebellar hemisphere and contralateral thalamus in the normal subjects, and between the supplementary motor area (SMA) and the contralateral putamen, consistent with prior knowledge. Interestingly, in P_{pre} subjects, the left cerebellum now connects with the right SMA, and the right SMA <——> left putamen connection is lost. Also, there are now bilateral primary motor cortex (M1) <——> putamen connections seen in the P_{pre} group, presumably as a compensatory mechanism. After medication (P_{post}), the left SMA < ——> Left thalamus connection is restored back to be normal. These results are consistent with the profound benefit imparted by L-dopa observed clinically.

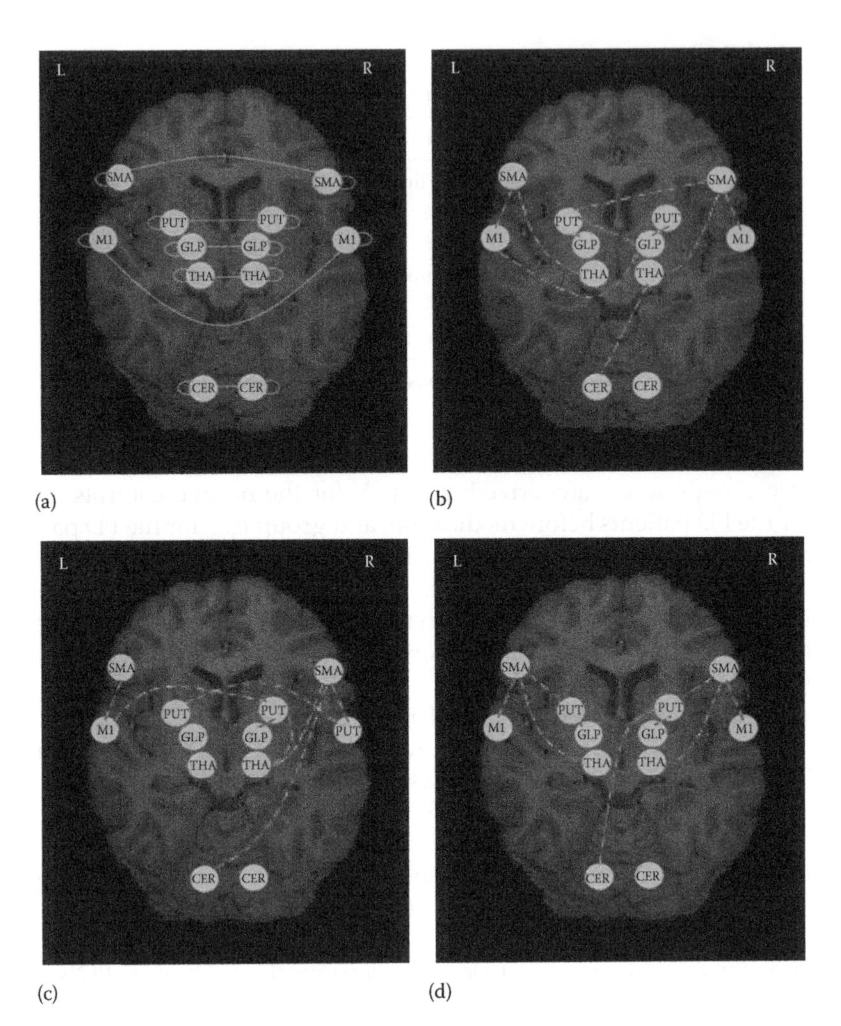

FIGURE 21.4
Connectivity differences between Parkinson's disease subjects and normal controls. (a) The predefined connectivity for three groups. (b) Learned brain connectivity for the normal group (group N). (c) Learned brain connectivity for the PD group before medication (group P_{pre}). (d) Learned brain connectivity for the PD group after medication (group P_{post}). Here "L" and "R" refer to the left and right sides, respectively.

21.4 Further Work

21.4.1 Connectivity Network with a Greater Number of ROIs

Current network-learning methods for graphical models can only reliably recover structures of dozens of nodes from data of a practical sample size, (typically several hundred time points in fMRI studies). Designing algorithms that are able to recover networks of hundreds or thousands of vertices will be of great potential interest.

Incorporating anatomic connectivity inferred from other imaging modalities, for example, DTI, as prior knowledge into functional network is a promising approach for improvement. When inferring brain connectivity from fMRI data, we are interested in a network that is associated with a certain cognition task. Since anatomic connection is the physical "infrastructure" for signal passing between brain regions, the signal passing network associated with a task will not "exceed" that defined by the anatomic connectivity. Using anatomic connectivity as the prior knowledge can reduce the search space of structure-learning algorithms, and hopefully lead to more accurate results.

Multi-scaling is another possible approach for improvement. Without geometric information on brain regions, graphical models simply treat ROIs as discrete vertices in graphs. However, if a connectivity network is composed of more ROIs, the ROIs become smaller and nearer, and their geometric relationship can potentially be exploited: smaller ROIs in the same functional region can be grouped together and viewed as one big ROI if the network is studied at a coarser level. Learning the network at a refined level can be based on the network at a coarse level. Multi-scaling has been widely used in image processing for acceleration, but its full potential in functional network learning has not been exploited.

21.4.2 Incrementally Updatable and Hierarchical Group Analysis

Incremental update means that group-inference results can be summarized as summary statistics, and forwarded for merging with the newly collected data. In research practice, experimental data are usually collected incrementally. For example, at first data of 80 subjects were collected, and half a year later 20 more are added. In this case, if the group analysis must start over with the newly added subjects, but not update incrementally, it may need to repeat the computation involving the previous 80 subjects, which might be cumbersome. Incremental updates can keep group analysis at the same pace with the data collection process, without starting over whenever new data are added, especially useful for the collection procedure of large subject pools.

Developing hierarchical group analysis methods for graphical models is a challenging and open problem. Subjects typically can be categorized into a hierarchical structure according to their epidemiology information. In this case, group analysis is not only to summarize information in one group of subjects, or just to contrast two groups, but to model diversity among subjects in a hierarchical manner. Hierarchical linear models have been developed for decades, but graphical models largely complicate the problem because graphical models not only involve coefficient parameters as linear models do, but also need to determine network structures. Conventional hierarchical linear models can be regarded as simplified cases with predefined network structures.

21.4.3 Perturbation Analysis

Perturbation analysis is important in understanding the contribution of an ROI or a pathway in a cognition task. The importance of a connectivity component, for instance, a vertex, an edge or a subgraph, can be revealed by removing it from the network and comparing the post-perturbation network with the original one. Difference between healthy subjects can be regarded as "unimportant" perturbation, while pathological difference between normal subjects and patients can be regarded as "important bad" perturbation, and effective medication can be regarded as "important good" perturbation. With perturbation analysis, the role of changes can be revealed, providing insight on network dynamics. Due to its high computational demand, perturbation analysis is still a challenging and important problem.

References

1. Ogawa S, Lee TM, Kay AR, Tank DW. Brain magnetic-resonance-imaging with contrast dependent on blood oxygenation. *Proceedings of the National Academy of Sciences of the United States of America*. December 1990;87(24):9868–9872.
2. Chang C, Cunningham JP, Glover GH. Influence of heart rate on the BOLD signal: The cardiac response function. *Neuroimage*. February 1, 2009;44(3):857–869.
3. Birn RM, Smith MA, Jones TB, Bandettini PA. The respiration response function: The temporal dynamics of fMRI signal fluctuations related to changes in respiration. *Neuroimage*. April 1, 2008;40(2):644–654.
4. Goebel R, Roebroeck A, Kim D-S, Formisano E. Investigating directed cortical interactions in time-resolved fMRI data using vector autoregressive modeling and Granger causality mapping. *Magnetic Resonance Imaging*. December 2003;21(10):1251–1261.
5. Greicius MD, Supekar K, Menon V, Dougherty RF. Resting-state functional connectivity reflects structural connectivity in the default mode network. *Cerebral Cortex*. January 2009;19(1):72–78.
6. Damoiseaux JS, Greicius MD. Greater than the sum of its parts: A review of studies combining structural connectivity and resting-state functional connectivity. *Brain Structure and Function*. October 2009;213(6):525–533.
7. Le Bihan D. Looking into the functional architecture of the brain with diffusion MRI. *Nature Reviews Neuroscience*. June 2003;4(6):469–480.
8. Friston KJ. Statistical parametric mapping and other analyses of functional imaging data. In: Toga AW, Mazziotta JC, eds. *Brain Mapping, the Methods*, 1st edn. San Diego, CA: Academic Press; 1996. pp. 363–396.
9. Ng B, Abugharbieh R, Huang X, McKeown MJ. Spatial characterization of FMRI activation maps using invariant 3-D moment descriptors. *IEEE Transactions on Medical Imaging*. 2009;28(2):261.
10. Friston KJ. Functional and effective connectivity: A review. *Brain Connectivity*. June 1, 2011;1(1):13–36.

11. Cao J, Worsley K. The geometry of correlation fields with an application to functional connectivity of the brain. *The Annals of Applied Probability*. 1999;9(4):1021–1057.
12. Sato JR, Takahashi DY, Arcuri SM, Sameshima K, Morettin PA, Baccalá LA. Frequency domain connectivity identification: An application of partial directed coherence in fMRI. *Human Brain Mapping*. 2009;30(2):452–461.
13. Sun FT, Miller LM, D'Esposito M. Measuring interregional functional connectivity using coherence and partial coherence analyses of fMRI data. *Neuroimage*. February 2004;21(2):647–658.
14. Shannon CE. A mathematical theory of communication. *SIGMOBILE Mobile Computing and Communications Review*. 2001;5:3–55.
15. Harrison L, Penny WD, Friston K. Multivariate autoregressive modeling of fMRI time series. *Neuroimage*. August 2003;19(4):1477–1491.
16. McLntosh AR, Gonzalez-Lima F. Structural equation modeling and its application to network analysis in functional brain imaging. *Human Brain Mapping*. 1994;2(1–2):2–22.
17. Friston KJ, Harrison L, Penny W. Dynamic causal modelling. *Neuroimage*. August 2003;19(4):1273–1302.
18. McKeown MJ, Jung T-P, Makeig S, Brown G, Kindermann SS, Lee T-W et al. Spatially independent activity patterns in functional MRI data during the Stroop color-naming task. *Proceedings of the National Academy of Sciences*. February 3, 1998;95(3):803–810.
19. Chen X, Jane Wang Z, McKeown MJ. A Bayesian Lasso via reversible-jump MCMC. *Signal Processing*. August 2011;91(8):1920–1932.
20. Varoquaux G, Gramfort A, Poline JB, Thirion B. Brain covariance selection: Better individual functional connectivity models using population prior. *Advances in Neural Information Processing Systems*. 2010;23:2334–2342.
21. Li J, Wang ZJ. Controlling the false discovery rate of the association/causality structure learned with the PC algorithm. *Journal of Machine Learning Research*. 2009;10:475–514.
22. Li J, Wang ZJ, Palmer SJ, McKeown MJ. Dynamic Bayesian network modeling of fMRI: A comparison of group-analysis methods. *Neuroimage*. June 2008;41(2):398–407.
23. Rajapakse JC, Zhou J. Learning effective brain connectivity with dynamic Bayesian networks. *Neuroimage*. September 1, 2007;37(3):749–760.
24. Biswal B, Yetkin FZ, Haughton VM, Hyde JS. Functional connectivity in the motor cortex of resting human brain using echo-planar MRI. *Magnetic Resonance in Medicine*. October 1995;34(4):537–541.
25. Greicius MD, Krasnow B, Reiss AL, Menon V. Functional connectivity in the resting brain: A network analysis of the default mode hypothesis. *Proceedings of the National Academy of Sciences of the United States of America*. January 7, 2003;100(1):253–258.
26. Andrews-Hanna JR, Snyder AZ, Vincent JL, Lustig C, Head D, Raichle ME et al. Disruption of large-scale brain systems in advanced aging. *Neuron*. December 6, 2007;56(5):924–935.
27. Cordes D, Haughton VM, Arfanakis K, Wendt GJ, Turski PA, Moritz CH et al. Mapping functionally related regions of brain with functional connectivity MR imaging. *American Journal of Neuroradiology*. October 2000;21(9):1636–1644.

28. Kaminski M. Determination of transmission patterns in multichannel data. *Philosophical Transactions of the Royal Society B: Biological Sciences*. May 29, 2005;360(1457):947–952.
29. Marrelec G, Krainik A, Duffau H, Pélégrini-Issac M, Lehéricy S, Doyon J et al. Partial correlation for functional brain interactivity investigation in functional MRI. *Neuroimage*. August 1, 2006;32(1):228–237.
30. Patel RS, Bowman FD, Rilling JK. A Bayesian approach to determining connectivity of the human brain. *Human Brain Mapping*. 2006;27(3):267–276.
31. McKeown MJ, Makeig S, Brown GG, Jung TP, Kindermann SS, Bell AJ et al. Analysis of fMRI data by blind separation into independent spatial components. *Human Brain Mapping*. 1998;6(3):160–188.
32. Esposito F, Aragri A, Pesaresi I, Cirillo S, Tedeschi G, Marciano E et al. Independent component model of the default-mode brain function: Combining individual-level and population-level analyses in resting-state fMRI. *Magnetic Resonance Imaging*. September 2008;26(7):905–913.
33. Damoiseaux JS, Rombouts SARB, Barkhof F, Scheltens P, Stam CJ, Smith SM et al. Consistent resting-state networks across healthy subjects. *Proceedings of the National Academy of Sciences of the United States of America*. September 12, 2006;103(37):13848–13853.
34. Greicius MD, Srivastava G, Reiss AL, Menon V. Default-mode network activity distinguishes Alzheimer's disease from healthy aging: Evidence from functional MRI. *Proceedings of the National Academy of Sciences of the United States of America*. March 30, 2004;101(13):4637–4642.
35. Greicius MD, Flores BH, Menon V, Glover GH, Solvason HB, Kenna H et al. Resting-state functional connectivity in major depression: Abnormally increased contributions from subgenual cingulate cortex and thalamus. *Biological Psychiatry*. September 1, 2007;62(5):429–437.
36. Jafri MJ, Pearlson GD, Stevens M, Calhoun VD. A method for functional network connectivity among spatially independent resting-state components in schizophrenia. *Neuroimage*. February 15, 2008;39(4):1666–1681.
37. Qi Z, Wu X, Wang Z, Zhang N, Dong H, Yao L et al. Impairment and compensation coexist in amnestic MCI default mode network. *Neuroimage*. March 2010;50(1):48–55.
38. Vanhaudenhuyse A, Noirhomme Q, Tshibanda LJF, Bruno MA, Boveroux P, Schnakers C et al. Default network connectivity reflects the level of consciousness in non-communicative brain-damaged patients. *Brain*. January 2010;133:161–171.
39. Fischer H, Hennig J. Neural network-based analysis of MR time series. *Magnetic Resonance in Medicine*. 1999;41(1):124–131.
40. Peltier SJ, Polk TA, Noll DC. Detecting low-frequency functional connectivity in fMRI using a self-organizing map (SOM) algorithm. *Human Brain Mapping*. 2003;20(4):220–226.
41. Baumgartner R, Windischberger C, Moser E. Quantification in functional magnetic resonance imaging: Fuzzy clustering vs. correlation analysis. *Magnetic Resonance Imaging*. 1998;16(2):115–125.
42. Cordes D, Haughton V, Carew JD, Arfanakis K, Maravilla K. Hierarchical clustering to measure connectivity in fMRI resting-state data. *Magnetic Resonance Imaging*. May 2002;20(4):305–317.

43. Heller R, Stanley D, Yekutieli D, Rubin N, Benjamini Y. Cluster-based analysis of FMRI data. *Neuroimage*. November 2006;33(2):599–608.
44. Penny W, Ghahramani Z, Friston K. Bilinear dynamical systems. *Philosophical Transactions of the Royal Society B*. May 2005;360(1457):983–993.
45. Chaminade T, Fonlupt P. Changes of effective connectivity between the lateral and medial parts of the prefrontal cortex during a visual task. *European Journal of Neuroscience*. 2003;18(3):675–679.
46. Honey GD, Fu CHY, Kim J, Brammer MJ, Croudace TJ, Suckling J et al. Effects of verbal working memory load on corticocortical connectivity modeled by path analysis of functional magnetic resonance imaging data. *Neuroimage*. 2002;17(2):573–582.
47. Bressler SL, Ding M, Yang W. Investigation of cooperative cortical dynamics by multivariate autoregressive modeling of event-related local field potentials. *Neurocomputing*. June 1999:625–631.
48. Buchel C, Coull JT, Friston KJ. The predictive value of changes in effective connectivity for human learning. *Science*. March 5, 1999;283(5407):1538–1541.
49. Friston KJ, Buechel C, Fink GR, Morris J, Rolls E, Dolan RJ. Psychophysiological and modulatory interactions in neuroimaging. *Neuroimage*. October 1997;6(3):218–229.
50. Huang S, Li J, Sun L, Ye J, Fleisher A, Wu T et al. Learning brain connectivity of Alzheimer's disease by sparse inverse covariance estimation. *Neuroimage*. 2010;50:935–949.
51. Shimizu S, Hoyer PO, Hyvärinen A, Kerminen A. A linear non-Gaussian acyclic model for causal discovery. *Journal of Machine Learning Research*. 2006;7:2003–2030.
52. Smith SM, Miller KL, Salimi-Khorshidi G, Webster M, Beckmann CF, Nichols TE et al. Network modelling methods for FMRI. *Neuroimage*. January 15, 2011;54(2):875–891.
53. Ramsey JD, Hanson SJ, Glymour C. Multi-subject search correctly identifies causal connections and most causal directions in the DCM models of the Smith et al. simulation study. *Neuroimage*. October 1, 2011;58(3):838–848.
54. Deleus F, Van Hulle MM. A connectivity-based method for defining regions-of-interest in fMRI data. *IEEE Transactions on Image Processing*. 2009;18(8):1760–1771.
55. Stephan KE, Penny WD, Daunizeau J, Moran RJ, Friston KJ. Bayesian model selection for group studies. *Neuroimage*. July 15, 2009;46(4):1004–1017.
56. Gates KM, Molenaar PCM. Group search algorithm recovers effective connectivity maps for individuals in homogeneous and heterogeneous samples. *Neuroimage*. October 15, 2012;63(1):310–319.
57. Ramsey J, Hanson S, Hanson C, Halchenko Y, Poldrack R, Glymour C. Six problems for causal inference from fMRI. *Neuroimage*. 2010;49(2):1545–1558.
58. Ramsey J, Zhang J, Spirtes P, eds. Adjacency-faithfulness and conservative causal inference. *Proceedings of the 22nd Annual Conference on Uncertainty in Artificial Intelligence*, Cambridge, MA; 2006.
59. Zheng X, Rajapakse JC. Learning functional structure from fMR images. *Neuroimage*. July 15, 2006;31(4):1601–1613.
60. Spirtes P, Glymour C, Scheines R. *Causation, Prediction, and Search*. Cambridge, MA: The MIT Press; 2001.
61. Lauritzen SL. *Graphical Models*. Oxford, U.K.: Oxford University Press; 1996.

62. Heckerman D, Geiger D, Chickering DM. Learning Bayesian networks: The combination of knowledge and statistical data. *Machine Learning*. 1995;20(3):197–243.
63. Palmer SJ, Ng B, Abugharbieh R, Eigenraam L, McKeown MJ. Motor reserve and novel area recruitment: Amplitude and spatial characteristics of compensation in Parkinson's disease. *European Journal of Neuroscience*. 2009;29(11):2187–2196.
64. Churchland P, Terrence, S. *The Computational Brain*. Cambridge, MA: MIT Press; 1992.

22

Medical Image Registration: A Review

Lisa Tang and Ghassan Hamarneh

CONTENTS

22.1 Introduction to Medical Image Registration

With the advent of different medical imaging modalities, clinicians can now perform diagnosis in a minimally invasive manner. Nevertheless, each modality currently only provides particular types of information. CT and MRI, for example, provide *structural information* of our body (e.g., an anatomical map), but do not convey *functional information** (i.e., relating to metabolic functions of an imaged organ). Conversely, modalities such as PET and SPECT capture functional but not structural information. This motivates image registration, the task of bringing two medical images into spatial alignment. Once images are aligned, information from different modalities can then be integrated and examined as a whole.

MIR indeed is a crucial step for many image analysis problems where information from different sources are combined and examined. The images involved may come from the same modality but captured at different times, or from multiple modalities and/or at different times, for example, as introduced earlier. Some general applications of MIR are listed in the following; more can be found in [18,38,89,160].

- *Multi-temporal image analysis.* Images of the same subject are acquired at different times and/or under different physical conditions. Registration of these *intra-subject* images allows us to monitor disease progression, or to detect and/or quantify changes in an anatomy. Registration of intra-operative (between preoperative and postoperative) images can also enhance surgical procedures.
- *Multi-subject image analysis.* Also known as (aka) intersubject registration; images of different subjects are registered for deformation-based morphology [8,114]. The subjects may also be of different species (e.g., human brain to chimpanzee brain for "asymmetry analysis" [7]).
- *Multimodality image fusion.* As introduced earlier, images acquired from different modalities are aligned to create image fusions that facilitate clinical diagnosis [44,113]. Some clinical applications may also involve matching images of different dimensionality.
- *Construction and use of atlases.* Group of images that are acquired from different sites and/or at different times are aligned simultaneously. This allows for statistical analysis of anatomical shapes [146], creation of probabilistic segmentations [139], brain functional localization [43], etc.
- *Dynamic image sequence analysis.* This analysis involves dynamic stacking static images that were acquired at different time steps form *dynamic image sequences*, which are typically used to capture and quantify motion of an anatomy, for example, respiratory/cardiac cycle of the lungs/heart. One example application is the fusion of 2D+time ultrasound with 2D+time MRI to measure the distensibility of carotid arteries [87].

* An analogy would be a topographic map of a city, which does not indicate the functional aspect of the city, i.e., information conveying population growth or utility of a location (residential vs. commercial).

As we will see later, most of the proposed techniques for the aforementioned problems involve performing the following tasks: (1) characterizing the registration solution, which may involve a set of constraints imposed on the desired solution; (2) defining a list of criteria to quantify the optimality of a candidate solution; and (3) devising a strategy to find the desired solution. In the coming sections, we will individually examine the essential components central to each task. Let us first review the preliminaries to MIR.

22.2 Preliminaries

22.2.1 Image Representation and Imaging Terminologies

A digital image I gives a discrete representation of a continuous signal that measures physical quantities (e.g., light in photographs, or photon densities in an x-ray image). Following conventions of [56,92], the term *physical* or *world* coordinate is used to describe an arbitrary location in a continuous domain and the term *image* coordinate is used to index a particular cell on the image grid. We also refer to the data value at a grid center as its intensity value ("pixel" or "voxel" are also used in the context of a 2D or 3D domain, respectively) and denote a physical coordinate as $\mathbf{x}$ (in 2D, $\mathbf{x} = [x_1\ x_2]$ represents the position of a point in the x and y dimensions).

Typically, structural images acquired from invisible light medical imaging modalities are scalar functions, that is, $I: \Omega \mapsto \mathbb{R}^r$, $r = 1$. For color images, or visible light medical images in general (e.g., dermoscopy, wound care imageries, etc.), I is a vector-valued function with $r = 3$. When I is an image sequence that is acquired to capture dynamic information occurring over time which is typically created by stacking static images to form 2D + time or 3D + time data, one may instead denote as $I: \mathbb{R}^d \times \mathbb{R} \mapsto \mathbb{R}^r$.

22.2.2 Data Interpolation

When one wants to query for intensity values at locations between cell centers (e.g., one third position of a cell), one must approximate these values from the original data. This is known as image interpolation or image resampling, and is usually done in a process where a continuous function is fitted to the original set of data points.

Common interpolation methods include nearest neighbor, linear (bi- or tri-), polynomial (cubic, quintic, etc.), and windowed sinc. Other less commonly used ones include Lagrange and Gaussian interpolation, as well as methods based on the Fourier and Wavelet transforms. For details on the latter methods proposed, see the survey of [72].

The simplest interpolation method is nearest neighbor (aka piecewise constant or proximal interpolation) where the intensity value of $\mathbf{x}$ is determined as the intensity value of its closest grid location. However, being a discontinuous function, nearest neighbor is not normally used for registration.

The more commonly used interpolant is linear interpolation (bilinear and trilinear for 2D and 3D, respectively). In 2D, it is defined as

$$\mathcal{I}_{linear}(\mathbf{x}, I) = (1-\epsilon_1)(1-\epsilon_2)I(i_1, i_2) + \epsilon_1(1-\epsilon_2)I(i_1+1, i_2)$$
$$+(1-\epsilon_1)\epsilon_2 I(i_1, i_2+1) + \epsilon_1\epsilon_2 I(i_1+1, i_2+1)$$

where i_1 and ε_1 are, respectively, the integer and remainder parts of $\mathbf{x}_1$ (similarly defined for those of $\mathbf{x}_2$). Unlike nearest neighbor, it generates spatially continuous intensity values. However, it is not differentiable at the center of grid cells, which is a problem when a gradient-based optimization technique is used to optimize the registration objective (see Section 22.3.4.2). It also causes spatially varying, low-pass filtering of the images [38], which can be a problem when computing similarity metrics on image pairs.

From the perspective of signal processing theory, the most ideal interpolant is one that uses the sine function because it has the fullest potential to recover the continuous signal underlying I [56,72]. For reasons of practicality, the sine function is only approximated with a Window function w. In 2D, a windowed sine interpolant is defined as

$$\mathcal{I}_{sinc}(\mathbf{x}, I) = \sum_{p=\lfloor x_1 \rfloor+1-m}^{\lfloor x_1 \rfloor+m} \sum_{q=\lfloor x_2 \rfloor+1-m}^{\lfloor x_2 \rfloor+m} I(\mathbf{x})K(x_1-p)K(x_2-q), \tag{22.1}$$

where m is the radius or spatial extent of w and

$$K(t) = w(t)\,\mathrm{sinc}(t) \tag{22.2}$$

Nevertheless, the windowed sine interpolant has several trade-offs in its utilization, for example, large m, while being more precise, will require longer computation times [144].

Spline interpolation may arguably give optimal balance between computation time and precision of interpolation [73]. The spline interpolant involves representing the continuous form of I by a set of spline basis functions β_j and computing a set of expansion coefficients c_j of β_j such that the resulting basis function interpolates the values of I at grid locations. The set of coefficients are computed by solving a linear system of equations, which is most efficiently done by recursive digital filtering. Intensity at $\mathbf{x}$ is computed by multiplying the B-spline coefficients with shifted B-spline kernels within a small support region of the requested position; in 2D, this is done as

$$\mathcal{I}_{spline}(\mathbf{x}, I) = \sum_j c_j \beta^{(d)}(\mathbf{x} - \mathbf{x}_j) \tag{22.3}$$

where j indices the pixel coordinates within the support region of the spline window.

22.2.3 Spatial (Geometric) Transformations

In image registration, a spatial transform plays the crucial role of representing registration solutions in a precise manner. Mathematically, it may be defined as $\mathbf{U}: \Omega \mapsto \mathbb{R}^d, \Omega \subseteq \mathbb{R}^d$, which describes a mapping between the coordinate system of one image and that of another image.

22.2.3.1 Terminologies

A mapping **U** is said to be *smooth* if all of its partial derivatives of up to a certain order exist and are continuous. It is said to be *bijective* if it is a one-to-one mapping. It is said to be *invertible* (aka homeomorphic or topology-preserving) if it is bijective and continuous, and that its inverse is also continuous. Note that smoothness of **U** is not equivalent to invertibility, that is, invertibility does not require differentiability of **U**. When **U** is both smooth and invertible, it is called *diffeomorphic*.

Some of these properties of **U** are highly desirable; after all, properties of **U** affect the appearance of the image to which it is applied. Generally, smoothness of **U** is desired as it directly affects the smoothness of the image on which it is applied to. Applying a random warp (vector field with random displacements) to an image, for instance, will destroy the *morphology* (level sets) of an image [34].

Bijectivity, on the other hand, may or may not be required. Bijectivity of **U** ensures that the folding of space does not happen as multiple points cannot be mapped into the same point. In some cases, however, the folding of space should be allowed. This occurs, for example, in intra-operative image registration where tearing and resection of tissues can occur in the postoperative image. Even for registration of intersubject brain images, the topology of the brain, the shape of the ventricles, the number and shape of the sulci may vary significantly from one subject to another [114] such that bijectivity should not be strictly enforced.

Invertibility also may or may not be required. It is needed when one needs to map individual images from a set to a common space. It is also needed in order to allow for decomposition of the transformation Jacobian, which is often done in registrations of diffusion tensor images to reorient tensor values at each location after deforming an image.

Lastly, we desire a diffeomorphic transform when we need to maintain the (non-) connectivities between neighboring anatomical structures after registration [12]. Note that a diffeomorphic transform that matches arbitrary number of points can always be found (proof given in [63]).

22.2.3.2 Spatial Transformation Models

Spatial transformation models (transforms) are often used to parameterize **U**, thereby allowing **U** to be compactly represented. The ones commonly used for image registration are categorized as either linear or nonlinear.

Linear transforms include translation, rotation, rigid, affine, and perspective. A rigid transform allows for translations and rotations only; an affine transform allows for rigid as well as scale and skew (distortions that preserve parallelism of lines); a projective transform allows for all distortions that preserve colinearity (lines remain straight after distortion).

Compared to linear models, nonlinear transformation models allow for more localized deformations. Common nonlinear models include those that are based on radial basis functions (RBF). Under these models, a set of points $\mathcal{P} = \{\mathbf{x}_i\}$, for example, manually identified landmarks in Ω, is used to extrapolate $\mathbf{U}$:

$$\mathbf{U}(\mathbf{x}) = H_{m-1}(\mathbf{x}) + \sum_i c_i \beta(\mathbf{x} - \mathcal{P}_i). \tag{22.4}$$

where

H_{m-1} ($\mathbf{x}$) is a polynomial of degree $m - 1$
β is a basis function
c_i are coefficients to be solved in a system of equations

Note that $\mathcal{P}$ may be irregularly located in Ω. Different forms of $\beta(r)$ results in different types of spline [124]. For example, with $r = \|\mathbf{x} - \mathcal{P}_i\|$, defining the RBF as $\beta(r) = r^2 \ln r$ and $\beta(r) = r$ gives the thin-plate-spline (TPS) in 2D and 3D, respectively. With $(r^2 + a^2)$, where $0 < b < 1$ and a is a user-parameter that increases the amount of smoothing, one gets the multi-quadric splines (MQ). Other choices are described in [52,124].

Many RBFs like TPS and MQ give global support, that is, one control point has a global effect on $\mathbf{U}$ (i.e., deformations in one image region affects the entire domain). This property may cause problems as large and inconsistent displacements in $\mathcal{P}$ can lead to singularities in the systems of equation to be solved. Furthermore, the size of $\mathcal{P}$ directly affects computational complexity of solving the system.* In these regards, basis functions with local support are highly desired.

One of such locally supported functions is the basis-spline (B-spline). When extended to a multivariate function via the use of tensor products, it gives the free-form deformation (FFD) model. Under FFDs, the image domain is divided into a lattice whose intersections represent the positions of a set of control points $\mathbf{B}$ such that translating the control points induces local deformations. Performing registration is then reduced to finding the optimal translations for each control point [46], or the optimal values of the set of B-spline coefficients [107], or the divergence† and curl of $\mathbf{U}$ at control

* Malsch et al. [84] proposed to partition the image domain and employ separate TPS models for each partition.

† This parameterization approach was shown to be more capable of avoiding grid folding more than uniform B-spline parameterization [54].

point [54]. The DOFs of FFD is thus determined by the size of this set of parameters. The spacing between control points determines the scale of local deformations, for example, lowering the spacing allows for more localized motion. In 2D, it has the form:

$$\mathbf{U}(\mathbf{x}) = \sum_{l=0}^{n}\sum_{m=0}^{n} \beta_{l,n}(r)\beta_{m,n}(s)\mathbf{B}_{i+l,j+m}. \tag{22.5}$$

where
$\beta_{\cdot,\cdot}$ is a basis function of degree n
(r, s) are lattice coordinates relative to $\mathbf{x}$

Despite some of its deficiencies [126], FFD is a popular choice in many applications due to its ease of implementation, low computational complexity, and the superior properties of B-spline functions (continuity and locality).

Another popular choice is to represent $\mathbf{U}$ with finite elements [102,122], which involves discretizing Ω with mesh elements, each with a set of meshes. To obtain values of $\mathbf{U}$ at points within a mesh element e, one then uses linear interpolation:

$$\mathbf{U}(\mathbf{x}) = \sum_{l=0}^{n} B^e\left(q_l^e\right)\left[\mathbf{U}\left(q_1^e\right)\mathbf{U}\left(q_1^e\right)\cdots\mathbf{U}\left(q_n^e\right)\right]^T. \tag{22.6}$$

where
B^e is a basis function for e
q_n^e is the nth node of mesh e

Lastly, parameterization of $\mathbf{U}$ has also been done with discrete Fourier transformation (DFT) [27] and discrete cosine transformation (DCT) [52], which involve globally supported basis functions (as they operate in frequency domain). Usually, only low-frequency basis functions are used for parameterization, which by construction gives inherently smooth regularized solutions [159]. Gefen [42] also modeled $\mathbf{U}$ with wavelet expansion and showed that its use can give more accurate registrations than use of TPS. However, the complexity of the wavelet approach depends on the size of $\mathbf{U}$.

In summary, linear models and nonlinear models differ in terms of their DOFs, which in turn determine the different types of geometric distortions they induce on images. In some applications, linear models are sufficient to describe misalignment between two images, for example, registration of brain images. In others, both linear and nonlinear models are needed to correct for global and local misalignment between two images. There are also cases where misalignment between two images involve both articulation (e.g., of rigid bones) and elastic deformations (e.g., soft tissues surrounding rigid bones) such that the use of a global nonlinear transform will give elastic deformations everywhere, or conversely, the use of a global linear transform will fail to align elastic tissues. In these cases, a common strategy is to partition the image domain into different regions, determine a local (non) linear transform for each region, and employ a scheme to spatially fuse the piecewise local transforms [100]. Another approach is the use of polyrigid models [3]. Details on the aforementioned linear and nonlinear models can also be found in focused reviews [52,144] or general surveys [18,44,77,83].

22.3 Fundamentals of Image Registration

22.3.1 Mathematical Formulation of Image Registration

Let there be two images F and M. The goal of registering F and M is to find a spatial transformation $\mathbf{U}$ that optimally maps points in F to the corresponding points in M.* Optimality of $\mathbf{U}$ is measured by a set of criteria of two types: *regularization* (aka internal forces), which measures the regularity of $\mathbf{U}$, and *data dissimilarity* (aka external forces), which measures how well $\mathbf{U}$ relate F and M based on image data. Details on these terms will be surveyed in the following subsections. Once a set of criteria has been defined for a specific registration problem, the optimal $\mathbf{U}$ can then be found via minimization of a cost function C that contains a weighted sum of two or more of these criteria, for example,

$$\mathcal{C}(\mathbf{U}) = \alpha_i \mathcal{D}_i(F, M \circ \mathbf{U}) + \alpha_{i+1} \mathcal{R}_{i+1}(\mathbf{U}) + \cdots \tag{22.7}$$

where

$\mathcal{D}_i$ denotes a data dissimilarity term
$\mathcal{R}_{i+1}$ denotes a regularization term
α_i is a weight on the ith term

22.3.2 Regularization

Solution of an image registration problem, $\mathbf{U}$, is generally not unique.† To regularize $\mathbf{U}$, we may restrict the type of $\mathbf{U}$ to a certain space, such as the space of affine transformations or the space of diffeomorphisms (Section 22.4.1). We may also explicitly enforce constraints on the behavior of $\mathbf{U}$ based on our

* Following conventions of [56], the image which we apply $\mathbf{U}$ is called the moving image and the target image which M is matched to is called the *fixed image*. The literature also refers to the fixed and moving images as the target and source images, or the reference and floating images, respectively.

† One intuition is that registration involves solving for a vector-valued mapping (i.e., $T(\mathbf{x})$) based on scalar information $\mathcal{D}$, thus leading to an under-determined problem [92]. Consider too an example where two images are completely homogeneous in grey values, then any arbitrary transformation would be a valid solution even though many are undesired ones.

prior knowledge about the expected solution. For example, when registering x-ray breast images, one might impose incompressibility constraints in tissue deformation [104]. When registering images containing rigid bodies (e.g., bones [129]) one might impose local rigidity constraints to disallow elastic deformations in rigid regions.

In Section 22.2.3, we have outlined the desired properties of **U**, including smoothness and invertibility, arguably two of the most important ones. We next present how these may be achieved. Other regularization constraints proposed in the literature are highlighted in Section 22.3.2.4.

22.3.2.1 Smoothness Regularization: Implicit Approaches

A common approach to regularization is to constrain the behavior of **U** via use of a spatial transformation model. As discussed previously, these models impose constraints on the type of geometric distortions they induce on images. For example, rigid transforms preserve all internal angles and distances, perspective transforms only preserve colinearity (points that lie on a line remain so after they have undergone transformation), while non-linear transforms allow for localized deformations.

Solving **U** using parametric transforms generally simplifies the optimization of the objective function (Equation 1.7). Rather than solving for **U** at each spatial location in Ω (c.f. Section 22.4.1), one only needs to resolve the transform's parameters, effectively reducing the size of the solution space (in case of linear transforms, the dimension of the search space is equal to the number of transform parameters). Furthermore, the inherent smoothness properties provided by a given spatial transformation model ensure some regularity of **U**. Thus, in many older works, for example [82], no explicit regularization was done. However, in more recent works, invertibility has become an important criterion [29]. As all of the aforementioned parametric models do not enforce invertibility, additional regularization terms are needed; these will be examined in Section 22.3.2.4.1.

22.3.2.2 Smoothness Regularization: Explicit Approaches

When no parameterization of **U** is used, any point may be displaced arbitrarily, which would lead to an unrealistic warp of M. To ensure smoothness of **U**, as well as other desired properties of **U** outlined in Section 22.2.3, explicit energy terms are needed ($\mathcal{R}_{i+1}$ in Equation 22.7). We now examine in detail some of the common ones found in the literature.

22.3.2.2.1 Regularizers Motivated from Basis Functions

Regularization terms in Equation 22.7 are typically defined based on an application-dependent Sobolev semi-norm (bilinear form) [90]:

$$\mathcal{R}_{norm}(\mathbf{U}) = \left\| \mathbf{L}[\mathbf{U}] \right\|_{L_2(\Omega)^q} d\mathbf{x} \tag{22.8}$$

where **L** is a differential operator on **U**.

One regularizer proposed by Horn and Schunk [53] is the diffusion regularizer, which penalizes variations in the gradients of **U**. In 2D, it is defined as

$$\mathcal{R}_{diff}(\mathbf{U}) = \frac{1}{2}\int_{\mathbf{x}\in\Omega}\sum_{i}^{d}\|\nabla \mathbf{U}_i(\mathbf{x})\|^2 dx \tag{22.9}$$

where $\mathbf{U}_i(x)$ denotes the ith component of **U** at **x**. This definition relates closely to the classical Tikhonov regularization for *ill-posed* problems. When instead an L^1 norm is used, such regularization falls in the class of total variation (TV) methods [10,41], leading to a TV regularizer that is defined as

$$\mathcal{R}_{TV}(\mathbf{U}) = \int_{\mathbf{x}\in\Omega}\sum_{i}^{d}|\nabla \mathbf{U}_i(\mathbf{x})|\, dx \tag{22.10}$$

which allows for discontinuities in **U**. However, as it is non-differentiable at zero, researchers have proposed to replace $|x|$ with $\sqrt{x^2+k^2}$ where value of k affects amount of discontinuities allowed.

Note that the aforementioned regularizers also penalize linear transformations [92]. To address this, the curvature regularization (aka bi-harmonic) [20,92] was proposed:

$$\mathcal{R}_{curv}(\mathbf{U}) = \frac{1}{2}\int_{\mathbf{x}\in\Omega}\sum_{i}^{d}\Delta U_i(\mathbf{x})^2 dx \tag{22.11}$$

which approximates the mean curvature of the surface of **U**. By taking the Laplacian of **U** instead of its gradient, linear transforms result in zero smoothness cost. Furthermore, as the null-space of the operator has infinite dimension and contains all harmonic vector fields, use of this regularizer ensures smooth deformations [92].

Another common regularizer is based on the thin-plate-spline model (Section 22.2.3):

$$\mathcal{R}_{thin\text{-}plate}(\mathbf{U}) = \frac{1}{2}\int_{\mathbf{x}\in\Omega}\left(\frac{\partial^2\mathbf{U}(\mathbf{x})}{\partial^2 x_1} + 2\frac{\partial^2\mathbf{U}(\mathbf{x})}{\partial x_1 \partial x_2} + \frac{\partial^2\mathbf{U}(\mathbf{x})}{\partial^2 x_2}\right)d\mathbf{x}. \tag{22.12}$$

Note that this regularization term is zero for any affine transformations and therefore penalizes only non-affine transformations.

In [22], Chen and Suter proposed a generalization of the scalar thin-plate splines (TPS) called the *m-order regularization*:

$$\mathcal{R}_{div+curl}(\mathbf{U}) = \frac{1}{2}\int_{\Omega} \beta_1 \,|\nabla^m \, div\mathbf{U}|^2 + \beta_2 \,|\nabla^m \, curl\mathbf{U}|^2 \, d\mathbf{x} \tag{22.13}$$

where the divergence and curl operators are used to impose incompressible flow and irrotational flow, respectively, and β_1 and β_2 weight the respective components. According to [22], this regularizer is applicable in tagged cardiac MRI registration that require the solenoidal component of **U** to be divergence-free.

22.3.2.2.2 *Regularizers Based on Continuum Mechanics*

The linear-elastic and fluid-flow models are two classic physics-based regularization methods [52,92,103].* Both models borrow concepts from continuum mechanics to describe the amount and type of forces exerted on a medium and model image deformations as responses to forces acting on either a elastic body or viscous fluid.

Elastic regularization is based on the theory of linear elasticity, which states that the body forces and "surface stresses" of a linear elastic material sum to zero when the material is in equilibrium, thus giving rise to the following *Navier-Cauchy linear elastic partial differential equation (PDE)* [52]:

$$\mu\nabla^2\mathbf{U}(x) + (\mu+\lambda)\nabla(\nabla\cdot\mathbf{U})^2 + \mathbf{f}(\mathbf{x}) = 0 \tag{22.14}$$

where **f** denotes the body force; μ and λ are the Lame parameters, which can be interpreted in terms of Young's modulus and Poisson's ratio†, and are used to describe the relationship between stress and strain (internal forces that bring the material into equilibrium). The first term is associated to the incompressibility of the material [15,52]. The second term allows for expansion (dilation) or contraction of the material [17,124]. Formulating MIR with this PDE, **f** is modeled as matching forces based on image similarity (more details in Section 22.3.3) and the remaining terms are seen as internal forces that regularize **U**. Accordingly, elastic regularization can be formulated as [92]

$$\mathcal{R}_{elastic}(\mathbf{U}) = \frac{1}{2}\int_{\Omega} \mu\nabla^2\mathbf{U}(x) + (\mu+\lambda)\nabla(\nabla\cdot\mathbf{U})^2 d\mathbf{x}. \tag{22.15}$$

The derivation of Equation 22.14 assumes that the stress and strain are linearly related, thus assuming that the elastic properties of the material is homogeneously by isotropic [52]. This is generally not true in many biological

* Note Zikic et al. [159] categorize them as parametric via use of particular basis functions.
† Poisson's ratio is the ratio between lateral shrinking and longitudinal stretching. A value of 0.5 means the material is incompressible.

materials. In practical terms, the linear relationship assumes that the restoring forces increase monotonically with strain [29], thus penalizing large deformations. Additionally, ignoring the second-order derivatives of **U** would lead to large numerical errors for large deformations [52].

To address these problems, Christensen et al. [29] proposed a viscous-fluid-based model, which views a deforming image as a highly viscous (sticky and thick) fluid. Specifically, this model is built on several physical laws that govern conservation of mass, energy, linear and angular momentum. We will highlight two of the most relevant. The first is the *constitutive equations* [52], which state that the stress tensor Σ_{stress} is linearly related to the *rate of deformation tensor* **D** as follows:

$$\Sigma_{stress} = -pI + \lambda_f \,\text{trace}(\mathbf{D})\mathbf{I} + 2\mu_f \mathbf{D} \tag{22.16}$$

where **D** is expressed as

$$\mathbf{D} = \frac{1}{2}\left(\nabla\mathbf{V} + (\nabla\mathbf{V})^T\right) \tag{22.17}$$

with **V** being the *instantaneous velocity* of **U**.

The second is the *conservation of linear momentum*, which states that

$$\nabla\rho + \mathbf{f} = \rho\frac{\partial\mathbf{V}}{\partial t} + \eta\mathbf{V} \tag{22.18}$$

where η is a parameter that allows for "arbitrary creation or destruction of mass." Combining Equations 22.16 through 22.18 results in the *Navier–Stokes–Duhem equation*:

$$\rho\frac{\partial\mathbf{V}}{\partial t} = \mathbf{f} - \nabla p + \mu_f\nabla^2\mathbf{U}(x) + \left(\mu_f + \lambda_f\right)\nabla(\nabla\cdot\mathbf{V})^2 - \eta\mathbf{V} \tag{22.19}$$

and assuming negligible change in ρ and ignoring* the inertial terms ($\rho\frac{\partial\mathbf{V}}{\partial t}$ and $\eta\mathbf{V}$) yield the following *Navier–Stokes PDE*:

$$\mu\nabla^2\mathbf{V}(x) + (\mu+\lambda)\nabla(\nabla\cdot\mathbf{V})^2 + \mathbf{f}(\mathbf{x}) = 0 \tag{22.20}$$

Remarkably, Equation 22.20 is similar to Equation 22.14 except that this PDE operates on **V** instead of **U**. Intuitively, operating on **V** gives a more accurate calculation of the displacement gradients (second-order gradients of **U** are not ignored) leading to a more realistic model of material deformations.

* Valid for slow flow rates.

22.3.2.3 Constraints for Invertibility

None of the aforementioned regularization methods detailed so far ensures invertibility of **U**. A simple approach to ensure invertibility is to add a regularization term to the registration objective function (Equation 22.7) that penalizes cases when $|D| < 0.3$ [106,126].

$$R_{invertibility}(U) = |D(x) - 1|. \tag{22.21}$$

In [62], Chen and Lee showed that FFD based on cubic B-splines is locally injective over Ω if the maximum displacement of control points is limited to 0.40δ, where δ is the spacing of control point grid. This means that invertibility of **U** can also be ensured by constraining the maximum displacement of control points of an FFD lattice (e.g., control points on a lattice with 20 mm spacing should be constrained to translate no more than 8 mm). However, this restriction becomes impractical when one needs to recover large deformations (i.e., >8 mm). A better alternative may be the polyrigid/polyaffine transformation models proposed in [3].

22.3.2.4 Other Regularization Constraints

22.3.2.4.1 *Inverse Consistency*

Related to the invertibility constraint is *inverse consistency*, which demands that registration results be consistent or *symmetric*: if $\mathbf{U}_2$ is the inverse transform of **U**, that is, $M \circ \mathbf{U} = F$ and $F \circ \mathbf{U}_2 = M$, then $\mathbf{U}_2 \circ \mathbf{U} = Id$ [8,27]. Inverse consistency is important when the end-goal of MIR is to transfer some quantitation from one image to another (e.g., segmentation via registration) such that results are invariant to choice of the reference image.

Theoretically, the process of finding an invertible transform is inherently symmetric; however, this is practically not true if an asymmetric data similarity cost is used [11]. One solution is to symmetrize the data term as done in [5,11,111], or to add a regularization that penalizes the residue between **U** and its inverse [26], for example,

$$\mathcal{R}_{inv\text{-}consistency} = \sum_{\mathbf{x} \in \Omega} \left\| \mathbf{U}^{-1}(\mathbf{x}) - \mathbf{U}(\mathbf{x}) \right\|^2. \tag{22.22}$$

22.3.2.4.2 *Volume-Preservation*

One may require **U** to be volume-preserving (VP), which is desired* when F and M are known to be capturing the same anatomy that does not change in size. An interesting formulation of VP is based on the *optimal mass transport*

* Conversely, it should not be enforced when, for example, matching two different perspective projections of an object, or matching MRI to PET where image intensities are assumed to predict mass density.

problem of Monge-Kantorovich [127]. A simpler alternative is the incompressibility constraint:

$$\mathcal{R}_j(\mathbf{U}) = \int_{\Omega} | \, log(D(\mathbf{x})) \, | \, d\mathbf{x}. \tag{22.23}$$

Other forms include $\int_{\Omega} | \, D(\mathbf{x}) - 1 \, | \, d\mathbf{x}$ and those given in [104].

22.3.2.4.3 Local Rigidity

When registering images containing rigid objects (e.g., bones, hard tissues), one demands that these hard objects remain rigid after registration. One can do this by enforcing additional local rigidity constraints over selected regions; for example, in 2D [91]:

$$\mathcal{R}_{rigidity}(\mathbf{U}) = \sum_{\mathbf{x} \in \Omega} \left(\sum_{i,j=1}^{2} \left\| \frac{\partial^2 \mathbf{U}(\mathbf{x})}{\partial x_i \partial x_j} \right\|^2 + \left\| J^T(\mathbf{x}) J(\mathbf{x}) - Id \right\|^2 + \left\| D(\mathbf{x}) - 1 \right\|^2 \right) \tag{22.24}$$

where each aforementioned term measures local deviations from area-preservation (linearity), angle-preservation (orthogonality), and VP, respectively. Another [48] formulates local rigidity as hard constraints. Pennec et al. [98] also measured local rigidity with $trace((J^T J - Id)^2)$.

22.3.3 Data (Dis)Similarity

The sources of information that the data (dis)similarity term is built upon can generally be classified as *model-based* or *intensity-based* [2,92].

Model-based information is often provided in forms of landmark correspondences identified by clinical experts or manual segmentation of the input images. Examples include [30] where Chui et al. used the *Robust Point Matching algorithm* to match manually extracted sulcal points for brain image registration, and [25,63] where researchers have matched landmarks via diffeomorphic warps. Fuzzy or hard segmentation labels, too, can be matched using the cardinality match or Kappa statistics similarity metrics [56].

Due to the need for human intervention, accuracy of model-driven registration highly depends on the correctness of landmark correspondences or segmentation. When employing landmark correspondences, the identified landmarks should be spatially distributed [56,92]. When using segmentation, one needs to ensure that interpolation artifacts would not corrupt evaluations of image similarity. Consequently, pure use of model-based data terms is quite restrictive.

In contrast to model-based information, intensity-based information is derived entirely from the input images. This information may be based on raw image intensities or features extracted from the intensities.

In mono-modality registration, data similarity terms using information about raw intensities usually assume that the similarity between F and M is related to their intensity distributions. Assuming both F and M contain the same anatomy and $\mathbf{U}$ represents an optimal alignment, their intensity relationship becomes:

$$F = M \circ \mathbf{U} + \tau + \zeta \tag{22.25}$$

where
τ is zero-mean Gaussian noise
ζ is an intensity correction field

Essentially, Equation 22.25 states that after applying $\mathbf{U}$ to M, M and F only differ by an intensity mapping with an additive noise.

For some applications, e.g., CT-CT intra-subject image registrations, we may assume $\zeta = 0$, thus allowing for the use of the Sum of Squared Difference (SSD) metric. If instead, intensities of F and M are linearly related, then one may use the (normalized) cross-correlation (CC) metric [5,144], the coefficient of variation [143], or those based on the normalized intensity gradients [49].

The assumption that ζ is zero or linear can easily be violated when images are corrupted by spatially-varying or "nonstationary" intensity distortions, for example, MR images. Real-world images are also often corrupted by spatially varying distortions, occlusions, illumination inhomogeneity, and reflectance artifacts [77,95].

To address intensity inhomogeneities, measures based on information theory may be used. These include joint entropy [92], entropy correlation coefficient [40], cumulative residual entropy [138], normalized mutual information (NMI) [2] and its numerous other extensions, for example, regional MI, alpha-MI, generalized survival exponential entropy–based MI (GSEE-MI) [75,78,81,107,137]. As these measures belong to a subclass of the divergence measures, which measures distances between distributions [101], they can be more robust against intensity inhomogeneities. However, while they are shown to be effective for many multimodal registration problems [2], these measures still assume global spatial dependencies between corresponding pixels. To better capture nonstationary dependencies, authors have proposed to compute these measures locally, that is, average over a small pixel neighborhood, based on the observation that spatially varying intensity distortion is constant within a small pixel neighborhood. Examples include local CC and regional MI.* More explicit ways include performing intensity correction and registration simultaneously or removing ζ (via some training procedures, etc.) before metric

* MI have several "weaknesses"; one being sensitivity to interpolation artifacts [38], another being its "globality" [5], that is, it examines joint and marginal intensity distributions only. One variant of MI addresses this problem to estimate MI locally. However, one requires a large number of samples to estimate MI; conversely, as locality in the MI estimate increases, its statistical reliability decreases. Thus, MI and local MI may be arguably less efficient and less robust than CC [5].

evaluation [37,93,95]. Alternatively, Droske and Rumpf [34] proposed to match image level sets using a Mumford-Shah formulation for registration.

Data dissimilarity terms built on feature-based information generally involve extracting and using particular type of information from the images for registration. Edge-based features include edgeness, ridges, local curvature extrema, corners, crest, and extremal lines [17,83,96,149]. Others that examine the texture of image regions include Gabor [16] and grey level cooccurrence matrix (GLCM) [16], "braintons" [75], and Laws' texture coefficients [60]. Features that analyze and capture the structural properties in images include vesselness [21], local structure [116], and moment invariants [39,151]. Those adopted from the computer vision community include SIFT [23], SURF [157], and shape context [1,128].

In the past decade, researchers began to adopt a hybrid approach where multiple sources of information are combined to improve existing metrics. One of the most prominent examples includes the HAMMER algorithm (Section 22.4.2), which combines segmentation-based information and geometric moment invariants (GMI), which are calculated in neighborhoods of different sizes. Avants et al. also [6] proposed the idea of integrating different image metrics and sources for DTI-registration (e.g., combining a correlation-based metric on T1 and FA images). Their software, too, allows users to combine multiple image metrics as well as landmarks. Hartkens et al. [50] also combined user-provided surface information with NMI for nonrigid registration of brain images, while Cao et al. [21] combined a feature-based measure with a conventional intensity-based measure for lung image registration. Table 22.1 highlights various other hybrid approaches.

22.3.4 Solution-Search

If the registration objective contains only one data term (e.g., Equation 22.7 contains D only), it may be possible to perform registration analytically, as we shall see in the next section. However, as Section 22.3.2 explained, regularization constraints are generally needed, meaning that Equation 22.7 involves more than one energy term. Optimization methods are therefore needed to minimize a potentially complex and nonlinear objective function. Section 22.3.4.2 will highlight some of the optimization methods that have been devised for MIR.

22.3.4.1 Closed-Form Solutions for Image Registration

Most closed-form solutions for image registration operate on ordered point sets (point sets with known correspondences) that have been extracted from the images to be registered. As discussed in Section 22.2.3.2, if point correspondences are available, one may compute a nonlinear transform directly via TPS interpolation. It is also possible to resolve the parameters of a rigid transform between two ordered point sets in closed form. According to [36],

TABLE 22.1

Hybrid Approaches to Construction of Data Dissimilarity Terms

Features + other sources
An edge-based alignment metric + MI + the SURF descriptor [157] for lung image registration [149]
NCC + local phase + image gradients for 3D echocardiography [158]
Geometric moments + segmentation information for various images [111,145]
Vesselness + 1 of 3 similarity metrics [21]
Shannon divergence on brainton + GSEE-MI +, etc. [75]
Landmarks + other sources
Kullback-Leibler + global MI + landmarks correspondences for brain images [65]
SSD + landmark correspondences, combined with spatially varying weight [7]
Intensity similarities + rough landmark correspondences [9]
Landmarks + point-wise MI for brain MRI and functional MRI (brain tumor) [80]
Airway bifurcations and vascular landmarks + sum of squared local tissue volume difference [155]
Landmarks + 1 of many metrics, implemented in a MRF-based optimization approach [47]
Surfaces + other sources
Points + surfaces + image metric [50]
Surface registration refined by voxel-based registration [102]
SAVOR [45]: performs nonrigid diffeomorphic registration using MR image intensity, subcortical binary segmentation, and volumetric cortical segmentations

the earliest method for closed-form rigid registration is based on minimization of a least squares (LS) error* (known as fiducial registration error if the point correspondences originate from fiducial markers or expert-identified landmarks) and is solved via SVD. Its later variants include different formulations of the Eigen system describing the point correspondences. Three of such are described in [36].

For affine motions based on least squares, regularization is needed to prevent all points in one point set to be mapped to a single point of the other set (as this would give an LS error of zero) [35]. For this, one may employ the Laplacian operator (i.e., $L = a\nabla^2 + cI$ where a, b are constants) for regularization and allow for inexact point correspondences by assuming Gaussian distributed noise around each. This led to the following Bayesian energy minimization as proposed in [28]:

$$\arg\min_{\mathbf{U},A,\mathbf{t}} \sum_{i=1}^{m} \left[(A\mathcal{P}_i' + \mathbf{t}) - (\mathcal{P}_i - \mathbf{U}(\mathcal{P}_i))\right] \sigma_i \left[(A\mathcal{P}_i' + \mathbf{t}) - (\mathcal{P}_i - \mathbf{U}(\mathcal{P}_i))\right] + \left\|L\,\mathbf{U}\right\|^2. \quad (22.26)$$

* Least squares error may be defined as $\sum_m \|\mathcal{P}_m - \hat{R}\mathcal{P}_m' - \hat{T}\|$, where $(\mathcal{P}_m, \mathcal{P}_m')$ denotes the mth pair of corresponding points in point sets $\mathcal{P}$ and $\mathcal{P}'$ and $\hat{R}$ and $\hat{T}$ are, respectively, the candidate rotation and translation components.

where

m is the number of correspondences

A, $\mathbf{t}$ are respectively the affine matrix and translation vector to be estimated

σ_i is the covariance matrix that represents the spatial uncertainty of the ith pair of correspondence

By inverting the system matrix of the following system of linear equations

$$(A\mathcal{P}_i' + \mathbf{t}) = \mathcal{P}_i\left(\sum_{j=1}^{m} K(\mathcal{P}_i, \mathcal{P}_j)\beta_j - \sigma_i\beta_i\right),$$

$$\sum_{i=1}^{m} \beta_i \mathcal{P}_i' = 0, \quad \sum_{i=1}^{m} \beta_i = 0,$$

one attains the optimal values for A, $\mathbf{t}$, and β_i (a vector of weights to be calculated for each correspondence), with $K(x, y) = k\ Id$ being a $d \times d$ matrix where $k \propto e^{\|x-y\|}$. The optimal transformation is then computed as $U(\mathbf{x}_i) = \sum_{i=1}^{m} K(\mathbf{x}, \mathbf{x}_i)\beta_i$. Another closed-form approach for affine registration is to reduce the affine problem to that of the orthogonal case [51]. However, the accuracy of this approach is questioned in [35].

When correspondences between extracted point sets are not known, rough estimation of the translation and rotation parameters can be done independently by first computing the displacement vector that would align the centroids of the point sets and then resolving for the rotation parameters via principal axis transform [18]. Spectral algorithms may also be used to solve rigid registrations, for example [58]. These algorithms operate by encoding point sets as affinity matrices and performing eigen-decomposition on these matrices. However, for spectral methods to work robustly, the eigenstructure of the affinity matrix must be rich in order to provide discriminating features for proper determination of point correspondences. Otherwise, optimization techniques such as the iterative closest point algorithm and its variants may be better alternatives [35].

Affine transformations may be solved in closed form using properties of the Fourier transform [125]. Furthermore, if $\mathbf{U}$ can be described by a finite set of displacement vectors, one may also solve for $\mathbf{U}$ by formulating a linear assignment problem (LAP) [130]. Recently, Westin et al. [64] also developed the *polynomial expansion* framework that was designed to solve global and local linear registrations analytically.

22.3.4.2 Iterative Optimization

A common approach taken by many optimization methods is to view the solution space as a 1D landscape, known as *functional landscape*, where a position on the landscape corresponds to a candidate solution and elevation corresponds to the optimality of a solution, that is, C. One then takes a step from the current to a new position on a path that would lead to the location of lowest elevation in the landscape. This can be mathematically expressed as

$$\mathbf{P}_{k+1} = \mathbf{P}_k + a_k \mathbf{d}_k \quad (22.27)$$

where $\mathbf{P}_k$ denotes a vector of transform parameters obtained in the current iteration, $\mathbf{P}_{k+1}$ corresponds to a candidate vector to be examined next, $\mathbf{d}_k$ is the direction of the next step, and a_k is the size of the step.

There are many strategies for determining a_k. For example, it can simply be set to a fixed value. It can also be defined using a decaying function of k or a *line search* scheme where for every k, one tries to minimize C along the search direction $\mathbf{d}_k$:

$$a_k = \arg\min_a C(\mathbf{P}_k + a\mathbf{d}_k) \quad (22.28)$$

This can be costly as each iteration involves multiple evaluations of C, so inexact searches are usually used [67].

How one exactly computes $\mathbf{d}_k$ and a_k differentiates different optimization techniques from one and another. We next present two general classes of optimization methods: gradient-based and gradient-free. The former computes $\mathbf{d}_k$ based on the gradient of the cost function, typically by examining the cost function with regard to changes in transformation parameters. Conversely, gradient-free approaches operate on the cost function directly. The next subsections present the details of each.

22.3.4.2.1 Gradient-Based

Gradient-based methods may be *deterministic* or *stochastic*, where the difference lies in the use of a deterministic or random process during the entire optimization process [67].

Deterministic techniques that have been employed for image registration include gradient descent, quasi-Newton, nonlinear conjugate gradient, etc. These are common in that the search direction $\mathbf{d}_k$ is based on the derivative of the cost function with respect to the parameters. Under gradient descent techniques, the search direction is updated as

$$\mathbf{P}_{k+1} = \mathbf{P}_k - a_k \frac{\partial C}{\partial \mathbf{P}_k} \quad (22.29)$$

Klein et al. [66] explored two variants of the gradient descent method. One applies a decaying function on k for a_k; the other uses an inexact line search routine called "More-Thuente" to determine a_k.

Quasi-Newton (QN) methods may have "better theoretical convergence properties than gradient descent," thanks to the use of second-order information [66]. They are based on the Newton–Raphson algorithm, which is given by

$$\mathbf{P}_{k+1} = \mathbf{P}_k - \left[H(\mathbf{P}_k)\right]^{-1} \frac{\partial C}{\partial \mathbf{P}_k} \quad (22.30)$$

where $H(\mathbf{P}_k)$ is the Hessian matrix of the cost function. However, calculation of the inverse of the computation is expensive, so QN methods approximate the Hessian, for example, $L = [H(\mathbf{P}_k)]^{-1}$. Calculation of L_k can be done in several ways, including Symmetric-Rank-1, Davidon–Fletcher–Powell, Broyden–Fletcher–Goldfarb–Shanno (BFGS), and the Limited memory BFGS, which is a variant of BFGS that avoids the need to store L in memory. According to [67], BFGS is very efficient in many applications.

The nonlinear conjugate gradient (NCG) method is based on the linear conjugate gradient method, which was first designed for solving a system of linear equations. NCG is an extension that is suitable for minimizing nonlinear functions. Under NCG, the search direction is defined as a linear combination of the cost gradient and the previous search direction, that is,

$$\mathbf{d}_k = -\frac{\partial C}{\partial \mathbf{P}_k} + \beta_k \mathbf{d}_{k-1} \quad (22.31)$$

where the scalar β_k affects the convergence property and can be computed using different equations (see examples in [67]).

All of the aforementioned methods require an exact calculation or close approximation of the first- or second-order functional gradient. This can be very computational expensive. To address this, approximation of the cost gradient using stochastic sampling is done. The Kiefer–Wolfowitz method, for instance, uses finite difference to approximate the gradient, that is, $\tilde{g} \approx \partial C / \partial \mathbf{P}_k$ [67]. This method requires $2N$ evaluations of the cost function for each k where N is the dimensionality of the solution space. The simultaneous perturbation method may also be used, which only performs two evaluations of the cost function, independent of N. The Robbins-Monro algorithm approximates the functional derivative using only a small subset of voxels randomly selected in every iteration. It was shown to give best compromise between convergence and accuracy over all aforementioned methods (both deterministic and stochastic ones), as evaluated in various registration experiments [67]. For further details on the aforementioned or other gradient-based optimization algorithms, we refer readers to Klein's thesis [67].

22.3.4.2.2 *Gradient-Free*

Next, we present different classes of gradient-free approaches, many of which make few or no assumptions about the problem being optimized and can conduct searches in a very large solution space. Often, some form of stochastic procedures is used in these methods.

22.3.4.2.2.1 Search Approaches

Powell's method (Powell's conjugate gradient descent method) is one commonly used gradient-free, search-based optimization technique. It finds the optimum in a d-dimensional solution space with a series of 1D searches. For example, finding the optimal parameters for an affine 3D transformation usually starts with a 1D search along lateral translation direction, then along the cranial–caudal translation direction, and followed by a rotation about each of the three axes. The choice of ordering influences the success of optimization and many have adopted heuristics to select the most suitable order.

A related algorithm is the simplex algorithm (aka Nelder-Mead simplex, downhill simplex, or amoeba method). Following the analogy of [144], one can see simplex as a creature with $n + 1$ feet crawling in an n-D search space. One of its feet stands on the initial parameter estimate while the rest are randomly placed. At each iteration, the cost function is evaluated at each foot position and the foot with the worst cost is moved to improve its cost. Some specific rules are designed to guide which foot to move next and by how much.

Other examples of search-based algorithms include simulated annealing, DIRECT [135], and Tabu search.

22.3.4.2.2.2 Evolutionary Algorithms

A representative class of metaheuristics is evolutionary algorithms (EAs). EAs are based on the principle of natural selection, that is, the fittest survives. The solution space is seen as a dynamic population. A candidate solution plays the role of an individual in a population and its function cost corresponds to its fitness value of surviving in the population. Evolution of the population then occurs after repeated applications of reproduction (aka inheritance), mutation, recombination (aka crossover), and selection. Many methods based on this idea have been proposed in the literature; an extensive review is given in [31].

One popular choice is the evolutionary strategy (ES). In each iteration, "children" are created via mutation of the population (its "parents"). A mutation is a random vector generated from a multidimensional normal distribution. The covariance matrix of the distribution is adapted in each iteration: it is increased if the new population consists of better individuals or decreased otherwise. The population then reduces back to its original size by keeping only the fittest. A variant is the one-plus-one ES [115], where both the population size and the number of children generated are equal to 1; this variant was used by Chillet et al. [24] to affinely register 3D models of brain and liver vasculatures (encoded with an inverted distance map) to CT or MR images.

A relatively new class of algorithms for multimodal function optimization is those based on artificial immune systems (AIS), which are inspired by theoretical immunologic models. They are capable of locating the global minimum of a function as well as a large number of strong local optima by automatically adjusting the population size and combining local with global searches. Recently, Delibasis et al. [33] applied it to solve for a spatial matching between point correspondences identified in medical images.

Of all EA methods, the current state-of-the-art is probably the covariance matrix adaptation method [67]. It has several phases. In the reproduction phase, a set of m search directions ($\mathbf{d}_k$) is generated from a normal distribution whose covariance matrix is computed to favor search directions that were successful in previous iterations (and m is a user-defined parameter controlling the population size). In the selection phase, the cost function is evaluated for each search direction using Equation 22.28, and the l best search directions are selected. Finally, in the recombination phase, a weighted average of the l selected trial directions is computed to generate the next set of search directions. Recommendations for the choice of m and l based on n are given in [67].

22.3.4.2.2.3 Discrete Optimization

Discrete optimization operates by searching for the best solution out of a discrete set of solutions. When applied to image registration, they usually involve modeling the deformation $\mathbf{U}$ as a random field and recasting registration of F and M as a Markov Random Field (MRF) energy minimization problem [14,46,71,109], which formulates the statistical dependence between F and M as

$$p(F, \mathbf{U} \mid M) = p(F \mid \mathbf{U}, M)p(\mathbf{U}). \tag{22.32}$$

where the conditional probability $p(F|\mathbf{U}, M)$ is the *data likelihood* that models dependence between intensity values between the two images and the prior probability $p(\mathbf{U})$ is the *prior* that favors plausible deformations. As we will see shortly, these probabilities essentially correspond to data dissimilarity and regularization terms of a typical registration objective given in Equation 22.7. Assume F is conditionally independent of $\mathbf{U}$ and M, the first term in the preceding becomes

$$p(F \mid \mathbf{U}, M) = \prod_{\mathbf{x} \in \Omega} p(F(\mathbf{x}) \mid \mathbf{U}(\mathbf{x}), M(\mathbf{x})). \tag{22.33}$$

Assuming further that $F(\mathbf{x})$ only depends on $M \circ \mathbf{U}(\mathbf{x})$ and not the rest of the values in M, one obtains

$$p(F(\mathbf{x}) \mid \mathbf{U}(\mathbf{x}), M(\mathbf{x})) = p(F(\mathbf{x}) \mid M \circ \mathbf{U}(\mathbf{x})) \tag{22.34}$$

We will now see how use of discrete optimization can simplify maximum *a posterior* (MAP) inference. Let the spatial coordinates of M be a graph G (whose set of nodes V correspond to $\mathbf{x} \in \Omega$ and edges $(\mathbf{x}_i, \mathbf{x}_j) \in \varepsilon$ encode neighborhood relationships, for example, 4-connectivity in 2D, etc.). In the simplest form, the data likelihood can model pixel-wise dependence between F and M with a Gaussian noise model, that is,

$$p(F \mid M \circ \mathbf{U})) = \prod_{\mathbf{x}_i \in V} p(F(\mathbf{x}_i) \mid M(\mathbf{U}(\mathbf{x}_i))) = \prod_{\mathbf{x}_i \in V} e^{-0.5(F(\mathbf{x}_i)-M(\mathbf{U}(\mathbf{x}_i)))^2}; \quad (22.35)$$

and the prior can be encoded as a regularization penalty that examines the interactions between displacements of two neighboring points $\mathbf{x}_i$, $\mathbf{x}_j$, that is,

$$p(\mathbf{U}) = \prod_{(\mathbf{x}_i,\mathbf{x}_j)\in\varepsilon} \left\| \mathbf{U}(\mathbf{x}_i) - \mathbf{U}(\mathbf{x}_j) \right\|^2 \quad (22.36)$$

Performing MAP inference can now be done as minimization of the following:

$$E(\mathbf{U}) = -p(F, \mathbf{U} \mid M) = -\prod_{\mathbf{x}_i \in V} p(F(\mathbf{x}_i) \mid M(\mathbf{U}(\mathbf{x}_i)))p(\mathbf{U}(\mathbf{x}_i)) \quad (22.37)$$

which leads to minimization of the following MRF energy:

$$-\log(E(\mathbf{U})) = \sum_{\mathbf{x}_i \in V} \underbrace{(F(\mathbf{x}_i) - M(\mathbf{U}(\mathbf{x}_i)))^2}_{\theta_i} + \sum_{(\mathbf{x}_i,\mathbf{x}_j)\in\varepsilon} \underbrace{\left\| \mathbf{U}(\mathbf{x}_i) - \mathbf{U}(\mathbf{x}_j) \right\|^2}_{\theta_{ij}} \quad (22.38)$$

where the first summation over *unary* potentials θ_i is identical to the SSD metric (Section 22.3.3) and the second summation over *pair-wise* potentials θ_{ij} resembles a Laplacian regularizer on $\mathbf{U}$ (Section 22.3.2). Note that this MRF energy contains potentials of *first-order** and that the maximum clique of its corresponding graph is size of 2.

MAP inference with MRF as given earlier can be solved with different classes of algorithms; these include *linear programming, message passing,* and *graph-cut.* Algorithms for solving integer linear programs include cutting-plane method, branch-and-bound, branch-and-cut, etc. Message passing algorithms include Belief Propagation, Tree Reweighted Message Passing (TRMP), and Max-Product algorithms. Graph-cut algorithms include α-*expansion* and α-β *swap* [14].

Different matching problems have adopted a discrete optimization approach [14,70]. For MIR, one of the earliest applications of MRF energy minimization include [121] where Tang and Chung resolved $\mathbf{U}$ at every

* Modeling $\mathbf{U}$ with pair-wise relationships may be sufficient; however, some problems may require terms of higher-order energies, e.g., surface registration [156] or point (hypergraph) matching [88], which would require more elaborate techniques, e.g., [57] to optimize.

location via minimization of Equation 22.38. Lombaert et al. [79] later extended [120] by incorporating the use of landmarks for deformable registration of coronary angiograms.

To control deformations more effectively, Glocker et al. [46] parameterized **U** with FFD transformation model and applied MAP inference to solve for the optimal translations of the control points of the B-spline model. To minimize their objective function, they employed Fast-PD [69] (stands for Fast Primal-Dual approximation), which works by solving the energy minimization problem by a series of graph-cut computations. The process reuses the primal and dual solutions of the linear programming relaxation of the energy minimization problem obtained in the previous iteration, thus effectively decreases run time; see review [47] for details. Their method was widely accepted due to its high efficiency and public availability made possible through their software called DROP, which contains implementations of various image similarity measures. Kwon et al. [71] also used the FFD framework for 2D registration, but proposed the following higher-order regularization term that does not penalize global transformations like Equation 22.38 would:

$$\sum_{(\mathbf{x}_i,\mathbf{x}_j,\mathbf{x}_k)\in\varepsilon}\sum_{d=\{1,2\}}\left\|-\mathbf{U}_d(\mathbf{x}_i)+2\mathbf{U}(\mathbf{x}_j)-\mathbf{U}_d(\mathbf{x}_k)\right\|^2 \tag{22.39}$$

where $\mathbf{U}_d$ denotes the dth component of **U**. The higher-order MRF energy was then converted to a pair-wise one, so it could be solved with TRWP.

In [109], Shekhovtsov et al. performed MRF-based matching using a linear programming relaxation technique. However, they proposed to decompose a 2D deformation model into x and y components so that the problem could be solved with sequential TRMP. Lastly, Lempitsky et al. [74] proposed the fusion-move approach to combine discrete optimization with continuously valued solutions for 2D image registration, effectively avoiding local minima while avoiding problems due to discretizing the deformation space. Specifically, candidate flow fields ($\hat{\mathbf{U}}$) were generated using the method of [53]. Each candidate flow was seen as a "proposal" label $l \in \mathcal{L}$, which was then iteratively examined and fused in a moving-making approach. MRF-energy minimization in each iteration was performed with the Quadratic Pseudo Boolean Optimization algorithm [68]. In regularizing **U**, the authors proposed the following pair-wise potentials:

$$\theta_{ij} = \sum_{d=1,2}\rho\left(\frac{\mathbf{U}_d(\mathbf{x}_i)-\mathbf{U}_d(\mathbf{x}_i)}{\left\|\mathbf{x}_i-\mathbf{x}_j\right\|}\right) \tag{22.40}$$

where $\rho(x)=\log\left(1+\frac{1}{2b^2}x^2\right)$ and b is a constant. Fused solutions were then refined with gradient-based optimization.

22.4 Two Commonly Adopted Registration Paradigms

22.4.1 Demons Algorithm and Its Extensions

To better understand thermodynamic equilibrium, the Scottish physicist James Maxwell [123] came up with the concept of *demons* as door guards that selectively allow "hot" molecules through. Adopting this concept, Thirion [123] proposed to formulate registration as a process where demons locally push voxels into or out of object boundaries to allow for image matching, by treating images as isointensity contours and computing demon forces to push these contours in their normal direction to encourage image alignment. The orientation and magnitude of the forces are derived from the *instantaneous optical flow equation* [56], which assumes that M is the deformed F, whose deformations $\delta\mathbf{x}$ are due to local motions that occur after δt, that is,

$$M(\mathbf{x}) = F(\mathbf{x} + \delta\mathbf{x}, t + \delta t)$$

$$F(\mathbf{x}, t) - F(\mathbf{x} + \delta\mathbf{x}, t + \delta t) = 0$$

where the last equation is known as the brightness constancy assumption. Using linear approximation, we get,

$$F(\mathbf{x} + \delta\mathbf{x}, t + \delta t) = F(\mathbf{x}, t) + \frac{\partial F}{\partial \mathbf{x}}\delta\mathbf{x} + \frac{\partial F}{\partial t}\delta t$$

$$0 = \frac{\partial F}{\partial \mathbf{x}}\frac{\delta\mathbf{x}}{\delta t} + \frac{\partial F}{\partial t}\delta t \quad \text{(known as the optical flow constraint)}$$

$$-\frac{\partial F}{\partial t}\delta t = \nabla F \frac{\delta\mathbf{x}}{\delta \mathbf{t}}$$

$$-(M(\mathbf{x}) - F(\mathbf{x})) = \nabla F(\mathbf{x})\mathbf{V}(\mathbf{x})$$

The method alternates between computing the image forces and performing regularization on the motion field using Gaussian smoothing. This simple scheme is computationally efficient. In [97], Pennec et al. showed that the Demons algorithm can be seen as an approximation of a second-order gradient-descent on the SSD metric. Bro-Nielsen [17] also showed that applying the Gaussian filter instead of the real linear elastic filter is an approximation of the fluid model; see details in Section 22.3.2.2.2. Surveys of the state of art of optical flow algorithms are given in [10,135] and the references therein; the latter in particular also provides public data sets and methodologies for evaluation of optical flow algorithms. [75] also provides ground truth motion fields of real-world videos.

Since its conception, different extensions of Thirion's Demons algorithm were proposed; these include extension of the image forces to be based on cross-correlation [5] or phase [59]. In [19], Cachier et al. proposed the PASHA algorithm, which incorporates the addition of a random field C (essentially, an "intermediate transformation" [85]) to allow for inexact point correspondences as given by **U**. This leads to minimization of the following energy

$$E(\mathbf{C},\mathbf{U}) = \frac{1}{\gamma_n}\|F - M \circ \mathbf{C}\|^2 + \frac{1}{\gamma_c}\|\mathbf{C} - \mathbf{U}\|^2 + \frac{1}{\gamma}\|\nabla \mathbf{U}\|^2 \tag{22.41}$$

where the first term measures the quality of the field of point correspondence **C** between the deformed M and F based on a data dissimilarity measure; γ_n accounts for the noise in the image, γ_c accounts for spatial uncertainty on the correspondences, and γ controls the amount of regularization on **U**. Rather than solving the complex energy minimization in Equation 22.46, Cachier et al. alternatively optimize Equation 22.46 with regard to **C** given **U**, and then optimize Equation 22.46 with regard to **U** given **C**. The second optimization has a closed-form solution when the regularization is quadratic and uniform, and can be obtained by setting it as the convolution of **C** by a Gaussian kernel [132].

To solve the first optimization, that is, minimize $\frac{1}{\gamma_n}\|F - M \circ \mathbf{C}\|^2 + \frac{1}{\gamma_c}\|\mathbf{C} - \mathbf{U}\|^2$ with regard to **C** given **U**, one linearly approximates the data (first) term for every pixel as

$$\psi(\mathbf{x}) = F(\mathbf{x})M(\mathbf{x}) \circ \mathbf{U} \circ \mathbf{C}(\mathbf{x}) \approx F(\mathbf{x}) - M(\mathbf{x}) \circ \mathbf{U}(\mathbf{x}) + J(\mathbf{x}) \circ \mathbf{C}(\mathbf{x}) \tag{22.42}$$

where $\mathbf{J}(\mathbf{x}) = -\nabla M^\circ\, \mathbf{U}(\mathbf{x})$. This allows one to rewrite the first optimization as

$$E(\mathbf{C} \mid \mathbf{U}) \approx \frac{1}{2|\Omega|}\sum_{\mathbf{x}\in\Omega}\left\| \begin{bmatrix} F - M \circ \mathbf{U}(\mathbf{x}) \\ 0 \end{bmatrix} + \begin{bmatrix} J(\mathbf{x}) \\ \frac{\gamma_n(\mathbf{x})}{\gamma_c} I \end{bmatrix} \mathbf{C}(\mathbf{x}) \right\|^2 \tag{22.43}$$

where $\gamma_n(\mathbf{x}) = |F(\mathbf{x}) - M \circ \mathbf{C}(\mathbf{x})|$ is a local estimation of noise that is used. Since the approximation for each pixel is independent of each other, one instead solves a simple system for each pixel using the following normal equation:

$$\left(J^T(\mathbf{x})J(\mathbf{x}) + \frac{\gamma_n}{\gamma_c} I\right)\mathbf{C}(\mathbf{x}) = -(F(\mathbf{x}) - M \circ \mathbf{U}(\mathbf{x}))J^T(\mathbf{x}) \tag{22.44}$$

$$\mathbf{C}(\mathbf{x}, F, M) = -\frac{F(\mathbf{x}) - M \circ \mathbf{U}(\mathbf{x})}{\|J(\mathbf{x})\|^2 + \frac{\gamma_n^2(\mathbf{x})}{\gamma_c^2}} J^T \tag{22.45}$$

All of this leads to the following modified demons algorithm:

1. Start with an initial estimate of $\hat{\mathbf{U}}$.
2. Compute an update for $\mathbf{C}$ as given in Equation 22.45.
3. To enforce fluid-like regularization (Section 22.3.2.2), set $\mathbf{C} \leftarrow K * \mathbf{C}$, where K is generally a Gaussian kernel.* Otherwise, skip this step.
4. Compute $\mathbf{C} \leftarrow \mathbf{U} + \mathbf{C}$.
   - To enforce elastic-like regularization, compute $\mathbf{U} \leftarrow K * \mathbf{C}$.
   - Otherwise, compute $\mathbf{U} \leftarrow \mathbf{C}$.

where the convolution of $\mathbf{C}$ with a Gaussian kernel is an efficient approximation to elastic or fluid regularization.

To spatially adapt regularization, Stefanescu et al. [114] smoothed $\mathbf{U}$ using a variable Gaussian kernel whose size depended on a scalar field that encoded the expected amount of deformations as estimated from prior segmentation of the anatomy. The authors also filtered $\mathbf{C}$ after each iteration according to a measure that is based on local intensity variance and gradient.

Yet another extension is called the Log Domain Diffeomorphic Demons (LDDD), which was proposed by Vercauteren et al. [131,133] to constrain $\mathbf{U}$ to reside in the space of diffeomorphisms $Diff(\Omega)$. Recall that when performing an n-dimensional optimization search in $\mathcal{R}^n$, the iterative step used in Newton's optimization method is of the form $\mathbf{P}_{k+1} = \mathbf{P}_k + a_k\mathbf{d}_k$, where $\mathbf{d}_k$ is the descent direction, which lies in the tangent space of $\mathcal{R}^n$. However, when we move from $\mathcal{R}^n$ to the space of diffeomorphisms (i.e., smooth manifold), the descent direction that lies in the tangent space at $\mathbf{P}_k$ is different from the tangent spaces of any other point. To map a point on the tangent space of $\mathbf{P}_k$ to a point on the manifold, one thus employs the *exponential map* operation, that is, $exp : \mathbf{V} \mapsto Diff(\Omega)$. Accordingly, parameterizing $\mathbf{U}$ and $\mathbf{C}$ by stationary velocity fields $\mathbf{V}$ and $\mathbf{V}_C$ via exponential map, that is, $\mathbf{U} = exp(\mathbf{V})$, $\mathbf{C} = exp(\mathbf{V}_C)$, one now minimizes the log-domain version of Equation 22.46 with regard to $\mathbf{V}_C$:

$$E(\mathbf{C},\mathbf{U}) = \frac{1}{\gamma_n}\left\|F - M \circ exp(\mathbf{V}_C)\right\|^2 + \frac{1}{\gamma_c}\left\|log\left(exp(-\mathbf{V}) \circ exp(\mathbf{V}_C)\right)\right\|^2 + \frac{1}{\gamma}\left\|\nabla \mathbf{V}\right\|^2. \tag{22.46}$$

The iterative update in the aforementioned step 4 becomes

$$\mathbf{C} \leftarrow \mathbf{U} \circ exp(\delta v), \tag{22.47}$$

* The Gaussian kernel is an estimate to the Green's kernel for the linear operator $\boldsymbol{L}$ for the fluid model [17].

where δv is a small update velocity field that was derived in [84] as $\frac{F(\mathbf{x}) - M \circ \mathbf{U}(\mathbf{x})}{\|J(\mathbf{x})\|^2 + \frac{\gamma_n^2}{\gamma_c^2}} J(\mathbf{x})$.

To further enforce inverse consistency (Section 22.3.2.4.1), Vercauteren et al. further proposed a symmetric version of LDDD [132]. Other extensions of the demons algorithm include multichannel diffeomorphic demons [99]; *spherical demons*, which operates on the sphere for surface registration [153]; as well as *iLogDemons* [85], which incorporates elasticity and incompressibility constraints.

22.4.2 Block-Matching and HAMMER

Block-matching (BM) is an efficient scheme for isolating and selecting salient parts of an image to drive the process of registration. Usually, a BM algorithm consists of the following steps that iterate until certain convergence criterion is met. First, blocks of image pixels computed from F and M are matched according to their similarity scores and displacement vectors are computed from the best matches. Second, displacements of all voxels in the entire domain are then computed from the matches via interpolation, for example, using TPS [84] or a Gaussian weighting scheme [111]. Then, regularization on $\mathbf{U}$ is subsequently enforced via Gaussian filtering on $\mathbf{U}$.

This matching method was originally proposed for video sequence* matching but has been successfully used for MIR since its first adoption in [32] by Collins et al. Some of its successful applications include elastic registration of neck-and-head, paraspine, and prostate images as performed in [84] (where blocks belonging to detected anatomical landmarks were matched based on local cross-correlation), as well as piecewise affine registration of biological images (e.g., histological sections, autoradiographs, cryosections, etc.) as performed in [100] (where robust least square was used to estimate the local transformations associated to each matching pair).

Both demons framework and BM algorithms are quite similar in that both perform regularization of $\mathbf{U}$ via post-filtering (i.e., apply filtering on update of $\mathbf{U}$ per iteration). However, rather than computing the data forces for all voxels, BM involves an additional/explicit step of preselecting salient regions only in which data forces are calculated for matching. Unlike the original demons algorithm, BM algorithms are also capable of recovering large deformations thanks to the strategy of matching blocks of different scales hierarchically (e.g., start with large blocks, then reduce their sizes in subsequent stages). For example, it was shown capable of recovering rotations up to 28° in a study involving rat brain sections (see details in [100]).

* A survey of BM algorithms applied to video sequence matching is given in [55].

One frequently cited BM-based registration framework is called HAMMER, which stands for Hierarchical Attribute Matching Mechanism for Elastic Registration. First proposed in [111] by Shen and Daatziko, important voxels or *driving voxels* are selected based on the *attribute vectors* (AV) (Section 22.3.3) computed from each voxel. The selection of voxels and their influences on the current estimate of **U** are hierarchically determined. Specifically, in the early stages of registration, the similarity between two driving voxels is determined based on the weighted sum of the AV similarities in the subvolumes that encompass each of the two voxels. A match occurs only when the integrated similarity is beyond a user-defined threshold, which is initially set to a high value. The threshold as well as the size of the subvolumes are then decreased at a user-defined rate in each iteration.

Some of the later extensions and applications of HAMMER include the following:

- Optimize parameters that define the set of AVs [145]
- Use of local spatial intensity histogram [110] or wavelets [150] to define the AVs
- Incorporation of landmark correspondences for matching, and replace Gaussian with TPS for interpolation of U [147]
- Groupwise registration using HAMMER [148]

22.5 Evaluating Registrations

A logical start on evaluating registration results is to examine them subjectively. For images of the same modality, one can visualize the results by examining the registered images individually on separate windows with a linked cursor ("split window validation" [86]), examining their checkerboard or difference image, or viewing them jointly with alpha-bending.* For images of different modality, the alpha-bending is the most common tool as all image characteristics in both images can be displayed simultaneously [119]. Furthermore, if segmentation of one or both of the two images is available, one can also overlay the segmentation contours of one image on top of the image to be matched to; see [16] for examples.

In general, due to limits of subjective evaluation,† qualitative approaches are also guided with quantitative evaluation.

* A checkerboard image is created by combining parts of two images in an interlaced manner. A difference image is computed as pixel-wise subtraction between them. In a color overlay, one image is displayed in gray-scale and another is overlaid on top with a color-scale.

† Wong et al. [142] studied limits of visual detection of misregistrations in PET-to-MRI brain registrations.

Quantitative evaluation of a registration technique may be broadly classified in terms of the nature of the experiment: synthetic or real.

Validation based on synthetic data is generally conducted in a controlled setting, usually performed to evaluate particular components of a registration method. For example, when performing registration with gradient-based optimization, one often examines the "smoothness" and the capture range of the metric (by computing the metric between two images with regard to a range of misalignments). One may also examine the *robustness* of an algorithm, that is, registration results remain reasonable even in extreme scenarios. A formalized protocol involving the aforementioned measures on a given image pair with known "ground truth" registration is given in [112]. One may also examine an algorithms' performance by studying its inverse consistency and transitivity (i.e., registration results transfer, e.g., propagating registration result from A to B to C to B, and then back to A should yield zero in the composition of all obtained deformations maps [13]).

For mono-modality registration, where one assumes that M is a deformed version of F with additive noise corruption, one common practice to quantify registration quality is to compute the mean square error (MSE) between the deformed moving image and fixed image. MSE may further be used to compute the Peak-Signal-to-Noise Ratio (PSNR), which equals to $10 log (I_{max}^2 / MSE)$, where I_{max} is the maximum intensity between two images. The PSNE is a common measure to evaluate image reconstruction quality, and has been used in [134] to measure registration accuracy. However, an implausible warp may always produce zero MSE. Generally, examining intensity values alone does not evaluate the plausibility and quality of **U** and so both MSE and PSNR should not be the sole quantitative measurements.

A more direct and arguably more complete evaluation approach is to examine an algorithm's ability to recover known misalignments. Starting with a pair of images, subjectively judged as registered, one then applies random misalignments and/or warps to one of the pair [61]. Then, a registration error metric (e.g., angular error or mean Euclidean distance) is computed to measure the residual between the obtained transform and the known transform (that misaligned the original pair) within the region of interest. Countless works, for example [10,46,71,109,117,118,121,140,157–159], have adopted this approach.

Validation based on real data involves performing actual image registrations and employing additional data, ideally provided by clinical experts, to quantify accuracy of the obtained results. For example, if landmark correspondences can be acquired, one may compute the target registration error (TRE) [141]. However, in order for precise assessment, it is important that the correspondences are as spatially distributed as possible, especially for results obtained from deformable registration. If segmentation of the images is available, one can also apply the warp to the segmentation and compute overlap measures* between the target and deformed segmentations. The latter approach is known as morphology-based evaluation [5,66].

Several issues evolve around landmark-based and morphology-based evaluations. First, acquiring additional validation data is difficult and laborious. For instance, landmarking 3D images (usually using 2D displays) is a challenging task due to limited sense of spatial context and depth. And when one or both of the input images are functional data (e.g., PET or SPECT, where anatomical boundaries cannot be reliably segmented automatically or manually), it becomes extremely difficult [136], if not impossible. Second, low interoperator variability† of the segmentation/landmarking results cannot always be guaranteed. To reduce variability that may be introduced by operator's biases and experience, protocols have been established for segmentation, for example, brain images [66], and for identification of anatomical landmark, for example, femur [108]. Third, morphology-based evaluation does not provide a complete insight of registration accuracy. One problem is that quantitation cannot be expressed in a unit that is relevant for task of registration (e.g., distance in mm). Another is that it focuses only on segmented regions and disregards unsegmented structures as well as the regions within or surrounding the segmented regions.

Due to the aforementioned factors, evaluation based on segmentations are usually regarded as "bronze" evaluation, suggesting that one should interpret with caution.

In the MIR literature, numerous researchers have performed comparative studies to evaluate a set of registration algorithms that have been designed for different problems. The earliest work is the Nonrigid Image Registration Evaluation Project (NIREP), which standardized a set of benchmarks and metrics (overlap, variance of image intensities in a population of registered images,‡ transitivity, and inverse consistency). The initial phase of NIREP involved a data set of 16 labeled brain images. Other comparative analyses include study of [128], where Urschler et al. compared six deformable registration algorithms with an experiment involving synthetic deformations applied to two pairs CT lung images.

Recently in 2010, Murphy et al. [94] have set up a web-based public platform called EMPIRE to allow for a "fair and meaningful comparison of registration algorithms applied to thoracic CT data." The website allows developers to download image pairs and upload the corresponding registrations they obtained for evaluation as performed by the website. Four sets of measures are then later reported on the site, which include those based on alignment of

* For example, Jaccard index, Dice coefficient (related to Jaccard index), Tanimoto coefficient (extended Jaccard index), etc.

† Some publications report registration accuracy in relation to interoperator variability (e.g., registration error is comparable to interoperator variability). One example is [136] for evaluation of whole-body CT-PET registrations.

‡ As mentioned earlier, intensity-based measures do not reflect accuracy [66,154] as they assume that the same imaging protocol was used in all acquisitions).

lung boundary and fissures. As of October 2011, over 30 teams have submitted their results to this website.

For brain image, comparative studies in which at least three nonlinear algorithms were compared on whole-brain registrations that have been completed in the last decade include:

- Studies by Hellier et al. (see [66]): A series of publications that compared five nonlinear methods that focused on registration accuracy in the cortical areas. Two sets of evaluation measures were used, including segmentation overlap, deformation quality (amount of singularities present), and symmetric hausdorff distance between segmented sulci surfaces. The authors also examined results in terms of shape similarity using PCA.
- Study by Yassa and Stark [152]: Six algorithms and two semi-automated methods were compared based on two measures (overlap and "measure of blur" similar to intensity variance) computed in the cortical areas. They remarked that their evaluation involving 18 images could not affirm that the nonrigid methods they examined can perform better in registering major cortical sulci.
- Study of Klein et al. [66]: the study compared 14 registration algorithms using four data sets by performing over 2168 image registrations per algorithm and computing various evaluation measures and performing statistical analyses on the obtained evaluation measures. Top-ranking methods were SyN (Symmetric Normalization) [8]), ART*, IRTK [105], and SPM's DARTEL toolbox [4], with the first two giving top scores in all tests and for all label sets. SyN is recommended if high accuracy is desired; ROMEO† and Diffeomorhpic Demons [131] are more efficient alternatives for time-sensitive tasks.

22.6 Conclusions

This chapter reviewed the various techniques proposed for MIR, as well as those relating to registration evaluation. In summary, key tasks of a registration procedure include (1) characterization of the registration solution Section 22.3.2; (2) definition of one or a set of criterion to quantify the optimality of the solution (Section 22.3.3; and (3) design and employment of an optimization strategy for finding the solution (Section 22.3.4.2). As there is

* Automatic Registration Toolbox. http://www.nitrc.org/projects/art/

† http://ralyx.inria.fr/2006/Raweb/visages/uid27.html

currently no generic registration algorithm that works in all applications, choices of data (dis)similarity terms and optimization strategy remain highly problem-specific.

There remain many open questions in MIR. For instance, MIR is challenged by the need for tuning a potentially long list of parameters that relate to optimization and feature extraction settings, as well as the weights between data dissimilarity and regularization terms. Furthermore, when operated on large images (e.g., vector- or tensor-valued images and dynamic sequences), MIR is burdened by heavy computation and memory demands. Future research works that aim at addressing these issues are therefore both interesting and important.

References

1. O. Acosta, J. Fripp, A. Rueda, D. Xiao, E. Bonner, P. Bourgeat, and O. Salvado. 3D shape context surface registration for cortical mapping. In *Proceedings of the 2010 IEEE International Conference on Biomedical Imaging: From Nano to Macro, ISBI'10*, IEEE Press, Piscataway, NJ, 2010, pp. 1021–1024.
2. A. Pluim, J. Maintz, and M. Viergever. Mutual information based registration of medical images: A survey. *IEEE Transactions on Medical Imaging*, 21:61–75, 2003.
3. V. Arsigny, X. Pennec, and N. Ayache. Polyrigid and polyaffine transformations: A new class of diffeomorphisms for locally rigid or affine registration. In *Proceedings of MICCAI*, Springer, Heidelberg, Germany, 2003, pp. 829–837.
4. J. Ashburner. A fast diffeomorphic image registration algorithm. *Neuroimage*, 38(1):95–113, 2007.
5. B. Avants, C. Epstein, M. Grossman, and J. Gee. Symmetric diffeomorphic image registration with cross-correlation: Evaluating automated labeling of elderly and neurodegenerative brain. *Medical Image Analysis*, 12(1):26–41, 2008. Special Issue on The Third International Workshop on Biomedical Image Registration—WBIR 2006.
6. B. Avants, N. Tustison, and G. Song. Advanced normalization tools: V1.0. *Insight Journal*, http://hdl.handle.net/10380/3113, July 2009.
7. B. B. Avants, P. T. Schoenemann, and J. C. Gee. Lagrangian frame diffeomorphic image registration: Morphometric comparison of human and chimpanzee cortex. *Medical Image Analysis*, 10(3):397–412, 2006. Special Issue on The Second International Workshop on Biomedical Image Registration (WBIR'03).
8. B. B. Avants, N. J. Tustison, G. Song, P. A. Cook, A. Klein, and J. C. Gee. A reproducible evaluation of ANTS similarity metric performance in brain image registration. *Neuroimage*, 54(3):2033–2044, 2011.
9. A. Azar, C. Xu, X. Pennec, and N. Ayache. An interactive hybrid non-rigid registration framework for 3D medical images. In *Third IEEE International Symposium on Biomedical Imaging: Nano to Macro, 2006*, Paris, France, April 2006, pp. 824–827.
10. S. Baker, D. Scharstein, J. P. Lewis, S. Roth, M. J. Black, and R. Szeliski. A database and evaluation methodology for optical flow. *International Journal of Computer Vision*, 92(1):1–31, 2011.

11. M. Beg and A. Khan. Symmetric data attachment terms for large deformation image registration. *IEEE Transactions on Medical Imaging*, 26(9):1179–1189, September 2007.
12. M. F. Beg, M. I. Miller, A. Trouvé, and L. Younes. Computing large deformation metric mappings via geodesic flows of diffeomorphisms. *International Journal of Computer Vision*, 61:139–157, February 2005.
13. E. T. Bender and W. A. Tomé. The utilization of consistency metrics for error analysis in deformable image registration. *Physics in Medicine and Biology*, 54(18):5561, 2009.
14. Y. Boykov and V. Kolmogorov. An experimental comparison of min-cut/max-flow algorithms for energy minimization in vision. In *Proceedings of the Third International Workshop on Energy Minimization Methods in Computer Vision and Pattern Recognition, EMMCVPR '01*, Springer-Verlag, London, U.K., 2001, pp. 359–374.
15. M. Bro-Nielsen. Medical image registration and surgery simulation. PhD thesis, IMM-DTU, 1996.
16. M. Bro-Nielsen. Rigid registration of CT, MR and cryosection images using a GLCM framework. In J. Troccaz, E. Grimson, and R. Mösges, eds., *CVRMed-MRCAS'97*, vol. 1205 of *Lecture Notes in Computer Science*, Springer, Berlin, Germany, 1997, pp. 171–180. 10.1007/BFb0029236.
17. M. Bro-Nielsen and C. Gramkow. Fast fluid registration of medical images. In *Proc. Visualization in Biomedical Computing (VBC'96)*, September, Springer Lecture Notes in Computer Science, Hamburg, Germany, vol. 1131, 1996, pp. 267–276.
18. L. G. Brown. A survey of image registration techniques. *ACM Computing Surveys*, 24:325–376, December 1992.
19. P. Cachier, J.-F. Mangin, X. Pennec, D. Rivière, D. Papadopoulos-Orfanos, J. Régis, and N. Ayache. Multisubject non-rigid registration of brain MRI using intensity and geometric features. In *Proceedings of MICCAI*, Utrecht, the Netherlands, 2001, pp. 734–742.
20. N. D. Cahill, J. A. Noble, and D. J. Hawkes. A demons algorithm for image registration with locally adaptive regularization. In *Proceedings of the 12th International Conference on Medical Image Computing and Computer-Assisted Intervention: Part I, MICCAI '09*, Springer-Verlag, Berlin, Germany, 2009, pp. 574–581.
21. K. Cao, K. Ding, G. E. Christensen, M. L. Raghavan, R. E. Amelon, and J. M. Reinhardt. Unifying vascular information in intensity-based nonrigid lung CT registration. In *WBIR*, Springer-Verlag, Berlin, Germany, 2010, pp. 1–12.
22. F. Chen and D. Suter. Image coordinate transformation based on div-curl vector splines. In *ICPR'98*, IEEE Computer Society Press, Brisbane, QLD, Australia, 1998, pp. 518–520.
23. W. Cheung and G. Hamarneh. n-sift: n-dimensional scale invariant feature transform. *IEEE Transactions on Image Processing*, 18(9):2012–2021, September 2009.
24. D. Chillet, J. Jomier, D. Cool, and S. Aylward. Vascular atlas formation using a vessel-to-image affine registration method. In R. Ellis and T. Peters, eds., *Medical Image Computing and Computer-Assisted Intervention—MICCAI 2003*, vol. 2878 of *Lecture Notes in Computer Science*, Springer, Berlin, Germany, 2003, pp. 335–342.
25. G. Christensen, P. Yin, M. W. Vannier, K. Chao, J. F. Dempsey, and J. Williamson. Large-deformation image registration using fluid landmarks. In *Fourth IEEE Southwest Symposium on Image Analysis and Interpretation*, 2000, pp. 269–273.
26. G. E. Christensen and H. J. Johnson. Consistent image registration. *IEEE Transaction on Medical Imaging*, 20(7):568–582, 2001.

27. G. E. Christensen and H. J. Johnson. Invertibility and transitivity analysis for nonrigid image registration. *Journal of Electronic Imaging*, 12(1):106–117, 2003.
28. G. E. Christensen, S. C. Joshi, M. I. Miller, and S. Member. Volumetric transformation of brain anatomy. *IEEE Transactions on Medical Imaging*, 16:864–877, 1997.
29. G. E. Christensen, R. D. Rabbitt, and M. I. Miller. 3D brain mapping using a deformable neuroanatomy. *Physics in Medicine and Biology*, 39(3):609–618, 1994.
30. H. Chui, J. Rambo, J. Duncan, R. Schultz, and A. Rangarajan. Registration of cortical anatomical structures via robust 3D point matching. In A. Kuba, M. áamal, and A. Todd-Pokropek, eds., *Information Processing in Medical Imaging*, vol. 1613 of *Lecture Notes in Computer Science*, 1999, pp. 168–181.
31. C. A. C. Coello. A comprehensive survey of evolutionary-based multiobjective optimization techniques. *Knowledge and Information Systems*, 1:269–308, 1998.
32. D. L. Collins and A. C. Evans. ANIMAL: Validation and applications of nonlinear registration-based segmentation. *IJPRAI*, 11(8):1271–1294, 1997.
33. K. K. Delibasis, P. A. Asvestas, and G. K. Matsopoulos. Automatic point correspondence using an artificial immune system optimization technique for medical image registration. *Computerized Medical Imaging and Graphics* (in press).
34. M. Droske and M. Rumpf. Multiscale joint segmentation and registration of image morphology. *IEEE Transactions on Pattern Analysis and Machine Intelligence*, 29(12):2181–2194, December 2007.
35. S. Du, N. Zheng, S. Ying, and J. Liu. Affine iterative closest point algorithm for point set registration. *Pattern Recognition Letters*, 31(9):791–799, 2010.
36. D. Eggert, A. Lorusso, and R. Fisher. Estimating 3-d rigid body transformations: a comparison of four major algorithms. *Machine Vision and Applications*, 9:272–290, 1997. 10.1007/s001380050048.
37. A. El-Baz, A. Farag, G. Gimel'farb, and A. Abdel-Hakim. Image alignment using learning prior appearance model. In *2006 IEEE International Conference on Image Processing*, Atlanta, Georgia, October 2006, pp. 341–344.
38. J. M. Fitzpatrick, D. L. G. Hill, and C. R. Maurer. Image registration. In J. M. Fitzpatrick and M. Sonka, eds., *Handbook of Medical Imaging, Vol. 2. Medical Image Processing and Analysis (SPIE Press Book)*. Academic Press, Inc., San Diego, CA, 2000, pp. 449–506.
39. J. Flusser. Moment invariants in image analysis. In *Proceedings of the International Conference on Computer Science. ICCS'06*, Reading, U.K., 2006, pp. 196–201.
40. F. Maes, A. Collignon, D. Vandermeulen, G. Marchal, and P. Suetens. Multimodality image registration by maximization of mutual information. *IEEE Transaction on Medical Imaging*, 16:187–189, 1997.
41. C. Frohn-Schauf, S. Henn, and K. Witsch. Multigrid based total variation image registration. *Computing and Visualization in Science*, 11:101–113, 2008. 10.1007/s00791-007-0060-2.
42. S. Gefen, O. Tretiak, and J. Nissanov. Elastic 3D alignment of rat brain histological images. *IEEE Transactions on Medical Imaging*, 22(11):1480–1489, November 2003.
43. A. Gholipour, N. Kehtarnavaz, S. Yousefi, K. Gopinath, and R. Briggs. Symmetric deformable image registration via optimization of information theoretic measures. *Image and Vision Computing*, 28(6):965–975, 2010.
44. A. Gholipour, N. D. Kehtarnavaz, R. W. Briggs, M. Devous, and K. S. Gopinath. Brain functional localization: A survey of image registration techniques. *IEEE Transaction Medical Imaging*, 26(4):427–451, 2007.

45. E. Gibson, A. Khan, and M. Beg. A combined surface and volumetric registration (SAVOR) framework to study cortical biomarkers and volumetric imaging data. In G.-Z. Yang, D. Hawkes, D. Rueckert, A. Noble, and C. Taylor, eds., *Medical Image Computing and Computer-Assisted Intervention MICCAI 2009*, vol. 5761 of *Lecture Notes in Computer Science*. Springer, Berlin, Germany, 2009, pp. 713–720.
46. B. Glocker, N. Komodakis, G. Tziritas, N. Navab, and N. Paragios. Dense image registration through mrfs and efficient linear programming. *Medical Image Analysis*, 12(6):731–741, 2008.
47. B. Glocker, A. Sotiras, N. Komodakis, and N. Paragios. Deformable medical image registration: Setting the state of the art with discrete methods. *Annual Review of Biomedical Engineering*, 13(1):null, 2011.
48. E. Haber, S. Heldmann, and J. Modersitzki. A computational framework for image-based constrained registration. *Linear Algebra and its Applications*, 431(3–4):459–470, 2009. Special Issue in honor of Henk van der Vorst.
49. E. Haber and J. Modersitzki. Intensity gradient based registration and fusion of multi-modal images. *Methods of Information in Medicine*, 46(3):292–299, 2007.
50. T. Hartkens, D. L. G. Hill, A. D. Castellano-Smith, D. J. Hawkes, C. R. Maurer, Jr., A. J. Martin, W. A. Hall, H. Liu, and C. L. Truwit. Using points and surfaces to improve voxel-based non-rigid registration. In *Proceedings of the Fifth International Conference on Medical Image Computing and Computer-Assisted Intervention-Part II*, MICCAI '02, Springer-Verlag, London, U.K., 2002, pp. 565–572.
51. J. Ho, M.-H. Yang, A. Rangarajan, and B. Vemuri. A new affine registration algorithm for matching 2d point sets. In *Applications of Computer Vision, 2007. WACV '07. IEEE Workshop on*, Austin, TX, February 2007, p. 25.
52. M. Holden. A review of geometric transformations for nonrigid body registration. *IEEE Transactions on Medical Imaging*, 27(1):111–128, January 2008.
53. B. K. P. Horn and B. G. Schunck. "Determining optical flow": A retrospective. *Artificial Intelligence*, 59(1–2):81–87, 1993.
54. H.-Y. Hsiao, H. mei Chen, T.-H. Lin, C.-Y. Hsieh, M.-Y. Chu, G. Liao, and H. Zhong. A new parametric nonrigid image registration method based on Helmholtz's theorem. In J. M. Reinhardt and J. P. W. Pluim, eds., *Medical Imaging 2008: Image Processing*, vol. 6914, 2008, pp. 69142W. SPIE.
55. Y.-W. Huang, C.-Y. Chen, C.-H. Tsai, C.-F. Shen, and L.-G. Chen. Survey on block matching motion estimation algorithms and architectures with new results. *The Journal of VLSI Signal Processing*, 42:297–320, 2006.
56. L. Ibanez, W. Schroeder, L. Ng, and J. Cates. *The ITK Software Guide*. Kitware, Inc. ISBN 1-930934-15-7, http://www.itk.org/ItkSoftwareGuide.pdf, 2nd edn., 2005.
57. H. Ishikawa. Higher-order clique reduction in binary graph cut. In *IEEE Conference on Computer Vision and Pattern Recognition*, Miami, FL, 2009, pp. 20–25.
58. V. Jain and H. Zhang. Robust 3d shape correspondence in the spectral domain. In *IEEE International Conference on Shape Modeling and Applications, 2006. SMI 2006*, Washington, DC, June 2006, p. 19.
59. G. Janssens, L. Jacques, J. O. de Xivry, X. Geets, and B. M. Macq. Diffeomorphic registration of images with variable contrast enhancement. *International Journal of Biomedical Imaging*, 2011, http://dx.doi.org/10.1155/2011/891585, 2011.
60. A. Jarc, P. Rogelj, and S. Kovacic. Texture feature based image registration. In *Biomedical Imaging: From Nano to Macro, 2007. ISBI 2007. 4th IEEE International Symposium on*, Arlington, VA, April 2007, pp. 17–20.

61. P. Jassi, G. Hamarneh, and L. Tang. Simulation of ground-truth validation data via physically- and statistically-based warps. In *Proceedings of MICCAI*, Springer, Heidelberg, Germany, 2008, pp. 459–467.
62. J. Chen, S. Caputlu-Wilson, H. Shi, J. Galt, T. Faber, and E. Garcia. Automated quality control of emission-transmission misalignment for attenuation correction in myocardial perfusion imaging with SPECT-CT. *Journal of Nuclear Cardiology*, 13(1):43–49, January–February 2006.
63. S. Joshi and M. Miller. Landmark matching via large deformation diffeomorphisms. *IEEE Transactions on Image Processing*, 9(8):1357–1370, August 2000.
64. Y. jun Wang, G. Farnebäck, and C.-F. Westin. Multi-affine registration using local polynomial expansion. *Journal of Zhejiang University—Science C*, 11(7):495–503, 2010.
65. J.-S. Kim, J.-M. Lee, Y.-H. Lee, J.-S. Kim, I.-Y. Kim, and S. I. Kim. Intensity based affine registration including feature similarity for spatial normalization. *Computers in Biology and Medicine*, 32(5):389–402, 2002.
66. A. Klein, J. Andersson, B. A. Ardekani, J. Ashburner, B. Avants, M.-C. Chiang, G. E. Christensen et al. Evaluation of 14 nonlinear deformation algorithms applied to human brain MRI registration. *Neuroimage*, 46(3):786–802, 2009.
67. S. Klein. Optimisation methods for medical image registration, PhD thesis, Image Sciences Institute, UMC Utrecht, the Netherlands, 2008.
68. V. Kolmogorov and C. Rother. Minimizing nonsubmodular functions with graph cuts-a review. *IEEE Transactions on Pattern Analysis and Machine Intelligence*, 29(7):1274–1279, July 2007.
69. N. Komodakis, G. Tziritas, and N. Paragios. Performance vs. computational efficiency for optimizing single and dynamic MRFs: Setting the state of the art with primal-dual strategies. *Computer Vision and Image Understanding*, 112(1):14–29, 2008. Special Issue on Discrete Optimization in Computer Vision.
70. M. P. Kumar, P. H. S. Torr, and A. Zisserman. Learning layered motion segmentation of video. In *Proceedings of the Tenth IEEE International Conference on Computer Vision (ICCV'05) Volume 1–Volume 01*, IEEE Computer Society. Washington, DC, 2005, pp. 33–40.
71. D. Kwon, K. J. Lee, I. D. Yun, and S. U. Lee. Nonrigid image registration using dynamic higher-order MRF model. In ECCV '08: *Proceedings of the 10th European Conference on Computer Vision*, Springer-Verlag, Berlin, Germany, 2008, pp. 373–386.
72. T. Lehmann, C. Gonner, and K. Spitzer. Survey: Interpolation methods in medical image processing. *IEEE Transactions on Medical Imaging*, 18(11):1049–1075, November 1999.
73. T. Lehmann, C. Gonner, and K. Spitzer. Addendum: B-spline interpolation in medical image processing. *IEEE Transactions on Medical Imaging*, 20(7):660–665, July 2001.
74. V. Lempitsky, C. Rother, S. Roth, and A. Blake. Fusion moves for Markov random field optimization. *IEEE Transactions on Pattern Analysis and Machine Intelligence*, 32:1392–1405, 2010.
75. S. Liao and A. Chung. A novel multi-layer framework for non-rigid image registration. In *2010 IEEE International Symposium on Biomedical Imaging: From Nano to Macro*, Varna, Bulgaria, April 2010, pp. 344–347.
76. C. Liu, W. Freeman, E. Adelson, and Y. Weiss. Human-assisted motion annotation. In *IEEE Conference on Computer Vision and Pattern Recognition, 2008. CVPR 2008*, Anchorage, Alaska, June 2008, pp. 1–8.

77. W. Liu and E. Ribeiro. A survey on image-based continuum-body motion estimation. *Image and Vision Computing* (in press).
78. D. Loeckx, P. Slagmolen, F. Maes, D. Vandermeulen, and P. Suetens. Nonrigid image registration using conditional mutual information. *IEEE Transactions on Medical Imaging*, 29(1):19–29, January 2010.
79. H. Lombaert, Y. Sun, and F. Cheriet. Landmark-based non-rigid registration via graph cuts. In *International Conference on Image Analysis and Recognition*. Springer-Verlag, August 2007.
80. H. Lu, P. Cattin, and M. Reyes. A hybrid multimodal non-rigid registration of MR images based on diffeomorphic demons. In *Engineering in Medicine and Biology Society (EMBC), 2010 Annual International Conference of the IEEE*, vol. 31, San-Diego, CA, September 2010, pp. 5951–5954.
81. H. Luan, F. Qi, Z. Xue, L. Chen, and D. Shen. Multimodality image registration by maximization of quantitative-qualitative measure of mutual information. *Pattern Recognition*, 41(1):285–298, 2008.
82. Maes, D. Vandermeulen, and P. Suetens. Medical image registration using mutual information. *Proceedings of the IEEE*, 91(10):1699–1722, October 2003.
83. J. Maintz and M. A. Viergever. A survey of medical image registration. *Medical Image Analysis*, 2(1):1–36, 1998.
84. U. Malsch, C. Thieke, P. E. Huber, and R. Bendl. An enhanced block matching algorithm for fast elastic registration in adaptive radiotherapy. *Physics in Medicine and Biology*, 51(19):4789, 2006.
85. T. Mansi, X. Pennec, M. Sermesant, H. Delingette, and N. Ayache. ILogDemons: A demons-based registration algorithm for tracking incompressible elastic biological tissues. *International Journal of Computer Vision*, 92:92–111, March 2011.
86. D. Mattes, D. Haynor, H. Vesselle, and T. Lewellen. PET-CT image registration in the chest using free-form deformations. *IEEE Transaction on Medical Imaging*, 22(1):120–128, January 2003.
87. C. Metz, S. Klein, M. Schaap, T. van Walsum, and W. J. Niessen. Nonrigid registration of dynamic medical imaging data using nD + t B-splines and a groupwise optimization approach. *Medical Image Analysis*, 15(2):238–249, 2011.
88. H. Mirzaalian, G. Hamarneh, and T. Lee. Graph-based approach to skin mole matching incorporating template-normalized coordinates. In *IEEE Conference on Computer Vision and Pattern Recognition, (CVPR)*, Miami, FL, June 2009, pp. 2152–2159.
89. T. Makela, P. Clarysse, O. Sipil, N. Pauna, Q. C. Pham, T. Katila, and I. E. Magnin. A review of cardiac image registration methods. *IEEE Transaction on Medical Imaging*, 21(9):1011–1021, 2002.
90. J. Modersitzki. *Numerical Methods for Image Registration*. Oxford University Press, New York, 2004.
91. J. Modersitzki. Flirt with rigidity: Image registration with a local non-rigidity penalty. *International Journal of Computer Vision*, 76:153–163, 2008.
92. J. Modersitzki. *FAIR: Flexible Algorithms for Image Registration*. SIAM, Philadelphia, PA, 2009.
93. J. Modersitzki and S. Wirtz. Combining homogenization and registration. In *Biomedical Image Registration*, vol. 4057. Springer, Berlin, Germany, 2006, pp. 257–263.
94. K. Murphy, B. van Ginneken, J. M. Reinhardt, S. Kabus, K. Ding, X. Deng, and J. P. Pluim. Evaluation of methods for pulmonary image registration: The EMPIRE10 study. In *Medical Image Analysis for the Clinic—A Grand Challenge*, 14(4):527–5382, 2010.

95. A. Myronenko and X. Song. Intensity-based image registration by minimizing residual complexity. *IEEE Transactions on Medical Imaging*, 29(11):1882–1891, November 2010.
96. X. Pennec, N. Ayache, and J.-P. Thirion. Landmark-based registration using differential geometric features. In I. N. Bankman and PhD, eds., *Handbook of Medical Image Processing and Analysis*, 2nd edn. Academic Press, Burlington, VT, 2009, pp. 499–513.
97. X. Pennec, P. Cachier, and N. Ayache. Understanding the "demon's algorithm": 3D non-rigid registration by gradient descent. In *Proceedings of MICCAI*, Cambridge, U.K., 1999, pp. 597–605.
98. X. Pennec, R. Stefanescu, V. Arsigny, P. Fillard, and N. Ayache. Riemannian elasticity: A statistical regularization framework for non-linear registration. In *Proceeding of MICCAI (2)*, Palm Springs, CA, 2005, pp. 943–950.
99. J.-M. Peyrat, H. Delingette, M. Sermesant, C. Xu, and N. Ayache. Registration of 4d cardiac CT sequences under trajectory constraints with multichannel diffeomorphic demons. *IEEE Transactions on Medical Imaging*, 29(7):1351–1368, July 2010.
100. A. Pitiot, E. Bardinet, P. M. Thompson, and G. Malandain. Piecewise affine registration of biological images for volume reconstruction. *Medical Image Analysis*, 10(3):465–483, 2006.
101. J. P. W. Pluim, J. B. A. Maintz, and M. A. Viergever. f-information measures in medical image registration. *Transactions on Medical Imaging*, 23(12):1508–1516, 2004.
102. G. Postelnicu, L. Zollei, and B. Fischl. Combined volumetric and surface registration. *IEEE Transactions on Medical Imaging*, 28(4):508–522, April 2009.
103. P. Risholm, E. Samset, and W. Wells. Bayesian estimation of deformation and elastic parameters in non-rigid registration. *Biomedical Image Registration*, 6204:104–115, 07 2010.
104. T. Rohlfing, C. R. M. Jr., D. A. Bluemke, and M. A. Jacobs. Volume-preserving nonrigid registration of MR breast images using free-form deformation with an incompressibility constraint. *IEEE Transactions on Medical Imaging*, 22:730–741, 2003.
105. D. Rueckert. IRTK: *Image Registration Toolkit*. Visual Information Processing Group, Imperial College, London, U.K., 2009.
106. D. Rueckert, P. Aljabar, R. A. Heckemann, J. Hajnal, and A. Hammers. Diffeomorphic registration using B-splines. In *9th International Conference on Medical Image Computing and Computer-Assisted Intervention (MICCAI 2006)*, Copenhagen, Denmark, October 2006.
107. D. Rueckert, L. I. Sonoda, C. Hayes, D. L. G. Hill, M. O. Leach, and D. J. Hawkes. Non-rigid registration using free-form deformations: Application to breast MR images. *IEEE Transaction on Medical Imaging*, 18(8):712–721, 1999.
108. C. Seiler, X. Pennec, L. Ritacco, and M. Reyes. Femur specific polyaffine model to regularize the log-domain demons registration. In *SPIE Medical Imaging (Image Processing)*, Orlando, FL, February 2011.
109. A. Shekhovtsov, I. Kovtun, and V. Hlavac. Efficient MRF deformation model for non-rigid image matching. *Computer Vision Image Understanding*, 112(1):91–99, 2008.
110. D. Shen. Image registration by hierarchical matching of local spatial intensity histograms. Lecture Notes in Computer Science, In *Proceeding of MICCAI*, 3216, 2004, pp. 582–590.
111. D. Shen and C. Davatzikos. HAMMER: Hierarchical attribute matching mechanism for elastic registration. *IEEE Transaction on Medical Imaging*, 21:1421–1439, 2002.
112. D. Skerl, B. Likar, and F. Pernus. A protocol for evaluation of similarity measures for non-rigid registration. *Medical Image Analysis*, 12(1):42–54, 2008.

113. P. Slomka and R. Baum. Multimodality image registration with software: state-of-the-art. *European Journal of Nuclear Medicine and Molecular Imaging*, 36:44–55, 2009.
114. R. Stefanescu, X. Pennec, and N. Ayache. Grid powered nonlinear image registration with locally adaptive regularization. *Medical Image Analysis*, 8(3):325–342, 2004. Medical Image Computing and Computer-Assisted Intervention—MICCAI 2003.
115. M. Styner and G. Gerig. Evaluation of 2D/3D bias correction with 1 + 1 ES-optimization. Technical report, BIWI, ETH Zurich, Switzerland, 2000.
116. E. Suarez, C.-F. Westin, E. Rovaris, and J. Ruiz-Alzola. Nonrigid registration using regularized matching weighted by local structure. In *Fifth International Conference on Medical Image Computing and Computer-Assisted Intervention (MICCAI'02)*, Tokyo, Japan, 2003.
117. L. Tang and G. Hamarneh. Efficient search and adaptive image similarity for multi-modal deformable registration using random walks. In *Proceedings of MICCAI*, Springer, Nagoya, Japan, 2013, pp. 1–8.
118. L. Tang, G. Hamarneh, and R. Abugharbieh. Reliability-driven, spatially-adaptive regularization for deformable registration. In *WBIR'10*, Springer, Heidelberg, Germany, July 2010, pp. 173–185.
119. L. Tang, G. Hamarneh, and A. Celler. Validation of mutual information-based registration of CT and bone SPECT images in dual-isotope studies. *Computer Methods and Programs in Biomedicine*, 92:173–185, November 2008.
120. T. Tang and A. Chung. Non-rigid image registration using graph-cuts. In *MICCAI (1)*, Springer Verlag, Berlin, Germany, 2007, pp. 916–924.
121. L. Tang, A. Hero, and G. Hamarneh. Locally-adaptive similarity metric for deformable medical image registration. In *Proceedings of Biomedical Imaging: From Nano to Macro (ISBI'12)*, Barcelona, Spain, May 2012, pp. 728–731.
122. N. Tatsuya. A two-dimensional, finite element analysis of vasogenic brain edema. *Neurologia Medico-Chirurgica*, 30(1):1–9, 1990.
123. J.-P. Thirion. Image matching as a diffusion process: An analogy with Maxwell's demons. *Medical Image Analysis*, 2(3):243–260, 1998.
124. P. M. Thompson and A. W. Toga. Warping strategies for intersubject registration. In I. N. Bankman PhD, ed., *Handbook of Medical Image Processing and Analysis*, 2nd edn. Academic Press, Burlington, VT, 2009, pp. 643–673.
125. P. K. Turaga, A. Veeraraghavan, and R. Chellappa. Unsupervised view and rate invariant clustering of video sequences. *Computer Vision and Image Understanding*, 113:353–371, 2009.
126. N. J. Tustison, B. B. Avants, and J. C. Gee. Directly manipulated free-form deformation image registration. *IEEE Transactions on Image Processing*, 18(3):624–635, 2009.
127. T. ur Rehman, E. Haber, G. Pryor, J. Melonakos, and A. Tannenbaum. 3d nonrigid registration via optimal mass transport on the GPU. *Medical Image Analysis*, 13(6):931–940, 2009. Includes Special Section on Computational Biomechanics for Medicine.
128. M. Urschler, J. Bauer, H. Ditt, and H. Bischof. SIFT and shape context for feature-based nonlinear registration of thoracic CT images. In *CVAMIA'06*, 2006, pp. 73–84.
129. M. van de Giessen, G. Streekstra, S. Strackee, M. Maas, K. Grimbergen, L. van Vliet, and F. Vos. Constrained registration of the wrist joint. *IEEE Transactions on Medical Imaging*, 28(12):1861–1869, December 2009.
130. O. van Kaick, H. Zhang, G. Hamarneh, and D. Cohen-Or. A survey on shape correspondence. In *Proceedings of Eurographics State-of-the-Art Report*, 2010, pp. 1–23.

131. T. Vercauteren, X. Pennec, A. Perchant, and N. Ayache. Non-parametric diffeomorphic image registration with the demons algorithm. In N. Ayache, S. Ourselin, and A. Maeder, eds., *Medical Image Computing and Computer Assisted Intervention (MICCAI'07)*, vol. 4792/2007 of *Lecture Notes in Computer Science*, Brisbane, QLD, Australia, 2007, pp. 319–326.
132. T. Vercauteren, X. Pennec, A. Perchant, and N. Ayache. Symmetric log-domain diffeomorphic registration: A demons-based approach. In *Medical Image Computing and Computer Assisted Intervention*, New York, 2008.
133. T. Vercauteren, X. Pennec, A. Perchant, and N. Ayache. Diffeomorphic demons: Efficient non-parametric image registration. *Neuroimage*, 45, 2009.
134. S. Vishnukumar and M. Wilscy. A comparative study on adaptive local image registration methods. In *2009 International Conference on Signal Processing Systems*, May 2009, pp. 140–145.
135. M. Wachowiak and T. Peters. High-performance medical image registration using new optimization techniques. *IEEE Transactions on Information Technology in Biomedicine*, 10(2):344–353, April 2006.
136. V. Walimbe and R. Shekhar. Automatic elastic image registration by interpolation of 3D rotations and translations from discrete rigid-body transformations. *Medical Image Analysis*, 10(6):899–914, 2006.
137. J. Walters-Williams and Y. Li. Estimation of mutual information: A survey. In P. Wen, Y. Li, L. Polkowski, Y. Yao, S. Tsumoto, and G. Wang, eds., *Rough Sets and Knowledge Technology*, vol. 5589 of *Lecture Notes in Computer Science*. Springer, Berlin, Germany, 2009, pp. 389–396.
138. F. Wang, B. C. Vemuri, M. Rao, and Y. Chen. A new and robust information theoretic measure and its application to image alignment. In *IPMI*, Ambleside, U.K., 2003, pp. 388–400.
139. Q. Wang, L. Chen, P.-T. Yap, G. Wu, and D. Shen. Groupwise registration based on hierarchical image clustering and atlas synthesis. *Human Brain Mapping*, 31(8):1128–1140, 2010.
140. J. Weickert, A. Bruhn, T. Brox, and N. Papenberg. A survey on variational optic flow methods for small displacements. In H.-G. Bock, F. Hoog, A. Friedman, A. Gupta, H. Neunzert, W. R. Pulley-blank, T. Rusten, F. Santosa, A.-K. Tornberg, V. Capasso, R. Mattheij, H. Neunzert, O. Scherzer, and O. Scherzer, eds., *Mathematical Models for Registration and Applications to Medical Imaging*, volume 10 of *Mathematics in Industry*. Springer, Berlin, Germany, 2006, pp. 103–136.
141. J. West. Comparison and evaluation of retrospective intermodality brain image registration techniques. *Journal of Computer Assisted Tomography*, 21:554–566, 1997.
142. J. Wong, C. Studholme, D. Hawkes, and M. Maisey. Evaluation of the limits of visual detection of image misregistration in a brain fluorine-18 FDG PET-MRI study. *European Journal of Nuclear Medicine*, 24(6):642–650, June 1997.
143. R. Woods, J. Mazziotta, and S. Cherry. MRI-PET registration with automated algorithm. *Journal of Computer Assisted Tomography*, 17:536–546, 1993.
144. R. P. Woods. Within-modality registration using intensity-based cost functions. In I. N. Bankman and PhD, eds., *Handbook of Medical Image Processing and Analysis*, 2nd edn. Academic Press, Burlington, VT, 2009, pp. 605–611.
145. G. Wu, F. Qi, and D. Shen. Learning best features and deformation statistics for hierarchical registration of MR brain images. In *IPMI'07: Proceedings of the 20th International Conference on Information Processing in Medical Imaging*, Springer-Verlag, Berlin, Germany, 2007, pp. 160–171.

146. G. Wu, Q. Wang, H. Jia, and D. Shen. Feature-based groupwise registration by hierarchical anatomical correspondence detection. *Human Brain Mapping*, 2011.
147. G. Wu, P.-T. Yap, M. Kim, and D. Shen. TPS-HAMMER: Improving HAMMER registration algorithm by soft correspondence matching and thin-plate splines based deformation interpolation. *Neuroimage*, 49(3):2225–2233, 2010.
148. G. Wu, P.-T. Yap, Q. Wang, and D. Shen. Groupwise registration from exemplar to group mean: Extending hammer to groupwise registration. In *ISBI*, Paris, France, 2010, pp. 396–399.
149. X. Han. Feature-constrained nonlinear registration of lung CT images. *In MICCAI 2010 Grand Challenges in Medical Image Analysis*: Evaluation of Methods for Pulmonary Image Registration (EMPIRE 10), Beijing, China, 2010.
150. Z. Xue, D. Shen, and C. Davatzikos. Correspondence detection using wavelet-based attribute vectors. In *Proceedings of MICCAI (2)*, Springer-Verlag, Berlin, Germany, 2003, pp. 762–770.
151. X. Yang, W. Birkfellner, and P. Niederer. A similarity measure based on Tchebichef moments for 2D/3D medical image registration. *International Congress Series*, 1268:153–158, 2004. CARS 2004—Computer Assisted Radiology and Surgery. Proceedings of the 18th International Congress and Exhibition.
152. M. A. Yassa and C. E. Stark. A quantitative evaluation of cross-participant registration techniques for MRI studies of the medial temporal lobe. *Neuroimage*, 44(2):319–327, 2009.
153. B. T. T. Yeo, M. R. Sabuncu, T. Vercauteren, N. Ayache, B. Fischl, and P. Golland. Spherical demons: Fast diffeomorphic landmark-free surface registration. *IEEE Transaction on Medicine Imaging*, 29(3):650–668, 2010.
154. L. S. Yin, L. Tang, G. Hamarneh, B. Gill, A. Celler, S. Shcherbinin, T. F. Fua, A. Thompson, M. Liu, C. Duzenli, F. Sheehan, and V. Moiseenko. Complexity and accuracy of image registration methods in SPECT-guided radiation therapy. *Physics in Medicine and Biology*, 55(1):237, 2010.
155. Y. Yin, E. A. Hoffman, K. Ding, J. M. Reinhardt, and C.-L. Lin. A cubic B-spline-based hybrid registration of lung CT images for a dynamic airway geometric model with large deformation. *Physics in Medicine and Biology*, 56(1):203–218, September 2011.
156. Y. Zeng, C. Wang, Y. Wang, X. Gu, D. Samaras, and N. Paragios. Dense non-rigid surface registration using high-order graph matching. In *2010 IEEE Conference on Computer Vision and Pattern Recognition (CVPR)*, June 2010, pp. 382–389.
157. G. Zhao, L. Chen, and G. Chen. A speeded-up local descriptor for dense stereo matching. In *2009 16th IEEE International Conference on Image Processing (ICIP)*, Cairo, Egypt, November 2009, pp. 2101–2104.
158. X. Zhuang, C. Yao, Y. Ma, D. Hawkes, G. Penney, and S. Ourselin. Registration-based propagation for whole heart segmentation from compounded 3D echocardiography. In *2010 IEEE International Symposium on Biomedical Imaging: From Nano to Macro*, April 2010, pp. 1093–1096.
159. D. Zikic, A. Kamen, and N. Navab. Unifying characterization of deformable registration methods based on the inherent parametrization. In *WBIR'10*, 2010, pp. 161–172.
160. B. Zitova and J. Flusser. Image registration methods: A survey. *Image and Vision Computing*, 21(11):977–1000, October 2003.

23

Medical Image Segmentation: Energy Minimization and Deformable Models

Chris McIntosh and Ghassan Hamarneh

CONTENTS

This chapter surveys the field of energy minimization as it applies to medical image segmentation (MIS). MIS remains a daunting task but one whose solution will allow for the automatic extraction of important structures, organs, and diagnostic features from medical images, with applications to computer-aided diagnosis, statistical shape analysis, and medical image visualization. Several classifications of segmentation techniques exist, including edge-, pixel-, and region-based techniques, in addition to clustering, and graph-theoretic approaches (Pham et al., 2000; Robb, 2000; Sonka and Fitzpatrick, 2000; Yoo, 2004). However, the unreliability of traditional, purely pixel-based methods in the face of shape variation and noise has caused recent trends (Pham et al., 2000) to focus on incorporating prior knowledge about the location, intensity, and shape of the target anatomy (Hamarneh et al., 2001). One type of approach that has been of particular interest to meeting these requirements is that of energy minimization methods due to their inherent ability to allow multiple competing goals to be considered.

In energy minimization methods, a function evaluates the goodness of a segmentation for a particular image, and the minimization of that function yields the segmentation of the image (Figure 23.1). Though highly variant in nature, the application of energy minimization methods to MIS is commonly built on five primary building blocks: (i) the energy function, (ii) the

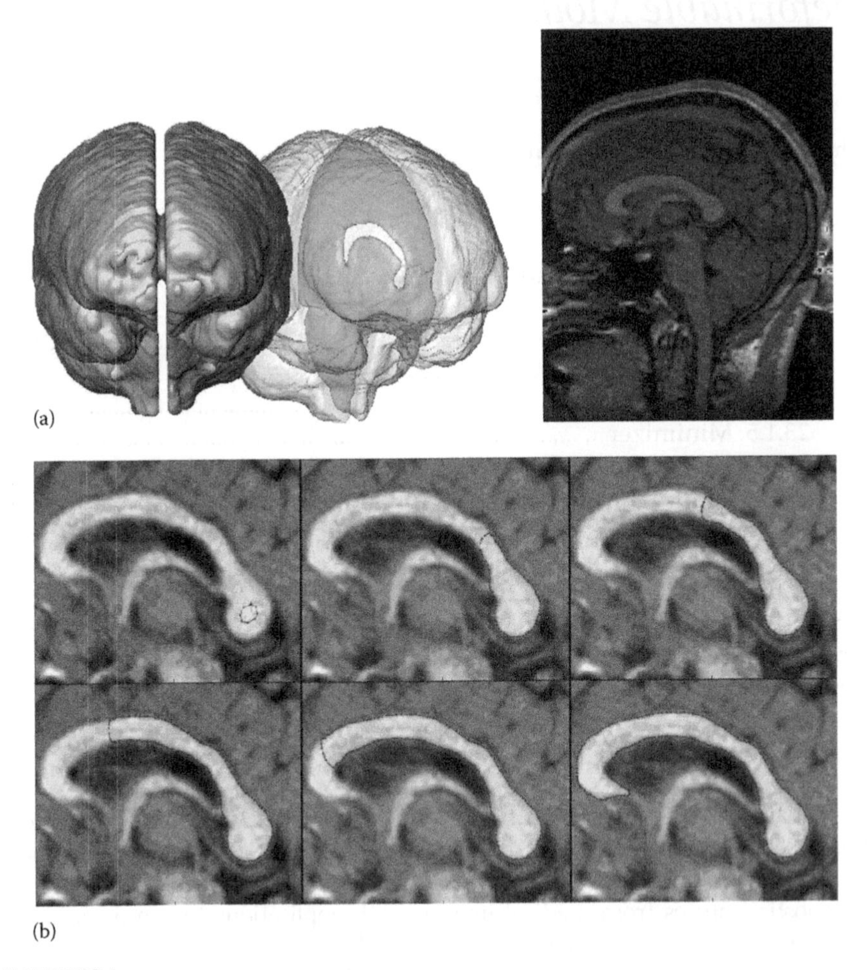

FIGURE 23.1
A corpus callosum (CC). (a) The CC is the band of nerve fiber tissue connecting the left and right hemispheres of the brain. (b) An energy minimization segmentation process. A shape model with progressively lower energy (left to right), showing a minimum of the energy function in bottom right. (Adapted from McIntosh, C. and Hamarneh, G., Vessel crawlers: 3D physically-based deformable organisms for vasculature segmentation and analysis, *IEEE Computer Society Conference on Computer Vision and Pattern Recognition, 2006*, Vol. 1, pp. 1084–1091, 2006a; Hamarneh, G. and McIntosh, C., Physically and statistically based deformable models for medical image analysis (Chapter 11), *Deformable Models: Biomedical and Clinical Applications*, pp. 335–386, 2007.)

segmentation representation, (iii) the image representation, (iv) the training data, (v) and the minimizer. In what follows we provide an overview of each building block, and the major works therein over the past few decades.*

23.1 Definitions and Foundations

We first give an outline of the energy minimization for MIS process. We define a medical image I_i and its corresponding segmentation (i.e., pixel labels) S_i, each having N pixels. Then $\mathbf{I} = \{I_1, I_2, \ldots, I_N\}$ and $\mathbf{S} = \{S_1, S_2, \ldots, S_N\}$ are sets of images and their corresponding ground-truth segmentations. In a slight abuse of the notation, we occasionally omit the subscript i from I_i and S_i for clarity and instead use I and S.

The first step in any energy minimization problem is the identification of the form of the energy function. In the next section, we will briefly group some popular energy terms into three main categories: boundary, region, and shape. Boundary terms are concerned primarily with the object boundary, region terms with the region inside or outside the object, and shape terms with the shape of the object. Other energy terms include spatial constraints on multipart objects, for example, containment (or layering), exclusion, or the number of labels (Delong et al., 2012; Nosrati and Hamarneh, 2013; Ulen et al., 2013; Yazdanpanah et al., 2011). Here we use these groupings to build a general energy functional. It may be convex or non-convex, as can the shape space over which it will be minimized. A general form is $E(S, I, \mathbf{w}) = w_1 \times boundary(S, I) + w_2 \times region(S, I) + w_3 \times shape\,Prior(S)$, with free parameter $\mathbf{w} = [w_1, w_2, w_3]$. Depending on the value of $\mathbf{w}$, minima of E tend toward best satisfying the boundary, region, or shape terms. We note that the boundary and region terms are often referred to as external terms, since they involve cues external to the shape model, while the shape prior is deemed an internal term. More generally, we can write

$$E(S, I, \mathbf{w}) = w_1 J_1(S, I) + w_2 J_2(S, I) + \cdots + w_n J_n(S, I), \tag{23.1}$$

where

J_i is the ith energy term
$\mathbf{w} = [w_1, \ldots, w_n]$ are the weights

We note here that depending on the nature of S, E may be called an energy function or an energy functional, with the latter indicating S itself is a function.

The segmentation problem is to solve

$$S^* = \arg\min_S E(S \mid I, \mathbf{w}), \tag{23.2}$$

which involves choosing a $\mathbf{w}$ and, depending on the nature of the energy functional, may also require training appearance and/or shape priors using

* This chapter summarizes and builds upon McIntosh (2012).

I and/or **S** and setting an initialization. A gradient-descent-based solver is typically used but combinatorial approaches have also been explored for discretized versions of the problem (Boykov and Kolmogorov, 2003) (see Section 23.5 for details). In either case, non-convexity, or supermodularity for combinatorial problems, can be quite problematic. There is no guarantee that another solution does not exist that better minimizes the energy and thus is potentially a better segmentation (Figure 23.2). Ideally both functional and shape spaces are convex, guaranteeing globally optimal solutions. However, whether convex or not the ground-truth segmentation, S_i, for image, I_i is not guaranteed to be an optima (local or global) of $E(S \mid I_i, \mathbf{w})$. The goal, in general, is to build an $E(S \mid I_i, \mathbf{w})$ such that S^* is as close as possible to S_i under some definition of closeness [e.g., (Dice, 1945; Jaccard, 1901; Tanimoto, 1957)].

One of the earliest developed, and perhaps most recognized, examples of energy minimization methods being applied to image segmentation is that of deformable models. Deformable models for MIS gained popularity since the 1987 introduction of *snakes* by Terzopoulos et al. (Kass et al., 1987; Terzopoulos, 1987). At a high level, energy-minimizing deformable models work by deforming a user provided initial shape to fit to a target structure in a medical image. Shape-changing deformations result from the minimization, with respect to the shape, of a cost function measuring how plausible is the shape model and how well it aligns with the boundaries of the target anatomy in the image. Since the shape model itself is most commonly represented by a function, the cost function is often termed an energy functional and its gradient is derived using methods from variational calculus. The shape deformations are therefore typically simulated by solving an initial value problem using gradient-descent optimization algorithms (Elsgolc, 1962, G2). One further development, "Deformable organisms," uses artificial life modeling techniques to augment the energy-minimizing deformable models with models of cognition, behaviors, and sensory input (Hamarneh et al., 2001, 2009).

We now begin our more in-depth discussion of the different components of the energy minimization for MIS process.

23.1.1 Energy Function

Since snakes paved the way, there have been numerous papers attempting to increase accuracy by contributing novel energy terms, each designed to address a particular class of images or alleviate a particular problem with the original terms. As there have been far too many proposals to survey them all, here we focus on the key terms, many of which themselves have spawned numerous new approaches to MIS or changed the way we think about the problem at hand. As much of the initial development focused on external energy terms, namely, on the boundary and region terms that deal with the image processing itself, we begin our discussion there.

One of the most fundamental problems noted with snakes relates to their boundary capture range. If placed near a strong edge in an image, the

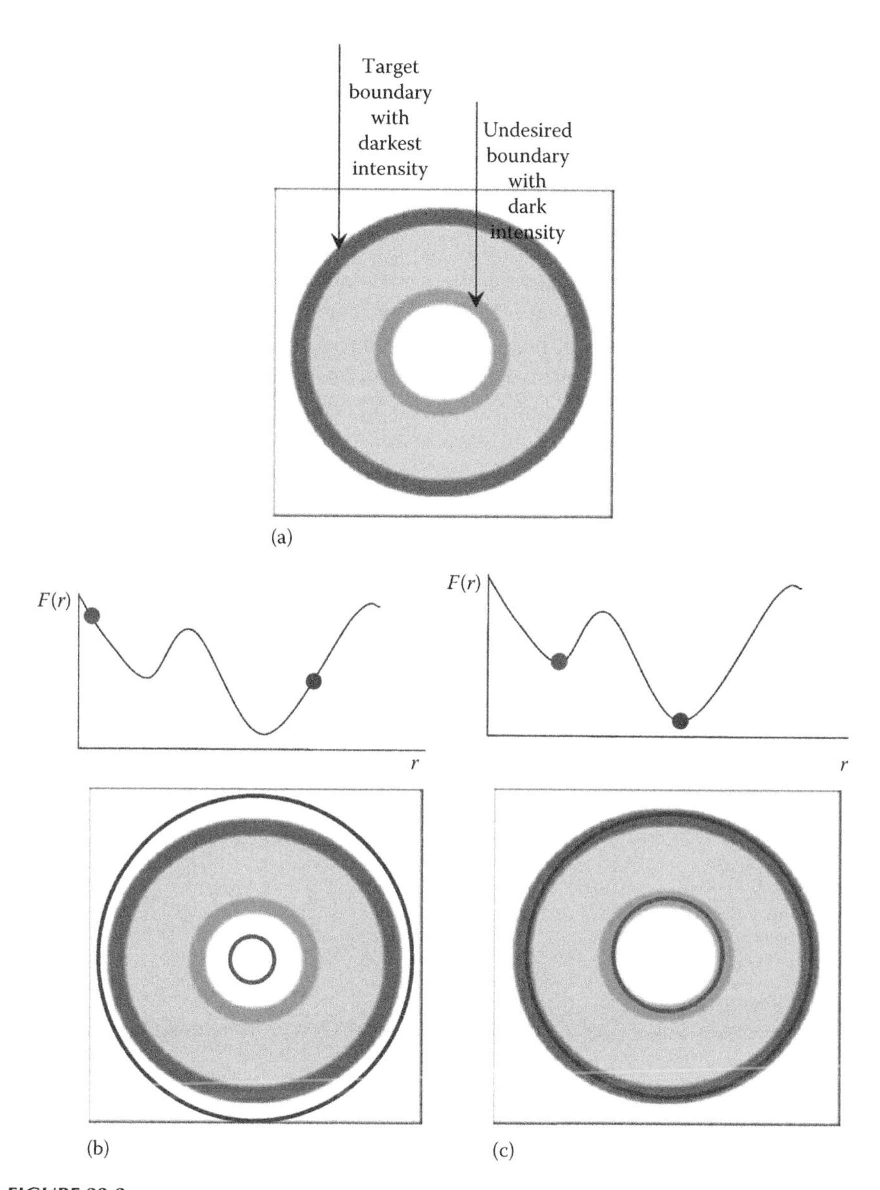

FIGURE 23.2
Synthetic example of single parameter deformable model with local minima. The circular deformable model's only parameter is its radius r. The energy function $F(r)$ is minimal for the circle with darkest average intensity. The input image is shown in (a) with the darkest outmost boundary representing the global minima. In (b) two deformable models are initialized. In (c) after gradient descent, each model has moved to the nearest minima. (Reprinted from McIntosh, C., Energy functionals for medical image segmentation: Choices and consequences, PhD dissertation, Simon Fraser University, Burnaby, BC, Canada, Copyright 2011. With permission.)

contour would quickly *snap* to the edge, but if initialized farther away the influence of the edge would not reach the contour. There are, in fact, numerous causes of this problem relating to not only the external energy terms (Caselles et al., 1997; Cohen, 1991; Xu and Prince, 1998) but also the way the segmentation was originally represented (an explicit contour—see Section 23.1.2 for details) and the minimization technique being used.

An early attempt to rectify the aforementioned problem was by adding a deflation or inflation force to the contour that would attempt to shrink/grow it toward edges (Cohen, 1991). Rather than rely on a constant force, gradient vector flows extend the influence of edges off into homogeneous regions, thus increasing the capture range (Xu and Prince, 1998). Geodesic active contours (GACs) were similarly developed by both Caselles et al. (1997) and Yezzi et al. (1997). The approach of Caselles et al. formulated a deformable model optimization problem as that of finding the optimal path in a Riemannian space. Termed GACs, these popular deformable models work by minimizing curve length, where length is measured as the geodesic distance on a Riemannian manifold defined via an edge indicator function. The shortest curve is then, by definition, the curve along the edges of the image, and thus GACs automatically shrink the curve to the edges. GACs have become the canonical example of boundary-based deformable models.

However, what about objects whose boundaries blurred due to their inherent nature or noise, like the example in Figure 23.3? In these cases the intensity statistics of the areas both inside and outside the contour can be used to attempt to divide the image into maximally separated regions. The approaches

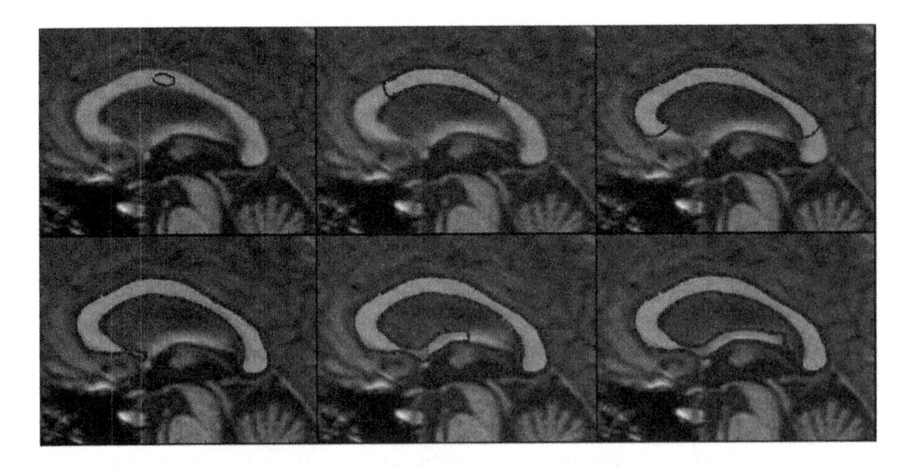

FIGURE 23.3
An energy minimization segmentation process. A shape model with progressively lower energy (left to right), showing a minima of the energy function in bottom right. Notice the leaking into neighboring structures that occurs as a result of weak edge strength. (Reprinted from McIntosh, C. and Hamarneh, G., 2006b, Genetic algorithm driven statistically deformed models for medical image segmentation, *ACM Workshop on Medical Applications of Genetic and Evolutionary Computation Workshop (MedGEC), in Conjunction with the Genetic and Evolutionary Computation Conference (GECCO)*, Seattle, WA, 2006b. With permission.)

of Chan et al. and of Yezzi et al. (Chan and Vese, 2001; Tsai et al., 2001) were similarly developed. Both are modeled after the Mumford–Shah functional wherein images are approximated by piecewise-smooth functions (Mumford and Shah, 1989). The approach of Chan and Vese is referred to as active contours without edges (ACWE) and has become a popular example of energy minimization for image segmentation. However, in their initial formulations, both methods approximate images by piecewise-constant functions, that is, objects are assumed to have a constant intensity. When objects are noisy, or their intensity changes gradually, a piecewise-smooth approximation is better suited, and thus an extension to piecewise-smooth functions was developed (Chan et al., 2007).

When both boundary- and region-derived statistics are not enough, shape-based terms are used. Shape terms provide resilience to false boundaries by heavily penalizing the implausible shape configurations that the false boundaries imply. The most basic terms attempt to achieve boundary smoothness by minimizing curve length, segmentation area, or curvature. More advanced terms compare the shape of the segmentation to some prior model of the shape in an effort to minimize their dissimilarity (Cootes et al., 1992, 1995, 2001; Cootes and Taylor, 1997; Cremers et al., 2001, 2002, 2008; Dambreville et al., 2006; Etyngier et al., 2007; Leventon et al., 2000; Paragios et al., 2006; Pohl et al., 2007; Vu and Manjunath, 2008; Warfield et al., 2000) (see Section 23.1.4 for discussions).

There has even been work trying to combine edge terms, piecewise-constant terms, and shape terms into a single formulation (Bresson et al., 2006). However, with so many competing assumptions about the object and the image inherent in each formulation, some level of trade-off must be established in the resulting functional (i.e., the weights, **w**, must be set).

If not appropriately set, the weights **w** can cause significant error. In fact, our results demonstrated that optimizing the weights has dramatic effects, reducing error in large data sets by as much as 60% (McIntosh and Hamarneh, 2007). However, optimizing the weights by hand for even a single image can be a long and tedious task, with no real guarantee of obtaining the correct segmentation. As such, there has been a number of works that seek to automatically set the weights (Anguelov et al., 2005; Finley and Joachims, 2008; Kolmogorov et al., 2007; McIntosh and Hamarneh, 2007, 2009; Szummer et al., 2008; Taskar et al., 2005 Rao et al., 2009, 2010).

Instead of *guessing* the optimal weights, suppose we write a function $\gamma(\mathbf{w}|I_j, S_j)$ evaluating how well weight **w** works for a given image-segmentation pair (I_j, S_j), such that a parameter is deemed better when it causes S^* to approach S_j, that is, the minimum of E to be the correct segmentation. Given S_j, we could then calculate the ideal weights for a particular image I_j by solving $\mathbf{w}^* = \arg\min_{\mathbf{w}} \gamma(\mathbf{w} \mid I_j, S_j)$. It is important that γ itself be convex or globally solvable in **w**. If γ was not globally solvable, uncertainty would remain in that another $\mathbf{w}^*$ may better minimize γ and thus better segment the image. Similarly, γ cannot contain free parameters, else those parameters would themselves introduce uncertainty, as was the case in McIntosh and Hamarneh (2007).

Recent advances in maximum margin estimation allow for weight optimization wherein the parameters of the objective function are sought such that the highest scoring structures (in our case segmentations) are as close as possible to the ground truth (Anguelov et al., 2005; Finley and Joachims, 2008; Szummer et al., 2008; Taskar et al., 2005). These methods do however assume that a single set of weights works for the entire test set of images, whereas different images can easily require different weights. In contrast, Kolmogorov et al. seek an optimal parameter on a per-image basis. Given a parameter range, their method simultaneously solves the objective function for a set of parameters that bound how the parameters influence the solution (Kolmogorov et al., 2007). Each solution is then treated as a potential segmentation. They propose a number of heuristics, including user intervention, to select the best segmentation from a set of potential ones.

A related topic is that of multiobjective optimization, where methods try to jointly minimize the terms, without weighting them. As the topic relates more to minimizing energy functionals, we defer its discussion until Section 23.1.5.

23.1.2 Segmentation Representation

As expected, all functions must have domains and so in turn energy functionals must have domains. That domain is the space of possible segmentations of the image or shapes. There are many different ways to represent the underlying segmentation, and that choice in turn impacts the image and shape terms that can be readily evaluated. For example, a shape using closed contours pairs most readily with region statistics. Here we summarize some of the most prevalent shape representations in use.

Naturally, we start with the representation originally detailed by Terzopolous et al., namely the explicit contour model (Kass et al., 1987; Terzopoulos, 1987). The contour is defined as an explicit, parameterized function of its arc length. Integrating along the arc-length integrates along the contour and allows for the evaluation of both internal and external terms (see Appendix B for details). These methods were extended to represent not only contours but surfaces and volumes (Cohen et al., 1992; Cohen and Cohen, 1993; Delingette et al. 1992; McInerney and Terzopoulos, 1995a; Miller et al., 1991; Staib and Duncan, 1992b; Whitaker, 1994). Other explicit models were introduced relating to spring-mass systems, where each boundary point is a mass, connected to other masses via springs (McInerney and Terzopoulos, 1996; Montagnat et al., 2001). There can be other types of masses, namely, internal nodes and medial nodes (Pizer et al., 2003). However, in general these aforementioned works are constrained to a fixed topology, which posed problems in many applications.

There have been two major directions of work to address topological adaptability. The first direction focused on novel ways to automatically re-parameterize the contour or surface enabling the evolution into complex geometries (McInerney and Terzopoulos, 1995b,c, 1999, 2000). Initially developed in 2D, *T-snakes* work by subdividing the image domain into a Freudenthal triangulation (McInerney and Terzopoulos, 1995b,c). At each deformation step, the intersections of the contours with the triangular grids are found, and a set of rules is followed to determine if the contour point should be split or merged. This work was later extended to 3D, with the introduction of *T-surfaces* (McInerney and Terzopoulos, 1999, 2000). Delingette simultaneously developed a somewhat related approach where simplex meshes are used to model the shape, and topological changes are performed "semi-automatically with an automatic detection of topological problems, but a manual validation of all operations" (Delingette, 1999). However, in both cases the explicit re-parameterization of the contour or surface can be costly and may not generalize well to even higher dimensions.

In contrast, implicit contours were built from the ground up to handle changes in topology (Caselles et al., 1997; Osher and Paragios, 2003; Osher and Sethian, 1988; Sethian, 1996). In these representations, the boundary is implied by a given function, instead of explicitly defined. Implicit representations are built around the signed distance function (SDF), where the object boundary is defined as the zero level set of the function. Integrating over the domain of the SDF implicitly integrates over the contour and thus allows for the evaluation of both internal and external terms, as before. Changes in topology are handled internally by the shape representation and require no re-parameterization of the model (Figure 23.4). Similarly, other functions can be used to implicitly represent shapes including characteristic functions (Tsai et al., 2004) and probability maps (Cremers et al., 2008).

More recently, graph methods have emerged where the segmentation is represented by the assignment of labels to a graph (Barrett and Mortensen, 1997; Boykov and Funka-Lea, 2006; Boykov and Jolly, 2001; Boykov and Kolmogorov, 2003; Falcão et al., 1998; Falcão and Udupa, 1997; Grady, 2006; Mortensen and Barrett, 1998; Shi and Malik, 2000). Pixels are represented as vertices in the graphs, and edges between pixel–vertices represent a connectivity neighborhood. Energy functionals, often called cost functions in graph-based works, are expressed as sums over the vertices and their neighborhoods that vary as a function of how each vertex is labeled.

Finally, our discussion thus far has been limited to that of a single object class, but it is also possible to represent multiple object classes at one time, using so-called multi-class shape representations that are built on both implicit (Paragios and Deriche, 2000, 2002; Samson et al., 2000; Vese and Chan, 2001; Zhao et al., 1996) and graph-based representations (Boykov and Jolly, 2001; Grady and Funka-Lea, 2004). Implicit shape models are adapted to multiple classes by defining multiple implicit functions, where the combinations and intersections of the functions denote which class is being represented.

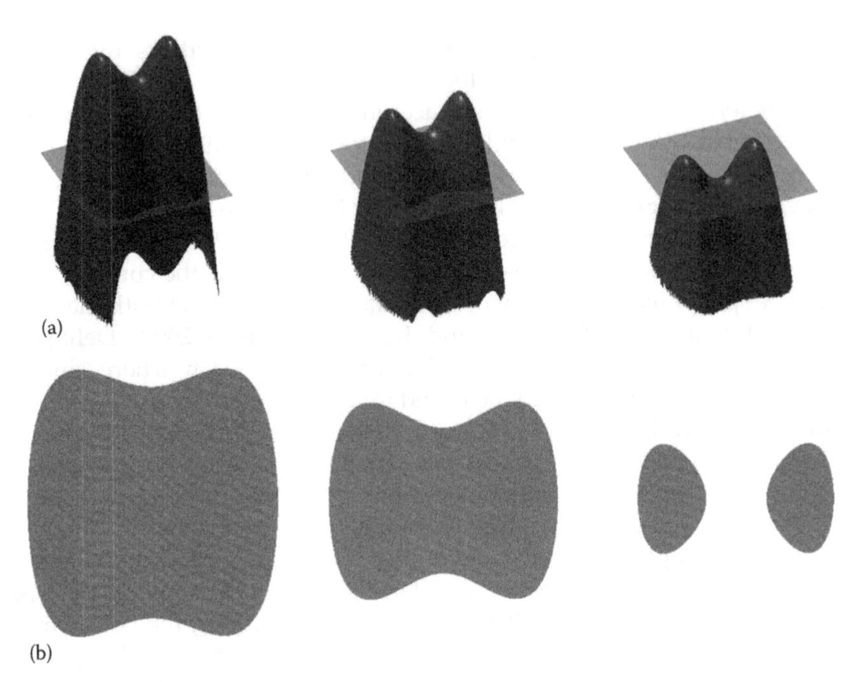

FIGURE 23.4
An exemplar SDF as a shape representation. (a) An SDF undergoing a simple deformation from left to right. (b) The zero level sets of the SDFs, displaying the segmentation each SDF represents. Notice how the topology automatically changes without any re-parameterization. (Reprinted from McIntosh, C., Energy functionals for medical image segmentation: Choices and consequences, PhD dissertation, Simon Fraser University, Burnaby, BC, Canada, Copyright 2011. With permission.)

Graph methods, however, are extended by increasing the number of possible labels for each vertex.

In the end, the choice of representation is ultimately determined by the desired segmentation task. As such, it is worth noting that specialized models can exist, where a shape representation is designed specifically for one type of anatomy. A popular example is that of tubular-branching objects, namely, vessels, where cylindrical models can be built and deformed (McIntosh and Hamarneh, 2006; O'Donnell et al., 1994).

23.1.3 Image Representation

As might be expected, how the image is represented will impact how the terms of the energy functional can be evaluated on it. In fact, early on there were many contributions demonstrating how existing techniques could be modified to fit higher dimensional data, vector-valued data (Ballerini, 2001; Chan et al., 2000; Sapiro, 1996; Shah, 1996; Shi et al., 2008), tensor data (Nand et al., 2011; Rousson et al., 2004; Wang and Vemuri, 2004; Weldeselassie and Hamarneh, 2007) (Chapter 19), texture-heavy images (Paragios and Deriche, 2000, 2002), or even temporal data (Saad et al., 2008 Hamarneh et al., 2004; Tang et al., 2012; Rana et al., 2009; Ng et al., 2012), as opposed to the static, 2D, grayscale images early algorithms were presented on. Furthermore, numerous methods have been adapted to handle the intricacies of specific medical imaging modalities including magnetic resonance images and ultrasound. As an example, we talk briefly about some of the issues inherent to angiography in Section 23.2.

23.1.4 Training Data

A substantial amount of knowledge is often available about anatomical structures of interest—characteristic shape, position, orientation, symmetry, relationship to neighboring structures, associated landmarks, etc. and about plausible image intensity characteristics, subject to natural biological variability or the presence of pathology. Once collected, the training data typically come into the energy functional in the place of shape priors (Cootes et al., 1995), but appearance priors (Cootes et al., 2001) have also been developed (see Heimann and Meinzer (2009) for a complete survey). As shape priors have been a particular area of interest in the field, here we discuss some of the key works.

In many applications, prior knowledge about object shape variability is available or can be obtained by studying a training set of shape examples. This knowledge restricts the space of allowable deformations to a learned shape space that approximates the space of anatomically feasible shapes (Cootes and Taylor, 1997; Cootes et al., 1992, 1995, 2001; Cremers et al., 2001, 2002, 2008; Dambreville et al., 2006; Etyngier et al., 2007; Leventon et al., 2000; Paragios et al., 2006; Pohl et al., 2007; Vu and Manjunath, 2008; Warfield et al., 2000). One of the most notable works in this area is that of Cootes et al., where they introduced and refined active shape models (ASM) (Cootes et al., 1992, 1993, 1995). In ASM, principal component analysis (PCA) is calculated over a set of landmark points extracted from training shapes. The resulting principal components are used to construct a point distribution model (PDM) and an allowable shape domain (ASD). In a natural extension to their previous work, Cootes et al. modify their method to include image intensity statistics (Cootes et al., 2001). Staib and Duncan constrained the deformable models in Fourier space by conforming to probability distributions of the parameters of an elliptic Fourier decomposition of the boundary (Staib and Duncan, 1992a). Statistical prior shape knowledge was also incorporated in implicit, level set-based deformable models. Leventon et al. introduced statistical shape priors by using PCA to capture the main modes of variation of the level set representation (Leventon et al., 2000). However, as Pohl et al. point out, level sets do not form a vector space and hence more accurate shape statistics could be captured by transforming the shapes into a vector space using the logarithm of odds before performing PCA (Pohl et al., 2007).

Though simpler to optimize than their nonlinear counterparts, linear models of shape deformation may not always adequately represent the variance observed in real data. Linear shape models assume the data lies on a linear manifold, but shapes often lie on nonlinear manifolds where the manifold's properties are not accurately captured by linear statistics (Fletcher et al., 2004). For example, try fitting an ellipse to an "S"-like shape space. In order to include the entire letter, extraneous white space (nonvalid shapes) must also be included. Nonlinear shape models have been introduced to address this problem (Cootes and Taylor, 1997; Cremers, 2008; Cremers et al., 2001, 2002; Dambreville et al., 2006; Etyngier et al., 2007; Fletcher et al., 2004; Sozou et al., 1995).

However, we argue that the problem with linear statistics, as described earlier, is not necessarily due to the application of a linear model to nonlinear data but rather because of the implicit nature in which the statistics are applied (McIntosh and Hamarneh, 2011). By *implicit* we mean the statistics attempt to model variation in the shape, rather than variation in the parameters governing the deformations themselves. Note that we are not referring to the shape representation being implicit or explicit but instead whether the deformations are implicitly or explicitly studied. Implicit shape statistics result from the majority of previous deformable model approaches adopting a boundary-based shape representation, aside from a few exceptions (Pizer, 2003). As a consequence, the statistics are calculated using boundary models of the shape instead of models representing the interior and skeletal topology of the structures. Studying the underlying structural changes of a shape allows deformations that were previously non-convex to be decomposed into linear models. We refer to these as *explicit* shape statistics since they are calculated over the very parameters responsible for varying the object's shape. Consider an object represented by a single pixel. Different images of the object show the pixel moving around in a circle. A nonlinear function is required to describe the pixel's motion and hence no linear statistics can capture the motion adequately as long as it is the object's x, y position being studied. However, once decomposed into a function of sin and cos, the underlying parameter that controls the objects variability is linear in its variation, and hence linear statistics will yield greatly improved shape statistics. The same argument carries forward, albeit more complexly, to a more complex object. A simple bending of a shape's medial axis is a linear deformation under the appropriate representation, as it is simply a rotation of some of the medial nodes. However, the bending is a highly non-convex deformation once embedded in the image domain, as either an implicit shape (Pohl et al., 2007) or an explicit boundary-based model (Cootes et al., 1995).

23.1.5 Minimizer

Once the image and shape representations have been set and the problem formulated, all that is left is to solve the minimization process and obtain the resulting segmentation. Though it may sound simple enough, this area has been a major focus of criticism of energy minimization-based methods over the years, and as such has recently become one of the most focused areas for research. Specifically, in their original inception, many of the aforementioned methods were plagued by problems of local minima and sensitivity to initialization. Here we describe the changes and revelations in the field on this topic over the past decade. For a broader review, the interested reader is referred to the following representative, but far from comprehensive, list: Cremers et al., 2011; Grady and Polimeni, 2010; Kolmogorov and Zabin, 2004; and Nikolova et al., 2006.

Energy functional minimization can be carried out in a variety of ways. One solution is to perform explicit differentiation under the Euler–Lagrange equations, where each new application with a modified energy functional must be accompanied by one such derivation (Caselles et al., 1997; Chan and Vese, 2001; Kass et al., 1987; Terzopoulos, 1987) (see Appendix A for details). The result is a set of equations, which, if satisfied, guarantee a stationary point of the energy functional. The solution is then obtained through a gradient-descent process where the change in the shape model (with respect to an artificial time variable) is equated to the Euler–Lagrange equation, that is, the deformable models come to rest when the equations are satisfied (Kass et al., 1987; Terzopoulos, 1987). There are, however, two main drawbacks with this approach. Firstly, performing gradient descent in the presence of image noise can lead to instability in the deformations over time (Sundaramoorthi et al., 2007). Secondly, as the number of dependent variables (shape, location, scale, orientation, etc.) increases so does the complexity of the search space, which often increases the number of local minima and requires the calculation of an increasingly large number of derivatives.

On the issue of stability, there has been work by Sundaramoorthi et al. on reformulating the gradient flow using Sobolev-type inner products, which induce favorable regularity prosperities into the flow, thus bringing smoother deformations over time (Sundaramoorthi et al., 2007). Recently Bar and Sapiro introduced a Newton-type method built on more generalized norms than the L^2 norm, obtained by equating the Euler–Lagrange constraint to an artificial time variable (Bar and Sapiro, 2009). They include the Sobolev norm and demonstrate improved results over L^2 norms, with faster convergence and more accuracy in the presence of noise. Unless used on convex problems, however, these methods are still prone to local minima.

If local minima do not suffice, global minima must be sought, and as such there have been a number of recent approaches to obtaining the global optima of energy functionals (Andrews et al., 2011b; Appleton and Talbot, 2006; Barrett and Mortensen, 1997; Boykov and Funka-Lea, 2006; Boykov and Kolmogorov, 2003; Bresson et al., 2007; Cremers et al., 2008; Falcão and Udupa, 1997; Falcão et al., 1998; Grady, 2006; Li and Yezzi, 2006; Mortensen and Barrett, 1998; Nikolova et al., 2006). There are three primary directions toward this goal: min-paths, min-cuts, and convex approximations.

Min-path techniques are formulated on the basis of Dijkstra's algorithm for finding the shortest path in an undirected graph with nonnegative edge weights. They were first presented for 2D interactive segmentation (Barrett and Mortensen, 1997; Falcão and Udupa, 1997), extended to 3D (Falcão and Udupa, 1997), and later specialized to 4D for vessel segmentation (Li and Yezzi, 2006; Poon et al., 2008, kawahara et al., 2013, Wink et al., 2004) and 6D (or higher) for spinal cord segmentation (Kawahara et al., 2013).

Graph cuts were demonstrated as a global minimization technique for a popular energy functional (Caselles et al., 1997), as a special case of computing a geodesic on a Riemannian space whose metric is computed from the image (Boykov and Kolmogorov, 2003). However, graph cuts have been shown to apply only to a restricted class of energy functionals that are submodular (Kolmogorov and Zabin, 2004), and their solutions are discrete approximations to the continuous formulations whose accuracy is dependent on the resolution of the approximating graph (Boykov and Kolmogorov, 2003). Naturally, as that resolution increases so does their running time. Random walkers were developed in a similar nature, solving image segmentation as a graph problem wherein the global optimum is obtained to a particular cost function (Grady, 2006). In fact, graph cuts and random walkers have been shown to be specific instantiations of a single framework (Sinop and Grady, 2007).

Another line of work has come from the relaxation of the underlying shape model from a non-convex space to a convex one, thereby defining convex energy functionals that can then be minimized instead of their non-convex counterparts. This convex relaxation work, which began in 2004 with a simple restricted class of functionals (Nikolova et al., 2006), was later extended to a broader class in Bresson et al. (2007), and then a similar work appeared in 2008 with the addition of a shape prior (Cremers et al., 2008). However, restrictions still exist in that the functionals and the shape spaces they are optimized over must be convex when defined over the relaxed space and that the relaxed shape space must itself be convex.

Though not guaranteed to find global optima, genetic algorithms (GAs) have also been applied to the minimization of energy functionals (Ballerini, 1998, 2001; Fan et al., 2002; Ibáñez et al., 2009; MacEachern and Manku, 1998; McIntosh and Hamarneh, 2011; Tohka, 2001). At a high level, GAs work by performing many simultaneous local searches, each individually optimizing the energy functional via a random walk in the search space. At the end of the process, the search that yielded the lowest value for the energy functional is adopted as the segmentation.

Ballerini extends the classical active contour models, developed by Terzopoulos (1987), by using GA to directly minimize the standard energy functional (Ballerini, 1998). Members of the GA population are hypothetical shape configurations, represented by their explicit contour locations. The method was later extended to color images by using one image term per color channel (Ballerini, 2001). MacEachern and Manku presented a similar method using a binary representation of the contour (MacEachern and Manku, 1998). Similarly, Tohka presented simplex meshes paired with image-based energies, minimized via a hybrid GA-greedy approach, and applied the technique to the segmentation of 3D medical images (Tohka, 2001). Fan et al. also developed a GA method for an explicit active contour but describe their method using Fourier descriptors and employ parallel GAs to speed up minimization (Fan et al., 2002). A different shape representation, known as topological active nets, is used by Ibáñez et al. to enable the segmentation of objects with unknown topologies or even multiple objects in the same scene (Ibáñez et al., 2009). However, aside from simple boundary smoothness constraints, all of these methods are based on classical active contour models or their variants without incorporating prior shape knowledge, making them prone to latching to erroneous edges and ill-equipped to handle gaps in object boundaries (as was discussed in Section 23.1.1).

In Hill and Taylor (1992), GAs were used with statistically based ASMs, where the parameter space consists of possible ranges of values for the pose and shape parameters of the model. The objective function to be maximized reflects the similarity between the gray levels related to the object in the search stage and those found from training. Additional works use convex, implicit, global shape statistics assuming a Gaussian distribution around a mean shape (Ghosh and Mitchell, 2006; Mignotte and Meunier, 1999; Ruff et al., 1999). Mignotte and Meunier (1999) incorporate prior shape information by defining the mean as a circular deformable template, while Ruff et al. (1999) use a PDM for occluded shape reconstruction, and Ghosh and Mitchell (2006) use a level set shape representation and a learned mean from training data. Although these techniques apply GAs to produce generations of plausible populations of shapes, the statistically based deformations are convex and may not offer the required flexibility to accommodate for nonlinear shape deformations. In McIntosh and Hamarneh (2011), we address this problem by using GAs to optimize statically based deformations that explicitly study the underlying shape variations, thus reducing the problem with linear shape statistics described in Section 23.1.4.

A somewhat related direction is that of multiobjective optimization, where each term is simultaneously optimized rather than optimizing a linear sum of the terms (23.1). Hence, in multiobjective optimization no weights are provided to combine the competing terms of the functional, instead a solution is sought for which no term can be improved without sacrificing another (Collette and Siarry, 2002). That solution is known as a *Pareto optimal* solution (Collette and Siarry, 2002). The set of Pareto optimal solutions for a given problem is known as the *Pareto front*, and there is no preference among them unless a ranking is provided between the objectives. Nakib et al. recently used multiobjective optimization to determine the parameters for a thresholding algorithm for image segmentation (Nakib et al., 2010). Hanning et al. present an approach using a piecewise approximation of the image similar to the Mumford–Shah model (Mumford and Shah, 1989) using multiobjective optimization to decide the trade-off between the number of segments and the image approximation error (Hanning et al., 2006). It is interesting to note that, for convex functions, if the ground truth lies on the Pareto front, then by definition a set of weights must exist that causes the optima to be the ground truth for the linear sum of terms formulation (23.1) (Geoffrion, 1968). In other words, a linear sum of terms model exists that yields the same segmentation as the multiobjective model, which is far more challenging to optimize. However, the necessary weights needed to achieve the desired segmentation are unknown. Weight optimization attempts to determine these weights (Section 23.1.1).

23.1.6 Related Methods

Though our discussion thus far has been focused on energy minimization methods for image segmentation that follow the prescribed building blocks, there are a few bodies of related research that follow a different path. Here we briefly detail two of them.

23.1.6.1 Relation to Bayesian Methods

Throughout this chapter, we have examined numerous methods built on energy functionals of the form $E(S|I,\mathbf{w}) = w_1 \times internal(S) + w_2 \times external(S,I)$. However, an interesting parallel to probabilistic approaches can be observed with a few simple assumptions and manipulations of this general form. Here we demonstrate that this model is actually equivalent to performing image segmentation via Bayesian inference. First, we restate the segmentation problem as

$$S^* = \arg\max_{S} P(S \mid I, \mathbf{w}), \tag{23.3}$$

where

$$P(S \mid I, \mathbf{w}) = \frac{P(I \mid S, \mathbf{w}) P(S \mid \mathbf{w})}{P(I \mid \mathbf{w})}. \tag{23.4}$$

Maximizing (23.3) is equivalent to minimizing its negative logarithm

$$S^* = \arg\min_{S} \, (-\log P(S \mid I, \mathbf{w})), \tag{23.5}$$

where = $-\log P(I|S, \mathbf{w}) - \log P(S|\mathbf{w})$, and the denominator has been removed as it has no consequence on the minimization. Now suppose we model our probabilities as

$$P(S \mid \mathbf{w}) = e^{-w_1 \times internal(S)}$$

$$P(S \mid I, \mathbf{w}) = e^{-w_2 \times external(S,I)} \tag{23.6}$$

then substituting back into (23.5) yields

$$S^* = \arg\min_S (-\log P(S \mid I, \mathbf{w})) = w_1 \times internal(S) + w_2 \times external(S, I), \qquad (23.7)$$

as before. From here it becomes possible to examine many of the approaches previously cited and see what independence assumptions they are making from a Bayesian standpoint.

23.1.6.2 Relation to Segmentation via Registration

Thus far we have discussed methods where the dependent function of the energy functional is one describing the shape of the current segmentation. However, a related field exists where the dependent function is instead a spatial transformation describing how one or more images are related. This is the field known as image registration, and it involves finding a mapping from the spatial coordinates of one image to another, identifying which pixels in a source image map to which pixels in a target image. Image registration can be used for segmentation when the mapping from a novel image is found to a training image with a known segmentation, since that mapping effectively labels the novel image. As this field is far too large and diverse to cover here, we refer the interested reader to Maintz and Viergever (1998), Zitová and Flusser (2003), and Chapter 22, for a complete survey. We do note, however, that many of the same issues discussed in this chapter, that is, global versus local optima and setting the weights for the energy functional, are also important problems in registration.

23.2 MIS via Energy Minimization

As already noted in this chapter, energy minimization methods have been applied to a wide variety of MIS problems. Two popular application domains are those of cardiac images and vascular images. As an example of how energy minimization can be applied to MIS, here we briefly discuss some key works relating to vascular segmentation. Though a complete survey of energy minimization for MIS does not exist, the interested reader is referred to McInerney and Terzopoulos (1996) for a related survey of MIS using deformable models.

One structure of particular interest in the diagnosis and understanding of many diseases is vasculature (Bullitt et al., 2003). Vessel segmentation remains an interesting application area of energy minimization methods because of its unique challenges. Firstly, the topology is complex, and as such many of the already mentioned topology-adaptive shape models were first demonstrated in their application to vessel segmentation (McInerney and Terzopoulos, 2000). Secondly, the vessels are often of very low contrast motivating advances in image terms (Frangi et al., 1999; Vasilevskiy and Siddiqi, 2002; Wink et al., 2001, 2004). For example, Vasilevskiy and Siddiqi built flux-maximizing geometric flows based on the observation that the gradient vector field near a vessel should be orthogonal to the vessel (Vasilevskiy and Siddiqi, 2002). They define a flux-maximizing geometric flow as one for which the inward normals of the underlying curve align with the direction of the gradient vector field. Near vessels the gradient vector field points inward toward the vessel centerline, and thus maximizing the flux will align the boundary of the segmentation to the boundary of the vessel. The last major challenge in vessel segmentation is that the vessels can be very thin, pushing the boundaries of numerical stability in many techniques and motivating new methods with increased stability to thin structures (Lorigo et al., 2001; Sundaramoorthi et al., 2007). For example, Lorigo et al. modify GAC to deform along the medial axis of a tubular shape, as opposed to its surface (Lorigo et al., 2001). For a complete survey of vessel segmentation techniques, the reader is referred to Lesage et al. (2009).

23.3 Discussion

Having briefly touched on each of the five primary building blocks of energy minimization methods, we conclude with a discussion of the main issues concerning their usage for MIS, namely, issues relating to how to build the energy functional, represent the segmentation, deal with the different imaging modalities themselves, train priors, and finally minimize the resulting system. We begin by touching briefly on issues relating to validating the method, as that is a fundamental step that we have not yet discussed.

Every MIS method must be validated. There are two main approaches to validate an MIS method, expert segmentations and synthetic data, each of which has its own inherit advantages and drawbacks. Expert segmentations can be time consuming and costly to obtain. Furthermore, expert segmentations suffer from both inter- and intraoperator variability because multiple experts, or even the same expert on different days, can obtain differing segmentations of the same object. Warfield et al. attempt to address problems with inter- and intraoperator variability through an expectation–maximization algorithm that weights different segmentations according to a variety of measurements and rules (Warfield et al., 2004). The key advantage to expert segmentations of real medical data is that the data is real and hence it demonstrates the applicability of the method to the problem at hand. Synthetic data can be created by either physical or computational phantoms. Physical phantoms are those physically constructed and then imaged in some manner, whereas computational phantoms are simulated using mathematical models designed to replicate human anatomy under specific imaging protocols (Cocosco et al., 1997; Hamarneh and Jassi, 2010). The main drawback in both cases is realism: segmenting a phantom well does not necessarily mean real data will also be segmented with high accuracy. The main advantage of synthetic data is certainty about the ground-truth segmentations. In Hamarneh et al. (2008), a hybrid method based on synthesizing deformation of a real data is presented. Whether real data or synthetic data, a measure of dissimilarity between the automatic segmentation and the ground truth must eventually be computed. Standard approaches for evaluating segmentation results given ground-truth segmentation include the Hausdorff distance, the Dice similarity coefficient (Dice, 1945), the Jaccard index (Jaccard, 1901), and the Tanimoto coefficient (Tanimoto, 1957).

Choosing the right energy functional for the given task is a crucial first step. Ideally, one hopes for strictly convex functions with their global minima lying at the correct segmentations for a given set of images, but this ideal scenario is rarely the case. The main challenges here are determining what energy terms could make a good functional and how to weight them in a such way as to best segment the images, In other words, one must appreciate the trade-off between the fidelity of the energy functional (how accurately it models the segmentation problems at hand) and its optimizability (how attainable is the functional's global optimum) (Hamarneh et al., 2011; Ulen et al., 2013).

There are often two main concerns when choosing the segmentation representation. Firstly, will it allow for the training of appropriate shape priors? Secondly, will it introduce problems in the minimization stage? Some shape representations do not form vector spaces, and thus performing statistics on them is difficult. Level sets (SDFs) are one such example, where a specific method for computing statistics over the resulting shape space was needed (Pohl et al., 2007). Level set-based representations can also cause problems in the minimization stage, as they do not form a convex space, and thus introduced non-convexity into the minimization problem (Cremers et al., 2008). However, they remain a popular technique due to the sub-pixel accuracy and automatic handling of topological changes. A more recent representation is the isometric log ratio (Changizi and Hamarneh, 2010) that has been used for segmentation of anatomical images (Andrews et al., 2011a,b).

Each imaging modality has its own inherent problem associated with it relating to different types of noise and spatial or temporal resolution of the structure of interest. As already exemplified with a brief treatment of blood vessel segmentation methods, each application can require customized energy functionals, segmentation representations, and optimization strategies.

When training priors, one is often forced to balance two competing goals: having the priors accurate versus having the resulting energy functional solvable. Linear statistics do not often fit the training data well but lead to easily solved energy functionals, whereas nonlinear statistics fit the data but are difficult to work with. The problem with linear statistics, however, is often not necessarily due to the application of a linear model to nonlinear data but rather due to the global aspect of the shape statistics themselves and due to the shape representation to which they are applied. Global shape statistics are those that model the shape variation of the entire shape simultaneously, that is, each shape is a single point in some high-dimensional space and the statistics (linear or not) describe some restricted set of that space. Many deformable model approaches adopt a boundary-based representation.

As a consequence the statistics are calculated using boundary models of the shape instead of models representing the interior and skeletal topology of the structures, leading to a loss of accuracy (McIntosh and Hamarneh, 2011).

The primary problem in minimization remains how to globally optimize an increasingly large set of energy functionals, as opposed to the restricted sets seen thus far (Bresson et al., 2007; Cremers et al., 2008; Kolmogorov and Zabin, 2004). Interestingly enough, even with global minima, existing functionals are not proving accurate enough, motivating the search for new energy terms or ways of building functionals that can better respect the image variability.

In short, developing MIS methods remains a daunting task from start to finish with numerous areas for research relating to each of the five building blocks commonly found in energy minimization techniques. However, with the ever-growing popularity of medical imaging for diagnosis and disease understanding, developing robust and automatic techniques for MIS remains an important goal. In order to reach that goal, we must continue to make breakthroughs in terms of accuracy and reliability of MIS methods. To do that, we need to consider where the bulk of the error and performance variability is coming from, so as to best focus our efforts there.

With fully automatic segmentation remaining an unsolved problem, user-based inspection and delineation of medical images is indispensable. This is trivial for scalar fields, but for complex manifold-valued images, even displaying these images requires careful considerations.

Realizing the challenge in achieving accurate yet fully automatic segmentation methods, recent methods opted to calculate the uncertainty in a resulting segmentation (e.g., via Shannon's entropy of probabilistic segmentation) and utilizing the uncertainty measures in active- and self-learning strategies (Andrews et al., 2011a,b; Changizi and Hamarneh; 2010, Saad et al., 2010a,b; Top et al., 2010, 2011).

Appendix 23.A: Euler–Lagrange

Starting with a function $\Phi(x)$, we build an energy functional

$$E(\Phi) = \int L\left(x, \Phi(x), \Phi_x(x), \Phi_{xx}(x)\right) \mathrm{d}x \tag{23.A.1}$$

where Φ_x is the first derivative of Φ with respect to x and Φ_{xx} the second. The functional E describes a desirable feature of Φ, in that it takes a high value when Φ is doing poorly and a low value when Φ is doing well. We know from Fermat's theorem that E obtains its extremum at a stationary point, that is, where the derivative is zero. To describe a stationary point one typically writes

$$\lim_{\tau \to 0} \frac{E(\Phi + \tau\omega) - E(\Phi)}{\tau} = 0 \tag{23.A.2}$$

where ω is a scalar, but here ω is a function, so this equation is known as the *first variation*, as opposed to fist derivative, of the functional. Developed in the 1750s, the Euler–Lagrange equation plays a fundamental role in variational calculus, defining a necessary condition under which the first variation tends to zero, and therefore the functional reaches a stationary point (Elsgolc, 1962). For a functional E as described earlier, the general form of the Euler–Lagrange differential equation is

$$\frac{\partial L}{\partial \Phi} - \frac{\partial}{\partial x}\left(\frac{\partial L}{\partial \Phi_x}\right) + \frac{\partial}{\partial x^2}\left(\frac{\partial L}{\partial \Phi_{xx}}\right) = 0 \qquad (23.A.3)$$

See Appendix B for an example of its application to energy minimization methods for MIS.

Appendix 23.B: Snakes: Details and Derivations

Classical deformable shape models (Terzopoulos, 1987) are represented as a 2D parametric contour $\mathbf{v}(s) = (x(s), y(s))$, where $s \in [0,1]$ traverses the contour (Figure 23.B.1) and $\mathbf{v}$ is deformed to fit to image data by minimizing an energy term ξ

$$\xi(\mathbf{v}(s)) = \alpha(\mathbf{v}(s)) + \beta(\mathbf{v}(s)), \qquad (23.B.1)$$

which depends on both the shape of the contour and the image data $I(x, y)$ reflected via the internal and external energy terms, $\alpha(\mathbf{v}(s))$ and $\beta(\mathbf{v}(s))$, respectively.

The internal energy term is given as

$$\alpha(\mathbf{v}(s)) = \int_0^1 \frac{1}{2}\left(w_1 \|\mathbf{v}_s\|^2 + w_2 \|\mathbf{v}_{ss}\|^2\right) ds, \qquad (23.B.2)$$

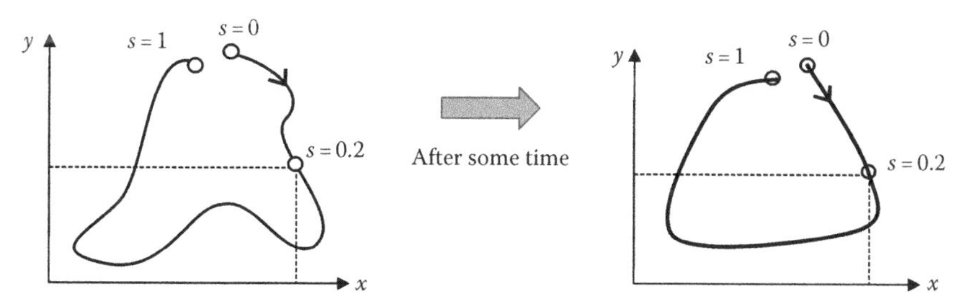

FIGURE 23.B.1
A parameterized contour undergoing a simple deformation (left to right). (Reprinted from McIntosh, C., Energy functionals for medical image segmentation: Choices and consequences, PhD dissertation, Simon Fraser University, Burnaby, BC, Canada, Copyright 2011. With permission.)

whereas the external energy term is given as

$$\beta(\mathbf{v}(s)) = \int_0^1 w_3(s)P(\mathbf{v}(s))ds. \tag{23.B.3}$$

The weighting functions w_1 and w_2 control the tension and flexibility of the contour, respectively, and w_3 controls the influence of image data. w_i's can depend on s but are typically set to different constants. For the contour to be attracted to image features, the function $P(\mathbf{v}(s))$ is designed such that it has minima where the features have maxima. For example, for the contour to be attracted to high-intensity changes (high gradients), we can choose

$$P(\mathbf{v}(s)) = P(x(s), y(s)) = -\left\|\nabla\left[G_\sigma * I(x(s), y(s))\right]\right\| \tag{23.B.4}$$

where $G_\sigma * I$ denotes the image convolved with a smoothing (e.g., Gaussian) filter with a parameter σ controlling the extent of the smoothing (e.g., variance of Gaussian).

The contour $\mathbf{v}$ that minimizes the energy ξ must, according to the calculus of variations (Elsgolc, 1962), satisfy the vector-valued partial differential (Euler–Lagrange) equation

$$\frac{\partial f}{\partial \mathbf{v}} - \frac{\partial}{\partial s}\left(\frac{\partial f}{\partial \mathbf{v}_s}\right) + \frac{\partial^2}{\partial s^2}\left(\frac{\partial f}{\partial \mathbf{v}_{ss}}\right) = 0 \tag{23.B.5}$$

where $f = \frac{1}{2}\left(w_1\left\|\mathbf{v}_s\right\|^2 + w_2\left\|\mathbf{v}_{ss}\right\|^2\right) - w_3\left\|\nabla(G_\sigma * I(\mathbf{v}))\right\|$. The first step in applying the Euler–Lagrange is to determine the partial derivatives as follows:

$$\frac{\partial f}{\partial \mathbf{v}} = -w_3\nabla\left\|\nabla(G_\sigma * I(\mathbf{v}))\right\|$$

$$\frac{\partial}{\partial s}\left(\frac{\partial f}{\partial \mathbf{v}_s}\right) = w_1\mathbf{v}_{ss} \tag{23.B.6}$$

$$\frac{\partial^2}{\partial s^2}\left(\frac{\partial f}{\partial \mathbf{v}_{ss}}\right) = w_2\mathbf{v}_{ssss}$$

Then substituting back into Equation 23.B.5, we get the final set of equations

$$-w_1\mathbf{v}_{ss} + w_2\mathbf{v}_{ssss} - w_3\nabla\left\|\nabla(G_\sigma * I)\right\| = 0 \tag{23.B.7}$$

where we make note that $\mathbf{v}(s) = (x(s), y(s))$, and so the earlier equation can be expanded to

$$-w_1 x_{ss} + w_2 x_{ssss} - w_3 \frac{\partial \|\nabla(G_\sigma * I)\|}{\partial x} = 0$$

$$-w_1 y_{ss} + w_2 y_{ssss} - w_3 \frac{\partial \|\nabla(G_\sigma * I)\|}{\partial y} = 0 \quad (23.B.8)$$

In order to solve the system of equations, we introduce an artificial time step, by equating the earlier equations to $-\partial\mathbf{v}/\partial t$. This yields a first-order iterative optimization method, though as outlined in this report other choices for optimization methods exist.

References

Andrews S, Hamarneh G, Yazdanpanah A, HajGhanbari B, and Reid D, 2011a, Probabilistic multi-shape segmentation of knee extensor and flexor muscles, in *Lecture Notes in Computer Science, Medical Image Computing and Computer-Assisted Intervention (MICCAI)*, Vol. 6893, pp. 651–658.

Andrews S, Hamarneh G, Yazdanpanah A, HajGhanbari B, and Reid WD, 2011a, Probabilistic multi-shape segmentation of knee extensor and flexor muscles, *Lecture Notes in Computer Science, Medical Image Computing and Computer-Assisted Intervention (MICCAI)*, pp. 651–658.

Andrews S, McIntosh C, and Hamarneh G, 2011b, Convex multi-region probabilistic segmentation with shape prior in the isometric log-ratio transformation space, *IEEE International Conference on Computer Vision (IEEE ICCV)*, Barcelona, Spain.

Andrews S, McIntosh C, and Hamarneh G, 2011b, Convex multi-region probabilistic segmentation with shape prior in the isometric Logratio transformation space, in *IEEE International Conference on Computer Vision (IEEE ICCV)*, pp. 2096–2103.

Anguelov D, Taskarf B, Chatalbashev V, Koller D, Gupta D, Heitz G, and Ng A, 2005, Discriminative learning of Markov random fields for segmentation of 3D scan data, *IEEE Computer Society Conference on Computer Vision and Pattern Recognition, 2005, CVPR 2005*, Vol. 2, pp. 169–176, San Diego, CA.

Appleton B and Talbot H, 2006, Globally minimal surfaces by continuous maximal flows, *IEEE Transactions on Pattern Analysis and Machine Intelligence* **28**, 106–118.

Ballerini L, 1998, Genetic snakes for medical image segmentation, *Proceedings of the SPIE Conference on Mathematical Modeling and Estimation Techniques in Computer Vision*, Vol. 3457, pp. 284–295, San Diego, CA.

Ballerini L, 2001, Genetic snakes for color images segmentation, in *Applications of Evolutionary Computing* (ed. Boers E), Vol. 2037 of Lecture notes in Computer Science, Springer, Berlin/Heidelberg, Germany, pp. 268–277.

Bar L and Sapiro G, 2009, Generalized Newton-type methods for energy formulations in image processing, *SIAM Journal of Imaging Science* **2**(2), 508–531.

Barrett WA and Mortensen EN, 1997, Interactive live-wire boundary extraction, *Medical Image Analysis* **1**(4), 331–341.

Boykov Y and Funka-Lea G, 2006, Graph cuts and efficient N-D image segmentation, *International Journal of Computer Vision (IJCV)* **70**(2), 109–131.

Boykov Y and Jolly MP, 2001, Interactive graph cuts for optimal boundary & region segmentation of objects in n-d images, *Proceedings of the Eighth IEEE International Conference on Computer Vision, 2001, ICCV 2001*, Vol. 1, pp. 105–112, Vancouver, BC, Canada.

Boykov Y and Kolmogorov V, 2003, Computing geodesics and minimal surfaces via graph cuts, *Proceedings of the Ninth IEEE International Conference on Computer Vision, 2003*, Vol. 1, pp. 26 –33, Nice, France.

Bresson X, Esedoḡlu S, Vandergheynst P, Thiran JP, and Osher S, 2007, Fast global minimization of the active contour/snake model, *Journal of Mathematical Imaging and Vision* **28**(2), 151–167.

Bresson X, Vandergheynst P, and Thiran JP, 2006, A variational model for object segmentation using boundary information and shape prior driven by the Mumford-Shah functional, *International Journal of Computer Vision* **68**(2), 145–162.

Bullitt E, Gerig G, Aylward S, Joshi S, Smith K, Ewend M, and Lin W, 2003, Vascular attributes and malignant brain tumors, *Medical Image Computing and Computer-Assisted Intervention—MICCAI 2003*, Vol. 2878 of Lecture notes in Computer Science, Springer, Berlin/Heidelberg, Germany, pp. 671–679.

Caselles V, Kimmel R, and Sapiro G, 1997, Geodesic active contours, *International Journal of Computer Vision* **22**, 61–79.

Changizi N and Hamarneh G, 2010, Probabilistic multi-shape representation using an isometric log-ratio mapping, in *Lecture Notes in Computer Science, Medical Image Computing and Computer-Assisted Intervention (MICCAI)*, Vol. 6363, pp. 563–570.

Changizi N and Hamarneh G, 2010, Probabilistic multi-shape representation using an isometric log-ratio mapping, *Lecture Notes in Computer Science, Medical Image Computing and Computer-Assisted Intervention (MICCAI)*, pp. Part III: 563–570.

Chan T, Moelich M, and Sandberg B, 2007, Some recent developments in variational image segmentation, *Image Processing Based on Partial Differential Equations*, Springer, Berlin, Germany, pp. 175–210.

Chan T and Vese L, 2001, Active contours without edges, *IEEE Transactions on Image Processing* **10**(2), 266–277.

Chan TF, Sandberg BY, and Vese LA, 2000, Active contours without edges for vector-valued images, *Journal of Visual Communication and Image Representation* **11**(2), 130–141.

Cocosco CA, Kollokian V, Kwan RKS, Pike GB, and Evans AC, 1997, Brainweb: Online interface to a 3D MRI simulated brain database, *NeuroImage* **5**, 425.

Cohen I, Cohen L, and Ayache N, 1992, Using deformable surfaces to segment 3-d images and infer differential structures, *CVGIP* **56**(2), 242–263.

Cohen L and Cohen I, 1993, Finite-element methods for active contour models and balloons for 2-d and 3-d images, *IEEE Transactions on Pattern Analysis and Machine Intelligence* **15**(11), 1131–1147.

Cohen LD, 1991, On active contour models and balloons, *Computer Vision Graphics and Image Processing: Image Understanding* **53**(2), 211–218.

Collette Y and Siarry P, 2002, *Multiobjective Optimization Principles and Case Studies*, Decision engineering, Springer, Berlin, Germany.

Cootes T, Taylor C, Cooper D, and Graham J, 1992, Training models of shape from sets of examples, *British Machine Vision Conference*, pp. 9–18. Springer-Verlag, Leeds, UK.

Cootes T, Taylor C, Hill A, and Halsam J, 1993, The use of active shape models for locating structures in medical images, *Proceedings of the 13th International Conference on Information Processing in Medical Imaging*, Flagstaff, AZ, pp. 33–47. Springer-Verlag.

Cootes T and Taylor C, 1997, A mixture model for representing shape variation, *British Machine Vision Conference*, pp. 110–119. Springer-Verlag, University of Essex, UK.

Cootes TF, Cooper D, Taylor CJ, and Graham J, 1995, Active shape models—their training and application, *Computer Vision and Image Understanding* **61**, 38–59.

Cootes TF, Edwards GJ, and Taylor CJ, 2001, Active appearance models, *IEEE Transactions on Pattern Analysis and Machine Intelligence* **23**(6), 681–685.

Cremers D, 2008, Nonlinear dynamical shape priors for level set segmentation, *Journal of Scientific Computing* **35**(2–3), 132–143.

Cremers D, Kohlberger T, and Schnörr C, 2001, Nonlinear shape statistics via kernel spaces, in *Pattern Recognition (Proc. DAGM)* (eds. Radig B and Florczyk S), Vol. 2191 of *LNCS*, pp. 269–276, Springer, Munich, Germany.

Cremers D, Kohlberger T, and Schnörr C, 2002, Nonlinear shape statistics in Mumford–Shah based segmentation, in *Computer Vision—ECCV 2002* (eds. Heyden A et al.), Vol. 2351 of *LNCS*, pp. 93–108, Springer, Copenhagen, Denmark.

Cremers D, Pock T, Kolev K, and Chambolle A, 2011, Convex relaxation techniques for segmentation, stereo and multiview reconstruction, *Advances in Markov Random Fields for Vision and Image Processing*, MIT Press, Cambridge, MA.

Cremers D, Schmidt F, and Barthel F, 2008, Shape priors in variational image segmentation: Convexity, lipschitz continuity and globally optimal solutions, *IEEE Conference on Computer Vision and Pattern Recognition, 2008, CVPR 2008*, pp. 1–6, Anchorage, AK.

Dambreville S, Rathi Y, and Tannen A, 2006, Shape-based approach to robust image segmentation using kernel PCA, *IEEE Computer Society Conference on Computer Vision and Pattern Recognition, 2006*, Vol. 1, pp. 977–984, New York.

Delingette H, 1999, General object reconstruction based on simplex meshes, *International Journal of Computer Vision* **32**, 111–146.

Delingette H, Hebert M, and Ikeuchi K, 1992, Shape representation and image segmentation using deformable surfaces, *Image Vision Computing* **10**(3), 132–144.

Delong A, Osokin A, Isack H, and Boykov Y, 2012, Fast approximate energy minimization with label costs, *International Journal of Computer Vision (IJCV)* **96**(1), 1–27.

Dice LR, 1945, Measures of the amount of ecologic association between species, *Ecology* **26**(3), 297–302.

Elsgolc L, 1962, *Calculus of Variations*, Pergamon Press Ltd., London, U.K.

Etyngier P, Segonne F, and Keriven R, 2007, Shape priors using manifold learning techniques, *ICCV 2007: IEEE 11th International Conference on Computer Vision, 2007*, pp. 1–8, Rio de Janeiro, Brazil.

Falcão AX, Udupa JK, Samarasekera S, Sharma S, Hirsch BE, and Lotufo RdA, 1998, User-steered image segmentation paradigms: Live wire and live lane, *Graphical Models and Image Processing* **60**(4), 233–260.

Falcão AX and Udupa JK, 1997, Segmentation of 3D objects using live wire, in *Medical Imaging 1997: Image Processing* (ed. Hanson KM), Vol. 3034, pp. 228–235. SPIE, Newport Beach, CA.

Fan Y, Jiang T, and Evans D, 2002, Volumetric segmentation of brain images using parallel genetic algorithms, *IEEE Transactions on Medical Imaging* **21**(8), 904–909.

Finley T and Joachims T, 2008, Training structural SVMs when exact inference is intractable, *ICML '08: Proceedings of the 25th International Conference on Machine Learning*, pp. 304–311, ACM, New York.

Fletcher P, Lu C, Pizer S, and Joshi S, 2004, Principal geodesic analysis for the study of nonlinear statistics of shape, *IEEE Transactions on Medical Imaging* **23**(8), 995–1005.

Frangi A, Niessen W, Hoogeveen R, van Walsum T, and Viergever M, 1999, Quantitation of vessel morphology from 3D MRA, in *Medical Image Computing and Computer-Assisted Intervention—MICCAI'99* (eds. Taylor C and Colchester A), Vol. 1679 of Lecture notes in Computer Science, pp. 358–367, Springer Berlin/Heidelberg, Germany.

Geoffrion AM, 1968, Proper efficiency and the theory of vector maximization, *Journal of Mathematics, Analysis and Applications* **22**(3), 618–630.

Ghosh P and Mitchell M, 2006, Segmentation of medical images using a genetic algorithm, *Proceedings of the 8th Annual Conference on Genetic and Evolutionary Computation, GECCO '06*, pp. 1171–1178, ACM, New York.

Grady L, 2006, Random walks for image segmentation, *IEEE Transactions on Pattern Analysis and Machine Intelligence* **28**(11), 1768–1783.

Grady L and Funka-Lea G, 2004, Multi-label image segmentation for medical applications based on graph-theoretic electrical potentials, *Computer Vision and Mathematical Methods in Medical and Biomedical Image Analysis*, pp. 230–245, Prague, Czech Republic.

Grady L and Polimeni JR, 2010, *Discrete Calculus: Applied Analysis on Graphs for Computational Science*, Springer, London, U.K.

Hamarneh G and Gustavsson T, 2004, Deformable spatio-temporal shape models: extending active shape models to 2D+time, *Journal of Image Vision Computing* **22**(6), 461–470.

Hamarneh G and Jassi P, 2010, Vascusynth: Simulating vascular trees for generating volumetric image data with ground truth segmentation and tree analysis, *Computerized Medical Imaging and Graphics* **34**(8), 605–616.

Hamarneh G, Jassi P, and Tang L, 2008, Simulation of ground-truth validation data via physically- and statistically-based warps, *Lecture Notes in Computer Science, Medical Image Computing and Computer-Assisted Intervention (MICCAI)*, pp. 459–467.

Hamarneh G, McInerney T, and Terzopoulos D, 2001, Deformable organisms for automatic medical image analysis, in *Medical Image Computing and Computer-Assisted Intervention—MICCAI 2001* (eds. Niessen W and Viergever M), Vol. 2208 of Lecture notes in Computer Science, pp. 66–76, Springer Berlin/Heidelberg, Germany.

Hamarneh G, McInerney T, and Terzopoulos D, 2001, Deformable organisms for automatic medical image analysis, *Lecture Notes in Computer Science, Medical Image Computing and Computer-Assisted Intervention (MICCAI)*, Vol. 2208, pp. 66–75.

Hamarneh G, McIntosh C, and Drew M, 2011, Perception-based visualization of manifold-valued medical images using distance-preserving dimensionality reduction, *IEEE Transactions on Medical Imaging*, **30**(7), 1314–1327.

Hamarneh G, McIntosh C, McInerney T, and Terzopoulos D, 2009, Deformable organisms: An artificial life framework for automated medical image analysis, in *Computational Intelligence in Medical Imaging: Techniques and Applications*, CRC Press, Boca Raton, FL, pp. 433–474.

Hamarneh G and McIntosh C, 2007, Physically and statistically based deformable models for medical image analysis (chapter 11), *Deformable Models: Biomedical and Clinical Applications*, pp. 335–386, Springer, New York, http://rd.springer.com/chapter/10.1007%2F978-0-387-68413-0_11.

Hanning T, Schöne R, and Pisinger G, 2006, Vector image segmentation by piecewise continuous approximation, *Journal of Mathematical Imaging and Vision* **25**, 5–23, 10.1007/s10851-005-4385-5.

Heimann T and Meinzer HP, 2009, Statistical shape models for 3D medical image segmentation: A review, *Medical Image Analysis* **13**(4), 543–563.

Hill A and Taylor C, 1992, Model-based image interpretation using genetic algorithms, *Image and Vision Computing* **10**(5), 295–300.

Ibáñez O, Barreira N, Santos J, and Penedo M, 2009, Genetic approaches for topological active nets optimization, *Pattern Recognition* **42**(5), 907–917.

Jaccard P, 1901, Étude comparative de la distribution florale dans une portion des alpes et des jura, *Bulletin del la Société Vaudoise des Sciences Naturelles* **37**, 547–579.

Kass M, Witkin A, and Terzopoulos D, 1987, Snakes: Active contour models, *International Journal of Computer Vision* **1(4)**, 321–331.

Kawahara J, McIntosh C, Tam R, and Hamarneh G, 2013, Globally optimal spinal cord segmentation using a minimal path in high dimensions, in *IEEE International Symposium on Biomedical Imaging*, pp. 836–839.

Kolmogorov V, Boykov Y, and Rother C, 2007, Applications of parametric maxflow in computer vision, *ICCV 2007: IEEE 11th International Conference on Computer Vision, 2007*, pp. 1–8, Rio de Janeiro, Brazil.

Kolmogorov V and Zabin R, 2004, What energy functions can be minimized via graph cuts? *IEEE Transactions on Pattern Analysis and Machine Intelligence* **26**(2), 147–159.

Lesage D, Angelini ED, Bloch I, and Funka-Lea G, 2009, A review of 3D vessel lumen segmentation techniques: Models, features and extraction schemes, *Medical Image Analysis* **13**(6), 819–845. Includes Special Section on *Computational Biomechanics for Medicine*.

Leventon M, Grimson W, and Faugeras O, 2000, Statistical shape influence in geodesic active contours, *Proceedings of the IEEE Conference on Computer Vision and Pattern Recognition, 2000*, Vol. 1, pp. 316–323, San Diego, CA.

Li H and Yezzi A, 2006, Vessels as 4d curves: Global minimal 4d paths to extract 3d tubular surfaces, *CVPRW '06: Conference on Computer Vision and Pattern Recognition Workshop, 2006*, pp. 82–82, NY, USA.

Lorigo LM, Faugeras OD, Grimson WEL, Keriven R, Kikinis R, Nabavi A, and Westin CF, 2001, Curves: Curve evolution for vessel segmentation, *Medical Image Analysis* **5**(3), 195–206.

MacEachern L and Manku T, 1998, Genetic algorithms for active contour optimization, *ISCAS '98: Proceedings of the 1998 IEEE International Symposium on Circuits and Systems, 1998*, Vol. 4, pp. 229 –232, Monterey, CA.

Maintz J and Viergever MA, 1998, A survey of medical image registration, *Medical Image Analysis* **2**(1), 1–36.

McInerney T and Terzopoulos D, 1995a, A dynamic finite element surface model for segmentation and tracking in multidimensional medical images with application to cardiac 4d image analysis, *Computerized Medical Imaging and Graphics* **19**(1), 69–83.

McInerney T and Terzopoulos D, 1995b, Medical image segmentation using topologically adaptable snakes, *Proceedings of the First International Conference on Computer Vision, Virtual Reality and Robotics in Medicine*, pp. 92–101, Springer-Verlag, London, U.K.

McInerney T and Terzopoulos D, 1995c, Topologically adaptable snakes, *Proceedings of the Fifth International Conference on Computer Vision, 1995*, pp. 840–845, Cambridge, MA.

McInerney T and Terzopoulos D, 1996, Deformable models in medical image analysis: A survey, *Medical Image Analysis* **1(2)**, 91–108.

McInerney T and Terzopoulos D, 1999, Topology adaptive deformable surfaces for medical image volume segmentation, *IEEE Transactions on Medical Imaging* **18**(10), 840–850.

McInerney T and Terzopoulos D, 2000, T-snakes: Topology adaptive snakes, *Medical Image Analysis* **4**(2), 73–91.

McIntosh C, 2011, Energy functionals for medical image segmentation: Choices and consequences, PhD dissertation, Simon Fraser University, Burnaby, BC, Canada.

McIntosh C and Hamarneh G, 2006a, Vessel crawlers: 3D physically-based deformable organisms for vasculature segmentation and analysis, *IEEE Computer Society Conference on Computer Vision and Pattern Recognition, 2006*, Vol. 1, pp. 1084–1091, New York.

McIntosh C and Hamarneh G, 2006b, Genetic algorithm driven statistically deformed models for medical image segmentation, *ACM Workshop on Medical Applications of Genetic and Evolutionary Computation Workshop (MedGEC), in conjunction with the Genetic and Evolutionary Computation Conference (GECCO)*, Seattle, WA.

McIntosh C and Hamarneh G, 2007, Is a single energy functional sufficient? Adaptive energy functionals and automatic initialization, in *Medical Image Computing and Computer-Assisted Intervention—MICCAI 2007* (eds. Ayache N, Ourselin S, and Maeder A), Vol. 4792 of Lecture notes in Computer Science, pp. 503–510, Springer Berlin/Heidelberg, Germany, 10.1007/978-3-540-75759-7_61.

McIntosh C and Hamarneh G, 2009, Optimal weights for convex functionals in medical image segmentation, *International Symposium on Visual Computing: Special Track on Optimization for Vision, Graphics and Medical Imaging: Theory and Applications (ISVC OVGMI)*, Vol. 5875-I, pp. 1079–1088, Las Vegas, Nevada, USA.

McIntosh C and Hamarneh G, 2012, Medial-based deformable models in non-convex shape-spaces for medical image segmentation using genetic algorithms, *IEEE Transactions on Medical Imaging*, on 31.1, 33–50.

Mignotte M and Meunier J, 1999, Deformable template and distribution mixture-based data modeling for the endocardial contour tracking in an echographic sequence, *IEEE Computer Society Conference on Computer Vision and Pattern Recognition, 1999*, Vol. 1, p. 230, Fort Collins, CO.

Miller JV, Breen DE, Lorensen WE, O'Bara RM, and Wozny MJ, 1991, Geometrically deformed models: A method for extracting closed geometric models form volume data, *SIGGRAPH Computer Graphics* **25**(4), 217–226.

Montagnat J, Delingette H, and Ayache N, 2001, A review of deformable surfaces: Topology, geometry and deformation, *Image and Vision Computing* **19**(14), 1023–1040.

Mortensen EN and Barrett WA, 1998, Interactive segmentation with intelligent scissors, *Graphical Models and Image Processing* **60**(5), 349–384.

Mumford D and Shah J, 1989, Optimal approximation by piecewise smooth functions and associated variational problems, *Communications on Pure Applied Mathematics* **42**, 577–685.

Nakib A, Oulhadj H, and Siarry P, 2010, Image thresholding based on pareto multiobjective optimization, *Engineering Applications of Artificial Intelligence* 23.3: 313–320.

Nand K, Abugharbieh R, Booth B, and Hamarneh G, 2011, Detecting structure in diffusion tensor MR images, *Lecture Notes in Computer Science, Medical Image Computing and Computer-Assisted Intervention (MICCAI)*, pp. 90–97.

Ng B, Hamarneh G, and Abugharbieh R, 2012, Modeling brain activation in fMRI using group MRF, *IEEE Transactions on Medical Imaging* **31**(5), 1113–1123.

Nikolova M, Esedoglu S, and Chan TF, 2006, Algorithms for finding global minimizers of image segmentation and denoising models, *SIAM Journal on Applied Mathematics* **66**(5), 1632–1648.

Nosrati M and Hamarneh G, 2013, Segmentation of cells with partial occlusion and part configuration constraint using evolutionary computation, *Lecture Notes in Computer Science, Medical Image Computing and Computer-Assisted Intervention (MICCAI)*.

O'Donnell T, Boult T, Fang XS, and Gupta A, 1994, The extruded generalized cylinder: A deformable model for object recovery, *Proceedings CVPR '94: IEEE Computer Society Conference on Computer Vision and Pattern Recognition, 1994*, pp. 174–181, Seattle, WA.

Osher S and Paragios N, 2003, *Geometric Level Set Methods in Imaging Vision and Graphics*, Springer Verlag, New York.

Osher S and Sethian JA, 1988, Fronts propagating with curvature dependent speed: Algorithms based on Hamilton-Jacobi formulations, *Journal of Computational Physics* **79**(1), 12–49.

Paragios N and Deriche R, 2000, Coupled geodesic active regions for image segmentation: A level set approach, in *Computer Vision—ECCV 2000* (eds. Vernon D), Vol. 1843 of Lecture notes in Computer Science, pp. 224–240, Springer Berlin/Heidelberg, Germany.

Paragios N and Deriche R, 2002, Geodesic active regions and level set methods for supervised texture segmentation, *International Journal of Computer Vision* **46**, 223–247.

Paragios N, Taron M, Huang X, Rousson M, and Metaxas D, 2006, On the representation of shapes using implicit functions, in *Statistics and Analysis of Shapes* (eds. Krim H and Yezzi A), Modeling and simulation in science, engineering and technology, Birkhäuser, Boston, MA, pp. 167–199.

Pham DL, Xu C, and Prince JL, 2000, A survey of current methods in medical image segmentation, *In Annual Review of Biomedical Engineering* **2**, 315–338.

Pizer SM, 2003, Guest editorial—Medial and medical: A good match for image analysis, *International Journal of Computer Vision* **55**(2–3), 79–84.

Pizer SM, Fletcher PT, Joshi SC, Thall A, Chen JZ, Fridman Y, Fritsch DS et al., 2003, Deformable M-Reps for 3d medical image segmentation, *International Journal of Computer Vision* **55**(2–3), 85–106.

Pohl KM, Fisher J, Bouix S, Shenton M, McCarley RW, Grimson WEL, Kikinis R, and Wells WM, 2007, Using the logarithm of odds to define a vector space on probabilistic atlases, *Medical Image Analysis* **11**(5), 465–477, Special issue on the *Ninth International Conference on Medical Image Computing and Computer-Assisted Interventions—MICCAI 2006*.

Poon M, Hamarneh G, and Abugharbieh R, 2008, Efficient interactive 3D livewire segmentation of objects with arbitrarily topologies, *Computerized Medical Imaging and Graphics* **32**(8), 639–650.

Rana M, Hamarneh G, and Wakeling J, 2009, Automated tracking of muscle fascicle orientation in B-mode ultrasound images, *Journal of Biomechanics* **42**(13), 2068–2073.

Rao J, Abugharbieh R, and Hamarneh G, 2010, Adaptive regularization for image segmentation using local image curvature cues, *European Conference on Computer Vision (ECCV)*, pp. 651–665.

Rao J, Hamarneh G, and Abugharbieh R, 2009, Adaptive contextual energy parameterization for automated image segmentation, in *Lecture Notes in Computer Science, International Symposium on Visual Computing: Special Track on Optimization for Vision, Graphics and Medical Imaging: Theory and Applications (ISVC OVGMI)*, Vol. 5875-I, pp. 1089–1100.

Robb RA, 2000, *Biomedical Imaging, Visualization, and Analysis*, Wiley-Liss Inc., New York.

Rousson M, Lenglet C, and Deriche R, 2004, Level set and region based surface propagation for diffusion tensor MRI segmentation, in *Computer Vision and Mathematical Methods in Medical and Biomedical Image Analysis* (eds. Sonka M, Kakadiaris IA, and Kybic J), Vol. 3117 of Lecture notes in Computer Science, pp. 123–134, Springer Berlin/Heidelberg, Germany, 10.1007/978-3-540-27816-0_11.

Ruff CF, Hughes SW, and Hawkes DJ, 1999, Volume estimation from sparse planar images using deformable models, *Image and Vision Computing* **17**(8), 559–565.

Saad A, Hamarneh G, and Moeller T, 2010a, Exploration and visualization of segmentation uncertainty using shape and appearance prior information, *IEEE Transactions on Visualization and Computer Graphics (Special Issue of the IEEE Visualization Conference 2010)* **16**(6), 1366–1375.

Saad A, Hamarneh G, Moeller T, and Smith B, 2008, Kinetic modeling based probabilistic segmentation for molecular images, *Lecture Notes in Computer Science, Medical Image Computing and Computer-Assisted Intervention (MICCAI)*, pp. 244–252.

Saad A, Moeller T, and Hamarneh G, 2010b, ProbExplorer: An uncertainty-based visual analysis tool for medical imaging, *Computer Graphics Forum (CGF) (Special Issue of the proceedings of Eurographics/IEEE-VGTC Symposium on Visualization 2010)* **29**(3), 1113–1122.

Samson C, Blanc-Féraud L, Aubert G, and Zerubia J, 2000, A level set model for image classification, *International Journal of Computer Vision* **40**, 187–197, 10.1023/A:1008183109594.

Sapiro G, 1996, Vector-valued active contours, *Proceedings CVPR '96: IEEE Computer Society Conference on Computer Vision and Pattern Recognition, 1996*, pp. 680–685, San Francisco, CA, USA.

Sethian J, 1996, *Level Set Methods: Evolving Interfaces in Geometry, Fluid Mechanics, Computer Vision and Materials Science*, Cambridge University Press, Cambridge, U.K.

Shah J, 1996, Curve evolution and segmentation functionals: Application to color images *Proceedings of the International Conference on Image Processing, 1996*, Vol. 1, pp. 461–464, Lausanne, Switzerland.

Shi J and Malik, J, 2000, Normalized cuts and image segmentation, *Pattern Analysis and Machine Intelligence, IEEE Transactions on 22.8*: 888–905.

Shi L, Funt B, and Hamarneh G, 2008, Quaternion color curvature, *Color Imaging (IS&T/SID CI)*, pp. 338–341, Portland, Oregon.
Sinop A and Grady L, 2007, A seeded image segmentation framework unifying graph cuts and random walker which yields a new algorithm, *ICCV 2007: IEEE 11th International Conference on Computer Vision, 2007*, pp. 1–8, Rio de Janeiro, Brazil.
Sonka M and Fitzpatrick J, 2000, *Handbook of Medical Imaging, Volume 2: Medical Image Processing and Analysis*, SPIE-International Society for Optical Engine, http://spie.org/X648.html?produce_id_=_831079
Sozou P, Cootes T, Taylor C, and Di Mauro E, 1995, Non-linear point distribution modelling using a multi-layer perceptron, *British Machine Vision Conference*, pp. 107–116, Birmingham, U.K.
Staib L and Duncan J, 1992a, Boundary finding with parametrically deformable models, *IEEE Transactions on Pattern Analysis and Machine Intelligence* **14**(11), 1061–1075.
Staib LH and Duncan JS, 1992b, Deformable Fourier models for surface finding in 3D images, *Proceedings of the Second Conference on Visualization in Biomedical Computing (VBC-92)*, Vol. 1808, pp. 90–104, SPIE, Bellingham, WA.
Sundaramoorthi G, Yezzi A, and Mennucci AC, 2007, Sobolev active contours, *International Journal of Computer Vision* **73**(3), 345–366.
Szummer M, Kohli P, and Hoiem D, 2008, Learning CRFs using graph cuts, in *Computer Vision—ECCV 2008* (eds. Forsyth D, Torr P, and Zisserman A), Vol. 5303 of Lecture notes in Computer Science, pp. 582–595, Springer Berlin/Heidelberg, Germany.
Tang L, Bressmann T, and Hamarneh G, 2012, Tongue contour tracking in dynamic ultrasound via higher-order MRFs and efficient fusion moves, *Medical Image Analysis* **16**(8), 1503–1520.
Tanimoto T, 1957, An elementary mathematical theory of classification and prediction, Technical report, IBM Internal Report.
Taskar B, Chatalbashev V, Koller D, and Guestrin C, 2005, Learning structured prediction models: A large margin approach, *ICML '05: Proceedings of the 22nd International Conference on Machine Learning*, pp. 896–903, ACM, New York.
Terzopoulos D, 1987, On matching deformable models to images, *Topical Meeting on Machine Vision, Technical Digest Series* **12**, 160–167.
Tohka J, 2001, Global optimization of deformable surface meshes based on genetic algorithms, *Proceedings of the 11th International Conference on Image Analysis and Processing, 2001*, pp. 459–464, Palermo, Italy.
Top A, Hamarneh G, and Abugharbieh R, 2010, Spotlight: Automated confidence-based user guidance for increasing efficiency in interactive 3D image segmentation, in *Medical Image Computing and Computer-Assisted Intervention Workshop on Medical Computer Vision (MICCAI MCV)*, Vol. 6533, pp. 204–213.
Top A, Hamarneh G, and Abugharbieh R, 2011, Active learning for interactive 3D image segmentation, in *Lecture Notes in Computer Science, Medical Image Computing and Computer-Assisted Intervention (MICCAI)*, Vol. 6893, pp. 603–610.
Tsai A, Wells W, Tempany C, Grimson E, and Willsky A, 2004, Mutual information in coupled multi-shape model for medical image segmentation, *Medical Image Analysis* **8**(4), 429–445.
Tsai A, Yezzi AJ, and Willsky A, 2001, Curve evolution implementation of the Mumford-Shah functional for image segmentation, denoising, interpolation, and magnification, *IEEE Transactions on Image Processing* **10**(8), 1169–1186.

Ulen J, Strandmark P, and Kahl F, 2013, An efficient optimization framework for multi-region segmentation based on Lagrangian duality, *IEEE Transactions on Medical Imaging* **32**(2), 178–188.

Vasilevskiy A and Siddiqi K, 2002, Flux maximizing geometric flows, *IEEE Transactions on Pattern Analysis and Machine Intelligence* **24**(12), 1565–1578.

Vese LA and Chan TF, 2001, A multiphase level set framework for image segmentation using the Mumford and Shah model, *International Journal of Computer Vision* **50**, 271–293.

Vu N and Manjunath B, 2008, Shape prior segmentation of multiple objects with graph cuts, *CVPR 2008: IEEE Conference on Computer Vision and Pattern Recognition, 2008*, pp. 1–8, Anchorage, AK.

Wang Z and Vemuri BC, 2004, Tensor field segmentation using region based active contour model, *Computer Vision—ECCV 2000*, Vol. 3024 of Lecture notes in Computer Science, pp. 304–315, Springer Berlin/Heidelberg, Germany.

Warfield S, Kaus M, Jolesz FA, and Kikinis R, 2000, Adaptive, template moderated, spatially varying statistical classification, *Medical Image Analysis* **4**(1), 43–55.

Warfield S, Zou K, and Wells W, 2004, Simultaneous truth and performance level estimation (staple): An algorithm for the validation of image segmentation, *IEEE Transactions on Medical Imaging* **23**(7), 903–921.

Weldeselassie Y and Hamarneh G, 2007, DT-MRI segmentation using graph cuts, *SPIE Medical Imaging*, Vol. 6512-1K, pp. 1–9, San Diego, California, USA.

Whitaker RT, 1994, Volumetric deformable models: Active blobs, *Visualization in Biomedical Computing* **2359**(1), 122–134.

Wink O, Niessen W, Verdonck B, and Viergever M, 2001, Vessel axis determination using wave front propagation analysis, in *Medical Image Computing and Computer-Assisted Intervention—MICCAI 2001* (eds. Niessen W and Viergever M), Vol. 2208 of Lecture notes in Computer Science, pp. 845–853, Springer, Berlin/Heidelberg, Germany, 10.1007/3-540-45468-3_101.

Wink O, Niessen WJ, and Viergever M, 2004, Multiscale vessel tracking, *IEEE Transactions on Medical Imaging* **23**(1), 130–133.

Xu C and Prince J, 1998, Snakes, shapes, and gradient vector flow, *IEEE Transactions on Image Processing* **7**(3), 359–369.

Yazdanpanah A, Hamarneh G, Smith B, and Sarunic M, 2011, Automated segmentation of intra-retinal layers from optical coherence tomography images using an active contour approach, *IEEE Transactions on Medical Imaging* **2**(30), 484–496.

Yezzi AJ, Kichenassamy S, Kumar A, Olver P, and Tannenbaum A, 1997, A geometric snake model for segmentation of medical imagery, *IEEE Transactions on Medical Imaging* **16**(2), 199–209.

Yoo TS, 2004, *Insight into Images: Principles and Practice for Segmentation, Registration, and Image Analysis*, A K Peters Ltd., Wellesey, MA.

Zhao HK, Chan T, Merriman B, and Osher S, 1996, A variational level set approach to multiphase motion, *Journal of Computational Physics* **127**(1), 179–195.

Zitová B and Flusser J, 2003, Image registration methods: A survey, *Image and Vision Computing* **21**(11), 977–1000.

Index

C

H

I

J

Q

R

S

Printed and bound by CPI Group (UK) Ltd, Croydon, CR0 4YY
18/10/2024
01776258-0002